T0186749

PROCEEDINGS
SEVENTH INTERNATIONAL CONGRESS
INTERNATIONAL ASSOCIATION OF ENGINEERING GEOLOGY
VOLUME 4

COMPTES-RENDUS
SEPTIEME CONGRES INTERNATIONAL
ASSOCIATION INTERNATIONALE DE GEOLOGIE DE L'INGENIEUR
VOLUME 4

Comptes-rendus Septième Congrès International Association Internationale de Géologie de l'Ingénieur

5–9 SEPTEMBRE 1994 / LISBOA / PORTUGAL

Rédacteurs
R.OLIVEIRA, L.F.RODRIGUES, A.G.COELHO & A.P.CUNHA
LNEC, Lisboa, Portugal

VOLUME 4
Thème 3 La géologie de l'ingénieur et la protection de l'environnement

A.A.BALKEMA / ROTTERDAM / BROOKFIELD / 1994

Proceedings
Seventh International Congress
International Association
of Engineering Geology

5–9 SEPTEMBER 1994 / LISBOA / PORTUGAL

Editors
R.OLIVEIRA, L.F.RODRIGUES, A.G.COELHO & A.P.CUNHA
LNEC, Lisboa, Portugal

VOLUME 4
Theme 3 Engineering geology and environmental protection

A.A.BALKEMA / ROTTERDAM / BROOKFIELD / 1994

INTERNATIONAL ASSOCIATION OF ENGINEERING GEOLOGY
ASSOCIATION INTERNATIONALE DE GEOLOGIE DE L'INGENIEUR

The publication of these proceedings has been partially funded by
La publication des comptes-rendus a été partiellement supportée par

Junta Nacional de Investigação Científica e Tecnológica (JNICT), Portugal
Fundação Calouste Gulbenkian, Portugal

The texts of the various papers in this volume were set individually by typists under the supervision of each of the authors concerned.

Les textes des divers articles dans ce volume ont été dactylographiés sous la supervision de chacun des auteurs concernés.

Complete set of six volumes / Collection complète de six volumes: ISBN 90 5410 503 8
Volume 1: ISBN 90 5410 504 6
Volume 2: ISBN 90 5410 505 4
Volume 3: ISBN 90 5410 506 2
Volume 4: ISBN 90 5410 507 0
Volume 5: ISBN 90 5410 508 9
Volume 6: ISBN 90 5410 509 7

Published by:
© 1994 A.A. Balkema, Postbus 1675, 3000 BR Rotterdam, Netherlands (Fax: +31.10.413.5947)
Distributed in the USA & Canada by: A.A. Balkema Publishers, Old Post Road, Brookfield, VT 05036, USA
(Fax: +1.802.276.3837)
Printed in the Netherlands

Publié par:
© 1994 A.A. Balkema, Postbus 1675, 3000 BR Rotterdam, Pays-Bas
Distribué aux USA & Canada par: A.A. Balkema Publishers, Old Post Road, Brookfield, VT 05036, USA
Imprimé aux Pays-Bas

SCHEME OF THE WORK
SCHÉMA DE L'OUVRAGE

Table of contents
Table des matières

XV

XVI

Conférence spéciale: Contributions de la géologie de l'ingénieur au stockage sûr des déchets ménagers et industriels toxiques: Exemples actuels et perspectives

Keynote lecture: Contributions of engineering geology to the safe storage of toxic domestic and industrial wastes: present situation and persopectives

B.Côme
ANTEA (groupe BRGM), Orléans, France

RESUME : Quelles que soient les améliorations techniques et réglementaires, les activités humaines de production et de consommation sont à l'origine de déchets plus ou moins toxiques, pour lesquels le stockage, dans le sol et le sous-sol, restera la voie d'élimination la plus compatible avec la protection de l'environnement. L'article examine les trois principales options de stockage (mise en décharge superficielle, dépôt en cavités souterraines, injection de liquides par puits profonds) et souligne à leur propos les domaines d'intervention et de compétences, actuels et envisageables, de la Géologie de l'Ingénieur. Il suggère, pour finir, des voies de développement souhaitables en la matière.

ABSTRACT : In spite of various technical and regulatory improvements, Man's activities like production and consumption still generate more or less toxic wastes, for which a safe storage, above ground or underground, will remain the most acceptable disposal solution. The paper reviews the three main storage alternatives (landfilling, underground cavern storage, deep well injection of liquids), and emphasizes the present and likely future contributions of Engineering Geology. To conclude, suggestions are made with a view to further developments and improvements in this field.

1 INTRODUCTION

C'est devenu un lieu commun que d'évoquer les "montagnes de déchets" qui résultent des activités humaines de production et de consommation. Les premières sont à l'origine de la majorité des déchets industriels (chimie, traitement des métaux, agro-alimentaire, etc.) ; quant aux secondes, la mise à dis-position de produits généralement emballés et/ou conditionnés a engendré l'augmentation du tonnage des déchets ménagers, à l'origine réduits à la fraction organique non-comestible de denrées alimentaires, et maintenant largement constitués d'emballages, bouteilles perdues, etc.

Bien que les nomenclatures et les méthodes de comptage diffèrent encore selon les pays, quelques chiffres suffisent à donner l'échelle du problème des déchets, comme présenté dans le tableau 1 concernant les pays considérés comme les plus industrialisés (EUROSTAT, 1992).

Dans les pays en développement, le problème des déchets est lié à la croissance démographique et à la concentration des populations en un petit nombre de mégalopoles et/ou de zones industrielles. Le tableau 2 rassemble à leur propos un certain nombre

de chiffres également significatifs (Campbell, 1993). On peut noter au passage la prédominance de résidus organiques dans les déchets ménagers de ces pays ; à titre de comparaison, la répartition des ordures ménagères en France (environ 1 kg/personne/jour) ne comporte qu'un quart de matières organiques en poids, pour un tiers d'emballages (verre, plastique, papier et carton) et 10 % de produits "divers". Parmi ces derniers se trouvent en particulier les résidus de peintures, piles usagées, ampoules électriques, produits de nettoyage, etc., qui sont à l'origine de la toxicité accrue des déchets ménagers des pays industrialisés (Carra et Cossu, 1990).

La notion de *toxicité* des déchets, intuitive mais difficile à quantifier, peut se définir de façon régle-mentaire ; c'est ainsi que la directive communautaire 67/548/CEE, sur les substances dangereuses, a réparti ces dernières en "substances nocives", "substances toxiques" et "substances très toxiques", selon les valeurs de dose létales, ou concentrations létales, dérivées d'essais toxicologiques sur les rats ou les lapins (CCE, 1967). Un grand nombre de produits de la chimie organique (phénols, solvants chlorés, etc.), et des composés de métaux lourds (cadmium, mercure, plomb, chrome hexavalent, etc.) font partie

Tableau 1. Production de certains déchets dans quelques pays de l'OCDE (en millions de tonnes) (EUROSTAT, 1992).

Pays	Année	Déchets ménagers	Déchets industriels	Déblais de démolition	Déchets miniers
Allemagne	1989	19,5	61,4	11,8	9,5
France	1989	17	50	-	100
	1993	20,5	-	-	-
Italie	1989	17,3	40	34,3	57
Royaume-Uni	1989	20	50	25	230
Etats-Unis	1989	208,7	760	31,3	1400
Canada	1989	16,4	61	1,5	10,5
Japon	1988	48,3	312	57,8	26

Tableau 2. Ratios de production de déchets ménagers dans quelques pays en développement et comparaison avec d'autres pays industrialisés (Campbell, 1993).

Pays	Production de déchets ménagers (kg/personne/jour)	Poids volumique des déchets ménagers (t/m^3)
Inde	0,25	0,25
Ghana	0,25	0,25
Egypte	0,30	0,24
Philippines	0,50	0,25
Malaisie	0,70	0,2
Europe	1,0	0,13
USA	1,25	0,10

des substances toxiques que l'on trouve dans des déchets industriels et aussi ménagers. Une quantité croissante de données concernant les effets de substances dangereuses pour la santé sont collectées dans différents pays (US-EPA, 1993).

Compte tenu du rythme de progression enregistré dans la production de déchets, et des problèmes de santé que ces derniers peuvent faire courir aux populations (outre la dégradation insidieuse de l'environnement en général), bon nombre de législations ont introduit des politiques volontaristes de gestion des déchets par *réduction* de leur production, *récupération* (notamment des emballages par tri sélectif) et *recyclage* (des métaux et plastiques par exemple). Il existe cependant des limites pratiques et économiques à ces techniques et procédés ; en outre, les déchets peuvent voir leurs composition et qualité varier au cours du temps, ce qui constitue un frein supplémentaire à leur traitement industriel à grande échelle. Pour toutes ces raisons, il subsiste et subsistera encore un certain temps, des résidus inutilisables pour lesquels la seule solution acceptable pour leur élimination sûre est celle de *la mise en dépôt dans ou sur le sol*, appelée par abus de langage "*stockage géologique*". En fait, le rôle du sol et du sous-sol en tant que réceptacles appropriés pour de nombreux déchets pouvant présenter un risque toxique est

reconnu depuis longtemps par la communauté des Sciences de la Terre (Innes Lumsden, 1992 ; Barrès et Côme, 1994).

C'est le propos du présent article d'illustrer la contribution de la Géologie de l'Ingénieur et des disciplines connexes (hydrogéologie, géotechnique, sciences des matériaux, géochimie, génie civil, etc.) à la gestion sûre des déchets toxiques, ménagers et industriels, par l'option du stockage ou confinement géologique. Les aspects liés à la collecte et au traitement des déchets eux-mêmes ne seront abordés qu'allusivement ; il en sera de même pour les aspects (fondamentaux) liés à la législation, à la sociologie, etc., davantage reflets des modes de vie et de l'évolution des sociétés, pour lesquels le lecteur intéressé pourra se reporter à d'autres ouvrages spécialisés (Barrère, 1992).

En vue du stockage des déchets, le milieu géologique a jusqu'à présent été utilisé de trois manières principales :

1. la mise en décharge (en surface du sol, ou à faible profondeur) de déchets solides, ménagers et industriels.

2. La mise en place de déchets solides dans des cavités souterraines, ou l'utilisation d'excavations existantes (minières, etc.), accessibles à l'homme.

3. L'injection, par puits profonds, de déchets liquides dans des horizons perméables et isolés des aquifères voisins.

Pour chacune de ces techniques, on envisagera d'abord les apports présents de la Géologie de l'Ingénieur, puis un certain nombre de perspectives possibles de développement.

2 LA MISE EN DECHARGE

2.1 *Généralités*

La mise en dépôt, par simple déversement dans une dépression superficielle des terrains, de déchets solides tels des ordures ménagères, des déblais de démolition, des biens d'équipement hors d'usage, est une pratique ancienne, qu'atteste l'archéologie par exemple. Ce mode d'élimination des déchets a évolué pour aboutir à la mise en décharge ordonnée actuelle, qui constitue le mode le plus répandu de stockage des résidus urbains et industriels. A titre d'exemple, 43 % des ordures ménagères produites en France ont été éliminées par mise en décharge contrôlée, pour 41 % traités par incinération avec ou sans récupération d'énergie. Compte tenu des tonnages extrêmement importants de déchets industriels (plusieurs centaines de millions de tonnes par an pour les pays développés), la décharge contrôlée de surface, aux prescriptions de plus en plus strictes, est encore le mode d'élimination majoritaire (pour des raisons de coût essentiellement), bien que les contraintes d'occupation de l'espace et l'opposition des riverains actuels ou potentiels ralentissent dans certains pays le rythme d'extension des installations existantes, ou la création d'installations nouvelles. Dans les pays de l'OCDE, les décharges pour déchets industriels peuvent se compter par dizaines, et par centaines dans le cas des déchets ménagers (Carra et Cossu, 1990).

Qu'elle accueille des déchets ménagers ou industriels, la décharge contrôlée actuelle - appelée parfois centre d'enfouissement technique - est conçue comme une "boîte" la plus étanche possible, restreignant l'arrivée et le départ de l'eau météorique ou souterraine (principal vecteur des éventuels polluants qu'on cherche à mettre hors de portée de l'homme et de son environnement) ; on applique ainsi le principe de gestion des déchets dénommé *"concentrer et confiner"*. Il est clair cependant qu'aucun confinement ne peut prétendre durer éternellement ; l'objectif pourra donc être de tolérer un relargage très faible de polluants, qui pourra ainsi être dilué naturellement, de telle façon que les critères de protection de la santé humaine (par exemple, qualité de l'eau pour l'alimentation) soient constamment vérifiés. Pour les déchets ménagers à fraction organique plus importante, la décharge jouera plutôt un rôle de *bioréacteur*, en produisant des liquides (jus de décharges) et des gaz (biogaz) jusqu'à minéralisation plus ou moins complète du stock.

En résumé, les réglementations et la pratique internationales s'accordent sur les *grands principes d'implantation et de fonctionnement d'un site de décharge* :

1. de préférence, asseoir l'ouvrage sur une formation argileuse ;

2. assurer ou renforcer l'imperméabilité de cette dernière ;

3. restreindre les arrivées d'eau de pluie ;

4. récupérer et traiter les éventuels lixiviats et/ou percolats ;

5. capter et traiter les gaz éventuellement produits (par la décomposition des ordures ménagères notamment).

De très nombreuses conférences, générales ou spécialisées, sont à l'origine d'utiles "états de l'art" sur cette question en évolution rapide (CBGI, 1985 ; AIH, 1988 ; Arnould, 1993). De même, un grand nombre de commissions techniques préparent des guides ou manuels dans lesquels les derniers progrès sont mis à la disposition des concepteurs et réalisateurs de ces ouvrages. Les actes des conférences "International Landfill Symposium", connus sous le nom générique SARDINIA, ainsi que ceux du premier symposium international GEOCONFINE (sur le confinement géologique des déchets toxiques) constituent aussi d'utiles références.

Bien que leur mise en oeuvre particulière puisse différer, il ne sera pas ici fait de distinction fondamentale entre les décharges pour déchets urbains et industriels ; de même, on ne consacrera pas de développement spécial aux *dépôts de résidus et/ou de stériles issus des exploitations minières*. En effet, pour ce dernier cas, le problème principal est d'éviter le "drainage minier acide", c'est-à-dire des lixiviats à pH très bas pouvant provenir de réactions complexes entre l'eau météorique et certains résidus tels les pyrites (Aubertin *et al.*, 1993) ; la solution passe alors par la création d'une structure de confinement voisine de celle d'un centre de stockage pour déchets ordinaires.

Enfin, l'expérience passée a montré que les dépôts d'ordures ménagères et ceux pour déchets industriels n'ont pas toujours été physiquement séparés. La majorité des spécifications techniques recommandent, actuellement, des critères et pratiques de confinement distincts pour ces deux types de déchets ; à l'intérieur des décharges pour déchets industriels, il est même recommandé de séparer les catégories de déchets dans des *alvéoles* distinctes. D'autres réglementations prévoient cependant la possibilité de la *co-déposition* des déchets, ménagers et industriels notamment (Greedy, 1993). L'argument majeur présenté en soutien de cette option est que certains produits organiques peuvent créer, dans la décharge, des conditions physico-chimiques favorables au piégeage de substances toxiques, par exemple des métaux lourds. L'efficacité réelle de ce mode de piégeage de polluants est, cependant, encore controversée, car difficile voire impossible à gérer sur sites réels.

2.2 *La construction et l'exploitation de décharges contrôlées ("sanitary landfills")*

Compte tenu de l'ancienneté de cette pratique, la mise en décharge fait l'objet de *prescriptions réglementaires*, d'ailleurs variables d'un pays à l'autre ; les législations correspondantes peuvent en effet être applicables au niveau de provinces ou national (Dörhöfer, 1993 ; CCME, 1991 ; US-EPA, 1989), ou même transnational (CCE, 1993). Les exigences de base pour le choix des sites correspondent généralement à des catégories de déchets ; c'est ainsi que les sites de décharge ou "d'enfouissement technique" en France sont répartis en trois "classes", respectivement dénommées III pour déchets inertes (déblais de démolition par exemple), II pour les ordures ménagères et/ou déchets industriels banals, et I pour les déchets industriels spéciaux ; à chaque classe correspond une valeur maximale de perméabilité du substratum argileux de ces décharges, et une exigence d'épaisseur minimale pour ce dernier. De plus en plus, les réglementations imposent le recours à des matériaux supplémentaires permettant d'assurer le drainage et l'étanchéité des décharges ; on assiste donc actuellement à la publication de documents ou manuels techniques permettant le choix et le calcul, par l'ingénieur de projet, des matériaux appropriés (Monjoie *et al.*, 1992) ; ces manuels aussi prennent en compte des cas particuliers comme l'aménagement d'excavations existantes (carrières, etc.) à priori adaptées.

En règle générale, les décharges doivent accueillir quelques milliers, voire dizaines de milliers de mètres cubes de déchets annuellement, d'où une emprise au sol non négligeable, de l'ordre de la dizaine ou centaine d'hectares. Cette exigence, combinée à d'autres dont celle d'imperméabilité du milieu-hôte, et à des considérations d'utilisations des terrains, introduit la nécessité d'un processus de choix de sites sur la base d'une *analyse multi-attributs formalisée* (Maquinay, 1985), dont la cartographie est l'expression finale la plus appropriée.

Une fois les sites candidats identifiés, la réglementation impose la *qualification* de ces derniers, au moyen notamment *d'essais de perméabilité* en place (et/ou au laboratoire sur échantillons). On ne s'étendra pas ici sur les difficultés (théoriques et pratiques) rencontrées dans ce type de mesure, où l'hétérogénéité plus ou moins marquée des formations testées, et leur degré de saturation plus ou moins élevé, doivent être pris en compte. La faiblesse des valeurs à mesurer (jusqu'à 10^{-9} m/s dans un grand nombre de réglementations) impose l'emploi d'appareillages particuliers. De plus, compte tenu du litage plus ou moins marqué des formations étudiées, l'anisotropie des perméabilités horizontale et verticale peut se révéler non négligeable ; c'est pour discriminer une telle particularité que des dispositifs tels le système "TSB" (Two-Stage-Borehole) ont été proposés et mis au point (Boutwell, 1993). Le TSB préconise deux mesures successives d'infiltration dans le même forage, avant et après approfondissement de ce dernier ; le changement de géométrie de l'écoulement permet d'accéder aux deux perméabilités, horizontale et verticale, par une modélisation appropriée des régimes d'écoulement correspondant aux deux étapes de l'essai (fig. 1).

Une tendance croissante des réglementations est de réserver au substratum naturel des décharges un rôle de barrière "passive" (et donc normalement non sollicité par des écoulements de fluides provenant de la décharge susjacente), tandis qu'un certain nombre de couches de matériaux rapportés au-dessus de l'horizon naturel constituent une *barrière "active"*, ou *"complexe d'étanchéité - drainage"* *artificiel*. Schématiquement, ce complexe est constitué d'un système de drains en fond de décharge, surmontant une ou plusieurs couches manufacturées à très faible perméabilité. Ces dernières sont constituées de *géomatériaux*

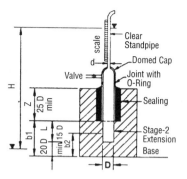

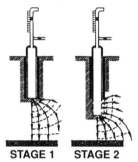

Fig. 1 : Système d'essai "TSB" et concept de la mesure (Boutwell, 1993).

mis en oeuvre selon les techniques du Génie Civil. Les plus communément préconisés et utilisés sont :

1. les sols (argileux) traités par procédés mécaniques (compactage) et/ou chimiques, par exemple des additifs appropriés (Bernhard, 1993) ;

2. les géocomposites, par exemple une couche mince d'argile bentonitique maintenue en sandwich entre deux couches textiles ;

3. les géomembranes ou nappes d'étanchéité en produits de synthèse (polyéthylène, feuille bitumineuse, etc.).

L'ensemble de ces éléments doit être calculé et mis en place, en fond et sur les flancs de la décharge, de manière à accepter un certain tassement sous l'effet du poids des déchets qui seront déversés, et de la compaction progressive du substratum, ainsi qu'une faible charge hydraulique lorsque des lixiviats (eau de percolation météorique, ou eau contenue dans les déchets eux-mêmes) viendront s'accumuler en fond de décharge avant d'être évacués par le système de drainage (Monjoie *et al.*, 1992 ; Gisbert, 1993). La figure 2 est un exemple de tels assemblages de géomatériaux à mettre en oeuvre en fond de décharge.

De manière à restreindre l'arrivée d'eau lors de la mise en place des déchets, l'exploitation des décharges se fait de plus en plus par alvéoles dédiées individuellement à un type de déchets spécifiques. Outre la rationalisation de la mise en décharge, l'avantage de

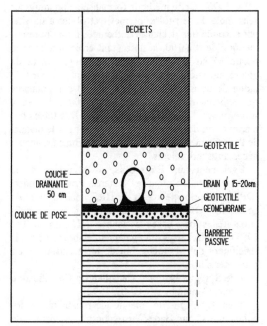

DECHETS

GEOTEXTILE

COUCHE
DRAINANTE
50 cm
○ ○ ○ ○ ○ ○

DRAIN ⌀ 15-20cm
GEOTEXTILE
GEOMEMBRANE
COUCHE DE POSE

BARRIERE
PASSIVE

Fig. 2 : Coupe type dans un dispositif d'étanchéité de fond de décharge (document BRGM).

cette pratique est de permettre l'installation de *toits mobiles*, se déplaçant au rythme du remplissage de l'alvéole (Raffin *et al.*, 1993).

Egalement pour minimiser la production de lixiviats ou jus de décharge, des calculs de *bilan hydrique* sont maintenant nécessaires dès le stade de conception d'une décharge. L'expérience a montré que la simple prise en compte de la pluie efficace n'est pas suffisante ; il s'agit donc de tenir compte des grandeurs caractéristiques du site, et des paramètres d'exploitation de l'ouvrage. Différents codes de calcul sur ordinateur permettent de simuler, pour des pas de temps généralement mensuels, la production d'effluents (ou lixiviats) à évacuer par le système de drainage, selon une équation du type :

$E = P - ETR + ED - I - S \pm R$, dans laquelle :

E est le volume d'effluents à évacuer (m^3)

P est l'eau de pluie,

ETR est l'évapotranspiration,

ED est l'eau apportée ou absorbée par les déchets,

I est l'infiltration dans le substratum,

S est la variation du stock d'eau libre dans les déchets,

± R est le ruissellement entrant ou sortant.

En France, le logiciel BHYDEC mis au point au BRGM (Sauter, 1988) est un exemple de tels outils, ainsi que le modèle HELP aux Etats-Unis (US-EPA, 1984).

Lorsque la quantité de déchets prévue a été mise en place, la décharge doit être munie d'une *couverture*, dont les objectifs peuvent varier selon le type des déchets et la durée envisagée (couverture temporaire ou permanente), mais peuvent généralement se regrouper comme suit (Chapuis, 1993) :

1. minimiser l'infiltration d'eau ;
2. maximiser le ruissellement des eaux météoriques ;
3. réduire l'émission des gaz éventuels vers l'atmosphère, ou maximiser leur récupération ;
4. offrir une barrière entre les déchets et des agents naturels perturbateurs, tels les racines de plantes et/ou les animaux fouisseurs.

Pour les dépôts de déchets miniers, il s'agit en outre de restreindre l'arrivée d'oxygène, de manière à éviter les "drainages miniers acides" déjà évoqués.

La mise en place de *couvertures multicouches*, incluant les mêmes éléments de base que les fonds de décharge (couches drainantes, sols compactés, géosynthétiques), est la solution communément recommandée en vue de la satisfaction des objectifs précédemment évoqués, dont celui de bilan hydrique approprié. La figure 3 présente un exemple de système multicouche ; on remarquera l'importance de la couverture végétale dans le bilan global d'évapotranspiration. Pour optimiser le ruissellement, une certaine pente est donnée à ces couvertures, en forme de dièdres ou de dômes aplatis ; il est donc souhaitable d'inclure, dans leur conception, des considérations relatives à l'érosion de leur surface, à l'infiltration des eaux, et à la stabilité mécanique de l'ensemble (Bonin et Aziz, 1992).

Au moins pour les déchets organiques, la *collecte et le traitement des "jus de décharge"* et du biogaz sont des impératifs pour l'exploitant de telles installations (AGHTM, 1990).

Un *système de surveillance*, comportant au moins quelques piézomètres, est demandé dans la plupart des réglementations actuelles en matière de décharges ; l'échantillonnage de l'eau à intervalle régulier est un moyen (tardif) de détecter une éventuelle fuite de lixiviats, et de prendre les mesures appropriées en vue de la protection des eaux souterraines voisines de la décharge. Cependant, une détection plus précoce de fuites éventuelles sous le système d'étanchéité est un objectif recherché, car permettant une intervention (réparation) plus précoce. De tels systèmes font, par exemple, appel à des mesures géoélectriques : contraste de résistivité, etc. (Furuichi *et al.*, 1993). Dans le même ordre d'idée, l'examen de *sites-pilotes* ou *d'installations expérimentales* permet d'accéder au comportement, en vraie grandeur et sur de longues durées, des géosynthétiques et autres systèmes d'étanchéité, et aussi des systèmes de drainage parfois sujets au colmatage par dépôts d'incrustations diverses (Düllmann *et al.*, 1993 ; Le Tellier *et al.*, 1993).

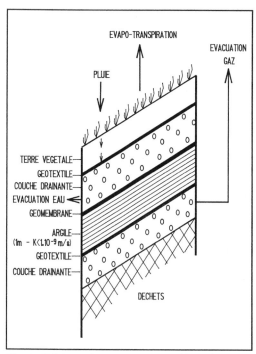

Fig. 3 : Schéma de couverture multicouche pour centre d'enfouissement de déchets (document BRGM).

2.3 La réhabilitation des décharges défectueuses

Certaines décharges anciennes pour déchets ménagers ou industriels toxiques sont maintenant à l'origine de pollutions de nappes phréatiques, par infiltration, dans le sous-sol, de lixiviats improprement ou pas du tout collectés. Ces décharges font partie des *sites pollués* recensés en nombre croissant, héritages d'activités industrielles ou urbaines peu soucieuses de l'environnement. Après les Etats-Unis dès les années 70, de nombreux pays se sont lancés dans des politiques de *réhabilitation* de tels sites (ENPC, 1993), dont le préalable est un *recensement* le plus complet possible des décharges défectueuses, dépôts incontrôlés de déchets, etc.

Compte tenu du nombre élevé de ces sites, et des budgets forcément limités disponibles pour leur réhabilitation, une étape souhaitable consiste en la *hiérarchisation* des sites pollués en fonction des risques plus ou moins grands qu'ils font courir à la santé des populations voisines, ou aux milieux naturels fragiles. Aux Etats-Unis, l'Agence de Protection de l'Environnement (US-EPA) a probablement été la première à élaborer un outil systématique de hiérarchisation des risques ou dangers liés à la pollution des sites, appelé HRS - Hazard Ranking System (US-EPA, 1990). Cette méthode formalisée, appliquée sur l'ensemble du territoire, permet d'attribuer à un site, par exemple une décharge défectueuse, une note numérique de 0 à 100, la note étant croissante avec le risque lié au site. Cette note est la résultante de quatre notations partielles concernant chacune le mode de transport (réel ou potentiel) des polluants respectivement par les eaux de surface, les eaux souterraines, l'air, et le sol lui-même. Des tables numériques ou des formules sont fournies par le manuel de l'utilisateur pour le choix des différentes grandeurs nécessaires aux calculs.

Egalement aux Etats-Unis, le Ministère de la Défense a mis au point le "Defense Priority Model" (DPM), de principe voisin du HRS, mais distinct de ce dernier ; DPM est utilisé surtout sur les sites militaires telles des bases aériennes (Hushon *et al.*, 1993). D'autres pays ont également élaboré des outils spécifiques dans le même but de hiérarchisation ; on peut citer ainsi :

1. le Système National Canadien de Classification des Lieux Contaminés (CCME, 1992) ;

2. en Allemagne, le "manuel de réhabilitation des points noirs" du Bade-Wurtemberg" (Ministère de l'Environnement du Bade-Wurtemberg, 1988), et le "guide des points noirs" appliqué en Bavière (Ministère bavarois de l'Intérieur, 1991).

On pourrait attendre de telles méthodes qu'elles fournissent la même hiérarchisation lorsqu'appliquées au même ensemble de sites pollués tels des décharges défectueuses. De récentes études comparatives ont cependant montré que ce n'est pas le cas ; cette différence des verdicts est le reflet des modes de notation et de calcul, extrêmement variables d'une méthode à l'autre (Hushon *et al.*, 1993 ; Côme *et al.*, 1993). Compte tenu des enjeux et de la taille du problème, par exemple à l'échelle de l'Union Européenne, un effort de rationalisation et d'homogénéisation en ce domaine constituerait sans doute un investissement vite rentabilisé, auquel la Géologie de l'Ingénieur devrait efficacement contribuer.

Lorsqu'il s'agit d'*évaluer le risque* pour la santé lié à un site pollué spécifique, une *modélisation détaillée* des processus de relargage des polluants, de leur transport par les eaux souterraines, et de leur ingestion par les populations voisines, peut être envisagée. Les doses de polluants ingérés sont converties en indices de danger pour la santé, ou en risques de cancer lorsque les données éco-toxicologiques nécessaires existent. La méthode MEPAS (Multimedia Environmental Pollutant Assessment System) a ainsi été mise au point par le Ministère de l'Energie américain dans le but d'évaluer les risques de sites de décharges défectueuses, et aussi de hiérarchiser de tels sites en vue de leur réhabilitation (Hushon *et al.*, 1993). L'intérêt de ce type d'outils détaillés est aussi de pouvoir tester, de façon prévisionnelle, l'efficacité de différentes techniques de réhabilitation en vue de la réduction du risque associé (Stephanatos *et al.*, 1989).

Une fois définis les objectifs de réhabilitation, les *techniques* les plus adaptées peuvent être choisies dans la vaste panoplie disponible et déjà testée. L'*enlèvement* des déchets les plus dangereux et leur transport vers un centre de stockage approprié, constituent la solution la plus élémentaire, qui n'est cependant pas toujours compatible avec la santé du personnel chargé de ces opérations. S'il s'agit de plus de forts tonnages, l'*amélioration du confinement* des déchets laissés en place devient alors l'option à privilégier. De nombreux exemples attestent de la faisabilité de ce type d'opérations (Ouvry *et al.*, 1993).

La *construction de parois étanches verticales*, ceinturant le stockage défectueux, et descendant jusqu'à une couche sous-jacente imperméable, est l'une des techniques les plus anciennes et les mieux étayées (SIMS, 1993). Plusieurs principes de base sont envisageables :

1. excavation du sol en place et remplissage par un matériau étanche (béton plastique, etc.) ;

2. déplacement (refoulement) du sol en place et incorporation d'un matériau étanche (coulis, palplanches) ;

3. réduction de la perméabilité du sol en place. L'injection, le "jet-grouting" et la congélation (cette dernière généralement temporaire) entrent dans cette catégorie.

Pour des barrières hydrauliques ordinaires, seules la perméabilité et la résistance mécanique entrent en ligne de compte. Dans le cas de lixiviats parfois agressifs, il faut également s'assurer de la pérennité des matériaux utilisés (exemple : éventuelle corrosion des palplanches, etc.) ; l'incorporation d'une membrane en polyéthylène à l'intérieur d'une paroi au coulis peut se révéler intéressante (Duplaine *et al.*, 1993).

D'autres développements visent à améliorer la capacité de rétention des matériaux, constitutifs des enceintes, vis-à-vis des polluants relargués par le site à confiner. C'est ainsi que des coulis spéciaux associent les mécanismes de piégeage d'ions métalliques, telles la précipitation et l'adsorption (Gouvenot *et al.*, 1993).

Des dispositifs d'étanchéité/drainage subhorizontaux ou sur pentes permettent en outre de reconstituer une *couverture étanche* sur des sites de stockage défectueux, éventuellement raccordée aux enceintes verticales évoquées précédemment. Les matériaux et techniques utilisés sont voisins de ceux déjà présentés à propos de la construction de centres de stockage sûrs ; c'est ainsi qu'on peut trouver de haut en bas de ces couvertures :

1. la couche de terre végétalisable ;

2. un géocomposite de drainage des eaux de surface ;

3. une géomembrane assurant l'étanchéité (par exemple une feuille de polyéthylène) ;

4. un géocomposite de drainage des gaz éventuels ;

5. une couche d'assise en sable ;

6. un géotextile pour la filtration et la séparation d'avec les déchets confinés.

L'avantage de ces structures composites peut résider dans leur facilité de mise en oeuvre, et de réaliser, à épaisseur réduite, la même performance en terme d'étanchéité qu'une épaisse couche de matériaux naturels. L'efficacité de ces couvertures multicomposantes dépend cependant très fortement de la qualité de leur dimensionnement et de leur mise en place.

Dans certains cas particulièrement favorables de résidus solides, situés non loin de matériaux argileux naturels appropriés, l'*encapsulage intégral ou "mise en tombeau étanche"* peut être envisagé (fig. 4).

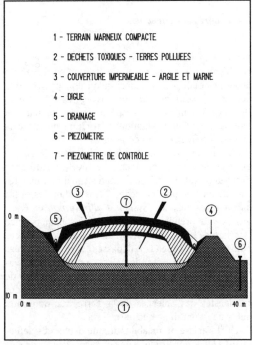

1 - TERRAIN MARNEUX COMPACTE

2 - DECHETS TOXIQUES - TERRES POLLUEES

3 - COUVERTURE IMPERMEABLE - ARGILE ET MARNE

4 - DIGUE

5 - DRAINAGE

6 - PIEZOMETRE

7 - PIEZOMETRE DE CONTROLE

Fig. 4 : Exemple de "mise en tombeau" de résidus toxiques dans l'Est de la France (Risler, 1993).

Un exemple de tel édifice a été réalisé dans l'Est de la France, montrant les étapes suivantes (Risler, 1993) :

1. fabrication d'une cuvette étanche destinée à recevoir les déchets et matériaux pollués (mise en forme du substratum, d'un drainage, de couches successives d'argile au dameur et au cylindre vibrant ; mesures de contrôle de perméabilité) ;

2. rassemblement sur la cuvette des déchets et produits pollués ;

3. fabrication d'une couverture étanche associant plusieurs couches d'argile ;

4. mise en place d'une couverture de terre végétale ;

5. mise en place de drains périphériques et de piézomètres de contrôle.

En outre, pour renforcer l'efficacité de ces dispositifs de confinement qualifiés de "passifs", il est parfois proposé d'installer des réseaux de forages créant, par pompages, un *confinement hydraulique* qui permet de bloquer d'éventuels panaches de polluants (Soyez, 1994) ; ces pompages peuvent aussi assurer un confinement provisoire avant la construction des parois verticales et couvertures précédemment décrites. Enfin, des phénomènes tels l'électro-osmose sont à la base de nouveaux concepts de confinement de polluants ioniques, dont la mise en oeuvre dépasse actuellement le stade du laboratoire (Yeung, 1993).

2.4 *Progrès envisageables*

Les réglementations environnementales font évoluer notablement le stockage de déchets en surface ou à faible profondeur ; c'est ainsi qu'on doit passer de la décharge (traditionnelle) à un ouvrage (ou "centre de stockage") soumis à des spécifications de plus en plus strictes. Cette mutation se traduira par un certain nombre d'exigences de progrès.

En ce qui concerne le *choix des sites*, le simple examen de la géologie locale se révèle insuffisant ; c'est à une évaluation "multicritères" qu'il faut maintenant de plus en plus procéder. Dans ce but, les outils informatisés tels les *Systèmes d'Information Géographiques* prendront de plus en plus d'importance (Dörhöfer, 1993). La valeur des conclusions de ces outils ne saura cependant dépasser celles des *données* introduites ; pour ces dernières, un effort de validation et d'assurance de leur qualité deviendra de plus en plus nécessaire. Loin de réduire le rôle des spécialistes des Sciences de la Terre, ces nouveaux outils ne feront donc que le renforcer.

Qu'il s'agisse de réhabilitation de décharges défectueuses ou de construction de nouveaux centres de stockage, les *géomatériaux* mis en oeuvre, naturels ou artificiellement améliorés, seront soumis à des exigences de qualité de plus en plus strictes. Outre leur *contrôle-qualité* à la production et lors de la mise en place, il faudra prévoir l'étude de leur évolution en terme de *vieillissement* et de longévité, notamment en présence de lixiviats souvent agressifs. Un vaste champ de recherche s'ouvre donc dans ce domaine, encore mal exploré ; là encore, la Géologie de l'Ingénieur possède les savoirs et les techniques nécessaires.

Egalement utilisable pour les nouveaux ouvrages, ou pour l'évaluation de travaux de réhabilitation, la *modélisation* détaillée des processus de relargage, de transport et d'ingestion des polluants est appelée à se développer ; elle implique une meilleure compréhension des mécanismes et des données d'entrée validées pour les calculs. En retour, la fiabilité des prévisions ne pourra être établie que par la *mesure* et le *contrôle*, sur le site, des principales grandeurs caractéristiques du fonctionnement du système (concentration d'espèces en solution, pH, etc.). La mise au point de *capteurs* stables et résistants aux agents agressifs devra être un objectif primordial des progrès en métrologie dans ce domaine.

Enfin, le *concept de décharge* lui-même est amené à évoluer ; plusieurs directions sont évoquées actuellement, et on se limitera ici à un petit nombre d'entre elles.

Reconnaissant qu'aucune barrière (naturelle ou artificielle) n'est absolument et indéfiniment étanche, certaines législations - dont celle de la France - préconisent en outre la réduction du caractère nuisible des déchets, par leur stabilisation (mécanique) et leur inertage (chimique) avant enfouissement. Dans ce concept, le déchet lui-même n'est d'ailleurs plus qu'un *résidu "ultime"*, sans réemploi technique et économiquement envisageable, tels une cendre d'incinération de déchet industriel, ou un résidu de dépollution (Militon, 1993). Il peut s'agir là d'une autre version du concept "multi-barrières" appliqué au stockage. La mise en place de déchets en blocs facilement manipulables, car de résistance mécanique adéquate, peut également favoriser leur *récupérabilité* ultérieure, au cas où une intervention deviendrait nécessaire sur le centre de stockage (au prix, éventuellement, de la destruction au moins partielle de la couverture).

Une autre option proposée est celle de rendre à la décharge moderne son rôle de réacteur ou de *bioréacteur*, en permettant à des déchets non traités d'évoluer jusqu'à leur point d'équilibre géochimique avec le milieu environnant (Joseph *et al.*, 1993 ; Marsily (de), 1993). Selon ce concept, il s'agirait essentiellement d'extraire les éléments mobiles des déchets, en utilisant les géomembranes et drains pour récupérer les lixiviats ; ces derniers seraient traités avant rejet dans le milieu naturel, jusqu'à innocuité des produits stockés. L'objection que soulève ce système est bien sûr la durée du traitement nécessaire, qui peut être très longue en cas de déchet à évolution lente.

Par opposition à ces développements de nature "technologique", car concernant les déchets avant stockage ou leur traitement en place, une autre solution à l'élimination des déchets dangereux consiste à *accroître l'importance de la barrière géologique*, par enfouissement en souterrain. Cette technique fait l'objet des paragraphes suivants.

3 LE STOCKAGE SOUTERRAIN DES DECHETS TOXIQUES ET DANGEREUX

3.1 *Généralités*

Compte tenu de l'emprise en surface nécessaire pour une décharge ou centre de stockage (y compris son périmètre de protection), l'implantation de tels ouvrages peut se révéler délicate, par exemple dans des régions à topographie accidentée. Par ailleurs, certains déchets industriels toxiques (par exemple, les sels résultant du traitement de surface des métaux) sont reconnus comme solubles, ce qui les rend peu aptes au stockage en surface. L'option du stockage de certains déchets solides dans des cavités ou galeries étanches, existantes ou spécialement créées dans les profondeurs du sous-sol, paraît donc compatible avec ces contraintes. Cependant, comme le coût du stockage en souterrain excède en général celui de la simple mise en décharge, cette option est à réserver, en première analyse, à des quantités pas trop importantes de déchets hautement toxiques, qui peuvent bénéficier ainsi d'une protection accrue de la part d'une barrière géologique de grande épaisseur. La toxicité de ces substances, concentrées et confinées, pose à son tour le problème de la sûreté (à long terme) de ce genre de stockage.

Ces considérations ont fait récemment l'objet des travaux et réflexions de la Commission n° 14 de l'Association Internationale de Géologie de l'Ingénieur (AIGI, 1989). La sûreté d'un stockage souterrain de déchets toxiques et dangereux serait obtenue si les conditions suivantes sont convenablement remplies :

1. choix judicieux du site de stockage : zone tectoniquement calme, à géologie simple, montrant des circulations d'eau souterraine très faibles voire nulles ;

2. emploi d'un système multi-barrières de confinement des déchets : par exemple conditionnement en fûts, remblayage des cavités de stockage après la phase d'exploitation, scellement des voies d'accès tels puits et/ou galeries inclinées ;

3. étude des scénarios envisageables pour le relargage de substances toxiques vers l'homme, en phase d'exploitation et après fermeture, et prise en compte des conclusions pour la sélection et la mise en oeuvre des barrières multiples évoquées précédemment.

Du point de vue géologique, les exigences concernant les sites potentiels de stockage sont donc, en particulier, la très faible perméabilité des formations susceptibles d'abriter les cavités ou galeries, et une résistance mécanique compatible avec la stabilité, à court et long terme, de ces dernières.

Les formations salines (sel en couche ou en dômes) étant, de par leur existence même, la preuve d'une absence quasi totale de circulations d'eaux

souterraines en leur voisinage, ce type de milieu-hôte potentiel a le plus souvent été considéré comme adéquat pour le stockage souterrain des déchets toxiques ; l'essentiel du rapport de la Commission AIGI n° 14 s'y réfère d'ailleurs explicitement. Cependant, s'il est souhaité de pouvoir récupérer les déchets à tout instant (donc de réaliser un entrepôt souterrain plus qu'une élimination définitive), des cavités en milieu rocheux très peu perméable (anhydrite, marnes, granite très peu fracturé) peuvent aussi être envisagées.

Si la mise en décharge (ou en centre de stockage de surface) est une pratique courante en matière d'élimination des déchets, l'utilisation de cavités souterraines, existantes ou spécialement créées, est beaucoup moins répandue ; il n'existe en effet que peu d'exemples d'installations en fonctionnement (Barrès *et al.*, 1991 ; ENPC, 1990).

3.2 *Exemples de réalisations et nouveaux concepts*

Comme évoqué plus haut, la *conversion*, en stockage, de *mines de sel existantes* semble l'option la plus compatible avec les exigences de sûreté et de coût. Un exemple caractéristique est celui de l'installation d'Herfa-Neurode, située dans les quartiers exploités de la mine Wintershall à Heringen-Werra (Land de Hesse) en Allemagne (Barrès *et al.*, 1991).

La formation salifère du Zechstein est, sur ce site, épaisse de 300 m ; la couverture comporte quatre couches d'argile d'une épaisseur cumulée d'environ 100 m. Les quartiers de la mine convertis en stockage pour déchets industriels toxiques sont situés à environ 800 m de profondeur. L'exploitation se fait par la méthode des chambres et piliers, sur 2,5 à 3 m de hauteur, avec des galeries de 15 à 20 m de large et un taux de dépilage moyen de 60 %, compatible avec la stabilité mécanique de l'édifice. Avant mise en place de déchets, le toit des galeries est purgé de ses éventuels blocs instables, puis boulonné.

Les déchets solides sont transportés sur le site sous forme de fûts d'acier de 200 litres, hermétiquement clos, et disposés sur palettes. Selon leurs caractéristiques, ils sont mis en place dans différentes parties du dépôt. Aucun produit explosif, instable, ou susceptible d'émettre des gaz, n'est accepté dans le stockage. Les groupes de produits acceptés sont les suivants : cyanures ; composés d'arsenic, de mercure ; résidus de galvanisation, de distillation ; condensateurs et transformateurs ayant contenu des PCB ; résidus d'incinération, de filtration, d'évaporation ; goudrons ; batteries ; substances médicamenteuses. Depuis sa mise en service en 1972, l'installation d'Herfa-Neurode a reçu en moyenne 40 000 t/an de ces types de déchets, strictement contrôlés à leur arrivée. Lorsqu'une zone de stockage est totalement remplie, elle est fermée par un cloison de parpaings. L'archivage complet de tous les fûts permet la reprise éventuelle de certains d'entre eux pour réutilisation de leur contenu (ce qui fut le cas pour 15 000 d'entre eux).

Après remplissage d'une zone de stockage (soit 5 ans d'exploitation), celle-ci est isolée des autres compartiments par une double muraille de briques, avec remplissage du vide intérieur au moyen de béton. Lorsque l'ensemble de l'installation sera rempli, elle pourra être convertie en dépôt définitif par rebouchage et scellement approprié des quatre puits d'accès.

Egalement en Allemagne, les autorités du Land de Rhénanie-Westphalie ont examiné la possibilité de *convertir des mines de charbon* de la Ruhr en dépôts pour résidus d'incinération, cendres volantes, de diverses origines industrielles (Striegel, 1993). Dans ce cas, les résidus seraient mis en place sous forme de pulpe susceptible de solidifier, assurant ainsi le remplissage et la stabilisation des vides existants. Une étude de pré-faisabilité a permis d'identifier les critères d'adéquation des mines en vue de leur conversion éventuelle en stockage :

1. profondeur minimale : 800 m ;
2. épontes du charbon à forte teneur en argile et faible perméabilité ;
3. distance minimale entre les sites de dépôt et les autres voies d'accès de la mine ;
4. repérage préalable de toutes les zones perméables et/ou aquifères ;
5. existence de propriétés de sorption des épontes vis à vis des polluants stockés ;
6. présence d'une épaisse couche de marnes (200-300 m) au-dessus des couches de charbon.

Trois mines de la Ruhr satisfaisant à ces critères font actuellement l'objet d'une demande d'autorisation en vue de leur conversion en stockage de tels résidus.

La *construction de cavités spécifiques au rocher*, par opposition à la conversion de vides existant, implique un investissement initial peut-être important, mais qui peut se révéler intéressant lorsqu'aucune autre solution de stockage en surface n'est possible, faute d'espace disponible par exemple, pour des déchets dangereux. Si un massif rocheux de qualité appropriée est disponible, l'option "cavités spécialement minées" est alors à considérer. C'est le cas de l'installation du Sörfjorden en Norvège (Aarvoll *et al.*, 1986).

Dans ce cas, une usine de zinc située près d'un fjord à rives escarpées, produit annuellement environ 65 000 m³, soit 70 000 t, de résidus solides (jarosite). Ces déchets autrefois déversés dans le fjord ont fini par créer un sérieux problème de pollution. A la demande des autorités, la société exploitante a donc étudié la possibilité de stocker ces résidus dans de grandes cavernes excavées directement sous le versant du fjord, haut d'environ 1 500 m à cet endroit.

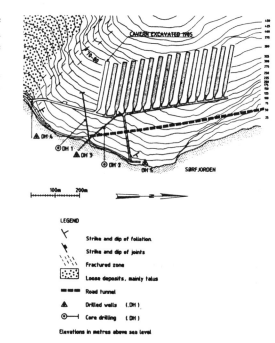

Fig. 5 : Implantation des cavernes de stockage à Sörfjorden, Norvège (Aarvoll *et al.*, 1986).

L'étude a abouti à la conception de cavernes de volume unitaire total 65 000 m³, longueur totale 211 m, section 17,5 x 23,5 m, séparées par un "pilier" de 20 m (fig. 5).

En préalable à l'excavation, les études géologiques ont comporté en particulier la réalisation et l'examen de trois forages, et des essais hydrauliques sur l'un des forages ; des mesures de contrainte in-situ ont suivi l'excavation de la première caverne. La roche s'est révélée partout de très bonne qualité (RQD ≥ 90 %, résistance uniaxiale 145 MPa) ; la faible perméabilité du massif limite à 15 m³/jour l'exhaure dans chaque caverne, même sans travaux d'injection.

De telles cavités ou cavernes pour résidus dangereux sont également étudiées en Autriche, pour les mêmes raisons de manque d'espace en surface dans les régions montagneuses, et de l'existence de roches dures (granites). Le concept proposé est celui de cavernes situées sous un flanc de montagne et au-dessus de la nappe phréatique, pouvant accueillir des déchets solides (avec emballage), moyennement toxiques, éventuellement récupérables ; les cavités doivent donc être accessibles, jusqu'à une décision de transformer l'installation en dépôt définitif, par remplissage des vides au moyen de béton. Les critères de choix des sites sont proposés comme suit (Neubauer, 1992) :

1. un bon drainage de surface, et de faibles précipitations ;

2. un recouvrement rocheux d'épaisseur inférieure à 400 m ;

3. une couche superficielle sédimentaire jouant le rôle de barrière pour une éventuelle rétention des polluants ;

4. une roche-hôte (granite) très peu perméable ($\leq 10^{-8}$ m/s), résistance à la compression > 50 MPa, module d'Young $\geq$ 20 GPa ;

5. une zone peu sujette à des influences sismiques.

Un tel système permettrait de garantir le confinement des polluants pendant 10 000 ans.

Dans le même ordre d'idées, le concept de *stockage en galeries sèches* et "à niveau" (c'est-à-dire sous un flanc de colline, sans puits d'accès vertical) a été récemment proposé en France (Cottez *et al.*, 1993). Il s'agit alors de réaliser les galeries au tunnelier, sous recouvrement moyen (de l'ordre de 200 m), dans des roches suffisamment compétentes et peu perméables pour ne pas nécessiter de soutènement : marnes, calcaires massifs, roches cristallines peu fracturées. L'utilisation du tunnelier implique de minimiser les démontages de la machine, ce qui peut se réaliser par un dépôt en spirale (fig. 6). Le coût de l'excavation devrait être alors notablement réduit, ce qui rendrait cette option intéressante par rapport à une décharge classique.

Enfin, l'enfouissement définitif de déchets (sous forme de coulis ou de granulés) dans des *cavités lessivées dans le sel* a également été proposé (Dusseault, 1993a). Là encore, l'utilisation de cavités existantes (par exemple, résultant de l'exploitation de sel par dissolution), ou la création de cavités spécifiques, sont envisageables. Dans les deux cas, le volume de stockage est le résultat de l'injection d'eau douce dans un forage, et de son pompage sous forme de saumure plus ou moins saturée. Lorsque la cavité a atteint sa taille finale, il serait possible de remplir la cavité de déchets pâteux ou granulaires, susceptibles de se solidifier en masse, et plus denses que la saumure, de telle façon que cette dernière soit progressivement remplacée par les déchets ; cette procédure évite la vidange totale de la cavité et assure une meilleure stabilité mécanique.

A la fin du remplissage, le puits d'injection pourrait être scellé définitivement par plusieurs bouchons de béton.

Par rapport aux critères de sûreté communément évoqués, en particulier ceux de la Commission de l'AIGI (AIGI, 1989), ce mode d'élimination semble cumuler un grand nombre de caractéristiques favorables ; en particulier, l'accès à la cavité se limite à un forage facile à obturer, ce qui minimise les perturbations de la barrière géologique. Paradoxalement, cette difficulté d'accéder aux déchets une fois mis en place, et donc la quasi-impossibilité de les récupérer,

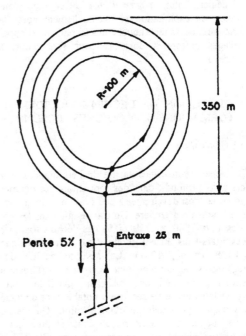

Fig. 6 : Concept de stockage souterrain à niveau et en spirale, sous flanc de montagne (Cottez *et al.*, 1993).

pourrait être la raison pour laquelle ce mode d'enfouissement reste souvent, au moins actuellement, au stade du concept.

3.3 *Progrès envisageables*

S'il est possible, comme on l'a vu dans la section précédente, de réhabiliter des décharges défectueuses (au prix de travaux plus ou moins longs et/ou coûteux), une telle opération apparaît beaucoup plus délicate, voire impossible, pour un dépôt souterrain de déchets toxiques, définitivement scellé, qui serait à l'origine d'une contamination notable de l'environnement, suite à une conception et/ou une réalisation incorrecte.

Les spécialistes de la mécanique des roches, de l'hydraulique souterraine, et de la géochimie, doivent donc associer leurs compétences pour mettre au point les *techniques d'études des sites* potentiels, qui permettront d'en vérifier l'adéquation par rapport aux critères de sûreté à court et long terme. Il s'agira également d'établir la conformité des travaux (excavation des stockages, mise en place des déchets, fermeture et, éventuellement, scellement définitif) avec les prescriptions en la matière, lorsqu'elles existent ; des *moyens de contrôle non-destructifs* seront à privilégier. Enfin, la *surveillance* de ce type d'ouvrages, en exploitation et éventuellement après la fermeture, devra faire appel à des techniques et appareillages appropriés, dont beaucoup sont encore à concevoir.

4 L'INJECTION DE DECHETS LIQUIDES EN FORMATIONS GEOLOGIQUES PROFONDES

4.1 *Généralités*

La technique de l'injection de déchets liquides, par des puits profonds, dans des formations géologiques poreuses, s'est développée à peu près en même temps que l'industrie pétrolière. On estime que dans les années 1970-1980, près de 60 % des déchets toxiques des Etats-Unis étaient éliminés par cette technique (Barrès *et al.*, 1991). La sévérité accrue des règlements a entraîné une certaine désaffection envers cette méthode ; mais, en parallèle, cette nouvelle exigence de sécurité a notablement diminué les risques potentiels de contamination. On assiste donc, depuis peu, à un regain d'intérêt pour cette option d'élimination, au moins aux Etats-Unis (Silliman, 1993). Dans d'autres pays, l'injection n'est autorisée que pour des déchets pour lesquels aucun autre mode d'élimination n'est techniquement ou économiquement envisageable.

Une de ses caractéristiques est en effet d'utiliser une barrière géologique la plus épaisse possible, à l'abri de laquelle le déchet liquide imprégnera les pores de la couche-hôte sélectionnée ; la possibilité de récupérer ces produits est donc pratiquement nulle.

Les *roches-hôtes* normalement envisageables pour cette technique sont les grès, les calcaires, les dolomies. Les zones adéquates doivent répondre aux critères hydrogéologiques et géologiques suivants :

1. la couche poreuse constituant le réservoir ne doit pas avoir valeur de ressource, ni contenir de l'eau susceptible d'être utilisée pour la consommation humaine, ni être utilisable pour le stockage d'hydrocarbures ou d'énergie géothermale ;

2. la roche réservoir (et ses fluides initialement inclus) doit être compatible chimiquement avec les déchets injectés ;

3. la superficie, le volume et la porosité de l'horizon-hôte doivent être compatibles avec le volume d'injection envisagé ;

4. la formation-hôte doit être suffisamment bien connue pour que le comportement des fluides injectés (répartition des pressions et débits) puisse être prévu ;

5. sa perméabilité doit être au moins égale à 10^{-6} m/s ;

6. la zone d'injection doit être limitée latéralement, à son toit, et à sa base, par des formations suffisamment épaisses et peu perméables pour constituer des barrières efficaces à la migration des déchets ;

7. la roche-réservoir doit être localisée dans une zone de faible activité sismique, et dépourvue de failles et/ou d'activité volcanique ;

8. s'il existe, le gradient hydraulique doit être connu pour que l'on puisse déterminer la direction d'écoulement.

Des critères plus quantitatifs ont été édictés par l'Agence américaine de Protection de l'Environnement (Silliman, 1993). En particulier, l'injection de déchets en formation poreuse profonde peut être autorisée si :

1. pendant 10 000 ans, les fluides injectés ne migrent pas hors de la zone de stockage, selon la verticale, ou latéralement dans l'horizon-hôte, jusqu'à un contact avec un aquifère destiné à l'alimentation en eau potable (c'est-à-dire contenant moins de 10 mg/l de solides dissous totaux) ; ou bien si,

2. les constituants des déchets ont perdu leur caractère dangereux (par exemple, par dégradation ou immobilisation des éléments toxiques dans l'horizon de stockage), avant qu'ils ne migrent jusqu'à une zone de contact avec un aquifère destiné à l'alimentation en eau potable.

Les *types de déchets* admissibles doivent posséder une viscosité suffisamment faible, et une teneur plutôt faible en solides sous forme de suspension. Il s'agit essentiellement de solutions acides ou alcalines diluées, de substances organiques (aldéhydes, alcools, phénols, glycols, etc.) et de déchets issus des champs pétroliers ou gaziers (saumures).

Les puits d'injection sont bien sûr l'équipement essentiel d'un tel ouvrage. Le débit souhaité pour l'injection est fonction du diamètre du puits. En règle générale, l'injection est réalisée par un tubage central descendant jusqu'au niveau de l'horizon-hôte ; l'espace annulaire entre ce tubage et le "casing" extérieur est scellé par un ou plusieurs obturateurs, l'étanchéité étant contrôlée par un fluide dont la pression est maintenue constante et enregistrée (fig. 7). En fin d'exploitation, le puits doit être obturé pour éviter toute remontée des fluides injectés vers des horizons perméables moins profonds.

Des déclenchements d'activité sismique accompagnant ou suivant l'injection massive et localisée de fluides ont été décrits en particulier aux Etats-Unis ; l'exemple de Denver dans les années 1960 (1 500 séismes de 1962 à 1969, magnitude d'environ 3) est parmi les plus connus (Grasso *et al.*, 1992). L'éventualité de telles conséquences doit être prise en compte au même titre que celle de fuites de déchet injecté, au-delà de l'horizon-réservoir.

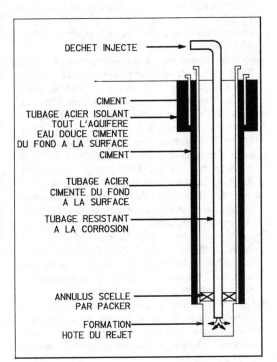

DECHET INJECTE

CIMENT
TUBAGE ACIER ISOLANT
TOUT L'AQUIFERE
EAU DOUCE CIMENTE
DU FOND A LA SURFACE
CIMENT

TUBAGE ACIER
CIMENTE DU FOND
A LA SURFACE

TUBAGE RESISTANT
A LA CORROSION

ANNULUS SCELLE
PAR PACKER

FORMATION
HOTE DU REJET

Fig. 7 : Schéma d'un puits d'injection de déchets liquides en formations profondes (Barrès *et al.*, 1991).

4.2 *Exemples actuels et nouveaux concepts*

Aux Etats-Unis, les puits d'injection dits "de classe 1" étaient, en 1990, au nombre de 185, la plupart d'entre eux situés près de la côte du Golfe du Mexique et dans la région des Grands Lacs. Dans les années 1980, on injectait jusqu'à 3,8 Mm³/an de déchets liquides dans la région du Golfe du Mexique (Barrès *et al.*, 1991 ; Silliman, 1993). Au Canada, les nappes profondes du Saskatchewan ont également reçu plusieurs milliards de mètres cubes d'effluents provenant de l'industrie pétrolière. En Europe, la majorité des puits en exploitation se trouve au nord de l'Allemagne (Barrès *et al.*, 1991).

Une variante de cette méthode a été récemment proposée et testée au Canada : il s'agit de l'injection de déchets faiblement toxiques (sous forme de suspension à grains fins) par *fracturation hydraulique*, dans des horizons poreux tels des grès moyennement profonds (entre 100 et 1 000 m), très perméables et bien isolés par des couches argileuses (Dusseault, 1993b). Plusieurs milliers de mètres cubes de sables légèrement chargés en pétrole ont ainsi été injectés dans l'Alberta entre 1989 et 1991, dans un grès de porosité 30 %, à 690 m de profondeur.

4.3 *Progrès envisageables*

Plus encore que pour les options de stockage précédemment décrites, tout développement dans ce domaine est conditionné par les progrès dans *l'évaluation fiable de la sûreté à long terme* du système constitué par le réservoir poreux et les déchets emmagasinés. C'est donc le domaine de l'hydrogéologie des réservoirs, qui peut s'appuyer sur les expériences et compétences pétrolière, gazière, et de l'énergie géothermique. Plus spécifiquement, les avancées souhaitables doivent concerner (Lawrence Berkeley Laboratory, 1994) :

1. les méthodes de caractérisation des formations-réservoirs profondes, du point de vue de leur géométrie, perméabilité, etc., ainsi que les techniques de surveillance des zones d'injection ;

2. la compréhension des phénomènes d'interaction entre roche-hôte (et fluides d'origine) et déchets injectés ;

3. la mise au point de matériaux résistants pour l'équipement des puits.

5 CONCLUSIONS GENERALES ET PERSPECTIVES

Initialement, le stockage des déchets par mise en décharge n'a que peu concerné les géologues et hydrogéologues, et seulement dans leurs domaines de compétences d'origine, en vue du choix de sites peu perméables et suffisamment éloignés des centres d'habitation.

L'évolution industrielle et démographique, la prise de conscience environnementale généralisée et les exigences légales et administratives, imposent maintenant une multitude d'interventions de la Géologie de l'Ingénieur en matière de stockage (provisoire ou définitif) des déchets. Cette dernière technique s'est en effet diversifiée : création et/ou réhabilitation de centres de stockage de surface pour déchets ménagers ou industriels, dépôts souterrains, et aussi injections profondes d'effluents industriels ; les Sciences de la Terre jouent un rôle fondamental lors des étapes suivantes de la vie de ces installations :

1. choix et qualification des sites ;

2. construction et exploitation des installations (y compris le contrôle de la qualité des matériaux) ;

3. contrôle et surveillance des impacts éventuels sur l'environnement ;

4. fermeture et abandon éventuel des ouvrages.

Dans une perspective de protection accrue de l'environnement à l'aide de systèmes multi-barrières de confinement des déchets, au fonctionnement optimisé en vue d'objectifs de performances de mieux en mieux spécifiés, la Géologie de l'Ingénieur peut fournir une vision intégrée des milieux concernés par les stockages (le sol, le sous-sol et les eaux souterraines). Pour maintenir et accroître le niveau de compétences nécessaires, des développements importants sont attendus dans les domaines suivants :

1. la cartographie et les systèmes d'information géographiques ;

2. les techniques de caractérisation des sites, selon diverses gammes de profondeur ;

3. le comportement des géomatériaux et son influence sur leur mise en oeuvre ;

4. l'étude du vieillissement et de la longévité des géomatériaux (agents stabilisateurs pour déchets, géosynthétiques pour barrières ouvragées, etc.) ;

5. les modélisations numériques couplées des éventuels transferts de polluants à partir des stockages ;

6. la télé-mesure et la surveillance (acquisition, traitement, restitution des grandeurs mesurées).

En s'appuyant sur ces développements théoriques et pratiques, la Géologie de l'Ingénieur pourra permettre une meilleure application des spécifications réglementaires en matière de stockage de déchets, lorsqu'elles existent ; dans les pays où de telles spécifications n'existent pas encore, il est à souhaiter que la Géologie de l'Ingénieur contribue à leur élaboration, sur des bases scientifiques et techniques les plus saines possibles.

REMERCIEMENTS

L'auteur remercie Michel Barrès (BRGM - Service Géologique National) pour ses conseils et la documentation fournie. Les opinions exprimées ici sont propres à l'auteur et ne coïncident pas nécessairement avec celles d'organismes nationaux ou internationaux.

REFERENCES

Aarvoll, A., Barbo, T.F., Hansen, R., Lövholt, J.V. 1986. Storage of industrial waste in large rock caverns. *Proc. Int. Symp. on large rock caverns*, Helsinki.

Arnould, M., Furuichi, T., Koide, H. (eds.) 1993. Management of hazardous and radioactive waste disposal sites. Papers presented at the International Symposium n° I-3-50, *29th International Geological Congress*, Kyoto, 1992. Engineering Geology, Special issue, vol. 34, n° 3-4, Elsevier.

Association Générale des Hygiénistes et Techniciens Municipaux (AGHTM) 1990. Les lixiviats de décharges : le point des connaissances en 1990. Rapport d'un groupe de travail. *Techniques, Sciences, Méthodes*, 6 : 289-314.

Association Internationale de Géologie de l'Ingénieur (AIGI) - Commission n° 14, 1989. Problems of underground disposal of waste. *Bulletin de l'AIGI n° 39*, Paris.

Association Internationale des Hydrogéologues (AIH) 1988. Actes du Colloque International "Hydrogéologie et sûreté des dépôts de déchets radioactifs et industriels toxiques", Orléans (France), Juin 1988. *Document du BRGM n° 161*, Orléans.

Aubertin, M., Chapuis, R.P., Bussière, B. & Aachib, M. 1993. Propriétés des résidus miniers utilisés comme matériaux de recouvrement pour limiter le drainage minier acide (DMA). *Proc. Geoconfine 93*, A.A. Balkema, Rotterdam, 1 : 299-308.

Barrère, M. (éd.) 1992. Terre, patrimoine commun. La science au service de l'environnement et du développement. *Editions "La Découverte"*, Paris.

Barrès, M., Côme, B., Lallemand-Barrès, A., Pauwels, H. 1991. L'utilisation du milieu souterrain pour l'élimination des déchets hautement toxiques : aspects techniques et institutionnels. *Rapport BRGM n° R 32900*, Orléans.

Barrès, M., Côme, B. 1994. Geoconfine 93, Les Sciences de la Terre au service du confinement des déchets toxiques. *Revue de l'Industrie Minérale, Mines et Carrières*, vol. 76, Janvier 1994 : 69-72.

Bernhard, C., Goussé, F., Matichard, Y. 1993. Utilisation des sols argileux : propriétés, mise en oeuvre, contrôle. *Proc. Geoconfine 93*, A.A. Balkema, Rotterdam, 1 : 141-146.

Bonin, H., Aziz, S. 1992. Approche multicritère pour la conception de couvertures de centres d'enfouissement technique de déchets. *Rapport du BRGM n° R 34850*, soumis au Ministère de l'Environnement, France.

Boutwell, G.P. 1993. Field permeability testing with TSB procedure. *Proc. Geoconfine 93*, A.A. Balkema, Rotterdam, 1 : 9-14.

Campbell, D.J.V. 1993. Waste management needs in developing countries. *Proc. Sardinia 93*, CISA, Cagliari (Italy), 2 : 1851-1866.

Carra, J.S., Cossu, R. (eds) 1990. International perspectives on municipal solid waste and sanitary landfilling. A report from the International Solid Wastes and Public Cleansing Association (ISWA). Academic Press, London.

Chapuis, R. 1993. Résumé et principales conclusions du thème 3: couverture et isolation de surface des stockages. *Proc. Geoconfine 93*, A.A. Balkema, Rotterdam, 2 : 729-732.

Côme, B., Lallemand-Barrès, A., Ricour, J., Martin, S. 1993. Applications comparatives de méthodes d'évaluation de risques liés aux sites pollués : premiers enseignements et perspectives. *Techniques, Sciences, Méthodes*, 9 : 447-451.

Comité Belge de Géologie de l'Ingénieur (CBGI) 1985. *Actes du Colloque National* "Problèmes de Géologie de l'Ingénieur en relation avec le stockage des déchets", Sart-Tilman, Liège (Belgique), Octobre 1985.

Conseil Canadien des Ministres de l'Environnement (CCME) 1991. Lignes directives nationales sur l'enfouissement des déchets dangereux. *Rapport CCME - WM/TRE-028F*.

Conseil Canadien des Ministres de l'Environnement (CCME) 1992. Programme national d'assainissement des lieux contaminés. *Rapport n° CCME EPC-CS39F*.

Conseil des Communautés Européennes 1967. Directive n° 67/548/CEE du 27 juin 1967 concernant le rapprochement des dispositions législatives, réglementaires et administratives relatives à la classification, l'emballage et l'étiquetage des substances dangereuses. *Journal Officiel des Communautés Européennes, n° L 196, 16 août 1967*.

Conseil des Communautés Européennes 1993. Proposition modifiée de directive du Conseil concernant la mise en décharge des déchets. *Journal Officiel des Communautés Européennes n° C212/33, 5 août 1993*.

Cottez, S., Piraud, J. 1993. Intérêt potentiel d'un stockage souterrain de déchets en galeries accessibles à niveau. *Proc. Geoconfine 93*, A.A. Balkema, Rotterdam, 1 : 493-498.

Dörhöfer, G. 1993. The role of natural barriers for the siting of landfills in Germany. *Proc. Geoconfine 93*, A.A. Balkema, Rotterdam, 1:39-45.

Düllmann, H., Eisele, B. 1993. The analysis of various landfill liners after 10 years exposure to leachate. *Proc. Geoconfine 93*, A.A. Balkema, Rotterdam, 1 : 177-182.

Duplaine, H., Esnault, A., Dufournet-Bourgeois, F. 1993. Le confinement de sites pollués. *Proc. Geoconfine 93*, A.A. Balkema, Rotterdam, 1 : 183-188.

Dusseault, M.B. 1993a. Solution cavern entombment of granular wastes. *Proc. Geoconfine 93*, A.A. Balkema, Rotterdam, 1 : 47-54.

Dusseault, M.B. 1993b. Slurry injection disposal of granular solid waste. *Proc. Geoconfine 93*, A.A. Balkema, Rotterdam, 1 : 511-517.

Ecole Nationale des Ponts et Chaussées (ENPC) 1990. Stockage en souterrain. *Actes des Journées d'études*, Novembre 1990. Presses de l'ENPC, Paris.

Ecole Nationale des Ponts et Chaussées (ENPC) 1993. *Actes du Colloque international* "Environnement et Géotechnique". Presses de l'ENPC, Paris.

EUROSTAT 1992. L'Europe en chiffres - 3ème édition. *Office des Publications Officielles des Communautés Européennes*, Luxembourg.

Furuichi, T. & Tanaka, M. 1993. Development of the detection system for leakage from the seepage control sheet of landfill disposal site. *Proc. Geoconfine 93*, A.A. Balkema, Rotterdam, 1:389-395.

Gisbert, T. 1993. Application des matériaux géosynthétiques à l'aménagement des CET de classe 1 : utilisation à ce jour et perspectives. *Proc. Geoconfine 93*, A.A. Balkema, Rotterdam, 1:201-206.

Gouvenot, D., Bouchelaghem, A. 1993. Barrières antipollution souterraines. *Proc. Geoconfine 93*, A.A. Balkema, Rotterdam, 1 : 207-212.

Grasso, J.R., Fourmaintraux, D., Maury, V. 1992. Le rôle des fluides dans les mécanismes d'instabilité de la croûte supérieure : l'exemple des exploitations d'hydrocarbures. *Bull. Soc. géol. France*, 1992, t 163, 1 : 27-36.

Greedy, D.R. 1993. Co-disposal in principle and practice - Lessons for developing nations. *Proc. Sardinia 93*, CISA, Cagliari (Italy), 2:1901-1910.

Hushon, J.M, Read, M.W., Morris, J.M., Zaragoza, L.V. 1993. Comparison of hazardous waste site ranking models. *Proc. 19th Annual Environmental Symposium of the American Defense Preparedness Association*, Albuquerque, N.M., USA.

Innes Lumsden, G. (ed.) 1992. Geology and the environment in Western Europe. A coordinated statement by the Western European Geological Surveys. *Clarendon Press*, Oxford (UK).

Joseph, J.B., Mather, J.D. 1993. Landfill: does current containment practice represent the best option ? *Proc. Sardinia 93*, CISA, Cagliari (Italy), 1 : 99-108.

Lawrence Berkeley Laboratory 1994. *Proceedings of an international symposium* "Scientific and engineering aspects of deep injection disposal of hazardous and industrial wastes", Berkeley, Mai 1994 (sous presse).

Le Tellier, I., Bernhard, C., Gourc, J.P., Matichard, Y. 1993. Expérimentation des systèmes d'étanchéité de centres de stockage de déchets à Montreuil-sur-Barse. *Proc. Geoconfine 93*, A.A. Balkema, Rotterdam, 1 : 225-230.

Maquinay, J.C. 1985. Schématique pour une prise de décision en matière de mise en dépôt de déchets urbains ou industriels. Colloque national "Problèmes de Géologie de l'Ingénieur en relation avec le stockage des déchets", Liège (Belgique), Octobre 1985. *Document du CBGI*, 1 : 1.22-1.29.

Marsily, G. (de) 1993. Résumé et conclusions du thème 1 : barrières géologiques naturelles. *Proc. Geoconfine 93*, A.A. Balkema, Rotterdam, 2 : 723-725.

Militon, C 1993. Nouveau concept français pour le stockage de déchets industriels spéciaux ultimes et stabilisés. *Proc. Geoconfine 93*, A.A. Balkema, Rotterdam, 1 : 547-550.

Ministère bavarois de l'Intérieur 1991. Altlasten-Leitfaden. *Document n° 14/91/07,* Munich.

Ministère de l'Environnement du Bade-Wurtemberg 1988. Altlasten - Handbuch. *Wasserwirtschaftsverwaltung,* Heft 18-19.

Monjoie, A., Rigo, J.M., Polo-Chiapolini, C. 1992. Vade-mecum pour la réalisation des systèmes d'étanchéité-drainage artificiels pour les sites d'enfouissement technique en Wallonie. *Publication de l'Université de Liège (Belgique).*

Neubauer, W.H. 1992. Eignung seichter Felskavernen in Hartgestein für Untertagedeponien. *Proc. XLI Geomechanik Kolloquium,* Salzburg, 1992.

Ouvry, J.F., Rouvreau, L., Faucon, V. 1993. Réhabilitation et mise en conformité de centres de stockages de déchets anciens. *Actes du Colloque Franco-Polonais "Géotechnique et Environnement",* Nancy, Nov. 1993 : 257-264.

Raffin, P., Leplat, L. 1993. Toits mobiles. *Proc. Geoconfine 93,* A.A. Balkema, Rotterdam, 1 : 331-336.

Risler, J.J. 1993. Le confinement ou la "mise en tombeau étanche", une technique adaptée à la neutralisation des dépôts de déchets industriels. *Techniques, Sciences, Méthodes,* 9 : 441-446.

Sauter, M. 1988. Mise au point d'un programme de calcul du bilan hydrique prévisionnel d'une décharge (BHYDEC). Actes du Colloque de l'AIH "Hydrogéologie et sûreté des dépôts de déchets radioactifs et industriels toxiques", Orléans, Juin 1988. *Document du BRGM n° 161,* vol. 1 : 587-595.

Silliman, J.M. 1993. The feasibility and cost of hazardous waste injection wells in the current regulatory climate. *Proc. 6th Annual Conference "Hazmat Central 93",* March 1993, Rosemont, Illinois, USA.

Société Internationale de Mécanique des Sols et des Travaux de Fondation (SIMS) - Comité Technique n° 8 1993. Geotechnics of landfill design and remedial works - Technical recommendations - GLR. *Second Edition-Ernst & John Publishers.*

Soyez, B. 1994. Le confinement : une nouvelle approche. A paraître dans le Bulletin de Liaison des Laboratoires des Ponts et Chaussées, Paris.

Stephanatos, B.N., Schuller, T.A. 1989. Selection of remedial alternatives at a hazardous waste site by use of quantitative risk assessment techniques. *Proc. 6th National Conf. Hazardous Wastes and Hazardous Materials,* New-Orleans (USA) : 70-75.

Striegel, K.H. 1993. Underground disposal in working coal mines of Northrhine - Westfalia, Germany. *Proc. Geoconfine 93,* A.A. Balkema, 1:569-577.

United States Environmental Protection Agency (US-EPA) 1984. The hydrologic evaluation of landfill performance (HELP) model. *User's guide, documents n° EPA/530-SW-84-009 et 010.*

United States Environmental Protection Agency (US-EPA) 1989. Requirements for hazardous waste landfill design, construction and closure. *Seminar publications EPA/625/4-89/022.*

United States Environmental Protection Agency (US-EPA) 1990. "Revised Hazard Ranking System - Final Rule". *Document n° PB91-100800.*

United States Environmental Protection Agency (EPA) 1993. Health effects assessment summary tables. Annual update. *Document ref. 540-R-93-058.*

Yeung, A.T. 1993. Waste confinement using electrokinetics. *Proc. Geoconfine 93,* A.A. Balkema, Rotterdam, 1:585-590.

3 Engineering geology and environmental protection
La géologie de l'ingénieur et la protection de l'environnement

Radon in soil gas in North-East England

Du radon dans le sol au nord-est d'Angleterre

A.G. Pickup & A.R. Selby
University of Durham, UK

Abstract.

Radon gas occurs naturally as three isotopes, 219,220 and 222, normally in very low concentrations. Radon gas and its metallic daughter products are hazardous to health due to their radioactive nature. The gas emanates from uranium, radium and thorium bearing strata, which are most abundant in igneous rocks. The gas is mobile, and migrates to ground surface both in gaseous form and in solution in water.

In the United Kingdom, the areas associated with the highest concentrations of radon are the south-west (Cornwall and Devon), and to a lesser extent parts of Northamptonshire and Derbyshire. However, other areas underlain by igneous rocks are of interest also, and this study relates to the northern Pennines in County Durham, and in particular the Weardale granite. Major fault zones in the area are of significance with respect to radon migration.

Initially a 5 km interval survey was conducted to identify localities of high radon counts using portable α-scintillator equipment. These localities were then surveyed in greater detail. Maximum values of total radon measured (with high repeatability) were 19600 Bq m^{-3}, compared with the average value for total radon of some 4300 Bq m^{-3} based upon readings in some 260 locations throughout the survey.

A study of the structural geology of the survey area identified some correlation between high radon levels and the major faulting around the edge of the Weardale granite.

Introduction

Radon is a naturally occurring inert radioactive gas with no smell, taste or colour. It is produced as three isotopes in the decay series of uranium and thorium. Radon-222 derives from uranium-238, has a half life of 3.83 days, and is of environmental concern because of its links with lung cancer. Radon-220 derives from thorium-232, and has a half life of 55 seconds. Radon 219 has a half life of only 3.9 s, and is therefore generally ignored in soil gas studies. Radon activity is expressed in bequerels (Bq), and gas concentration may be expressed in terms of Bq/m^3 of air. One Bq is equivalent to one disintegration per second.

Radon is hazardous to health due to its radioactive nature and its metallic daughter nucleides. The gas is highly mobile and can migrate both in gaseous form and in solution in water. A comprehensive review of the behaviour of radon in the geological domain is provided by Ball et al (1991) and aqueous migration is discussed by Heath(1991).

The occurrence of radon in the geological environment, and the consequential emission of the gas to atmosphere, is influenced by the concentrations of uranium and thorium in the strata, the mineralogy, the permeability of rocks and soils including faults and discontinuities and the nature of transport mechanism. The relation between radon and uranium mineralised zones was reported by Miller and Ostle(1973).

The National Radiological Protection Board conducted a survey of radon in dwellings throughout the UK, followed by more detailed surveys in areas of high observed levels. The detailed results are reported by Green et al. (1992), and these indicate that the strong majority of homes above the agreed action level of 200 $Bq.m^{-3}$ are in Cornwall and Devon. Somerset and Northampton also show results above the national average of 20 Bqm^{-3}.

These counties, however, are not the only areas in which uranium and thorium bearing deposits occur. Also the correlation between radon in soil gas and radon ingress into dwellings is not total. Further, a previous survey of radon in soil gas reported by Watson and Selby (1992) showed rapid variations over short distances, with very high anomalies in localised survey.

Consequently, a soil gas survey was undertaken in the County of Durham in the North East of England. This area is underlain in part by the Weardale granite, and both Weardale and Teesdale to the south have a history of mineral extraction, including tin and lead.

Radon detection.

Most procedures for identifying and quantifying radon depend upon its alpha emission. The equipment used in this survey is a zinc sulphide scintillation counter. When alpha particles interact with zinc sulphide, photons are produced. These pulses of light may be counted using a photomultiplier tube and a suitable electronic counter. The concentration of radon is broadly proportional to the counts per minute displayed digitally.

In the field, it is necessary to obtain a sample of soil gas in a consistent and repeatable manner which can then be analysed by the ZnS scintillation counter. This method of 'grab' sampling is affected by variables such as soil moisture and soil permeability, but was chosen in preference to the alternative method of burial and later retrieval of plate detectors, because of its speed and flexibility in a comparative survey such as this. The instrument used was the Scintrex RD200.

The technique adopted for gas sampling was to push a probe, comprising a tube and a close fitting central rod, some 700mm into the ground. The rod was withdrawn steadily, drawing soil gas into the space left within the tube. Connection was then made to the portable unit comprising a photomultiplier and counter, and a sample of the soil gas was drawn by hand pump into the measurement chamber. Counts per minute during the first three minutes elapsing after sampling were then made, and corrected for any small instrument background reading.

The values obtained over the first three minutes were used to detect the total radon observed. In addition, because radon-222 and radon-220 have different half lives, a method developed by Morse(1976) was used to detect the radon-222 and the radon-220 contents. A further parameter sometimes quoted is the 'flux ' or count1 to count3 ratio.

The survey area.
The area chosen for the initial, 5km density, survey was a rectangle defined crudely by towns at its corners of Haltwhistle(NW) to Newcastle(NE), and Appleby(SW) to Darlington(SE). The grid references defining the area are Eastings 3650 to 4300, and Northings 5150 to 5650. The survey area was thus very much centred upon the Weardale granite as shown in the geological map of the basement features of the area in Figure 1. Bott (1966) proposed that the Weardale granite is a post tectonic intrusion with respect to the Caledonian orogeny. Over much of the area Lower Palaeozoic slates are believed to intervene between the top of the granite and the overlying pre-carboniferous surface. The northern, western and southern margins of the Alston block are marked by major structural features of fault systems indicated in Figure 1. The Butterknowle fault is of some significance in this survey.

Some 200 measurements were taken during the main 5km survey, with a higher density of sampling in the Eastgate and Cotherstone areas which showed relatively high readings of radon-222. From the raw data in cpm, levels of total radon, of radon-222, of radon-220, (all in Bequerels/m^3) and of 'flux' C1/C3, were calculated, and the results were converted to graphical display in the form of coloured contour plots and of 'mountain range' diagrams, using the Uniras package.

Results.
The observations of radon will be discussed in terms of the four parameters of radon-222, radon-220, total radon, and of flux.

1) Radon-222.

The radon-222 levels in soil gas measured across the full survey area is indicated in figures 2 and 3. The highest level recorded was 19630 Bqm^{-3} near to the village of Cotherstone. This is atypical, in that of the many readings taken, 58% were lower than 1000 Bqm^{-3}, 6% were above 5000 Bqm^{-3}, and only 1% were above 10000 Bqm^{-3}.

The figures 2 and 3 show clearly that the levels of radon-222 across the area are generally low, but with some very high values observed locally. A comparison of figures 1 and 2 show some strong correlations between the high radon-222 zones with geological features. The high of 11000Bqm^{-3} observed near Rookhope (GR3935 5430) relates well to the main granitic cupola; the area is networked by several faults and veins.

A ridge of moderately high radon-222 running N-S correlates closely with the line of the Burtreeford Disturbance.

High levels in the location around GR 4200 5405 correlates with the eastern termination of the Deerness fault, and not far from the Cornsay cupola of the granite batholith. In addition, the area was heavily mined for coal, so that numerous subsurface voids and fracture paths are to be expected (Sibley, 1992). High levels to the NW of Hexham lie approximately over the Haydon Bridge mineral field. The anomaly near to Bowes, GR4000 5120 sits on the Stainmoor Summit fault line which penetrates the Namurian sediments.

The highest radon-222 value of 19630 Bqm^{-3} was found near Cotherstone at GR 4000 5199. The soil gas was sampled on two different days with a very close repeatability. This anomaly appears to be linked to the high radon-

222 values near Bowes. Bott et al (1972) suggested that high heat flow values obtained in nearby Woodland may be related to rising hydrothermal water near the Butterknowle fault system, which may in turn be responsible for rapid transport of radon.

2) Radon-220.

The levels of radon-220 observed across the survey area are displayed in figures 4 and 5. The highest level recorded was 8880 Bqm^{-3} at GR 4054 5650. Of the full set of readings, 12% were less than 500 Bqm^{-3}, while 14% were above 5000 Bqm^{-3}. On average the radon-220 values were nearly double those of radon-222. Some areas high in radon-222 were also high in radon-220, e.g. Cotherstone, Rookhope and Durham. However, areas high in radon-220 emerged which had not indicated high radon-222, around Darlington and Appleby. The latter correlates well with the Scordale granite cupola and the Cross Fell inlier. The high to the west of Darlington may be linked to fault hanging wall concentrations and in the west it occurs above the Middle coal Measures and upper carboniferous limestones. Workings, and numerous minor faults may be contributory.

As a whole the area exhibits a ridge of high radon-220 extending NE-SW from Appleby through Rookhope towards Newcastle. This follows one of the prevalent fault directions of the area, but is not clearly linked to a specific fault structure.

3) Total radon.

Total radon levels in soil gas are shown in figures 6 and 7. The highest (repeated) recording of 24820 Bqm^{-3} was near to Cotherstone, GR 4000 5199. Of all the observations, 12%

were below 1000 Bqm^{-3}, while 5% were above 10000 Bqm^{-3}.

Most of the western and northern areas recorded less than 4000 Bqm^{-3}, except for a few peaks and ridges around the Burfordtree Disturbance and the Pennine faults. Areas of high radon have already been discussed in the geological context.

A broader view shows that most of the high areas of total radon occur either over the central part of the granite pluton where it is closest to the surface, or around the edges and in particular where deep seated major faults lie.

4) Count 1 to Count 3 ratio.

Plots of this ratio are displayed in figures 8 and 9, to indicate areas of rapid gas flow. This parameter shows little correlation with the three other radon data sets. The highest value, of 13.5 occurred at GR 3751 5551. Other high values were found near Hexham, Alston, Rookhope and Brough.

The eastern area shows generally low ratios, and surprisingly there appears to be little correlation with major fault lines.

5) The anomaly west of Cotherstone.

After the primary survey had indicated a very high, repeatable, level of radon to the west of Cotherstone, the locality was surveyed more intensely. The results are shown in figures 10 11 and 12, which indicate that the high was very localised.

The radon-222 peak lies over near-horizontal sandstone, see figure 13, and ridges lead away west and SE, perhaps influenced by a fault dipping steeply to the SW.

The radon-220 shows a rather different

picture, in which a ridge of high runs from west to east, without a clear relation to any one stratum.

The total radon plot is the sum of those from the two isotopes. Past the faults to the north, the radon reduces rapidly due to the scavenging effect of the faults.

6) The Eastgate anomaly.
Because the primary survey indicated radon-222 in excess of 11000 Bqm^{-3}, a more detailed localised survey was made. Results are shown in figures 14 and 15.

The radon-222 plot suggests that the high is linked to other areas to the west and southwest, but no relation emerges with respect to the geology of the area. It seems more plausible that the current mineral quarrying and mining may be the precursor.

The radon-220 also shows a high, of some 7700 Bqm^{-3}, to the north of Eastgate. However the eastern side shows only low values of radon-220, perhaps because of a capping effect by the Upper Limestones (Namurian).

Conclusions.
One objective of this study was to investigate whether high levels of radon in soil gas may occur locally in areas generally considered to be of low radon risk.

The primary survey of an area centred on the Weardale granite showed generally fairly low levels of total radon, but a small number of anomalies were identified in which the soil gas contained very high levels of radon. More detailed investigation confirmed that such observations were repeatable on different days, but showed that the highest values were very localised. The

Weardale granite is known to lie at least 400m below ground surface, so that radon has a long and possibly tortuous route to the surface.

If the overlying rocks and soils were dense, unfissured and of low permeability, then radon in surficial soil gas could be expected to be very low. However, where localities of high radon were observed, there was some correlation with known major faults, or proposed hydrothermal rise, or other geological discontinuities.

More detailed surveys recorded by Pickup (1993) indicated that mine shafts and local faulting may channel radon to the surface, possibly with a scavenging of adjacent subsurface zones.

It seems probable, therefore, that in an area such as that studied here, there can be an broad view of low soil gas radon level, or of radon in dwellings, and yet, where particular geological conditions occur, rapid radon transport may yield, very locally, some significantly high levels of radon. The detailed nature of the pathways for radon transmission is not generally known.

Further soil gas radon surveys should be undertaken in areas where plutonic granites underlie, to establish whether radon anomalies exist elsewhere, similar to those identified here.

It would be desirable to locate long term detectors either below ground or in semi-sealed units at surface, in the Eastgate and Cotherstone areas, to try to identify long term radon gas doses.

References

Ball,T.K., Cameron, D.G., Colman, T.B. & Roberts, P.D. (1991) Behaviour of radon in the geological environment: a review. QJEG, 24, 169-182

Bott,M.H.P. (1966) Geophysical Investigations of the Northern Pennine Basement Rocks. Proc. Yorks, Geol.Soc. **V36-2-9**, 139-168

Bott, M.H.P., Johnson, G.A.L., Mansfield, J. & Wheildon, J. (1972) Terrestrial Heat Flow in North-East England. Geophys.J.R.Atron.Soc. **27**, 277-288

Heath M.J. (1991) Radon in the surface waters of southwest England and its bearing on uranium distribution, fault and fracture systems and human health. QJEG, 24, 183-189.

Miller, J.M. & Ostle,D. (1973) Radon measurement in uranium prospecting. In: 'Uranium Exploration Methods' Int. Atomic Energy Agency, Vienna 229-239.

Morse, R.H. (1976) Radon counters in uranium exploration. In: Exploration for Uranium ore deposits. Int.Atomic Energy Agency Symp. Vienna.

Pickup, A.G. (1993) Radon in soil gas in the Northern Pennines. MSc diss., Durham Univ.

Sibley,R.D. (1992) Gas migration studies at a disused mine to detect possible rock fracture patterns and the extent of underground working. Q.J.E.G.,**25**, 47-56.

Watson, J. & Selby, A.R. (1992) Some measurements of radon in south west England. Geotech. and Geolog.Eng., **10**, 95-115

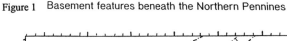

Figure 1   Basement features beneath the Northern Pennines

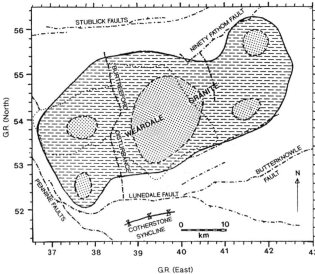

Figure 2   Radon-222 levels in soil gas around County Durham

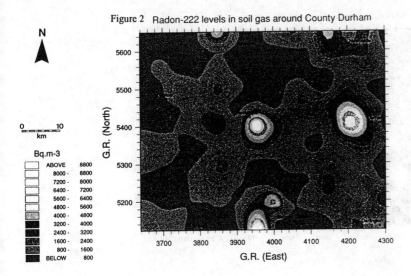

Figure 3   Radon-222 levels in soil gas around County Durham

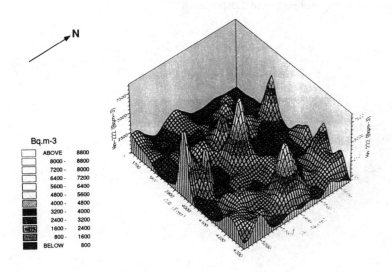

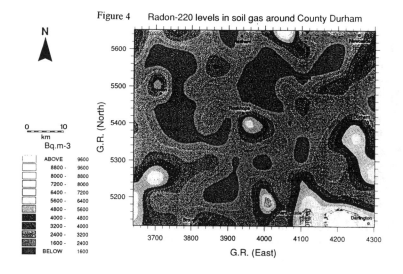

Figure 4   Radon-220 levels in soil gas around County Durham

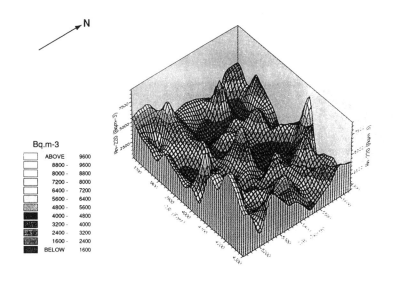

Figure 5   Radon-220 levels in soil gas around County Durham

Figure 6   Total radon levels in soil gas around County Durham

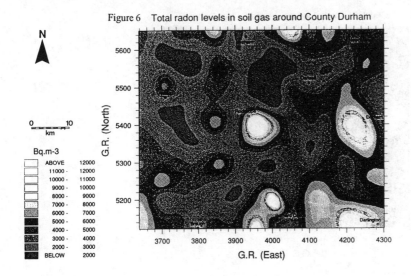

Figure 7   Total radon levels in soil gas around County Durham

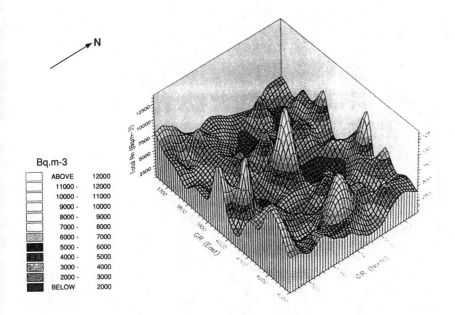

Figure 8   Count 1 / Count 3 values around County Durham

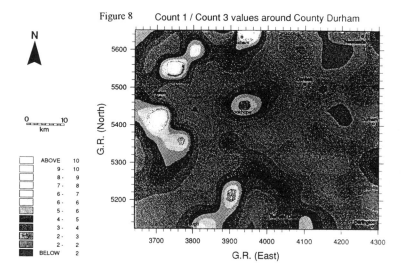

Figure 9   Count 1 / Count 3 values around County Durham

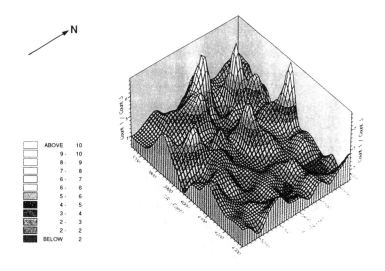

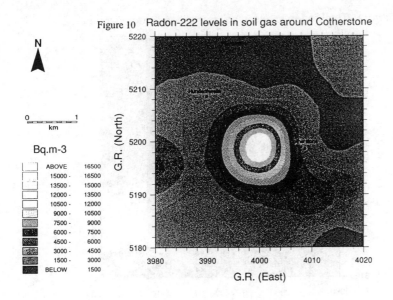

Figure 10   Radon-222 levels in soil gas around Cotherstone

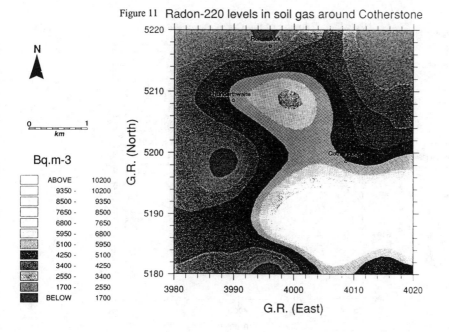

Figure 11   Radon-220 levels in soil gas around Cotherstone

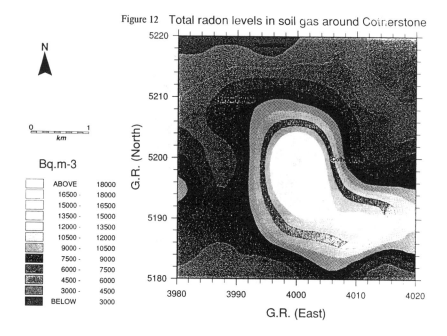

Figure 12   Total radon levels in soil gas around Cotherstone

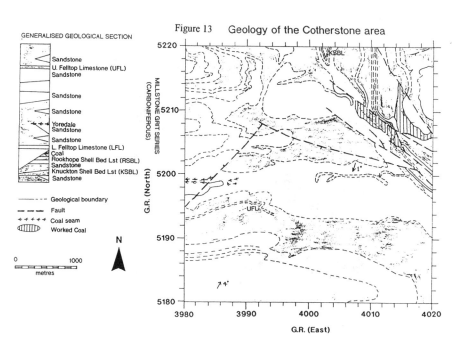

Figure 13   Geology of the Cotherstone area

2360

Figure 14 Radon-222 levels in soil gas around Eastgate

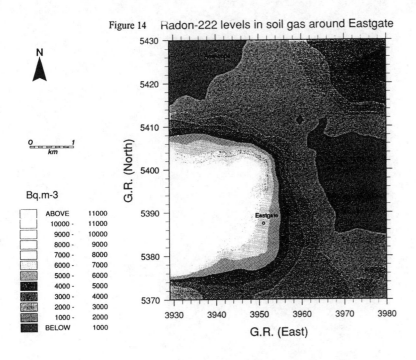

Figure 15 Radon-220 levels in soil gas around Eastgate

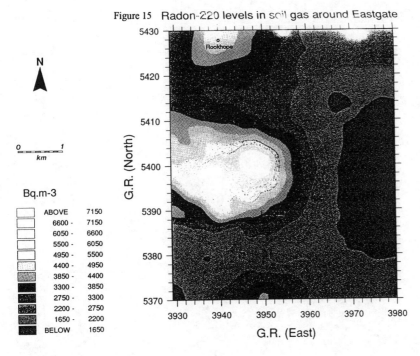

Hydraulic conductivity of a waste silty clay

La conductivité hydraulique d'une argile à plasticité moyenne pour un stockage de déchets urbains

Roberto Meriggi
Department of Georesources and Territory, University of Udine, Italy

Andrea Zagolin

ABSTRACT: The paper describes the results of an experimental research conducted on waste material deriving from the washing of aggregates in order to verify the feasibility of its use in the construction of an impermeable liner in an urban waste landfill. Some specimens of the material, a medium plasticity clay sedimented in an artificial lagoon, were compacted using the Standard AASHO Method and then tested with different laboratory methods to verify the variability of hydraulic conductivity under different boundary conditions and different hydraulic gradients. The research was carried out using both routine tests, such as oedometer and the falling head method, both using triaxial apparatus and Rowe's consolidation cell. Only the latter two methods resulted useful to the reproduction of water flow through the specimen under known conditions of effective confining pressure and under constant values of the hydraulic gradients. The values of permeability measured through the different methods were compared using a statistical method with those which may be obtained in situ.

RESUME': L'article décrit les résultats d'une recherche expéerimentale menée dans le but de vèrifier s'il est possible d'employer un matériau de rebut dans la construction de la couche imperméable de base d'une decharche municipale. Un èchantillonage de ce matériau - une argile de plasicité moyenne sédimentée dans un bassin articiel - a été rendu compact en utilisant la methode AASHO Standard et a ensuite été soumis à différent tests de laboratoire afin d'en vérifier la conductibilité hydraulique dans des conditions où les paramètres de depart et les pentes hydrauliques diffèrent. La recherque a ètè menée en utilisant des testes comme le test édométrique et le test à charge variable, ainsi qu'en utilisant l'appareil triaxial et la cellule de consolidation de Rowe. Seules ces deux dernières méthodes se sont révélées utiles pour reproduire le mouvement du filtrage de l'eau à travers les echatillons dans des conditions où la pression de la bordure et la pente hydraulique sont connues et constantes. Les degrés de perméabilité mesurés grace aux différents tests effecttués ont été ensuite comparés, par des méthodes statistiques, à ceux qui peuvent etre obtenus sur place.

1. INTRODUCTION

The main problem concerning the design of a compacted soil liner is to find the moisture content requirements, w, and dry density, γ_d, that the used clayey materials must meet to ensure the necessary low permeability. Once these requirements were identified, the designer must define the method of compaction and the compactive force necessary to achieve in situ the same results.

Although the method of in situ compaction is of great importance, the first step of the design must satisfy the difficult objective of estimating the variability of hydraulic conductivity under the different boundary conditions existing in situ and then to use laboratory tests which best reproduce these conditions. In fact, as pointed out by Olsen at al. (1985), the high values of hydraulic gradients used in the routine laboratory tests that measure permeability, such as constant or falling head methods, which are much higher than those existing in situ, can cause deviations from Darcy's law and substantial errors in the measurements because they can induce alterations in the solid skeleton of the specimen with a consequent increase in the suffusion phenomena.

Furthermore, the laboratory tests are usually performed on small specimens recreated with a known compactive effort using distilled water as permeant fluid and, what is more, they may be not representative of the actual hydraulic conductivity in situ which depends largely on the bonding and connection of each lift of the liner, by the type and weight of compactor used and on the nature of the leachate.

2. EXPERIMENTAL PROGRAM

Some experimental research was carried out in order to investigate the influence of test methods, hydraulic gradients and confining pressures on the measurement of permeability of a compacted medium plasticity clay with the following index properties: LL = 46, liquid limit; LP = 20, plastic limit, PI = 26, plasticity index; Gs = 27,6 kN/mc specific gravity, %<2μ = 46%, clay fraction. The main mineralogical compounds were quartz, calcite, dolomite, chlorite and illite. Eight specimens of clay, each with a different value of moisture content w, were compacted using the Standard AASHO Method to obtain the compaction curve shown in fig. 1 in

which w_{opt} is the Optimum Moisture Content (OMC) required to reach the maximum dry density, γ_{dmax}, for material compacted with the Standard compactive effort. This preliminary test is necessary because the hydraulic conductivity of specimens recreated with the same compactive effort depends greatly on their moisture content.

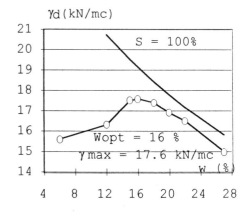

Fig.1 Compaction curve

In fact, according to Lambe (1958), the solid structure of the compacted specimens dry of OMC is dispersed and then more permeable of the specimens compacted wet of OMC, in which the arrangement of soil particles are almost horizontally oriented with a lower percentage of voids. Six of these specimens, each with a different moisture content, were tested for permeability using four different laboratory methods.

2.1 Oedometer test

The specimens, with a diameter Φ = 7,1 cm and a height H = 2,0 cm, were located in an oedometer apparatus between two porous stones and then loaded with three different vertical pressures:100 kPa, 200 kPa, 400 kPa. During each phase of loading, the coefficient of consolidation, c_v, was measured plotting the settlements of the specimen againsts the elapsed time.

By using the consolidation theory (Terzaghi, 1943), it was possible to measure the vertical coefficient of permeability with the following relation:

$$k = c_v \gamma_w m_v \qquad (1)$$

in which γ_w is the unit weight of water and m_v the coefficient of compressibility. For each specimen is was then possible to measure permeability for three different values of effective confining pressures.

2.2 Falling head permeability test.

After each stage of consolidation, the specimen was tested for permeability using the falling head method in the oedometer apparatus. The lower side of the specimen was connected to a glass burette filled with distilled water so as to start a seepage flow through it.
   Then the variation of the piezometric head was measured against time. By applying the known formula of the falling head method, it was possible to measure three different values of vertical permeability for each specimen at the same confining pressures used in theoedometer method.

In figures 2a and 2b the results obtained with both tests are shown and it can be seen that specimens dry of w_{opt} are more permeable than the wet ones but that the results obtained with the falling head method do not fall into line with those obtained through the oedometer tests and, for w drier than optimum, independent of the confining pressure. Other than to the possible water leakage along the walls of permeameter permeameter, this variance may also be due to the variability of hydraulic gradients during the test and to the fact that steady condition of flow through the specimen can never reached (Olsen, 1984).

2.3 Measurements of permeability with a triaxial apparatus.

The third method used to measure the permeability of compacted clay is not a routine test (Head, 1986) and was carried out using a triaxial cell, a pore pressure transducer, a volume change gauge and three pressure systems, one connected to the triaxial cell and the other two connected respectively to the upper and to the lower side of the specimen (fig.3).

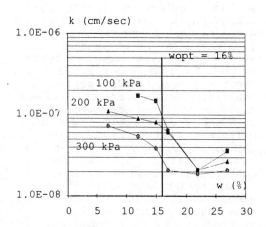

Fig. 2a. Permeability v.s.content moisture; Oedometer tests.

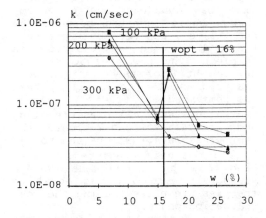

Fig. 2b. Permeability v.s.moisture content;Falling head tests.

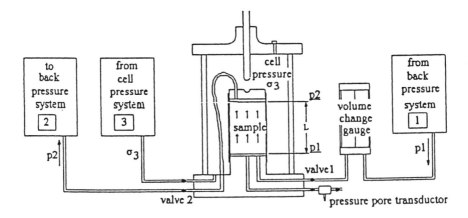

Fig 3. Triaxial apparatus for permeability test (Head, 1986)

Each sample, with a diameter Φ = 3,8 cm and a height H = 7,6 cm, were located in the triaxial cell between two porous discs and sealed with an impermeable elastic rubber membrane. After the saturation phase, the valves n. 1 and n. 2 were closed and the water pressure inside the cell was raised to a value of σ_3 by pressure system n. 3 while the pressure in the other two back pressure systems, n. 1 and n.2. were equalised to p. The specimen, subjected to an effective confining pressure of σ'_3 = σ_3 - p, was then consolidated by opening valve 1 and the water flow against time was measured by a volume change gauge inserted between the open valve 1 and the back pressure system n. 1. When consolidation was complete, the new area A and the new height H of the specimen were measured and the permeability test was commenced.

At a constant cell pressure of σ_3 and with valves 1 and 2 closed, the back pressure n.1 was raised to $p_1 <$ σ_3 and the back pressure n.2 lowered to $p_2 < p_1$ so that the average pressure, $(p_1+p_2)/2$, was equal to the pressure u used during the consolidation phase.

At the opening of the two valves, the hydraulic gradient, $i=(p_1-p_2)/H$, caused water to flow between the two ends of the specimen and

the cumulative value Q of water flowing through the specimen at the time t was measured with the volume change gauge and plotted in a graph. When the flow reached the steady condition, the mean rate of flow was calculated, q=dQ/dt. and then the coefficient of permeability was calculated using the equation:

$$k = \frac{q}{i\ A} \qquad (2)$$

This method allowed us to perform hydraulic conductivity tests under a known effective stress system acting on the specimens and with a wide range of constant hydraulic gradients. In the tests four values were used ranging from i = 20 to i = 200. In fig. 4 we can see the typical trend of permeability against the hydraulic gradients measured on three specimen with w = 20% and consolidated under three different isotropic pressures. It may be noted that, at constant values for moisture content, the permeability decreases as confining pressures increase while it remains almost constant in the case of increasing gradients. In Fig. 5 the average values of permeability are shown measured in relation to different moisture contents and consolidation pressures.

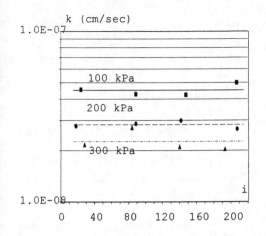

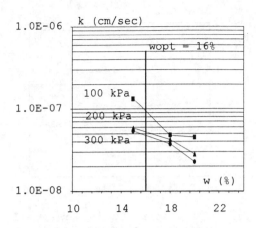

Fig. 4. Permeability in triaxial cell v.s. hydraulic gradient at W = 20 %.

Fig. 5. Permeability in triaxial cell v.s. moisture content for three consolidation pressures.

2.3. Measurements of permeability with Rowe cell

The setup of the Rowe cell and ancillary equipment used for permeability test is shown in fig. 6. The tests were only performed on one specimen, with a diameter Φ = 15,3 cm and a height H = 3,0 cm, compacted directly inside the cell with a moisture content w = 19.5%.

After the saturation phase, the diaphragm was loaded by applying pressure σ_d from the constant pressure system and the sample subjected to backpressure p from systems 1 and 2 connected respectively to the top and the base.

After the sample was consolidated at the effective stress of $\sigma' = \sigma_d - p$, the constant permeability test was commenced by adjusting the pressure difference across the sam-

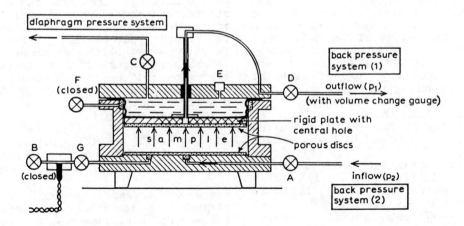

Fig. 6. Arrangement of Rowe cell for permeability test (Head, 1986)

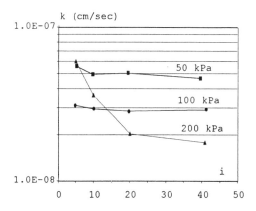

Fig. 7. Permeability in Rowe cell v.s. hydraulic gradient at W = 19.5%.

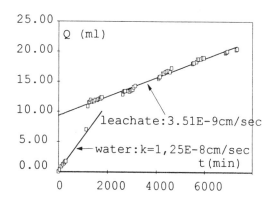

Fig. 8. Total flow of water and leachate v.s. time at the pressure p = 200 kPa.

ple following the same method used in the triaxial test method and measuring the cumulative flow Q of water through the specimen at the time t using two volume change gauges. The coefficient of permeability, computed with equation (2), was measured for three different consolidation pressures, 50 kPa, 100 kPa, 200 kPa, and for a wide range of hydraulic gradients. The results are shown in fig. 7.

A comparison between the results obtained throgh the Rowe and triaxial cells show that hydraulic conductivity remains almost constant at the same confining pressure and for high gradient values. As the tests were conducted starting with the higher values of hydraulic gradients, the lower values of permeability measured at 1 = 5 - 10 are probably due to loss of material happened during the seepage.

To evaluate the influence of the permeant fluid on permeability, at a pressure of 200 kPa, the deaired water, was replaced with a leachate characterized by the following compounds: suspended solids: 0.142 g/l, volatile:0.043 g/l, total dissolved solids: 8.18 g/l, azote in ammonia: 932 mgN/l and PH = 7,6.

The different hydraulic conductivities measured (fig. 8) are due to the different levels of viscosity and density of the two permeant fluids used in the test.

3. DISCUSSION OF THE RESULTS

The hydraulic conductivity measured through laboratory tests cannot be directly correlated with that achieved in situ because of the differences in the compactive effort used and the difference in the fabric of the soil.

Nevertheless a rough comparison between laboratory and in situ permeabilities, which is usefull in the first stage of the design of the liner, may be made using the statistical model developed by Benson et al. (1994). The model, based on the data collected from the construction reports of 67 landfills in North America, is a functional relationship between the values of in situ permeability, measured in laboratory tests on undisturbed specimens and the five most important variables in predicting hydraulic conductivity.

The relationship is:

$$\ln Kg = -18.35 + 894/P - 0.08PI + \\ -2.87Si + 0.32G^{0.5} + 0.02C + \\ + \varepsilon \qquad (3)$$

where Kg (cm/sec)= geometric mean permeability measured, P = compactor weight in kilonewtons, G = percent gravel, C = percent clay, Si = initial saturation and ε = a mean-zero Gaussian random error term.

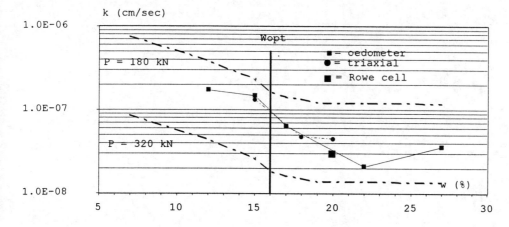

Fig. 8. Comparison between permeability measured in laboratory and its probable trend in situ achieved by varying the compactor weight.

Equation (3) was used to make a comparison between the values of hydraulic conductivity measured on specimens recreated in the laboratory with Standard compactive effort and those values which may be achieved in situ by varying the compactor weight, P, and correlating the initial saturation degree, Si, to moisture content w.

The probable trends of in situ permeabilities for P = 180 kN and P = 320 kN are reported in fig. 9 together with the values measured by oedometer, triaxial and Rowe cell tests at the confining pressure of σ = 100 kPa.

The values of hydraulic conductivity measured using these three methods concord fairly well but a heavy compactor, P = 320 kN, has to be used to achieve in situ the same order of magnitude since the permeability which it is possible to reach using a lighter compactor is 2 - 5 times higher.

REFERENCES

BENSON, C.H., ZHAI, H. AND WANG, X. 1994. " Estimating hydraulic conductivity of compacted clay liners", Jour. of. Geotechnical Engineering, Vol. 120, No. 2, 366-387.

HEAD, K. H. 1986. Manual of soil laboratory testing. Vol. 3. Pentech Press. London

LAMBE,T.W. 1958." The engineering behaviour of compacted clay", Jour. Soil Mech. Found. Division., ASCE, Vol. 84, No. SM 2, 1655-1 -35.

OLSEN,H. W., NICHOLS,R.W. and RICE T.L.1985. "Low gradient permeability measurements in a triaxial system", Geotechnique, Vol. 35, No. 2, 145-157.

TERZAGHI K. 1943. Theoretical soil mechanics. J. Wiley &S., New York.

Groundwater oil pollution and remedial technologies

Pollution des eaux souterraines par les hydrocarbures et mesures de correction

A. Bonoli, G. Brighenti, F. Ciancabilla & M. Pacciani
Mining Science Institute, Bologna, Italy

ABSTRACT: This paper briefly outlines the types of oil pollution, the techniques and criteria for choosing remedial methods. It also presents the data from a case history considered useful both in defining the guidelines according to the case histories method, and identifying the characteristic parameters and the various skills.

RESUME': Cet article présente sinthetiquement les modalitées de pollution par hydrocarbures, les techniques et les jugements de choix des méthodes de bonification. On présente aussi les données de un "case history" interessant pour la définition des "guide-lines" par la méthode des cas et pour l'individuation des paramètres charactéristiques et des differents compétences professionelles.

1 INTRODUCTION

Ground water is still one of the major sources of drinking water today. In Italy, over 85% of the water distributed by the aqueducts is groundwater. Protecting it from pollution is thus an important duty, especially in densely populated areas.

The spread of the use of petroleum products has led to an increased spread of oil pollution in groundwater, both crude and refined oils. This pollution follows special laws, since these liquids are not readily soluble with water and form a separate phase (non-aqueous phase liquid, NAPL).

The sources of greatest pollution are refineries, petroleum plants, oil pipelines and large storage deposits of crude oils or products. However, we must not neglect the pollution spread throughout the territory due to spills from the millions of tanks of heating fuel present in nearly every home, service stations, tipping of tank trucks in car accidents, and finally the small leaks of petroleum products from cars in circulation, aggravated by the uncontrolled disposal of used oil by many individuals.

We wish to draw attention to these cases in particular, because on the one hand they are the ones for which there is little or no protection and for which detection, and thus remedy, is late and often carried out by personnel inexperienced and unfamiliar with clean up methods, and on the other the authorities responsible for controlling the situation are unable to define a valid set of criteria for choosing clean up strategies. In addition, it is a question of working on extremely complex phenomena, the study of which requires a vast range of skills ranging from those of the biologist and toxicologist to those of the geologist, chemist and hydrogeologist. Within this framework, the figure of the petroleum engineer appears to be especially qualified, but he must complete his skills in the area of hydrogeology. In this regard, we feel it important to call the attention to the need for greater coordination among all experts in the field of underground fluid mechanics, in order to arrive at choosing the same parameters (indicated by the same symbols) and the same mathematical formulation to characterize the same phenomenon, even when it involves various sectors such as reservoir engineering, hydrogeology, agricultural sciences, etc.

In this note, we wish to point out that a valid method for approaching this type of problem is to define corrective action plans and guidelines prepared based on a definition of the characteristic parameters (such as type of pollutant, soil type, etc.). In this way, objective of reference data can be found for the local control authorities.

The processes of oil pollution are briefly outlined, as well as the techniques and criteria for choosing clean-up techniques. In addition, the data are presented from a "case history" considered useful both in defining guidelines according to the case histories method and in identifying characteristic parameters and the various skills necessary.

2 MECHANICS OF OIL POLLUTION

Hydrocarbons make up most of the non-aqueous phase liquids, and are distinguished in terms of density greater than (DNAPLs) or less than water (LNAPLs or "sinkers").

This paper considers only LNAPLs, which include crude oil, hydrocarbon fuels and many other products that we shall indicate here, generically, as oil. However, we should point out that DNAPLs also include highly toxic substances, such as halogenated hydrocarbons [Bruce, 1993], which are currently a serious problem for our aqueduct technicians [Raffaelli, 1993].

As mentioned, hydrocarbons have a very low water solubility rate; this decreases as the density increases and increases when passing from paraffins to aromatics (for example, approximately 61 ppm for butane, 1780 ppm for benzene[Testa, 1991], although oils are virtually insoluble, some are still sufficiently soluble to supply the water with a flavour and an odour that can render the water unsuitable for human consumption.

Following an oil spill, the oil migrates through the insaturated zone until it reaches the capillary frange and the aquifer; part of it remains trapped in the pores in a state of irreducible saturation. Thus if the volume of oil spilled is small, it cannot reach the

aquifer and in the retention zone a suspended oil body grows. If instead the spill is sufficiently large, the oil reaches the capillary fringe at the water table and forms a layer of increasing thickness. If the aquifer is in motion, the oil migrates in the same direction as the ground water until the threshold of residual saturation is reached or until it reaches a discharge point.

If we examine its distribution in the subsoil in detail, the oil is found: partly in the gaseous state, partly free, partly in residual saturation (adsorbed on the surface of grains of soil or trapped in capillary pores, depending on whether the oil is wetting or non-wetting), and partly dissolved in the water.

From a quantitative standpoint, this distribution depends on the conditions of the subsoil and the chemical and physical features of the oil, water and soil. The determining factor for this distribution, in any case, is the type of soil and its homogeneity and anisotropy characteristics.

3 CLEAN-UP TECHNIQUES

While in the case of refineries, petroleum chemical plants, etc., it is possible to design preventive actions to minimise pollution of groundwater, in the case we have examined of small, widespread incidents it is possible to take remedial measures only once the spill has occurred. In this regard, the rapidity of the intervention is essential: as time passes, the size of the polluted area increases, and thus the difficulty in cleaning up as well; on the other hand, effective measures require a sufficient knowledge of the geological, hydraulic and geotechnical features of the spill site and the exact boundaries of the polluted area. It is thus most appropriate for early, emergency measures to be followed immediately by a study of the state of the area and the types and extension of the pollution, making it possible to decide upon the most appropriate clean-up strategies. A vast bibliography exists on such remedies, and handbooks and suggestions for measures to take in various situations (taking into account the type of soil and the degree of clean-up required) have been proposed by the American Petroleum Institute [API, 1979], CONCAWE [1990], as well as many other specialists.

Among the main remedial measures, we recall: removal of the soil for treatment (incineration, dumping, etc.), removal of the mobile hydrocarbon through wells, drains or ditches, ventilation, biological treatement, containment (using physical barriers, controlling the movement of groundwater), vitrification and solidification of the polluted soil [Russel, 1992].

Each of these methods has its own fields of application and limitations; usually several methods are used in sequence to achieve the desired level of clean-up. However, we point out that many remedial systems will require years of operation before they achieve acceptable levels in the soil or groundwater. The intervention times may range from a few days or weeks for soil removing, through to many months or years for recovering free oil by wells or drains or for bioremediation.

4 HOW CLEAN IS CLEAN?

Protection of groundwater must certainly be one of the priority objectives, given the increasing scarcity of good-quality water; however, the problem here is how far to push underground and aquifer clean-up techniques in each case. Specialists hold varying opinions on this matter, and thus the regulations change from country to country [cfr. Concawe, 1990, Russel, 1992]. In densely populated countries such as those in Western Europe, it appears necessary to allow multiple use of the territory, meaning that the same place can serve for several functions, as well as be used for different functions over time (for a detailed examination of this problem see, for example, [Brighenti, 1992], [Bonoli, Brighenti, 1988].

This criterion would thus lead us to consider soil and subsoil as a single system, made up of a solid stage, the fluids that saturate it and by the living beings that populate it, and would make it necessary to keep such a system in a state allowing multiple use in the present and in the future. The protection of the territory should be aimed at protecting all of its potential, and not require separate consideration of maintaining the quality of its components, setting purity standards for each.

However, while this criterion is valid from a theoretical standpoint, it is difficult to apply in practice.

In the first place, our knowledge is not currently sufficient to allow sufficiently precise forecasts for the future; secondly, the need to make each site suitable for multiple use may often lead to unacceptable costs; thirdly, a set of global protection standards indicating only the final objectives is quite generic in its specific requirements, leaving a great deal of discretional power to local control organizations, often not especially competent and subject to political, economic and other pressures.

Sometimes in such situations the local control organizations have made requests for clean-up with extremely severe standards, or at the limit of the sensitivity of instruments; on the contrary, at other times cases of extreme permissiveness have been encountered.

It is therefore advisable, from a practical standpoint, to set general, strict standards valid throughout the country, and create a specialized team of regulation to control each operation, requiring that a "corrective action plan" be submitted for approval before action is taken, giving a detailed hydrogeological outline, description of the site and the spill, as well as the plan for the suggested clean-up operations, indicating the results to be achieved and the methods for testing afterwards to verify their achievement [cfr. Russel, 1992, Testa, 1991].

With these guaranties, it is our opinion that it is realistic to take into account -for clean-up purposes- the importance of the aquifers affected by the spill, their actual use and real insulation from aquifers used or which may be used in the future for human consumption, as well as the level of pollution present in soil before the spill. Only when all of these factors have been taken into account is it possible to have criteria that are both technically valid and economically feasible.

5 CASE HISTORY

Corrective action plans were developed at the *Istituto di Scienze Minerarie* at the University of Bologna on cases of pollution following diesel fuel or gasoline spills from tanks or due to road accidents. In this paper, we refer especially to the case of pollution due to fuel oil spills from a tank used for a power plant, and which at the time of the accident was no longer in use and was waiting to be emptied of the unused oil. This case is briefly described here; for a more detailed description, see a paper in printing [Bonoli, Pacciani, 1994].

The tank, having a volume of 300 cubic meters, is located in an industrial area near Bologna (Italy), in the northern part of the city (Fig. 1). In the autumn of 1991, for reasons yet to be identified, fuel oil leaked out of a bottom valve in the tank, for a spill estimated at approximately 200 cubic meters. The spilled oil reached the soil, percolating through cracks in the containment tub, and partly returned to the surface in a nearby canal (Navile Canal) close to the site of the accident. The owners took the first emergency measures, consisting of a series of boreholes and cone penetration tests that made it possible to determine the approximate extension of the pollution, and attempted to contain the polluted area and recover the oil by means of a trench and by installing a system with 47 well points.

Subsequently, the environmental control authority ordered that further action be taken to restore the original conditions of the site and avoid spreading the contamination to adjacent areas. Thus the *Istituto di Scienze Minerarie* was asked to outline the necessary clean-up operations. A corrective action plan was suggested in order to provide:

- a hydrogeological, geological and morphological description of the area, to assess the hazard related to the spill as accurately as possible;
- the chemical-physical parameters of oil, to assess the forms of pollution that could be present and thus the necessary corrective action;
- a definition of the distribution of the plume and its evolution over time, in order to determine the boundaries of the polluted area and choose the remedial measures;
- a description of the use of the site and its basic pollution level before the accident.

Figure 1 - AREA'S LOCATION

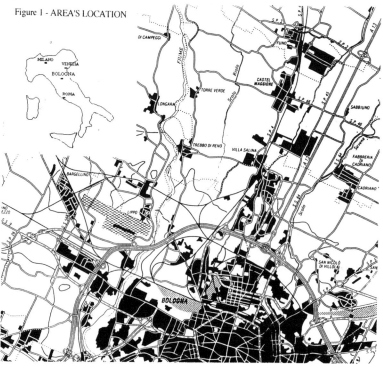

5.1 Geological and hydrogeological features of the area

The area studied is in the plains area outside Bologna. The morphology is generally plains set-up. In the immediate vicinity there is a water collection canal between the Reno River and the Idice River that conveys the water into the Reno.

In order to identify the hydrological cycle of the water involved in the study area, studies were gathered that had previously been carried out on a regional scale on the vulnera-bility of the aquifers [Ciabatti et al. 1992]. In addition, in situ tests were made both locally (in an area of approximately 50 x 50 metres) and on a wider scale (approxi-mately 500 x 500 meters). The in situ tests were aimed at studying the stratigraphic succession and the aquifers. Seven borings with continuous sampling were run, located as shown in Fig. 2. The borehols were drilled until they reached a clay layer at an average depth of 7 m from the land surface. Part of the samples were used for fluid characteriz-ation studies, while the others were subjected to petro-physic tests. Each well was then completed with a piezometer to check the groundwater levels and oil thicknesses. Fig. 3 shows the stratigraphic sections.

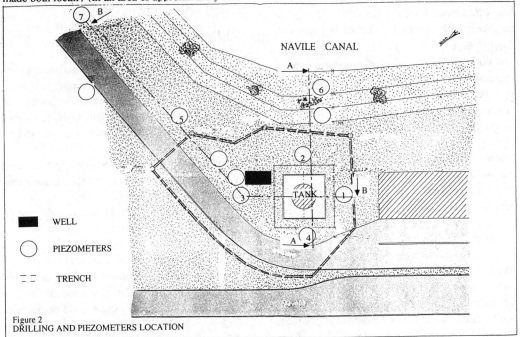

WELL

PIEZOMETERS

TRENCH

Figure 2
DRILLING AND PIEZOMETERS LOCATION

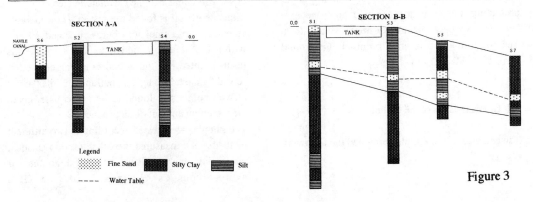

Figure 3

The soil layers were found to consist of packs of sandy silt and silty sand varying in thickness from 1 to 1.5 meters, crossed by thin strips of more or less clay-dense silts: the small fractions are sharply predominant over the sandy-silt ones. At a depth of approximately -6.5 meters, a more markedly silty-clay layer was found, well-consolidated, whose considerable thickness and homogeneity were pointed out by soundings made at deeper levels. Laboratory tests then showed that the horizons and layers dominated by silty-sand and sandy-silt fractions are characterized by modest permeability (K) ranging from 10^{-4} to 10^{-5} cm/s, while for those predominantly made up of silt and clay-silt, as found in the bed layer mentioned above, we can describe a nearly impermeable lithological unit, with $K<10^{-7}$ cm/s.

From a geological standpoint, the sector studied thus consists of flood deposits, locally characterized by a sharp spatial unevenness with horizons and layers often lens-shaped, in various positions based on the past deposits of the water bodies. The area directly involved by the fuel oil spill has also been broadly "readjusted" by man's intervention, such as the construction of the road and the tank.

The hydrogeology is adapted to these features: modest aquifers are found, often perched and not very extended in the most permeable horizons, mainly discontinuous, but isolated and fed by local precipitation. In the area near the tank there is a phreatic aquifer at a depth of approximately 3 meters from the surface, with a gradient of approximately 1% aimed at the canal (Fig. 2). This aquifer is not connected to the deep aquifers, more important from the standpoint of drinking water production thanks to the constant presence of a layer of grey clay, practically impermeable, from -6.5 m from the reference point of the ground surface (23.80 a.s.l.) to -13 m.

5.2 Characteristics of the fuel oil

The analyses of the samples provided the following results:

relative density (15°C/4°C)	0.837
distillation 150 °C	/ (% in volume)
distillation 210 °C	4 (% in volume)
distillation 250 °C	28 (% in volume)
distillation 350 °C	91 (% in volume)
slippage point	- 15 (°C)
flammability point	70 (°C)
viscosity (10 °C)	5,13
viscosity (20 °C)	4,43
viscosity (38 °C)	2,53
sulphur	0.84%
Water and sediments	traces n.r. < 0.05

5.3 Delineation of the contaminated area

In order to assess the entity of the pollution and its distribution over time, it was necessary to identify the plume of mobile hydrocarbon, the presence of residual hydrocarbon and determine the boundaries of the areas involved by pollution due to the soluble components.

In addition, this study intended to verify the amount of oil present in the soil, control its evolution over time and estimate the recovery time.

The mobile oil was assessed by measuring the oil thickness in the piezometers,.

The oil in a state of residual saturation was identified by analyzing soil samples.

In order to measure the actual thickness in the soil, given the exaggeration phenomenon in the piezometer, reference was made to the equations of De Pastrovic el al. [Concawe, 1979], while the bail down tests confirmed the calculated values.

Thus mobile oil is found to exist in a considerable amount, with a thickness between 2 and 10 cm, within the area surrounded by the trench. This result pointed out the actual containment action served by the trench against pollution, while the use of well points was found to be inadequate, given the minimum amount of oil produced.

The samples were tested according to two different methods: one measures the oil saturation as such, while the other measures the total amount of hydrocarbons as N-hexane with the GLC determination method as such. The tests were run on 14 cases, selected based on the qualitative indications gathered during the executive stage. Thus samples taken from the grey clay were selected, especially where small lenses of clay could be found and those for the various lithologies: brown silty clay, brown sandy clay, brown silty sand of medium-fine consistency. The two analysis techniques, as shown in table 3 (where the fuel oil is given in terms of weight/weight) pointed out sharply different results. Each sample contained a certain amount of oleous substances different from those spilled from the tank, indicating a generalized state of pollution existing before the accident and relative to the entire area studied. The spilled fuel oil is thus in a state of irreducible residual saturation in the unsaturated soil, in the area beneath the protective tub of the tank and in the subsoil interested by water table fluctuation, for the entire area bordered by the trench. In addition, it may be hypothesized that a small amount of oil is found in the capillary fringe. It was also pointed out how the movement of the groundwater, due to the use of the well points, led to an expansion of the phenomenon.

Finally, in the piezometer 5 (Fig. 2), oil was present only in traces, due to the adsorption by the soil of the easily volatile soluble elements carried by the water.

These tests thus made it possible to make an initial assessment of the pollution conditions created in the area and the functional response of the plants installed in the emergency measures taken, and to formulate a hypothesis on the most appropriate remedial measures.

In order to assess the quantity of hydrocarbons, further laboratory tests were made, measuring the porosity φ of those samples most representative of the lithologies present in the subsoil. The results are shown below:

1-3 Brown sandy silt	$\varphi = 39\%$
2-2 Brown-grey silty clay	$\varphi = 33\%$
3-4 Brown silty medium-fine sand	$\varphi = 41\%$
2-3 Brown-grey clay-based silt	$\varphi = 37\%$
3-3 Brown sandy silt	$\varphi = 37\%$
4-4 Brown-grey sandy silt fine sand	$\varphi = 38\%$

Displacement tests were then run on some of these. From the data gathered from the previous tests it was then possible to estimate the amount of oil in the soil, the amount of mobile -thus removable- oil, and the amount of residual oil, thus what remains in place. Based on these amounts, the clean-up techniques were chosen (Fig. 4).

It may be hypothesized that the mobile oil is just above the water table for an extension equal to that bordered by the trench, while the oil in a state of residual saturation is in the unsaturated soil beneath the containment basin, from the land surface to the groundwater surface, and in the soil beneath the mobile oil for a depth equal to that of the water table, and finally in the section between the trench and the canal reached by the oil before the emergency measures were taken.

From the data gathered from the above tests, and with an average residual saturation of 20% assessed, it was then possible to estimate the volumes of mobile hydrocarbon and residual *in situ* hydrocarbon:

Total V_{oil} *in situ* $= 32 \text{ m}^3$

Recoverable $V_{oil} = 20 \text{ m}^3$

V_{oil} that will remain *in situ* $= 12 \text{ m}^3$

6 CLEAN-UP PROPOSAL AND CONCLUSIONS

Based on the results of the tests and measurements described, it was found to be impossible to remove the polluted soil. This was due both to the excessive cost of such intervention, and because this would have required demolishing existing buildings and interrupting the road system. Thus the technique of recovering the mobile hydrocarbon was chosen, draining it through the trench and an appropriate number of wells placed accordingly. As far as the fuel oil in irreducible residual saturation is concerned, it is known that measures generally tend to modify its physical state, making the most volatile components evaporate and oxidizing and/or biodegrading the others. These effects are achieved by various methods, all of which require the presence of oxygen.

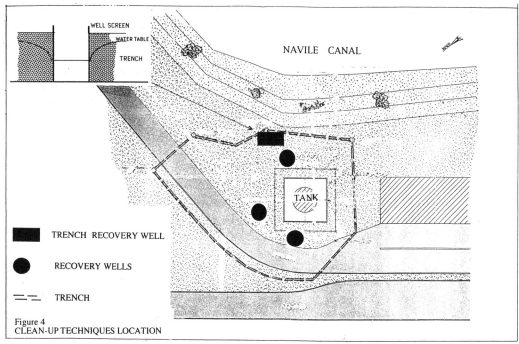

WELL SCREEN
WATER TABLE
TRENCH
NAVILE CANAL
TANK

▬ TRENCH RECOVERY WELL

● RECOVERY WELLS

‒‒‒ TRENCH

Figure 4
CLEAN-UP TECHNIQUES LOCATION

Techniques range from the simplest, moving the land with excavators so that the oil comes into contact with the air, thus causing the processes mentioned above, to more complex ones which require the installation of wells aeration systems and addition of bacteria and nutrients to the soil the achieve biological degradations of pollution. The decision to use either technique is based on the admissible value of residual hydrocarbon saturation as well as economic considerations.

In order to schedule the most appropriate clean-up techniques in this specific case, it was deemed necessary to carry out further studies in order to predict the recovery of free oil more precisely.

More specifically, well tests were run, carried out by means of a pilot well and aimed at determining the aquifer properties and the permeability of the soil, so as to size the lifting system, and oil capacity tests to estimate the intervention time and recoverable quantities.

For field testing, a test well was drilled near the piezometer n. 3 (cfr. fig. 2), as during the previous tests this point highlighted the greatest oil thickness and the highest recoverable oil capacity. It could therefore be deduced that in this area, together with the area just below the tank, there was the greatest concentration of oil. The well was built by a percussion method to a depth of 7 m. Inside the hole, a screen, measuring approximately 400 mm in diameter, with a gravel pack was placed at a depth of up to 5 m. Water and oil were then extracted in order to identify the best water rate, the oil recovery, and the permeability of the layers (Fig. 4).

Before carrying out the tests, it was necessary to wait several weeks for the system to stabilize and full operating conditions to be reached. The well test was then conduction in a stationary condition.

Permeability was found to be K = 5 x 10^{-4} cm/s,

while the oil recovery vs. time is shown over time in Fig. 5.

Figure 5 Oil Recovery vs. Time

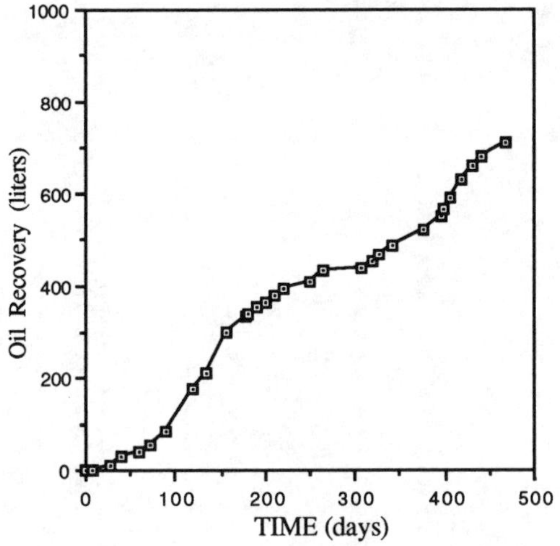

Based on the above results, it was possible to place 3 wells to complete the mobile hydrocarbon drainage system (see Fig. 4) and estimate the duration of the operation at approximately 3 years.

As far as further clean-up operations are concerned, it is felt that these need not be pushed to very high levels, given the initial state of the site and its use as an industrial area. It is probably sufficient to isolate the polluted groundwater using containment barriers anchored in the impermeable stratum, and make the surface soil impermeable.

However, where health authorities require, it is possible to examine ventilation and biodegradation measures. But these present considerable difficulties, and require a long time given the low permeability of some polluted strata.

REFERENCES

API, 1979. Underground movement of gasoline on groundwater and enhanced recovery by surfactants, API pubbl. N. 4317.

Bonoli A., Brighenti G., 1988. Uso multiplo del Territorio ed inquinamento delle falde acquifere sotterranee - Acque sotterranee, III, 29-33.

Bonoli A. Pacciani M., 1994. Un Case History di inquinamento da gasolio dels uolo e della falda: metodologia di indagine e tecniche di bonifica. In publishing

Brighenti G. 1992. Criteria for programming investigation on groundwater pollution. 2nd Int. Congr. Energy, Environment and Technological Investigation, Roma, 59-64.

Bruce L.C., 1993 - Refined Gasoline in the Subsurface, Am. Ass. Petr. Geol. Bull, February, 212-224.

Ciabatti M, Francavilla F, Giorgi G., 1992. Vulnerabilità degli acquiferi in rapporto all'attività estrattiva nella pianura bolognese, Ed. Grafis, Bologna.

Collins , 1986. Groundwater contaminants: Field methods, Proceeding ASTM Congress.

CONCAWE, 1979. Protection of groundwater oil pollution, rep 3/79, Bruxelles.

CONCAWE, 1990. European soil and groundwater legislation: implications for the oil refining industry, rep. 4/90. Bruxelles.

Raffaelli G., 1993. Distribuzione dei composti organo-alogenati negli acquiferi della Provincia di Bologna. Siti contaminati, procedure di controllo ambientale e di bonifica. Workshop, Bologna, 8 giugno.

Russel D.L., 1992. Remediation Manual for Petroleum Contaminated Sites, Technomic, Lancaster, USA..

Testa S.M., Winegardner D.L.,1991. Restoration of Petroleum-Contaminated Aquifers, C. Lewis Publ. Chelsea.

Yong R.N., Mohamed A.M.O., Warkentin B.P., 1992. Principles of contaminant transport in soils, Elsevier, Amsterdam.

Environmental studies related to solid waste disposal in the State of São Paulo, Brazil

Études sur l'environnement pour le stockage des déchets solides dans l'État de São Paulo, Brésil

Márcio Angelieri Cunha & Angelo José Consoni
Division of geology, Technological Research Institute of São Paulo State, Brazil

ABSTRACT: This paper briefly presents some activities developed in relation to solid waste disposal by Division of Geology – Digeo, of Technological Research Institute – IPT, in Sao Paulo State, Brazil. Basically three main activities have been carried out: studies to sanitary and industrial landfills installation; technical analyses of environmental impact studies, and impact assessment of contaminated sites due to inadequate disposal of solid waste.

RÉSUMÉ: Cet exposé présente quelques activités developpées par la Division Géologie – Digeo de l'Institut de Recherchers Technologiques – IPT de l'état de São Paulo, Brésil, en ce qui concerne les dépôts de déchets solides. La Digeo travaille sùr trois activités principales: des Études pour l'aménagement des dépôts sanitaires et industriels; des analyses techniques de l'impact de ces dépôts sur l'environement; évaluation de l'impact des sites contaminés par la déposition inadequate des déchets solides.

1 INTRODUCTION

The management of urban solid waste is a worldwide problem and is beginning to assume critical dimension also in Brazil, specially in Sao Paulo State, where is an area heavily populated and industrialized.

After decades of uncontrolled disposal of urban and industrial waste, the last years have been showing an increasing concern about management of such materials. In this way, it has been intensified the effort to the reduction of volume of generated waste, to the recycling of different used materials, to the application of new intermediary treatments (in order to reduce the harmfulness and volume to disposal) and, mainly a final adequate environmental destination. Nevertheless, this stage of such activities is still incipient.

The Technological Research Institute has been working to provide technical solutions for the problems described before. Through its Division of Geology – Digeo, several studies has been carried out with the aim fo the correct solid waste disposal, preventing pollution of surface and underground water resources. Among other Digeo's activities, the most important are those related to the installation of landfill (sanitary and industrial),

impact assessments due to improper disposal of solid waste, and studies related to disposal of contaminated river sediments.

2 LANDFILL INSTALLATION

Concerning to landfill installation (sanitaries and industrial) we have geologic-geotechnical and environmental studies for the selection of places suitable for such installations, and assessment of Environmental Impact Studies (EIS), Environmental Impact Reports (EIR) and Technical and Economic Viability Projects.

2.1 The site selection process

The process of selection of a site in order to install a landfill has been conducted as a decision process in three successive phases with increasing level of detail, involving technical criteria and political decisions. Each level has its characteristic geographic dimension, its own criteria of selection and degree of detail. Subsequently, the analyzed area will became smaller; the criteria, more specifics; and the suitable sites, less numerous, ending with the final indication.

In the first phase bibliographic information is collected about analyzed area, including qualitative and quantitative data about generated waste, several thematic maps (geological, soil use and occupation, environmental conservation unities, etc.). Scale of work is generally 1:50,000 or lesser. The criteria used in this phase area generic and of rapid application, such as maximal distance from sources of residues, legal restrictions, regional characteristics (geology, hydrogeology, geomorphology, etc.). Applying these criteria potential sub-areas are identified and prioritized, being other ones rejected.

In second phase, by the assessment of the sub-areas previously prioritized, more suitable sites are individualized to the landfill installation. Some aspects of physical environment, in greater or lesser degree, are assessed during this phase, depending on the adopted criteria and on the specific conditions of the analyzed site. The information collected in this phase is analyzed and interpreted, having at the ending a detailed description of the quality of each candidate place. This phase involves studies more detailed, with application of more specific criteria, generally in a scale of 1:10.000.

Finally applying criteria such as minimization of potential for environmental impacts generation, longer useful life, reduction of implementation and operational costs, reclamation of degraded areas, etc., are established the potential for each site. The result of these studies is a discription of recommended sites, with justification, and measures in order to the correct utilization of each site, so that the manager can define together with the environmental agency, the real site to the installation of landfill for disposal of solid waste which is proposed.

These studies normally attend solicitations of the municipalities from Sao Paulo State, through the Municipality Technique Assistance Programme, of Secretary of Science, Technology and Economic Developing of Sao Paulo State.

2.2 Technical analysis of aspects related to landfill installation

Such activities have as purpose the analyse of documents related to the legal process of installation of sanitary or industrial landfills, such as Environmental Impacts Study (EIS) and its respective Environmental Impact Report (EIR), and the project of landfill itself.

These works are concentrated on the assessment of characterization of physical environment (mainly in the geologic, geotechnic and hydrogeologic aspects), influence of such aspects in the project and in the environmental protection. They consist in the assessing of the available documents, and in carry out 'in situ' investigations in order to obtain complementary information and also inspections to verify execution of designed works (specially in cases of assessment and following of projects).

Along the analises of technical documents is utilized a physical environmental approach (developed by the Division of Technological Research Institute) to assess environmental impacts. In this approach the components of physical environment are identified which can be affected by the project and assessed if the technological processes included in the project are sufficiently characterized, in order to prevent present and future risks to the environmental quality of the region.

The projects are analyzed according to the usual technology applied in such situations, being pointed any inadequation that will be found.

It is discussed aspects of site selection (geologic, pedologic, hydrogeologic aspects, etc.), design of the landfill, proposed systems for treatment of effluents and monitoring, etc.

The analises carried out point deficiences and also suggest measures to correct them, preventing future problems, with consequent damage to the environment and society. Frequently, some complementary studies are also required.

Such studies are carried out mainly under solicitations of Public Ministry and Secretary of Environment of Sao Paulo State.

3 IMPACT ASSESSEMENT DUE TO INADEQUATE DISPOSAL OF SOLID WASTE

The studies of impact assessment due to inadequate disposal of waste involve characterization of waste and affected environment, with determination of pollution extension and level of contamination, and proposal of mitigatory measures to the detected problems.

The studies that are specially important for Sao Paulo City, are related to disposal of contaminated sediments, from dredging process carried out on Tiete and Pinheiros rivers, when they cross the metropolitan area. At the Sao Paulo State level, studies related to characterization of contaminated areas by improper disposal of urban and industrial waste are predominant.

3.1 Disposal of contaminated sediments

Such studies are in the earlier phases of developing. Nevertheless, data already collected clearly show the pollutant effect of industrial effluents (directly discharged to these rivers) over sediments transported by them.

In the sediments dredged from Tiete, it was found higher concentrations of heavy metals, mainly mercury, lead and, in smaller quantity, nickel and chromium. This fact increase in gravity when we observe the deficiencies in the management of these dredged material with disposal of in landfill for inerts, along the riversides (some of them, inclusive situated within environmental protection areas).

The solutions foreseeable to solve this problem, besides main measures in course to control effluent emission into these rivers, pass also by elimination of Tiete and Pinheiros rivers silting causes. Also, it must be identified some possibilities of utilization of this sediments, such as the decreasing quantities of materials to disposed in landfill.

3.2 Remediation activities at contaminated sites

Under this major title are included studies related to domestic and industrial waste contamination. These activities have been developed just partially, mainly in the pollution characterization phase. However, IPT intends to start working deeper these problems, also carrying out planning and implementation of remediation activities, as well as possible, by applying simpler and lower cost measures, within reach of most municipalities and enterprises in the Sao Paulo State.

These activities will be specially necessary with the increasing strictness of the action of Sao Paulo Environmental Agency to obligate municipalities and industries to promote the improvement of solid waste management.

Activities carried out so far include diagnostic studies related to contamination by inadequate disposal of municipal solid waste (in open dump disposal sites), industrial waste (foundry industries) and leaks from petrol station storage tanks.

4 FINAL CONSIDERATION

The management of solid waste in the Sao Paulo State is demanding immediate improvement in face of serious problems nowadays we are trying to overcome.

As a government instrument, Technological Research Institute has been carrying out several activities in this subject, mainly in helping Sao Paulo State's municipalities to solve their problems, by setting up orientations in a near and long term planning.

These activities, until now concentrated only in some aspects of the management, are intended to be considerably increasing in the coming years such as this improvement in Solid Waste Management.

REFERENCES

Cunha, M.A. s.d. Recuperação de áreas degradadas. Anais 1. Seminário Internacional de Gestão e Tecnologia de Tratamento de Resíduos, 1, 1991, São Paulo: sessão 2, debates. São Paulo.

Cunha, M.A., Augusto Jr., F. 1991. Resíduo industrial inerte como alternativa para tratamento de leito de estrada de terra. Anais do 2. Simpósio sobre Rejeito e Disposição de Resíduos, 2, 1991, Rio de Janeiro. Rio de Janeiro: ABMS/ABGE/CBGB.

Cunha, M.A., Benvenuto, C. 1992. Escorregamento em taludes de aterro sanitário na cidade de São Paulo, Brasil. Atas Simpósio Latino Americano sobre Riesgo Geológico Urbano, 1992, Pereira, Colombia.

Cunha, M.A., Benvenuto, C. 1992. Principais problemas geológico-geotécnicos relacionados com a disposição de resíduos na Região Metropolitana de São Paulo. Anais 1. Seminário sobre Problemas Geológicos e Geotécnicos na Região Metropolitana de São Paulo, 1, 1992. São Paulo: 133-145. São Paulo: ABAS/ABGE/SBG/SP.

Cunha, M.A., Consoni, A.J. 1993. Os estudos do meio físico na disposição de resíduos. São Paulo: IPT. (4º Curso de Geologia Aplicada a Problemas Ambientais).

Fornasari Filho, N. (Coord.). 1992. Alterações do meio físico decorrentes de obras de engenharia. São Paulo: IPT 165p. (Boletim, 61).

Instituto de Pesquisas Tecnológicas do Estado de São Paulo. Diagnóstico hidrogeológico e hidrogeoquímico do aqüífero livre na área instalação da Usina Termoelétrica de Paulínia. São Paulo: IPT, 1990. (IPT, Realtório 28.082/90).

Instituto de Pesquisas Tecnológicas do Estado de São Paulo. Avaliação do poten-

cial de alteração da qualidade da água
subterrânea na área da Mineração Horii,
Moji das Cruzes, SP. São Paulo: IPT.
(IPT, Relatório 29.253/91).

Instituto de Pesquisas Tecnológicas do Estado de São Paulo. 1991. Levantamento e
análise das causas do escorregamento de
massa de lixo no Aterro Sanitário Ban-
deirantes – AS-I, em Perus, São Paulo.
São Paulo: IPT. (IPT, Relatório,
29.596/91).

Instituto de Pesquisas Tecnológicas do Estado de São Paulo. 1992. Apoio técnico
a avaliação de estudos de impacto ambiental, relatórios de impacto ambiental e
planos de recuperação de áreas degradadas. São Paulo: IPT. (IPT, Relatório,
29.919/92).

Instituto de Pesquisas Tecnológicas do Estado de São Paulo. 1991. Apoio e as-
sistência técnica na área de hidrogeo-
logia como subsídio a implantação do
Terminal de Derivados de Ribeirão Preto,
SP. São Paulo: IPT. (IPT, Relatórios
29.638/91, 29.941/91, 30.253/91,
30.431/92 e 30.797/92.

Instituto de Pesquisas Tecnológicas do Estado de São Paulo. 1992. Avaliação da
contaminação por hidrocarboneto das
águas subterrâneas do aqüífero livre na
área localizada no Jardim São João, Gua-
rulhos, SP. São Paulo: IPT. (IPT, Relatório 30.502/92).

Instituto de Pesquisas Tecnológicas do Estado de São Paulo. 1992. Considerações
sobre a questão ambiental de locais e
formas de disposição de resíduos sólidos
no município de Iguape, SP. São Paulo:
IPT. (IPT, Relatório 30.997/92).

Instituto de Pesquisas Tecnológicas do Estado de São Paulo. 1993. Análise técnica do projeto do aterro sanitário do Sí-
tio São João. São Paulo: IPT. (IPT, Relatório 31.350/93).

Instituto de Pesquisas Tecnológicas do Estado de São Paulo. 1993. Visita técnica
a empresa Profundir S.A. – Produtos de
Aciaria e Fundição, município de Praia
Grande, SP. São Paulo: IPT. (IPT, Relatório 30.560/93).

Instituto de Pesquisas Tecnológicas do Estado de São Paulo. 1993. Estudos geológicos para escolha de local para instalação de aterro sanitário no município
de Guaratinguetá, SP. São Paulo: IPT.
(IPT, Relatório 31.794/93).

Silva, W.S. da, Fornasari Filho, N. 1992.
Unidades de conservação ambiental e
áreas correlatas no Estado de São Paulo.
São Paulo: IPT. 85p. (Boletim 63).

Permafrost as a matter for burial of highly concentrated industrial waste

Le permafrost – Un facteur concernant le stockage des déchets industriels à haute concentration

V.N.Borisov & S.V.Alexeev
Institute of the Earth's Crust, Irkutsk, Russia

ABSTRACT: It is accepted world-wide to bury sewage at deep horizons or into artificially produced underground capacities. However, on the Earth there are enormous areas of permafrost sequences, impenetrable for poorly mineralised waters, where waste sewage may be buried. Sewage of some industries achieves mineralization 200-400 g/l and do not freeze at known -5..-2 °C temperatures of permafrost zone. When interacting with permafrost, the texture-like ice melts releasing space on account of water transition from solid to liquid phase, solutions fill the cavities freely. The physico-chemical disequilibrium of ice and brine is the driving force of interaction. This approach to bury industrial sewage is theoretically substantiated and realised in one of mining companies of Russia. Discharge of technogenic brines is performed via free pouring into a surfacial infiltration collector or through bore-holes. The authors summarized available data and established new factors affecting interaction of brines with ice (permafrost). The factors reflect physico-chemical properties of gas, parameters of the environment and development of processes.

RESUME: Comme c'est admis dans la pratique modialement répandue, on entère les diversoirs dans les horizons profonds ou dans les récipients souterrains spéciaux. Pourtant il existe sur la terre des régions où les diversoirs peuvent être enterrés dans la substance de nature spécifique-less masses du permafrost. Pour les eaux faiblement minéralisées elles sont impénétrables. Des diversoirs liquides de certaines productions atteignet la minéralisation jusqu'à 200-400 gr/l et ne gèlent pas aux températures du zone kryolitique -5..-2 °C. A leur interaction avec le sol gelé, la glace qui cimente le permafrost fond, des espaces libres se créent grâce au passage de l'eau de la phase solide a la phase liquide, les solutions remplissent les vides. La force motrice de l'interaction est la non-equivalence physique et chimique de la glace et de la saumur. Le moyen examiné de l'enterrement des diversoirs est théorétiquement argumenté et mis en pratique dans une des enterprise de l'extraction des matières premières en Russie. Le diversoir des saumurs technogènes se fait par le versement libre dans l'accumulateur infiltrant superficiel ou dans les puits. Les auteurs ont classifié les faits déjà connus et ont relevé des facteurs nouveaux influensant l'interaction de la saumur et de la glace (le permafrost). Ces facteurs reflètent les carateristiques et chimiques des phases, les paramètres du milieu et du développement des processus.

1 INTRODUCTION

Diversity of natural environment, in which human activity takes place, has expanded recently on account of the cryolite zone, particularly in Russia, where permafrost occupies over half of the total area of the country. In East Siberia, like in other regions of Russia, some industries cause accumulation of liquid industrial waste harmful for the environment. The results presented in this paper were obtained by the authors who performed investigations aimed at solving ecological problems at the mining company in Eastern Siberia. They consist in substantiation of the possibility, if adequate conditions are available, to bury some types of industrial waste in permafrost sequences. As is known, under common conditions the industrial waste is pumped to deep horizons of the

geological section, where the rock cavities are filled with pressure (artesian) waters. This necessitates displacing underground water from a consuming collector, as well as to use elastic capacity of the bed. The principle of burial in permafrost is on the contrary based on other phenomena, e.g. physico-chemical interaction between ice and mineralised solution. This interaction refers to the knowledge still to be developed, and our article is focused on generalization of available results.

2 PROCESSES IN BRINE-ICE (PERMAFROST) SYSTEM AT NEGATIVE TEMPERATURES

One of the features of the East Siberia cryolite zone consists in practically entire absence of fresh subpermafrost waters and on the other hand,

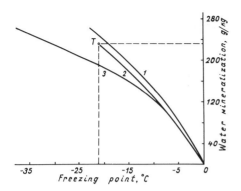

Fig. 1. Freezing temperature of some water solutions versus mineralization (degree of concentration). 1 - sea water, 2 - sodium chloride solution (reference data), T - cryoeutectic point of sodium chloride, 3 - ground water Na-Mg-Ca chloride brines of Eastern Siberia.

ubiquitous propagation of chloride brines with mineralization 100-300 g/l and more. Consequently, sewage produced by some mining and exploration companies show the same salinity. These are drainage waters of deep (iron ore, diamond extraction) quarries and mines, salt mines, petroleum-associated waters, etc. Their input may achieve many hundreds of cubic meters per hour.

Such brines, depending on mineralization and chemical composition, retain liquid and consequently chemically active state in a wide range of negative temperatures, e.g. to -22..-37 $^{\circ}$C and lower (Fig. 1), considering that they do not freeze at observed temperatures of intra-permafrost sequences (commonly -5..-2 $^{\circ}$C). This is why brines are not in equilibrium in contact with ice and melt it as the temperature of the environment is above the freezing point of liquid phase, which is well known to permafrost specialists.

Brines are able to transform ice to the liquid phase, to eliminate water resistance of permafrost rocks, their bound state and bearing capacities.

The literature review indicates that interaction of ice and mineralized solutions attracted many scientists. In particular, the rate of melting of ice specimen immersed in solution was repeatedly determined in experiments. Note that results of experiments are far from imitation of natural and many technogenic systems and in principle demonstrate the effect expected. The necessary and appropriate conditions, which define the possibility to realize the processes under examination, remain insufficiently revealed. To fully understand the problem the authors earlier (Pinneker et al. 1989) systematized the factors responsible for the processes in the equilibrated system: this was the first compilation of such type. Later these factors were supplemented, refined and represented herein as complete. The major conclusions were based on

the data of laboratory experiments and test hydrogeological works devoted to the study of water-bearing capacity, drainage of mining workings and burial of drainage brines with mineralization of 300 g/l and more in then northern region of Eastern Siberia (Table 1).

The first order factors imply the properties of brine, ice, the mechanism of their interaction at unequilibrated temperature, exo- and endothermal effects. The rest represent the refined value of stability of the rock structure after melting ice filling the pores and cracks. In contrast to rocks, loose and poorly lithified deposits are deformed, and liquid phase is saturated with finely dispersed material up to the floating earth state which slows down or arrests the process.

The role of chemical composition of the brine is in a large difference of values of cryoeutectic points for solutions of some salts. In water solutions the temperature of freezing may vary from -1.9 to -55 $^{\circ}$C, accordingly, from the point of calcite precipitation to cryoeutectic of calcium chloride. In solutions of complex composition the temperature of crystallization of some salts ranges. For instance, the cryohydrate of sodium chloride started to precipitate from Mg-Na-Ca chloride brine with mineralization 320 g/l, when the solution was cooled to -37 $^{\circ}$C, whereas having the solution of this salt only, precipitation takes place at -22 $^{\circ}$C.

The third group of factors was studied to a lesser extent and displays elements of some novelty. This first of all refers to the spatial ratio between the solid and liquid phases, conditions for density convection of water, zonation along the boundary of phases after intensity of ice melting. These factors control the main condition for ice melting in the brine, i.e. the possibility to remove desalinated solution from the zone of contact. Any other favourable prerequisites are insufficient for the process to proceed. If such a condition is not the case, the system is in equilibrium. At the same time, the outflow from the zone of contact depends on the position of liquid phase relative to ice: whether it is lower or higher than ice, or in the lateral contact. Note that in this study only closed systems are concerned, in which the water movement occurs via density convection, that is without mechanic filtration.

The first case (Fig. 2) is typical for permafrost-hydrogeological sections, when sub-permafrost waters are represented by negative temperature brines. On the contact between the phases there is always a layer of desalination, in which the density of solution is lower than that of the underlying brine, that is why convection is absent there. If the system is close and desalinated mixture does not flow from the zone of contact, its salinity is established near the point of cryoeutectic for the given temperature of the environment, and melting of ice ceases. The concentration diffusion at negative temperature sharply slows down and does

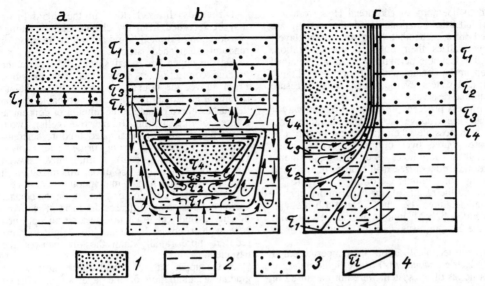

Fig. 2. Liquid-solid phase interaction in the systems: a - brine beneath ice, b - brine over ice, c - brine in lateral contact with ice (from experimental data) 1 - primary ice, 2 - brine, 3 - secondary ice, 4 - position of boundary (contact) between phases in time τ_i.

Table 1. Factors and results of brine-ice (permafrost) interaction

Brine properties	Ice properties	Environments characteristics	Process indices	Results
Temperature	Temperature	Spatial position of brine relative to ice (over, lower, lateral contact)	Density gradient in liquid phase	Complete or partial melting of ice
Mineralization	Mineralization	Area of contact	Zonation in liquid phase along contact with ice	Desalinisation of brine
Density	Volume, form and area of body surface	Degree of closure (current) of space of interacting	Endo- and exothermal effects (heat of melting and ice formation, dissolution and crystallization of salts	Liberation of space on account of phase transition of water
Chemical composition	Structure	Condition for density convection of liquid phase	Concentration diffusion	Formations of secondary saline ice
Volume	Inclusion: saline, mechanic gaseous	Possibility of ice floating up in the volume of interacting	Duration and rate of interaction	Increase of permeability of rocky permafrost
	Spatial orientation of body (horisontal, vertical, inclined)	Degree of lithificatin of hosting rocks and stability of cavities in rock (after ice melting)		Distortion of stability and bearing capacities of loose and poorly lithified rocks
		Permeability of rocks		Formation of technogenic inter-permafrost "taliks"

not lead to liberation of additional H_2O from ice. As a result, the components of the system are separated from each other by a transitional layer, which neutralizes their potential unequilibration. Due to this fact, the permafrost rocks are retained unaltered in the contact with subpermafrost highly concentrated waters even at high water pressure and unlimited reserves.

The process proceeds in quite a different way if the brine occurs above ice or has lateral contact with it. It is possible when the natural or artificial brines are poured over permafrost rocks: when pumped from bore-holes and filling in them, preparation of mineralized drill mud in the pits and desiccation of mining workings. Figure 2 demonstrates the results of our experiments performed at temperatures -12..-5 °C. Natural brine with mineralization 320 g/l poured in the vessels with ice on the bottom, was stained for visual observation.

Because of technical reasons the temperature and density (electric conductivity) of the liquid phase in different points of the system were not measured in detail.

The brine applied begins to freeze at a fairly strong cooling (-37 °C), therefore the process proceeded rapidly, its first evidence became evident during one hour. It appears that the brine penetrates through the contact of ice with the walls of vessel and solid inclusions (in one of the experiments the ice-clasts mixture was applied), that is melting of ice occurs not ubiquitously but selectively, i.e. the brine flows over ice surface (the factor reflecting the shape and area of ice surface). This is true because initially these were the forces of electrochemical nature (molecular attraction, wetting, sorption) that were active. They are followed by the effect of space liberation in transition H_2O in the liquid phase. Between ice and inert body there originates and rapidly expands a cavity subsequently filled by solution. Convection - floating of desalinated mixture occur over the ice surface, the brine descends along the walls of foreign bodies. In the meantime the heavy brine tends to move to the lower part of liberated space, where extremely unequilibrated with ice, it intensely melts its base.

Zonation is formed according to the degree of the system unequilibration from below upwards along the base and lateral walls of ice body in the liquid phase. The zone of the most "aggressive" contact occurs in the low point of interaction where the first portions of desalinated solutions emerge and the mixture starts ascending. In the latter the H_2O content constantly raises and floating of ice slows down; consequently distinguishing the zone of the process slowing down is possible. With a sufficient extensive movement of ascending liquid it desalinates to such an extent that loses ability to melt ice, which results in origination of the neutral contact zone.

Differently directed flows in the liquid phase may be recorded from the change of its staining, decoloration of diluted liquid as well as transport of randomly suspended particles.

On the upper horizontal surface of the ice the ordered convection is deteriorated and invisible. From above, the boundary lowered in parallel to the initial horizontal position. It is probable that on the surface there existed some amount of concentric centres of ice melting (convective cells), from which the streams of desalinated mixture surrounded by a descending strong brine, went down. Dissolution of ice occurred uniformly, slowed down with decreasing of solution mineralization in this part of the medium of interaction.

As a result of zonation origination on the contact between the interacting phases the form of ice from initially cylindrical gradually became an upturned cone with truncated basement; the height of cone reduced more rapidly than the diameter of the basement.

In the upper part of liquid phase there was a gradual accumulation of the layer of desalinated liquid with lighter coloration than lower-lying solution. In the end of experiments, which lasted 3-4 days, the layer of desalination froze.

In the experimental study of lateral interaction of the same solution and ice (analogy consists in pouring of brines in the bore-hole driven through permafrost rocks) along the vertical contact the liquid immediately started a regular ordered convective movement. The ascending desalinated mixture gradually terminates to affect ice, which melted more rapidly in the lower part per unit of time, than in the upper one. During a day the base of the ice body separated from the cylinder bottom, in which the experiment was conducted. The strong brine first of all dissolved the most immersed part of ice. This is why its base, at first tilted, gradually acquired the horizontal orientation and further ascended in parallel to this position.

The process ceased when the layer of desalinated mixture gradually accumulated from the side of ice, then became as thick as the rest of ice, and the horizontal component of density gradient disappeared in the liquid phase. The dissolved brine also froze, the secondary ice appeared in the system. Its salinity at temperature -5 °C was 75 g/l, which corresponds to the experimentally established dependence of freezing temperature of brines of similar complex composition on their mineralization. It has been revealed by experiments that brine also penetrates inside the ice body through different heterogeneities, e.g. crystallization, texture, gas bubbles, mechanic deformations which generally occur in ice.

For many industries the study of the processes of brine-ice interaction is greatly important, particularly when the production technology envisages work in permafrost, where brines come into contact with permafrost rocks: drilling bore-holes with application of washing solutions; test of hydrogeological works, interpretation of results; storage and discharge of related brines at oil and gas industries; desiccation of mining workings, discharge of drainage waters; exploration and development of salt deposits; construction of sewage collectors; chemical industries.

Theoretically, of particular significance is the role of brines in formation of cryolite zone, their influence on permafrost continuity, the depth of freezing and cooling, formation of gas hydrates.

3 SPECIFIC FEATURES OF PERMAFROST AS THE MATTER FOR SEWAGE BURIAL

The suitability of objects for burial of industrial waste is known to be evaluated from a number of parameters, e.g. permeability, depth of occurrence of consuming collector, hydraulic relation to adjacent blocks of rocks, isolation from the site of polluted water outburst onto the Earth surface and in the ground water horizons. In this respect permafrost is characterized by peculiar qualities preconditioned by the presence of specific substance (ice) in pores, an impact of phase transitions of water on the rock structure, local specific conditions of freezing and evolution of mountainous massif.

According to existing viewpoints, permafrost does not represent a continuous screen which would make the exchange between subpermafrost and suprapermafrost water impossible (Olovin 1993). As a rule, cryogenic sequences contain two (lower and upper) ice-saturated layers, between which the cavities in rocks are filled with gas rather than ice. Formation of such zonation may be exemplified by the region in the north of Eastern Siberia where permafrost formed on the Pliocene-Pleistocene boundary. In that period cooling of climate was followed by uplifting of land and lowering the world ocean level by 100-130 m. The boundary of land in Siberia moved 800-900 km northwards from the modern position. The climate became more continental, amount of liquid sediments and meteogenic recharge of ground water sharply reduced. This caused a profound non compensated draining of rock massifs propagating into the earth interior beneath erosional incisions, that is the aeration zone thickness noticeably increased. Under conditions of advancing cooling of the interior the depth of the liquid phase infiltration was limited by the near-surface interval and on average did not exceed the 30-50 m depth. From above there originated a continuous water-resistant ice-rich cover, below which the most of cavities in rocks remained ice-devoid (Fig. 3). Depending on the relative raggedness of the relief (at present on average 150-300 m) the thickness of the zone of ice-free collectors may achieve many dozens of meters. Below it is restrained by the

depth of regional paleodrainage, that is the level of the upper ancient boundary of water-saturated zone. The latter also froze with increasing the depth of cooling the geological cross-section (about 1000 m in local conditions). It appears that ice-free collectors may be preserved only inside watershed massifs, whereas in the river valleys the upper and lower horizons of continuous ice content merge.

Such a scheme of formation and structure of permafrost is supported by the facts from practice: by observations of ice-free cavities in drill core, intense consumption of drainage brines, drilling mud for bore-hole driving, output or pumping of air, etc.

Descriptions of such structure and formation of permafrost, presence of interrelated cavities are made by many Russian scientists (Bakulin 1958; Tsitovich 1973; Romanovsky 1977; et al.), the newest and most complete investigation of this problem is accomplished by B.A. Olovin (1993).

Deep drainage of mountainous massif at the stage of freezing is not the only reason for presence of cavities with no ice. The important role in displacing water from cavities belonged to underground free gases, their mass migration from deep horizons of sedimentary basins in Eastern Siberia refers to typical phenomena. The resultant upper ice-rich screen (together with lithological confining layer) contributed to accumulation of cooling beds of primarily methane and nitrogen-methane gases. The formation of permafrost was accompanied by a noticeable reduction decline of pressure in sub-permafrost acquifers, that intensified separation of gases in a free state dissolved in ground waters and increased their total outflow from the Earth interior. In the present environment, the high gas saturation of collectors in permafrost is reflected in numerous gas occurrences in the boré-holes, in some cases as large fountains. The depth of occurrence of pressure gas accumulations (from 30 to 150-200 m) approximately corresponds to the ice-poor part of the geological cross-section.

Thus, permafrost is not an obligatory fluid rest for gases except for the upper ice-rich layer and are accessible for filtration of mineralized solutions not freezing at natural temperature of the cross-section. For strong brines icy rock is not an obstacle as well if it is permeable in melt state. As is considered above, the system "brine-over-ice" is extremely disbalanced, in which the processes cause melting ice and convective redistribution of phases in accordance with revealed regularities and principle of self-organisation of unstable systems. Heavy brine will occupy the lowest position and its mixture with H_2O will float up together with the rest of ice separated from the walls of the capacity of interaction (vessel in tests or crack in rock) and will freeze as the secondary saline ice.

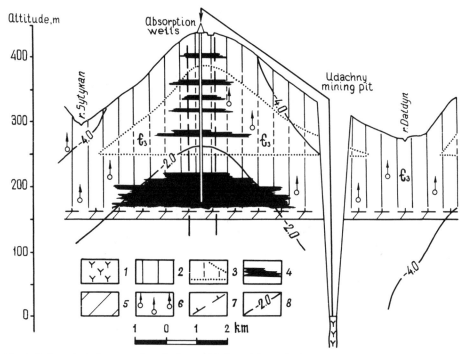

Fig. 3. Hydrogeological cross-section of polygon of drainage brine burial in permafrost. 1 - kimberlites, 2 - icy fluid-resistant sporadically gas-saturated interval of permafrost, 3 - potentially ice-devoid gas-saturated rocks, 4 - technogenic "talik" filled with drainage brines, 5 - sub-permafrost horizon of underground brines with negative temperature, 6 - local occurrences of sodium-hydrocarbon gases, 7 - lower boundary of permafrost, 8 - isotherms.

Analogous consequences will be also imprinted in the environment in which semi-open and open systems predominate. Due to repeated cycles of freezing and thawing rocks are subjected to intense cryogenic disintegration, which is traced for dozens of meters, in places up to 100-200 m, inside the permafrost massif. This results in higher permeability than in rocks not affected by cryogenesis. This is why, the brine placed on permafrost ground melting ice, will tend to the lowest part of the entire hydraulically bound space. With unlimited volume of brine and conservation in liquid phase of density gradient the process may in principle develop to the depth of wedging of permeable channels, that is to the lithological confining layer or the basement of the zone of exogenous fissuring. In this sense the role of tectonic fracturing is more understandable.

In this case the penetration of brine through pores and fissures filled with ice is concerned. After it achieves ice-free cavities there begins descending filtration to the lower ice-saturated permafrost layer, where disbalance of interacting phases occurs. This leads to formation of open or closed "talik" in the rock sequence retaining negative temperature and frozen state. The useful

capacity of the permafrost massif needed for sewage burial is defined not only by connecting cavities devoid of ice. The capacity reserve is also available as the space released in H_2O transition in the liquid phase, this is about 9% of ice volume in rocks. Besides, of real value is air porosity of natural ice, on average providing 3% (Saveliev 1980). In core samples from bore-holes the authors observed concentration of gas bubbles to 5-10%, which is not rarity for permafrost ice, particularly in the zones of tectonic fracturing. To sum this up, the useful capacity may achieve the volume sufficient for practical burial of sewage. This was confirmed by the industrial experience of Joint Stock Company "Diamonds of Russia-Sakha".

One of the diamond-mining quarries is located in the permafrost area, where the thickness of freezing stage amounts to 150-600 m, and zero isotherm is traced to the depth about 1100 m. The temperature of rocks at the depth of zero amplitude (15-20 m) varies from -2.8 to -7.8 °C. The zone of fresh ground waters is entirely frozen, sub-permafrost waters are represented by chloride sodium-magnesium-calcium brines, their minera-lization achieving 300-400 g/l. The geological cross-section is composed of sedimentary deposits of the Lower Palaeozoic age: intercalated dolomites, limestones, marls and their clay varieties of total thickness 2500 m. This sequence is disrupted by kimberlite pipes and trap intrusions (dolerites). The ground water influx in the quarry started at its depth about 110 m progressively increasing; at present the depth of the quarry is over 300 m, mineralization of drainage waters is over 300 g/l.

In the initial period of the quarry desiccation an attempt was made to store drainage waters in a pit 100 m deep. When making the pit, the layer of fissured carbonatites in places was destructed and cemented by ice. After the first attempt to fill the pit with drainage water (mineralization not over 40 g/l), it was found that on its bottom opened exogenous and tectonic fissures through which water is entirely absorbed. Thus, over 100000 m^3 of liquid with average debit 25-35 l/sec was discharged. Warm season and positive temperature of drainage waters favoured intensification of absorption. This caused rapid melting of ice-cement of permafrost and establishment of hydraulic relation with lower-lying gas-saturated cavities. In winter time mineralization of drainage waters raised to 170-200 g/l and their temperature dropped to -2..-5 °C, but absorption proceeded. In permafrost a series of technogenic acquifers within the depth range 30-180 m was formed (Fig. 3).

At the next stage the discharge of drainage waters went on via free pouring through bore-holes into the same massif. The test works established distribution of fissured absorbing collector within depth interval 35-220 m. The thickness of permeable beds (limestones) varies from 10-20 cm to 10-20 m , the technogenic concentrations of water are commonly formed on interbeds of clay carbonates. For eight years of burial polygon development in the permafrost massif about 3500000 m^3 of brine with mineralization 200-300 g/l was discharged. Part of water is supposed to return back to the quarry. However this does not hinder its further working out.

Distribution of technogenic water-bearing concentrations in the upper part of geological cross-section shows temporary character, with the tendency of their submerging into lower-lying beds. The ways for descending migration are not only lithologically permeable windows and tectonic fissuring but numerous observation bore-holes designated for degassing the massif of burial. Practically in all holes, after reaching technogenic waters, great ice corks are formed, evidently due to accumulation of desalinated mixture of drainage brines with melt ice in the upper part of the water column. Ice corks are the evidence of secondary ice which conserves from above the space occupied by absorbed drainage water.

Applying the procedure proposed successful burial of sewage is due to favourable circumstances: presence of appropriate permafrost-hydrogeological, structural-tectonic, hydrogeochemical features of the region. Considering these conditions some lines of evidence may be put forward for assessment of potential objects of burial and prospects of application of this procedure.

First of all, permafrost sequences should have an areal distribution, significant thickness (100-200 m) and position of the lower boundary 30-50 m relative to the bottom of the nearest valleys. Their temperature must be low (-2..-3 °C) and its existence should be predicted for some hundred years. These qualities may be best referred to permafrost of the northern type. Considering the volume of water capacity the massifs of sedimentary lithified (rocky) deposits, as well as coarse-grained sediments with isotropic permeability are preferable. For the procedure of pumping it is not expedient to choose the zones of tectonic fracturing which play the role of "large consuming bore-holes with open shaft". Permafrost is proposed for burial of relatively low-toxic sewage, the basic substance of which are natural brines or highly soluble salts (part of them enumerated above). Such solutions are common for natural environment, as they often reach rivers via artesian discharge (through springs).

The authors express deep gratitude to specialists of Stock Company "Diamonds of Russia-Sakha" whose help in testing experimental and theoretical studies was invaluable. Ms. Tatiana Bounaeva is thanked for translating the text into English.

REFERENCES

Bakulin, F.G. 1958. *Ice content and sediments in melting permafrost of Quaternary deposits of Vorkuta region.* Moscow: USSR Academy of Sci. Publ. House (in Russian).

Olovin, B.A. 1993. *Filtration permeability of permafrost.* Novosibirsk: Nauka (in Russian).

Pinneker, E.V., S.V. Alexeev, V.N. Borisov 1989. The interaction of brines and permafrost. *Proc. WRI-6 International Symposium:* 557-560. Rotterdam: Balkema.

Romanovskiy, N.N. 1977. *Formation of polygonal ore structures.* Novosibirsk: Nauka (in Russian).

Saveliev, B.A. 1980. *Structure and composition of natural ice.* Moscow: Moscow State Univ. Publ. House (in Russian).

Tsitovich, N.A. 1973. *Mechanic features of permafrost ground.* Moscow: Vysshaya Shkola (in Russian).

Geomorphic assessment for a waste disposal facility

Évaluation géomorphique pour un stockage de déchets

John D. Rockaway
University of Missouri-Rolla, Mo., USA

Rheta J. Smith
Jacobs Engineering Group, Weldon Spring Site Remedial Action Project, US Department of Energy, St. Charles, Mo., USA

ABSTRACT: A critical concern in the design of a hazardous waste disposal facility is the long-term geomorphic stability of the disposal site. Site location in a geomorphologically stable area is imperative to prevent the failure of retainment structures and to protect surface and groundwater quality. Evaluation of landscape stability is made through the study of both short- and long-term rates of geomorphic processes. Engineered facilities may provide enhanced protection of the disposal structure over short time periods (i.e. historic time); but a successful long-term containment (hundreds to thousands of years) requires sitting the facility at a location where geomorphological processes will not diminish its structural integrity. A geomorphological assessment provides an evaluation of the impact that geomorphic processes will have on site stability.

RÉSUMÉ: Un souci important pour la planification d'un dépôt de déchets hasardeux est la stabilité géomorphique a long terme du site. Il est essentiel de choisir le site dans une région géomorphologiquement stable pour empêcher la rupture des structures de retenue et pour protéger la qualité de l'eau de surface et de la nappe phréatique. L'évaluation de la stabilité du paysage se fait par l'étude, a court terme et a long terme, des taux des processus géomorphiques. Des dépôts étudiés peuvent offrir une protection améliorée de la structure a court terme (i.e. le temps historique); mais une retenue réussie a long terme (de plusieurs centaines à des milliers d'années) exige qu'on situe le dépôt sur un site où les processus géomorphologiques ne diminueront pas l'intégrité de la structure. Une évaluation géomorphologique fournit un jugement sur l'impact que des processus géomorphiques auront sur la stabilité du site.

1.0 INTRODUCTION

A critical concern in the design of a disposal facility is the long-term geomorphic stability of the disposal site. Site location in a geomorphologically stable area is imperative to prevent the failure of retainment structures and to protect surface and groundwater quality. Evaluation of landscape stability, in terms of historical long-term time periods, is made through the study of long- and short-term rates of geomorphic processes.

Engineering considerations may provide protection of the disposal structure over short time periods (i.e. historic time); however, successful long-term containment (hundreds to thousands of years) requires an extensive evaluation of regional and local geomorphic settings and processes. The objective of this investigation has been to assess the geomorphic stability of the disposal cell encapsulation system for the short-term (0 - 50 year), intermediate-term (50 - 200 year), and long-term (200 - 1000 year) durations.

2.0 GEOMORPHIC SETTING

The Weldon Spring Site Remedial Action Project (WSSRAP) has a unique gemorphological setting (Figure 1) in the mid-continent United States. It is located at the boundary between the Dissected Till Plain section of the Central Lowlands Physiographic Province and the Salem Plateau Section of the Ozark Plateaus Physiographic Province. This boundary is the southern limit of Pleistocene glaciation. The site also is situated across the divide separating the drainage basins of the Missouri and Mississippi River systems. The Weldon Spring Remedial Action Project thus, is located within a transition zone defined by the boundaries among a number of major geomorphic features.

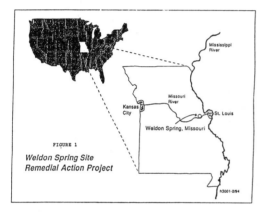

FIGURE 1

Weldon Spring Site
Remedial Action Project

The northern two-thirds of the study area which includes the chemical plant property, is in the Dissected Till Plains section (Figure 2). This area is characterized by moderately to slightly undulating topography developed upon the loess and glacial deposits which overlie the limestone bedrock. The area is moderately dissected with broad, flat drainage divides between the tributary streams. The southern one-third of the area lies on the northeastern flank of the Salem Plateau Section. This area is characterized by rugged topography, narrow, irregular drainage divides and is drained by many short streams with steep

gradients. The transition area between the two provinces is just south of the chemical plant and corresponds to a primary drainage basin divide.

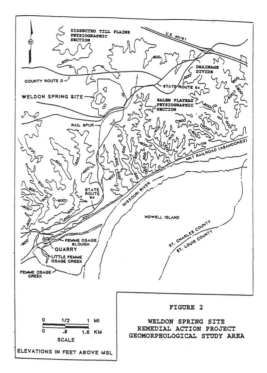

FIGURE 2

WELDON SPRING SITE
REMEDIAL ACTION PROJECT
GEOMORPHOLOGICAL STUDY AREA

The site itself is for the most part an upland area with rolling topography, becoming more rugged on the adjacent slopes. To the north, the topography in the Dardenne Creek watershed is gently rolling. This area of the site has been developed on Kansan age till and is overlain by loess deposits. Areas draining southward to the Missouri River are much more rugged and highly dissected. Glacial till is for all practical purposes absent south of the drainage divide.

Topographic elevations in the area range from a maximum of about 220 meters (msl) along the drainage divide to a minimum of about 135 meters (msl) along the Missouri River. In general, the relief from the Missouri River alluvial plain to the drainage divide is about 85 meters, which occurs over a distance of about three kilometers. To the north, the elevation drops to 145 meters (msl) along Dardenne Creek over a distance of about five kilometers.

The landform features at the site have developed as the result of the characteristic erosional and depositional processes associated with fluvial, glacial and eolian activity as well as in-situ weathering. The extent and nature of these processes is illustrated in the generalized stratigraphic column (Figure 3), which includes residuum over weathered bedrock, overlain by glacial deposits, which turn are overlain by loess. The present landforms at the Weldon Spring Site have developed primarily as a result of the erosion of these materials by surface waters.

3.0 ANALYSIS
3.1 LONG-TERM STABILITY

One of the major criteria for evaluating the suitability of a site for the long-term containment of hazardous wastes is to determine that the land surface (landforms) are stable. Landscapes which have been in existence for thousands of years represent stable conditions. Such landscapes suggest that major erosional and depositional processes have not altered the landscape. Although modern geomorphic processes may be modifying the old landscape, the presence of older landscapes suggests a stable environment for disposal structures.

Determining landscape age may be accomplished by a number of procedures (Longmire, 1985; Schumm, 1985). These include:

1) Radiometric dating of material within surficial deposits making up the landscape (i.e., organic carbon).
2) Correlating stratigraphic or geomorphic features with known dated sequences.
3) Determining the degree of soil-profile development on the surficial deposits (i.e., determining a chronosequence when parent materials, climate, vegetation, and topography are similar.

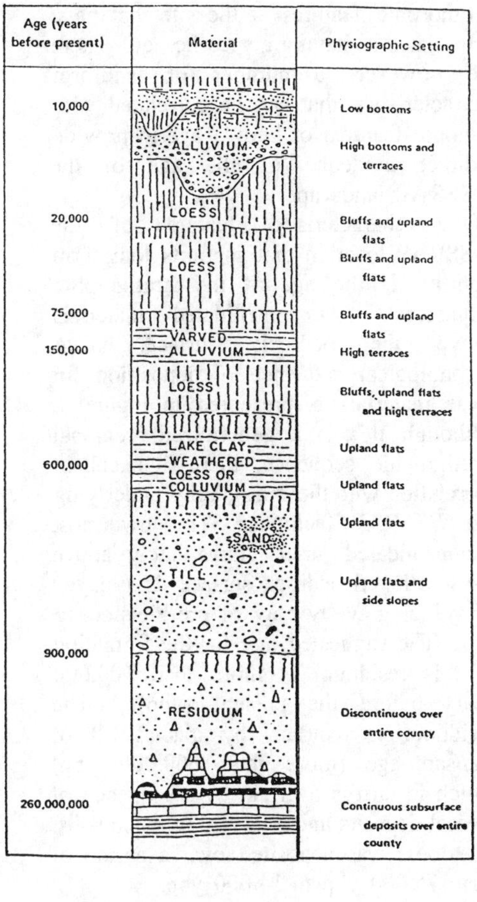

Age (years before present)	Material	Physiographic Setting
10,000	ALLUVIUM	Low bottoms
		High bottoms and terraces
20,000	LOESS	Bluffs and upland flats
	LOESS	Bluffs and upland flats
75,000		Bluffs and upland flats
150,000	VARVED ALLUVIUM	High terraces
	LOESS	Bluffs, upland flats and high terraces
	LAKE CLAY, WEATHERED	Upland flats
600,000	LOESS OR COLLUVIUM	Upland flats
		Upland flats
	SAND, TILL	Upland flats and side slopes
900,000		
	RESIDUUM	Discontinuous over entire county
260,000,000		Continuous subsurface deposits over entire county

FIGURE 3

Chronology of Surficial Material Deposition, WSSRAP Vicinity
(Adapted from Allen and Ward, 1977)

3.1.1 RADIOMETRIC DATING

No studies involving the radiometric dating of organics included within the soils or unconsolidated surficial materials were conducted for this investigation. Such studies would provide quantitative data which would permit a definite determination of the age of the WSSRAP landscape.

3.1.2 CORRELATION WITH DATED SEQUENCES

There are no specific stratigraphic or

geomorphic features at the site that have been dated as having a specific age. There are however, a number of landform characteristics that can be correlated with regional features of known age to provide indirect evidence of the age of the WSSRAP landscape.

One characteristic feature of the WSSRAP location that supports long-term stability is the age of the stratigraphic sequences of unconsolidated materials above the bedrock surface. A chronological sequence of deposition for these materials is presented in Figure 3. Although this is a generalized, regional stratigraphic sequence, there is excellent correlation with the stratigraphy underlying the WSSRAP facilities. The lowermost unconsolidated strata, immediately above the weathered bedrock surface, is a clayey gravel or gravelly clay of pre-Pleistocene age. The suggested age for the formation of this residuum is more than 900,000 years before the present (ybp). The residuum is overlain by glacial till of Kansan age (more than 600,000 ybp) which in turn is overlain by a sequence of loessal deposits and intervening paleo soils. The loess was deposited over a period of from 75,000 ybp to 14,000 ybp.

The chronology defined for the deposition of the unconsolidated materials upon which the present landscape has been developed indicates that these landforms and materials have maintained their stability since the most recent period of loessal depositions more than 14,000 years ago.

3.1.3 SOIL PROFILE DEVELOPMENT

Degree of soil profile development is a major criteria for assessing long-term stability of surfaces. Better developed profiles (those with thicker B horizons and well defined structure) indicate long-term stability. Although the soils in the vicinity of the chemical plant have been disturbed by cultural activity, the undisturbed soils surrounding the plant site exhibit the characteristics of a well developed B horizon.

The major soil series in this area are the Armster, Mexico and Weller (U.S. Dept. Ag., 1982). The parent material from which these soils have developed is loess, which was deposited during the latter stages of Pleistocene glaciation. Collectively, these soils are characterized as having a well developed, fine to medium, blocky, subangular structure. The boundaries of the B horizon are well defined and the thickness of the B horizon ranges from about 18 inches for the Mexico soils to 53 inches for the Weller. It is understood that without radiometric dating, the absolute age of a soil usually cannot be determined. However, established age for the deposition of the parent materials and the well developed and well defined characteristics of the B horizons indicate that the surficial materials of the upland regions of the WSSRAP location have been subject to a stable geomorphic environment for period of many thousands of years.

3.1.4 ARCHAEOLOGICAL ANALOGY

An indirect approach to evaluating the anticipated geomorphic stability of the disposal cell is through comparison with archaeological evidence. Archaeological data, particularly that obtained by the evaluation of archaeological mounds as analogs of engineered covers, can provide insight into the long-term effects of geomorphic processes on closure caps.

Archaeological mounds are analogs to hazardous waste disposal cells in that they also are constructed of soil embankments, are sometimes protected by rock covers, are similar in size, and are subject to the same erosional processes. The mounds were usually constructed as roughly four-sided or conical structures and typically were located on relatively flat topography. Thus the design and location concepts are similar to those of disposal cells. Since archaeological mounds are hundreds to thousands of years old their performance is a record of long-term geomorphic stability.

Monks Mound in Madison County, Illinois, is an excellent example of such a structure in the vicinity of the Weldon Spring site. Although the similarities in form, construction details and intent are very elemental, a comparison of the proposed configuration of the Weldon Spring disposal cell with that of Monks Mound indicates that the stability record of Monks Mound is an appropriate analog for the disposal cell. Monks Mound is a 30 meters high, rectangular platform-shaped structure with a base area of approximately 60,000 square meters. It is located on level topography, composed principally of clay and silty clay, and was constructed in stages, sometime during the period between 900 to 1150 AD. Monks Mound is considered to be in good condition and has not experienced much erosive damage.

3.2 GEOMORPHIC PROCESSES

The WSSRAP is located in an upland area of rolling topography in a humid, temperate climate. The principle geomorphic processes which have been active in this environment are rejuvenation, gully extension and/or sediment deposition in the drainage network; degradation, knickpoint migration and bank erosion in the existing stream channels; instability, retreat and mass movement of the surrounding slopes; and solution of the underlying limestone bedrock to form characteristic karst features. Although the surficial materials and landforms of the WSSRAP location are formed of wind deposited sediments, eolian processes are no longer active under present climatic conditions. Of the geomorphic processes that are active, rejuvenation and gully extension in the drainage network is the process with the greatest potential impact on geomorphic stability.

3.2.1 DRAINAGE NETWORK EXTENSION

Drainage network extension may be expressed in terms of the development of rills from overland flow and the headward extension of gullies in the outlying reaches of the drainage basin.

Analysis of areal photography of site conditions as they existed prior to landscape modification required for construction of the chemical plant (U.S. Dept. Ag., 1937) indicates that the surficial materials are susceptible to gully erosion. Prior to construction of the chemical plant, land use in the area now included within the WSSRAP was almost entirely agricultural. Interpretation of the photography illustrates that gully erosion occurred quite frequently in the upper reaches of the tributary draingeways in the open fields and active bank erosion was occurring along some of the defined stream channels. It should be noted however that agricultural practice in 1937 did not incorporate conservation practices that are designed to minimize erosive activity.

The visual historical record of the WSSRAP location is about 55 years (USDA 1937 aerial photography). Conceptually, gully development over the 55 year period could be measured to define a short-term rate of gully advancement. Unfortunately, present expressions of gully erosion, although reflective of active geomorphic processes at the Weldon Spring Site, are unreliable indications of the susceptibility of the surficial materials and landforms to short-term, let alone long-term, gully erosion. As noted, the landscape was extensively modified for the construction of the chemical plant facilities and for over 100 years prior to that, the area was used for agricultural purposes. It is to be expected that much of the existing surface erosion was initiated as a result of previous farming activity.

3.2.2 DEGRADATION, KNICKPOINT MIGRATION AND CHANNEL EROSION

The immediate area of the WSSRAP chemical plant and proposed borrow source area is drained by a number of ephemeral streams. South of the drainage divide upon which the site is located the streams drain directly to the Missouri River. The base level for these drainageways is the present elevation of the Missouri River flood plain. These streams are characterized as having rather steep gradients in their lower reaches with a concave upward stream profile. Although the concave profile suggests the potential for degradation and valley erosion up the channel, the stream channels are incised into the underlying bedrock throughout most of their course. In the upstream areas where the stream channels are underlain by unconsolidated surficial materials, the stream profiles are less steep and generally convex in form.

Channel erosion in the southward flowing streams will be controlled by the presence of the underlying bedrock. The down stream areas, although steeper, will be subject to little channel erosion. Lateral migration and valley side erosion along the ephemeral stream channels is minimal because of the relative straightness of the channels, the confining bedrock slopes and the relatively small ephemeral discharges. Erosion that does occur will be principally that associated with removing unconsolidated materials derived from slope wash or past depositional cycles from the channel basin.

Overall, the erosion of the bedrock channel itself can be expected to proceed at about the general rate associated with the regional rate of long-term denudation. There have been attempts to evaluate rates of erosion on a regional basis (Schumm, 1963, Young, 1969) but the resulting values (in terms of mm/1000 years) are subject to considerable deviation. There are a number of difficulties encountered in applying those theoretical rates to present landscapes and those limitations preclude their usefulness for site specific studies (Daniels and Hammer, 1992).

3.2.3 SLOPE INSTABILITY AND MASS MOVEMENT

The slopes in the vicinity of the WSSRAP facilities are quite variable. The slopes south of the upland upon which the facilities are located are steep and stream channels are deeply incised into the valley sides. Typically the upper parts of the slopes are convex and moderately steep, whereas the lower slopes are straight and steep. The slopes are composed of a variable thickness of residual soil and colluvium over the weathered bedrock surface. Bedrock outcrops are common on the steeper slopes near the stream channels. The slopes north of the uplands are much more gentle. Slopes are composed of unconsolidated materials, principally loess and glacial till. The slopes are broadly convex in form and do not steepen significantly toward the base.

Overall, mass movement is not a significant geomorphic process on the slopes on either side of the uplands. There is little evidence of past slope instability and that which has occurred has been limited to the lower parts of the steeper slopes on the south side. Here shallow movement of overlying colluvium and residual material has occurred on the top of the bedrock surface. This movement generally occurs as creep or very shallow slumps in the unconsolidated surficial materials and does not include movement or failure of the underlying bedrock. Continued small scale mass movements of this nature can be expected in the future, but since they are localized in extent, shallow and do not involve failure of the underlying bedrock, they will not affect the stability of the site.

3.2.4 KARST PROCESSES

The solution of the underlying limestone bedrock is an active geomorphological process in the region in which the WSSRAP is located. Although there are a number of caves, springs and sinkholes in the general vicinity of the WSSRAP, there are no apparent surface features at the chemical plant area that have resulted from the solution of the underlying limestone. All of the karst features identified are associated with losing streams and are located in areas where most or all of the loess and glacial materials overlying the limestone bedrock have been removed by erosion.

One of the principle concerns about the impact of karst processes on geomorphological stability arises from the potential detrimental effects that the void spaces resulting from solution activity would have on the integrity of the site. This consideration results from the possibility that the overburden or bedrock between a subsurface void and the surface facilities will not be of sufficient strength to support a structure. When this condition exists, subsidence or catastrophic collapse may occur. Evaluation of the potential for collapse to occur must include not only the near-term (immediate) potential but also the long-term (200 - 1000 year) potential since solution activity in limestone bedrock is a continuing process in a humid environment.

3.2.4.1 NEAR-TERM STABILITY

Interpretation of the boring logs obtained from the subsurface exploration program indicates that there are no major void spaces in the subsurface that could lead to subsidence or collapse. Evaluation of forty exploratory borings drilled in the vicinity of the site showed that only twenty-five percent of the borings exhibited voids and that those voids were small (Garstang, 1991). Ninety percent of the void space was within the upper ten feet of bedrock and no void space was encountered in the overlying unconsolidated materials. Two-thirds of the voids encountered had openings of less than one foot and all were partially or totally filled with clay. Later examination of a number of these boreholes with a downhole camera did not identify any voids with a significant opening or extensive lateral continuity.

3.2.4.2 LONG-TERM STABILITY

The solution activity which has resulted in the development of the karst topography and aquifer systems in the region in which the WSSRAP is located is a naturally occurring geomorphic process. This process may be expected to continue throughout the life of the project. The net result, independent of scale, will be an increase in the size, continuity and extent of the system of solution enlarged voids and conduits which currently exist.

Solution activity, although persistent in a humid climate, is a very slow process with respect to the actual development of karst features. Rates of limestone solution, like rates of other geomorphic processes, are subject to a wide variety of controls which vary over both time and location. Determination of these rates is limited by the incomplete record of factors which have influenced the process throughout geologic history. However, both laboratory tests and field studies of the rates of the limestone solution process have indicated that the maximum solution rate is roughly on the order of one millimeter per year (Palmer, 1984). This value must be considered as a very general estimate which at best, provides only a relative scale of magnitude for the rate of limestone solution. It nevertheless suggests that the rate of long-term development of solution features is insufficient to affect facility stability.

This conclusion is further supported by site-specific history of karst development.

It has been established that the development of karst features in the immediate vicinity of the WSSRAP was coincident to, if not previous to, the period of major down-cutting of the Missouri and Mississippi Rivers during Kansan times, 600,000 to 700,000 years ago (Meier, 1992). Thus, since solution activity in the intervening 600,000 years has not created a system of large voids and conduits, there is no reason to believe that such a system will develop in the next 200 to 1000 years.

4.0 GEOMORPHIC STABILITY AT THE WSSRAP
4.1 ANALYSIS

The analysis of existing geomorphic processes, historical geomorphological activity, site geology, soil stratigraphy and the archaeological record indicates that the Weldon Spring site is located in an area of excellent geomorphological stability. The only geomorphological process that could have a long-term detrimental effect at WSSRAP is fluvial erosion. Flooding is not possible at the upland location, wind erosion will not be a factor because of the vegetative cover that will develop naturally in a humid environment and, even though the adjacent southern slope is relatively steep, mass movement is not considered to be a significant concern because of the stability of the underlying bedrock strata. Solution of the underlying limestone bedrock has resulted in a network of enlarged bedding planes and fractures, but has not resulted in the development of subsurface openings of sufficient size to cause subsidence or collapse.

The fluvial erosive processes that may effect geomorphic stability include both overland/rill erosion, and gully erosion. The procedures implemented to prevent or mitigate the detrimental effects of erosion however must include consideration not only of processes acting on-site but off-site as well. This is a particularly significant consideration in investigating the effects of gully erosion since gullies may initiate elsewhere and migrate onto the site through headward erosion. As a result, the mitigating alternatives developed must include consideration of the regional geomorphic environment in addition to site specific conditions.

4.2 MITIGATING ALTERNATIVES

Given the upland setting of the Weldon Spring Site, headward erosion of the tributary streams toward the drainage divide is a natural geomorphic process. As a result, the most appropriate approaches toward assuring long-term geomorphic stability are to reduce the magnitude of the in-situ parameters that contribute to this process. As such, practices implemented to reduce the erosive energy developed on the landscape are more desirable for long-term stability than structures designed specifically to stop erosional activity. In addition to considerations of rock riprap protection and permanent vegetation, two additional concepts of erosion protection have been included in the site development plan for the disposal cell.

4.2.1 LANDSCAPE MODIFICATION

An effective approach to preventing gully development is to reduce the level of activity of the geomorphic factors that are responsible for the headward migration of the gullies. Specifically, a general equation for estimating the rate of gully advancement is $R = 1.5A^{.49} \, S^{.14} \, P^{.74P} \, E^{1.0}$ (Thompson, 1964). In this equation, R is the average rate of gully head advance (in feet); A is the drainage area (in acres); S is the slope of the approach channel (as a percentage); P is the sum of annual rainfall (in inches) equivalent to or greater than 0.5 in/hr; and E is the clay content (.005 mm or smaller) of the eroding soil profile, as a percentage of weight. Of these parameters the area (A) drained by the gully system and the slope (S) both are parameters that can be reduced to minimize potential gully development.

Area may be adjusted by landscape modification. Site topography may be adjusted so that areas contributing run-off to those streams considered to have a greater potential for gully erosion and advancement are reduced and the run-off directed toward those basins considered more stable. Landscape modification in this capacity principally requires slope grading to relocate the divides between adjacent drainage basins. Decisions regarding drainage basin modification must however, also consider the impact that increased run-off and channel length will have on the receiving drainage basins.

Effective landscape modification also should include reestablishment of preexisting stream patterns, stream gradients and drainage densities to the maximum extent possible. Natural drainage patterns have developed, over a period of many years, a state of dynamic equilibrium with the amount of sediment and the volume of water they can carry. When this equilibrium is disturbed, the stream attempts to adjust to the new conditions. Thus, increased water and sediment loads may result in stream channel erosion in the form of bed and bank erosion and headcutting or gully migration in the upper reaches. It has been shown (Barr, and Rockaway, 1980) that reestablishing the original stream regimen through terrain sculpturing is an effective procedure to avoid detrimental long-term erosional consequences of drainage basin modification.

Design of terrain sculpturing plan should be based on an analysis of preexisting undisturbed topographic conditions. At the WSSRAP location, this would include an evaluation of naturally established run-off area relationships, drainage density stream gradients and slope profiles from 1937 areal photography. Also, in the process of creating drainage channels, some effort should be taken to sculpture the land into concave slopes where possible. It has been established that concave slopes are the most stable and in near equilibrium with precipitation run-off that produces sheet erosion. Thus, the sculpturing of concave slopes within the watershed would tend to reduce the potential for erosive activity.

4.2.2 BASE LEVEL ADJUSTMENT

An additional option to retard long-term headward erosion is to limit downward erosion by raising the elevation of the local base level. This effectively reduces the regional stream gradients by creating a higher elevation at which stream equilibrium was attained, thus lowering stream velocity and erosional activity. Increasing the elevation of the local base level could be accomplished by using channel stabilization structures to control the channel gradient and elevation. Use of check dams and grade stabilization structures are the techniques most commonly used to accomplish this. Unlike revetments, which are used to protect the entire channel or its sides and bottom, check dams are designed to protect only the base, or bottom of the channel from erosion. These structures are placed across the channel at intervals along the alignment to inhibit physically, the moving water from eroding the bottom of the channel. Grade stabilization structures are installed to stop the advance or prevent the formation of gullies in natural waterways or other points of concentrated flow. The structures may be drop inlets, chutes or earth dams which are designed to have the capacity to pass the channelized flow anticipated for the drainage basin.

5.0 CONCLUSIONS

The analysis of existing geomorphological processes, historical geomorphological activity, site geology, soil stratigraphy and the archaeological record indicates that the

Weldon Spring site is located in an area of excellent geomorphic stability. The only geomorphological process that could have a long-term detrimental effect at WSSRAP is fluvial erosion. This includes both overland/rill erosion and gully erosion. Although it does not appear that the effects of fluvial erosion will be detrimental to the long-term performance or local environment at the Weldon Spring site, there are several procedures that can be implemented to mitigate the impact of future erosion.

The most effective mitigating alternative to be considered, in addition to rock riprap apron, permanent vegetative cover would appear to be modification of the post-closure terrain through terrain sculpturing and base level adjustment. Site topography could be adjusted so that the area contributing runoff to those drainage considered to have greater potential for gully erosion and advancement are reduced and the runoff directed toward those drainages considered less susceptible to erosive activity.

REFERENCES CITED

Allen, W. H. and Ward, R.A., 1977, Soil: The Resources of St. Charles County, Missouri: Land, Water and Minerals: Missouri Division of Geology and Land Survey, Dept. of Natural Resources.

Barr, D. J. and Rockaway, J. D., 1980, How to Decrease Erosion by Natural Terrain Sculpturing; in Weeds, Trees and Turf, January.

Daniels, R., B. and Hammer, R. D., 1992, Soil Geomorphology; John Wiley and Sons, Inc., New York.

Garstang, Mimi, 1991, Collapse Potential of the Defined Study Area at the Weldon Spring, Chemical Plant Site, Weldon Spring, Missouri; in Proceedings of the Geoscience Workshop, MK-Ferguson Company and Jacobs Engineering Group, DOE/OR/21548-197.

Longmire, P. A., et al, 1981, Geologic, Geochemical and Hydrological Criteria for disposal of hazardous wastes in New Mexico, in Wells, S.G., and Lambert, W. (eds) Environmental Geology and Hydrology in New Mexico: New Mexico Geological Society Special Publication 10 P. 93-102.

Meier, D. H., 1992, Pleistocene Fluctuations in Valley Levels at the Former Weldon Spring Ordnance Works Superficial Site, St. Charles County, Missouri, Unpubl.

Palmer, A. N., 1984, Geomorphic Interpretation of Karst Features; in Groundwater as a Geomorphic Agent, R. G. LaFleur, ed.; The Binghamton Symposia in Geomorphology: International Series, No. 13, Allen and Unwin.

Schumm, S. A., et al, 1982, Geomorphic Assessment of Uranium Mill Tailings, Disposal Sites: Uranium Mill Tailings Management, OECH Nuclear Energy Agency and U.S. Department of Energy, Colorado State University, October 1981.

Thompson, J. R., 1964, Quantative Effect of Watershed Variables on Rate of Gully Head Advancement: Transactions; Am. Soc. Ag. Engrs., V. 7, No. 1.

U.S. Dept. Agriculture, Soil Conservation Service, 1982, Soil Survey of St. Charles County, Missouri.

Young, Anthony, 1969, Present Rate of Land Erosion; Nature, Volume 224, Nov.

Geoenvironmental assessment of radioactive waste repository

Évaluation de géologie de l'environnement d'un site de stockage de déchets radioactifs

D. Evstatiev, P. Petrov, R. Angelova & D. Karastanev
Geotechnical Laboratory of the Bulgarian Academy of Sciences, Sofia, Bulgaria

ABSTRACT: The geoenvironment in the area of the only operating radioactive waste repository in Bulgaria has been analyzed. The repository is intended for storage of all kinds of low and medium level radioactive wastes with the exception of these from nuclear power production. The performed investigations prove that the 30 years of operation have not caused pollution of the geoenvironment. Meanwhile the existing complex geological setting does not provide prerequisites to rely only on the natural geological safety barriers.

RÉSUMÉ: La situation géoécologique du seul dépôt de stockage des déchets radioactifs en Bulgarie a été analysée. Le dépôt a pour but d'entasser toutes sortes de déchets de faible et de moyenne radioactivité excepté ceux de l'énergie nucléaire. Les explorations qu'on a fait prouvent qu'aprés une exploitation de trente ans du dépôt il n'y a pas de la pollution du géoenvironnement. En même temps les conditions géologiques compliquées ne permettent pas de compter seulement sur les barrières protectrices géologiques.

INTRODUCTION

Within the scope of the large-scale programme on radioactive waste management that has started recently, a special attention was paid to the only operating radioactive waste repository in Bulgaria. It is situated in the neighbourhood of the capital Sofia in the Lozen mountain region.

The repository is functioning since 1964 and is intended for storage of radioactive wastes discharged by medicine, stock-breeding, metallurgy, construction works control, research laboratories, etc. Radioactive wastes from nuclear power production are not stored in this place. The repository construction is of the shallow type and there are already about 300 m³ of low- and medium level radioactive wastes buried in it.

The main object of the investigations described in the present report is to analyze the geoenvironment in the area of the repository after almost 30 years of operation and to assess the hazard for contamination of the environment. In this respect geological, engineering geological, hydrogeological and radio-geological surveys were performed in the repository region and a system for monitoring and control of ground water, rocks and soils was founded. The investigations were in conformity with the criteria of the International Atomic Energy Agency (IAEA), with the experience in other countries and with their principles for the near-surface disposal of radioactive wastes (Stevens & Debuchananne 1976, Langer 1983, Operational Experience... 1985, Langer 1986, Morfeldt 1986, Nold & Lieb 1986, Quast et al. 1986, Chapman et al. 1987, Squires 1989).

GEOGRAPHICAL CHARACTERISTIC

The repository is located directly under the ridge of the Lozen mountain at an altitude of 920 m, the closest inhabited locality being at a distance of 4 km. The slope acclinal is 13-16 % in N-NE direction (Fig.1). The relief of the region is formed mainly by the plain denudation and gully erosion. There are two ravines with variable water outflow in the vicinity of the site, draining the shallow ground water and surface water running downwards the slope which could eventually transport the radioactive contaminants.

According to the data from the nearest meteorological stations the predominating wind direction throughout the year is W-NW with the exception of the fall season when the direction is SE-E. Precipitation values are very close to the average for the country. The absolute maximum daily amount of precipitation - 90.6 mm, was established in June 1969.

The presented information about the geographical conditions in the region show that there is no evidence for observed hazardous atmospheric phenomena. The repository site is not endangered by flooding and gully erosion.

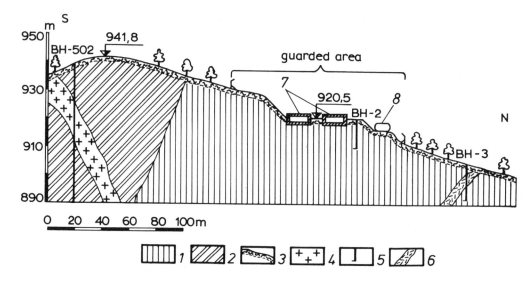

Fig.1. Cross section along the mountain slope in the area of the repository.
1 - Phyllite-schist formation; 2 - Diabase formation; 3 - Weathered zone; 4 - Magmatic rocks; 5 - Boreholes;
6 - Mylonitized phyllites; 7 - Reinforced concrete bunkers; 8 - Liquid waste tanks.

LITHOSTRATIGRAPHY

The investigated region of the Lozen mountain consists of Lower Paleozoic rocks, Upper Paleozoic sediments, Lower Triassic terrigenous deposits, Neogene sediments, Quaternary deposits and several magmatic bodies of granitoides and lamprophyre dikes (Fig.2).

Diabase formation. The oldest rocks in the region are of pre-Cambrian and Cambrian age /Pɛ-Є/ (Iliev and Katskov 1990). They are represented by dynamometamorphized, thin layered, diabasic, green and reddish tuffs; diabases; chlorite-sericitic-amphibole schists and laminated quartzose sandstones with arkose intercalations. The rocks from this formation are stratified or intersected by quartz lenses and veins with a thickness from 1 to dozens of centimeters. The formation is observed on the southern slope and along the ridge in the immediate proximity of the repository (Fig.2). Deep boreholes (750 m) pass through the rocks of this formation and lens-like bodies from plagiogranites and lamprophyre dikes have been established at various depth in them. The magmatic bodies are surrounded by halos of higher radioactivity.

Phyllite-schist formation. This formation builds up the terrain of the repository site as well as a wide strip to the North and Northeast near it (Fig.2). The formation thickness is about 300-500 m. It consists of thin layered, semicrystalline, clayey, quartz-sericitic phyllites. They are of greyish green to dark-grey colour, sometimes with glancing and finely undulated surfaces. Under weathering or tectonic processing they are transformed to ochre yellow or rusty brown clayey mylonites. The weathering zone of phyllites reaches 5-7 m. They have a fine laminated - parallel fabric and an aleurolite to micro-lepidoblastic texture. Fine flaky sericite, quartz and a little biotite predominate in their mineral composition. On some places thin (up to 1 m) intercalations of quartzites are established among the phyllites. Stratified and transversal quartz lenses and veins, several centimeters thick, are encountered in the whole formation. The phyllite schistosity is of E-W orientation (80°-90°) and steep inclination to the North (74°-80°). The Arenigian-Llanvirnian /2O$_{1-2}$/ age of this formation has been proved by palinologic and stratigraphic investigations.

Červenigrad formation. It consists of breccia-conglomerates and alternating conglomerates and sandstones with small aleurolite and argillite intercalations. The red colour is characteristic for the rocks. The age of the whole formation is Stephanian /če C$_2$s/ and its thickness is 550-1100 m.

Gabra formation. It is built up of sandy aleurolites, argillites and marlstones. Calcareous concretions, whole carbonate strips and carbonate cement have been established among the sandstones. The age is Upper Stephanian-Permian /gb C$_2$s - P$_1$/ and the thickness is 380-550 m.

Tarnava formation. The age is Lower Permian /tr P$_1$/. It is represented by unsorted breccia-conglomerates and gravelly sandstones. The thickness is 160-200 m and is distributed at the northern slope of the mountain.

Ravuljano formation. It is situated also at the northern slope of the mountain but is built up of sandstones with aleurolite and argillite intercalations. The thickness is 50-200 m and the age is Upper Permian /ra P$_2$/.

Rusamskidol member of Lozen formation. The Rusamskidol member /lo/r T$_1$/ consists of versi-

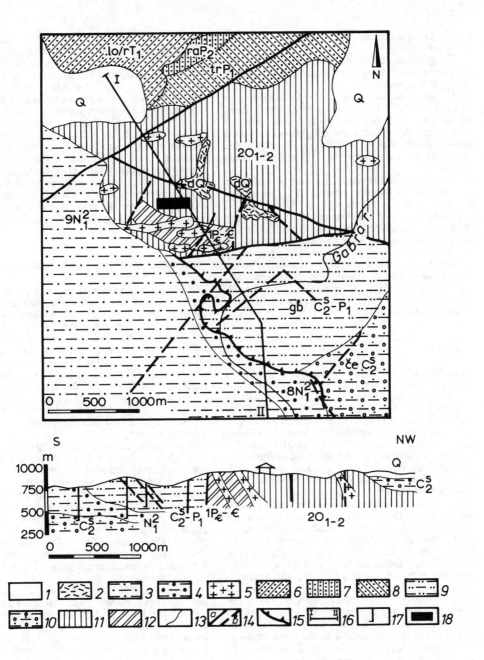

Fig.2. Geological map of the investigated area.
1 - Quaternary sediments; 2 - Talus deposits; 3 - Clayey sandstone and clay formation with Čukurovo coal;
4 - Conglomerate - sandstone - clay formation; 5 - Magmatic rocks; 6 - Rusamskidol member of Lozen
formation; 7 - Ravuljano formation; 8 - Tarnava formation; 9 - Gabra formation; 10 - Červenigrad formation;
11 - Phyllite - schist formation; 12 - Diabase formation; 13 - Lithostratigraphic boundary; 14 - Fault:
a) certain, b) supposed; 15 - Thrust; 16 - Profile line; 17 - Boreholes; 18 - Radioactive waste repository area.

2405

coloured quartzose sandstones with a total thickness of 240 m.

Conglomerate-sandstone-clay formation. The thickness is about 50 m and the age is determined to be Upper Miocene /8 N_1^2/.

Clayey sandstone and clay formation with Čukurovo coal. It is represented by clayey sandstones, sandy clays, fine-grained sandstones, 16 coal layers, plastic clays and conglomerates with a total thickness of about 200-250 m. The age of the sediments in this formation is Upper Miocene /9 N_1^2/.

Quaternary sediments. In the investigated region they are formed by talus deposits /d Q_{h-p} /, eluvial and eluvial-talus sediments /e Q, e-d Q/ and alluvial deposits /a Q_{p3}/. The quaternary sediments are not thick, with granulometry changing from rock pieces and gravel to clays and have a limited distribution.

Magmatic rocks. These bodies are observed to the south of the repository site and consist of biotite microgranite with a transfer to syenodirite and quartz-syeno-porphyry. Their texture is porphyritic, hypidiomorphic-granular. The magmatic rocks age is Upper Cretaceous /gδ K_2/.

TECTONICS

According to the first geostructural investigations the region belongs to the Čukino fault system, which is restricted between almost parallel faults - Čukurovo and Iazdirastovo (Dimitrov 1937). Three tectonic blocks have been determined in the area of the repository site - northern, southern and western one. The Iazdirastovo fault with SE-NW direction passing to the north of the site (Fig.2) forms the boundary between the northern and the southern block. A deep fault zone oriented SW-NE has been identified by gravity prospecting to the west of the site. It is on the border between the western block and the others. One of the faults of this zone can also be observed on the surface. Southeast of the site, a fault separates the Permo-Carboniferous from the Lower Paleozoic rocks (Fig.2). The Upper Carboniferous sediments have overslided on those from the Miocene and make up the allochthonous structure, 260-380 m thick, which is moved by gravity sliding. The thrust surface has been established by deep boreholes. Discontinuity of the magmatic granitoid bodies in depth has been observed and is obviously due to the thrust origin of the whole block. Some smaller faults are also encountered in the area around the site.

The Paleozoic rocks in the repository region are fissured and jointed mainly in four dominant sets: NNE (10°-20°), WNW (290°-300°), NE (40°-50°) and ENE (70°-80°). The fissures in the Ordovician phyllites correspond to the orientation of the near joint zones with almost E-W direction and a steep inclination to the North. They are sub parallel to schistosity.

Special measurements of recent vertical crust movements in the repository region have not been carried out. There are geological and geomorphological evidences for the young rising of the Lozen mountain (from Miocene till now). The permanent earthquake's epicenters in the Sofia seismic zone proves the late tectonic activity. According to the map of the recent vertical crust movements, the Lozen mountain has been rising at a constant velocity of 1-2 mm per annum.

It is obvious that the site of the radioactive waste repository is located in a very complex tectonic structure, which is the result of intensive tectonic activity that began as early as in the Paleozoic period continuing after the Neogene. In consequence, this massif has been greatly deformed, divided by faults and non-homogenous, whereas the rocks that make it up are broken, fissured and in some zones turned to mylonite.

ENGINEERING GEOLOGICAL SETTING

Engineering geological mapping of an area of 1 km² in scale 1:1000 was performed with the view of thorough investigation of the engineering geological conditions in the repository region. The data obtained from the six boreholes, 250-790 m deep, drilled during the period 1984-1988 and located at 80 to 1500 m around the site, were analyzed. Core drilling within the area of the site included three boreholes (BH-1, BH-2 and BH-3) with corresponding depths 15, 20 and 25 m. The rock and soil samples were subjected to determination of their physical and mechanical properties and cation exchange capacity.

The investigations provided the possibility to establish that the soil base of the repository is composed of the following soil varieties:

Artificial fill. It is made during the levelling of the site for the building of the different installation facilities. It has a maximum thickness of 1.5-1.8 m and consists of poorly graded pieces of weathered phyllite with sizes from 3-4 to 10-12 cm and fine-grained clayey-sandy filler. The artificial fill is insufficiently compacted and is susceptible to differential settlement under moistening. It possesses unequal bearing capacity.

Weathered phyllite (eluvium). The thickness of the weathered phyllite layer is 3.0-4.5 m. It is transformed to a thicker (5-8 m) deluvial-eluvial cover in N-NE direction along the slope. In the upper part of the weathered layer the clayey-silty fraction is predominant (up to 70 %) with inclusions of small (up to 3-4 cm) angular pieces of easily splitting phyllites. The quantity and sizes of the phyllite pieces increase in depth, while the content of the fine fractions - silt and clay, decreases. According to their grain size distribution and plasticity index I_p, the eluvial deposits are classified as sandy clays which turn from silty on the surface to gravelly ones in depth. The index properties of weathered phyllites change in the following ranges: bulk density ρ_n=1.98-2.06 g/cm³; water content W_n=14-18 %; dry density ρ_d=1.73-1.75 g/cm³; density of solid particles ρ_s=2.66-2.68 g/cm³; porosity n=34-36 % and void ratio e=0.52-0.56. The relative consistency C_r of the weathered phyllite fine-grained fraction is greater than 1, which means that it is with very stiff to hard consistency. The cation exchange capacity varies from 5 to 33 mg equ/100 g. The value of the bearing capacity of weathered phyllite is 0.3-0.4 MPa and this variety may be assumed to be a suitable foundation base for constructions of the type of the existing concrete bunkers and shafts.

Broken and fissured phyllites. They have been established at a depth of 4.5-5.5 m under the surface. The weathered phyllites turn gradually into fissured and broken, fine bedded, quartz-sericite ones. Phyllites are mylonitized and turned to clay in the fault zones. Broken and fissured phyllites have been established to the depth of 790 m. Among the very fissured phyllites single less fissured and stronger interlayers are encountered which have variable thickness in horizontal direction. The bedding of borehole core phyllite samples is with E-W line of strike and angle of dip from 55º-60º to 80º-85º to N-NE. The dominant joint set (Fig.3) has a line of strike in NNE (10º-20º) - SSW (190º-200º) direction. The density of solid particles of unweathered phyllites is ρ_s=2.72-2.73 g/cm³. Their

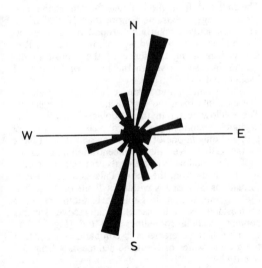

Fig. 3. Rose diagram of joints in the phyllite-schist formation.

bulk density varies in a wider range - from ρ_n=2.31-2.37 g/cm³ to ρ_n=2.63-2.68 g/cm³ since they are fissured to a different degree. This is also the reason for the rather variable unconfined compressive strength of water soaked samples, which changes from 1.3 MPa to 12-14 MPa. The strength of phyllites is largely decreased after periodical frost

action which is also proved by the thick weathered layer. The bearing capacity of the strongly fissured phyllites is 0.5 MPa and of the less fissured ones - 1.5 MPa. They have safe bearing capacity and are a suitable base for foundation of all kinds of constructions.

No landslides and rock falls were established in the repository region which could be an eventual hazard. There are no conditions for the evolution of such processes except for the case of deep excavations or slope cutting. There are certain and supposed faults around the repository area (Fig.2). The nearest fault is at about 100 m to the North of the site and is a part of the deep and wide Iazdirastovo fault zone. In the process of investigations it was crossed by one of the boreholes in the interval between 4 and 14.5 m. In this borehole phyllites are strongly mylonitized and ground to clay at some places. Their colour is changed to ochre and dark brown. The angle of bedding is greater - from 70º-75º to almost vertical. According to the seismic zonation of Bulgaria over a period of 1000 years (Bulgarian Building Code ... 1987) the repository site is located on the boundary between a region with seismic intensity of VIII degree and a region with seismic intensity of IX degree on the MSK-64 scale. The situated in the proximity tectonic faults with recent earthquake activity as well as the importance of the repository are the reason to assume the IX degree of seismic intensity and the seismic constant K_c=0.27. This constant could be corrected in eventual microseismic mapping of the site. The radioactive waste repository is not exposed to danger of the destructive action of landslides, rockfalls, karstification, gully erosion, subsidence or significant settlement of the soil base. The only hazard to the safety of the repository is the high seismicity. The strongly fissured phyllites do not allow the rock massif to be considered as homogenous in regard to its physical and mechanical behaviour.

HYDROGEOLOGICAL SETTING

The Paleozoic rocks in the region represent a low water bearing and low permeable formation. Their permeability is higher in the tectonic strongly fissured and faulted zones and in the upper layer of the secondary weathered fissuration. The ground water in the Paleozoic rocks is shallow circulated in rock fissures and is recharged only by precipitation. The schistosity of the Ordovician phyllites determines their anisotropic water permeability. They are a confining bed in transversal direction to the stratification. Their water permeability increases in the fissured and faulted zones and parallel to schistosity. The fine-banded phyllites, inclining steeply to the North enable the precipitation water to infiltrate into the massif. The permeated water is drained at the intersections of schistosity with tectonic faults and cracks and with the erosion forms of the river-ravine system. The surface water outflow in the gullies is not permanent and depends on precipitation, running dry during the hot seasons. The only larger spring which drains the formation of phyllite-schists is situated at an elevation of about 850 m to the North from the repository and is captured for its water supply. The maximum debit is about 0.2 l/s. The water has a hydrocarbonate-sulphate-sodium-calcium composition with a total salts content 0.08 g/l and a slightly acidic reaction (pH=5.6).

There are several small springs along the valley of the Gabra river that originate from the Carboniferous-Permian terrigenous complex, their waters being mainly hydrocarbonate-calcium-magnesium.

A regular aquifer has not been formed in the Ordovician phyllites where are the repository construction foundations. A significant phreatic zone exists in the rock massif along the watershed above the site which infiltrates a part of the precipitation waters. The piezometric boreholes drilled inside the site area prove the variable filtration characteristics of the phyllites and the unstable water table of the shallow unconfined ground water at a different depth from the surface - from 6-7 m to 15-16 m. A full loss of the drilling fluid occurs when passing through strongly fissured phyllites or quartz veins. In order to assess the filtration properties of phyllites a field pumping-in permeability test has been carried out in the boreholes drilled in the repository site. The values of the determined coefficient of permeability K_f are from 0.06 m/d to 0.7 m/d.

The chemical composition of the water samples from the boreholes is sulphate-hydrocarbonate-sodium-calcium with a total salts content 0.136 g/l.

The hydrogeological setting of the radioactive waste repository is not very favourable for constructions of this type. The established water levels below the depth of 6-7 m have to be taken under consideration in an eventual enlargement of the repository. The tectonically fissured and faulted massifs represent one of the most difficult for prediction and modelling continua concerning the flow of the ground water and the migration of radionuclides.

RADIOGEOLOGICAL SETTING

A characteristic feature of the region around the repository is the significant diversity of natural radioactivity of the rocks and of the radiohydrogeological anomalies.

The rocks from the Lower Paleozoic phyllite-schist formation have gamma radiation of 22-30 µR/h. The values for the quartzose sandstones and diabase rocks are lower - 14-18 µR/h and 6-10 µR/h respectively. The young Paleozoic terrigenous formations of sandstones and aleurolites with conglomerate intercalations have a natural gamma intensity of 22-26 µR/h. Along the whole ridge (including the southwestern corner of the repository site) and on its southern slopes the natural

radioactive intensity is more than 40 µR/h. Several gamma-anomalies have been established here - from 100 to 140 µR/h, which are due to the uranium ore-bearing bodies (Fig.4). The higher natural background is determined by the magmatic bodies from microgranites and lamprophyre dikes which are incorporated in the Paleozoic rocks. A series of smaller but numerous sections with intensity of 45 to 75 µR/h with single maxima of up to 120 µR/h are connected with the influence of the dike rocks, distributed among the phyllitoid formation to the North, Northeast and East from the repository (Fig.4).

The performed gamma-ray log of the piezometric boreholes (BH 1, BH 2 and BH 3) in the site shows that the weathered and fissured phyllites to the depth of 7-13 m have gamma radiation of 18-20 µR/h, while the fresh and less fissured phyllites beneath have gamma radiation of 12-14 µR/h. Radiation of various intensity is superposed on the natural geological radioactive background from the concrete bunkers and shafts for storage of the different kinds of waste. The radiation over the plutonium waste shaft is 466 µR/h; near the gamma radioactive waste shaft - 23 µR/h; near the bunker with biological wastes - 27 µR/h; near the concrete wastes - 20 µR/h; near the bunker with solid wastes - 29 µR/h; near the metal containers for waste transport - 65 to 350 µR/h and near the fluid waste tanks - 40 µR/h. The radiation from the building for radioactive isotopes storage is extreme - more than 10 000 µR/h. The radioactive background around the administrative and laboratory buildings is low - 16-18 µR/h. Everywhere along the fence of the repository the gamma intensity is within the limits of the natural background - 18-23 µR/h.

The ground water forms its content of radioactive elements from the ingredient rocks themselves. The shallow water in the Ordovician phyllites has negligible quantities of radioactive elements - 0.1-0.9 µg/l and only in one of the springs on the right bank of the Gabra river their concentration reaches 15 µg/l. The radioactive elements content in spring water drained from the young Paleozoic terrigenous formations falls usually within the limits of 0.1 to 7.4 µg/l. Some radiohydrogeological anomalies have been established in this case, the radioactive elements content reaching sometimes 420 µg/l (Fig.4). The quantity of these elements is comparatively small in the ground water of Neogene sediments - from 0.2 to 6 µg/l.

The radioactivity log of the rocks, the gamma-ray log of the boreholes on the site and the laboratory radiochemical analyses of the water sources in the region have not detected radioactive contamination in the rock massif and in the ground water.

CONCLUSION

The results from the performed surveys and investigations prove that the 30 year long operation of the repository for low- and medium radioactive wastes

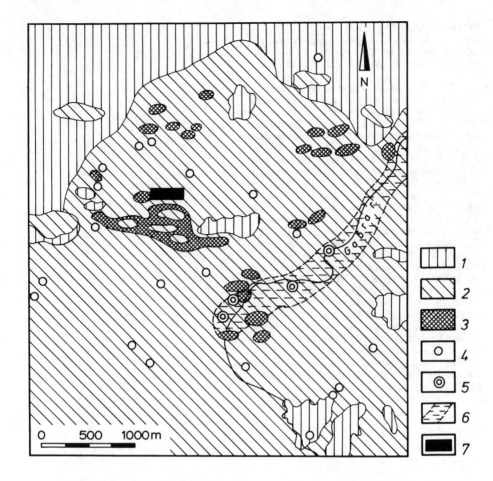

Fig.4. Map of the natural rock radioactivity.
1 - Rock gamma rays up to 20 μR/h; 2 - Rock gamma rays from 20 to 40 μR/h; 3 - Rock gamma rays over 40 μR/h; 4 - Spring with radioactive elements content up to 10 μg/l; 5 - Spring with radioactive elements content over 10 μg/l; 6 - Radio-hydrogeological anomaly; 7 - Repository area.

has not caused pollution of the geoenvironment in the Lozen mountain. It has been established that the higher radiation in some neighbouring zones is due to uranium ore-bearing bodies, i.e. to natural reasons.

Meanwhile the existing geological setting does not provide prerequisites to rely only on the natural geological safety barriers. It is necessary to improve the quality of the artificial engineering barriers and to refine the system for environmental monitoring and control.

After the building of the National repository for low- and medium radioactive wastes from the nuclear power industry it will be convenient to store there the wastes from the investigated repository.

REFERENCES

Bulgarian Building Code for Construction in Seismic Areas. 1987. Sofia, EC "Building and Architecture", 67 pp.
Chapman, N., I. Mckinley & M. Hill 1987. The Geological Disposal of Nuclear Waste. John Wiley & Sons, Chichester, 280 pp.
Dimitrov, S. 1937. Notes on the Geology and Petrography of the Lozen mountain. I. Stratigraphy and Tectonics. Annual Book of the Sofia Univ., II Phys.-Math. Dept., No 3, Natural Science, 163-218 (in Bulg.).
Iliev, K. & N. Katskov 1990. Geological Map of Bulgaria. Page Ihtiman in scale 1:100 000,

Military Topography Service.

Langer, M. 1983. La geologie de l'ingenieur peut elle aider a resoudre le probleme du stockage des dechets radioactifs. *Bull. IAEG*, No 28, Paris, 5-16.

Langer, M. 1986. Main activities of engineering geologists in the field of radioactive waste disposal. *Bull. IAEG*, No 34, Paris, 25-38.

Morfeldt, C. 1986. Swedish primary formations (Granites and Gneisses) as Hostrock for nuclear waste deposits. *Bull. IAEG*, No 34, Paris, 45-50.

Nold, A. & R. Lieb 1986. The Grimsel test site - part of the disposal project for radioactive waste in Switzerland. *Bull. IAEG*, No 34, Paris, 51-58.

Operational Experience in Shallow Ground Disposal of Radioactive wastes. 1985. Technical Reports Series, No 253. Vienna, IAEA, 88 pp.

Quast, P., E. Hawickenbrauck & M. Schmidt 1986. Engineering geological and safety technological aspects for the final disposal of in situ - consolidated radioactive waste in hard rock and salt formations. *Bull. IAEG*, No 34, Paris, 73-86.

Squires, D. 1989. Waste Disposal in Shallow Land Burial. In: *IAEA International Training Course - Management of radioactive Wastes with Regard to Radioisotope Application.* Karlsruhe Nuclear Research Center, 18 Sept.-13 Oct., 1-25.

Stevens, P. & G. Debuchananne 1976. Problems in Shallow Land Disposal of Solid Low-Level Radioactive Waste in the United States. *Bull. IAEG*, No 14, 161-171.

Using volcanic deposits among difficult limestone terrain as sanitary landfill sites in rural areas

Décharge controlée dans des couches volcaniques d'un massif calcaire complexe dans une région agricole

M. R. Khawlie & R. Nasser
American University of Beirut, Lebanon

ABSTRACT: Open dumping of solid wastes over karstified limestone rugged terrain in rural areas is leading to serious environmental problems. Previous attempts failed to locate sanitary landfill sites. New geoenvironmental investigations located 4 such sites in the typical rural Metn-Kesserwan area covering 18 villages-towns, over 59 km^2 and a population of about 100 thousand. General studies reflect suitability of area while specific parameters delineate potentials of landfilling capabilities. The four sites, about 1km^2 each, consist of three in volcanic tuffs and one silty shales. The area is typically a faulted U.Jurassic lst., mountainous, with deep valleys and rather steep slopes, one major stream, a deep water table and a little soil cover. Land-use varies among the little towns and villages, being mostly residential, agricultural or unexploited. The surficial cover of the proposed sites shows acceptable values of permeability 0.79-3.5cm/min, plasticity index 3.87-9.49%, particle density 2.43-2.82g/ml, organic content 1.72-2.22%, clay content 31.96-36.96%, and varying beteen a clay loam to a sandy clay loam with adequate attenuation capacity. The significant capability analysis parameters are weighted and rated to arrive at a priority ranking of the four sites.

RÉSUMÉ: Les sites de disposition des déchets solides sur des terrains calcaires Karstifiés, soulèvent un problème enviromemental critique. Tous les tentatives précédentes ont échouées pour localiser des sites hygiéniques convenables. De nouvelles études ont permis de localiser 4 Sites dans le secteur rural du Metn-Kesserwan, couvrant 59km^2 de surface et où 100 mille habitants sont distribués sur 18 villages et villes. L'étude comporte 4 sites, chacun de 1km^2, dont 3 dans des terrains volcaniques et 1 dans un terrain argileux appartenan au Jurassique supérieur. La région est montagneuse avec des vallées profonds, trés faillées, avec un coursd'eau principal, et une nappe phréatique profonde. A la surface, les roches présentent des valeurs acceptables de perméabilité, 0.79-3.5cm/min., l'indice de plasticité, 3.87-9.49%, la densité, 2.43-2.82 g/ml, le contenu organique, 1.72-2.22%, l'argile, 31.96-36.96%. Les paramètres significatifs ont été évalués dans chaque région pour donner la priorité au meilleur site.

1 INTRODUCTION

The problem of solid waste disposal has already proved its universality. This is a particularly notable issue in developing countries(Schertenleib & Meyer, 1991), and specifically in their rural areas. There, open dumping in river or dry stream courses and other improper waste disposal practices, i.e. burning, littering, etc. are increasing and turning into a real environmental menace. The conditions in rural Lebanon are not different, on the contrary they are worse because of the previous sixteen years of internal strife where daily affairs went on unchecked. If this was a matter of municipal services or human attitude only, it could be resolved by regulations and control. But the problem is linked to the natural setting of the site, where disposal is to take place, and its geoenvironmental character.

Lebanon is a mountainous area that is overwhelmingly made up of limestones and heavily faulted. Add to this the intense precipitation befalling the mountains during winter, makes finding suitable places for landfilling a compounded problem. In fact, a major study was conducted on Lebanon as part of a master plan for solid waste and waste water management (CDR, 1982). It surveyed current practices describing them as inadequate in terms of storage, collection and final disposal of solid wastes. The disposal is carried out in open dumps sited for convenience without regard to environmental or public health factors. Sanitary landfilling was recommended as the most feasible solution, however, rural areas in the mountains were regarded as unsuitable. Although the study did not carry on detailed field work it hinted to the possibility of finding few suitable "discrete patches".

This investigation aimed at finding such suitable patches, and was carried out in a typical mountainous rural area, Metn-Kesserwan(Fig.1), covering 59km^2 and including various sized towns-villages with a population of around 110,000$\pm$ There are other studies related to the subject matter, though they do not utilize the geoenvironmental conditions necessary for the current research. A survey on the needs for rehabilitation and development of Lebanon was done (Hariri Fdn.,1987) which produced maps showing suitability indices for implementing projects relating to solid waste services. Khawlie et al.,(1991) worked on the capital Beirut producing a preliminary waste disposal capability map delineating potential sanitary landfill sites for the area. A similar approach covering socio-economic, environmental and technical factors but with emphasis on rural conditions and a difficult rugged limesone terrain is covered here. This requires relating disposal sites to existing land-use pattern, natural resources, accessibility in service as well as the geology, surficial processes and materials properties to assure minimal, or controllable, environmental inpacts from the proposed landfills(FAO,1990; Canter, 1991; Jessberger,1990).

2. CURRENT STATUS
2.1 Geography and climate

The Metn-Kesserwan area covered in this study lies between N 35° 40'to N 35° 45' latitude and E 33° 55'to E34°

longitude making 59km^2. It is an area in Mount Lebanon, similar to equivalent areas along the eastern Mediterranean. Typically, it consists of a scattered number of villages, some often grow into towns without proper planning, erratically spread over a rugged terrain. Eighteen villages and/ or towns of population varying from a few thousand to 30 thousand each are located(Fig.1) at mountain shoulders, near valleys or across local plateaux.

It is a mountainous terrain cut by some deep valleys, with highest elevation of 1200m at Feytroun and lowest of 300m in El-Kalb river valley. There are pronounced slopes of over 40° and cliffs, with local relief ranging between moderate to high in the limestones, anywhere between 300-800m, and relatively low, less than 200m, in volcanic deposits.

The climate is the regime prevailing in central Mount Lebanon zone. Average yearly precipitation is around 900-1000mm, though most of it falls in short intervals between November to March, more or less torrential. The intense precipitation pattern causes a high rate of surficial washing and downward movement of leached components. In the dry period,on the other hand, with approximately 50% evaporation, the migration is strongly reversed.

Two questionnaires were distributed to people, the first asking information on current disposal practices of solid waste, the second on characterization of the domestic wastes. From the 325 questionnaires, 70% of the population said that the collection is individualised, improperly done and disposed off in three or more common open dump sites, after being put temporarily in hundreds of makeshift sites. Wastes are burnt in 20% of the latter, and often open burning takes place in the large final disposal sites. All surveyed population complained of annoying smells, horrible sceneries, increasing insects and rodents. Field observations noted barren spots and deteriorated soils in and around the disposal sites.

Characteristically,the average composition of waste consists of 24% paper, 4% plastics, 4% textiles, 5%glass, 5%metal and 40-50% organics. The average per capita production is about 0.75kg per day.

2.2 Socioeconomic aspects
The study revealed three major open dump sites, in Achqout valley, along

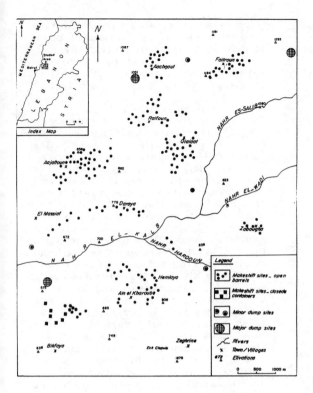

Fig. 1 Location and geographic features of study area.

Feytroun-Faraya highway and the third between Mar Boutros and Beit Chebab. A major site is defined as one having permanency, of area not less than few hundred m^2 and receiving more than 10 tons of wastes daily. The area is littered with many other sites of open dumps which are minor or temporary. During the crisis in Lebanon, because of no essential control, any little site could grow larger to become an environmental problem.

The littered and major sites, located near urbanized or in agricultural land, are obviously the result of the NIMBY attitude (Not In My Back Yard). They overlap with all sorts of urban and agricultural everyday operations. The major ones serve communities within 1-3km distance. The year-round main income is from agriculture, but also some major construction materials' quarries are available and contribute their industrial wastes directly into river beds. Forests, or what is left of them, are small scattered patches often with a spot or more of littered garbage showing signs of fires. The important natural resource is water, both surface and subsurface

though the latter is deep. Pollution of this water is inevitably a common feature.

2.3 Environmental impact of existing open dump sites

Over and above the terrible smells, sceneries, insects, etc. the burning of wastes pollutes the air and causes fires. Toxic materials in the ash is affecting the soil and vegetation. Along valley slopes, in rivers and streams, polluted water is a direct threat. The indirect threat is due to the fact that these sites lie over a karstified terrain therefore polluting underground water. This was confirmed in several instances where Jeita, a subsurface water system(Bakic, 1972), that is famous along the coastal stretch of El-Kalb river (few tens of km west from study area) showed obvious pollution that can be easily traced back to the study area garbage and construction materials wastes.

This ground water pollution was evaluated using a modified DRASTIC scheme (Canter et al.,1988). It takes into consideration the following parameters as crucial to enhancing or preventing the pollution: Depth to water table, Recharge

2413

rate, Aquifer media, Soil media, Topography i.e. slope, Impact of vadose zone, and Conductivity (hydraulic) of the aquifer. All three major dump sites showed values indicating pollution potentials are high.

3 AREA SUITABILITY FOR SANITARY LANDFILLS

In order to find spots where sanitary landfill sites could be located, the whole area was studied in terms of the factors deciding this potential, i.e. social performance, technical performance and management.

3.1 Social performance
The landuse patterns described above give an idea about the whole area. One major element deciding categorically the unfittness of a spot is urban extent. Therefore, all stretches covered by urban encroachment are automatically eliminated (Fig.1). This leaves other general features that have to be noted like extent of valleys, agriculture, forests,economic activities, streams, etc. If the urban centers (including villages) are spread erratically, as is the case,which would reflect on population distribution, then it is essential to decide on pivotal areas for potential sites as they would serve a common purpose. That

purpose would cover securing sanitary landfill sites for the clusters of population, at feasible distance away and of course have minimal or no environmental impacts noted previously with the terrible on-going practice at existing sites.

3.2 Technical performance
This covers the general area geology (Fig.2), landforms, hydrogeology and engineering geological character. The rock formations start with the Kesserwan lst. (M.Jurassic, Bathonian-Callovian), composed mainly of limestones and dolomitised lst., covers about 60% of area, highly fractured, medium to thick bedded or massive, with thickness of around 1000m. It is an excellent aquifer. It is followed stratigraphically upward by the Bhanness Complex(U.Jurassic, Kimmeridgian) composed- in the area - mainly of a variety of volcanics i.e. tuffs, agglomerates and basalts, with localised intercalations of clastic lst. and marls. It covers 19% of the study area, is weak, variably weathered, shows mostly flow-type or pyroclastic-type deposits, variably cut by calcite veins, with thickness between 50-100m. The other formations are minor in extent (Fig.2) and therefore not significant, unless one of them is in vicinity of the sanitary landfill sites

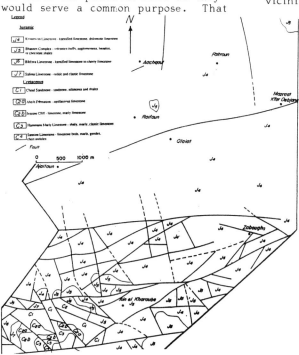

Fig. 2 Geological map and structures.

to be proposed. This applies to the Bikfaya lst. (U.Jurassic, Portlandian) covering 4% of area, also lst. and dolomitised lst.with some thin marl horisons, bedded to massive, fractured and karstified, with thickness of about 65m.

The area is thus essentially a karstified carbonate mountain mass with scattered hills or plateaux of volcanic cover. The terrain is dissected by deep valleys and numerous primary and secondary drainage of a dendritic pattern. The major basin is El-Kalb river and its main branch Es-Salib stream valley. Few isolated hills show over mountain or slope shoulders. Where the volcanics dominate, the land surface is gently to moderately undulating. Valley slopes are steep, rough, rocky with gradients exceeding 40° and sometimes with overhanging cliffs. Erosion of soil cover has left many slopes bare and rugged. Valley courses or beds do not contain appreciable amounts of fluvial deposits, rather, they are colluvial and man-dumped. Torrential rains and episodic turbulent high velocity ($>$ 50m/s) regimes wash materials downstream.

The major perrenial river is El-Kalb and its upstream continuation Es-Salib river. During summer, as water use becomes excessive, the two rivers thin down noticeably. Other minor intermittent or ephemeral rivers run in Aa'chqout and Hardoun valleys, while smaller streamlets branch here and there. Few minor springs are available in the vicinity of Bikfaya and result from localized perched water table, productive only in the replenishment period (Bakic, 1972). outside the area, about 3km westward however, are two important springs, Jeita and Kashkoush, that make part of the hydrological system covering the area. They both issue from the Kesserwan lst.mentioned above. At several instances, the Water Authority in Lebanon indicated these springs were polluted because of operations going on in area of study. The water table in the area under investigation is deep ranging between 500m to actual level of El-Kalb river, with seasonal fluctuations varying between 5-15m ±. The hydraulic gradient of the water slopes down westward with the general natural topographic profile.

The two Jurassic formations are fractured, and the area is faulted (Fig.2) which affects water transmissivity and migration. Two extensive faults trending EW cross the whole area, while

minor faults abound. This, added to the precipitation, topography, relief, lithological character and man's interference lead to several spots being potentially unstable. Slope instability, especially that certain portions of the area contain quarries that use explosives, is sometimes problematic. Mass wasting deposits are common at valleys and break in slopes. As the open dumping itself takes place along steep slopes and its detrimental effects on surface cover, i.e. deterioration of soil and plants, is contributing to instability.

The management performance relates to how the local authorities are handling the solid waste operation in view of their final disposal. The proposed sanitary landfill sites have to have roads for accessibility, enough areal extent to contain whatever thousands of tons of waste per year, should be easily workable sites in preparation,excavation, design, etc. and of course should have proper lining and cover materials for preventing the wastes from polluting the environment.

All the preceding suitability procedures were applied for screening the land. The resultant outcome are 4 patches, distributed fairly decently around the area, that could serve as potential sites of sanitary landfills (Fig.3). Therefore, the next step is to find out their specific function as sanitary landfills, i.e. their sanitary landfilling capability.

4. LANDFILL CAPABILITY OF PROPOSED SITES

The difference between suitability and capability is in covering details, in smaller areas and may be investigating other factors/parameters that are purpose-specific, in this case sanitary landfilling. Thus the background covered by the suitability analysis above is sufficient for assessing the socioeconomic aspects, the landforms and surficial processes, the hydrogeology and structures as well as management. The four patches Feytroun, Abou Mizane, Ech-Chawie, Zabbougha, lie in convenient settings.

4.1 Technical factors

All sites except one lie in the Bhanness Complex volcanics (in Lebanon given symbol J_5), while the exception, Ech-Chawie, is in silty shales though occupying the same stratigraphic position. Table 1 gives the detailed lithology of the four sites. Emphasis should be made

Table 1. Lithology in areas of proposed landfill sites

| Site | General lithology | Detailed Lithology | | | | | |
		Structure	Surface material	Permeability cm/min	Attenuation components	Cemen-tation	Hard-ness
Feytroun	A greenish outcrop of J_5, 100m, thick, weathered	Massive volcanic deposits, intercalated with thin beds of shales and limestone	48.64% sand,18% silt,33% clay,sandy clay loam	3.01	.Organic content= 2.22%(by wt) .Clay type: Disordered kaolinites to illites	Moderately to poorly cemented	Weak
Zabbougha	A greyish brown J_5 volcanic outcrop, 80m thick, weathered	Massive volcanic deposits, with some limestone beds	52.64% sand,16% silt,31.36% clay,sandy clay loam	3.5	.Organic content= 1.97%(by wt) .Clay type: Disordered kaolinites to illites	Moderately to poorly cemented	Medium
Abou Mizane	A greyish to green J_5 volcanic outcrop, 50m thick, weathered	Massive volcanic deposits, with limestone aggregates	44.64% sand,22% silt,33.36% clay,sandy clay loam	1.1	.Organic content= 2.02%(by wt) .Clay type: Disordered kaolinites to illites	Poorly to moderately cemented	Medium
Ech-Chawie	A brown chocolate shale outcrop in J_5 30m thick	Thin chocolate shale beds with marl and limestone intercalations	39.04% sand,24% silt,33.96% clay,sandy clay loam	0.79	.Organic content = 1.72%(by wt) .Clay type: illites to disordered kaolinite	Poorly cemented	Weak

in the Table on the thickness, surface material permeability and attenuation components, (composition) as these are crucial for sanitary landfilling. In capability analysis of investigated purpose, geotechnical studies are also quite important for deciding on sanitary landfilling. Several aspects are included here starting with site capacity, i.e. the volume of landfill soil material compared to the expected volume of wastes. One ton of waste with necessary cover material occupies a volume of $2m^3$ without pulverization or compaction. The four sites show apparent capacities, what is exposed without further excavation,of (tons): 250,000, 160,000, 100,000 and 7500, for Feytroun, Zabbougha, Abou Mizane and Ech-Chawie, respectively. Other aspects are the presence of any physical

constraints, such as an operating quarry downslope of Zabbougha site, the stream bed downslope of Abou Mizane, and potentially unstable steep slopes of the shales plus a nearby water pipe at Ech-Chawie. Cover and base materials for the landfilling operations are available, and excavation is easy to moderate.

Composite samples of the surficial materials at the four sites were obtained and tested at the laboratory. Table 2 shows the major items that are important for projecting the behavior of the landfill, its engineering design and operation. Furthermore, a simulated leachate test was done, and attenuation capacity was approached by noting the amounts of clays and organic matter in Table 2 (Knight,1987; Bagchi,1990).

Table 2. Some important geotechnical properties

Site	Consistency%				Density g/ml	Grain Size %			Organic Matter%
	L.L.	P.L.	P.I.			Sand	silt	clay	
Feytroun	30.1	20.606	9.49	2.43		48.64	18	33.0	2.22
Zabbougha	26.5	22.63	3.87	2.46		52.64	16	31.36	1.97
Abou Mizane	22.0	14.4	7.6	2.82		44.64	22	33.36	2.02
Ech-Chawie	24.4	17.93	6.47	2.61		39.04	24	36.96	1.72

The chemistry of the simulated leachate was compared to a clean fresh water source just for reference to give an idea about possible contamination types. The relatively high amounts of clays and organic matter in the four sites imply a good attenuation capacity. The volume of soil that must be available for attenuation was calculated by using (Bagchi, 1990): V=R.A.H where

V is volume of the soil needed.
R is a reduction factor varying between 1/1.2-1/1.3 accounting for soil fabric.
A is base area of landfill
H is average depth of unsaturated zone beneath the landfill.

No doubt, these parameters are very significant in soil behavior notably for the purpose of landfilling. The soil volume implies the potential of contaminants getting attenuated before some critical distance-may be to water table is reached.
The calculated values came up to be (m^3) 8300, 6640, 6640 and 830 for Feytroun, Zabbougha, Abou Mizane and Ech-Chawie, respectively.

Those parameters considered crucial in a final decision on siting and proper functioning of a landfill site, therefore its capability, are given weights of significance relative to each other ranging from 3 to 10, Table 3, and each is

Table 3. Capability analysis of the four proposed landfill sites

Site*	Land use (10)	Slope (5)	Vegetation (3)	Drainage (8)	Depth W.T. (10)	Surface geology (8)	Thickness of formation (6)	Compactness (6)	Geotechnical (10)	Excavation (6)	Haulage distance (5)	Sum weighted rating
E	3 Partially to sparse	3 Gentle	3 Sparse	2 3 drainage lines	2 perched	2 Clay loam	2 Moderate	3 Loose	4 **	2 Easy to moderate	3 <5km	203
A	3 Unexploited	3 Gentle to moderate	3 Sparse	3 1drainage line	3 Deep	2 Clay loam	3 Thick	2 Loose to moderate	3	3 Easy	2 <10km	201
F	3 Unexploited	3 Gentle to moderate	3 Sparse	3 1drainage line	3 Deep	1 Sandy clay loam	3 Thick	2 Loose to moderate	2	3 Easy	2 <10km	188
Z	3 Unexploited	2 moderate to steep	3 Sparse	3 1drainage line	3 Deep	1 Sandy clay loam	3 Thick	2 Poor to moderate	1	2 moderate	1 ≈10km	168

* E= Ech-Chawie, A=Abou Mizane, F=Feytroun, Z=Zabbougha
** Geotechnical properties include consistency limits, permeability and density

rated on its own between 1 to 3 to reflect its value (Khawlie et al.,1991).The
following examples clarify the approach:

Parameter	Weight	Rate
Landuse	10	Dense= 1
		Moderate= 2
		Sparse= 3
Slope	5	Steep 20°= 1
		Moderate 10°-20°=2
		Gentle 10°= 3

4.2 Solid waste management

The proposed spots lying at the J_5 stratigraphic position (whether occupied by
volcanics or shales) proved capable of
serving sanitary landfill sites. This is
an important finding not only for the
area of concern, but also for the whole
Lebanese as well as other eastern Mediterranean mountains because their geology resembles what is exposed here.

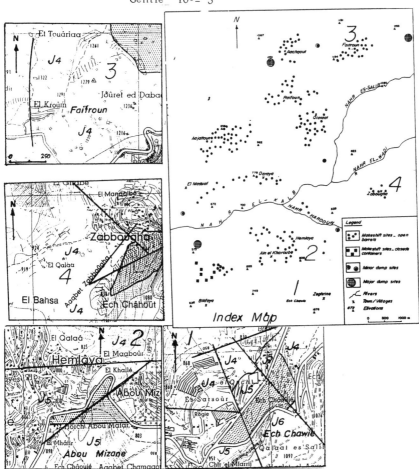

Fig. 3 Capability of proposed sites

The higher the weight the more difficult it is to modify that parameter for
preparing and efficient functioning of
the engineered landfill site. Then, as
shown in Table 3, the weighted rating
values (obtained by multiplying for each
parameter its weight by its rate and
summed up together) are calculated to
show what are the capability of the four
sites. Fig.3 shows the four sites and
their indicated capabilities.

Proper waste management schemes need
planning of operations from the initial
collection, through transport and final
disposal in adequate sites. In the study
area, or for that matter somewhere else
in rural Lebanon, it is not only necessary to find suitable landfill sites but
to have a complete solid waste handling
plan.

REFERENCES

Bagchi,A. 1990 Design,construction and monitoring of sanitary landfills. John Wiley and sons,N.Y.,281 p.

Bakic,M.,1972 Jeita:The famous karst spring of Lebanon. United Nations Development Program(UNDP),Ministry of Hydraulic and Electric Resources, Republic of Lebanon,378p.

Canter,L.1991 Environmental impact assessments for hazardous-waste landfills. J.Urban Planning and Development, ASCE,117:2:59 -74.

Canter,L.Knox,R. and D.Fairchild,1988 Ground water quality protection.Lewis Publishers,Inc.,Michigan,562p.

CDR,1982 Master plan for solid wastes management, The national Waste Management Plan.Vol.IV.Council for Development and Reconstruction,372p.

FAO,1990 Lebanon land cover map, at 1:50,000. Sheet 6. CDR, Ministere des Travaux Publiques,Republique Libanaise.

Hariri Fdn.1987 Sewers,septic tanks and solid waste activities. Lebanon at present and its needs for rehabilitation and development, Final Summary Report pp.42-58.

Jessberger,H.1990 Some technological aspects of waste disposal.Proc.6th International Congress, International Association of Engineering Geology, Amsterdam, 6-10 Aug.1990,pp.131-139.

Khawlie,M.; Fattouh,S. & Jeha, Z.1991 General assessment of land inputs for producing a preliminary waste disposal capability map for Greater Beirut area, 4th Jordanian Geological Conference,Amman,4-7 Nov.1991,14p.

Knight,M.1987 Role of geology in siting and design of waste disposal landfill for minimal environmental impact. Proc.Land Plan III,Symposium on Role of Geology in urban Development,Hong Kong, 15-20Dec. 1986, 22p.

Acknowledgment: The authors extend sincere appreciation to the Lebanese National Council for Scientific Research as it gives   continuous support for the Geology Department.

Waste as a high risk 'geotechnical' unit

Les déchets solides vue comme une unité géotechnique hasardeuse

Wilmar T. de Barros & Ricardo N. d'Orsi
GEORIO Foundation, Rio de Janeiro, Brazil

ABSTRACT: The disordered occupation of the steep slopes of the City of Rio de Janeiro has caused serious man-induced landslides. A very peculiar type of mass movement in these areas are the wasteslides, that has caused severe damages and loss of lives. The GEORIO Foundation, together with the County Cleaning Company, are trying to solve this problem carrying out a great number of measures that include since the cartography and studies of these waste disposals until their complete or partial removals and even the execution of works for their stability.
Although these waste disposals have artificial origin and a nonnatural composition, their features and dynamic behavior led the GEORIO technicians to treat them, when mapping or building retaining works, as a high risk "geotechnical" unit. These features and behaviors are here presented, as well as some cases of wasteslides and the remedial measures carried out.

RÉSUMÉ: L'occupation desordonnée sur les pentes des montagnes de la ville de Rio de Janeiro a provoquée de graves glissements. Un type particulier de ces mouvements de masses sont les glissements de déchets qui ont comme conséquences de serieux problèmes matériels et très souvent des pertes de vies humaines. La Fondation GEORIO avec l'appui de la Compagnie Municipale de Nettoyage Urbaine, a essayé de resoudre le problème utilisant la cartographie, l'étude des dépôts de déchets, le transport des déchets (integral ou parcial) a travers des méthodes différentes, mais aussi par l'éxecution d'ouvrages de contention.
Malgré le côté antropique et la composition non-naturelle de tels dépôts de déchets, leur caractéristiques et comportement dinamique ont amené les techniciens de la Fondation GEORIO à traiter ces dépôts comme des unités géotechniques de grand risque. Ces caractéristiques et comportements sont ainsi représentés comme quelques cas de glissement de déchets et les solutions nécessaires et preventives adoptées.

1 INTRODUCTION

Rio de Janeiro County has an area of 1,356 km^2, where 64% correspond to hilssides base and fluviomarine plains and the remains presents a moved relief that embodies isolated hills and three mountainous massives: Tijuca Massive, Pedra Branca Massive and Gericinó Massive (Heine, 1986). Its population is of 5,341,000 inhabitants, 17,7% of which live in slum (*favelas*) areas (FIBGE, 1992). It is estimated that today there are 537 *favelas* in all the county, 45% of them located at hillsides (IPLAN, 1993). The origin of these *favelas* at hillsides goes back to the end of the last century when, after the end of slavery, many groups of Afro-brazilian, homeless, found in the abrupt hillsides of Tijuca Massive its home solution. With the quick development of the city, and since there was no intervention on these areas, *favelas* followed growing and from small locations became great *favelas* complex.

The decades of 50, 60 and 70 were decisive for the consolidation of these areas. At the same time of a great migration inland-urban centers that has occured in this period, a very precarious urban substructure (opening of access path, installation of electrical network and water supply) was being installed, creating more favourable conditions for living. As a consequence, the searching for these areas was very intense and the population of these areas had a exponential growth. But it grew also the degradation processes of hillsides, starting with deforestation, excavations and indiscriminate embankments and waste disposal on the hillside, in many cases on the natural drainages and on the more declivous areas.

The population impoverishment (mainly from the decade of 70 on), the lack of an urban planning on these regions and the very high birth rate there registered made it chronic the habitation conditions. The high suscepti- bility to diseases due to the lack of basic sanitation was aggravated to the risks of landslides on the cut screes and of disseminated embakments. The pluviometric events of 1966, 1967 and 1968, when hundreds of people died due to hillsides slidings, confirm the high risk condition of these places. It has to be said that from mid 70s the causes of the mass movements on hillsides of Rio de Janeiro County become basicaly of anthropic origin.

Of the several kinds mass movements - with fatal consequences - registered in hillsides *favelas*, the sliding of waste disposal directed GEORIO Foundation (county agency responsible for geotechnical problems management in Rio de Janeiro hillsides) attention. These slidings are occurring with an increasingly frequency, dislocanting masses with enormous volumes (they can bury many houses) and covering large areas. Between December 1983 and February 1994, GEORIO Foundation has registered tens of slidings with different intensities, at waste disposal, with a great material damage and the loss of 58 lifes.

The recommeded measures by GEORIO technical department for mitigate the accident risks deriving from slidings in waste disposal vary according to the geologic-topographic characteristics of the hillsides, of the ocupational density of the encircling area of the disposals and the handiness of access to the site. The interdition of the risk area is always determined (although rarely it is accomplished by dwellers) and the option for the simple waste collecting is always a priority. For this, the COMLURB (County Company of Urban Sanitation) uses different methods according to local conditions. Among these methods for waste collecting there are: execution of wasteduct (permanent structure made of reinforced concrete), metallic drip, telepheric, manual transport with baskets and/or plastic canvas, use of micro-tractors and mules. But the ocupational density and the reincidence of waste disposal - in the same place - makes it very difficult the removal/collecting option.

Some attempts for using local workforce for resolve troubles related to the waste have produced results in some communities. These are the cases of the Community Sweeper Program (training people from the community for the daily and localized service of transferring the waste to municipal mechanized collecting points) and of the Differenciated Collecting Program (sorting out of glass, plastic and paper), performed by the dwellers, with a lucrative return for the community.

There are cases, although, in which restraint works are necessary, mainly when waste disposal are uphill of large and declivous screes. In these cases mixed solutions of waste partial removal and disposal restraint are the most technically and economically feasible.

2 FIELD OBSERVATIONS

To deal with the risks related to these waste disposals at hillsides, the technical departament of GEORIO Foundation responsible for the risk areas survey (scale between 1:250 and 1:2,000) and for indicating solutions started dealing with these disposals as "geotechnical unities".

The difficulty in finding samples for laboratory test (extremely heterogeneous material) and for adapting the current equipemts for testing the kind of material found at waste disposals allowed only - up to now - the obtainment of data deriving from field observations. These observations revealed, for the most of the disposals, the occurrence of peculiary geotechnical features where point out the very low resistence to shearing, very high plasticity, high porosity and low permeability. The analysis of the constituent material and the comparison of the

more recent disposals with the oldest waste disposals (up to 15 years) makes one perceive on those ones significative increasing of percentual plasticity (mainly plastic bags from supermarkets), what contributes to increase, especially, the capacity of these disposals accumulating "liquid bags", randomly disposed on the disposals (therefore the low permeability) and diminishing the internal friction. Other compounds commonly found on disposals correspond to works remains (timber, masonry, wall tiles fragments, etc) and earthy material deriving from excavations. The whole characteristics at- tributes to the movements a similar behavior of a mudflow, due to the great speed and to the low viscosity of the material.

3 PRINCIPAL ACCIDENTS

Every year Civil Defence of Rio de Janeiro County registers several accidentes associated to landslidings at hillsides waste disposals, and of these the majority corresponds to the movements of small masses that obstruct streets, piping, etc, causing many troubles in city's life.
Among the principal accidents, two of them are pointed out due to its catastrophic consequences: accidentes at Pavão-Pavãozinho *Favela* and Santa Marta Hill (see Table 1).

3.1 Wasteslide at Pavão-Pavãozinho *Favela*

The hillside where there is the Pavão-Pavãozinho *Favela* (Copacabana district, south of the city) is geologically constituted of augen gneiss (Hembold, 1965), showing a mainly rectilineal topography, with a 45° inclination on the area where it occurred the accident. The instable mass, whose waste volume was of 1,600 m³, was

over a thin soil coating (40cm thickness) that covered all the underlying rocky massive and caused a 150m long and 20cm width fissure (figure 1). In its trajectory, the mass reached and demolished the four reinforced cemente columns of the 20,000 liter water deposit, increasing the destructive effects of the sliding. As consequences, 22 houses were demolished and 19 people died buried. It is attributed to the development of positive subpressions in the waste/rock contact the principal factor deriving the movement, when it was registered a pluviometric rate of 40,8mm in 24 hours (INM,1993).

Table 1. Principal cases of waste-slides occurred between 1983 and 1994 at hillsides Favelas of Rio de Janeiro County.

FAVELA / DATE	CONSEQUENCES*
Coroa - 03/23/83	4 D, 2 BD
Pavão-Pavãozinho - 12/25/83	19 D, 22 HD
Santa Marta - 02/12/88	12 D, 23 HD
Candelária - 02/12/88	3D, 2HD
Páu da Bandeira - 06/16/89	6 D, 6 HD
Nova Pixumas - 01/05/92	2 D, 1 HD
Prazeres - 11/07/92	SO
Rocinha - 11/27/92	2 D, 6 HD
São Carlos - 03/12/93	9 D, 7 HD

* D=deaths; HD=houses demolished; BD=bulding damage; SO= street obstruction

As remediable solution, it was constructed a reinforced concrete wasteduct for removing the house waste at the highest portion of the hillside. It was constructed also a impact wall at half hillside. This wall was made of reinforced concrete and was solderesd with lead to the rock with countreforts. The objective of the wall was to protect the downhill houses against new slidings of the remaing soil coating (figure 2).

3.2 Wasteslide at Santa Marta *Favela*

The accident at Santa Marta *Favela* (Dona Marta Hill, Botafogo district), although foreseen with reasonable antecedence, caused 12 deaths and the demolishion of 23 houses (figure 3). Located in an hillside area with 35° mean inclination and over a altered leptynite gneiss rocky massive (Hembold, 1967), the waste disposal there existing had accumulated - in its 50 years (IPLAN, 1993) - a volume of 1,800m³, concentrated in a "waste tongue" with 90cm mean thickness (at some points up to 3m), 20cm width and 100m long (figure 4). It is important to say that the local hydrologic conditions were very unfavourable to the disposal stability, since besides then fissure conditions of the rocky massive make it possible the presence of water in some hillside points, therewas a great contribution derived from used water from houses encircling the disposal. Uphill the disposal there were about 90 houses, the majority made of masonry and some made of wood; there were 450 people. The geologic-geotechnical survey (scale 1:250), performed on that *favela* in January 1987, through a accord GEORIO Foundation and Federal University of Rio de Janeiro Geology Department indicated the high risk of

Figure 1. Wasteslide at Pavão-Pavãozinho *Favela*

Figure 3. Aerial view of Santa
Marta *Favela*, before wasteslide.

Figure 2. Pavão-Pavãozinho *Fav.*:
stabilization work and wasteduct.

Figure 4. Wasteslide at Santa Marta *Favela*.

accidents and recommended the immediate interdiction of the houses and arrangements to solve the problem. GEORIO Foundation interdicted 25 houses that were on the influence area of the probable sliding. But the dwellers, although notified, refused to leave the area. After some studies with COMLURB, it was decided to repressed the waste since it was impossible to remove it. The initial project for restraint had a reinforced concrete impact wall, with counterforts, soldered with lead to the rock. But in this initial phase of cleaning up the terrain for implementation of the wall (January 1988), it occurred a partial instability of the disposal base, what caused the death of a worker. As a security measure for following up the work, it was decided to constrain the disposal through two metallic tissue of high resistence, superposed, hold on extremities (uphill and downhill) by reinforced concrete binding beams soldered to the rock. The heavy rain on February 1988 with 785,2mm (INM, 1993) in the first days of the month, increased the weight of the waste deposit in such a way that the metallic tissue was breaken up and the waste disposal slided.

From the accident on, there were a modification in the initial project. Then it was indicated the removal of the small amount of residual material of the disposal and the construction of restraint curtains anchored, echeloned (figure 5), aiming at the same time to create, on that wanting community, plain areas for new houses, avoiding this way the formation of a new waste disposal.

4 CONCLUSIONS

Before the complexity, proportions

Figure 5. Stabilization woks at Santa Marta *Favela*.

(number of cases and involved volumes) and the approach of GEORIO Foundation's geotechnicians, the problem of waste disposal sliding in Rio de Janeiro County constitutes, probably, an unique case in the whole world. Its origin goes back to tens of years and the solutions for definitive eradication of the associated risk of this "mass movement" embody several and lenghty activities that should include intensive programs of basic and environmental education up to very heavy financial investments for *favelas'* urbanization, with sewerage system, superficial drainage and access vias that allow the waste removal. While these measures are not adopted, it increases the accident risks and the work of mapping and reducing the accidents potential has shown a better result when waste disposal are considered as short comprising geotechnical unities. The exigence of field observations (especially as to the movement processes) and detailed characterizations certainly will lead to the costs reduction and to optimization of the implemented services in the elimination of accidents risks.

ACKNOWLEDGMENTS

The authors are grateful to Mr.Ari Maciel photographer of GEORIO Foudation for all the photos and to geographer Antonia Rodrigues de Brito (GEORIO Foundation) for her help in the preparation of this paper.

REFERENCES

Fundação Instituto Brasileiro de Geografia e Estatística (FIBGE) 1992. Censo demográfico 1991: resultados preliminares. Ministério da Economia e Planejamento, Rio de Janeiro.

Heine, U.H.R. 1986. 20 anos de Geotécnica - Características das Encostas do Rio de Janeiro. Revista SEAERJ (Sociedade dos Engenheiros e Arquitetos do Estado do Rio de Janeiro), ano XVI, n° 20. Dezembro. pp 7-45.

Hembold, R 1965. Mapa geológico do Estado da Guanabara, escala 1:50.000. Ministério das Minas e Energia - DNPM. Divisão de Geologia e Mineralogia.

Instituto de Planejamento do Município do Rio de Janeiro (IPLANRIO) 1993. Favelas Cariocas: alguns dados estatísticos.

Instituto Nacional de Meteorologia do Ministério da Agricultura (INM) 1993. Normais Climatológilógicas. Rio de Janeiro.

Proposal for the utilization of solid waste by marble industries

Proposition sur l'application des déchets solides de l'industrie du marbre

María Beatriz P.de De Maio & Fabio Luna
National Institute of Industrial Technology, Argentina

ABSTRACT: During marble manufacturing process, certain waste (sludge) is produced as a result of cutting of stone material. Sludge creates a problem for manufacturers who must look for its disposal at high cost. This paper tends to find some possibility for the utilization of waste, attempting to minimize the environmental impact caused by the need of a place for its disposal as well as to make profitable use of it. The EPA instructions were taken as a reference, so as to satisfy international requirements about industrial waste disposal. The study was developed as an analysis of the potential degree of contamination this waste may cause. At the second stage tests were carried out so as to determine the effect of stabilization / solidification that is produced by means of mixing sludge solids with cement. Lastly, with the final product obtained, the idea is to determine its possible utilization as a construction element.

RÉSUMÉ: Le procès d'industrialisation du marbre produit un résidu après la taille et le polissage du matériel. Celui-là constitue un problème pour les producteurs qui doivent chercher des alternatives pour la disposition, et cela signifie un coût important. Ce travail cherche quelques possibilités d'emploi du résidu pour éviter l'impact sur l'environnement dans un lieu de décharge et la viabilité de sa ré-utilisation. A fin de répondre aux réquisitions internationales on a pris comme référence les directives émanés de l'EPA. On a analysé le degré potentiel de contamination que ces résidus peuvent provoquer. Dans une seconde étape, on a fait les essais qui ont permis de déterminer l'effet de stabilisation / solidification qui se produit quand on mélange le contenu de solides du barre avec le cément. Finalement, avec le produit final, déterminer sa possible utilisation dans l'industrie.

INTRODUCTION

The increasingly use of stones by the construction industry generates a greater production of waste.

This fact represents a problem to be solved at the lower cost possible.

From this point of view, industrialists must answer questions involving such aspects as:

If resulting waste is contaminating, where to derive and dispose it and, eventually, think about the possibility of its recycling as another industrial product re-inserting it into the trade market.

Studies and tests that were carried out, try to give a satisfactory answer to the above mentioned requirements, without raising costs managed at present by industrial development.

With this purpose, papers were oriented according to the US. Environmental Protection Agency (EPA) statements on the processing of solid waste that is considered potentially hazardous.

Following this outline, procedures were used towards the stabilization of the possibly contaminant elements and their solidification for avoiding its migration. One of the procedures commonly used, is that which is based on lime and cement chemical. The mixture of waste derived from marble pla te (and granites in lower proportion) sawing activity with cement, produce a matrix whose physicochemical characteristics reduce the mobility of potentially hazardous elements.

The study fullfilled up to present included analysis to determine if the produced waste should be considered as contaminant; tests on mixtures with cement to obtain stabilization / solidification and test of the produced product towards its possible re-utilization.

1.-Characterization of waste

During this first stage of study, two sludges were selected

from different factories:
* Sludge F : produce of the sawing of marble plates with low proportion of granite origin waste.
* Sludge K: produce of the manufacture of floor tiles with dolomite-calcite compound.

1 a) *Chemical Analysis*

Table I presents chemical analysis corresponding to both sludges.

TABLE I

Oxide contents	Sludge F (%)	Sludge K (%)
Si_2O	50.8	13.8
CaO	12.6	32.0
Na_2O	3.0	0.4
K_2O	1.6	0.2
Fe_2O_3	15.0	4.9
Al_2O_3	10.5	3.0
MgO	0.3	12.5
Calcination loss (1000 °C)	5.6	32.8

1 b) *Determination of the acidity degree*

This determination was done according to method 423 that was published in " Standard Methods for the Examination of Water and Wastewater "

	Sludge F	Sludge K
pH value	11.6	7.8

2.- *Mineralogical Analysis*

Due to the fineness of the sample, the determination was done by X Ray diffraction, using Philips equipment, with Cu radiation, Ni filter, under conditions: 40 Kv, 20 mA, 600 mm/h speed and 1000 scale. Results were as follows:

	Sludge F	Sludge K
Main compounds	Quartz Feldspar Calcite	Calcite Dolomite
Secondary comp.	Chlorite Illite	Clay

Although through XRD analysis, during which samples used are sieved passing through N°. 325 (45- μm) sieve there is no detection of presence of any iron compound, it must be explained that sludge F contains fragments of iron scourings used while sawing the material; the latter are detected in the retention over N°. 200 (75-μ m) sieve and through chemical analysis (see Table I). It is estimated that the amount of iron scourings in Sludge F arrives to 5 % of the whole sample.

3.- *Physical Tests*

Physical tests are carried out for the characterization and comparison of the studied material, before and after the stabilization and solidification procedure.
This data also provides basic information about waste processing, and about the possibility of costs estimation in relation to such process and handling.
Among the physical tests most commonly used, the Index Property Testing stands out, being carried out for obtaining the general characteristics of the material, taken as a reference during operational processes (i.e. pumpability). Tests are performed using a sample with neither stabilization nor solidification.
The following determinations where obtained from both samples:

3.1 *Particle Size Grading,* done according to ASTM D 422-63 (1990) Standard, by current-gauging (Fig. 1)
From these results on, it is possible to predict with enough certainty, from a quantitative point of view, a comprehensive range of characteristics, including aspects related to waste permeability and compressibility.

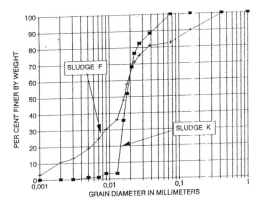

	Sludge F	Sludge K
Texture	Loam silt	Silt
Uniformity	9.5	1.4
Concavity	1.9	1.0

3.2 *Atterberg limits grading*, done according to ASTM D 4318-84 Standard, allows the definition of the physical characteristics of a material as a function of its water contents

	Sludge F	Sludge K
Liquid limit (%)	37	35
Plastic limit (%)	28	26
Plasticity index (%)	9	9

3.3 *Moisture Contents:* Determination of percentage Determination of total and volatile solids was done according to Standard Methods for Examination of Water and Wastewater (1985)

	Sludge F	Sludge K
Moisture (%)	60.0	46.5
Total solids (%)	40.0	3.5
Volatile solids (%)	0.25	8.2

4.- *Characterization of stabilized waste*

The method chosen for stabilizing sludge was to add portland cement to the waste (see Table II), using a water/cement relation (w/c) of 0.33.

The mixture was submitted to the compressing test stated by ASTM D 558-82 (1990) and D 698-78 (1990) standards, so arriving to maximum compression and density.

This determination is quite helpful for waste used as soil to be re-compressed once stabilized, without having the problem of an inadequate compactation in case materials are too dry or too wet.

TABLE II
Chemical composition of portland cement

Loss due to calcination	(%)	2.21
Insoluble waste	(%)	1.58
Sulfur trioxide(SO_3)	(%)	2.33
Magnesium oxide (MgO)	(%)	0.79
Sulphurs (S_2—)	(%)	0.03
Chlorides(Cl^-)	(%)	0.01
Silicon dioxide(SiO_2)	(%)	21.9
Calcium oxide(CaO)	(%)	61.9
Aluminum oxide(Al_2O_3)	(%)	5.04
Iron oxide(Fe_2O_3)	(%)	4.03
Sodium oxide(Na_2O)	(%)	0.09
Potassium oxide(K_2O)	(%)	0.86
Free Calcium oxide	(%)	0.60
Total alkalines	(%)	0.66

4.1 *Characterization:* Samples prepared with waste and cement were cured along 28 days and tested subsequently so as to determine their density, water absorption and strength characteristics. Table III shows resulting values

TABLE III

	Sludge F	Sludge K
Density (Kg / m^3)	1260	1035
Water absorption (%)	39.8	51.6
Strength testing (Kg / cm^2)	19.0	22.4
Strength testing (psi)	276	318

Next, photographs are shown about sludge samples already stabilized with cement and their aspect after compression testing

Sludge F

Sludge K

4.2 Mineralogical Composition

New analysis were carried out by X Ray diffraction applied to sludge already stabilized, in which although no change in their composition was detected, there appears an evident change in the bond between crystals presenting maximum values in both diffractographs a variation in their development and amplitude.

4.3 Wetting and drying test

This determination was done according to ASTM D 559-89 Standard so as to evaluate the resistance of the stabilization of the mixture waste / cement. The test was carried out along 13 cycles being results as follows:

	Sludge F	Sludge K
Weight loss (%)	0.8	4.6
Volume difference (%)	0	0

Tubes in both samples showed no changes in relation to physical damage, maintaining their exterior aspect integrity.

5.- Evaluation of results

The granulometric characteristics of both samples show the high fineness of wastes; nevertheless the one named F presents a lower steep indicating a better gradation of its particles, while sample K has a pronounced steep showing very close grain sizes.

This way, each sludge is identified as loam silt and silt, respectively.

Values obtained through Atterberg limits, for mixtures waste / cement, show that measured plastic limits are placed between those considered as typical ranges: between 20 and 50 %.

Along future testing, the proportion of cement to be added may vary, so as to obtain stabilized waste that should be submitted to different tensions, so modifying its compressibility, swelling, contraction and cut resistance properties.

Densities obtained on stabilized samples are placed at media values and allow to obtain data for calculation of the volume of material to be handled in case of being delivered to a disposal site, roads, cut work and back filling in construction, etc.

Strength obtained in stabilized waste rise much above minimum value demanded by EPA, which is 50 psi.

Finally, the performance of samples submitted to wetting an drying cycles for proving the stability of cemented waste, show low loss of material and very good conservation of their physical integrity, arriving to the conclusion that an adequate degree of stabilization has been achieved.

Obviating the freezing and defrosting test has to do with the ambient characteristics of the influence range of possible utilization of the produced waste, not existing at it freezing media temperatures.

6.- Usage possibilities

Taking into consideration that an average factory produces 1400 ton / year waste, at a cost of u$s 35000, obtaining no benefit of it because it is delivered to final disposal sites, it becomes important to find some way of recycling.

According to this preliminary study, it may be considered that sludge produced as a residue of marble factories (including granite sawing) has no hazardous or contaminant elements.

The applied and presented method allows to assume that mixture with cement in both samples would act as a stabilizing physical means for contaminating heavy elements (Pb, Cr, Ca, etc.), composing the sludge cake, in which the high fineness of material is one of its most important properties.

Besides, the addition of a certain proportion of sand would offer the possibility of producing articulated pavement (at present under development) or as sub-base for road construction too.

The re-insertion of this sludge cake (case F) into the industrial production system focuses other uses such as the correction of agricultural soil acidity values .

It must be noted too that sludge F, whose composition includes carbonates and silicates, has a correspondence with Italian methods, which demonstrate that these dewatered sludge while forming a 30 cm layer, acts as " absorbing " of heavy metals lixiviated by water (Frisa Morandini & Verga 1990)

REFERENCES

U.S. Environmental Protection Agency. 1989. Stabilization / Solidification of CERCLA and RCRA Wastes. EPA / 625 / 6-89 /022. May 1989

Sludge KU.S. Environmental Protection Agency. 1985. Standard Methods for the Examination of Water and Wastewater. 16th Ed

American Standard Testing Materials, Annual Book of ASTM Standards, Volume 04.08, 1991

Frisa Morandini, A. & Verga, G. 1990. Problemi connesi con lo smaltimiento dei residui di lavorazione delle pietre ornamentali. Bolletino della Associazione Mineraria Subalpina. Anno XXVII, 1-2, marzo-giunio

Simulation of pollutant-emission by using geoelectric surveys of an artificially introduced salt tracer

La simulation de l'émission de matière polluante: L'application d'un traceur de sel et la vérification de sa migration à l'aide de méthodes géophysiques

W. F. H. Kollmann
Geological Survey Austria, Department of Hydrogeology, Vienna, Austria

J. W. Meyer
Vienna, Austria

R. Supper
Institut für Meteorologie und Geophysik, Universität Wien, Austria

ABSTRACT: Pollutant-Immission and their simulation have to develop an innovative method for the determination of actual flow paths and velocities of a plume. Using geoelectric surveys for the detection of artificially introduced salttracers, the method supplements ex-situ and conventional tracer hydrology, which permits only point determinations of direct (bee-line) data in aquifers. Geoelectrical mapping was carried out, using a starlike grid for depth-specific contouring and subsequent automation in a data logger over an extendable probe network. The injection of the salt causes a reduction of the specific electric resistivity afterwards. The actual subterranean flowpath and -velocity is thus readily determined from the surface without drilling. Assuming the scenario of contamination having occurred, latent hazard potentials and reaction times required for the implementation of counteractive measures are identified, thus providing pertinent environmentally relevant information packages. These are of potential interest to waste management facilities and drinking water supply systems as well as their areas of influence and protection.

RÉSUMÉ: Les diffusions de matières polluantes dans les eaux souterraines ne se conforment pas toujours aux regularités hydrauliques. Les phases insolubles avec un poids specifique superieur à celui de l'eau, ont une dynamique d'écoulement particulière, qui est determinée par la pente et le relief de l'aquiclude et non pas par la pente du niveau phréatique. Une nouvelle méthode de marquage a été combinée avec une verification géophysique. La simulation d'une infiltration de matière polluante se base sur la reduction de la résistance électrique provoquée par l'eau salée. La migration d'un traceur de sel est prouvée par les sondages de WENNER qui ont été repetés plusieures fois. Ce teste peut servir comme modèle pour la diffusion d'autre matière polluante.

1 INTRODUCTION

1.1 Basic Principles

On the one hand the transmission of contaminants to aquifers occurs at discrete points, abruptly through accidents on transportation routes, in general undetected and continuously in populated areas and on commercial and industrial sites. On the other hand, these phenomena occur areally and latently, e. g. by improperly managed agricultural operations, undetected leakages from waste water treatment plants, as well as old waste repositories.

The dissemination of contaminants in groundwater does not al ways follow laws governing hydrologic systems. Those insoluble phases with a higher specific gravity than water have their own flow dynamics, which does not depend on the groundwater surface gradient, rather they are governed by the dip and relief of the aquitard (Fig.1).

1.2 Practical demand

The identification of pollution emitters, as well as the location of future potential emitters, should include the site orientation of actual, potential or simulated pollution plumes. A precise recognition of actual flow paths and flow rates is a prerequisite to resolving conflicts of interest affected by groundwater protection considerations and include appropriate water rights. The possible optimisation of water protection areas thus realised could reduce the extent of these areas and on the other hand minimize the extent of compensatory payments claimed.

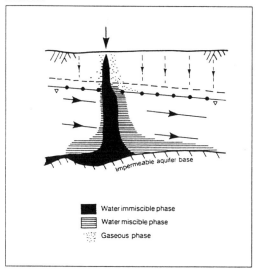

Fig.1: Seepage of contaminants and migration of specifically heavier phases does not drift along the groundwater table, but rather follows the relief of the groundwater aquitard and further following its inclination (from DOWNING & WILKINSON, 1991).

2 Innovation

A method was developed to allow the actual determination of in-situ migration by the geoelectric detection of artificially introduced salt tracers, as a combination with conventional tracer methods, which in former times only permits the identification of point like locations in direct line-of-sight (KOLLMANN et al., 1992).

Using geoelectric mapping, which was improved to yield depth specific information and finally automated via a data collector over an extendable probe network, the actual contaminant distribution in the underground can then be simulated readily from the surface without involving any environmental contamination and drilling. Naturally, any changes in flow rates or directions (meander dynamics) that means the real flow path can also be recorded.

Thus, as a result of contamination, any latent risk potential and the reaction time available to implement counter measures become readily apparent, offering environmentally relevant information for drinking water supply facilities.

From this point of view, a contaminant scenario (e. g. of chlorinated hydrocarbons) or the tracing of groundwater, polluted for example by over-fertilisation with liquid manure, with its specifically heavier saline solution, is comparable with the physical properties of the water in which the table salt is dissolved.

In a general clean-up program it is insufficient to know only the flow direction of pure uncontaminated water, it is also essential to have the capability of prognosticating the effective contaminant distribution. In the final analysis, it is irrelevant whether there is coincidence with the direction of groundwater flow or whether the specifically heavier contaminant follows the relief of the aquitard and drifts down-dip.

A further objective of the geoelectric tracer experiments is to optimise the quantity of the injected matter (i. e. minimisation of the quantity of salt added, as well as the postflushing required), due consideration being given to the necessity for causing least possible hydrologic changes and pollution. In practice, the alternative available, either intermittent or continuous injection, should be selected with the criterion of avoiding an infiltration cone in the saturated zone with an unintentional gradient in the groundwater gauge level and a consequent rise in flow velocity.

Based on numerous pumping trials, using monitoring drillings, conventional hydrochemical analysis could be employed, which is equally necessary for the minimization of the injection-quantity.

3 METHODOLOGY AND EVALUATION OF THE SALT INFILTRATION METHOD

3.1 Preliminary Measurements and Planning

In principle, no prior knowledge of flow direction and velocity and soil parameters is necessary for the application of this method. Nevertheless, such information can considerably ease the planning and implementation of the survey.

For example, knowledge of the approximate flow direction can facilitate a more precise alignment of the scanner grid, although it has also been shown that most of the available data are too approximate and inexact and that flow direction can locally diverge considerably from the general trend. If one were to rely on such indications, any incorrect alignment of the scanner grid can detract from the accuracy of a series of data measured. In most cases therefore, a pre-orientation of the survey grid should be avoided and the stellar array of the configuration of points measured, as described in the next chapter, for which no prior information is required, should be selected.

In order to carry out geoelectric mapping a certain spacing of the external electrodes must be selected, which is dependent not only on the depth of the aquifer but also on the vertical resistance distribution underground. To do so, it is necessary before beginning with the actual survey, to carry out several geoelectric (Schlumberger) depth probe measurements (soundings) within the planned survey area. Model calculations are used to arrive at an approximate distribution of the vertical resistance. It is assumed, that at 10° C the resistivity of water decreases from an original value of 21 ohm.m to 7,5 ohm.m through salt addition (this corresponds to about 1,000 mg NaCl/l, the maximum concentration expected) it is possible to calculate the porosity value of an aquifer with a pure water resistivity of 210 ohm.m, using the equation

$$(x,y) = (0.62 * R_W / R_S)^{(1/2.15)}$$

R_W ...Resistivity of pure water R_S ...Resistivity of aquifer

and the resistivity value of the aquifer. To obtain the resistivity value subsequent to salt injection, the porosity previously calculated at a water resistivity of 7,5 ohm.m, is inserted in the equation:

$$R_S (x,y) = R_W * (0.62)^{(-2.15)}$$

Using the model calculation, a theoretical value of resistivity distribution, before and after salt addition, for each of the survey configurations (mostly Wenner), can be calculated. Subtracting the curves from each other, one can derive the external electrode spacing giving the maximum change. In addition, it is possible to estimate the anticipated value.

3.2 Injection of the Salt

Originally about 200 kg NaCl were dissolved in 1,000 l of water. However, it turned out that at a depth of 2 to 10 m, a reduction to 60 kg still produced a sufficiently measurable effect. These empirically derived values are area specific and pertain particularly the Quaternary sediments of the Leibnitz Field.

The best method for injection proved to be via an tapping drill-hole which is completed at top of the aquifer without penetration so that only the overlying cover is intersected. A "complete" well with filter and swamp pipe built into the aquitard proved to be unsuitable, as the salt solution could not immediately migrate through the filter slots into the groundwater conductor.

If no overburden is in place and the groundwater conductor occurs at shallow depths (2 to 4 m), injection can be effected via a trench. This type of injection can be considered as a model case of polluting matter penetrating groundwater (e. g. fertilizer dressing, liquid chemical transport, leaking fluid container, derelict dump).

3.3 Application of Geoelectrics

The geoelectric method (using the direct current process), in particular the Wenner configuration, has so far proven to be the best method for detecting a change in resistance in the groundwater conductor caused by adding salt, as the method is universally applicable and has already been tested in numerous field evaluation programs.

A disadvantage of the method is that one must work with an electrode grid in which electrode probes must remain buried in soil for several days; external factors such as rain water can change the resistance at electrode surfaces and wrongly register a measurable effect. Practice has also shown that electrodes can often disappear, accidentally or by deliberate action. If one electrode is missing, it is in effect eliminating four points of measurement. On the other hand a reintroduction into the ground is often the cause of large errors in heterogeneous underground conditions.

3.4 Grid Modifications

Initially, measurement were made using a rectangular grid. However, whereever the flow direction was not known accurately, this did not prove to be effective. It is for this reason that the stellar array was designed, for which no information on flow direction is necessary. The basic grid for this configuration consists of a circle with a radius of 20 m, which as soon as the approximate flow direction becomes evident from the concurrent evaluation of field measurements, can be enlarged in the sector corresponding to ground water flow direction. A prerequisite is instrumentation using a PC suitable for field operations and a battery-driven plotter.

3.5 Presentation of Data

A suitable presentation of data on the salt migration process is to prepare incremental plots, using data from a blank run carried our prior to salt injection. In this, a drop in resistance signifying a higher salt concentration by tracing in groundwater is used as the maximum in the difference plots.

It was shown to be advisable to carry out at least two blank runs, as single point errors can recur again and again.

4 Case Example Bachsdorf

A tracer infiltration experiment was carried out in the area of the Haslacher Au, near Leibnitz (southern Styria, refer to Fig.2) in October 1992. Apart from method development, the experiment was undertaken to obtain answers to problems pertaining to water management. A test well was situated in the area of the power plant of Gralla (Mur-River hydroelectric project) by the Drinking-Water Supply-Association "Leibnitzer Feld" (VB-H in Fig.2). This well was drilled to obtain acceptable drinking water in the future, in terms of quality and quantity, from bank-filtered river-water of the Auwald area. As both quality and quantity were found to be acceptable, it became necessary to define the boundaries of an additional

protected area. In this connection, a sewage treatment plant, located approximately 300 m upstream, presented a possible problem. An accidental release scenario caused by leakages from the collecting main or settlement basins was to be simulated.

Local geological-geophysical stratification underground, deduced from drill-hole profiles and Schlumberger surveys, shows that it consists of a highly permeable aquifer at a depth of 3 to 6 m (k value approximately 1.6×10^{-2} m/s, specific electrical resistance: 500 ohm.m), a vadose zone of coarse gravel with high resistance values in the unsaturated zone (1,000 to 3,000 ohm.m) and underlain by an impermeable base of sands and clays having a low resistance value (about 200 ohm.m).

The blank measurements were carried out first to assess the "as is" situation of the particular area, including any pollution that had possibly already occurred. However, this possibility was discounted after the background measurements were taken and which could then be used as the baseline for the artificial simulation of pollutant migration by tracing a leakage in the area between the sewage treatment plant and the locality of Brunnenfeld.

A total of 60 kg of table salt dissolved in water were injected in the centre of the stellar array of the Wenner configuration on Oct. 22 1992 between 8.30 and 9.45 a.m.

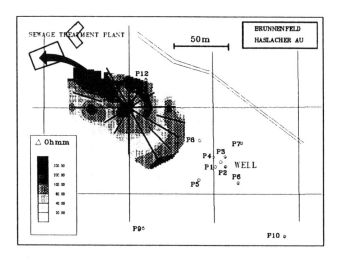

Fig.2 Sites of the the Drinking-Water-Well and Sewage-Water-Treatment, intermediately the Stellar Array Gridding of Geoelectric Soundings Network. Change of Resistance (actual value minus background) after two weeks indicating the Flow- Direction (arrows)

Fig.3 shows the change in resistance after one day. Apart from an infiltration cone in the area of the point of injection, no distinct flow direction is as yet apparent.

In Fig.4, showing the change in resistance after six days, a pronounced flow direction toward the NW is already apparent.

Two weeks after injection (Fig.2) the salt can be seen to have already drifted away from the point of injection to the NW. A minor quantity of salt still remains at the point of injection.

It is evident from the results obtained, that the major portion of the salt solution migrated toward the Northwest to the river Mur, following the relief of the aquitard. The distance-velocity of the main flow direction of the salt amounted to approximately 6.6 m/d. A pollutant having similar physical properties to the salt would therefore, if infiltration of sewage plume were to take place in the area between the treatment plant and at a reasonable clearance from the drinking water supply well, present an insignificant hazard to the drinking water supply.

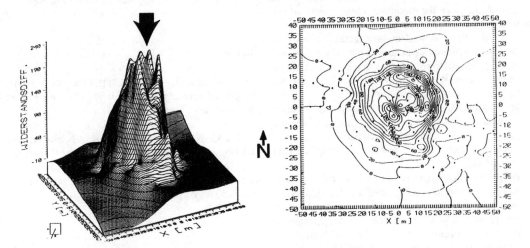

Fig.3   Change of Resistance (actual value minus background) after one day

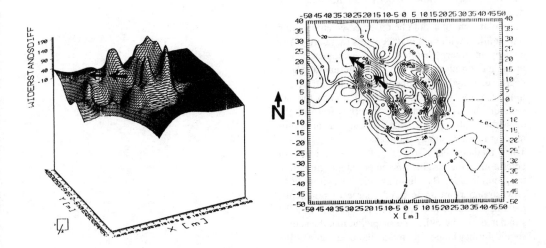

Fig.4   Change of Resistance (actual value minus background) after six days

2439

5 Critique and Conclusions

A critical review of the field tests carried out comes to the conclusion that, with the use of the geoelectric determination of a salt infiltration a method was developed that is simple and for certain investigations effective. Whether there is coincidence between the specifically heavier organic chemical solvents remains to be determined by columnar tests in the laboratory. Further practical experience will be gathered from the application of this method in surveys of closed-down waste repositories and the pollutant plumes emitted by these.

Flow directions as well as propagation velocity of the salt cloud could be measured with the use of this method. In contrast to other tracer methods, the determination of actual pollutant flow paths forms the basis for further calculations.

For this reason real flow velocities are determined rather than distance-velocities. Furthermore, data on porosity and hydraulic conductivity can be calculated from the resistivity mapping derived therefrom.

Possible areas of application particularly pertain to aquifers, at shallow to intermediate depths, that are overlain by beds of gravel. Where groundwater occurs at depths of more than 15 m, changes in resistance, occasioned by the infiltration of representative quantities of salt (50-150 kg NaCl), should not be expected to differ significantly from those caused by natural background variations. For greater depths, only an indication of flow direction can be deduced; flow velocity can no longer be determined.

Particular caution should be observed when injecting into aquifers with high clay contents (aquitard). Any change in hydro-chemical water parameters can bring about a change in clay structure and a related decrease in hydraulic conductivity. This effect can impede or even prevent the introduction of the salt solution and its dissemination and thus frustrate the evaluation of a tracer experiment on questions relating to water supply services.

This method is also an effective means of determining the originators of pollutant immission to groundwater. By defining the real flow path downstream from an inferred source of pollutants, the dispersion route of the immission can be simulated, so that either the suspicion that the source should further be considered as the originator of the pollution of a nearby well or not can be confirmed or rejected.

Finally, the enormous cost savings of the method, compared with the standard tracer methods, through savings on circular test drilling should be pointed out and which could surely be further enhanced by automatisation of the survey series. Requirements for environmental compatibility are met by using non-toxic, non-radioactive and non-carcinogenic table salt tracer, which can be employed below the rooting layer of the soil and which disperses rapidly in the saturated zone by dilution after some days.

References

DOWNING,R.A. & W.B.WILKINSON (1991): Applied Groundwater Hydrology. - 340 p., Ocford Science Publ., New York.

KOLLMANN,W.H. et al. (1992): Nachweis des tatsächlichen unterirdischen Fließweges und der Fließzeit durch geoelektrische Kartierung von eingebrachtem Salztracer. - Ber.FA Hydrogeol., Geol.B.-A., 67 p., Wien.

The significance of organic soils in the evaluation of power plant ash deposits influence on groundwater

L'importance des sols organiques pour l'évaluation de l'influence des cendres des centrales thermiques sur les eaux souterraines

E. Myślińska, E. Hoffmann & A. Stepień
Institute of Hydrogeology & Engineering Geology, Department of Geology, Warsaw University, Poland

ABSTRACT: It has been stated that in searching an adequate site for ash deposits from power plants, the most suitable areas are those covered by organic soils (peat and peat-earth muck). Organic soils, especially with a high rate of decomposition have high sorption abilities for some components of these soils (sulphates, magnesium, heavy metals: Pb, Zn, Cu, etc.). Furthermore it has been proved that loading these soils with waste causes their condensation, which considerably decrease their filtration.

RESUME: On a trouvé, que les régions de sols organiques (les tourbes et les tourbes décomposées) sont des places les plus convenables pour l'émmagasinement de cendres produites par les centrales éléctriques. Les sols organiques, sourtout après leur décomposition importante, sorbent facilement des constituants compris dans les cendres (les sulfates, magnésium, les métaux lourds: Pb, Zn, Cu, etc.). En outre, les sols organiques chargés avec les cendres deviennent plus condensés et leur filtration diminuent.

1 DISCUSSION AND OBSERVATIONS

The presented paper is a result of a research conducted for selecting the most suitable location for an ash deposit site from a power plant in south - western Poland (E. Myślińska et al., 1990). Field works have shown that from the economical point of view (distance from power plant, presence of infertile land) the most suitable localities for ash deposit sites are areas covered by organic soil (mainly peat and its upper, weathered layer - peat-earth muck). The peats represent a reed-sedge type. They are characterised by a light and medium decomposition and a roasting loss between 80 - 90%. Some of the upper parts of peat are overlain by peat-earth muck, mucky peat or by surface soil. Sandy peats, with an admixture of organic matter (mainly quartz, less commonly carbonate sand) often occur in the lower parts of many profiles. In the evaluation of peats and other organic soils for the purpose of groundwater protection from the influence of power plant ash (wet ash) the following items were examined:

1. ability of organic soils to retain polluted covering water;

2. changes in permeability of organic layers due to their silting-up by ash particles as well as peat consolidation under waste load.

The chemical analysis and pH of water of a composition close to the one of covering water (J. Pachowski et al., 1976) filtrated through samples of peat and peat-earth muck was tested (Tab.1) to estimate the possibilities of various ions retention by peat and peat-earth muck.

The examination of water extract from peat showed that a hydrochemical background has to be taken into consideration when evaluating the influence of ash

Table 1. Changes in the chemical composition of model water
filtrated through samples of natural peat and peat-earth muck

Type of sample	I_{om}[*] (%)	Chemical composition of water mg/dm^3								pH
		SO_4^{2-}	Cl^-	HCO_3^-	Ca^{2+}	Mg^{2+}	Na^+	K^+		
model water: cove-ring water	–	280.2	312.0	73.2	162.2	4.9	77.0	32.0		6.9
peat	78.5	288.2	390.0	61.0	220.3	7.3	77.0	23.0		5.2
peat	89.8	280.2	354.5	61.0	212.3	7.3	73.0	26.0		5.8
peat	90.8	184.1	354.5	61.0	158.2	7.3	74.0	24.0		5.7
peat	90.8	184.1	354.5	61.0	158.2	7.3	72.0	22.0		5.8
peat-earth muck	61.2	176.1	106.4	48.0	212.0	7.3	15.5	2.0		5.8
peat water	–	16.0	53.2	317.4	86.1	10.9	27.0	3.0		7.6

[*] – Content of organic matter

deposit sites on ground water (Tab.1). The variable chemical composition of the obtained filtrates results from two different causes: different ability of ion retention by organic soil samples and a different primary pollution of natural samples. The data analysis shows that peat-earth muck, that is only the upper layer of peat with a high grade of organic matter decomposition has considerable sorption abilities for most ions -especially SO_4^{2-}, Cl^-, Na^+, K^+. The reduction of the SO_4^{2-} content down to 176,1 mg/dm^3 and the Cl^- content down to 106,0 mg/dm^3 gives values lower than the limits established in Polish Standards for drinking water- SO_4^{2-} up to 200 mg/dm^3 and up to 300 mg/dm^3 Cl^-. Peats reveal sorption abilities mainly for sulphates. The maximum sulphate sorption abilities of natural peat and peat-earth muck reach up to 12-16 mg per 100g of sample in favorable conditions (lack of this ion in the tested sample). Peats did not reveal sorption abilities for Cl^-, and this ion can be sometimes filtrated from peat by covering water, which leads to Cl^- concentration in the filtrate after filtering covering water through peat samples. The model testing of peat (content of 50% of highly decomposed organic matter) and quartz sand mixtures, through which solutions of an approximately similar content (100 mg/dm^3) of various ions (Tab.2) were filtered, shows that these mixtures have a sorption ability for Mg^{2+} and SO_4^{2-} ions, which increases with the percentage of peat content in these mixtures. Higher sorption abilities of

Table 2. Ion contents in filtrates (mg/dm³)

Solution[*]	Sand ——— Peat	Ca^{2+}	Mg^{2+}	Cl^-	SO_4^{2-}
1	1:0	2.4	1.4	21.3	11.2
	20:1	15.8	1.9	14.2	12.8
	10:1	31.7	3.8	14.2	22.4
	5:1	51.6	8.7	14.2	38.4
	2:1	69.9	8.3	14.2	25.6
	1:1	194.7	6.3	21.3	115.2
2	1:0	–	77.5	272.9	–
	20:1	–	64.0	262.3	–
	10:1	–	42.9	255.2	–
	5:1	–	28.9	262.3	–
	2:1	–	33.1	262.2	–
	1:1	–	14.6	233.9	–
		3			**4**
3;4	1:0	65.0	–	–	28.8
	20:1	38.1	–	–	28.8
	10:1	26.9	–	–	32.2
	5:1	57.1	–	–	32.2
	2:1	45.7	–	–	19.2
	1:1	84.2	–	–	25.6

[*] Solution 1 – H_2O dest.; pH 6.7
Solution 2 – 100 mg/dm³ Mg^{2+} and
320 mg/dm³ Cl^-; pH 5.57
Solution 3 – 100 mg/dm³ SO_4^{2-}; pH 2.95
Solution 4 – 100 mg/dm³ Ca^{2+}; pH 6.79

Table 3. Absorption of heavy metals by various soils (mg/1000g soils)

Soil	Solutions		
	20 mg/dm³ Pb	16 mg/dm³ Cu	16 mg/dm³ Zn
Peat–ear– th muck	400.0	316.2	316.8
Peat	386.0	314.4	314.0
Surface soil	391.1	304.4	231.2
Sand with $CaCO_3$	396.0	317.8	316.6
Quartz sand	384.0	273.4	64.0

mixtures (with a considerably lower organic matter content) in comparison with natural samples are caused by a higher degree of organic matter decomposition in mixtures than in natural samples. For chosen samples of peat, peat-earth muck and surface soil, as well as carbonate and quartz sand, the static batch method (E.Osmeda-Ernst & S.Witczak, 1991) was used in order to test sorption abilities for heavy metals. Table 3 shows sorption for lead, copper and zinc from model solutions. According to collected data all samples excluding quartz sand have sorption abilities for heavy metals from model solutions, especially for zinc. For example, a 1 m² surface of a 10 cm thick layer of peat-earth muck is able to refine 1628 dm³ of a lead solution (concentration of 20 mg/m³), and the same peat layer is able to refine 556 dm³ of the same solution. The amount of a similar zinc solution is 2336 dm³

for the same layer of peat-earth muck and 800 dm³ for a layer of peat. These are estimated values, enabling the comparison of sorption abilities for various organic soils (E.Myślińska et al., 1993).

For chosen peat and peat-earth muck samples compressibility was also tested. As was expected, its values are high (E.Myślińska, 1988), the lowest to be observed in peat-earth muck (approximately 0,352 MPa within the load range of 0 - 0,025 MPa). A considerable increase of volumetric density of samples, which results in reducing the permeability of laden soils, was also noticed. The coefficient of filtration for organic soil samples was tested with different methods, including the modelling of conditions occurring after loading soil with waste (water with ash) and filtration of covering water within the profile. The permeability of the following samples was tested:

a) natural samples filtered with covering water, without suspended matter;

b) natural samples filtered with covering water with ash suspension (200g of ash in 1 dm³ of water);

c) samples after compressibility tests in edometers, filtered with covering water without suspended matter;

d) model samples filtered with various solutions.

The results of these tests are shown in Tables 4 and 5.

2443

Table 4. Coefficient of filtration peat and peat-earth muck

Type and depth of sample (m)	I_{om}* (%)	Type of filtred solution	k_{10} (m/s)
peat n(1) 0.6	78.5	cov. water	5.80×10^{-5}
peat n(1) 0.5–1.0	87.8	cov. water	3.82×10^{-5}
peat n(1) 0.5–1.0	89.8	cov. water	2.76×10^{-5}
peat n(1) 1.9	90.8	cov. water	6.44×10^{-5}
peat n(1) 4.5–5.0	79.3	cov. water	1.10×10^{-5}
peat n(1) 0.9	84.5	covering water + ash	9.03×10^{-6}
peat n(1) 0.5–1.0	89.8	covering water + ash	5.41×10^{-6}
peat e(2) 1.6–2.0	90.8	cov. water	1.79×10^{-6}
peat–earth muck n(1) 0.6–0.8	61.2	cov. water	1.14×10^{-4}
peat–earth muck n(1) 0.6–0.8	61.2	covering water + ash	1.06×10^{-5}
peat–earth muck e(2) 0.6–0.8	61.2	cov. water	2.13×10^{-5}

* – Content of organic mater;
n(1) – peat and peat–earth muck in natural conditions;
e(2) – peat and peat–earth muck after compressibility tests in edometers.

The comparison of data from Tab.4 shows that in natural conditions the uppermost layers of peat and peat-earth muck have the highest filtration velocity of covering water, which results from their increased concentration in the lower parts of profiles. Another evidence comes from data showing a lower filtration velocity for samples compressed in edometers. Water with ash suspension used as a filtrating solution in a clear way decreases the filtration velocity . This is a result of the fact that ash particles silt-up a peat layer, making it less permeable. Organic soils is interactive with solution of different salts, which is proved by the fact that the value k_{10} is decreasing in samples with a higher content of this substance (above 2% I_{om}; table 5).

Table 5. Coefficient of filtration for model samples with various solution

Solution	Sand / Peat	I_{om}* (%)	k_{10} (m/s)
$H_2O_{dest.}$			1.65×10^{-4}
$KHSO_4$	1:0	0.0	2.53×10^{-4}
$MgCl_2$			2.53×10^{-4}
$(CH_3COO)_2Ca$			2.42×10^{-4}
$H_2O_{dest.}$			1.56×10^{-4}
$KHSO_4$	20:1	2.3	1.65×10^{-4}
$MgCl_2$			1.63×10^{-4}
$(CH_3COO)_2Ca$			1.59×10^{-4}
$H_2O_{dest.}$	10:1	4.5	7.86×10^{-5}
$H_2O_{dest.}$			6.48×10^{-5}
$KHSO_4$	5:1	8.2	4.08×10^{-5}
$MgCl_2$			3.18×10^{-5}
$(CH_3COO)_2Ca$			2.56×10^{-5}
$H_2O_{dest.}$	2:1	16.5	5.92×10^{-5}
$H_2O_{dest.}$			6.80×10^{-6}
$KHSO_4$	1:1	24.7	5.94×10^{-6}
$MgCl_2$			4.99×10^{-6}
$(CH_3COO)_2Ca$			4.78×10^{-6}

* – Content of organic matter

2 CONCLUSION

1. Organic soils can in certain conditions be a basement for industrial waste deposit sites, in the tested example power plant ash deposit sites and wastes polluted with heavy metals.

2. Peats and peat-earth muck have sorption abilities for some of the inorganic contents of wastes, especially sulphates and magnesium, as well as lead, copper and zinc, and in the case of peat-earth muck also other ions.

3. In the evaluation of waste deposition on organic soils from the point of noxious contents retention within soil layers, the main role is played by the rate of decomposition of organic particles and in the second place by its percentage in soils.

4. Ion sorption from solutions will take place in the first stage of deposit sites starting, when the soil has not yet been compressed.

5. With continuing waste

loading of organic soils, their compression will be observed, which in turn will cause a decrease in their permeability.

6. At the same time a continuing decrease in permeability of soils will follow as a result of their sealing with suspensions (in the tested example with ashes) carried by water infiltrating the soils.

7. In an evaluation of carried off water influence on peat water, the content of SO_4^{2-} and Cl^- ions in peat samples is very important, as peats will not be an efficient filter for these ions in wastes.

REFERENCES

Hoffmann, E. & Myślińska, E. & Stępień, A. 1991. The isolation properties of peats underlying wet ash deposits. Współczesne problemy hydrogeologii 48, SGGW -AR Warszawa:17-21 (in polish).

Myślińska, E. 1988. Influence of loading on changes of structure of organic sediment at Komarno, Biała Podlaska District. Przegląd Geologiczny 11: 630 - 633 (in polish).

Myślińska, E. & Falkowska, E. & Hoffmann, E. & Stępień, A. 1993. Lithology of soils in the Supraśl River valley (E Poland) and their ability to hold pollution. Geological Quarterly, vol. 37, No 3: 467 - 484.

Osmęda-Ernst, E. & Witczak, S. 1991. Some problems related to laboratory investigetions of parameters of heavy metals migration in ground-water. Zeszyty naukowe AGH Sozologia i Sozotechnika 31: 9-18 (in polish).

Pachowski, J. 1976. Volatile ashes and their aplication in road construction. WKiŁ Warszawa: 90 (in polish).

Geotechnical problems of sanitary landfill waste disposal site of Bandung City, Indonesia

Problèmes géotechniques d'un terrain de décharge de matières solides à la ville de Bandung en Indonésie

K. Sampurno
Department of Geology, Institute of Technology, Bandung, Indonesia

ABSTRACT : Since the last 25 years the city of Bandung, Indonesia, developed very rapidly. The number of people also almost doubled, reaching $\pm$ 2 million people. As the consequent the volume of domestic waste also increase and reach about 3 thousand tons or 6 thousand cubic meter daily. Old open dumping waste desposal is not recomended any more, and new sanitary landfill than has been proposed. Systematic screening of the area under general as well as specific geological criteria has been done in an effort to select potentially suitable sites for landfill solid waste desposal.

RESUME : Depuis les 25 dernières années la Ville de Bandung en Indonésie, s' est développée très rapidement. La Population aussi a presque doublé, atteignant $\pm$ 2 millions d'habitants. En consequence le volume de déchets domestiques augmente aussi et atteint environ 3 mille tonnes ou 6 mille mètres cubes par Jour. Le Vieux système de dépotoir n' est plus recommandé, et donc une nouvelle décharge sanitaire a été proposée. Un référé systematique de la région a été efectué, tant sur un critère général que du point de vue spécifiquement géologique, dans le but de sélectionner des sites potentiels convenables pour un terrain de decharge de matières solides.

1 INTRODUCTION

Bandung city, the capitol of West Java province, Indonesia, is located on the southern foot of the Tangkuban Prahu volcano (2200 m), a Quarternary volcanic mountain complex.

Its slope gradually passes into the Bandung Plain, a former lake plain (650 m). About 25 years ago the Bandung city has developed open dumping waste disposal site, but by the time many problems concerning the environment have arized : smoke, smell, flies, and also spring and ground water polution.

Bandung city, with the surrounding districts, accomodate about 2 millions inhabitans, produce about 3000 ton or 6000 cubic meter of solid waste daily.

New solid waste disposal sites of Bandung city, especially the sanitary landfill type, than has been proposed. The geology of Bandung and the surroundings area, polution problems of the old open dumping disposal sites, geotechnical and hydrogeolgical data of the new solid waste disposal sites will be put forward in the following discussions.

2 GEOLOGICAL AND ENVIRONMENTAL SETTING

Bandung city and its surrounding area is located on the southern foot of Tangkuban Prahu (220 m) volcano, a Quarternary vol canic mountain complex which basically extends in an east-westerly direction.

The slope gradually passes into the Bandung Plain, a former lake bed (650 m a.s.l).

The volcanics consist of volcanic breccia, tuffs, laharic deposits, and lava flow, while the lake beds consist of sandstones, clay, some organic deposits, and coarse grained sand and clay layes of lake deltaic deposits.

Those materials rest on Tertiary sediments such as limestones, quartz sandstones, claystones, marl, greywackes, and turbidite breccia and sandstones which are exposed to the west of Bandung city.

The volcanics, especially tufts, have coeficien of permeability about. 10^{-5} – 10^{-4} cm/sec. Many springs can befound at elevation between 750 – 1000 m a.s.l.

The Bandung basin is drained by the

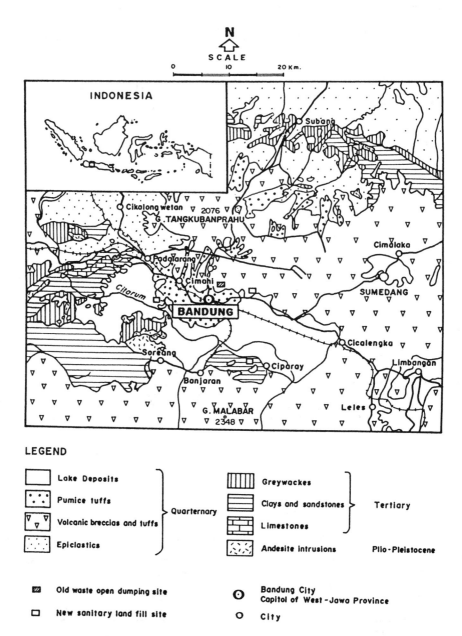

LEGEND

▭	Lake Deposits		▥	Greywockes
⠂⠂	Pumice tuffs		▤	Clays and sondstones
▽▽	Volcanic breccios and tuffs	Quarternary	▨	Limestones
⠂⠂	Epiclostics		▧	Andesite intrusions

Quarternary

Greywockes
Clays and sondstones } Tertiary
Limestones

Andesite intrusions   Pilo-Pleistocene

▨ Old waste open dumping site   ⊙ Bandung City
                                        Capitol of West-Jowa Province

▭ New sanitary land fill site   ○ City

FIG.I : GEOLOGICAL SKETCH MAP OF BANDUNG AREA
INDONESIA

Citarum river, one of Java's largest streams.

The lake deposits, such as clays, have very low permeability (k = 10^{-6} - 10^{-7} cm/sec) while the coarse sediments, especially the lake deltaic deposits, have higher permeability.

The Northern part of Bandung city is located on the volcanics deposits, while the southern part on the lake deposits (Fig.1)

The Bandung areas is at this moment one of the most populous conurbation of Indonesia. The Bandung municipality had 145.000 inhabitants at the outbreack of the 2nd world war but at present accomodate more than 2 million people. Originally,the economic base of the city was agriculture, trade, education, and administration, but in the last 25 years industrial development has assumed an increasingly leading role. The majority of industries, however, are textiles and garments, pharmacy; shoes manufacture, leather and plastic processing industries, and pulp are another economic mainstay.

The agricultural potensial of the Bandung areas is quite considerable. In the last 25 years, however, there has been a shift in landuse in the hilly land, namely from perennial crops to annual crops planted on dryland fields.

The lands have by now been fully developed, and locally there are even indication of over-exploitation.

On the botton of the Bandung basin natural condition favour the cultivation of irrigated rice. However, large areas of agricultural land in the basin bottom are lost every year to development of industries, human settlements, and infra structures. The development not only affecting the low lands areas but also the hilly land of north Bandung.

There are some quarries on the Bandung areas such as trass for pozzolan bricks (North of Bandung city), andesite quarries (South of Bandung), and limestone quarries (west of Bandung).
One of the former quarry sites of north Bandung has been used as solid waste open dumping site. This site is located on permeable volcanic materials which later on gave rise polution problems to the surrounding areas.

3 THE PROBLEMS OF WASTE DISPOSAL

The total ammount of waste produced by the Bandung conurbation is not known exactly. Some estimation come to rough figure available of the solid waste generated daily between 5700-6800 cu.m.

Part of this amount is disposed of throungh private efforts, while an average of about 4100 cu.m per day is collected by neighbour-hood services. This solid waste consists mainly of referse from house holds and markets; the general composition is as follow :

Table 1. Solid waste composition in Bandung
(%)

1. Garbage (Organic waste)	79.59
2. Non-Organic waste	
1) Papers	8.03
2) Plastics	6.37
3) Glasses	1.86
4) Metals	1.50
5) Textiles	0.75
6) Woods	0.74
7) Rubbers	0.37
8) Batteries	0.18
9) Others	0.61
	100.00

The water content is mostly 50 %, and composstable matter as high as 79 % , so making incineration a less-than-ideal option.

In the past, waste was disposed along embankments of roads and even rivers, or in abandoned quarries and sand pits.

It seemed to be the most efficient or cheapest way of removing the unsightly matter from the house or street sides.

The danger of polution of surface and ground water alike was often ignored. Reclamation was, as a rule, left to nature, admittedly with surprisingly good result, given by the tropical climatic conditions. The reclamation is, however, mostly a superficial one. As these is no watertight cover on the accumulated waste material, rain water infiltrates the waste body in large quantities, and the leachate thus generated forms a major pollution hazard to the suroundings, i.e. dug wells, springs and creek.

Until 1990, the city of Bandung operated one such disposal site, located in an abandoned quarry in the northern part of the city. The area consists of lapilli tuffs, with medium permeability (k = 10^{-5} - 10^{-4} cm/sec) (Fig-2).

The site has now been closed and covered with a thin layer of soil. The bottom and the side of the pit had, however, not been preconditioned, i.e., lined with an impermeable layer before use. Therefore, a considerable seepage of leachate is still discernable.

4 SANITARY LANDFILL WASTE DISPOSAL OF BANDUNG

Untill very recently, open dumping of domestic waste was commonplace as final disposal in Indonesia. Due to

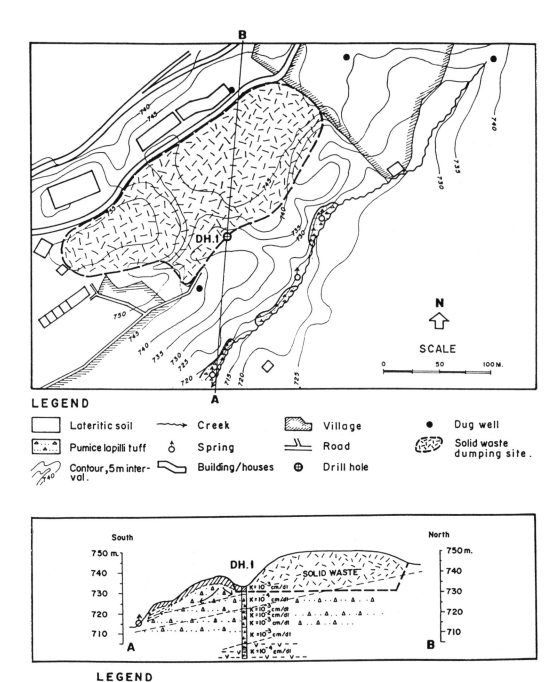

LEGEND

Lateritic soil		→ Creek		Village		●	Dug well
Pumice lapilli tuff		Ŏ Spring		Road			Solid waste dumping site.
Contour, 5 m interval.	740	Building/houses		⊕	Drill hole		

FIG. 2   GEOLOGICAL OUTCROP MAP AND CROSS SECTION
SOLID WASTE OPEN DUMPING SITE OF NORTH BANDUNG
INDONESIA

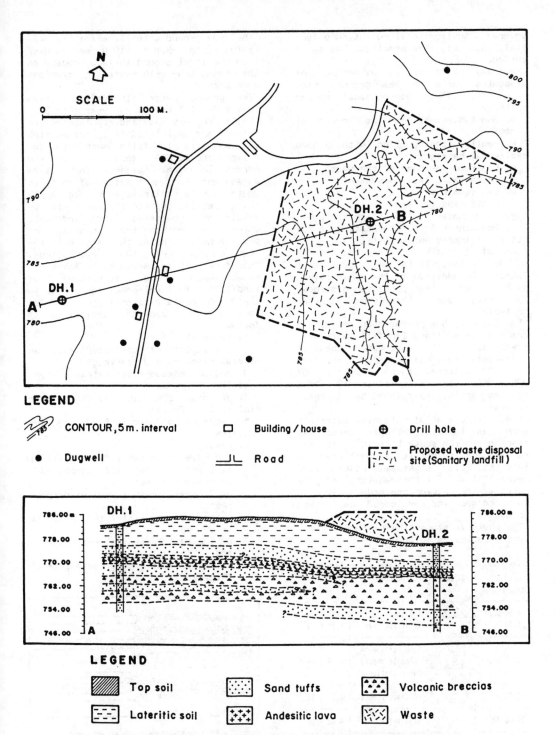

LEGEND

~~785~~ CONTOUR, 5 m. interval	▢ Building / house	⊕ Drill hole
● Dugwell	⌐⌐ Road	Proposed waste disposal site (Sanitary landfill)

LEGEND

▨ Top soil	▨ Sand tuffs	▨ Volcanic breccias
▨ Lateritic soil	▨ Andesitic lava	▨ Waste

FIG.3   TOPOGRAPHIC SKETCH MAP OF NEW SANITARY LANDFILL WASTE DISPOSAL OF BANDUNG, INDONESIA

severe environmental pollution potential, however, this practice has to be curbed.

Sanitary landfill is a method of disposing waste on the ground which should fullfill the geotechnical design criteria :

   a. Short distance from the source of waste;

   b. Gently sloping land with a good geological barrier;

   c. Liner on the base of the waste;

   d. Leachate collection system;

   e. Gas venting;

   f. Monitoring wells;

   g. Cover material and, if possible;

   h. Geotechnical instrumentation to monitor settlement and stability.

One of the main consideration of the siting of landfill is the geological factor, in addition to the other items, such as waste transportation distance, aesthetics and other environmental criteria.

Bandung city has prepared Sukamiskin as one of the sanitary landfill waste disposal sites; it is located about 4 kilometers east of Bandung city. The area 20 hectares, at the elevation of 725 - 800 meter a.s.l. sloping 15 - 25 % to the south, and flanked by 2 small rivers.

The area consist of volcanic materials such as tuffs, volcanic breccias, dipping to the south $5° - 10°$. At the surface the materials are weathered to lateritic clay sometimes reach 7.00 meter thick. The permeability of the rocks varies between the orde of 10^{-3} to 10^{-5} cm/sec and considered as rather permeable.

The physical properties of these materials are as follows :

DEPTH (m)	SIMBOL	LITHOLOGY	PERMEABILITY (cm/sec)
0.00-0.30		Top soil	
0.30-7.00		Lateritic clay	1.51×10^{-4}
7.00-17.00		Sand tuffs with volcanic brecia intercalation	5.80×10^{-5}
17.00-20.00		Volcanic breccia	2.20×10^{-4}
20.00-22.50		Sand tuff	4.63×10^{-4}
22.50-25.75		Volcanic breccia	2.48×10^{-4}
25.75-30.00		Sand tuff	1.41×10^{-5}

The free ground water level of the area varies from -3,8 to -10.0 meter deep from the local ground surface, depend on the topography with hydraulic gradient more than 2 % .

The ground water fluctuation varies from 1.0 - 2.5 meters per year (Fig.3).

There are very detailed regulations on methods and quality of the construction of sanitary landfills such as for example the US-EPA regulations or the German Technical Guidance for waste management. But, as a matter of fact, it would not be realistic to try to apply these regulations in its entirely in developing countries like Indonesia which have neither the financial nor the technical means to cary out the requirements of the regulations. Therefore, it is necessary to adapt the establishment and techniques of landfilling to local conditions, i.e. to apply appropriate technologies. The construction of sanitary landfill is as follow :

   1. bed rock

   2. two layers of compacted soil as mineral liner each 30 cm thick

   3. bottom bamboo-mats instead of geotextiles

   4. drainage layer (30 cm), coarse material (ø 20 - 50 mm)

   5. upper bambbo-mats instead of geotextiles

   6. waste

   7. top cover of a sanitary landfill :

      a. layer of normal soil compacted to compensate setting

      b. 2 layers (each 25 cm) of compacted clay

      c. coarse material for drainage (± 30 cm)

      d. top soil layer (1m) not compacted

      e. vegetation: grass, bushes, etc. (Fig.4)

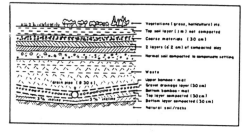

FIG.4 : LEACHATE COLLECTION AND TOP COVER OF SANITARY LAND FILL

5 CONCLUSION

Until recent years, the wastes generation has been increased rapidly accompanied with economic growth. Municipal goverment find it difficult to dispose of the ever in creasing of solid waste without causing air, ground and water pollution or wrecking the municipal budget. Many cities in Indonesia, have already faced to this phenomena since about ten years ago. Bandung city as an example produces about 6000 cu.m of domestic waste per day.

The problem now is how to dispose the amount of waste in a good manner so that it can reduce the pollution inpact. The sanitary landfill, which has been adapted to the establishment and techniques to local conditions, i. e. to apply appropriate technologies (low — cost) is necessary.

Attention should be paid to a good preparation in selecting landfill disposal sites, rather than to cope up with the environmental damage caused by groundwater pollution due to improper siting of sanitary landfill disposal of solid waste.

REFERENCES

Owens, I. S. & Khera, R. P. 1990. Geotechnology of Waste Management. Cambridge : Butter worth.

Sampurno 1990. Geology of Bandung (Indonesia) and its significance for the development of Bandung city. Proc. sixth Int. Congr. IAEG. Amsterdam : Balkema.

Sampurno 1991. Geology and City Planning. Kota. 4 : 23 - 27. Jakarta : BKS-AKSI.

Wiriosudarmo, S. 1992. Site selection and sanitary landfill disposal of waste. Waste and sustainable development: 113 - 127. Jakarta : Goethe Inst. & BPPT & Unesco.

La géologie et le régime juridique des sites contaminés au Brésil
Geology and law concerning contaminated sites in Brazil

P.A. Leme Machado
UNESP, Instituto de Biociências, Departamento de Ecologia, Rio Claro, São Paulo, Brazil

J.de O.Campos
UNESP, Instituto de Geociências e Ciências Exatas, Departamento de Geologia Aplicada, São Paulo, Brazil

RESUMÉ: La realité physique du territoire brésilien, souvent caracterisé par un environnement tropical, présente aspects bizarres assez importantes pour normaliser les sites pour la déposition des déchets industriels dangereux. Cette realité, que dans la plupart des situations signifie une enormité des risques dans ce que concerne à la pollution de l'environnement, doit être considerée correctement pour établir la législation spécifique. Néamoins, chez nous, on n'observe qu' une approche très générique de la realité, aggravée par la multiciplicité d'attributions et par les structures conservatrices et inefficaces des départements governamentaux responsables par la fiscalisation et l'administration de ce sujet. La considération des aspects géologiques comme facteurs pour guider la législation de l'environnement c'est une necessité qui ne peut être ajournée chez nous.

ABSTRACT: The brazilian physical reality, almost distinguished by a tropical environment, presents unusual aspects of great significance to a legal normalization of industrial dangerous waste deposition. This reality, which many times means enormous risks to the environment' s pollution, must be properly considerate in the improvement of a specific environmental legislation. Nevertheless, we can observe among us, a generical approach of the geological reality, aggravated by the multiplicity of atributions and by the conservative and inefficients administratives structures of the governamental organisms responsables by the fiscalization and managements of the matter. The annexation of the geological aspects as guiding factors of the environmental legislation is a imperative undelayable among us.

1 LES CONDICIONNENTS GÉOLOGIQUES REPRESEN-TÉS PAR LES PROFILS D'ALTÉRATION DES ROCHES

Les condicionnements géologiques dans les sites pour la déposition des déchets industriels au Brésil, révélent une extreme complexité à cause des différences existentes dans la nature, ce qui vient démontrer la grande difficulté pour établir les critères de sûreté, dans ce que concerne à la circulation des polluants dans le sous-sol.

On peut affirmer que les principaux problèmes attendus pour definir la viabilité des sites de déposition, sont liés aux profils de décomposition des roches et aux minéraux résultants. Pour bien comprendre les profils d'altération on peut se baser dans les idées de Deere et Miller (1971), et même sur les sugestions de Showers (1988). Récemment, The Geological Society Engineering Group Working Party Report: Tropical Residual Soils, a proposé une concéption des horizons d'altération, en ajoutant la échèle des degrés de décomposition des masses rocheuses de la ISRM (Figure 1). Chez nous, Vargas (1985) a montré les types des profils possibles d'occurrence en plusieurs régions brésiliènnes (Figure 2), dans lesqueles ont peut envisager une realité physique assez problématique. Pastore (1992) a resumé quelques importantes sugestions des profils d'altération, an ajoutant ces propres idées (Quadre 1). Aux complications mentionées, ont peut accroître la irregularité du sommet de la roche saine, dû aux temperatures elevées et à la intensité des pluies, étant les hydrolises doublées on triplées pour chaque $10^{\circ}C$ d'élevation de temperature dans un profil en granites et gneiss (Döbereiner et Porto, 1990). Ainsi, l'épaisseur du profil d'altération peut être de 30m ou 40m dans les gnaiss du Precambrien brésilien, mais il peut être de plus de 100m dans les mines de graphite à Pedra Azul au

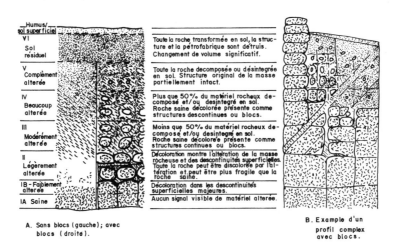

Toute la roche transformée en sol, la structure et la pétrofabrique sont détruis. Changement de volume significatif.

Toute la roche decomposée ou désintegrée en sol. Structure original de la masse partiellement intact.

Plus que 50% du matériel rocheux decomposé et/ou desintegré en sol. Roche saine décolorée présente comme structures descontinues ou blocs.

Moins que 50% du matériel rocheux decomposé et/ou desintegré en sol. Roche saine décolorée présente comme structures continues ou blocs.

Décoloration montre l'altération de la masse rocheuse et des descontinuités superficielles. Toute la roche peut être discolorée par l'altération et peut être plus fragile que la roche saine.

Décoloration dans les descontinuités superficielles majeures.

Aucun signal visible de matériel altérée.

A. Sans blocs (gauche); avec blocs (droite).

B. Example d'un profil complex avec blocs.

Figure 1 – Profils d'altération idealisés (d'après The Geological Society Engineering Group Working Party Report : Tropical Residual Soils, 1990, modifié).

nord-est de Minas Gerais (Cella et al, 1989). Les minéraux originés dans les procédés de la décomposition des roches sont aussi très variables, selon l'intensité des pluies, les pentes topographiques et la roche original (Döbereiner et Porto (1990). En outre, les différents minéraux d'argile dans les profils ont importance dans le niveau de contamination du sol par les déchets. Renaut (1993) nous rende compte que la présence des déchets alumineaux comme les cyanures, fluorures, chlorures, métaux lourds et hydrocarbures poly cycliques aromatiques (HPA), dans certaines concentracions atteignent 650 mg/l, entraine les modifications dans les minéraux argileux, comme la réduction des surfaces totale et interne et la capacité à gonfler. Évidemment, la compréhension de la mise en scène dès où peut arriver les changes chimiques circulation des liquides, dans le sous-sol, c'est fondamental pour le contrôle des risques entourés. De cette façon, le concept de barrière géologique (Dörhöfer, 1993), lequel repose sur une compréhension du système total déchet-revêtement-géologie, peut être un outil très importante pour la mise en place du contrôle mentionné. Une barrière géologique c'est une occurrence de roche à faible permeabilité, consolidée ou non consolidée, de plusieurs mètres d'épaisseur, qui aie un haut potenciel de rétention des contaminants, qui s'étendre au-delà des limites du terrassement, et qui soit raisonablement homogène. Si ces exigences ne sont pas attendues, le matériel naturel doit être replacé par une couche d'argile compactée avec un coefficient de permeabilité $k \leq 10^{-7}$ cm/s. Cette homogeneité peut être

observeée parfois, dans certaines conditions. Marchi et al (1993) décrivent un site pour une déposition, où une couche de latosol rouge sombre colluvial de la décomposition des basaltes, a une épaisseur de 10m, étant le coefficient k du 10^{-3} jusque le 10^{-4} cm/s dans l'état naturel, et 10^{-6} après compactage dans le laboratoire. Dans ce cas, la nappe phréatique est à une profondeur de 28. Cette situation configure une barrière géologique convenable pour la déposition des déchets industriels, car au dessous du latosol rouge sombre il y a 28m de roche alterée et seulement en bas de celui-ci, la roche basaltique fissurée très permeable.

2 DÉFINITION JURIDIQUE DES SITES CONTAMINÉS

Il n'existe pas une définition jurifique de site contaminé. Il y a l'emploi du terme "aire degradée" dans le sens large qui peut aussi servir à comprendre le site contaminé. Pour construire une définition dans l'avenir, il faut considérer deux points: premièrement, le produit déposé n'a pas être traité préalablement; deuxièmement, les précautions pour la conservation des ressources naturelles n'ont pas été mis au point. Nous croyons qu'un site pourra être consideré comme contaminé, ou par l'action clandestine, ou par l'action autorisée, mais négligente.

Le Brésil a commencé à se préocuper de la contamination des sols et du sols-sol par les dépôts des déchets industriels, depuis la fin de 1990, avec las découverte des dépôts clandestins. Surtout dans la région "Baixada Santista", fortement indus

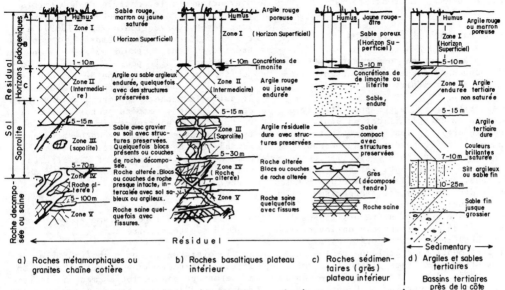

Figure 2. Quatre types principaux de profils d'altération dans le Brésil Central (d'après Vargas, 1985, modifié).

Deere et Patton (1971)		de Mello (1971)	Vargas (1974)			Woole (1985)		Pastore (1992)	
I - Sol résiduel	IA Horizon "A"	Sol mûr	**Niveaux pédologiques**	A		Sol superficiel	**Sol résiduel**	**Sol résiduel ou transporté**	Horizon organique (1)
	IB Horizon "B"			B	Sol résiduel mûr	Sol mûr			Horizon de sol latéritique (2)
	IC Horizon "C" (saprolite)	Sol résiduel ou saprolite		C	Horizon II Inter mediaire	Sol saproli-tique	**Résiduel**	Horizon de sol saprolitique (3)
II - Roche alterée	IIA Transition du sol saprolitique/roche alterée		**Sols résiduels** / **Saprolite**	Horizon III Saprolite		Saprolite		Horizon de saprolite (4)
	IIB Roche partiellement alterée	Roche alterée	**décomposée**	Horizon IV Roche alterée		Roche alterée		Horizon de roche assez alterée (5)
								Horizon de roche alterée (6)
III	Roche saine	Roche saine	**Roche? ou**	Horizon V Roche saine		Roche saine		Horizon de roche saine (7)

Quadre 1. Comparaison entre quelques propositions des profils d'altération au Brésil (d'après Pastore, 1992, modifié).

trialisée dans l' État de São Paulo.

3 LES MÉCANISMES DE PRÉVENTION

3.1 La planification concernant l'ouverture de dépôts de déchets et leur localisation sur le territoire

Les autorités compétentes pour établir un programme ou une planification sont, au niveau fédéral: l' Ibama - Institut Brésilien de l' Environnement et des Ressources Naturelles Renouvelables, avec le concours du Conama - Conseil National de l'Environnement; et aux niveaux des états - membres et des communes - les organismes chargés de l'environnement.

Le Conama à décidé, à travers la Résolution 006/1988, de promouvoir un inventaire des déchets industriels gérés au pays.

Il faut mentionner quelques règles des Constitutions des États, lesquelles ont été élaborées dans l' année 1989, après la votation de la Constitution Fédérale de 1988. Quelques états ont prévu l' obligation des pouvoirs publics de contrôler l' usage du sol pour des objectifs du stockage de matériaux qui risquent d'avoir des effects dommageables pour la vie ou por l' environnement (Amazonas, Ceará, Goiás); l' État du Maranhão prévoit zonage du territoire et l'établissement des directives générales pour les dépôts des déchets solides humaines, des égouts domestiques et industriels; Minas Gerais et Pará obligent aussi le cadastre préalable des produits; Paraná et Pernambuco obligent la récupération des aires dégradées; selon les états de Piaui, Rio de Janeiro, Rio Grande do Sul et Sergipe, l' État et les communes doivent coopérer pour le traitement des déchets urbains et industriels. Quelques états ont interdit l' introduction des déchets dans leurs territoires, en provenance d' autres états ou d' ailleurs (Amazonas), ou l' introduction dans le milieu ambient des substances qui peuvent occasionner le cancer (Goiás, Rio Grande do Sul). D' autres états conditionnent le stockage de produits dans leurs territoires, à la preuve de que l' usage du produit est permis dans l' état d' origine (Mato Grosso, Rondônia).

Les lois fédérales de "directives sur le budget" de 1992 - loi 8.447, et de 1993 - loi 8.694 - dans le secteur des programmes, ont constaté l' obligation "d' installation de dépôts pour la destination définitive des déchets radioatifs".

Cependant, la question des dépôts de déchets dangereux et des sites contaminés, n' est pas traitée concrètement dans les programmes. Toutefois, considérant que la

Convention sur le Contrôle de Mouvements Transfrontières des Déchets Dangereux et leur dépôt (Convention de Bâle/1989), a entré en vigueur le 39 décembre 1992 pour le Brésil, on peut soutenir qu' il y a une obligation d' agir pour l' administration publique selon le principe de la convention de "l' administration environnementale saine de déchets dangereux ou d' autres déchets", à qui compète prendre des mesures pratiques pour garantir que les déchets dangereux et d' autres déchets soinet administrés de façon à proteger la santé humaine et l' environnement des effects nuisibles qui puissent provoquér les déchets.

3.2 Le droit de l' Urbanisme et de l' amenagement du territoire

Par le biais de la Loi 6.766/1989 (loi fédérale du parcellement du sol urbain), on de permettra pas le lotissement ou le remembrement "dans les terrains qui auraient été remblayés avec des matériaux nuisibles à la santé publique, sans qu' ils aient été préalablement assainis. Il s' agit de procéder à une analyse sanitaire des élements employés dans le remblai, pour que l' on sache s' ils sont ou non nuisibles à la santé. Tant que l' on n' aura pas écarté les préjudices directs ou potenciels d' un lieu, le permis de lotissement ne sera pas ocotroyé".

Cette norme juridique n' est pas destinée exclusivement aux dépôts de déchets, mais en assurant un parcellement urbain sans ordures, elle contribue à localiser et nettoyer les décharges clandestines.

4 RÈGLES APPLICABLES AUX SITES INDUSTRIELS EN ACTIVITÉ IN CE QUI CONCERNE LEURS DÉCHETS DE PRODUCTION ET LEUR STOCKAGE SUR LES LIEUX DE PRODUCTION: EXIGENCE D'ÉTUDE D'IMPACT

Il y a l' exigence d' étude d' impact pour les décharges sanitaires, processus et destination finale de déchets toxiques et dangereux. C' est une norme générale fédérale, applicable à tout le territoire brésilien. Donc, les Etats-membres seulement peuvent exiger plus et non moins - c'est la compétence supplémentaire. D'autre côté, la Constitution Fédérale de 1988 (article 225, 1, IV exige l' étude d' impact pour toutes les activités qui peuvent causer du dommage significatif à l'environnement. L'étude d' impact sera exigée de la même façon dans les hypothèses du stockage sur les lieux de production ou ailleurs. L' autorisation environnementale est une obligation des Etats-membres. Mais, le gou vernement fédéral peut aussi instituer une autorisation pour ce genre d'activité. Il y a déjà l'autorisation fédérale octroyée par l'Ibama, pour importer des déchets dan gereux, mais elle n'est pas explicite pour les décharges. Les communes peuvent aussi créer un système d'autorisation pour le dé pôt de déchets. Il n'y a pas de règles applicables aux sites exclusivement consa crés au stockage des déchets industriels et ménagers.

L'année de 1993, de Conseil National de L'Environnement a voté la Résolution 5 du 5 août, publiée dans le Journal Officiel - "Diário Oficial da União" - du 31 août 1993 -p.12297. Cette résolution prévoit un "plano de gerenciamento de resíduos sóli dos" (plan d'aménagement de déchets soli des). Il est obligatoire pour les déchets produits dans les ports, aéroports, gares de chemins de fer et routières et dans les établissements de santé. Chaque administra tion des établissements mentionnés doit présenter de plan d'aménagement des déchets solides à l'organisme compétent de l'envi ronnement et de la santé. L'implantation du système de traitement et de destination finale de ces déchets solides seront sou mises à l'autorisation administrative. Cette autorisation dépend de trois condi tions: a) élimination des caractéristi ques de danger ou déchet; b) préservation des ressources naturelles; c) obéissance aux standards de qualité environnementale et de santé publique. Cette résolution veut éviter la naissance d'un site contami né. Si toutes les exigences sont accom plies, le site du dépôt ne sera pas conta miné.

Le Brésil n' a pas fait une réglémenta tion fédérale pour l'aménagement de déche ts industriels dangereux. La compétence pour établir des règles environnementales est concurrente entre le gouvernement cen tral ("União") et les états-fédérés (art. 24 de la Constitution Fédérale). Il y a des Etats qui ont fait des normes réglées ou non discrétionnaires sur les déchets.

5 DROIT A L'INFORMATION ET DROITS DE RE-COURS DES CITOYENS ET DES ASSOCIATIONS EN CE QUI CONCERNE LES CHOIX DES SITES DE DE-POTS DE DECHETS.

Le droit à l'information est présent en tout le droit de l'environnement brésilien. Il n'y a pas de droit spécial à l'informa tion en ce qui concerne les choix des si tes de dépôts de déchets; Il a une base constitutionnelle (art. 5, XXXIII de la Constitution Fédérale de 1988). Le droit à l'information existe dans la Loi de Politi

que Nationale de L'Environnement pour de mander les résultats des analyses. Il y a aussi le devoir du Pouvoir Public d'infor mer sur les sujets environnementaux et de publier les demandes d'autorisation et l' acte administratif d'autorisation. Il faut aussi mettre en relief le droit à l'infor mation pendant la procédure de l'étude d' impact, y compris la réalisation de l'audi ence publique.

Le citoyens ont le droit de recours au Pouvoir Judiciaire par le biais de l'Acti on Populaire". Aussi la Constitution Fédé rale a mis un but environnemental pour cette action, c'est l'article 5, LXXIII qui a l'objectif d'annuler l'acte domma geable au patrimoine public, à la moralité administrative, à l'environnement et au patrimoine historique et culturel.

Les associations n'ont pas le droit à l'action populaire", mais elles ont le droit à l'"Action Civile Publique". Cette action a l'objectif d' obliger l'accom plissement de faire ou de ne pas faire ou d'exiger la réparation du dommage à l'envi ronnement. Il faut rappeler que la Conven tion de Bâle (en vigueur au Brésil) oblige le pays à assurer la disponibilité des ins tallations adéquates pour le depôt de dé chets, qui doivent être installées, dans la mesure du son possible, dans leur terri toire. Elle oblige, aussi, les autorités responsables pour l'administration de dé chets dangereux d'éviter la pollution de ces déchets ou minimiser les consequences de la pollution pour la santé humaine et l'environnement. L'application de ces rè gles doivent arriver sur tous les déchets dangereux, pour éviter la fraude et le men songe de se dire que les déchets dangereux produits et deposés dans le même pays ne sont pas inclus dans la convention. Le Bré sil a demandé des mesures plus sévères dans le sense de réduire la quantité et le contenu toxique de déchets dangereux et d' assurer sa destination finale adéquate du point de vue de l'administration environne mentale saine et voisine du lieu de la production.

Ces deux moyens d'actions en justice ne visent pas exclusivement au stockage ou aux depôts de déchets.

Les procédures de planification générale et de choix de sites destinés à être plus ou moins contaminés sont accompagnées d' études scientifiques (géologiques, chimi ques, pédologiques et agronomiques), dans le cas où l'étude préalable d'impact envi ronnementale est obligatoire. Il y a la possibilité de la contre -expertise ou par décision de l'Administration ou pendant l'Audience Publique, à la demande des asso ciations, des citoyens ou du Ministère Public.

La responsabilité sans faute est la règle générale par le dommage environnemental.

Il y a la responsabilité du propriétaire de terrain en cas de location du terrain, si le locataire l'utilise pour deverser des déchets. La vente d'un terrain où il y a déjà un dépôt de déchets n'exonère pas le vendeur de sa résponsabilité pour le dommage à l'environnement.

La mise en cause de la responsabilité est faite par l'action d'indemnisation ordinaire promue par la victime directe ou par de biais de l'Action Civile Publique. Le constat de l'état de lieux peut être fait à travers la procédure de "l'enquête civile" (avant de l'Action Civile Publique), par l'Action "Cautelar" ou pendant l'Action Civile Publique, par le biais de l'expertise sous le contrôle du juge et des parties.

6 LES MÉCANISMES DE RESTAURATION DES LIEUX, DE RÉPARATION DES DOMMAGES ET DE RÉPRESSION

6.1 Cas de contamination d'un site et de son voisinage toujours exploité par un industriel

6.1.1 Mécanismes juridiques de restauration des lieux

La Constitution Fédérale a prévu que "quiconque exploite des ressources minérales est obligée à récupérer l'environnement degradé, d'accord avec la solution technique exigée par l'organisme public compétent, selon la forme de la loi" (art. 225, 2). La Constitution a, donc, inclu un principe de récupération des aires dégradées, bien qu'elle ait parlé des aires minérales dans ce paragraphe. C'est pour montrer la première directive du législateur constitutionell. Cependant, la Constitution Fédérale continue dans la même direction juridique où est prévu que l'obligation de réparer le dommage causé à l'environnement est separé des sanctions imposées aux personnes qui causent ce dommage.

La Loi de Politique Nationale de L'Environnement oblige le pollueur de récupérer les dommages causés ou de l'indemnisé. Cette loi a défini le pollueur comme "la personne physique ou morale, de droit public ou privé, responsable, directe ou indirectement, par l'activité qui cause la dégradation de l'environnement". Il y a aussi dans la Loi de Politique Nationale de l'Environnement le principe de que les aires degradées doivent être récupérées.

Nous avons des règles explicites sur la responsabilité de faire la récupération des sites contaminés dans les Etats de la Fédération: Maranhão, Mato Grosso, Pará et Rio Grande do Sul, toujours en faisant l'application de la norme générale fédérale de la responsabilité sans faute.

6.1.2 Pouvoirs de l'Administration et des Tiers

L'Ibama - organisme fédéral compétent pour contrôler la pollution - peut infliger des amendes sur les personnes physiques ou morales- qui causent pollution de quelque nature, de façon à provoquer des dommages à la santé ou ménaces au bien-être. L'amende pourra être réduite jusqu' à dix pour cent du total, si le responsable signe un accord avec l'Administration pour restaurer le lieu endommagé. L'accord doit être approuvé par le Conseil National de L'Environnement.

Les tiers pourront utiliser le droit de pétition à l'Administration.

L'Administration publique environnementale (fédérale, de l'état ou de la commune) pourra s'utiliser de l'"Action Civile Publique" pour obliger l'industriel de ne pas déposer les déchets dangereux; pour obliger à faire un dépot selon les normes juridiques et scientifiques et scientifiques ou pour obliger a enlever les déchets; dans ce cas-la, le juge imposera des astreintes sur le défendeur.

Les tiers - s'ils sont groupé dans une association, ils ont les mêmes droits à l'"Action Civile Publique" que l'Administration. le Ministère Public a aussi le même droit d'action. Si les tiers sont isolés ou aiment agir en tant qu'individus, ils ont droit à l'action "comminatoire" ou à l'action d'indemnisation du Code de Procédure Civile, mais ils n'ont pas droit à l'Action Civile Publique.

L'Administration Publique doit communiquer au Ministère Public les événements liés aux déchets dangereux pour la poursuite pénale. La disposition des déchets de façon contraire aux normes fédérales peut configurer le crime de mise en danger, puni d'un an à trois ans de prison. Les crimes pratiqués à travers l'activité industrielle ou par le biais de transport, pendant la nuit ou pendant les jours fériés sont punis de deux à six ans de prison. Les autorités qui s'ommetent dans la promotion de mesures pour empêcher les comportements ci-dessus, sont soumises aux mêmes sanctions que les auteurs directs.

2460

6.1.3 Obligations imposées en cas de cessation de l'activité

Les normes juridiques ne sont pas claires dans le sens d'indiquer le comportement de l'industriel ou de l'entreprise qui a produit le déchet, dans le cas de cessation de l'activité. Il faut, donc, faire appel aux règles de contrôle des activités polluantes - concrètement - aux normes sur le suivi des activités polluantes. Le suivi ou de "monitoring" fait partie de l'obligation d'informer à l'Administration de l'environnement. La Loi de Politique Nationale de l'Environnement prévoit "la garantie de la transmission des renseignements rélatifs à l'environnement". Si l'industriel ou l'entreprise ne transmet pas le renseignement, l'Administration est obligée à le produire, dit la loi mentionnée.

6.2 Cas de contamination provenant d'un site abandonne ou dont l'exploitant ancien est inconnu

6.2.1 Mécanismes de restauration des lieux

Las mêmes mentionnés dans l'alinea 6.1.1.

6.2.2. Pouvoirs de l'Administration et des tiers

Las mêmes mentionnés dans l'alinea 6.1.2.

6.2.3. Rapports avec le propriétaire non exploitant

Le propriétaire du terrain où se trouve le site contaminé en est responsable, même s'il n'est pas le producteur des déchets dangereux. La Convention de Bâle considère producteur des déchets la personne qui possède les déchets, dans le cas où le producteur est inconnu. Si l'acheteur d'un terrain ne connaît pas l'existence d'un site contaminé à l'intérieur de la propriété, il peut entamer une procédure contre le vendeur, avec le fondement du defaut occulte (Côde Civil Brésilien).

7 RESPONSABILITES CIVILES APPLICABLES AUX CAS 6.1 ET 6.2 -DESSUS

Au Brésil, la justice administrative et la justice civile appartient au Pouvoir Judiciaire. Il n'y a pas de tribunaux administratifs qui peuvent décider avec la force de la chose jugée. Dans l'Action Civile Publique l'argent payé en tant que condamnation, ne sera pas mis à la disposition de l'auteur de l'action, mais sera versé au "Fond de Droit Diffus" fédéral ou de l'

état. Ce fonds-là est chargé de mettre en ceuvre la restauration des lieux endommagés. Il n'y a pas de loi obligeant l'industriel et le transporteur de déchets dangereux de faire des contrats d'assurance dans les normes fédérales. Heureusement, il en y a l'exigence innovatrice dans l'Etat du Pará. Cependant, la responsabilité civile sans faute pour dommage à l'environnement n'a pas de plafond, donc le dommage doit être integralément réparé.

4 CONCLUSIONS

Dans le territoire brésilien nous pouvons envisager, d'une façon resumée, les aspects suivants de la géologie comme condicionnements pour la déposition des déchets industriels:
1-Dépots de colluvions sableux cénozoiques d'origine tropical étendus et épaisses (10 m à peu près, ou même plus, quelquefois) poreux, permeables et potencielle ment instables (collapsibles);
2-Zones de cisaillement et des failles dans les formations géologiques anciènnes (Precambrien) ou dans les coulées de basaltes, avec de plans profondément alterés, où la conductibilité hydraulique est souvent très haute;
3-Zones de roches decomposées épaisses de plus de 150 m dans les domaines des forêts tropicaux ou dans les chaines cotières, lesquelles ont une grande compressibilité, importantes horizons de percolation d'eau et occurrence d'étendus dépôts de gros blocs, qui ont un modèle géometrique embrouilé de dificille compréhension, même pour les méthodes géophisiques;
4-Roches sédimentaires tendres, disposées en 70% du territoire brésilien, dès que argileuses susceptibles d'une intense et rapide désagrégation et "slaking", dûe aux variations de la humidité et des cycles d'humectation/séchage, outre cela d'être goflants;
5-Le phénomène de la latérisation, responsable par les procédés dans lesquels les changes cationiques influencent d'une façon encore peu claire, le comportement géotechnique et le modèle structurel de quelques types de sols;
6-Présence des nappes phréatiques suspendus de corrélations dificiles avec les nappes profondes;
7-Rares connaissances disponibles, par rapport aux formulations des théories scientifiques et des travaux pratiques dans le domaine des polluants dans les millieux poreux non saturés et fissurés, outre cela les impropres et suspects méthodes pour les contrôles de la pollution de l'environnement à travers des nappes phréatiques

southerrains et aussi des ruisselements vers les réservoirs superficiels.

Dans ce que concerne aux aspects juridiques, ont peut voir qu' il n' y a pas que de références géneriques à la realité géologique: "Les procedures de planification générale et de choix de sites destinés à être plus au moins contaminées sont acompagnées d' études scientifiques (géologiques, chimiques et agronomiques) dans le cas où l'étude préalable d' impact environnementale est obligatoire"; l' expérience montre que, dans la plupart des cas, les considérations du domaine géologique manquent de perspicacité et précision.

Nous croyons que les questions relatives à la contamination de l'environnement, nettement les sources d'eau subterraines et superficiels par la déposition des déchets, seront proprement engagées au Brésil par la compréhension de la realité physique décrite, par la considération des effects de chaque condicionnement géologique et de tous les phénomènes physique-chimiques entourés dans les procédés et la circulation des fluides dans le sous-sol, dans ce que concerne à son degré d' importance, et par la mise en évidence de tous ces considérations, d' une façon convenable dans l' esprit de la loi spécifique.

REFERENCES

Aguiar Dias, J. 1987. Responsabilidade Civil. Ed. Forense, vol. II, 8ª edição, Rio de Janeiro.

Cella, P.R., Döbereiner, L., Nunes, N. 1989. Study of the slope stability of the Pedra Azul mines from the "National de grafite". 2nd Simposium EPUSP of Rock Mechanics applied to mining. São Paulo (in portuguese).

Deere, D.U. & Patton, F.D. 1971. Slope stability in residual soils. Proc. 4th Pan-American Conference on Soil Mechanics and Foundation Engineering, Puerto Rico, VI, 87-170.

Döbereiner, L. & Porto, C.G. 1990. Considerations on the Weathering of Gneissic Rocks. Proc. The Engineering Geology of Weak Rocks. University of Leeds, p. 228-240.

Dörhöfer, G. 1993. The role of natural geological barriers for the siting of landfills in Germany. Proc. Géoconfine 93, Géologie et Confinement des Déchets Toxiques, Montpellier, p. 39-45.

Journal Officiel (Diário Oficial da União). 1993. Déclaration des Réserves du Gouvernement Brésilien à la Convention de Bâle (1989). 20 Juillet, Brasilia.

Leme Machado, P.A. 1981. La loi brésilienne relative à l' utilisation du sol urbain, à l' environnement et à la qualité de la vie. In Rivista Trimestrale di Diritto Pubblico. Milano A. Giuffrè Ed. vol. 1, p. 225-272.

Marchi, A.J., Campos, J. de O., Prada, J.M.M., Löschl Filho, C., Weiss, J.L. 1993. La méthode RIA pour le projet du centre de traitement de déchets industriels de Piracicaba, Brésil. Proc. Géoconfine 93, Géologie et Confinement des Déchets Toxiques, Montpellier. Posters Theme 5, 15 p.

Pastore, E. 1992. Maciços de solos saprolíticos como fundação de barragens de concreto gravidade. Tese de Doutorado apresentada à USP - São Carlos, Departamento de Geotecnia, inédito. 290 p.

Renault, J.F. 1993. Contaminations du sous-sol par des déchets alumineus. Proc. Géoconfine 93, Géologie et Confinement des Déchets Toxiques. Montpellier, Vol. 1, p. 105-107.

Showers, G.F. 1988. Foundation problems in residual soils. Int. Conf. on Eng. Problems of Regional Soils, Beijing, China, 154-171.

The Geological Society Engineering Group Working Party Report: Tropical Residual Soils. The Quarterly Journal of Engineering Geology, Vol. 23, 101 p.

Vargas, M. 1985. The concept of Tropical Soils. 1st Intern. Conf. on Geomech. in Tropical Lateritic and Saprolitic Soils. ISSMFE, Brasilia. Proc. Vol. 3, p. 101-134.

Technological characterization of piedmontese clays for waste disposal impermeabilization

Caractérisation des argiles du piedmont pour l'utilisation dans l'imperméabilisation des sites de stockage de déchets

G. Barisone, G. Bottino & R. Crivellari
Dipartamento Georisorce e Territorio Corso Duca degla Abruzzi, Torino, Italy

ABSTRACT: The parameters considered for the choice of clays to use for waste disposal impermeabilization are generally limited to permeability and to geotechnical characteristics, in order to avoid stability problems (and consequently impermeabilization problems) caused by the assestment of clayey beds. The here described work took the main outcropping geological clayey Formations in the Piedmontese territory (North-Western Italy) into account and made a detailed study of all the characteristics of some importance, in order to determine the aptitude of a clay for use in waste disposal realization. About 40 samples (representing 14 quarried areas) were collected and analyzed; the following determinations were made: Atterberg's limits (at different Ph values), granulometric distribution, permeability on natural and reconstituted samples (by mean of a large-dimension permeameter), shear strength, mineralogy, clayey minerals and their percentage (by mean of X-ray diffraction), CaCO3 percentage and cationic exchange capacity. The results obtained from this study permitted a good characterization of the analysed clays; in particular, it was possible to distinguish between the formations whose clays are "normal" clays, suitable only for disposals of "non-dangerous" wastes, and those whose very good mineralogical characteristics make them also suitable for deposits of toxic and noxious wastes.

RESUME': Les facteurs techniques que l'on considére dans le choix d'une argile à user pour la realisation de couches imperméables pour les dépôts de déchets sont en general la permeabilité et les caractéristiques geotéchniques, dans le but d'eviter problémes de stabilité (et, en consequence, d'impermeabilisation) causés par les mouvements des couches argileuses. Le travail qu'on presente ici a comporté l' examen des principales formations argileuses qui affleurent en Piemont (Nord-Ouest de l'Italie), avec la determination de toutes les caractéristiques d'une certaine importance pour l'evaluation de leur aptitude a etre utilisées dans les dépôts de déchets. En particulier, une quarantaine d'echantillons (en representance de 14 zones d'extraction) ont eté recoltés et etudiés, en executant les determinations suivantes: limites d'Atterberg (a différents valeurs du pH), distribution granulometrique, permeabilité des echantillons naturels et reconstitués (au moyen d'un permeamétre realisé à l'occasion), resistence au cisaillement, mineralogie (particuliérement pour les mineraux des argiles, par diffractométrie au rayons X), calcimétrie, capacité d'echange cationique. Les résultats obtenus ont permis une bonne caracterization des argiles etudiées; en particulier, il a eté possible de distinguer parmis les formations argileuses "normales", utilisables seulement pour les dépôts de déchets non toxiques, de celles de meilleure qualite, indiquées aussi pour les dépôts de déchets dangereux ou toxiques.

1 INTRODUCTION

A series of laws concerning waste disposal in Italy according to European regulations were brought into force almost ten years ago. The law at present in force on this matter is Law DPR 915/82, that classifies the wastes according to their dangerosity and gives rules for the carring out of the related disposals. Another law, Law 441/87, concerns the arrangement of reclamation projects for polluted areas and the drawing up of a national plan for the detection of areas fit for waste disposal.

However, the parameters required by Italianaws to define the aptitude of different natural materials to use for waste disposal impermeabilization only refer to permeability and geotechnical characteristics even though other countries had proved the inadequateness of such parameters to assure the absence of a leachate flow through the lining.

The here described study, which concerns the aptitude of Piedmontese clays for use in waste disposal impermeabilization, took not only the hydraulic and geothecnical characteristics into consideration, but also some other parameters which, although rarely considered, are of great importance in influencing the efficiency of an "impermeable" clayey barrier; among these, the most important are the mineralogical composition, the cationic exchange capacity and the pH of leachate, which may greatly influence the geotechnical characteristics.

By considering the large scale homogeneity of the argillaceous sediments which form the different Piedmontese formations, the study was based on a systematic sampling with the purpose of obtaining the parameters which characterize each formation.

2 GEOLOGY OF STUDIED AREAS

From a geological point of view, the Piedmontese area can approximately be divided into three zones: the alpine zone, where methamorphic rocks largely outcrop ; the plain zone, formed by continental sediments of the quaternary age; the hilly zone of the Langhe and Monferrato, with large outcrops of Cenozoic sedimentary rocks (Fig.1). The most important argillaceous formations outcropping in the plains are connected to highly weathered paleosols, dated Mindel and (partially) Riss, whilst in the hills clayey marls and clays of the miocenic and pliocenic age outcrop.

The study was centered on four representative areas, choosen among those where the clays more largely outcrop:

1. The South-West portion of the Turin province, called "Altopiano di Poirino", formed mainly by pleistocenic loams and clays lying on older sediments of Villafranchian facies. Samples 1-8 were collected in this zone at different heights, in fluvial deposits about 8 m thick.

2. The North-west zone of the Langhe hills, 10 km South-West of the town of Alba; in this zone the "Marne di S. Agata fossili", a marly-clayey-conglomeratic formation, is the only formation that extensively outcrops having important levels of gray clays, often showing high plasticity. Samples 9-10-11 were collected in this area.

3. The Northern Astigiano zone, characterized by three main argillaceous formations: sample 12

comes from the "Argille di Lugagnano" formation, of lower Pliocene; sample 13 is a clay collected in lower Villafranchian, a transition facies with alternating layers of sands and clays; samples 14 and 15 come from the medium-upper Pliocene ("Sabbie di Valle Andona" formation).

4. The High Vercellese plane (the so called "Barragia"): here are widely outcropping sedimentary formations of transitional and continental-alluvial facies, often highly weathered. Samples 16, 17 and 18 were collected from the top of these formations, whose ages vary from upper Pliocene to lower Pleistocene.

3 LABORATORY TESTS

All the tests carried out on the different argillaceous formations were in order to determine the average values of the most important parameters to be used for the project of an impermeable layer in controlled waste disposals.

In addition to the usual geotechnical factors, the mineralogical composition and the cationic exchange capacity were also investigated for this purpose; a detailed description of the tests and of the results obtained is given in the following section.

3.1 *Atterberg limits*

These limits were determined using the well known techniques, but at three different pH : 7 (as usual), 3 and 11, using as the liquid phase distilled water, diluted acetic acid and ammonia respectively; this was done in order to operate in conditions similar to those that a clay might find if directly interested by leachates. The so obtained values are shown in Tab. 1.

Examining these results, it can be noted that:

1. At pH 11, the only significant variation of WL is shown by the "Altopiano di Poirino" clays, with an increase of about 21%;

2. At pH 3, only the "Argille di Lugagnano" and the "Sabbie di Valle Andona" clays have a WL reduction (on average 10%); 3. At pH 11, the WP index grows by about 32% for the "Altopiano di Poirino", 16% for the "Argille di Lugagnano" and 41% for "Baraggia" clays;

4. At pH 3 only 12 samples were tested for the WP index; all the clays showed a reduction of this parameter, with percentages varying from 4% to 17%.

2464

Figure 1. Piedmont, geological units

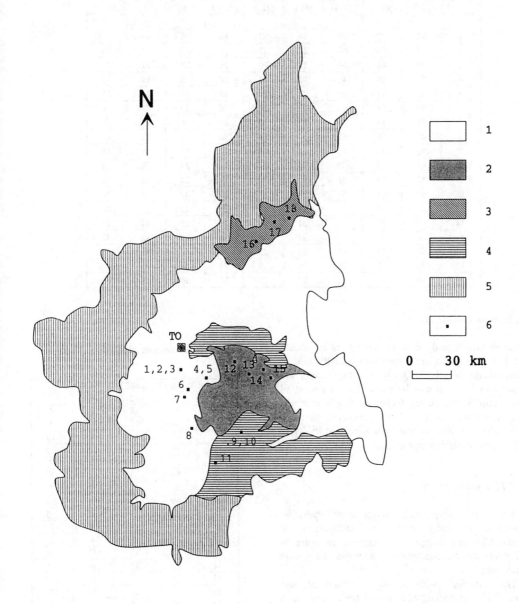

1, Olocene and Pleistocene; 2, Pliocene; 3, Pliocene; 4, Miocene;
5, Alpin Arc; 6, Sampling points

Table 1. Atterberg limits

SAMPLE	WL % ph 7	WL % pH 3	WL % pH 11	WP % pH 7	WP % pH 3	WP % pH 11	IP % pH 7	IP % pH 3	IP % pH 11
1	82	77	95	36	37	38	46	40	57
2	54	53	66	27	33	38	27	20	28
3	40	41	49	21	20	28	19	21	21
4	38	38	41	20	24	28	18	14	13
5	45	42	49	22	21	32	23	21	17
6	55	54	65	28	26	36	27	28	29
7	31	33	38	22	20	30	9	13	8
8	41	46	63	24	23	30	17	23	33
9	41	40	40	24	21	22	17	19	18
10	47	34	50	24	27	29	23	7	21
11	42	38	52	20	20	24	22	18	28
12	49	42	51	27	24	27	22	18	24
13	38	35	36	26	25	26	12	10	10
14	60	52	60	31	28	36	29	24	24
15	62	57	69	29	25	36	33	32	33
16	65	65	68	35	29	40	30	36	28
17	70	68	70	29	29	34	41	39	36
18	44	43	48	28	26	31	16	17	17

3.2 Granulometric analysis

Granulometric analysis were carried out by means of wet sieving for the dimensions >0.040 mm, of decantation for the <0.040 mm; the results are shown in Tab.2. It may be noted that, in almost all the samples, the clayey fraction clearly prevails; the uniformity coefficient C (D60/D10 ratio) is normally < 5, also indicating a good uniformity in the granulometric composition.

3.3 Permeability tests

These tests were made at constant hydraulic charges, operating both on undisturbed or remoulded and compressed samples, in order to simulate the true conditions of an impermealization clay layer.

The samples, with a 42mm diameter and previously saturated, were tested for 6 hours at a pressure of 0.12 MPa (approximatively 12m of water column); the Ki (undisturbed samples permeability coefficient) values obtained are all regular for clayey materials, and are summarized in Tab. 3.

Table 2. Granulometric analysis

SAMPLE	% CLAY	% SILT	% SAND	C
1	88	12		3,33
2	80	18	2	4
3	50	38	12	5
4	48	46	6	10
5	59	31	10	2,5
6	73	24	3	2
7	56	33	11	5
8	50	42	8	5
9	69	17	14	2,5
10	83	13	4	2,5
11	69	23	8	2
12	60	36	4	2
13	35	42	23	33
14	51	43	6	15
15	75	23	2	1,25
16	31	21	48	150
17	56	28	16	6,67
18	43	20	37	50

The Kr (remoulded samples permeability coefficient) values showed, on the contrary, marked differences between the tested clays, the Kr/Ki ratio varying from >1 (75% of the samples) to <1 for the clays (the best consolidated ones) with, the lowest Ki (10^-10 m/s). for the undisturbed samples. The so obtained values are shown in Fig. 2.

These results confirm the extreme behaviour variability of the argillaceous formations, because of their granulometric composition and mineralogy (clay mineral contents). It was not possible to find a satisfactory correlation between these factors, also because of the too reduced number of tested formations; it must be noted, however, that several authors state the possibility of easily reaching, with artificial compactation, a permeability of 10^-10 m/s for remoulded clay layers: the results obtained in this research suggest a more prudent approach to this theme, too dependent on parameters that are yet not well known.

All the "Altopiano di Poirino" clays seem suitable for cat. I disposals, as far as the limits required by Italian laws (DPR 915/82) are concerned, whilst their characteristics hardly reach the values required for cats. IIc and III disposals. Similar conclusions can be drawn for "Argille di Lugagnano" and "Sabbie di Valle Andona" clays; the "Baraggia" clays, on the contrary, are unsuitable for waste disposal impermeabilization, their permeability characteristics being lower than those required for cat. I deposits.

Table 3. Permeability tests

SAMPLE	Ki	Kr	Kr/Ki
	[m/s]	[m/s]	[m/s]
1	7.5*10^-8	6.1*10^-9	0,08
2	***	1.42*10^-10	***
4	***	1.96*10^-9	***
5	4.8*10^-8	2.9*10^-10	0,006
6	***	1.46*10^-10	***
7	2.57*10^-10	8.64*10^-9	33,6
11	***	1.57*10^-9	***
12	2.3*10^-8	***	***
13	1.92*10^-8	3.23*10^-8	1,68
14	2.4*10^-8	***	***
15	2.73*10^-10	2.44*10^-9	8,94
16	2.5*10^-8	1.72*10^-8	0,69
17	3.33*10^-8	3.16*10^-9	0,095
18	7.3*10^-7	3.41*10^-8	0,047

3.4 Shear tests

The shear tests were only carried out in the laboratory, using the Casagrande apparatus, in UU conditions (unconsolidated undrained), on remoulded samples. The friction angle and the cohesion which result are shown in Tab. 4 , and all fall in the normal variability interval corresponding to medium-high plasticity clays.

Table 4. Shear tests results

SAMPLE	c'	Peak	Residue
	[kPa]	[°]	[°]
1	98	34	20
2	25	20	17
5	93	28	20
7	55	28	24
9	45	20	17
11	60	24	18
12	65	26	18
13	53	28	22
14	35	23	16
16	40	27	19
17	30	24	16
18	35	30	26

3.5 Mineralogical analysis

This was carried out using a Cu-k-alfa radiation at an X-ray diffractometer, this tecnique being the best for clay mineral determination. Three specimens were analyzed for each sample, all concerning the < 0.002mm granulometric fraction: one simply oriented, another previously saturated with etilenglicol at 60 C for 24 hours, the last after 2 hours of heating at 500 C. The semi- quantitative results are shown in Tab. 5.

It may be noted that "Altopiano di Poirino", "Argille di Lugagnano" e "Sabbie di Valle Andona" clays have very similar mineralogical compositions, with prevailing illite and kaolinite, a good presence of quartz and calcite and rare feldspars. The "Marne di S. Agata fossili" are mainly composed of illite, kaolinite, chlorite, smectite, quartz and carbonates; the "Baraggia" clays have largely prevailing kaolinite and illite, and also abundant quartz.

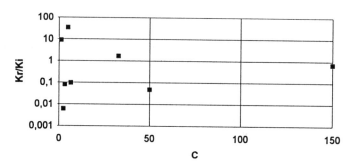

Figure 2a. Correlation between Kr/Ki and C

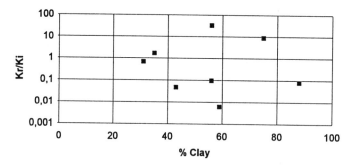

Figure 2b. Correlation between Kr/Ki and clay

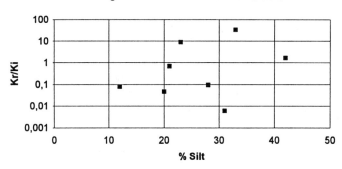

Figure 2c. Correlation between Kr/Ki and silt

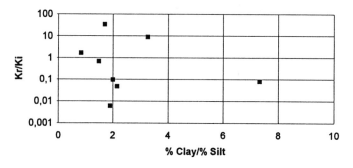

Figure 2d. Correlation between Kr/Ki and clay-silt ratio

Table 5. Mineralogical characteristics

STRATIGRAPHY		SAMPLE	FORMATION	MINERALOGY							
				I	K	C	S	V	QZ	F	CA
OLOCENE		4		50	-	-	-	50	+++	(+)	-
PLEISTOCENE	UPP.	1	Clays of "Altopiano di Poirino"	25	55	-	20	-	+++	++	(+)
		2		25	50	-	25	-	+++	+	(+)
		3		45	25	-	30	-	++	+	-
		5		45	55	-	-	-	+++	+	-
		6		40	15	-	-	45	+++	(+)	-
		7		35	45	20	-	-	+++	+	-
	MED.	8		-	60	-	40	-	++	(+)	-
	LOW.	18	Clays of "Baraggia"	40	60	-	-	-	+++	+	-
PLIOCENE	VILLAFR.	12		60	40	-	-	-	+++	+	++
		13	"Argille di Lugagnano"	100	-	-	-	-	+++	+	+
		14	"Sabbie di Valle Andona"	40	40	20	-	-	++	(+)	++
		15		40	45	15	-	-	+++	+	++
		16	Clays of	30	70	-	-	-	++	(+)	-
		17	"Baraggia"	50	50	-	-	-	+++	-	
MIOCENE	TORTON.	11	Marly clays of "Marne	30	40	20	10	-	++	(+)	+++
	MED.	9	di S. Agata Fossili"	40	35	15	10	-	+++	(+)	+++
		10		45	40	10	5	-	+++	(+)	+++

I=Illite; K=Kaolinite; C=Chlorite; S=Smectite; V=Vermiculite; QZ= Quartz; F= Feldspats; CA=Carbonates

+++ very abundant; ++ abundant; + scarce; (+) traces.

In the samples that had shown a high carbonates content, a quantitative evaluation of these minerals was carned out, by means of calcimetric tests; the results are included in Tab. 6.

3.6 Cationic exchange capacity

Another important function required of the clays used for impermeable layers in waste deposits is the capacity to act as a chemical filter, fixing the noxious ions in their cristalline reticule: this capacity is well shown by the cationic exchange capacity (CSC) test.

This test was carried out using the barium chloride and triethanelamine method, operating at pH 8.1. The standard procedure was followed, changing only the shaking time with a mechanized shaker: 5 minutes shaking was used. The CSC was obtained from the formula:

$$CSC = 1/P * [250 - N/M * 10 * (25 + B - A)] \ [meq/100g]$$

where P = sample weight; N,M ml of EDTA used for CSC determination and for the blank test; B-A g of water retained from the sample during the test.

After this, measurements were also carried out with an atomic absorption spectrophotometer, in order to obtain the CSC for K, Mg and Ca ions; the results obtained are summarized in Tab.6 .

The CSC values are medium-low for almost all the samples, considering that for clay minerals the average values normally range from 5 meq/100g for kaolinite to 55 meq/100g for bentonite: almost half of the specimens analyzed show CSC values varying between 16 meq/100g and 35 meq/100g, whilst the other half ranges from 5 meq/100g to 12 meq/100g. The cationic exchange capacity appears to be correlated to the mineralogical composition, kaolinite, chlorite and calcite being the minerals which characterize the formations with low CSC values.

2469

Table 6. CSC, results obtained

SAMPLE	A [g]	B [g]	H2O B-A [g]	E.D.T.A. [ml]	CSC [meq/100 g]	K+ [meq/100 g]	Mg++ [meq/100 g]	Ca++ [meq/100 g]	CaCO3 [%]
1	22.69	24.43	1.74	13.8	35.00	1.25	5.99	27.76	5.5
2	22.46	23.96	1.50	15.4	25.46	0.428	3.92	21.11	9.45
3	22.22	23.44	1.22	16.9	16.92	0.140	4.54	12.24	6.3
4	21.91	23.16	1.25	16.9	16.80	0.140	2.48	14.18	***
5	21.85	23.24	1.39	16.9	16.22	0.109	4.54	11.57	***
6	22.75	24.26	1.51	15.5	24.78	0.453	4.97	19.36	***
7	22.75	23.95	1.20	17.65	12.21	0.128	3.72	7.36	***
8	22.15	23.53	1.38	15.8	23.34	0.192	5.16	17.99	***
9	22.34	23.69	1.35	17.6	11.89	0.441	1.24	10.2	26.77
10	22.31	23.97	1.66	17.5	10.56	0.447	1.86	8.25	27.56
11	22.54	23.89	1.35	18.1	8.67	0.505	3.72	4.45	31.5
12	22.60	24.13	1.53	18.2	7.23	0.352	3.93	2.94	19.3
13	22.54	23.92	1.38	18.2	7.90	0.255	0.62	7.02	11.8
14	22.36	24.12	1.76	20.0	6.17	0.486	0.62	5.04	16.5
15	22.17	24.01	1.84	18.3	5.20	0.217	1.65	3.33	19.05
16	22.51	24.13	1.62	18.1	7.48	***	***	***	***
17	22.87	24.33	1.46	16.6	17.87	0.140	7.37	10.36	***
18	22.59	23.97	1.38	16.7	17.55	0.128	4.75	12.67	***

Fig. 3 Correlation between CSC and clay content

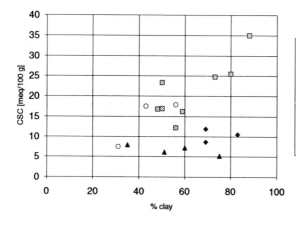

⊠ Olocene and Pleistocene, Altopiano di Poirino clays

◆ Miocene, S. Agata Fossili marly clays

▲ Pliocene, Argille di Lugagnano and Sabbie di Valle Andona clays

○ Pliocene and Pleistocene, Baraggia clays

A correlation was also for searched between CSC and the granulometric fraction < 0.002mm; as it is shown in Fig.3 , many samples appear to be placed on a straight line. An exception being the six mio-pliocenic samples from the Langhe and Astigiano formations, whose mineralogic composition is at the origin of the much lower CSC values.

4 CONCLUSIONS

The here described work concerns the definition of some uncommon parameters of argillaceous formations largely outcropping in Piedmont, in order to evaluate their appropriatness for use in waste deposit impermeabilization.

The tests carried out have shown the variability of some parameters, such as the Atterberg limits and permeability, in correspondence to the changing of pH, i.e. of the chemistry of the leachate.

The more recent studies have shown the importance of an impermeabilization which is not simply inactive, but also able to react with noxious substances by fixing them in its interior; a good parameter for this is the CSC, which is related to the percentage and the minerals of the argillaceous fraction.

As a final conclusion, the studied clayey formations can be classified as follows:

1. The "Altopiano di Poirino" quaternary clays are suitable for use in the impermeabilization of urban waste deposits (cat. I for the Italian law). These clays could also be used for deposits of toxic wastes of not too high toxicity, because of their medium-high CSC; it must be noted, however, that the permeability of these clays is rather high (at the law limits), and it will be therefore necessary to take great care in the compactation operations.

2. The "Marne di S. Agata fossili" miocenic clays are only useful for urban waste deposits, because of their high permeability and medium-low CSC due to a high carbonates content.

3. The "Argille di Lugagnano" and "Sabbie di Valle Andona" pliocenic clays have fields of use very similar to miocenic clays, as they have similar technical characteristics.

4. The "Baraggia" plio-pleistocenic clays normally have a low clayey fraction (about 50%); this fact negatively influences both the permeability and CSC, sometimes pushing one or another of these parameters under the limits permitted by law, and making these materials unsuitable for waste deposit impermeabilization.

REFERENCES

Bottino G., Grassi G., Stafferi L., 1988. Le argille del Bacino Terziario piemontese - aspetti genetici e caratteristiche tecnologiche. Boll. AMS, XXV, 2-3.

Esposito A.P., Ferrante L., 1990. Sull'impiego delle argille costipate per la impermeabilizzazione dei fondi delle discariche controllate. Rivista It. di Geotecnica, 2.

Gavasci R., Prestininzi A., Tiberio A., Sirini P., 1989. Studio dei siti per discariche: variazioni delle caratteristiche chimico- fisiche indotte dall'interazione fra terreni e percolati. Proc. *Suolosottosuolo*, Congresso Int. di Geoingegneria, Torino.

Ishikawa H., Amemiya K., Yusa Y., Sasaki N., 1989. Comparison of fundamental properties of japanese bentonites as buffer material for waste disposal. Proc. 9th Int. Clay Conference, Strasbourg.

Wagner J.F., 1991. A double mineral base liner for waste disposal sites. Proc. 7th Euroclay Conference, Dresden.

Characteristics of a clayey soil from Kyongjukun, with emphasis on its potential as a hydraulic barrier material

Caractéristiques d'un sol argileux de Kyongjukun concernant son usage comme barrière hydraulique

Daesuk Han
Korea Institute of Geology, Mining and Materials, Korea

ABSTRACT: A clayey soil was collected from the wastes of a bentonite mine located at Yangnammyon, Kyongjukun. The soil was tested for its specific gravity, natural moisture content, grain-size distribution, Atterberg limits, free swell, linear shrinkage, compaction characteristics, permeability, shear strength, mineralogy, petrography, and chemistry. The test procedures and soil characterization were discussed.

The soil was classified as SC according to the Unified Soil Classification System, containing 27% fines by weight. The main inorganic constituents of the fines were Si and Al; the oxides of these two elements comprised 78% of the composition. X-ray diffraction analysis and scanning electron microscopy revealed that the fines were composed of Ca-montmorillonite, quartz, plagioclase, muscovite, illite, mordenite, and hornblende. The clay fraction was as much as half of the fines, giving rise to the relatively high values of free swell in water and liquid limit, i.e. 1.98 ml/g and 63.0%, respectively.

The hydraulic conductivity of the soil varied from 1.0×10^{-8} to 1.0×10^{-9} m/s for the specimens prepared under different compactive efforts, indicating that the drainage characteristics are practically impervious.

Based on the test results, the soil may be utilized as a hydraulic barrier material for the disposal of urban and industrial wastes.

RESUME: Le sol argileux a été relevé à partir des déblais d'une mine de bentonite située à Yangnammyon, Kyongjukun. Le sol a été analysé pour le poids spécifique, la teneur en eau naturelle, granulométrie, limites d'Atterberg, gonflement libre, contraction linéaire, caractère du compactage, perméabilité, résistance au cisaillement, minéralogie, pétrographie, et chimie. Les procedures de l'essai et le caractére du sol ont été discutés.

Le sol a été classifié comme SC d'après le Système de Classification des Sols Unifiés, contenant 27% en poids des grains fins. Les principaux constituents inorganiques des grains sont Si et Al; les oxides de ces deux éléments se composent 78% de la composition totale. L'analyse à balayage a révélé que les grains fins ont été composés de la Ca-montmorillonite, le quartz, le plagioclase, la muscovite, l'illite, la mordenite, et la hornblende. La fraction argileuse est autant que la moitié des grains fins, donnant lieu à la valeur relativement haute de gonflement libre en eau et limite de liquidité, i.e. 1,98 ml/g et 63,0% respectivement.

La conductivité hydraulique du sol se varie entre $1,0 \times 10^{-8}$ m/s à $1,0 \times 10^{-9}$ m/s pour les échantillons préparés sous différents efforts compactifs. Ceci indique que les caractéristiques de drainage sont practiquement imperméables.

Basé sur les résultats de l'essai, le sol peut être utilisé comme un matériau d'une barrière hydraulique pour la disposition des déchets urbans et industriels.

1 INTRODUCTION

The Ministry of Environment, the Republic of Korea has reported that the urban areas produced about 75,000 metric tons of solid waste each day in 1992, whilst the firms generated about 48,000 metric tons of industrial waste daily. Approximately 44% of the industrial waste was the hazardous one

from 20,427 sources. At the present time, the Republic of Korea is confronted with a big problem that the existing controlled landfill facilities are too short to receive all the wastes, thus attributing to the improper waste disposal which can lead to the environmental problems such as soil and water contamination. An example of the improper waste-disposal sites in the Republic of Korea is shown in Figure 1. The open dump of the figure is in the city of Pohang located in the southeastern part of the Republic of Korea. A stream is running along the north side of the dump situated about 500 meters west of the east coast. A project to construct a lot of additional controlled landfill facilities is believed to require one of the most expensive environmental expenditures that the government must pay.

Figure 1. Open dump, Yeonamdong, Pohang, 1990.

Clayey soils, which have low hydraulic conductivity, can be chosen in designing a waste-disposal facility to restrict the flow of water as well as the migration of pollutants. About 130 kg of clayey soil was collected from the waste pile of a bentonite mine, Yangnammyon, Kyongjukun, which is located in the southeastern part of the Republic of Korea. The clayey soil was observed to occur between the bentonite deposits which had been formed from the volcanic ash and tuffaceous breccia of Tertiary age. A laboratory investigation on the soil was undertaken during the period from December, 1992 to July, 1993 for the following purposes: (1) to find out its physical, engineering, mineralogical, and chemical characteristics, (2) to evaluate its utilization potential as a hydraulic barrier material, and (3) to recommend further research on it.

2 LABORATORY INVESTIGATION

Throughout this and subsequent chapters, the term, "waste material" refers to the clayey soil collected from Kyongjukun. A series of laboratory tests was carried out on the waste material, including natural moisture content, specific gravity, grain-size dis-

tribution, Atterberg limits, free swell, linear shrinkage, compaction characteristics, permeability, shear strength, mineralogy, petrography, pH, cation exchange capacity, and chemical composition.

2.1 Sample preparation

The whole waste material collected was first oven dried at 60ºC. In order to have all kinds of laboratory tests be performed on the samples with a very similar grain-size distribution, the oven-dried material was then divided into 20 group samples by using a riffle box, each weighing about 6 kg. Unless stated otherwise, a riffle box was used whenever necessary during the course of the subsequent laboratory testing.

2.2 Physical characteristics

Five jar samples were collected especially for the determination of the natural moisture content of the waste material. When determined in accordance with ASTM D 2216, the moisture content ranged from 11.9 to 12.8%, the mean value being 12.5%.

A representative sample of about 100 g was obtained by quartering a group sample for its specific gravity determination. The particles larger than 0.420 mm were broken down to pass through a No. 40 sieve. The determination was conducted in accordance with ASTM D 854, the specific gravity being 2.71.

The determination of Atterberg limits for the waste material was conducted in accordance with ASTM D 423 and D 424. The liquid limit was determined at 63%; the plastic limit at 29.4%.

A representative sample of about 500 g was obtained by quartering a group sample of the waste material. Of the sample, only the particles passing a No. 40 sieve were subjected to the linear shrinkage determination by the method of BS 1377:1975, Test 5. The linear shrinkage is computed as a percentage of the original length of the specimen from the following equation (1).

$$LS = (1 - Ld/Lo) \times 100\% \ldots\ldots\ldots\ldots (1)$$

where Lo = original length (140 mm); Ld = length of dry specimen. Using the equation, the linear shrinkage was determined at 21%.

The grain-size distribution for the waste material was determined in accordance with ASTM D 422. A group sample of the waste material was wet-sieved using 1½" (38.1 mm), 3/4" (19.0 mm), 3/8" (9.5 mm), No. 4 (4.75 mm), No. 8 (2.36 mm), No. 40 (0.420 mm), No. 60 (0.250 mm), No. 100 (0.150 mm), and

No. 200 (0.075 mm) sieves. The particles finer than 0.075 mm were tested following a hydrometer method. A grain-size distribution curve (Figure 2) was constructed using the above two test results. As illustrated in the figure, the waste material comprises 21% gravel-size particles, 52% sand-size particles, and 27% fines.

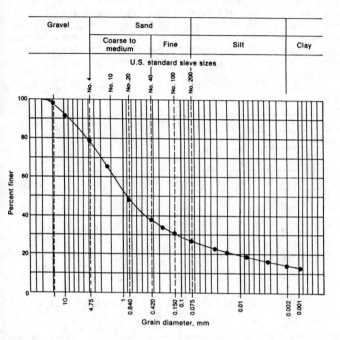

Figure 2. Grain-size distribution curve for the waste material.

Free swell test was undertaken following the method by Gibbs and Holtz (1956). The dry particles passing a No. 40 sieve was first placed loosely in a 25 ml cylinder up to the 10 ml mark. The weight of the particles in the cylinder was 10.23 g. They were then poured slowly and steadily into the 50 ml glass measuring cylinder filled with 50 ml distilled water. "Free swell" is defined as the change in volume of the dry sample expressed as a percentage of its original volume. The value was determined at 103%, being 1.98 when expressed as ml/g.

2.3 Engineering characteristics

The moisture-density relationship for the waste material was determined following the procedure specified in ASTM D 698, "Standard Test Methods for Moisture-Density Relations of Soils Using 2.5 kg Rammer and 304.8 mm Drop" and using a 101.7 mm diameter mold. A diagram was constructed illustrating the relationship between moisture content and dry density (Figure 3). According-ing to the figure, the maximum dry density

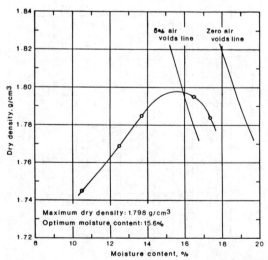

Figure 3. Moisture-density relationship for the waste material.

and optimum moisture content are 1.798 g/cm^3 and 15.6%, respectively.

Five group samples of the waste material were compacted over a range of moisture contents with different compactive efforts using a compaction mold, 101 mm diameter and 116 mm long. The compacted specimens were tested to obtain their hydraulic conductivity following a generally accepted procedure of falling head permeability test. The hydraulic conductivities determined at 20°C are listed in Table 1. Besides the above tests, a direct falling head test in a oedometer consolidation cell was carried out for the particles passing a No. 10 sieve. The compacted soil in the same mold as for the previous tests was used to prepare a cutting ring specimen, 63 mm diameter and 20 mm high. The determined value is also listed in Table 1. Using the above data, a diagram (Figure 4) was constructed, illustrating the relationship between dry density and permeability.

Table 1. Hydraulic conductivity for the waste material.

Specimen number	Specimen preparation		Moisture content, %	Dry density, g/cm^3	Test method	k_{20}, m/s
	Compactive effort					
	No. of layers	Blows/layer				
1*	3	10	10.7	1.575	Falling head test in compaction mold	1.0×10^{-8}
2*	3	15	9.8	1.621	– do –	5.2×10^{-9}
3*	3	20	10.2	1.661	– do –	2.8×10^{-9}
4*	3	25	7.6	1.698	– do –	1.8×10^{-9}
5*	3	30	7.5	1.734	– do –	1.0×10^{-9}
6**	3	25	16.3	1.527	Falling head test in oedometer consolidation cell	1.5×10^{-11}

* For the whole waste material.
** For the particles passing a No. 10 sieve.

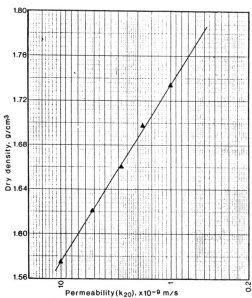

Figure 4. Dry density versus permeability for the waste material.

The measurement of the shear strength of the waste material was accomplished by means of direct shear tests. An ELE motorized machine having a 60 mm shearbox was used to define the shear strength parameters of the particles passing a No. 10 sieve. The specific gravity of the particles were determined at 2.72. In order to have the shearbox tests be performed on the specimens with a very similar density, they were prepared by the following manner. A representative sample of about 3,000 g was obtained by quartering the particles finer than 2.00 mm. The sample was well mixed with the 480 g water and then divided into three portions. Each portion was placed into a compaction mold of 101 mm diameter and compacted under the effort of 25 blows by a 2.5 kg rammer. The dry density of the three specimens ranged from 1.526 to 1.528 g/cm^3 with the moisture content of 16.3%. When the normal stresses of 36, 61, and 88 kN/m^2 were applied to the specimens, the corresponding maximum shear stresses at failure were 81, 91, and 112 kN/m^2. The loading rate for all the tests was 1.06

mm/min. The shear strength parameters, cohesion and internal friction angle were 62 kN/m² and 29°, respectively. The test results are graphically presented in Figure 5.

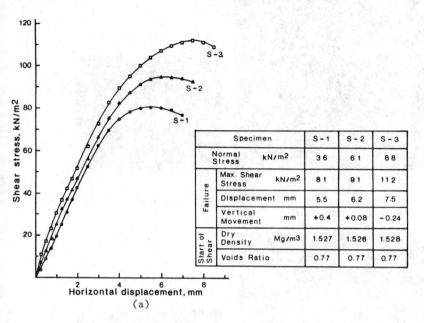

Specimen		S-1	S-2	S-3
Normal Stress kN/m²		3 6	6 1	8 8
Max. Shear Stress kN/m²		8 1	9 1	11 2
Displacement mm		5.5	6.2	7.5
Vertical Movement mm		+0.4	+0.08	-0.24
Dry Density Mg/m³		1.527	1.528	1.528
Voids Ratio		0.77	0.77	0.77

(a)

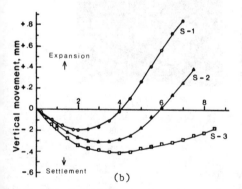

(b)

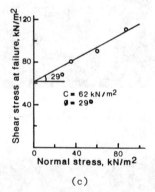

C = 62 kN/m²
ϕ = 29°

(c)

Figure 5. Graphical results of the shearbox test for the particles passing a No. 10 sieve: (a) stress-displacement, (b) vertical movement, and (c) Coulomb envelope.

2.4 Mineralogy and petrography

Several fine-grained chunks of the waste material were subjected to the mineralogical analysis by SEM and EDX. For X-ray analysis, the particles passing a No. 40 sieve were divided into two portions, one passing and the other retained on a No. 200 sieve; the latter was ground to pass through a No. 100 sieve. The common minerals identified by the two analyses were Ca-montmorillonite and plagioclase. Figure 6 shows a typical texture of Ca-montmorillonite aggregates. In addition to the above two minerals, the presence of muscovite and hornblende was revealed by the scanning electron micrographs (Figures 7 and 8), whilst that of quartz, illite, and mordenite by X-ray powder diffraction patterns (Figure 9).

The coarse grains of the waste material are generally angular to subangular in shape. The thin sections of them were examined under a polarizing microscope, the major rock types being tuff, shale, rhyolite, feldspar porphyry, and tuffaceous sandstone.

2477

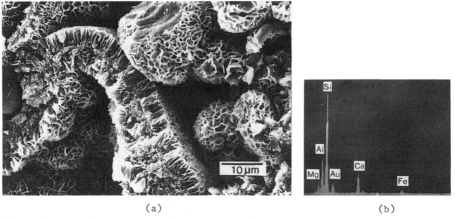

(a)   (b)

Figure 6. Scanning electron micrograph showing a typical texture of Ca-montmorillonite (a) and EDX pattern of the montmorillonite (b).

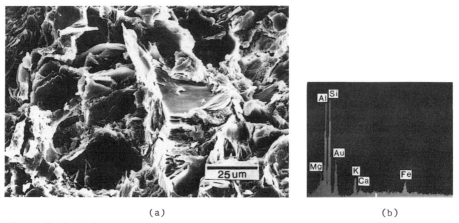

(a)   (b)

Figure 7. Scanning electron micrograph showing muscovite flakes (a) and EDX pattern
of the muscovite (b).

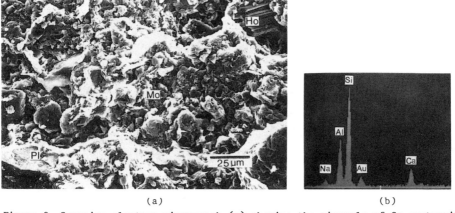

(a)   (b)

Figure 8. Scanning electron micrograph (a) showing the minerals of Ca-montmorillonite (Mo), plagioclase (Pl), and hornblende (Ho) and EDX pattern of the plagioclase (b).

2478

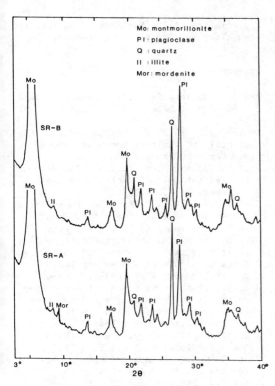

Figure 9. X-ray powder diffraction patterns
for the particles finer than 0.075 mm (SR-A)
and the particles passing a No. 40 sieve
but retained on a No. 200 sieve (SR-B).

2.5 Chemical characteristics

The determination of pH was conducted in
accordance with the Electrometric Method,
BS 1377:1975, Test 11(A). A representative
sample of about 50 g was obtained by riffl-
ing the particles passing a No. 4 sieve.
The sample was crushed to pass through a
3.35 mm sieve. A 30 g of the prepared sample was placed in a 100 mm beaker and 75 ml
distilled water was added to it. The mixture was allowed to stand overnight and then
measured for its pH, the value being 8.8.

Seven chemical constituents were determined for two kinds of particles, one passing a No. 200 sieve and the other passing
a No. 40 sieve but retained on a No. 200
sieve. The determination of SiO_2, Al_2O_3,
and Fe_2O_3 was made by colorimetric methods,
whilst that of CaO, MgO, Na_2O, and K_2O by
a spectrochemical method. The contents of
the chemical constituents are presented in
Table 2.

Besides the above tests, the cation exchange capacity for the particles finer
than 0.420 mm was determined at 70.1 meq/
100 g.

Table 2. Chemical composition of the waste
material.

Chemical constituent	Content, %	
	Sample 1*	Sample 2**
SiO_2	59.80	61.62
Al_2O_3	18.53	18.21
Fe_2O_3	4.57	4.03
CaO	3.31	2.86
MgO	3.06	2.80
Na_2O	1.60	1.93
K_2O	1.02	1.34

* For the particles finer than 0.075 mm.
** For the particles passing a No. 40 sieve
but retained on a No. 200 sieve.

3 DISCUSSION

Based on the test results of the grain-size
distribution and Atterberg limits, the whole
waste material containing the 13% clay is
classified as SC according to the Unified
Soil Classification System; the particles
passing a No. 40 sieve as CH.

The two X-ray diffraction patterns (Figure
9), one for the particles passing a No. 200
sieve and the other for the particles passing a No. 40 but retained on a No. 200 sieve,
reveal the same minerals except the mordenite in the former.

Of the fines, the clay fraction is as
much as about 50%. The clay particles are
assumed to be composed mostly of Ca-montmorillonite considering that the particles
passing a No. 40 sieve possess the cation
exchange capacity of 70.1 meq/100 g and the
free swell value of 103% (19.8 ml/g) together with the linear shrinkage of 21%.
The above assumption is also supported by
the pH value of 8.8. According to the investigation on the Korean bentonites of
good quality, the cation exchange capacity
ranges from 80.0 to 97.0 meq/100 g; the pH
value from 8.5 to 10.5.

The permeability of the waste material
appears to depend upon its clay content and
the degree of compaction. The compacted
specimens of the waste material having the
dry density from 1.575 to 1.734 g/cm^3 range
from 1.0×10^{-8} to 1.0×10^{-9} m/s (Table 1).
The waste material with such permeability
is characterized by very low degree of permeability according to the permeability
classification of Terzagi and Peck (1967).
The hydraulic conductivity test on the specimens of higher density were not carried
out because the measurement by the falling
head method in a compaction mold is virtually impossible. For such specimens, however, their permeability may be predicted
using the diagram shown in Figure 4. For

2479

instance, when the waste material with the natural moisture content of 12.5% is compacted by the method described in the paragraph 2.3, the dry density of about 1.77 g/cm3 can be acquired according to the moisture-density curve (Figure 3). Using Figure 4, the hydraulic conductivity is about 8.2 x 10-10 m/s, the compacted material being practically impermeable.

The permeability of the particles passing a No. 10 sieve is as low as 1.5 x 10-11 m/s, suggesting the possibility that the permeability of the waste material may be lowered up to less than 1.0 x 10-11 m/s if the material is compacted properly.

The shear strength parameters ($\emptyset$ = 29°, C = 62 kN/m2) are only for the particles finer than 2.00 mm. For the whole waste material, the internal friction angle is assumed to be higher than 29°; the cohesion to be about the same as or slightly lower than 62 kN/m2.

4 CONCLUSION

Secure landfills, which are lined with an impermeable material such as a clay, are the ultimate disposal for most wastes. They can accomodate not only a broad variety of waste but also the residue from other treatment methods, such as chemical treatment and incineration.

Based on the test results, the clay mineral (Ca-montmorillonite) must control the properties of the waste material. For that reason, the hydraulic conductivity of the waste material is relatively low comparing with that of the soils containing clay minerals, such as kaolinite and illite. The waste material, which has a dry density lower than 1.575 g/cm3 and the corresponding permeability higher than 1.0 x 10-8 m/s, can not be used as a hydraulic barrier material for secure landfills but as a cover material for sanitary landfills. However, the waste material compacted to have a dry density higher than 1.734 g/cm3 may be used as a hydraulic barrier material for secure landfills since its permeability is lower than 1.0 x 10-9 m/s.

Because the solute and solvent phases of waste leachate may affect the permeability of the waste material, it is recommended that the permeability test on the waste material be performed with permeants other than water to evaluate the material for containment in relation to the urban and industrial wastes.

5 ACKNOWLEDGEMENT

The writer wishes to express his sincere appreciation to the Ministry of Science and Technology for its financial support.

REFERENCES

American Society for Testing and Materials 1980. Annual book of ASTM standards, Part 19.
Bowles, J.E. 1978. Engineering properties of soil and their measurement, 2nd ed. New York: McGraw-Hill.
British Standard Institution 1975. Methods of test for soils for civil engineering purposes, BS 1377: 1975.
Gibbs, H.J. & W.G. Holtz 1956. Engineering properties of expansive clays. Trans. Am.Soc.Civ.Eng. Vol.121, No.1, Paper 2814.
Han, D., et al 1990. Environmental geology for the control of subsurface pollution. Korea Institute of Energy and Resources Research Report KE-90-(B)-7, p.44-60.
Han, D., et al 1993. Capability of clayey materials for providing a hydraulic barrier to pollutants, with special emphasis on its improvement (part 1). Korea Institute of Geology, Mining & Materials Research Report KR-93(T)-5, p.23-76.
Head, K.H. 1980. Manual of soil laboratory testing Vol.1. London: Pentech.
Head, K.H. 1982. Manual of soil laboratory testing Vol.2. London: Pentech.
Keller, E.A. 1982. Environmental geology, 3rd ed., p.271-281. Columbus: Charles E. Merrill.
Terzaghi, K. & R.B. Peck 1967. Soil mechanics in engineering practice. New York: Wiley.

Methodology of specific engineering geological mapping for selection of sites for waste disposal

Méthodologie de cartographie géotechnique spéciale pour la sélection de sites pour le stockage de déchets

L. V. Zuquette, O. J. Pejon & O. Sinelli
USP/FF CLRP/Ribeirão Preto, Brazil

N. Gandolfi
USP/ESSC /São Carlos, Brazil

ABSTRACT: In countries with extense territories and that present social-economic problems it is necessary the assessment of the conditions of the geological environment (waters, relieves, rocks and unconsolidated materials) for disposal of the waste produced in residences, industries, laboratories, hospitals and other. Several studies were elaborated in some regions of Brazil and now we are proposing an attribute group and levels of attributes that must be considered in the studies of selection of sites for waste disposal in sanitary landfills, lagoons, cesspool, septic tanks, irrigation and spreading. The selection of the attributes and their levels was based in the general conditions of the tropical regions and in the limitations that occur more frequently and that are mapped in the process of engineering geological mapping. The attributes that must be evaluated in the process are: rocks substrate depth, unconsolidated materials (texture, permeability, mineralogy), water level, chemical and physical characteristics of the waters, declivity, landforms and cation exchange capacity. The principal objective of this work is to orientate the elaboration of the special engineering geological mapping for waste disposal at scales between 1:100,000 and 1:25,000.

RESUMÈ: Dans les pays avec grand extension territorial et que présentent des problèmes sociaux et économiques est nécessaire l'évaluation de la condition du environment géologique (eaux, relief, roches et materiaux non consolidés) pour le stockage de déchets produites en résidences, industries, laboratoires, hôpitals et autres. Plusieurs études ont été elaboré dans quelques regions du Brésil et maintenant nous proposont un group d'attributs et niveaux d'attributs que doivent être utilizé dans les études pour la selection des sites pour stockage de déchets dans les decharges controllées, logons, fosses noires, irrigation et similaire. La selection des attributs et leurs niveaux ont été basés dans de conditions générales des regions tropicales et dans la limitation la plus souvent et que sont étudiés en cartographie géotechnique. Les attributs que doivent être evalués dans le procès de cartographie sont: profondité du substratum, materiaux non consolidés (texture, perméabilité, mineralogie), produndité de la nappe fréatique, caractéristiques chimique et fisique des eaux, déclivité, *landforms* et capacité d'exchange cationique. Le principal objetif de ce travail est d'orienter l'eraboration d'une cartographie géotechnique especial pour le stockage de déchets en écheles entre 1:100.000 et 1:25.000.

1. INTRODUCTION

Only few cities in Brazil have adequate sites for waste disposal that were selected by studies elaborated for selection and choice of the most eficient technological mechanisms in the protection of the environment. Due to these conditions, we are proposing a preliminary procedure to the development of studies of the components of the geological environemnts for selections of geologically adequate sites.

In countries with extense territories and low social-economic conditions it is necessary the development of low cost works. These works must consider one group of attributes and basic procedures for its obtention.

2. METHODOLOGY

The works for preliminary selection of the sites must be elaborated by specific engineering geological mapping at between 1:100.000 and 1:25,000 scales.

This work must consider the attributes and levels of attributes and forms of disposal (sanitary landfills, cesspool, septic tank, irrigation, lagoon and spreading).

The attributes that must be considered in these works are presented in Zuquette (1993) and Souza (1992) that are:

Rock Substrate
1. Litology
2. Depth
3. Descontinuities (fractures, strutural lines, folds)

Unconsolidated Material
1. Texture
2. Profile variations
3. Mineralogy
4. Presence of rock blocks
5. pH and ΔpH
6. Salinity
7. Cation exchange capacity or anionic exchange capacity
8. Compressibility conditions
9. Colapsivity/expansibility
10. Erodibility index
11. Retardation factor
12. Compaction conditions

Water
1. Ground water level
2. Ground water flow direction
3. Overland flow
4. Infiltration (permeability coefficient)
5. Recharge area
6. Well and spring localization
7. Drenability conditions

Geologic Process
1. Erosion
2. Landslides

Relief
1. Declivity
2. Landform
3. Surface water boundary
4. Wet zones
5. Flood zones

Climatic Condition
1. Wind direction
2. Evapotranspiration

3. Rainfall
4. Water balance

These attributes must be obtained by mechanisms that are adopted by technical associations and by proceding presented in Zuquette (1993) and Zuquette & Gandolfi (1990) and the user must to obey the sequency proposal in the Table 1. The principal manners of waste disposal that are considered in this work are: sanitary landfill, lagoon, cesspool, septic tank, irrigation and spreading.

The table 2 shows the attributes and levels of the attributes that permit the definition and delimitation of the units that are classified as:

Favorable - the natural attributes of the area present levels adequate for waste disposal.

Moderate - some natural attributes of the area present levels not adequate for waste disposal. The correction is possible with low costs and common technological mechanisms.

Severe - More than 50% of the natural attributes present levels not adequate for waste disposal. There is a necessity of special technological mechanisms for correction.

Restrictive - the natural attributes of the area present levels not adequate for waste disposal. It is necessary very special technological mechanisms and high costs for correction of the limitations. The occupation can produce intense environmental impacts.

3. CONCLUSION

The use of the metodology, attributes and other mechanisms permit the execution of works of the especific-purpose engineering geological mapping for selection of the sites for waste disposal in large regions with low costs.

These documents obtained by these works present the ranking of sites that must be studied in details before the instalation of: sanitary landfills, lagoons, cesspools, septic tanks, irrigation and spreading areas.

BIBLIOGRAPHY

SOUZA,N.C.de - 1992 - Mapeamento Geotécnico regional da Folha de Aguaí: com base na compartimentação das formas de relevo e perfis de tipos de alteração. São Carlos/SP. 2 volumes (Mestrado-EESC/USP).

Table 1 - Sequency for obtention of the attributes.

Attributes			More adequate scales
Floodzone Wet zone Landslide features Depth of the bedrock Ground water level Drenability condition Declivity Lithology or lithologic group	Basic unit of control of the fotointerpretation and field works Landform Variation of the profile of the unconsolidated materials Erosion features Textures of the unconsolidated materials	Block of rock	1:100,000
Recharge area PH/ΔPH Mineralogy Cation exchange capacity Overland flow Drainage basins	Descontinuities Welland spring localization Water balance	Rainfall Evapotranspiration Wind direction	1:50,000
Refardation factor Coeficient of permeability Ground inter flow direction Erodibility index	Colapsivity potential Compressibility conditions Salinity	Compactation conditions	1:25,000 or more large
Fundamental attributes	Secundary attributes	Complementary attributes	

Table 2 - Principals attributes and levels that must be used in the definition and delimitation of the homogeneous units.

Components	Classe Atribute	SANITARY LANDFILL			
		FAVORABLE	MODERATE	SEVERE	RESTRICTIVE
ROCK SUBSTRATE	(1) Lithology			Sandstone	Limestone
				Aquifers	Aquifers
	(2) Depth (m)	> 15	5 - 10	< 5	< 3
	(3) Descontinuities			Very fractured	Very fractured
	(4) Texture	Clayey sand	Sandyclay	Sandy	Very sand
UNCONSOLIDATED MATERIALS	(5) Variation of the profile	Heterogeneous	Heterogeneous	Homogeneous	Homogeneous
	(6) Mineralogy	Mineral type 2 x 1	Mineral type 1 x 1	Inert minerals	Inert minerals
	(7) Rock blocks	Few and small	Few and small	Very and small	Very and large
	(8) pH/ΔpH[*]	> 4/negative	> 4/negative	> 5/negative	< 4/positive
	(9) Salinity (mhos/cm)	< 16	< 16	> 16	High
	(10) C.E.C.[**] (Meq/100g)	> 15	5 - 15	< 5	< 2
	(11) Compressibility conditions	Not	Not	Occur in the surface bed	Occur in the surface bed
	(12) Collapsivity	Not occur	In surface bed (2m)	In surface bed (4m)	In surface bed (6m)
	(13) Erodibility index	Low	Low	High	Very high
	(14) Retardation factor	High	Intermediate	Low	Low
	(15) Compactation condition	Good	Good	Inadequate	Inadequate
WATER	(16) Ground water level (m)	> 10	> 6	< 4	< 2m
	(17) Ground water flow direction	1	1	2 or 3	> 3
	(18) Overland flow	Laminar	Laminar (low)	Laminar (high)	Concentrate
	(19) Permeability coeficient (cm/s)	10^{-4}	10^{-3} - 10^{-4}	> 10^{-3}	Very high (> 10^{-2})
	(20) Recharge area	Not	Not	Not	Occur
	(21) Distance of the well and spring (m)	> 500	400	> 300	< 300
	(22) Drenability condition	Good	Good	Inadequate	Inadequate
PROCESS (FEATURES)	(23) Erosion	Not	Not	Potential intensity	High potential
	(24) Landslide	Not	Not	Potential	Occur
RELIEF	(25) Declivity (%)	2 - 5	> 5 < 2	> 15	> 20
	(26) Landform	Flat slopes		Ingreme slopes	Very ingreme slopes
				Flood zones	Flood zones
	(27) Boundary between drainage basins	(> 200m) Far (200 m)	(> 200m) Far (100m)	Near	Coincidental
	(28) Wet zone	Not	Not	Not	Occur
	(29) Flood zone	Not	Not	Return time > 20 years	Return time < 20 years
CLIMATIC CHARACTERISTICS	(30) Evapo-transpiration	High	Intermediate	Low	Very low
	(31) Wind direction				Toward urban area
	(32) Rainfall (mm)			> 2000 (mm)/year	> 3000 (mm)/year

(*) ΔpH = pH H_2O - pH KCl
(**) C.E.C. - cationic exchange capacity
(———) Not very important
( ) Variable

TABLE 2 (cont.)

Components	Classe / Atribute	LAGOON			
		FAVORABLE	MODERATE	SEVERE	RESTRICTIVE
ROCK SUBSTRATE	(1) Lithology			Sandstone Aquifers	Limestone Aquifers
	(2) Depht (m)	> 10	8 - 4	< 4	< 2
	(3) Descontinuities	Few	Few	Very fractured	Very fractured and open fractures
	(4) Texture	Sand clay	Clayey sand	Sand	Sand
UNCONSOLIDATED MATERIALS	(5) Variation of the profile	Heterogeneous	Heterogeneous	Homogeneous	Homogeneous
	(6) Mineralogy	Without minerals 2:1 type	Withouth minerals 2:1 type	Inert minerals	Inert minerals
	(7) Rock blocks	Depth	Few and small	A loto in lowdepth	A lot in surface
	(8) pH/ΔpH[*]	> 4/negative	> 4/negative	> 4/negative	< 4/positive
	(9) Salinity (mhos/cm)	< 16	< 16	> 16	> 16
	(10) C.E.C.[**] (Meq/100g)	> 15	5 - 15	< 5	< 2
	(11) Compressibility conditions	Not occur	Not occur	Occur in surface bed	Occur in surface bed
	(12) Collapsivity	-----	-----	-----	
	(13) Erodibility index	Low	Low	High	Very high
	(14) Retardation factor	-----	-----	-----	-----
	(15) Compactation condition	Good	Intermediate	Inadequate	Minerals
WATER	(16) Ground water level (m)	> 7	> 4	< 4	< 2m
	(17) Ground water flow direction	1	1	2 or 3	Several
	(18) Overland flow			High	Concentrate
	(19) Permeability coeficient (cm/s)	< 10^{-4} > 10^{-5}	10^{-5} - 10^{-4}	> 10^{-3}	> 10^{-2}
	(20) Recharge area	Not	Not	Not	Occur
	(21) Distances of the well and spring (m)				< 300
	(22) Drenability condition	Good	Good	Bad	Bad
PROCESS (FEATURES)	(23) Erosion	Not	Not	Potential	Occur
	(24) Landslide	-----	-----	Potential	Occur
RELIEF	(25) Declivity (%)	< 2	< 5	< 7	> 7
	(26) Landform			Steep slopes and wet and flood zones	
	(27) Boundary between drainage basins	Far (> 200m)	Far (100m)	Near	Coincidental
	(28) Wet zone	Not	Not	Not	Occur
	(29) Flood zone	Not	Not	Occur	Occur
CLIMATIC	(30) Evapo-transpiration	High	High	High	Low
	(31) Wind direction				Toward urban area
	(32) Rainfall (mm)				Rainfall (> 1500 mm/year)

(*) ΔpH = pH H_2O - pH KCl
(**) C.E.C. - cationic exchange capacity
(———) Not very important
( ) Variable

TABLE 2 (cont.)

Components	Classe / Atribute	CESSPOLL			
		FAVORABLE	MODERATE	SEVERE	RESTRICTIVE
ROCK SUBSTRATE	(1) Lithology				
	(2) Depht (m)	> 8	< 6	< 5	< 4
	(3) Descontinuities	-----	-----	-----	-----
UNCONSOLIDATED MATERIALS	(4) Texture	Moderate	Moderate	Sandy	Sandy or clayey
	(5) Variation of the profile	Heterogeneous	Heterogeneous	Homogeneous	Homogeneous
	(6) Mineralogy	Clay minerals type 2:1	Clay minerals type 1:1	Inert minerals	Inert minerals
	(7) Rock blocks	Depth > 5	Depth < 3	Depth < 2	Occur in surface
	(8) pH/ΔpH(*)	> 4	> 4	> 5	< 4
	(9) Salinity (mhos/cm)	< 16	< 16	> 16	> 16
	(10) C.E.C.(**) (Meq/100g)	> 15	5 - 15	< 5	< 2
	(11) Compressibility conditions	-----	-----	-----	-----
	(12) Collapsivity	-----	-----	-----	-----
	(13) Erodibility index	-----	-----	-----	-----
	(14) Retardation factor	High	High	Low	Low
	(15) Compactation condition	-----	-----	-----	-----
WATER	(16) Ground water level (m)	> 8	> 4	< 3	< 2
	(17) Ground water flow direction	1	1	2	> 3
	(18) Overland flow	-----	-----	-----	Concentrate
	(19) Permeability coeficient (cm/s)	10^{-4} (near)	10^{-4} - 10^{-3}	10^{-3} - 10^{-2}	> 10^{-2}
	(20) Recharge area	Not	Not	Occur	Occur
	(21) Distance of the well and spring (m) (m)	> 50	> 30	< 30	< 30
	(22) Drenability condition	Good	Good	Inadequate	Inadequate
PROCESS (FEATURES)	(23) Erosion	-----	-----	-----	High
	(24) Landslide	Not			
RELIEF	(25) Declivity (%)	< 8	< 12	12 - 15	> 15
	(26) Landform			Wet and flood zones	
	(27) Boundary between drainage basins				
	(28) Wet zone	Not	Not	Occur	Occur
	(29) Flood zone	Not	Not	Occur	Occur
CLIMATIC CHARACTERISTICS	(30) Evapo-transpiration	High	Moderate	Low	Low
	(31) Wind direction	-----	-----	-----	-----
	(32) Rainfall (mm)	-----	-----	-----	-----

(*) ΔpH = pH H_2O - pH KCl
(**) C.E.C. - cationic exchange capacity
(-----) Not very important
( ) Variable

TABLE 2 (cont.)

Components	Classe / Atribute	SEPTIC TANK			
		FAVORABLE	MODERATE	SEVERE	RESTRICTIVE
ROCK SUBSTRATE	(1) Lithology			Sandstone	Limestone
	(2) Depht (m)	> 4	< 3	< 1	< 6.3
	(3) Descontinuities	—	—	—	—
	(4) Texture	Moderate	Moderate	Sandy or clayey	Sandy or clayey
UNCONSOLIDATED MATERIALS	(5) Variation of the profile	—	—	—	—
	(6) Mineralogy	Clay minerals			Inert minerals
	(7) Rock blocks	Depth > 3	Depth < 2	In surface	
	(8) pH/ΔpH[*]	> 4	> 4	> 5	< 4
	(9) Salinity (mhos/cm)	< 16	< 16	> 16	High
	(10) C.E.C.[**] (Meq/100g)	> 15	5 - 15	< 5	< 2
	(11) Compressibility conditions	—	—	—	—
	(12) Collapsivity	—	—	—	—
	(13) Erodibility index	—	—	—	—
	(14) Retardation factor	High	Moderate	Low	Low
	(15) Compactation Conditions	—	—	—	—
WATER	(16) Ground water level (m)	< 3	> 3	< 3	< 2
	(17) Ground water flow direction	1	1	2 or 3	> 3
	(18) Overland flow				Concentrate
	(19) Permeability coeficient (cm/s)	10^{-4} - 10^{-3}	< 10^{-3}	10^{-3} - 10^{-2}	> 10^{-2}
	(20) Recharge area	Not	Not	Occur	Occur
	(21) Distances of the well and spring (m)	> 50	> 30	< 30	< 30
	(22) Drenability condition	Good	Good	Inadequate	Inadequate
PROCESS (FEATURES)	(23) Erosion	—	—	—	Very high
	(24) Landslide	Not	Potential		
RELIEF	(25) Declivity (%)	< 8	8 - 15	> 15	> 15
	(26) Landform		Wet and flooding zones		
	(27) Boundary between drainage basins	—	—	—	—
	(28) Wet zone	Not	Not	Not	Occur
	(29) Flood zone	Not	Not	Not	Occur
CLIMATIC CHARACTERISTICS	(30) Evapo-transpiration	High	Moderate	Low	Low
	(31) Wind direction	—	—	—	—
	(32) Rainfall (mm)	—	—	—	—

(*) ΔpH = pH H_2O - pH KCl
(**) C.E.C. - cationic exchange capacity
(———) Not very important
( ) Variable

TABLE 2 (cont.)

Components	Classe / Attribute	IRRIGATION/SPREADING	
		FAVORABLE	RESTRICTIVE
ROCK SUBSTRATE	(1) Lithology		Limestone
	(2) Depth (m)	> 5	< 1
	(3) Descontinuities		
UNCONSOLIDATED MATERIALS	(4) Texture	Intermediate	Clayey sand
	(5) Variation of the profile	Heterogeneous	Heterogeneous
	(6) Mineralogy	Minerals 2:1 type	Inert minerals
	(7) Rock blocks	-----	-----
	(8) pH/ΔpH[*]	> 5/negative	< 4/positive
	(9) Salinity (Mhos/cm)	< 16	> 16
	(10) C.E.C.[**] (Meq/100g)	> 10	< 2
	(11) Compressibility conditions	-----	-----
	(12) Collapsivity	-----	-----
	(13) Erodibility index	Low	Intermediate and high
	(14) Reatardation factor	High	Low
	(15) Compactation condition	-----	-----
WATER	(16) Ground water level (m)	> 5	< 2
	(17) Ground water flux direction	-----	-----
	(18) Overland flow	Low	Concentrate
	(19) Permeability coeficient (cm/s)	10^{-3}(near)	< 10^{-2}
	(20) Recharge area	Not	Occur
	(21) Distance of the well and spring (m)	> 300	< 300
	(22) Drenability condition	Good	Inadequate
PROCESS FEATURES	(23) Erosion	Low potential	Intermediate and high potential
	(24) Landslide	Not	
RELIEF	(25) Declivity (%)	< 12	> 12
	(26) Landform	Flat slope, sever slope, wet and flood zones	
	(27) Boundary between drainage basins	-----	-----
	(28) Wet zone	Not	Occur
	(29) Flood zone	Return time < 20 years	Return time < 5 years
CLIMATIC	(30) Evapo-transpiration	High	Low
	(31) Wind direction	-----	-----
	(32) Rainfall (mm)		Rainfall (> 1500 mm/year)

(*) ΔpH = pH H_2O - pH KCl
(**) C.E.C. - cationic exchange capacity
(-----) Not very important
( ) Variable

2488

ZUQUETTE, L.V. & GANDOLFI, N. - 1990 - Engineering Geological Mapping: A Methodological Proposition. Geociências, São Paulo, 9:55-66.

ZUQUETTE, L.V. 7 GANDOLFI, N. - 1991 - Problems and rules to select the landfill waste disposal sites Brazil. Symposium International sur la Geologie Urbaine, pp. 300-309, Sfax, Tunisie.

ZUQUETTE, L.V. - 1993 - Importância do Mapeamento Geotécnico no uso e ocupação do meio físico: Fundamentos e Guia para Elaboração. 2 volumes - São Carlos/SP (Tese Livre Docência - EESC/USP).

Waste disposal on the mine tipped clayey fill – Engineering geological approach

Stockage des déchets en couches argileuses – Une approximation de la géologie de l'ingénieur

J. Herštus, J. Mühldorf, K. F. Kloss & J. F. Štastný
AGE Co. Ltd, Praha, Czech Republic

ABSTRACT: Penetration methods are shown to be suitable for assessment of properties of the tipped clayey spoil fill used as a site for the waste disposals.
Une brève aperçu sur des méthodes de pénétration, qui permettent d'apprécier les caractères des couches antropogènes argileuses, utilisées comme les dépositoires des déchêts.

1 INTRODUCTION

Brown coal has been extracted from open pit mines in northwestern Bohemia already for decades. The areal extent of mine tipped spoil exceeds 100 km² (Fig.1 - black).

Fig. 1. Tipped spoil fill areas

Tipped spoils are composed of fragments of claystones which form the overburden of the coal seams. The spoil is being deposited without compaction. The thickness of these tipped spoils usually reaches 50 m, exceptionally as much as 150 m.

Tipped spoils were exposed to the large land slides, particularly in places where they were deposited on the less bearing alluvial ground. The largest landslide involved almost 140 million m³ of ground.

Large density of the population in the Czech Republic causes major problems in selection of suitable sites for waste disposal. With respect to limited possibility of using the tipped spoil sites as built-up areas, the sites may be used for deposition of waste from coal burning power plants, municipal wastes of large towns neighbouring the tipped spoils.

Prior to designing a waste disposal site, it is necessary to assess the stability of the dump's body, hydrogeology and to estimate the body's settlement. At the same time, it is necessary to consider peculiarities of the filling ground behaviour.

2 BASIC DATA ON PROPERTIES OF THE SPOIL GROUND

The claystones which consitute the roof of the brown coal seams show W_L values ranging between 70% and 100% and those of IP between 40% and 50% (Herštus 1990). The claystones are often overconsolidated, slightly consolidated and show numerous fractures. Natural moisture is close to W_p. When stripping overburden, the originated spoil is composed of these clay-stones. A typical grain-size curve of such spoil is shown in Fig.2.

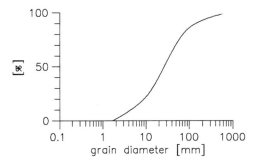

Fig. 2. Tipped spoil granulometry

Following the spoil deposition, there are continuous gaps between rock fragments, and the spoil behaves as rockfill of weak rock. The gaps between fragments become smaller with increasing load, and the structure of the spoil changes from granular soil to cohesive soil. A strong compression of deposited materials takes place during this process. Fig.3 shows values of settlement of tipped spoils vs. their height.

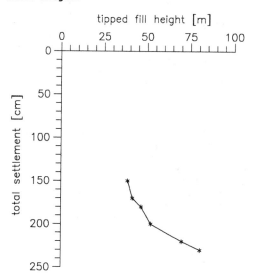

Fig. 3. Settlement curve of tipped claystone spoil fill

Change of shear strenght also occurs in the course of these structural alterations (cf.Fig.4). Provided the spoil shows a granular structure, then it exhibits a relatively large angle of internal fric-

tion. Along with the extinction of continuous gaps, an induced pore pressure originates resulting in a difference between effective and total envelope of the shear strenght. (Feda et al. 1994).

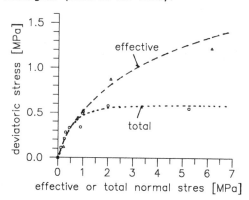

Fig. 4. Strength envelopes (total and effective) of fragmentary clayey material

Infiltration of atmospheric water or groundwater through the spoil while it shows continuous gaps, result in cosiderable changes in the character and behaviour of the spoil. Due to overconsolidation, the claystone fragments strongly absorb water. The surface of the fragments softens in relatively short period of time (Figs. 5 and 6).

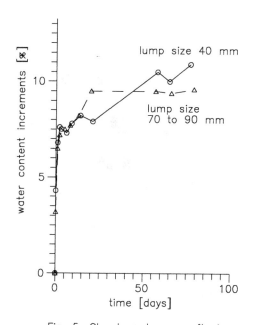

Fig. 5. Claystone lumps softening

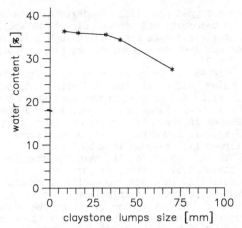

Fig. 6. Relationship between natural water content and claystone lumps size

The spoil is altered fast into a stiff and soft clay in which "swimm" hard remnants of original fragments. Effective values of shear strenght of such altered spoil correspond well to those obtained from shear test of normally consolidated clay prepared by disintegration of fragments (clayslurry) of a claystone (Fig.7).

In case of more lithificated clay, preserved layers of granular character can be enclosed in layers altered into clay. This may lead to the origin of confined groundwater horizons inside the tipped spoil.

Superficial less loaded layers of the spoil maintain continuos gaps and consequently show high permeability. The thickness of these layers depends on the degree of lithification of claystone fragments.

3 METHODS OF EXPLORATION OF WASTE DISPOSAL SITES LOCATED ON THE TIPPED SPOILS

3.1 Data for the stability assessment of the tipped spoil body

The behavior of claystone spoil and its regularities described in parag 2 were verified by the laboratory investigation involving laboratory tests of artificially prepared spoil in triaxial device on the specimens having diameter 10 to 38 cm. Triaxial and direct shear box tests were done on the undisturbed specimens taken in boreholes. Since the undisturbed specimens contain random remnants of the claystone fragments, the determination of representative values of shear strenght needs a great number of tests. This procedure, however, is not useful for routine site investigation.

Consequently, application of a static penetration test with a measurement of pore pressure was verified to be applied for routine tests. The comparison of results of static penetration with those obtained by the laboratory tests proved the employment of the former method to be reliable. The resistance maxima measured at the penetration cone were found to be represen-

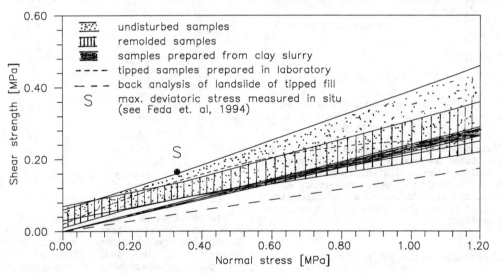

Fig. 7. Review of efective shear strength test results on tipped clayey spoil fill

ting undrained strenght of the claystone fragment remnants and the minima represent the strenght of the clay matrix.

A possibility of measurement of pore pressure values using a transducer located behind the static penetration cone was tested at the same time. Detailed verification using the finite element method showed negligible deformation of measured values through the cone disturbance of the ground. A comparison of the pore pressure values obtained in this way with those obtained by permanently fitted piezometer proved this method to be correct (Fig.8), at least for the relatively more permeable layers.

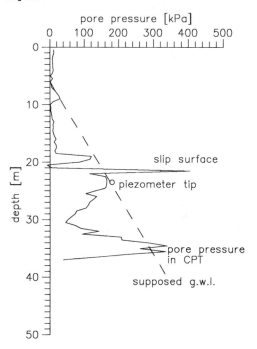

Fig. 8. Pore pressure changes with depth bellow tipped fill surface
- example 1

This procedure also showes that, at the same time, horizons having properties of an aquifer can be distinguished from the layers exhibiting properties of an aquiclude, and also indicates a contingent unconfined groundwater level in a superficial position.

3.2 Data for settlement prediction

Relatively accurate prognosis of a settlement due to the weight of deposited waste is necessary when locating waste disposal

on a tipped spoilfill. Fundamental data on the deformational properties of the spoil materials can be obtained through analysis of surveyed settlement of the tipped spoil (Fig.3).

Tipped spoils up to 50 m thick show the medium value of the oedometric modulus E_{ocd}: 19 MPa; higher tipped spoilfills exhibit many times greater modulus for the deeper layers. From point of view of the disposal foundation, the upper layers of tipped fill play an important role.

These layers still maintain granular, intersticial structure of the tipped spoilfill. In order to verify deformational properties of this layer, an experiment in situ was done in which the state of stress and deformability in a block of 3 x 3 x 3 m were measured (Feda et al. 1994). The deformational modulus of this layer was found to be usually ranging between 5 and 14 MPa.

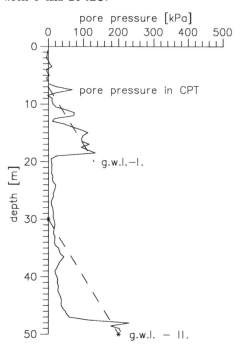

Fig. 9. Pore pressure changes with depth bellow tipped fill surface
- example 2

Data obtained in this way were verified using the loading tests in a borehole. Established deformational moduli fully correspond to the results obtained from a large scale experiment. A comparison of measured moduli with cone resistance measured during the static penetration test showed ratio E/QST in range of 2 to 6.

3.3 Assessment of permeability

In order to assess the necessarily degree of safety of the waste disposal site located on a tipped spoilfill, it is essential to know the permeability of the fill. The measurement of permeability in a borehole was strongly deformed by the presence of a remolded layer which originated along the borehole walls. Therefore, in order to perform reliable measurements of permeability, we applied filters which were driven in a desirable depth.

Permeability of the tipped clayey spoilfill was found to be changing considerably with the depth. The granular character of the spoil with continuous gaps was found to be prevailing down to the depth of 15 m. The permeability of this layer is usually higher than E-5 m/s. Lower layers, in which continuous gaps are missing, show permeability usually lower than E-8 m/s.

4 PRINCIPLES OF A DESIGNING WASTE DISPOSAL SITES LOCATED ON THE TIPPED SPOIL

First of all, it is necessary to assess the stability of the tipped spoil body at the load produced by deposited waste. The stability calculation must take into consideration the verified confined aquifers and really existing conditions of dissipation of pore pressures.

The waste disposal on a tipped spoilfill should be oriented in a direction allowing drains to have an inclination following the greater calculated settlement.

It is useful to remove few meters of the superficial layer which is strongly compressible and very permeable. Consequently, the subgrade settlement as well as the permeability will be considerably reduced whereas the capacity of the waste disposal site will increase.

5 CONCLUSION

Location of the waste disposal sites on the tipped clayes spoilfills of open pit mines deserves an assessment of stability, settlement and permeability. The application of the static penetration with the measurement of a pore pressure and the measurement of a permeability using penetration infiltration needles proved to be useful when seeking the entry data for a disposal design.

REFERENCES

Feda,J., Herštus,J., Herle,I., Šťastný, J. Landfills of the waste clayey material. 1994 XIII ICSMFE New Delhi, p.1623-1628

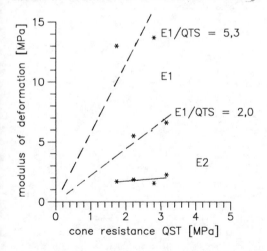

Fig. 10. Relationship between initial modulus of deformation E1, residual modulus E2 on the cone penetration resistance in statik cone penetration test

Restoration of derelict and contaminated land: Some case histories

Restauration des terrains abandonnés et contaminés: Quelques exemples

F.G. Bell
Department of Geology and Applied Geology, University of Natal, Durban, South Africa

A.W. Bell
White Young Consulting Group, Leeds, UK

ABSTRACT: The term 'derelict' is frequently used to describe land which has been spoiled by the extraction of minerals or by industrial operations. Generally the area is left in an unsightly state and is of little use without undergoing some rehabilitation. Not only is derelict land a wasted resource but it also has a blighting effect on the surrounding area and can deter new development. Hence its restoration is highly desirable and a significant contribution to the economy can be made by bringing derelict land back into worthwhile use. The use to which derelict land is put should suit the needs of the surrounding area and be compatible with other forms of land use. Restoring a site to a condition which is well integrated into its surroundings upgrades the character of the environment beyond the confines of the site.

Spoil heaps represent the most notable form of dereliction associated with subsurface coal mining. Two old colliery spoil heaps at Wharncliffe-Woodmoor near Barnsley, South Yorkshire, were restored. Part of the site was developed for a bakery and the rest of the site became a golf course. A site investigation preceded the restoration work. The latter is essentially a large scale exercise in earthmoving to achieve an acceptable contoured landform within the site which relates to the topography of the adjoining land. Unfortunately the work encountered problems with hot spots. These were normally around 600°C but at some locations temperatures were as high as 900°C. They were dealt with in a number of ways. After levelling, the installation of drainage and harrowing the waste, the spoil was then treated with lime and covered with soil. Fertilizers were applied prior to seeding and planting.

A sand and gravel pit adjacent to a canal and to the immediate west of Brighouse, West Yorkshire, was converted to a marina. The site lies in the bottom of the Calder valley and has a high water table. In fact, the sand and gravel workings were flooded and most extraction took place below water. Two berms separated the workings into three more or less equal areas. The flooded workings were converted into a lake by removal of the berms, the material in the berms being used to form islands in the lake. The western side of the site is significantly higher than the eastern side and so regrading had to be carried out by means of cut and fill. At the same time the slopes around the lake were regraded to fall more gently to the lakeside. Access was provided from the lake to the adjacent canal. The site was landscaped and planted, and various facilities provided.

An old foundry site in Leeds, West Yorkshire, was to be rehabilitated for light industrial development. The desk study indicated that various problems could be associated with the site, notably subsurface voids and both ground and groundwater contamination. The site exploration confirmed the present of old cellars and foundations, and revealed some two to four metres of made ground. This overlay sand and gravel which rested on Coal Measures strata. The water table tended to occur at 3.5 to 4.0 m beneath the surface. Samples of made ground and groundwater were analysed to determine whether or not they were contaminated. The concentration of sulphate was very high in some rubble fill. Surprisingly for an old industrial area, the amounts of lead, cadmium, zinc, arsenic, copper, nickel and zinc were below levels likely to present problems. Elevated concentrations of magnesium occurred at certain locations and this together with small fragments of asbestos sheeting could represent a dust problem on reworking. The rehabilitation involved bulk excavation and re-use of the rubble fill, the bearing capacity of which was enhanced by vibrocompaction.

RESUME: Le terme "abandonné" est souvent employé pour décrire un terrain endommagé par des extractions de minéraux ou par des operations industrielles. En genéral, le terrain est dans état lamentable et est inusable sans avoir été restauré. Non seulement une perte économique, le terrain abandonné a aussi un effet peu esthétique sur l'environnement, ce qui peu décourager de futures dévelopments. C'est pourquoi que la restauration est trés désirable et est une importante contribution pour l'économie. Cette dernière est possible par la transformation d'un terrain abandonné en un terrain utilisable. L'emploi de ce dernier devrait être compatible aux besoins de la région et aussi compatible avec les differentes formes d'utilisation de terrains.

La restauration d'un site qui s'adapte bien a l'environnement amèliore l'aspect de la region au-delà des limites de cette région. L'amencollement de débris représent la forme la plus comune d'abandon associé aux mines sous- terraines de charbon. Deux vieux amencollements debris carbonifères à Wharncliffe-Woodmoor prés de Barnsley, South Yorkshire, ont été restauré. Une partie du terrain fut developpée pour la contruction d'une boulangerie et le reste en un terrain de golf. La restauration fut precede par une étude du terrain. Les travaux de restaurataion consistèrent pricipalement de deplacements de terre sur une grande echelle defacon a produire une formation de terraian conformément a la topographie de la région. Malheureusement des problèmes se sont presenté durant ces travaux, dûs aux "pointschad". Ceux-ci étoient généralement de l'ordre de 600°C, mais a certains "points" etait aussi elevée que 900°C. Ces problèmes furent résolus au moyen de differentes méthodes. Après le miuellement des installations de drainages et hersage des débris, ceux-ci furent traité a la cheux et recouvert de terre. Des fertilisants furent employé avant la semence et le plantage.

Une fosse de sable et graviers adjacent á un canal a l'ouest de Brighouse, West Yorkshire, fut convertie en une "marina". L'endroit se trauve au fond du Calder Valley. Cette vallée possède un haut niveau d'eau sous-terraine. En fait l'étendue de sable et gravier etait innondée et la plupart des travaux d'extraction forent fait sous l'eau. Deux bermes diviserent le système en trois etendues plus ou mains égales. La partie innondée fut convertie en un lac par l'enlèvement des bermes. Le materiel des bermes fut emplouyée pour former de îles dans le lac. Le terrain de la partie occidentale etant plus élevé que le terrain de la partie orientale du site, on a dû modifier le gradient des emb pentes par remlissages et talutages, conjoinlement les rampes autour du lac furent modifées de façon a descendre moins rapidement vers les bords du lac. Un passage fut crée entre le lac et le canal avoisinant. Le site fut gardiné et planté de même que plusieurs facilités se sont établis.

L'emplacement d'une ancienne fonderie etait considéré pour être réhabilité afin d'être developé pour l'éstablissement d'industries legères. L'étude du projet au bureau indiqua que le terrain pourrait présenter divers problèmes notamment des cavité sous-terrains et contamination de l'eau de sous-sol. L'exploration du terrain confirme la presenee de vieux celliers et fondations et revela la presence de deux a quatre metres de terre travaillé. Sous cette couverture de sable et de gravier, qui elle-même est placée sûre une couche carbonifère setrouve l'eau sous-terrain dont le niveau est a 3,5 jusqúa 4 metres au-dessous de la surface. Les échantillions de terre et d'eau provenant du sous-sol furent analysés pour détérminer la possibilité de contamination. La concentration de sulphates etail trés forte dans certains amas de débris. Tres surprenant pour une anciene location industriel, les quantités de plomb, cadmium, zinc, arsenic, cuivre, nickel furent au-dessous d'un niveau considéré dangereux. De fortes concentrations de magnésium furent detectés a certains endroits et ceci conjointment avec la découverte de petits fragments de feuille d'asbeste pourait causer un problème de poussieres pendant le travail de réhabilitation. La réhabilitation consistait en excavations massifs et le re-emploi des débris pour le remplissage. La capacité de support fut augmenté par vibrocompaction.

1 INTRODUCTION

The term 'derelict' is commonly used to describe land which has been spoiled by the extraction of minerals, by industrial operations or as a result of general neglect and so is incapable of beneficial use without treatment. Usually the area is left in an unsightly state. Ironically much of the landscape of Britain had been deliberated and skilfully improved by professional effort prior to the Industrial Revolution while the latter laid waste large areas of land and the creation of derelict land has continued from then until the present day. Fortunately attitudes towards derelict land have changed and legislation has been enacted in the form of planning acts to limit its development and to facilitate its restoration. Furthermore the development of large earthmoving equipment has meant derelict features such as spoil heaps can be readily regraded. In the United Kingdom some of the worst dereliction is associated with old mineral workings. These may take the form of old open pits and quarries, and of waste dumps and tailings ponds.

Not only is derelict land a wasted resource but it also has a blighting effect on the surrounding area and can deter new development. Hence its restoration is highly desirable and a significant contribution can be made to the economy by bringing derelict land back into worthwhile use. One of the most important factors influencing the cost of restoration is frequently the volume of waste which has to be removed and the haulage distances involved. Hence the cheapest schemes are usually those in which the quantities of 'cut and fill' are approximately equal so that the movement of materials is restricted to the site and thereby kept to a minimum. Restoration of the original ground level is not essential in such schemes. Depending on ease of access, clay pits and quarries may be filled with domestic refuse and then landscaped. On the other hand abandoned sand and gravel pits, especially if flooded, may be unsuitable for such disposal as this could lead to pollution of water supply. The use to which derelict land is put should suit the needs of the surrounding area and be compatible with other forms of land use. Restoring a site to a condition which is well integrated into its surroundings upgrades the character of the environment beyond the confines of the site.

As recreation plays a major role in the life of a community, derelict areas have frequently been reclaimed for such a purpose. Parklands, playing fields and other sporting facilities, marinas and such like often have arisen from areas of former dereliction. Industrial estates are also often located on restored land.

The restoration of a derelict area requires a preliminary reconnaissance of the site, followed by a site investigation. Aerial photographs may prove of value, especially for large sites which contain spoil heaps, tailing lagoons and associated mine buildings; or for large open pit workings. The investigation provides essential input for the design of remedial measures. Derelict land may also present hazards, disposal of industrial wastes may have contaminated land, in some cases so badly that earth has to be removed; spoil heaps associated with coal mining may be burning; abandoned mine workings may have old shafts which are unfilled and shallow subsurface workings may be unrecorded. Details relating to such hazards should be determined during the site investigation.

After a derelict site has been levelled, regraded or filled, the actual surface still needs restoring. This is not so important if the area is to be built over (e.g. if it is to be used for an industrial estate), as it is if it is to be used for amenity or recreational purposes. In the case where buildings are to be erected, however, the ground must be adequately compacted so that they are not subjected to adverse settlement. In the case where the land is to be used for amenity or recreational purposes, then soil fertility must be restored so that the land can be grassed and trees planted. This involves the application of fertilizers, laying of top-soil (where available) or substitute materials and seeding (frequently by hydraulic methods). Adequate subsoil drainage will also need to be installed.

2 CASE HISTORY 1

Some of the worst dereliction in the United Kingdom has been associated with mining activities, especially of coal since this has been the most extensively worked mineral. Spoil heaps represent the most notable form of dereliction associated with subsurface coal mining. They are particularly conspicuous and can be difficult to rehabilitate into the landscape. Furthermore there has been large scale closure of collieries in recent years. Such dereliction contributes towards urban blight in that it tends to influence population migration and deters the establishment of new industry. In order to counteract such trends, affected areas need to be restored to sufficiently high standards to create a pleasant environment.

Spoil heaps formed of coarse colliery discard frequently are high and of conical shape in South Yorkshire. Their configuration depends upon the type of equipment used in their construction and the sequence of tipping the waste. The shape, aspect and height of a spoil heap affects the intensity of exposure, the amount of surface erosion which occurs, the moisture content in its surface layers and its stability. The mineralogical composition of coarse discard from different mines obviously varies but pyrite frequently occurs in the shales and coaly material is present in spoil heaps. When pyrite weathers it gives rise to the formation of sulphuric acid, along with ferrous and ferric sulphates and ferric hydroxide which gives rise to acidic conditions in the weathered material. Such conditions do not promote the growth of vegetation. To support vegetation a spoil heap should have a stable surface in which roots can become established, must be non-toxic, and contain an adequate and available supply of nutrients. The uppermost slopes of a spoil heap frequently are devoid of near-surface moisture. Hence spoil heaps are generally barren of vegetation and represent ugly blemishes on the landscape. Their restoration involves special treatment.

Two old colliery spoils heaps at Wharncliffe-Woodmoor near Barnsley, South Yorkshire, are provided as examples of reclamation to improve the local amenity and to provide a site for light industrial development. The site consisted of two distinct areas, namely, a western area comprising 19 ha of agricultural land, and the area containing the spoil heaps occurred on the eastern part. The latter covered 29.5 ha and contained two spoil heaps, one 45 m in height, the other 25 m high, and a large area occupied by old tailings ponds, as well as pithead buildings and stock grounds (Fig. 1). The smaller heap and pithead buildings were separated from the main heap and tailings ponds by a canal which carried compensation water from a local reservoir to the river Dearne. As a consequence the canal had to be diverted prior to reclamation work commencing.

The larger spoil heap had an estimated volume of 3 000 000 m³ and consisted mainly of shale, some of which had been burnt. The smaller spoil heap to the north of the canal was composed of similar material and its volume was less than 1 000 000 m³. Spoil had been spread over other areas and had been tipped over some of the tailings ponds. However, there was no evidence of burning in the latter spoil.

The dominant mineral in the unburnt shaley material was quartz. Illite was the most important clay mineral, with kaolinite usually averaging less than 5 or 10%. Mica and chlorite tended to occur in similar amounts to that of kaolinite. Feldspars, sulphates, pyrite and carbonate occurred in trace amounts.

The coarse discard of which spoil heaps are made, contains other material in addition to shale. As run-of-mine material it reflects the various rock types which are extracted during mining operations. Hence it contains various amounts of coal which have not been separated during the preparation process and in old tips there may be appreciable proportions of coal. Those discards with relatively high coal contents frequently are burnt in part or are still burning. This is a result of spontaneous combustion of carbonaceous material, frequently aggravated by the oxidation of pyrite. It can be regarded as an atmospheric oxidation (exothermic) process in which self-heating occurs. The moisture content and grading of spoil are important in this respect. An increase in free moisture at relatively low temperatures increases the rate of spontaneous heating. In material of large size air can cause heat to be dissipated whilst in fine material the air remains trapped which means that burning ceases when the supply of oxygen is consumed. Consequently ideal conditions for spontaneous combustion exist when the grading is intermediate between these two extremes. Hot spots may develop under such conditions. What is more the rate of oxidation generally increases as the specific surface of the particles increases.

The moisture content of the spoil material concerned tended to increase with increasing content of fines. It was also influenced by the permeability of the material, the topography and the climatic conditions. Generally it fell within the range 6% to 13%. The bulk density of the unburnt spoil showed a wide variation, with most of the material having values between 1.5 and 2.5 Mg/m³. Low densities were mainly a function of low specific gravities. These obviously were influenced by the relative proportions of mudrock, sandstone and coal in the waste. In particular, the higher the coal content, the lower was

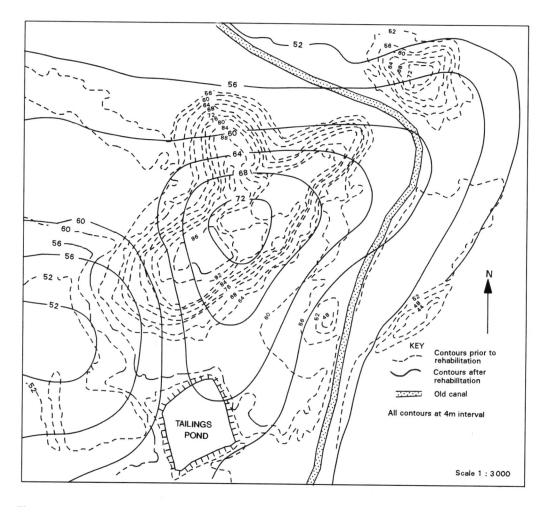

Fig. 1. Lay-out of spoil heaps showing contours prior to and after rehabilitation.

the specific gravity. As far as the particle size distribution of the coarse discard was concerned, it fell mostly within the sand range although significant proportions of gravel and cobble size were present. At placement, coarse discard consists mainly of gravel and cobble size but subsequent weathering reduces particle size. However, once buried within a tip, coarse discard undergoes little further reduction in size.

Although the pyrite content of a spoil heap is in trace amount, nonetheless its presence can have a significant effect. As mentioned above, pyrite breaks down rapidly under the influence of weathering to produce, amongst other things, sulphuric acid. Oxidation of pyrite within tip waste depends on access of air which, in turn, is influenced by particle size distribution, moisture content and degree of compaction. However, although oxidation products may be formed, they may be neutralized by alkaline materials in the waste material. If this is not the case and waste material is acidic, then it will probably be deficient in plant nutrients and so would require careful

treatment if vegetation is to be established, notably dozing with lime. In fact there was little vegetation cover over the tipped areas, that which did occur being found at the base of spoil heaps. However, the results of pH tests on the waste material gave values around 6.5 to 6.8. Hence the material was only very slightly acid.

The site investigation included the production of topographic plans of the site from aerial photographs and sinking boreholes in the old tailings heaps and spoil heaps. The tailings ponds appeared to be quite stable, so that their level could be raised by covering with waste material during the reclamation process.

Reclamation of a colliery spoil heap is essentially a large scale exercise in earthmoving and the coarse discard is frequently spread over a larger area than that occupied by the heap (Fig. 1). The first part of the reclamation work involved stripping over 1 000 000 m³ in order to reduce the height of the spoil heaps. The purpose of the levelling was to achieve a pleasant contoured landform within the site, related to the topography of the adjoining land. This was brought about not only by reducing the height of the spoil heaps, but also by regrading and rounding off the steeper slopes. The other areas such as those occupied by the pithead buildings, as well as the tailings ponds, were buried by spoil. These areas were also sloped. In this way a gently sloping ridge, valley and hill topography were created on the site.

This work unfortunately encountered problems with hot spots. These were normally around 600°C but at some locations temperatures were as high as 900°C. The heat often caused engines on the scrapers to cease and rescue of both operator and machine often proved hazardous. In addition the tyres on scrapers frequently melted. Where levelling was still in progress and hot spots were present, it was decided to only remove a layer of 300 mm in thickness and then compact the waste with a vibratory roller (Fig. 2). This reduced the temperature by about 50% after two hours. Levelling could then commence again. When hot spots were found in areas where the tip material had been lowered to its finished level, 450 mm of clay was spread and compacted on top of the hot spot.

(a)

(b)

Fig. 2. (a) Levelling of spoil over a "hot spot".
(b) Compacting hot material with a vibratory roller.

Then a layer of shale was spread and compacted over the clay. Boreholes were sunk into these hot spots to determine whether or not they were cooling. Those spots where the temperatures had not dropped within a year were enclosed by an injected curtain wall of pulverized fuel ash which extended down to original ground level. Spontaneous combustion may give rise to subsurface cavities in the spoil, the roofs of which may be incapable of supporting a person. Fortunately no such cavities were met with on this site.

Drains, in the form of open ditches containing pipes and filled with gravel, were constructed after levelling was completed and led into existing streams or culverts (Fig. 3). An open ditch was also placed around the site to catch any surface run-off.

The next operation involved harrowing the waste, after which lime was applied at 7.5 tonnes per hectare. Once liming was complete, subsoil and then topsoil were applied. This was obtained from the agricultural land on the western part of the site. Fertilizers were spread over the ground prior to seeding.

Part of the site was developed for a large bakery and most of the rest was used as a golf course. The latter was mainly grassed, although a large number of trees were planted on the course. Trees were also planted to enhance the character of the landscape, to reduce erosion on steeper slopes and to screen parts of the site. In all, some 20 000 trees were planted over a three year period. The area now represents a source of employment and a popular local amenity rather than an unpleasant eyesore.

3 CASE HISTORY 2

The extraction of sand and gravel deposits in low lying areas along the flanks of rivers frequently means that the workings eventually extend beneath the water table. On restoration it is not necessary to fill the flooded pits completely. Partial filling and landscaping can convert such sites into recreational areas offering such facilities as

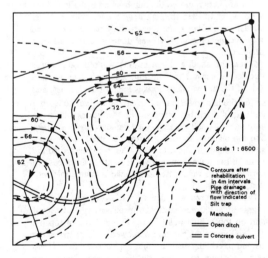

Fig. 3. Drainage for rehabilitated spoil heap.

sailing, fishing and other water sports. It is necessary to carry out a thorough survey of flooded workings, with soundings being taken so that accurate plans and sections can be prepared. The resultant report then forms the basis of the design of the measures involved in rehabilitation.

Old sand and gravel workings to the west of Brighouse in West Yorkshire were rehabilitated. The site was approximately 16.3 ha in extent. It was about 1.2 km long and 0.4 km wide, at its widest point. The site had been an active sand and gravel pit located in the valley of the river Calder, containing a main road, a canal, and railway line. The canal was closed for navigation but was still in a reasonable condition. Approximately 200 000 tonnes of sand and gravel were extracted annually from the workings. Work commenced at the eastern end of the site, where there was between 150 and 230 mm of topsoil, and 2 to 2.5 m of subsoil to remove, before the sand and gravel could be excavated. As the site has extended westwards, the subsoil increased up to 6 m in thickness, which meant that excavation became deeper. Normally if there is more than 1 m³ of overburden to remove in order to win 1 m³ of sand and gravel it is left alone, as it is unprofitable. However, as the site was in an area where much development was taking place there was a ready sale of the material for aggregate locally.

The overburden on the site was removed by a scraper, towed behind a dozer. Ground conditions were too wet and heavy for self-propelled scrapers to operate. A dragline was used to excavate the sand and gravel, loading the material directly into dump trucks. As the site lies in the bottom of a valley, there is a high water table and much of the excavation was below water. To reduce on the amount of pumping needed two berms were constructed across the workings from north to south, splitting the site into three approximately equal parts (Fig. 4). It was only necessary to keep the water level down in the active pit.

The silt-clay fraction which was removed from the sand and gravel by washing was piped into a settling pond. A series of filters removed the fine fraction and allowed the wash water to pass through a pipe back into the river Calder, in a clean state. Occasionally the settling pond required deepening. This was done with the aid of a dragline.

When the sand and gravel was worked out, it was decided to convert the site into a marina. The western side of the site was substantially higher than the east and so the site had to be graded, excavation being done on a cut and fill basis by scrapers. The two berms which ran across the flooded workings were removed to form a lake and the material from the berms was used to construct several islands in the lake. The lake is connected to the canal in the valley bottom by a lock at its south eastern corner and the water level in the lake is compatible with that of the canal.

The whole area was landscaped. After the site was regraded, a 150 mm layer of subsoil was spread over it and on top of this was laid a 150 mm layer of topsoil. The soil was ploughed and harrowed at least twice, the second pass being made at right angles to the first. Grass seed was then sown over the whole area. Trees were planted alongside the main road adjoining the site and

Fig. 4. Berms in sand and gravel workings.

around the lake, and flower beds and small shrubs set out along the sides of the site roads and around the marina. Grass was sown and trees planted on the artificial islands in the lake. The lake and islands provide a sanctuary for wild life, and the lake was stocked with fish.

A road system was constructed to cross the development area. Access to the marina was provided at the north east and north west corners of the site from the major road in the area.

About half way along the northern bank of the site, adjacent to the main road, were some disused farm buildings. As they were in a good state of repair, they were incorporated into the scheme by converting them into a boathouse and restaurant. These buildings command an excellent view over the marina. A cafe and snackbar were built alongside. An access road was laid from the boathouse to the edge of the lake, and a launching area and jetty were provided at the lakeside.

A large area on the northern bank was paved and grassed, to provide a caravan park. All the usual amenities, such as a site shop, washing and toilet facilities, a launderette and refuse disposals areas were made available. Services, such as water supply, electricity and sewage disposal were installed.

In the south west corner of the site an outdoor swimming pool was constructed, and alongside it there is a paddling pool. Children are catered for by the provision of an adventure playground.

4 CASE HISTORY 3

From the early part of this century light industrial foundry work has taken place in the southern part of the city of Leeds. Over the years various plots of land were developed to suit the needs of the factories existing at the time. With the advancement of casting and machining technologies, as well as economic changes generally, many of these buildings became redundant and some were subsequently demolished, often only to ground level. An additional problem associated with many of these sites is that of ground contamination. The latter need not only present a hazard to the construction of foundations but it can also mean hazardous conditions for operatives to work in. Hence it is necessary to establish whether or not precautionary measures are required when such sites are redeveloped.

One such old foundry site required rehabilitation in order to make it available for light industrial development. Accordingly a desk study was undertaken in order to establish as accurately as possible the previous usage of the site, as well as to gain some data regarding the site conditions. It revealed a plan of the last layout of buildings on the site, indicating areas where manufacturing took place. These areas represented areas of potential underground hazards in that man-made obstructions, and both ground and groundwater contamination, may be present.

The site exploration programme consisted of a number of boreholes, sunk by light cable and tool rig, to a maximum depth of 12.5 m below ground level. These were supplemented by the excavation of trenches and pits (Fig. 5). The subsurface exploration confirmed the presence of a number of underground obstructions (Fig. 6) and in fact some of the boreholes had to be sunk from the bases of pits dug through rubble and old foundations. The exploration revealed some 2 to 4 m of made-ground overlying a layer of sand with gravel, the maximum thickness of which was 6 m. The made-ground consisted of ashes, cinders, slag, foundry sand and brick rubble. Depending on the position within the site either mudstone or siltstone, belonging to the Coal Measures, occurred beneath the sand with gravel. Both rock types were highly weathered. Standard penetration tests were carried out within the sand with gravel, indicating N-values increasing from as low as 8 (loosely packed) to 32 (densely packed). The degree of compaction tended to increase with depth and most of the sand with gravel was

Fig. 6. Old foundations and cellars revealed by site investigation.

medium dense (angle of friction 35° to 40°). Groundwater levels were recorded in boreholes at the beginning and end of each working day. The water table tended to occur at 3.5 or 4.0 m below the surface. Samples of made-ground and groundwater were taken for chemical testing to determine the degree of contamination.

The pH values obtained form testing the made-ground ranged from 4.2 to 9.1, with most materials being alkaline. Total sulphate concentrations were very high in some samples of rubble fill and would warrant the use of sulphate resistant cement in foundations. More sophisticated chemical testing was undertaken to determine the concentrations of lead, cadmium, arsenic, copper, nickel, zinc and chloride across the site since their levels could be a cause of concern in relation to industrial redevelopment or landscaping (Table 1). In general, the concentrations of these materials within made-ground were below levels likely to represent problems. For an old industrial area their occurrence in most cases was surprisingly only in trace amounts. However, elevated concentrations of magnesium were found in certain locations on site (e.g. over 0.9% at two locations). These were due to the fact that part of the site was used for a munitions factory during the Second World War. Reworking the site dust from such material could cause irritation to the eyes. Accordingly it was recommended that eye protection should be worn and that good eye wash facilities be available on site. Concentrations of phenol high enough to permeate plastic water service pipes were present in some foundry sands. Any piping involved in subsequent redevelopment which would come in contact with such deposits therefore would have to be either phenol resistant (e.g. made of uPVC) or laid in trenches backfilled with clean sand. Minor oily

Fig. 5. Excavation of a trench for the site investigation of derelict land.

Table 1. Amount of contaminants present in and pH values of samples.

Sample Description	Total lead	Total cadmium	Total arsenic	Total magnesium	Total copper	Total nickel	Total zinc	Total cyanide	Cl	Total sulphate as SO₃	Phenol	Toluene extract	pH
	mg/kg	mg/kg	mg/kg	mg/kg	mg/kg	mg/kg	mg/kg	mg/kg	mg/kg	mg/kg	mg/kg	(%)	
1. Soils													
Grey soily slag fill	45	0.6	7.2	1481	78	45	27		18	270	<1.0	0.224	8.49
Black ashy cindery fill	15	0.3	4.7	1014	83	46	14		28	1541	<1.0	0.096	6.18
Black sooty fill	3	0.4	3.7	1441	7	9	26		<1	577	<1.0	0.104	8.39
Red clayey silt	90	0.4	8.1	2148	33	36	110		26	521	<1.0	0.312	7.22
Trench spoil, grey rubbly gravelly fill	83	0.2	10.3	1901	72	44	91		<1	19507	<1.0	0.152	6.44
Trench spoil, black gravelly sooty fill	33	0.6	10.5	1788	53	27	51		<1	11856	<1.0	0.048	4.22
Brown sandy silty fill	127	0.3	12.1	9810	111	31	120		<1	1098	<1.0	0.248	8.16
Trench spoil, black oily silty fill	16	0.4	0.4	1527	14	13	79		29	448	<1.0	0.688	7.78
Trench spoil, black ashy cinder fill	198	0.2	7.6	1448	68	37	79		18	443	<1.0	0.288	7.89
Blue-grey silty fill	283	0.7	13.6	310	34	120	74		<1	1276	<1.0	0.104	7.47
Red-brown slag fill	77	0.5	11.3	1148	94	39	34		16	361	<1.0	0.016	6.59
Black sandy sooty fill	5	0.2	3.1	1509	6	5	23		26	8738	<1.0	0.016	7.55
Black oily seepage									46	294	<1.0	36%	6.92

Table 1 continued

Buff sand	4	0.4	2.3	1424	3	3	11		28	233	<1.0	0.048	8.42
Trench spoil, black ashy sooty fill	64	0.8	2.9	6215	87	24	478		<1	933	<1.0	0.160	9.07
Trench spoil, black ashy sooty fill	11	0.3	2.4	1516	17	10	79		21	375	<1.0	0.012	8.16
Trench spoil, dark brown sandy fill	13	0.5	1.3	576	57	22	65		21	310	11.7	0.224	7.59
Trench spoil, black sooty fill	16	0.3	0.7	1524	38	13	95		21	557	<1.0	0.200	8.16
Buff sand	3	0.4	0.3	617	9	5	44		18	688	<1.0	0.064	7.93
Black ashy cinder fill	36	0.3	4.5	9390	63	41	70		<1	753	<1.0	0.224	8.82
Trench spoil, crushed, red brick/fire brick	144	0.5	11.3	1670	81	36	102		19	832	<1.0	0.152	7.89
White, blue-green limey material	41	2.4	3.2	2330	3380	163	823	<0.1	2420	343	<1.0	0.720	7.53
2. Water	mg/l	mg/l	mg/l	mg/l	mg/l	mg/l	mg/l	mg/l	mg/l	mg/l	mg/l	%	
Groundwater	<0.01	<0.01	<0.01	30	0.01	0.02	0.01		34	80	<0.1		7.44
Water in brick chamber	<0.01	<0.01	<0.01	29	0.01	0.02	0.01		18	835	<0.01		9.30
Water from brick chamber	<0.01	<0.01	<0.01	333	0.01	0.02	0.06		14	570	<0.01		7.77

contamination was exposed at two locations in trenches but the volumes were not large enough to be significant. Small fragments of asbestos (chrysotile) sheeting were scattered across the site and within the fill. Although this type of material is low risk and does not generally release dangerous dust, it would be better to remove it rather than crush and recompact it with fill during redevelopment.

A scheme for rehabilitation of the site was then developed which involved bulk excavation of the material on site. The depth of excavation across the site would have to be varied in relation to the location of known obstructions. Specifications for the handling and selection of the excavated ground, and its re-use as compacted backfill, incorporating the above recommendations, were also included. Re-using the made-ground as rubble fill and compacting it in lifts would cost £200 000. However, although this would rehabilitate the site as a landscaped open space amenity it would not necessarily be suitable for subsequent industrial development. It therefore was further recommended that if the site was to be used for such development, then the most economic method of enhancing the ground carrying capacity after backfilling would be by using vibrocompaction. Vibrocompaction would cost £500 000.

5 CONCLUSIONS

Derelict land refers to that which has been spoiled by the extraction of mineral deposits or by industrial operations. As such it is a wasted resource, as well as having a blighting effect on the surrounding area. Accordingly the restoration of such areas is of importance if the economic well being of the associated community is to prosper.

In the United Kingdom much dereliction is associated with old collieries. The most notable surface expressions of dereliction due to past coal mining are spoil heaps and tailings ponds. Two old colliery spoils heaps near Barnsley, South Yorkshire, were reclaimed to provide local amenity and a site for development of light industry. Regrading spoil heaps so that their appearance becomes aesthetically acceptable involves large scale earthmoving operations. These can be made difficult if the spoil is undergoing spontaneous combustion. This means that hot spots are present on the heap and in the case considered, temperatures up to 900°C were recorded. Once a spoil heap has been regraded, a system of drainage has to be installed. Then the spoil in those areas which are to be vegetated has to be improved by the addition of fertilizer, prior to seeding or planting.

Old sand and gravel pits are frequently flooded due to the presence of high water tables. This was the case with some workings near Brighouse in West Yorkshire. Commonly such sites are converted into recreational areas which involve various water sports. In this way, the pits only need partial filling and landscaping.

One of the problems with old industrial areas is that the ground may be contaminated which presents problems when such an area is to be rehabilitated. Such a site occurred in Leeds, West Yorkshire. Hence the site investigation entailed samples of made-ground and groundwater being taken for chemical analysis to ascertain the degree of contamination. Fortunately the concentrations of toxic constituents were generally below those levels likely to cause problems. However, there were some elevated concentrations of magnesium and phenol present in certain locations which would necessitate particular precautions being taken during site operations. In addition small fragments of asbestos sheeting were found on site and they needed to be removed. If the site was to be reclaimed for building development, then it would have to be bulk excavated, the material then being recompacted and, lastly, its load carrying capacity improved by vibrocompaction.

Correlative and regressive methods of mapping of metallurgical industry wastes

Méthodes de corrélation et régression pour la cartographie des déchets de l'industrie métallurgique

L. K. Govorova & A. B. Lolaev

Norilsk Industrial Institute, Russia

ABSTRACT: Mapping of technogene dispersion of metals in soils on the sites of the industrial enterprises is a labor-consuming and expensive process especially in the case with rare and dispersed chemical elements. Correlative and regressive method is highly suitable for approximate semi-quantative estimation of the technogene origin of microcomponents. This method was widely used during last decade for testing concomitant metals of ore deposits.

RESUME: Le levé et établissement des cartes thématiques de la diffusion technogéne des métals en sols sur les territoires des enterprises de production etc. exigent beaucoup de travail; c'est un processus couteux, surtout pour les éléments chimiques rares et diffusés. A l'appreciation approximative semi-quantitative de la technogenése des microcomposants la méthode de corrélation-régression fait ses preuves qui ort été obtenu l'expansion visible aux épreures du minerais sur les métals-satellites en derniéres périodes déccennales.

1 INTRODUCTION

Before planning the mapping of the contamination of the industrial sites it is necessary to have the preliminary data about wasteness or wastelessness of the technology. It is generally known that irrevocable losses of chemical elements on the stage of extraction are estimated by means of the material and metallurgical balance accounts. If the accounts are disbalanced then it is necessary to study the reasons of disbalance of the accounts income and expenditure. The best way to determine the loss of the pecious component is the stanistic calculation of balance.

Full analysis of metallurgical balance accounts is carried out at ten-thirteen stages of the calculation algoblock of about 40 algorithms. Application of the above mentioned analysis in the subsystem of the industrial accounting of metals is possible only when special digital massifs are available besides the data of the analytical and technical monitoring of an industrial enterprise. That is why for application at industrial enterprises the method of multistep control of quality (accuracy) and balance of accounts is converted into expressive form of monostep algorithm.

2 CORRELATIVE-REGRESSIVE METHOD

The necessary income data are composed of the temporal and immediate selections of the technical and analytical control:

a) masses of the income and expenditure items of the balance accounts M ;

b) concentration of metals in the items of the balance accounts C ;

c) correlative faults of the mass measuring Vb ;

d) relative faults of the determination of metal concentration Van ;

In the course of calculation the necessary values are found:

a) masses of each item of balance ac-

2507

counts, kg:

$$L = M * C \quad ; \tag{1}$$

b) actual disbalance, kg :

$$E_{fact} = \sum_{i=1}^{k} L_{in} - \sum_{i=1}^{k} L_{exp} \quad ; \tag{2}$$

c) absolute metrological faults of each item of balance accounts, kg :

$$\blacktriangle L = V_{st} * L * 10^{-2} \quad ; \tag{3}$$

d) limit of the absolute metrological fault of balance accounts, kg :

$$E_{lim} = \sum_{i=1}^{k} (\blacktriangle L)_{in} + \sum_{i=1}^{k} (\blacktriangle L)_{exp} \quad ; \tag{4}$$

e) dispersion of balance accounts equation, kg :

$$E_m^2 = \sum_{i=1}^{k} (\blacktriangle L)_{in}^2 + \sum_{i=1}^{k} (\blacktriangle L)_{exp}^2 \tag{5}$$

f) balance accounts metrological fault, kg :

$$E_m^2 = \sqrt{\sum_{i=1}^{k} (\blacktriangle L)_{in}^2 + \sum_{i=1}^{k} (\blacktriangle L)_{exp}^2} \quad ; \tag{6}$$

g) amount of uncontrolled total loss of metal, kg :

$$E_{un} = E_{fact} - E_m \quad ; \tag{7}$$

h) experimental value of Fisher's criterium in comparison with table value:

$$F_e = E_{fact}^2 : E_m^2 \geqslant F_{tab} \quad ; \tag{8}$$

f) experimental value of Student's criterium in comparison with table value:

$$T_e = E_{fact} : E_m \leqslant T_{tab} \quad ; \tag{9}$$

Expression: $T_e < T_{tab}$ tells the proper use of metallurgical balance accounts, while the inequation of $T_e > T_{tab}$ shows the inadequate one.

Expression: $F_e < F_{tab}$ points out that the loss of metal from the technological system is not considerable, and the reverse inequation of $F_e > F_{tab}$ shows the presence of the canals of losses and considerable wastes with the applied technology.

Expression: $F_e \gg F_{tab}$ signifies considerable escape of precious metals while their processing as well as disturbance of environment. Therefore it is necessary to study the environment after the use of metallurgical balance accounts and their statistic analysis.

For example, platinum balance at one of the enterprises has the result:

$$F_{fact} = 3.9^2/4.0^2 = 0.95 \approx 1 < F_{tab} = 2.8 \quad ;$$

It means that there is no loss of metal with the technological system.

Osmium balance at the same enterprise has the result:

$$F_{fact} = 57^2/13^2 = 19 > F_{tab} = 2.8 \quad ;$$

It means that there is considerable escape of metal.

At last, the result of the elementary sulpher balance is:

$$F_{fact} = 75^2/3^2 = 50625 \gg F_{tab} = 1.76 \quad ;$$

This signifies huragan loss of sulpher with applied technology.

These results correlate with data of the service of the dust and gas ventilation and purification of Norilsk mining-metallurgical company which show that in the eighties every year in the atmosphere polluted 2.4 mln tonnes of sulpher dioxide (about 1.2 mln tonnes of the elementary sulpher) and dozens of thousands tonnes of dust, much chlorine, nitrogen and others. Chemical composition of the effluent from different industries are dependent on the kind of the industry. With acid rains and technological wastes on the surface while snow melting contaminants penetrate into the seasonly melting layer. Hence, at the end of the summer the active layer has a concentration of pollutants.

In winter, the active layer freezes trapping the pollutants within itself. Eventually, after a number of freezing and thawing cycles, these pollutants start influencing the properties of the permafrost. Poluektov (1982) has presented data which show high concentration of sulphate ions SO_4^{2-} present in the pore water in active layer of soil taken from the Norilsk region of Russia. A growing concern about the problems associated with the saline of permafrost has developed because of its detrimental impact on the perfomance of foundations. The salts in the pore water affect the behavior of frosen soils.

3 MATERIALS AND TEST METHOD

The fine-grained soil used in this study was a silty clay obtained from a test site at The University of Calgary campus. The soil has been investigated for its properties without any salts by Joshi & Wijeweera (1990). Three types of salts, namely, sodium chloride, sodium bicarbonate and sodium sulphate, were considered to represent industrial wastes produced in Norilsk region of Russia. The salt concentration was varied between 0 to 150 parts per thousand (ppt). The tests were conducted at temperatures varying between +20°C to -20°C.

The physical properties such as Atterberg's limits of the soil with varying salt concentrations were determined according to ASTM D4318-84 (1984). The temperature at which the pore water started to freeze was determined by a thermocouple placed within the specimen. The temperature of the sample was recorded every 15 seconds.

The electrical conductivity of soil was measured by a SYBRON/Barnstead PM-70CB conductivity bridge with YSI 3417 series conductivity cells. The soil water extractions of soluble salts for pH and electrical conductivity were prepared in a 1:2 ratio, as recommended by Jackson (1958). Preparation of soil samples for compressive strength tests involved mixing soil with salt solutions prepared by adding predetermined quantities of different salt to dis-

tilled water. The water content of the soil water system was maintained at 14 % which was close to the plastic limit of the soil. Compressive strength of the soil samples was determined by testing samples made by compressing soil water mixtures in plastic molds 2.5 cm in diameter, 5 cm high. The samples were then stored in an air tight container for several days to allow them to equilibrate. After equilibration the samples were tamped to a constant density of 1.9 gm/cm^3 and placed in a cold chamber at -12°C to freeze.

Constant strain uniaxial compression tests were conducted to determine the short term strength of the frozen soils. The tests were conducted using a closed loop control MTS machine. A temperature controlled cabinet using liquid CO_2 was mounted with the MTS machine to mantain the required temperature during the test. The temperature close to the specimen was monitored using a thermocouple and was found to vary no more than $\pm$ 0.3°C. A constant deformation rate of 1.7×10^{-2} mms^{-1} was then applied axially on the specimen until a maximum of 30 percent axial strain was reached. Axial load and deformation were recorded by a microprocessor base data display in the MTS machine. At the end of the test the deformed shape of the specimen was noted. A total of more than 150 samples were tested.

4 TEST RESULTS AND DISCUSSION

The data physical properties of the soil with varying salt content are presented in Table 1. The data show that the liquid limit generally decreases with introduction of salt in the soil water system. This decrease in liquid limit is more pronounced for sulphates and carbonates than for chlorides. The plastic limit of all the soils increased with the addition of salt. The plasticity index tends to decrease for sulphates and carbonates but shows no specific trend with chlorides.

As the salt content increases, the pH decreases marginally for sulphate and significantly for chloride. Presence of sodium bicarbonate cause an increase in the

2509

soil pH from 8.25 to about 10.5.

The electrical conductivity increases significantly with increase in salt content. Presence of sulphates tends to increase the electrical conductivity for soil but at a decreasing rate as the salt concentration increases. The conductivity increase due to the presence of bicarbonate is lower than that due to either chloride or sulphate salts.

The effect of salt concentration of sodium chloride on the freezing temperature of the soil is much more pronounced than the effect of the other two salts. The freezing temperature is lowered from $-0.1°C$ to $-11.1°C$ for 150 ppt of sodium chloride. However, when sodium sulphate was mixed with soil, the temperature of freezing decreased only from $-0.1°C$ to $-1.1°C$ for 150 ppt. This depression of the freezing point was significant only up to a salt concentration of 50 ppt, beyond which the effect on freezing was rather minimal.

The average peak strength for frozen soil under different salt concentration and temperature conditions are presented in Table 1. The peak strength data indicate that the strength decreases with both increasing salt concentration and temperature. The compressive strength of natural soil at a temperature of 20°C was 0.45 MPa and incresed to 13.2 MPa as temperature decreased to $-10°C$, and further increased to 17.5 MPa as the temperature was lowered to $-20°C$. Addition of salt significantly loweres the compressive strength of the soil. Salt level of 50 ppt of sodium chloride at a temperature of $-20°C$ has the same compressive strength as of non-saline soil at a temperature of 20°C.

5 CONCLUSIONS

Before planning the mapping of contamination of the industrial sites in is necessary to have preliminary data about wasteness or wastelessness of the metallurgical technology. The losses of chemical elements on the stage of extraction are estimated by means of the material and me-

tallurgical balance accounts.

Metallurgical balance accounts are the system of the mathematical equations, which connect natural and formal factors and are used for finding the links in technological process. The dependence between the metallurgical balance accounts is found with the banks of regressive equations on the basis of data of analytic control of technological process. Volumes of losses of metals and pollutants which are found on the basis of correlative and regressive method used for mapping of metallurgical production wastes.

Mapping of metallurgical production wastes is important not only for environment, but also for engineering geology purposes esspecially in permafrost regions. The geocryological conditions in permafrost are significantly affected during the economic development of these regions. Artificial salting is one of the effects of antropogenic activities.

The physical properties of the soil change with an increase in the salt content. The nature of these changes depends on the type of salt present. Different salts have different effects on the pH and electrical conductivity of soils.

The peak strength of frozen soils decreases with an increase in temperature and salt concentration. Sodium chloride has the maximum effect on the compressive strength of the soil. However, at a temperature of $-20°C$ the influence of all types of salts is similar.

Accurate prediction of the behaviour of the frozen soils having different salts can be made only if the type of salt present in the soil is known, which may be done on the basis of metallurgical balance account analyse.

REFERENCES

American Society of Testing and Materials (ASTM). 1984. Annual Book of ASTM Standards,section 4,vol.04.08, D4318–84
Jackson,M.L. 1958. Soil chemical analysis. Prentice-Hall inc.,Englewood Cliffs,N.J.
Joshi,R.C. and Wijeweera,H. 1990. Post peak axial compressive strength and deforma-

Table 1. Physical and mechanical properties of soil with and without salts

Properties	Without salt	NaCl				Na$_2$SO$_4$				NaHCO$_3$		
		25	50	100	150	25	50	100	150	25	50	100
Physical properties:												
Liquid limit (%)	31.2	32.6	30.5	31.7	32.0	29.7	30.0	30.0	28.5	30.8	30.2	32.1
Plastic limit(%)	14.4	14.3	15.1	13.2	14.6	15.1	15.1	16.4	16.0	15.4	15.3	14.5
Plasticity index (%)	16.8	18.3	15.4	18.5	17.4	14.6	14.9	13.6	12.5	15.4	14.9	17.6
pH (1:2 H$_2$O)	8.24	8.17	8.10	7.81	7.78	8.18	8.12	8.08	8.02	9.66	10.2	10.5
Conductivity (mmho/cm) (1:2 H$_2$O)	0.47	7.60	11.1	21.5	33.8	5.73	10.9	21.7	28.6	2.22	5.35	10.7
Temperature of start of freezing (°C)	−0.1	−1.2	−3.1	−7.1	−11.1	−0.2	−0.9	−0.9	−1.1	−0.2	−0.2	−2.3
Compressive strength (MPa):												
+ 20°C	0.45	0.45	0.40	0.40	0.42	0.50	0.45	0.40	0.40	0.68	0.48	0.20
− 5 °C	1.75	0.82	0.55	0.49	0.37	1.60	1.35	1.05	0.46	0.66	0.62	0.50
− 10°C	13.2	3.70	2.30	0.85	0.65	9.70	7.70	6.80	6.20	6.50	6.20	5.30
− 20°C	17.5	2.40	1.05	0.95	0.75	5.10	3.70	2.90	1.16	3.20	1.46	0.54

tion behaviour of fine-grained frozen
soil. Proc. 5th Canadian Permafrost
Conf.: 317-325.
Poluektov,V.E. 1982. Foundation construction in permafrost. Leningrad: Stroyiz-
dat (in Russian).

Finegrained soils of Slovakia as geobarriers for waste repositories

Les sols fines de Slovaquie en tant que barrières pour le stockage des déchets

V. Letko & M. Hrašna
Comenius University, Bratislava, Slovakia

ABSTRACT: A majority of waste repositories in Slovakia is situated on geological environment which does not fulfill requirements of its permeability. However the sorption properties of finegrained soils can also stop or substantially decrease the migration of pollutants from waste disposals. A preliminary classification of soils on the slovak territory as possible geobarriers is proposed.

RÉSUMÉ: La majorité des décharges est située dans l'environnement géologique qui ne correspond pas aux demandes concernant leur épaisseur et/ou leur perméabilité. Cependant, les qualités de la sorbtion des sols à grains fins peut arreter ou éssentiellement limiter la migration des composés nuisibles provenant des décharges. La classification préliminaire des sols présentant les géobarrières possibles du territoire slovaque est proposée.

1 INTRODUCTION

One of the crucial moments in waste disposal siting is the assessment of the pollution threating to geological environment, especially to groundwaters. The migration of contaminants from waste disposal sites can be prevented by using either costly, artificial impervious membranes or the insulation properties of rock materials in their natural or processed states. According to existing national standards and recommendations two heading properties of such materials are their permeability and thickness. Depending on the type and the category of wastes the minimum coefficient of permeability prescribed ranges usually from 10^{-8} to even 10^{-11} m^{-1} and the minimum thickness of the stratum ranges from 3 to 10 m. There exist not many places in which geological conditions fulfill these parametres, however, various impermeable sealing elements can be constructed.

A recent inventory of waste disposals in Slovakia (Letko et al 1993) shows that most of repositories are indiscriminate ladfills composed mainly of typical municipal and building wastes. For these wastes containing harmful components in low concentrations the prescribed barrier parametres seem to be rather severe ones.

The transport of waste water particles acts more due to their diffusion and migration in soils than due to percolation caused by hydraulic forces. Thus, the ion exchange and sorption properties of the soil material can provide a sufficient prevention against transport of pollutants into groundwater, eventhough if it does not possess above mentioned values of permeability or thickness, and can serve as geobarrier.

2 SORPTION PROPERTIES DETERMINATION

The most favourable sorption properties can be found in finegrained soils which contain clay minerals, especially smectites. On the Slovak territory (fig. 1) there can be found such soils

Fig. 1 Finegrained soils of the Slovak territory: 1 - Quaternary loess loams and loesses, 2 - Neogene clayey deposits, 3 - Paleogene shales and claystones

mainly among wheathered paleogene shales, neogene clayey-silty sediments and various types of Quaternary deposites. For their preliminary classification as possible geobarriers material series of less costly laboratory batch tests were performed using solutions with relatively high concentrations of following metals: Cd, Co, Cu, Mn, Ni, Pb and Zn. The contact time between the soil and the solution for each initial concentration (Co) was 24 hours, the ratio between the sample weight (P) and the solution volume (V) was 1:100. Samples were taken gradually over the period of time and solute concentration changes (Ci) were determined by AAS-3 Hitachi Z-800.

The calculation of the sorption capacity (N, mg.g^{-1}) is based on the equation

$$N = \frac{C_o - C_i}{P} V$$

The mineral composition and the cation exchange capacity were determined, as well.

3 TEST RESULTS AND DISCUSSION

The following samples of lithological soil types were tested: loess to loess loam (Nr. 1,8,9)

eluvium of Paleogene shales (Nr. 4,5), Neogene clays to silty clays (Nr. 2,7,10). Thier sorption capacity and cation exchange capacity values are shown in Table 1.

It is known that sorption capacity increases in accordance with increased concentrations of the solution and contact time, and, gradually tends to reach a limit which can be used as the maximum sorption capacity for a particular ion. The sequence of cation sorption capacity changes with the solution concentration, however. Czurda and Wagner (1991) state a totally different cation sorption capacity sequence for low concentrations (Cd<Pb<Zn<Cu<Cr) and for high concentrations (Cr<Cd=Cu<Zn<Pb). Using average values for all of samples tested, the sequence for low concentrations (from 40 to 60 mg.l^{-1}) is Mn=Cd<Co<Cu=Zn<Ni<Pb and for high concentrations (from about 200 to 1000 mg.l^{-1}) it is Co<Mn<Ni<Zn<<Cu=Cd<Pb. All types of soils we have tested showed almost equal values for the two lowest capacities for Mn and Cd (fig. 2). There were, however, great differences between the two highest sorption capacities for Pb and Cu (fig. 3) as seen particularly in samples which represent Paleogene, Neogene and some loess sediments. This can be explained in terms of their mineral composition which

2514

Table 1. Sorption capacity (N) and cation exchange capacity (E) of samples

Nr. of sample	N (g.mg⁻¹)							E $\frac{mg.eq}{100g}$
	Cd	Co	Cu	Mn	Ni	Pb	Zn	
1	14,4	10,1	35,0	8,4	15,7	64,2	10,1	17
2	12,2	5,8	5,2	7,8	9,0	11,5	5,8	11
3	14,6	8,5	14,6	9,0	14,5	18,1	8,5	22
4	12,2	7,0	53,0	6,8	9,0	80,8	7,0	8
5	10,4	4,0	11,6	8,8	10,5	19,8	4,0	7
6	13,7	8,0	13,8	8,6	13,5	16,6	8,0	20
7	18,8	7,8	55,6	11,8	13,0	44,5	7,8	28
8	9,4	6,0	52,6	6,6	7,5	82,4	6,0	8
9	13,0	6,8	18,2	12,5	14,1	25,8	6,8	21
10	16,0	20,0	80,0	9,1	7,1	150,0	20,0	27

includes carbonates (calcite, dolomite) in addition to a higher content of clay minerals (Table 2).

The sequence of maximum sorption capacities for Paleogene shales which contain clay minerals as well as calcite, dolomite and chlorite is Mn=Co<Ni=Cd<Zn<<Cu<<Pb (fig. 4.). This corresponds to the results of Czurda and Wagner (1988) who found a similar sequence in Neogene clay: Cd<Ni<Zn, Cu<Pb.

Tab. 2 Mineral composition (%) of Paleogene shales (Nr. 4) and loess loam (Nr. 9)

Nr. of sample	cal-cite	chlo-rite	dolo-mite	illi-te	micro-cline	plagio-clase	quartz	smec-tites	gyp-sum
4	7,6	4,6	8,7	25,3	4,9	8,3	37,6	0,6	2,4
9	-	-	-	25,7	6,2	9,3	51,9	6,9	-

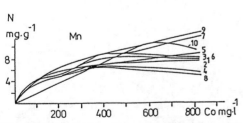

Fig. 2 Cd sorption isotherms for soil types tested

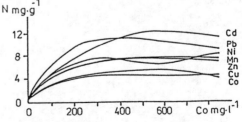

Fig. 5 Sorption isotherms of metals tested in Quaternary loams

2515

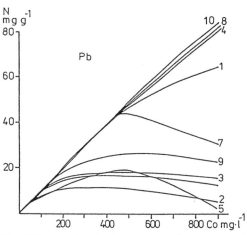

Fig. 3 Pb sorption isotherms for soil types tested

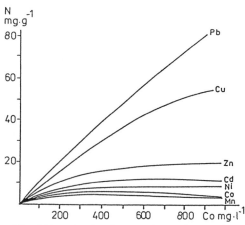

Fig. 4 Sorption isotherms of metals tested in Paleogene shales

4 CONCLUSIONS

Finegrained clayey sediments of Paleogene and Neogene possess very good sorption properties and are suitable materials for natural geobarriers. Values of their sorption capacities range to several tens mg.g^{-1} for the majority of metals tested (fig. 4), Quaternary deposites like deluvial loams (fig. 5) and loess loams possess low sorption properties with sorption capacities reached only several mg.g^{-1}. They are of low value for utilization as geobarriers. The majority of other Quaternary deposites possess medium sorption properties with sorption capacities reached about ten mg.g^{-1}. They are moderately suitable and require geotechnical processing to qualify as barrier material.

In a locality selected, however, a more detailed investigation must be done using the more precise determination of sorption capacity by column tests with actual compostition of waste waters.

REFERENCES

Czurda, K.A. & Wagner, P. 1988. Transfer and sorption of heavy metals in clayey barrier rocks. Schrift, Angew. Geol., Band 4: 225-245.

Czurda, K.A. & Wagner, P. 1991. Cation transport and retardation process in view of toxic waste depositon problems. Eng.Geol., Vol. 30: 103-113

Letko, V., Hrašna, M. & Holzer, R. 1992. Waste disposal problems in Slovakia inventory, land suitability maps and soil material as geobarriers. In.: Entsorgung efahrlicher Abfalle, Collegium Hungaricum, Wien: 71-76.

Final closure plan for Nanticoke GS ash landfill

Plan d'enfouissement des cendres de charbon de la centrale de Nanticoke

R.J.Heystee & M.R.Jensen
Ontario Hydro, Toronto, Ont., Canada

ABSTRACT: A final closure plan for the 18 million m³ Nanticoke TGS coal ash landfill has been developed to mitigate potential long-term impacts to the local surface and ground water bodies. The cover materials would maintain a balance between infiltration into the landfill and the current leachate exfiltration rate through the clay base of the landfill. There would be no leachate discharge through the sides of the landfill and thus no requirement for the post-closure treatment of leachate. The cover and its vegetation will also protect the landfill from effects of wind and water erosion, and will improve the visual appearance of the site. This paper describes the hydrogeologic conditions at the landfill site and predictions of long-term leachate migration. This information formed the basis for the proposed final closure plan.

RÉSUMÉ : Les auteurs ont développé un plan qui permettrait de condamner de façon définitive le site d'enfouissement de la centrale thermique de Nanticoke, qui contient 18 million de m³ de cendres de charbon, de façon à atténuer ses effets possibles à long terme sur les eaux de surface et les eaux souterraines. Le plan prévoit la mise en place de matériau de couverture permettant d'assurer un équilibre entre l'infiltration dans le site et le taux actuel d'exfiltration des lixiviats au niveau de la base argileuse du site. Comme on ne s'attend à aucune exfiltration latérale, aucune méthode de traitement des lixiviats n'a donc été prévue. Le matériau de couverture et sa végétation protègeront le site des effets de l'érosion éolienne et hydrique, en plus de rehausser son apparence visuelle. L'article décrit les conditions hydrogéologiques qui prévalent au site d'enfouissement et analyse différents modèles de migration à long terme des lixiviats. Les renseignements présentés constituent le fondement du plan proposé pour le site d'enfouissement.

1 INTRODUCTION

Ontario Hydro's 4000 MW Nanticoke Generating Station (GS) is coal-fired and has historically produced approximately 600,000 tonnes of coal ash each year. The station is located on the north shore of Lake Erie in the Province of Ontario (Figure 1). The coal ash is being placed in a 76 Ha landfill site which is the largest coal ash landfill operated by Ontario Hydro. The landfill has a design capacity of 18 million tonnes and a proposed design height of 45 m at its centre. The landfill is founded on a clay stratum with an average thickness of 6 m.

clay stratum with an average thickness of 6 m.

Starting in 1973, the landfill site was operated as an ash lagoon with 12 m clay containment dykes. Since the early 1980's the ash has been sluiced to the landfill, dewatered in cells, and then placed with a high water content in the ash disposal area. Beginning in 1992, the "dry" fly ash (with a 15% water content) has been transported by truck into the landfill site from nearby ash storage silos. The landfill will likely reach its design capacity in the early 2000's and at that time final cover materials will be placed on the landfill.

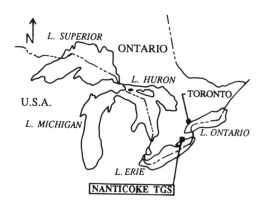

Figure 1 Location plan

2 COAL ASH CHARACTERISTICS

Coal ash is predominantly comprised of noncombustible material that is produced by the burning of coal. Coal ash is generally made up of 85% fly ash and 15% bottom ash. Fly ash is a heterogenous mixture of highly vitreous spherical particles, called cenospheres, crystalline matter and unburnt carbon. It is a fine powdery material that looks like grey talcum powder, and is removed from the flue gas by electrostatic precipitators. The larger and heavier ash particles either fall to the bottom of the boiler or are deposited on the furnace walls or heat exchangers; ultimately they are removed either in a molten state or in granular form. The molten materials are referred to as boiler slag while the solid granules are called bottom ash.

2.1 Solid coal fly ash composition

The chemical composition of the ash is a function of the type of coal that is burned, the extent to which it is prepared before being burned and the operating conditions of the station. The bulk chemical analyses of Ontario Hydro fly ash shows that the ash is primarily composed of silicon, aluminum, iron and calcium in their oxide forms as well as unburnt carbon which can be as high as 15%. The fly ash also contains small but significant concentrations of trace elements such as arsenic and selenium.

The fly ash is classified as a non-hazardous (non-registerable, non-leachate toxic) solid waste in accordance with the Ontario Ministry of the Environment and Energy (MOEE) Regulation 347 leachate extraction test procedures. According to the regulations that govern waste disposal activities in the province, this solid waste must be placed in a landfill or contained in some suitable manner.

The average concentrations of several elements in coal fly ash are at levels that are considered to be phytotoxic (Johnston and Eagleson, 1991). Therefore the Nanticoke GS coal ash landfill would have to be covered with a suitably thick layer of inert material before the site could be vegetated and opened for public use.

2.2 Leachate

Studies of Ontario Hydro coal ash leachate show that the concentrations of the various solutes fall within a predictable range of values (Johnston and Eagleson, 1991). For example, the porewater (interstitial water) in dry bituminous fly ash landfill sites is generally characterized by alkaline pH and elevated concentrations of calcium, sulphate and boron.

Target contaminants in leachate are often used in hydrogeologic impact assessments at Ontario Hydro coal ash landfills. Target contaminants have relatively high concentrations with respect to Ontario Drinking Water Objectives (ODWO), and are also potentially toxic to humans and/or mobile in the subsurface environment. Johnston and Eagleson (1991) have identified the following four target contaminants in the leachate of Ontario Hydro bituminous coal ash landfills: sulphate, boron, arsenic and selenium.

3 HYDROGEOLOGIC CONDITIONS

The station property is situated in the physiographic region known as the Haldimand Clay Plain. The clay at the station site is described as a hard to stiff, brown to grey and is fractured at shallow depths. Before the construction of the landfill, the clay stratum

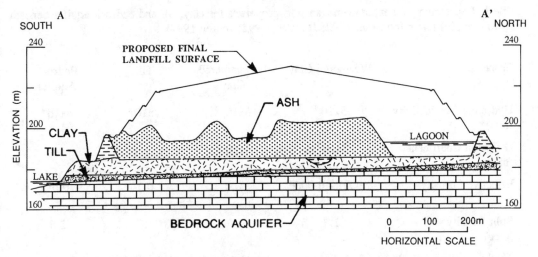

Figure 2 Geologic section A-A' through landfill

thickness beneath the landfill ranged between 5 and 10 m. After the excavation of approximately 1.2 million m³ of weathered clay to construct the containment dykes, the clay stratum thickness was reduced to between 3 and 9 m with an average thickness of 6 m.

At the base of the clay formation is a till unit which is better graded, harder and less plastic than the overlying clay. The till is underlain by a Devonian cherty limestones. The upper 5 to 10 m of these limestone formations is considered to be the aquifer at the landfill site (Figure 2).

3.1 *Properties of clay stratum*

The clay stratum is the primary barrier to the downward migration of coal ash leachate. It is divided into the upper weathered and fractured clay unit, and the lower unweathered and intact clay unit. Approximately 50% of the soil consists of clay minerals with illite being the predominant clay mineral. The upper fractured clay has mottled brown and grey colour. In the vicinity of the landfill the fractures extend up to 5 m below ground surface and the presence of plant roots at this depth would suggest that the fractures are continuous from ground surface (Golder Associates, 1988). The average thickness of the weathered clay beneath the landfill is approximately 3 m. The fracture

porosity of the weathered clay is estimated to be in the range of 0.0001 and 0.0006 (Heystee, 1992).

Due to the presence of fractures in the upper weathered silty clay, this unit is more permeable than lower unweathered and unfractured silty clay. In situ measurements in the fractured clay yield a bulk hydraulic conductivity value of 5×10^{-9} m/s. This value is consistent with hydraulic conductivity data obtained for the fractured Sarnia area clay (D'Austous et al, 1989). There is direct evidence from hydraulic gradient measurements beneath the landfill that the upper fractured clay is more permeable than the underlying unweathered clay. The gradients in the upper weathered clay are in the range 0.2 to 0.6 and in the lower unweathered clay they are in the range of 0.9 to 1.7 (Vorauer, 1989).

The lower unweathered and largely intact silty clay is distinguished from the upper fractured silty clay by its grey colour and by the reduced number of fractures. The average thickness of the unweathered clay beneath the ash landfill site is 2.5 m. The hydraulic conductivity of this unit is estimated to be 5×10^{-10} m/s.

Other contaminant transport properties of the clay stratum as well as the till and bedrock aquifer units are summarized in Table 1.

Table 1. Representative contaminant transport properties for clay, till and bedrock aquifer beneath coal ash landfill (after Johnston, 1992; Heystee, 1992; Jensen 1994)

Property	Weathered Clay	Unweathered Clay	Till	Bedrock Aquifer
Hydraulic Conductivity (m/s)	5×10^{-9}	5×10^{-10}	1×10^{-8}	4×10^{-6}
Diffusion Coefficent (m²/s)	3×10^{-10}	3×10^{-10}	-	-
Effective Porosity	0.0001 - 0.0006	0.4	0.3	0.05
Bulk Density (g/cc)	1.7	1.7	1.7	2.5
Thickness (data range) (m)	3 (1.3 - 4.5)	2.5 (0.4 - 5.2)	2 (0 - 3.9)	5 to 10
Retardation Factor Arsenic	4000	-	-	-
Retardation Factor Selenium	19	-	-	-

3.2 Existing ground water flow patterns

On a regional scale, the ground water flow in the bedrock aquifer is in a southeasterly direction and the aquifer water ultimately discharges into Lake Erie. Due to 10 to 12 m of water-table mounding within the 76 Ha landfill site, there is a local disturbance to the regional pattern of ground water flow within the bedrock aquifer (Jensen, 1992). Approximately 23 million litres of leachate is flowing downwards through the base of the landfill into the aquifer each year. In order to achieve a balance between this downward flow and flow in the aquifer, the vertical and horizontal gradients in the aquifer must increase. This, in turn, leads to a mounding of the potentiometric surface within the aquifer. As a result, aquifer water flow is radially divergent from the centre of landfill (Figure 3).

4 LEACHATE MIGRATION

If the current ground water flow conditions are maintained over the long-term, then at some point in time leachate will migrate in a radially divergent pattern away from the landfill. This leachate migration process would likely lead to a small zone of slightly contaminated aquifer water in the immediate vicinity of the landfill. The ultimate size of the zone and the contaminant concentrations in the zone will depend on a number of factors; eg the source concentration of the contaminants in the landfill, the period of time over which they are leached from the ash, the amount of dilution that will occur due to mixing with clean aquifer water and clean water recharging the aquifer.

4.1 Present location of leachate plume

Vorauer (1990) detected coal ash leachate within the overburden beneath the landfill to a depth of 4.5 metre below the ash/clay interface. Vorauer has postulated that the elevated concentrations of arsenic and boron at the 4.5 m depth may have been due to contamination of the piezometer sampling zone with fly ash during their installation. An alternative explanation is that the fractures in the weathered zone have allowed the relatively rapid migration of these

2520

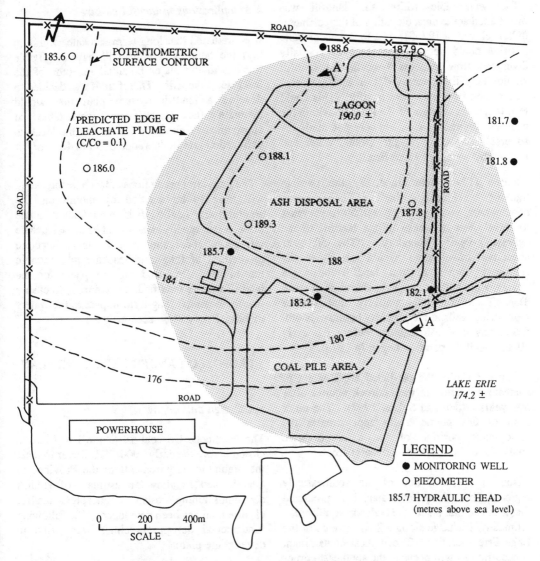

Figure 3 Ground water flow and leachate migration in bedrock aquifer

contaminants to the interface between the weathered and unweathered clay (average depth of 3 m) over a period of several years. Then the leachate slowly migrated into the unweathered clay during the next 1 to 2 decades.

These observations of leachate migration are consistent with the results of the on-going ground water monitoring program which has found no evidence of off-site leachate migration in the aquifer.

4.2 *Assessment of long-term leachate migration*

Allan and Sykes (1992) and Jensen (1994) carried out a numerical modelling studies to predict potential long-term coal ash leachate migration patterns in the bedrock aquifer. The ground water flow models were calibrated to an extensive hydraulic head data set for the bedrock aquifer.

The water-table within the landfill was maintained at the crest elevation of the perimeter dykes (elevation 197 m). If a lower water-table position could be achieved, then the radially divergent flow patterns beneath the landfill would likely disappear and "clean aquifer water" would then flow under the landfill. This could lead to greater dilution of the leachate within the bedrock aquifer and a lower potential for adversely impacting off-site ground water and the nearby waters of Lake Erie.

Jensen (1994) determined, by the particle tracking technique, that a conservative contaminant would take between 50 to 95 years to travel from the base of the landfill to the nearest property boundary. The 50 year advective travel time would occur at the southeast corner of the landfill near the embayment on the Lake Erie shoreline (Figure 3). Therefore it is unlikely that aquifer monitoring wells at the property boundaries will detect any coal ash leachate before the early 2000s (landfill operation began in 1973).

Figure 3 shows the predicted areal extent of leachate migration in the bedrock aquifer after 500 years (Allen and Sykes, 1992). The outer limit of this plume likely approximates the steady-state position of the a conservative contaminant (ie boron or sulphate) front.

Due to the influence of the southeasterly regional flow of aquifer water, it is predicted that leachate that enters the bedrock aquifer may ultimately discharge along a 2 Km stretch of the Lake Erie shoreline. The highest contaminant concentrations will occur at the southeast corner of the landfill where ground water/leachate discharge is focusing on an embayment. The contaminant concentrations in the discharging ground water will be smaller in either direction along the shoreline. It is anticipated that the quality of this discharging leachate will meet provincial surface water quality guidelines and that it would be considered acceptable for release into Lake Erie. Extensive mixing of the discharging ground water and the nearshore waters of Lake Erie would further reduce the concentration of the contaminants and their potential impact on lake water quality.

4.3 *Implications to landfill closure*

It is predicted that long-term leachate migration into the bedrock aquifer will have negligible impacts to the major potential receptor of the leachate, Lake Erie. Therefore it was decided to develop a landfill closure plan that would maintain the post-closure water-table at approximately elevation 197 m and the leachate exfiltration rate in the range of 30 to 40 mm/yr.

The long-term predictions of leachate migration will have to be verified by monitoring the aquifer water quality. It is our intention to use these water quality data to calibrate a leachate migration model and to make more accurate predictions of long-term leachate migration in the future. Proposed closure plans for the landfill will be modified, as required, to ensure that the potential long-term impact to the waters of Lake Erie will be acceptable.

5 PROPOSED LANDFILL CLOSURE PLAN

5.1 *Design philosophy*

The Nanticoke GS coal ash landfill will not be closed before the early 2000s. However to meet the regulatory requirements of the Province of Ontario and to allow an estimate of landfill decommissioning costs, a conceptual landfill closure plan has been developed. The following fundamental design guidelines were used to develop the plan:

● the landfill must be environmentally acceptable,

● the landfill must be structurally sound,

● the landfill should have a passive design and require little long-term maintenance, and

● post-closure leachate handling and treatment should not be required.

To meet these design guidelines it was determined that the water-table within the landfill should not rise above its present maximum elevation of 197 m. A higher long-term water table position could lead to slope instabilities (Lau and Chan, 1992). The potential for unacceptable impacts to off-site ground and surface water would also increase if the water-table were allowed to rise above elevation 197 m.

So that there would be no potential need for post-closure treatment of leachate, a balance between water infiltration though the landfill cover and current leachate exfiltration rate through the clay base would have to be achieved. Under these water balance conditions all soluble components of the coal ash will be forced to exfiltrate through the landfill base and to be naturally assimilated back into the subsurface environment. This type of landfill design is known as a "natural attenuation landfill". An imbalance between cover infiltration and base exfiltration would lead to excessive leachate mounding within the landfill. The net effect would be the discharge of leachate though the sides of the landfill to the perimeter ditch. If the water in the perimeter ditch does not meet water quality standards for discharge into Lake Erie, then it would have to be treated prior to discharge.

5.2 Final cover materials

Figure 4 illustrates the proposed final cover for the Nanticoke GS coal ash landfill. The compacted clay would form the main hydraulic barrier to water infiltration into the ash and would have a hydraulic conductivity of 1×10^{-9} m/s. Water balance calculations using the HELP model (Schroeder et al, 1984) indicate that this combination of cover materials will limit long-term infiltration to approximately to 4% of the mean annual precipitation or 34 mm/yr (Heystee, 1992). This infiltration rate is similar to the current leachate exfiltration rate through the landfill base.

The 450,000 m³ of clay that would be required to construct the cover will likely be

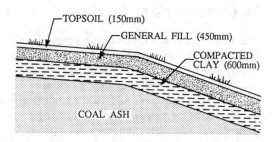

Figure 4 Proposed landfill cover materials

obtained in the immediate vicinity of the landfill site and would be placed on the landfill using standard construction procedures. However if clay is found to be unavailable or too costly, then alternate hydraulic barrier materials such as a geosynthetic clay liner (Daniel, 1991) or a stabilized fly ash material will be considered.

The general fill material will be required to protect the clay barrier from the adverse effects of freeze/thaw and desiccation processes, and will provide a medium for plant growth. Since a variety of clean soils could be used for this layer, it is anticipated that the 330,000 m³ of general fill material will be obtained from a various local sources.

The cover will also require 110,000 m³ of topsoil and will be vegetated to prevent soil erosion, and to promote evapotranspiration water losses. It is anticipated that the ultimate end use of the site will be public parkland. The cover design and the vegetation will be selected so that it is compatible with this end use and to minimize the visual impact.

A network of drainage ditches will be constructed on the final cover to handle storm water run-off from the landfill surface and to avoid severe erosion damage to the landfill cover materials. It is predicted that the run-off from the landfill could result in a peak flow rate of approximately 3 to 4 m³/s in the perimeter ditches. All ditches will be lined to prevent erosion damage that would otherwise occur during rain storm events.

5.3 Contingency plans

The long-term landfill monitoring program will include the visual inspection of the landfill cover, and the collection of water and leachate samples. Ground water quality monitoring on the north, east and south sides of the landfill will be conducted to detect significant changes in ground water quality that may be related to leachate migration. Should the on-site aquifer monitoring program indicate that off-site leachate migration will occur, then a contingency plan would be implemented. The preferred contingency plan would be to the establish a contaminant attenuation zone with restricted land use.

An alternative contingency plan would be to lower the water-table in the mound and thus reduce the leachate exfiltration rate through the landfill base. This could be achieved by two different methods. The first method would involve installing an internal well system in the ash mound and reducing the landfill water-table. The pumped leachate would likely have to be treated before discharge to Lake Erie. The second method would involve installing a tighter cover on the landfill. The water infiltration rate would become smaller than the exfiltration rate through the landfill base. This, in turn, would lead to a lowering of the landfill water-table.

6 SUMMARY

The construction of a low permeability cover on the Nanticoke GS coal ash landfill will reduce the water infiltration rate so that it is approximately equal to the current leachate exfiltration rate through the landfill base. Under these water balance conditions all soluble components of the coal ash will be forced to exfiltrate through the landfill base and to be naturally assimilated back into the subsurface environment. This type of landfill design, known as a "natural attenuation landfill", would make it unnecessary to collect and treat leachate in the form of leachate discharges through the landfill sides. A balance between water infiltration and exfiltration, will also ensure that the water-table within the landfill does not rise

above elevation 197 m or the top elevation of the existing containment dykes. This will make the landfill sideslopes safe against sloughing-type slope failures during the post-closure period.

Leachate exfiltration through the thick clay formation into the bedrock aquifer has lead to a mounding of the aquifer potentiometric surface and a radially divergent ground water flow pattern beneath the landfill. These ground water flow patterns will continue during the post-closure period and will likely lead to the development of a leachate plume in the immediate vicinity of the landfill. Leachate from the landfill would ultimately discharge into Lake Erie via the bedrock aquifer. It is anticipated that contaminant concentrations in this discharge water will meet provincial surface water quality guidelines.

Ground water quality monitoring will be conducted over the next two to three decades to detect any changes in aquifer water quality in the vicinity of the landfill. The water quality data will be used to calibrate a leachate migration model and to make more accurate predictions of future leachate migration. The results of these analyses will confirm the suitability of the aforementioned closure plan or lead to the modification of this closure plan.

REFERENCES

Allan, R.E. & J.F. Sykes 1992. Nanticoke TGS coal ash landfill, analysis of ground water flow and contaminant transport. Ontario Hydro Report 92045.

Daniel, D.E 1991. Geosynthetic clay liners. *Geotechnical News* 9: 28-33.

D'Austous, A.Y. et al 1989. Fracture effects in the shallow groundwater zone in weathered Sarnia-area clay. *Canadian Geotechnical Journal* 26: 43-56.

Dayal, R. 1991. Personal communication

Golder Associates Ltd 1988. FGD waste disposal at Nanticoke GS, preliminary hydrogeological investigations. Ontario Hydro Report 88184.

Heystee, R.J. 1992. Nanticoke TGS coal ash landfill, final closure plan. Ontario Hydro Report 92023.

Jensen, M.R. 1992. Nanticoke TGS coal ash landfill, ground water quality impact study. Ontario Hydro Report 92037.

Jensen, M.R. 1994. Nanticoke GS coal ash landfill, 1993 hydrogeology field program and numerical flow simulation, Ontario Hydro Report FBU-25200-3

Johnston H.M. & K.E. Eagleson 1991. Chemical characteristics of Ontario Hydro coal fly ash: a review. Ontario Hydro Research Division Report 89-155-K.

Johnston, H.M. 1992. Personal communication.

Lau, K.C. and H.T. Chan 1992. Long-term slope stability of the ash mound at Nanticoke G.S. Ontario Hydro Research Division Report S92-9-K.

Schroeder, P.R., et al 1984. The Hydrological Evaluation of Landfill Performance (HELP) model. US EPA Office of Solid Waste and Emergency Response Report EPA/530-SW-84-009: Washington D.C.

Vorauer, A.G. 1989. Groundwater study at Nanticoke ash lagoon, part 1: physical hydrogeology. Ontario Hydro Rep 89-185-K.

Vorauer, A. 1990. Groundwater study at Nanticoke ash lagoon, part 2: chemical analysis of local groundwaters. Ontario Hydro Report 90-214-K.

Réalisation des barrières étanches composées de sols perméables traités à la bentonite pour le fond et la couverture des centres de stockage de déchets

Construction of impervious barriers by adding bentonite to permeable soils for waste disposal

J. Cavalcante Rocha & G. Didier
INSA, Laboratoire Géotechnique-Bât. 304, Villeurbanne, France

RÉSUMÉ: Devant la difficulté de trouver des formations argileuses pour y implanter des sites destinés au stockage des déchets conformément à législation (perméabilité K, inférieure à $1*10^{-9}$ m/s) des solutions utilisant des outils géotechniques permettent de répondre à ces critères, à partir du traitement de sols perméables à la bentonite.

Nous présentons ici des résultats d'essais en laboratoire sur des mélanges sol-bentonite qui, après compactage, présenteront une perméabilité à l'eau inférieure à $1*10^{-9}$ m/s, ainsi que des résultats d'essais de perméabilité avec un lixiviat provenant d'e deux sites d'enfouissement téchnique de classe II.

Enfin nous analysons la performance hydraulique des barrières qui pourront être réalisées, pour le fond et la couverture de décharges, à partir de ces sols traités, selon les résultats obtenus des essais de perméabilité à l'eau et au lixiviat.

ABSTRACT: In order to overcome the difficulty of constructing a controlled landfill with an adequate geological formation (K < $1*10^{-09}$ m/s, French specicication), the solutions implement a geotechnical method of adding bentonite to a permeable soil, in order to improve the hydraulic conductivity. The results are presented of laboratory permeability tets on compacted sand-bentonite mixtures, thet were permeated by two different leachates. Finally we analyze the performance of a proposed barrier (liner and cover), the tests consisted of a soil mixture saturated with a tap water, and then percolated with two lechates of domestics landfills.

INTRODUCTION

La circulaire ministérielle relative aux dispositifs de stockage de déchets ultimes inertés établit les caractéristiques géologiques et hydrogéologiques du site et introduit deux niveaux de sécurité. De bas en haut nous trouverons:

- le niveau de sécurité passif représenté par une formation géologique naturelle ou par un sol dont les caractéristiques ont été améliorées et présentant sur une hauteur continue un coefficient de perméabilité égal ou inférieur à $1*10^{-09}$ m/s;

- le niveau de sécurité actif constitué d'un matériaux drainant ayant un coefficient de perméabilité égal ou supérieur à $1*10^{-4}$ m/s reposant sur une géomembrane.

Face aux difficultés pour répondre naturellement à ces contraintes réglementaires, basées essentiellement sur la valeur du coefficient de perméabilité et sur l'homogénéité du substratum, il peur être envisagé l'amélioration des caractéristiques d'un sol par l'utilisation de la bentonite.

Cette étude concerne l'évaluation en laboratoire de la performance hydraulique d'une barrière passive composée de sable-bentonite et la méthodologie utilisée pour recommander un mélange compacté donnant, suivant la réglementation, une perméabilité, K, inférieure à $1*10^{-9}$ m/s.

ETUDE DE SABLES TRAITES A LA BENTONITE

Pour atteindre les objectifs d'une barrière passive, une approche faisant état des connaissances et des procédures existantes (techniques de compactage, méthodes d'essais, prélèvement des échantillons, etc.) doit être initialement établie, celle-ci devra s'appuyer sur des mesures fiables.

Nous présentons ci-après les étapes qui amèneront au choix d'un mélange idéal de sable-bentonite et sur la conduite d'essais qui permettront d'évaluer la performance hydraulique de ce mélange:

a) une investigation préliminaire des propriétés des matériaux utilisés: la granulométrie du sable, le gonflement de la bentonite ou des fines argileuses éventuellement présentes dans le sable;

b) l'étude de la compatibilité chimique de la bentonite exposée au lixiviat, dans le cas d'un centre de stockage déjà existante ou à un lixiviat synthétique ayant des caractéristiques similaires à celles qui seront produites par les déchets devant être stockés;

c) compactage du sable à l'énergie Proctor normal, (ou modifié) conforme à l'essai ASTM, pour établir la teneur en eau optimale (Wopt), conférant au sable la densité sèche maximale;

d) préparation et stockage des mélanges, dosés à différents pourcentages de bentonite, et humidifiés à des teneurs en eau correspondant d'une part, à la teneur en eau optimale et d'autre part au coté humide de l'optimum du sable utilisé (ex.: Wopt+2%, Wopt+3%);

e) compactage des mélanges et essais de perméabilité à l'eau afin de définir le pourcentage optimal de bentonite, le couple teneur en eau-densité sèche et l'énergie de compactage qui confère au mélange un coefficient de perméabilité, $K < 1*10^{-09}$ m/s;

f) réaliser un essai de perméabilité au lixiviat sur l'échantillon donnant un K à l'eau inférieur à $1*10^{-09}$ m/s afin de assurer qu'il n'y aura pas changement important de la perméabilité.

Chapuis (1990), Marcotte et Mardon (1993) ont montré que les sables contenant des fines confèrent au mélange sable-bentonite une perméabilité plus faible que celle des mélanges constitués par des sables propres. En ce qui concerne les étapes (a) et (b), lorsque nous sommes en présence de plusieurs types de sables et de bentonites, les mélanges se feront plutôt avec les sables ayant le plus fort pourcentage de fine

($\% < 80\mu$) et les bentonites présentant d'une part un plus grand indice de gonflement à l'eau et d'autre part une faible inhibition du gonflement au lixiviat.

Lorsqu'un mélange de sable-bentonite compacté à l'énergie Proctor normal ne donne pas le coefficient de perméabilité escompté; on peut envisager d'augmenter l'énergie de compactage et donc la densité pour conférer au mélange la perméabilité désirée. Cette solution sera problablement plus économique que celle consistant à augmenter le pourcentage de bentonite dans le mélange. Le choix de l'énergie de compactage est un critère économique qui dépendra aussi des engins de compactage disponibles sur le chantier.

Le compactage se fera à la teneur en eau optimale ou de préférence du coté humide de l'optimum. En effet, Cavalcante Rocha et Didier (1993) ont constaté que la bentonite presente dans les éprouvettes compactées du coté humide est plus stable et chimiquement plus résistante qui celles compactées du coté sec de l'optimum.

Pour que les recommandations citées plus haut puissent être employées efficacement il est essentiel de maîtriser les paramètres qui influencent les mesures de perméabilité en laboratoire. Pour ce faire, nous observerons les règles suivantes:

- éviter l'utilisation des gradients hydrauliques excessifs, un gradient hydraulique maximum de 100 peut être admis à condition qu'il soit augmenté par paliers;

- il faut s'assurer que la perméabilité est mesurée à la saturation de l'échantillon. Le volume de liquide passant à travers l'échantillon doit être équivalant à 4 ou 6 fois le volume des vides. Notons qu'il est possible d'accélérer la saturation par application d'une contre-pression, Daniel et al. (1984);

- pour chaque gradient, les mesures doivent être effectuées lorsque le régime permanent est atteint;

- il faut effectuer des mesures de la température pendant l'essai afin de corriger la perméabilité en fonction de la variation de viscosité.

METHODE DE MESURE

Les essais de perméabilité ont été réalisés dans des perméamètres à parois rigides, constitués d'une bague acier en inoxydable de diamètre 9.0 cm et de de 4.0 cm hauteur, fermée par deux embases en PVC. Les éprouvettes sont compactées à une teneur en eau correspondant à la teneur en eau du sable seul compacté à l'énergie Proctor normal.

La saturation de ces échantillons se fait sous une faible charge hydraulique de 5 cm. Les essais sont réalisés à charge constante, avec un gradient hydraulique variant graduellement, par palier de 25, jusqu'a une valeur maximale de i=100.

Pour la mesure du gonflement unidirectionnel on a utilisé des échantillons compactés à une densité sèche de 1.40, sans aucune contrainte appliquée.

BENTONITE

La perméabilité d'un mélange constitué de sable-bentonite dépend surtout de la structure de la bentonite dans le mélange. A l'état sec une particule de montmorillonite ressemble à un livre fermé, composé de plusieurs feuillets maintenus ensemble par les forces de Van der Waals et par les cations. Chaque feuillet a une déficience de charge au niveau de la structure cristalline mais la neutralité de l'ensemble est assurée par la présence de cations éloignés de la surface des feuillets. Lorsque l'eau et la bentonite sèche sont mélangées, l'eau rentre dans les particules de montmorillonite pour hydrater la surface des feuillets et les cations.

Afin d'étudier le comportement des mélanges suivant les bentonites utilisées, nous avons réalisé deux études. Dans la première nous avons utilisé une bentonite activée, (Na)Activée 1, alors que pour la deuxième étude nous avons utilisé en plus deux bentonites de natures différentes: une activée, (Na)Activée 2 et la Wyoming, qui est une montmorillonite sodique naturelle. Leurs caractéristiques sont décrites dans le Tableau I, où nous donnons les valeurs du gonflement unidirectionnel et les limites d'Atterbeg lors de la saturation à l'eau du réseau et à deux lixiviats provenant d'un site de stockage de classe II.

Tableau 01: Caractéristiques de la bentonite utilisée.

Bentonite	Eau	Lixiviat L2	Lixiviat L1
WYOMING	LL= 530 LP= 197 G=305%	LL= 365 LP= 123 G=222%	LL= 360 LP= 128 G=164%
(Na) ACTIVEE 1	LL= 282 LP= 121 G=276%	LL= 258 LP= 94 G=96%	LL= 241 LP= 39 G=113%
(Na) ACTIVEE 2	LL= 328 LP= 84 G=187%	LL= 320 LP= 50 G=144%	LL= 241 LP= 39 G=65%

LL: limite de liquidité;
LP: limite de plasticité;
G: le gonflement à une densité sèche de 1.40.

LE SABLE

Nous avons utilisé trois sables dont les analyses granulométriques sont présentées figure 01:

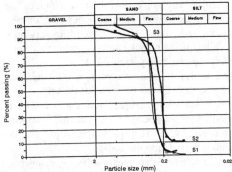

Figure 01: Analyse Granulométrique
S1: sable propre mal gradué; Sm;
S2: sable limoneux (10% des fines;80μ), SL;
S3: sable propre bien gradué (2% fines;<80μ);Sb

Les sables S2 et S1 correspondent à des matériaux prélevés in-situ à u
ne teneur en eau naturelle de 12% et 4%, respectivement. Le sable S3 provient d'une carrière de sable destiné à la fonderie.

La figure 02 donne la courbe de compactage à l'énergie Proctor normal de ces sables. Le sable S2, donne des densités plus élevées vu la quantité des fines existantes. Les couples teneur en eau optimales-densité sèche maximale relevés sont : sable S2 (10%; 1.80), sable S1 (10%; 1.65) et le sable S3 (13%; 1.624).

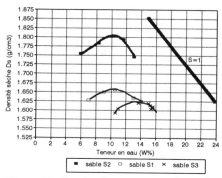

Figure 02: Courbe de compactage des trois sables étudiés.

LE LIXIVIAT

Il est très important de réaliser des essais de perméabilité en employant, comme liquides de percolation les lixiviats habituellement rencontrés, puisqu'on travaille avec des matériaux argileux (bentonite). Il est connu que les produits suivants peuvent augmenter la perméabilité:
- les acides et base: en grandes quantités par dissolution des certains constituants du sol;
- les liquides inorganiques non acides (sels en solution) en affectant la double couche électronique; par exemple l'effet d'un sel de calcium sur une bentonite sodique;
- les composés organiques (solvants, essences), les effets deviennent significatifs lorsque ce composées sont présents en fortes concentrations.
Nous avons utilisé deux lixiviats provenants de centres de stockage de classe II, dont les analyses sont présentées dans le Tableau 02.

Tableau 02: Caractéristiques des lixiviats utilisés.

	EAU	L2	L1
pH	7.14	7.59	7.78
Cond.(μS/cm)	434	7308	5863
TA(méq/l)	0	0	0
TAC(méq/l)	3.6	60.2	117.5
Mg(mg/l)	6.3	73.4	41.3
Ca^{+2}(mg/l)	123.7	72.22	107.6
Cl^-(mg/l)	30.75	1280	4830
NO_3^-(mg/l)	7.52	0.72	0.55
PO_4^{3-}(mg/l)	0	0	18
SO_4^-(mg/l)	22.14	26.57	9.4
Na^+(mg/l)	8.68	413.8	2160
NH_4^+(mg/l)	0	544.6	2060
K^+(mg/l)	1.79	331.8	896
DCO(mgO/l)	0	552.21	2050

LES MELANGES

Les échantillons étudiés sont réalisés en mélangeant manuellement un poids P_s, de sable sec, un poids P_b de bentonite et en ajoutant la quantité d'eau relative à la teneur en eau, choisie après l'étape de compactage du sable seul. Par la suite ces mélanges sont stockés pendant une période de 24h, avant le compactage. Le pourcentage de bentonite, B est défini par:

$$B = \frac{P_b}{P_s} * 100$$

Le sable S2 a été dosé à 2, 4, 5 et 6 % de bentonite, (Na)Activée 1 et à une teneur en eau de 12 % (Wopt+2), correspondant à la teneur en eau naturelle de ce sable. Le choix de cette teneur en eau est d'éviter une opération en plus qui pourrait être l'ajustement de la teneur en eau insitu par séchage du sable constituant la barrière étanche.

Les mélanges faits avec les sables S1 et S3 ont été réalisés à une teneur en eau optimale des ces sables, respectivement de 10% et 12% et avec les pourcentages de 2, 3, 4, 5 et 6 % de la bentonite sodique activée, (Na)Activée 1.

Le Tableau 03 donne les caractéristiques des échantillons obtenus:

Le Tableau 03 donne les caractéristiques des échantillons obtenus:

Tableau 03: Caractéristiques des échantillons

Sable	B (%)	W (%)	Ds	Keau (m/s)
	0	10.77	1.65	4.6E-06
	2	11.50	1.659	8.6E-09
S1	3	10.93	1.717	1.26E-10
	4	10.95	1.732	1.06E-10
	5	11.47	1.759	8.22E-11
	6	10.97	1.786	2.92E-11
	0	11.68	1.782	2.5E-08
	2	12.53	1.839	9E-11
S2	4	12.4	1.85	2.5E-11
	5	12.37	1.874	2.2E-11
	6	12.16	1.881	4.7E-11
	0	12.97	1.624	6.2E-06
	2	12.73	1.627	2.5E-09
S3	3	12.65	1.655	1.5E-10
	4	13.38	1.685	1E-10
	5	12.56	1.709	1.2E-10
	6	12.49	1.716	1.6E-10

B: pourcentage de bentonite;

W: teneur en eau;

Ds: densité sèche;

Sr: degré de saturation.

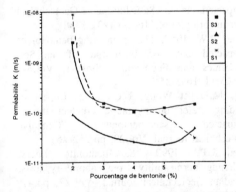

Figure 03: Variation de la perméabilité en fonction du pourcentage de bentonite.

Nous constatons que, pour tous les sables utilisés la perméabilité décroît avec le pourcentage de bentonite. Pour les sables S1 et S3 nous obtenons une perméabilité superieure à 1E-09 m/s pour un pourcentage de bentonite de 2%. Pour ce même pourcentage, dans le cas du mélange avec le sable

S2, la perméabilité est de l'ordre de 1E-10 m/s. Ceci est du à la presence de 10% d'élements inférieures à 80 μ.

Afin d'étudier l'influence du type de bentonite sur la perméabilité, le sable S3 a été dosé avec les bentonites (Na)Activée 1, (Na)Activée 2 et la Wyoming, à un même pourcentage, de 4%, et à une teneur en eau de 13% (teneur en eau optimale du sable obtenu au Proctor normal). Les caractéristiques des échantillons obtenus, sont données dans le tableau 04. Après avoir mesuré la perméabilité à l'eau du réseau, l'eau est remplacée par les lixiviats (L1 et L2). Ensuite nous suivons l'évolution de la perméabilité au lixiviat jusqu'à stabilisation. Les résultats obtenus sont synthétisés sur l'histogramme figure 04.

Tableau 04: Caractéristiques des échantillons:

Bentonite	W (%)	Ds g/cm	Keau (m/s)	KL1 (m/s)	KL2 (m/s)
(Na) Activé 1	12.85	1.684	1E-10	4E-10	***
(Na) Activé 2	12.96	1.660	3E-09	2E-09	***
Wyoming	12.67	1.697	4E-10	1.6E-10	***
(Na) Activé 1	12.88	1.678	1E-10	***	2.5E-10
(Na) Activé 2	12.70	1.680	3E-09	***	8.5E-10
Wyoming	12.98	1.703	4E-10	***	1.5E-10

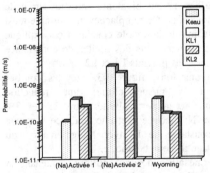

Figure 04: Variation du coefficient de perméabilité après remplacement d'eau par les lixiviats L1 et L2, respectivement.

Nous remarquons, figure 04, que seules les mélanges contenant les bentonites (Na)Activée 1 et Wyoming permettent d'avoir un coefficient de perméabilité, K, inférieur à $1*10^{-9}$ m/s. Ces deux bentonites presentent, tableau 01, les plus forts gonflement et limite de plasticité lorsqu'elles sont hydratées avec l'eau du réseau.

Malgré que la performance, K < 1E-09 m/s ne soit pas vérifiée pour le mélange constitué de la bentonite (Na)Activée 2, nous avons continué les mesures de perméabilité en présence des lixiviats L1 et L2. Nous constations une amélioration de la perméabilité aux deux lixiviats.

ANALYSE DES RESULTATS

Les résultats sur le Tableau 01 montrent que les limites de consistance ainsi que le gonflement determinés en utilisant les lixiviats L1 et L2 sont moins importants que ceux obtenus avec l'eau du réseau. La plus importante réduction est observée lors de la saturation avec le lixiviat L1, qui est le plus chargé en cations Ca et K. Le potassium ayant un rôle d'inhibiteur de gonflement justifie les réductions observées.

Foreman et Daniel (1986), Fernandez et Quigley (1985) ont montré que les propriétés de la double couche affecte la conductivité hydraulique et les limites de consistance des argiles. Quand une bentonite sodique est percolée par un liquide riche en Ca, ce dernier se substituera au sodium de cette bentonite, puisque l'affinité des cations bivalents prévaut sur celle des monovalents. Ce remplacement entraîne ainsi une réduction de la double couche. Ceci explique en partie les variations des limites de consistance réalisés avec les lixiviats L1 et L2.

Nous constatons, figure 03, que les courbes présentent la même allure: une réduction importante de la perméabilité entre 2% et 3% bentonite (Na)Activée 1 et une tendance vers une valeur constante de K avec l'augmentation du pourcentage de cette bentonite.

La perméabilité des échantillons contenant la bentonite (Na)Activée 2 et la Wyoming diminue lorsqu'ils sont percolés avec les lixiviats, figure 04. Ceci est sûrement dû au changement de la structure de la bentonite qui peut causer un déplacement et le colmatage des échantillons. Nous remarquons que, malgré le changement du coefficient de perméabilité en presence des lixiviats L1 et L2, celui ci reste toujours inférieur à 1E-09 m/s pour les mélanges ou furent utilisés les bentonites (Na)Activée 1 et Wyoming.

CONCLUSION

Bien qu'un faible pourcentage de bentonite (d'ordre de 2.5% pour les sables étudiés) ait donné le coefficient de perméabilité attendu, il est conseillé de réaliser des mélanges à un pourcentage plus élevé si l'homogénéité in-situ de l'ensemble risque de n'être pas assurée.

Dans la réalisation d'une couche étanche constituée de sable-bentonite, il doit être recherché la vulnérabilité à l'attaque des produits chimiques ou à des percolats de la bentonite employée. Cette qualification peut être assurée par des mesures du gonflement, des limites de consistance et de la perméabilité réalisées en laboratoire.

La performance hydraulique finale devra être contrôlée par de mesures de perméabilité sur des échantillons prélevés in-situ.

REFERENCE BIBLIOGRAPHIQUE

Benson, C.H; et Daniel, D.E. 1990. Influence of clod size and hydraulic conductivity of compacted clays. ASCE Journal of Geotechnical Engenineering, Vol.116; pp.1231-1248.

Lambe, T.W. 1958. The engineering behavior of compacted clay. ASCE Journal of the Soil Mechanics and Foundation Division, Vol.84, pp.1655-1_1655-35.

Haug, M.D. et Wong, L.C. 1992. Impact of molding water on hydraulic conductivity of compacted sand-bentonite. Canadian Geotechnical Journal, Vol.29, pp.253-262 .

Chapuis,R.P. 1990. Sand-bentonite liners: predicting permeability from laboratory tests. Canadian Geotechnical Journal, Vol.27, pp.47-57.

Cavalcante Rocha, J. et Didier, G. 1993. Laboratory permeability tests: evaluation of hydraulic conductivity of sand-liners. IV International Landfill Symposium, Sardinia 1993. CISA publisher. Proc.I, pp. 299-304.

Technogenesis influence on the frozen ground properties

Influence technogénique sur les propriétés des sols gelés

V.I.Grebenets & D.B.Fedoseev
Research Institute of Bases and Underground Structures, Norilsk, Russia

A.B.Lolaev
Norilsk Industrial Institute, Russia

ABSTRACT: Negative tecnogenic influences on the engineering - geocryological properties in Norilsk industrial zone are analyzed in this paper. Natural observations of dangerous cryogenic processes, technogenic salinity and thermo-regime were carried out. The aggravation of engineering geological conditions and permafrost were emphasized by the investigations. The evolution of the ground in different natural technogenic landscapes is shown. Some methods of improving ground properties have been worked out.

RÉSUMÉ: Cet article rend compte de l'analyse de l'influence négative de l'activité économique sur les propriétés géologiques dans la région au nord de la zone industrielle de Norilsk. On montre les résultats de l'observation du développement des processus criogeniques dangereux, du salinage et du changement de la température des terrains. On a constaté l'aggravation des conditions géotechniques des tendances la dégradations dans la congélation des sols sous l'influence de l'activité économique. On montre les modifications de la capacité portante des terrains dans des cadrer naturelles différents. On réfère des projets d'amélioration des propriétés géotechniques des terrains.

Big cities negatively influence on the natural-territorial complexes, climate, permafrost and ground conditions and other components of nature. What concerns the engineering-geological characteristics changes (bearing capacity, conductivity, etc). In addition to the global climatic heating tendency and to the prognosticated permafrost zone destroyment, technogenesis influence brings to the negative factors: mechanical, chemical and heating pollutions of the natural engineering-geocryological complexes.

Only in Siberia in the areas of large mining and metallurgical entertainments dozens millions cubic metres of different depozits are accumulated on the land surface, which have formed new, locally situated technogenic types of grounds with high level of the pollution. Erosion and cryopelitisation of the upper ground layers, non-prognosticated changes of the surface level due to the new depozits, freesing of the bottom ground layers, evolution of hydrogeological and hydrological conditions, glaciololgical genesis influence and other processes bring to the composite structures arising with poor texturical consolidation and number of frozen-melted layers, which makes them not-availabel for building purpose (Grebenets & Fedoseev, 1992). Poor northern vegetation exclude the traditional recultivation methods usage. Technogenic depozits occupy in Norilsk region 150 sq.km. and due to the wind, snow melting, rains, underground water migration they are the sources of environmental pollution. Different pollutants, first of all lightly dissolved salts, penetrate into the ground complexes and change engineering ground properties.

The rocks chemical composition changes also occur with the acid rains or ground pollution while the snow melting. For example, Norilsk plants throw out into the atmosphere 2 millions tons of sulphuric dioxide, dozens thousands tons of chlorine, nitrogen, dust, heavy metals, etc. While the ground salting, the bearing capasity and the temperature of the beginning of freezing decrease, thick layers of underground ice (10...12 m), the active layer thickness increases and permafrost temperature rises. In table 1 the results of on-location geochemical invistigations in different natural-technogenic landscapes are presented.

Technogenic and active layer grounds are salted. Most of pollutants are concentrated at the grounds surface and active layer bottom. Average depth of of the active layer is 0.4...0.7 metres in peats and clays and 1.8...2.5 metres in sandy grounds.

Heating influence of industrial centres and global climatic changes are also negative. For example, at the depths 20...60 metres permafrost temperature rised on 0.6°C at last 30 years (Grebenets, 1989), this fact testifies about the whole thermal field degradation, because the annual climatic changes don't reach that depths. Similar tendency makes stronger the degradation of local thermal fields in the upper ground layers under buildings. It must be noted, that the main part of buildings in the North were built in 50-80-s years according to the principal rule - preservation of permafrost during all the explotating period - the bearing capasity is being ensured by the freezing forces between the underground foundation and frozen grounds. Destroyment of permafrost doesn't only increases the buildings deformations, but, firstly, can lead to the ruins of buildings, secondly, can lead to environmental pollution - pollutants are conserved in the upper permafrost layers.

Many buildings are deformated or destroyed because of bad engineering-geocryological conditions: 35% of buildings in Dudinka, 30% - in Igarka, 10% - in Norilsk. Chemical and heating pollution brought to the formation of vast territories with high-temperature grounds (upper 10...15 metres) with permafrost temperature 0...-0.3°C. This horisonts are not stable and include much ice (25-40%). Dangerous engineering-geological processes occur: technogenic pollution rises, diagenes, thermocarst, erosion, etc. It is very difficult to use this grounds as the bases for building purposes - high-temperature permafrost exists, but the bearing capasity of permafrost is very small. Such territories, as a rule, occupy the nearest to the urban zone places and the process of town grouth is problematic. Special meliorations (Grebenets, 1990) improving grounds properties with the help of different autonomic seasonally-cooling constructions which utilise natural cold are put into practise. A lot of waste ground places

with very composite ground conditions exist and they are not used in building purposes.

Changes of engineering-geocryological ground properties are not equal in different landscapes. The main part of pollutants penetrates into the ground through the surface and active layer and is being accumulated in permafrost. The ammount of pollutants is different for territories of tundra and urban zone. On-location investigations proved that natural complexes are being developed in the conditions of the aggravation of the frozen grounds properties, so, the problem of analysing of the changes and working out of special technical methods of engineering-geological properties improvement is an actual task.

Technogenic landscapes with specific techno-geological and biological processes are being formed arround the cities.

In the conditions of piedmont territories with the uniform offensive of the town technogenic landscapes settle down concentrically. While the evolution of cryogenic, hydrogeological, technogenic and other processes go on, the surface appearance in close arrival to the urban zone is being submitted to technogenesis.

One of the biggest nothern cities Norilsk is an example of confirmation betweeen technogenesis and cryolitogenesis. Geographical conditions and man's activities change the orientation of the landscapes, but in the northern direction in the bounds of Valikovskaya limno-alluvial terrace they are quite clear.

There can be observed the following territorial complexes: 1) The forest-tundra zone in the bounds of Valikovskaya limno-alluvial terrace; 2) The urban zone surroundings with the serious destroyments of cryogenic complexes; 3) Mountain and piedmont landscapes, which are involved in industrial occupation; 4) The contemporary urban zone with complicated permafrost and ground conditions; 5) The territories of building mastering; 6) Agricultural landscapes; 7) Territories with the artificial lakes (system of waters refining); 8) Industrial zone; 9) Depozits of industrial waste materials; and some others.

Urban territory is the epicentre of antropogenic violatons. Twenty years of on-location observations over the geothermal conditions at 30-50 metres depths in the central part of the urban zone exhibited the growth of the temperatures for 0.3...0.5°C, meanwhile the similar trend of the average air temperature is absent. Heating of the upper layers of permafrost brings to the water saturation of rocks and the thickness of a seasonally melted

Table 1. Maintenace of different chemical elements in ground moisture, mg/l, Norilsk, Siberia.

Type of ground and place	Contents chemical elements, mg/l							
	Ca++	Mg++	Na+,K+	Cl-	SO4--	HCO3-	Ni	Cu
1. Wastes of nickel plant, depth 0.2 m, place 1	4430	1153	638	220	15682	622	-	-
2. Wastes of nickel plant, depth 0.2 m, place 2	1209	682	145	148	5593	342	-	-
3. Technogenic ground (metallurgical slag), depth 0.5 m, place 3	6881	45	900	238	17326	1514	1.30	1.80
4. Loam of active layer from urban zone, depth 2 m	184.5	86.8	36.2	57.7	1690	875	0.003	0.014
5. Sand of active layer from urban zone, depth 0.5 m	326	55.0	34.7	14.0	804	195	-	-
6. Marine loam of permafrost, depth 21.3 m	67	12	4.6	10	78	158	0.0019	0.014
7. Loam of active leer 50 km far from Norilsk, depth 1.55 m	38	16	39	4	118	-	0.01	0.04

layer increases.

Most of the undertakings are situated at the moutain slopes with average steepness 8...15°C. It was estimated that industrial undertakings are the local heating and water saturating lots. Friable artificial grounds are used in the constructions of bases. Water saturation of the artificial grounds brings to the solifluction beginning (fast solifluction is frequent). Different relief forms arise: banks, ridges and terrases.

Mechanical breaching and loosening of rocks in the time of communications and roads laying and other building works bring to activisation of errosion processes.

The road on the slopes, long constructions and different object change orohydrogeography of foothills. Temporary and permanent water flows arise. The flows repeat at the orientation of manly performed constructions.

Numerous unrational depozits of cultivated rocks and rubbish heaps dam the natural watercourses, that leads to the beginning of antropogenic screes, landslides and mud flows.

Problem of stabilising of natural-technogenic landscapes and frozen grounds properties improving can be salved in the present climatic nothern conditions with the help of special man's activities.

It is more expedient to improve the frozen grounds properties in two main directions: usage of different technical methods with orientation on landscapes evolution influence and construction of special underground systems utilising natural nothern cold resourses. Coomplex of technical methods has been worked out which enables to stop the negative tendencies and improve ground properties.

REFERENCES

Grebenets, V.I. & Fedoseev, D.B. 1992. Industrial influence on glacial process in mountains of circumpolar regions. In Annals of Glaciology, 16, 212-214.

Grebenets, V.I. 1989. Izmeneniye pod vliyaniyem zastroyki geokriologicheskoy obstanovki v g.Noril'ske [Permafrost changes under building constructions in Norilsk]. In Problemy inzenerno-geologicheskikh izyskaniy v kriolitozone [Problems of engineering-geological investigations in the permafrost zone]. Magadan, Gosstroy Russia, p.328-331. [In Russian].

Grebenets, V.I. 1990. Antifiltration curtains constructions with the natural cold utilization. In Proceedings of the 6th International EREG Gongress. Amsterdam, 6-10 August, 1990. A.A.Balkema (Rotterdam/Brookfield), p.1285-1287.

La méthode d'augmentation des propriétés écraniques des sols aux endroits du stockage de déchets toxiques industriels

The method of increasing screening ability of soils in the foundation of waste disposal sites

S. A. Lapitski, V. I. Sergeev, Z. P. Malachenko & M. E. Skvaletski
Laboratoire de la Protection du Milieu Géologique (LOGS), Faculté de Géologie, Université Lomonossov, Moscou, Russie

RESUME: Les auteurs ont étudié la possibilité d'augmentation des propriétés écraniques des sols de la zone d'aération à l'aide des solutions silicatées formant des hydrosilicagels. La capacité d'adsorption (la capacité de rétention) des hydrosilicagels a été étudiée par rapport aux métaux lords: Cu, Cd, Zn et d'autres. La liquidation des dislocations structurales dans des sols de la zone d'aération à l'aide des gels permet d'utiliser complétement leur capacité d'adsorption et d'augmenter l'effectivité écraniques des sols naturels.

ABSTRACT: The possibility to increase the sorption capacity of the aeration zone soils with the aid of gel-forming silicate solutions was studied. The sorption capacity of the gels was investigated with respect to heavy metals: Cu, Cd, Zn, etc. When structural disturbances of soils in the aeration zone are eliminated using gels, the soils manifast their potential sorption capacity that makes the soil layer to serve as a natural barrier.

1. INTRODUCTION

Parfois dans des régions industrielles où il est nécessaire de stocker des déchets toxiques il est impossible de trouver dans ces conditions géologiques un endroit satisfaisant pour la constuction des dépôts de déchets par example là où la zone d'aération est composée des roches sableuses, des loess ou des dépôts argileux dont l'épaisseur n'est pas grande.

Les auteurs ont examiné la possibilité d'augmentation des propriétés écraniques de tels sols à l'aide des solutions silicatées formant des hydrosilicagels.

On sait que le silicat de sodium (le verre soluble) forment avec certains produits chimiques le hydrosilicagel, lequel on peut utiliser pour le tamponage des pores du sol. Simultanément avec la diminution de la perméabilité du sol ce système "sol-hydrosilicagel" a la capacité d'adsorption des métaux lourds.

2. LA CAPACITE D'ADSORPTION DES HYDROSILICAGELS

On a étudié la capacité d'adsorption (la capacité de rétention) des hydrosilicagels de différentes modifications. Les essais de percolation ont été éffectués dans des conditions dynamiques où travers l'échantillon "sol-hydrosilicagel" on filtrait des solutions de différents métaux lourds et une phase liquide de déchets, provenant d'entreprises de l'idustrie de l'extraction d'or.

Comme les dursisseurs du silicat de sodium on a utilisé les substances suivantes: $Al_2(SO_4)_3$, $H_2C_2O_4$, $MgSO_4$, $CaCl_2$, $MgCl_2$, H_2SiF_6 et d'autres (Lapitski et al. 1992)

Au cour de l'action des sels solubles de Ca, Mg et Al avec le silicat de sodium se forment des hydrosilicagels avec les liens calsium-siloxan, magnesium-siloxan et aluminium-siloxan. Il en resulte la formation de l'hydrosilicagel dont le réseau est ajouré avec le liquide intramicellaire à l'intérieur. Cette phase liquide contient du sodium, du silicat, de l'alcali et de Ca, Mg et Al sels du dursisseur qui n'ont pas réagit.

La composition et la structure de ces hydrosilicagels sont favorables pour la future adsorption des cations des métaux lourds. La structure des hydrosilicagels est proche à la stucture des zéolithes.

Ces processus s'effectuent par l'interaction avec la surface solide d'hydrogel mais aussi qu'avec le liquide intramicellaire.

Dans le but d'obtenir des caracreristiques quantitatives d'adsorption de différents hydrosilicagels et d'étudier ces derniers comme des barrièrs géochimiques on a fait une série des essais de laboratoire.

La capacité d'adsorption de différents hydrosilicsgels a été examinée dans les conditions dynamiques (l'essai de percolation). Les solutions des sels de Cu, Zn, Ni, Cd, Mn, Co, Fe on été injectées (le gradient de filtration variait de 1 à 5) aux echantillons des gels. La concentration de métal en petits volumes percolés à travers une couche de gel a été determinée à l'aide du spectrophotomètre de l'absorption atomique "Hitachi Z-8000".

La figure 1 montre la courbe de retention qui est typique pour des essais. A la base de ces courbes on a calculé la quantité des ions de métaux retenues. La durée des essais a varié de quelques jours à 1,5 ans.

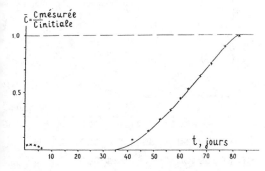

Figure 1. La courbe de reténtion de cadmium par le gel n°1. C_{in} de Cd en solution est égale à 43 mg/l.

On a été établi que la valeur de la capacité d'adsorption (N) dépend de deux facteurs essentiels: de la composition de solution et de la concentration (C). La valeur N pour tous les gels varie de 2 à 20 mg/cm^3. Si C<200 mg/l les valeurs se trouvent dans des limites 2-5 mg/cm^3.

Les tableaux 1 et 2 représentent les résultats principaux des essais de la capacité d'adsorption des différents gels par rapport aux métaux lords étudiés.

En effet la capacité d'adsorption des gels avec de différents dursisseurs ne varie plus qu'en 2-2,5 fois. La solution n° 1 avec le dursisseur complexe $H_2C_2O_2$ + $Al_2(SO_4)_3$ est préférable du point de vue technologique.

Tableau 1: La capacité d'adsorption des différents gels (avec des dursisseurs différents) pour le cuivre et le cadmium (C<50 mg/l).

n° de gel	Dursisseur de verre soluble	Capacité d'adsorption (N) mg/cm^3 pour des métaux	
		Cu	Cd
1	$H_2C_2O_4$ + $Al_2(SO_4)_3$	2,0	3,5
2	$H_2C_2O_4$ + $MgCl_2$ + $Al_2(SO_4)_3$	2,5	3,1
3	$H_2C_2O_4$ + $MgCl_2$	4,0	2,6
4	$CaCl_2$	4,4	2,8
5	H_2SiF_6	3,0	3,5

Tableau 2: La capacité d'adsorption de gel n°1 pour le cuivre, le cadmium. le zinc des solulions dont les concentratons sont différentes.

Elément adsorbat	Concentration étudiée mg/l	Capacité N mg/cm^3
Cu	24	2,0
	214	2,1
	700	2,2
	4500	17,5
Zn	25	2,3
	175	2,3
	250	2,4
	560	3,6
Cd	43	2,6
	75	2,5
	120	3,0
	440	4,9

La relation des composés
en 1 m^3 de la solution

solution A $(Na_2O \cdot nSiO_2)$; $\gamma = 1.19 - 1.13 \,\%/cm^3$; ℓ	solution B $(H_2C_2O_4 \cdot 2H_2O + + A\ell_2(SO_4)_3 \cdot 18H_2O)$; ℓ
625 — 833	375 — 167

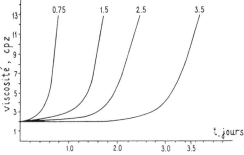

La durée de la formation
du gel

Le changement de la viscosité
de solution en dépendance
du temps de la formation du gel

Figure 2: La composition et les propriétés de la solution n°1

Le temps de formation de gel pour cette solution peut être controlé facilement et ne dépend pas de la température dans le domaine +5 ÷ +30°C. La figure 2 montre la composition de la solution recommamdé et le changement de sa viscosité dans le temps au cour de la formation de gel.

3. LE MODELE ELABORE DE SIMULATION MATHEMATIQUE DE LA DESCRIPTION DU PROCESSUS DE LA MIGRATION ET DE LA DETERMINATION DES PARAMETRES DE MIGRATION

Les courbes de retention obtenues pour des échantillons de gel donnent la possibilité de déterminer non seulement la valeur de capacité d'absorption mais aussi des paramètres de migration.
A partir des investigations on a

déterminé que pour les conditions de l'écran en gel ou en sable-gel est appliqué le modèle de simulation d'hydrodispersion. Cela est possible si les paramètres n et D se changent en n_e et D_e.

$$n_e \frac{dC}{dt} + v \frac{dC}{dx} = D_e \frac{d^2C}{dx^2}$$

$$n_e = n + K_d \quad ;$$

$$D_e = \frac{K_d^2 \, v^2}{\alpha \, n_e^2} \quad ,$$

où $K_d = N/C$

Donc pour les calcules pronostiques on peut utiliser la résolution suivante du modèle de microdispertion:

$$C = 0.5 \; erfc \; \xi \; ,$$

$$\xi = \frac{vt - n_e t}{2 \sqrt{D_e n_e t}} \; ,$$

où: n_e - la porosité effective, D_e- le coefficient effectif de dispersion, v - la vitesse de filtration, t - le temps de passage de solution, l - la distance.

La détermination des paramètres de migration offre aussi la possibilité de calculer l'épaisseur mininale de l'écran en gel ou en sable-gel si la possibilité écranique des sols est insuffisante. Pour cela des paramètres de migration sont déterminés pour le materiel qui sera utilisé pour la creation de l'écran. L'épaisseur de l'écran (m) est calculée selon la formule suivante (Sergeev et al. 1993):

$$m = \frac{1}{n_e} \; (vt + 2\xi \sqrt{D_e n_e t} \;)$$

C'est ainsi qu'on a déterminé l'épesseur de l'écran en sable-gel était pour le site de confinement des déchets de la fabrique de l'extraction de Sb.

La superficie de site du stockage de déchets était égale à 1000 m^3, le volume des déchets de 90 à 100 m^3/j. La phase liquide des déchets a contenu le cadmium dont la concentration est égale à 0,66 mg/l. La capacité d'adsorption de gel par rapport au cadmium a été determinée comme 2,5 mg/cm^3 et conformement pour l'écran en sable-gel comme 0,87 mg/cm^3.

La courbe de retention pour la solution contenante le cadmium a été obtenue pour l'échantillon sable-gel (figure 1). Les paramètres de migration ont été calculés sur l'ordinateur par la méthode grapho-analytique. Le coefficient de corrélation montrant l'applicabilité du modèle mathéma tique est égal à 0,98 (figure 3a).

Le calcul de l'épaisseur de l'écran montre que celle-ci ne doit pas être moins que 0,5 m. Dans ce cas le cadmium ne passera pas à travers l'écran au cours de tout le temps de l'exploitation du site de stockage. La répartition de la concentration de Cd en écran en sable-gel dans 20 ans est repro duite sur la figure 3b.

La seconde destination des solutions formant des gels c'est la liquidation des dislocations structurales dans des sols de la zone d'aération. Cela donne la possibilité d'utiliser la capacité potentielle d'adsorption des sols argileux ou des loess.

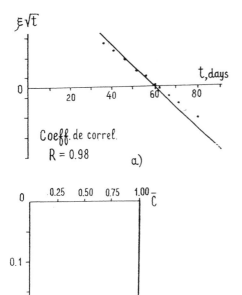

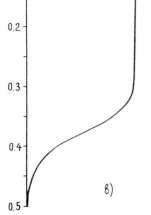

Figure 3: Les résultats de l'étude de la migration de Cd en mélange "sable-gel";
a) le graphique diagnostique de l'application du modèle d'hydrodispertion;
b) la répartition de la concentration relative de Cd en écran en sable-gel dont l'épaisseur est égale à 0,5 m après 20 ans de l'exploitation.

5. CONCLUSION

Les proprietés d'adsorption et de tamponage des hydrosilicagels nous a permis de les considérer comme le materiel pour l'augmentation de l'effectivité écranique des sols naturels. Dans ce cas le milieu hétérogène de la zone d'aération devient quasi homogène et des essais expérimentaux nous montrent que pour la simulation du transfer de masse on peut utiliser le model mathèmatique d'hydrodispertion.

6 REFERENCE

LapitskiS.A, Malachenko Z.P., Sergueev V.I. 1992, L'étude des silicagels pour leurs utilisation comme des adsorbents des métaux lourds.
La protection des eaux soutèrraines de la pollution. Edition MGU. p.47-53 (en russe).
Sergeev V.I., Shimko T.G., Kulechova M.L., Skvaletsky M.E. 1993, Procedure of investigating a subsoil layer as a geochemical barrier for heavy metals.
Géologie et confinement des déchets toxiques. GEOCONFINE 93/MONTPELLIER FRANCE. Edition BALKEMA, p.115-121.

Assessment of the screening capacity of the subsoil layer in the foundation of the sites intended for dumping toxic wastes

Évaluation de la capacité de rétention des couches du sous-sol dans les sites d'emplacement des décharges controlées de déchets toxiques

V. I. Sergeev, Ye. V. Petrova, M. L. Kuleshova & T. G. Shimko
Moscow State University, Russia

ABSTRACT: A study of clayey soils sorption capacity regarding heavy metals made it possible to work out the procedure for the determination of migration process parameters and on the basis of them to make a forecast of heavy metals trasport through out the aeration zone subsoil in time .The program MASSEX has been developed to solve the forecast problems in one-, two-, and three layer medium.

RESUME: L'étude de la capacité de rétention des sols argileux par rapport aux métaux lourds d'élaborer la méthode de détérmination des paramètres de migration et à la base de ces paramètres la méthode de la prévision du trasport des métaux lourds an fonction du temps dans la zone d'aération. Le programme "MASSEX" a été élaboré pour la résolution des problèmes pour des milieux à une seul, à deux ou à trois couches.

1 INTRODUCTION

The development of mining and mining-and-processing industry is associated with the need to dump the wastes. A study of the composition of these wastes indicates that, depending on the nature of production, they may contain harmful and toxic es like metals, cyanides, arsenic, antimony, etc.

In Russia as well as in some other countries, both solid and liquid wastes are dumped for storage. The ratio of the solid and liquid phases is usually 1:3. Thus, the dumping sites for the wastes of mining-and-processing plants, extracting non metals, gold, etc., are the sources of environmental pollution both during the period of tailing dumps operation and after they are preserved.

Ground water is that component of geological environment which is most susceptible to the polluting effect of industrial wastes. The usual mechanism of ground water pollution in areas of toxic wastes dumping is infiltration of the liquid phase through the foundation of the dump.

The only obstacle to pollutants infiltration is the layer of subsoil between the foundation of the dump and the aquifer surface (aeration zone). This natural screen may be represented by subsoils ranging from sands to heavy clays with a seepage factor from hundreds of meters per day to 0.001 m/day.

As we know, the clay component of dispersion soils may have a sorption capacity in respect of some chemical elements, in particular, of heavy metals. The intensity of the sorption process is dependent on the composition and structure of the sobsoil, the type of pollutants, their concentration and some other factors.

This paper deals with assessment of the subsoil layer in the aeration zone as a geochemical barrier on the way to migration of toxic pollutants. Such an assessment may be quantified in the form of a computed quantity of a pollutant (or of several pollutants) which may be retained by a particular subsoil layer at the sacrifice of the subsoil sorption capacity. On this basis, the question of the length of operation of a particular dumping site is resolved.

The study included the solution

of the following problems:

1. Determining the sorption capacity of the subsoils of varying grain size and mineral composition under static and dynamic conditions in respect of Pb, Cu, Zn, Co, Cd, Mn, Ni.

2. Development of a method to eliminate seepage non-uniformity in loamy subsoils of the aeration zone.

3. Choosing a mathematical model to describe the process of toxic elements migration in the subsoils of varying grain size. Determining migration parameters of the process of migration.

4. Solution of pollutant migration forecast problems in a single- and multi-layer subsoil. Elaborating the procedure of calculating a maximum permissible period of operation of a site for dumping toxic wastes and computing the thickness of a creen (if it is necessary), ruling out ground water pollution.

2 RESULTS OF STUDIES

The studies of the sorption capacity of clay subsoils of varying composition were conducted under static and dynamic conditions.

Static batch tests make it possible to determine maximum, or potential, sorption capacity of the subsoil in respect of a particular pollutant, as they use samples with a disturbed structure. The procedure of staging such experiments is described in [1]. Table 1 presents some data on the sorption capacity of clay subsoils in respect of some heavy metals.

The studies performed with 21 different types of dispersion subsoils in respect of the aforesaid seven heavy metals indicate that, from the standpoint of their adsorption capacity, heavy metals fall into two groups: (Pb, Cu, Zn) and (Cd, Co). Depending on the subsoil, absorption of metals in the first group may be 100 times different. The crucial factor of their absorption is the presence of carbonaceous minerals. This account for high values of N in loess subsoils, characterized by a high content of calcite. Apparently, when Pb, Cu, Zn interact with carbonaceous subsoils, the main processes are precipitation and co-precipitation of the hydroxyl complexes of these metals.

The elements of the latter group generally get sorbed worse than those of the former one. The fundamental processes here are physical adsorption and ionic exchange.

Experiments under static conditions, yielding results relatively fast, make it possible to assess, at the first stage of the studies, the sorption capacity of the subsoil. Building up a bank of data on the values of sorption capacities of the ion subsoils of varying grain size in respect of heavy metals will enable one in future to assess it by the data on mineral and chemical composition of the subsoil.

In natural conditions, the interaction of a polluted water flow with subsoils is different from the conditions of a static batch test. In subsoils with an undisturbed structure, the surface of particles, coming into contact with the solution, smaller than their combined surface, which is due to the aggregated condition of the subsoil, the presence of cohesive water and trapped air. All this preconditions the low value of active porosity despite the high value of total porosity. Therefore sorption capacity of the subsoil obtained in column tests under dynamic conditions is invariably lower than one obtained in batch tests for the same metal. It has been found through numerous experiments that the values of sorption capacity in dynamics and in statics differ 2 to 4 times.

Column tests with samples of undisturbed structure (Fig. 1), despite the long time they take, are essential not only for obtaining a more accurate value of sorption capacity. The breakthrough curves make it possible to determine the values of the process migration parameters: these allow us to forecast the changing concentrations of heavy metals in a subsoil layer time-wise.

An important prerequisite for a column test is maintaining a constant rate of solution filtration through the sample, which is achieved by using micro-pumps.

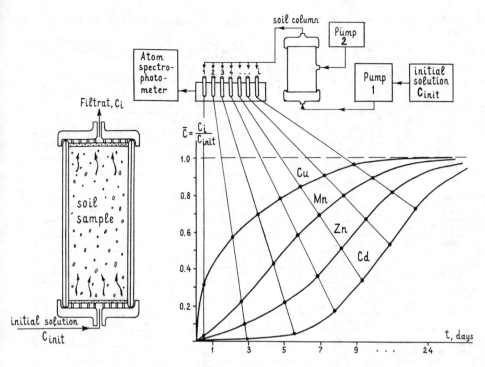

Fig. 1 The scheme of sorption capacity study in dynamic conditions

3 ELIMINATING SEEPAGE NON-UNIFORMITY OF SUBSOILS IN THE AERATION ZONE

The characteristic feature of the upper part of the aeration zone is structural non-uniformity of the subsoil which is due to plant root residues, burrows left by rodents, etc. These micro-disturbances of the structure result in that seepage non-uniform. The area of large seepage canals may account for just 1 - 5% of the subsoil area, but these become major conduits for pollutants transport. As its time of contact with a pollutant is too short, the subsoil cannot display its potential sorption capacity.

In this case, migration parameters, obtained in the subsoil sample, will be different from the process of pollutant transport in the aeration zone.

The notion of a "differentiated permeability of the subsoil" was introduced specifically for the characterization of seepage non-uniformity, and a method to determine this permeability was devised. The method is based on the use of a gel-form cate solution and on determining the discharge of the liquid through various structural openings and through the main body of the subsoil. A diagram of a field experiment is presented in Fig. 2.

The gist of the method consists in finding, at the first stage of the experiment, (a) the liquid discharge through the natural body of the subsoil and (b) the seepage factor K1. Then, macro-disturbance are eliminated from the seepage process with the aid of a silicate gel-forming solution with a viscosity close to that of water, the time of gel formation being under full control. At the third stage, the discharge of the liquid and the seepage factor K2 are determined for the same area with filled macro-disturbances.

The field studies performed at some project sites indicate that the discharge of the liquid through structural openings may amount to 90% of the total discharge through the subsoil (Table 2).

Table 1. Composition and sorption capacities of subsoils for heavy metals (static conditions)

N	Soil, age, genesis (location)★	Content of particles <0.01mm	Mineral composition					Cation exchang. mg-equiv per100g	Absorption capacity for metals, mg/l						
			mont-moril	hydro mica	kaoli nite	cal-cite	dolo-mite		Cu	Zn	Pb	Cd	Mn	Ni	Co
1	2	3	4	5	6	7	8	9	10	11	12	13	14	15	16
1	Loess-like loam $Q_{\overline{II}}$rs	48,2	1.6	2.9	-	8.3	5.7		108	88		10	4		
2	-"- $Q_{\overline{II}}gl$	44,1	7,3	7,8	0,5	11,8	8,1	30,9	130	82	200	11	6	5	8
3	-"- $Q_{\overline{IV}}sd$	38,1	6,0	6,8	1,5	10,6	6,3	26,2	100	50	130	10	6	6	7
4	Loess-like clay $Q_{\overline{II}}$	74,1	-	2,2	1,4	10,9	-	34,5	100	40	100	13	6	11	10
5	Clay loam MZ	46,6	0,2	0,3	0,4	0,9	-	12							
6	-"-	46,3	-	27,8	14,1	-	-	3							
7	Light loam MZ	26,1	-	-	26,7	-	-	2							
8	Heavy benton. clay P	95,6	91,4	-	-	7,9	-	90,5	19	10	72	35	10	16	10
9	Kaolinite clay P	82,79	-	-	97,6	-	-	14,7	5	2	15	4	2	4	1
10	Hydromica clay C	71,67	-	96,1	-	-	-	14,3	42	18	65	20	10	15	8
11	Clay loam N	45,1	-	33,0	-	3,8	9,9	27,0	80	40	160	13	9	7	20
12	Loess-like loam Q	31,21	5,4	9,5	-	2,5	3,6	17,0	35	15	64	14	8	16	10
13	Light clay N	52,64	6,9	25,7	1,5	-	-	11,0	5	7	11	12	8	9	6
14	Clay Q	72,09	3,5	3.1	0,8	-	-	22,0	15	13	18	15	9	15	8
15	Clay P	73,06	0,6	25,3	-	7,6	8,7	8,0	53	17	80	12	7	9	7
16	Light clay P	65,09	2,3	15,6	-	-	-	7,0	12	8	20	10	9	10	4
17	Loess light clay Q	54,45	3,9	2,9	0,5	-	-	20	14	9	17	14	9	14	8
18	Heavy clay N	99,27	0,4	21,3	-	10,6	14,2	28,0	56	26	44	19	12	13	8
19	Loess heavy clay Q	90,87	-	8,1	-	7,7	16,0	8,0	53	12	82	9	7	7	6
20	Loess clay loam Q	47,91	0,6	4,6	-	-	-	21,0	18	10	26	13	12	14	7
21	Soil		4,6	0,6	0,2	-	-	37,0	5	3	12	10	3	4	6

★ 1,2,3 - Central Asia; 4 - Kasakhstan; 5,6,7 - South Ural, t.Plast; 8 - Gorgia; 9 - Ukrain;10 - St.Petersburg; 11,12,13,14,15,16,17,18,19,20 - Slovakia; 21 - Perm.

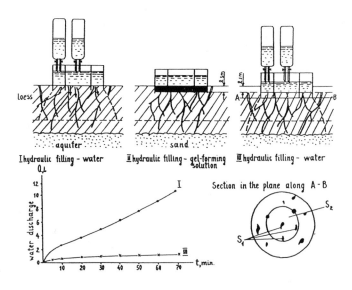

Fig. 2 The diagram of differential permeability determination

Table 2. Results of differential permeability determination

Soil, genesis	K1, m/day	K2, m/day	Discharge through macrodisturb
Loess-like loam a-d $Q_{\underline{\Pi}}$ ts	0.50	0.10	80%
-"-	0.35	0.09	75%
-"-	0.55	0.20	65%
Loess-like loam a-pr $Q_{\underline{IV}}$ sd	0.25	0.05	80%
Loess-like loam a-pr $Q_{\underline{III}}$ gl	0.40	0.20	50%
Pebbles with loamy filler a $Q_{\underline{IV}}$sd	4.00	0.20	95%

Gels that are used to fill the structural macro-disturbances exhibit not only a low permeability (seepage factor of 0.005 - 0.001 m/day), but a relatively high sorption capacity in respect of heavy metals.

Not all lithologically-different subsoils manifest seepage non-uniformity. For example, it is unlikely in heavy, solid clays and sands. However, structural disturbances are typical of the aeration zone made up of loess subsoils.

When seepage canals are filled with a gel-forming solution, the non-uniform environment of the aeration zone becomes quasi-uniform. In this case, mass transfer models may be used to describe it (Sergeev, Shimko et al.1993)

4 MATHEMATICAL SIMULATION OF THE PROCESS OF POLLUTANT TRANSPORT

To be able to simulate the processes of pollutant migration in an unsaturated porous medium, which is represented by the subsoils of the aeration zone, MASSEX program has been developed, designed for calculating the migration parameters and solving prediction problems.

When plotting the model, a combination of various physico-chemical processes is taken into account. In each particular case, depending on the natural features of the site, composition of migrating elements, the scale of space and time and nature of boundary conditions, there must be a specific model.

With the aid of the existing mathematical tools, using special mathematical functions, it is possible to calculate parameters of geomigration and sorption capacities in respect of a chemical element under study. The system of data processing well-known equations, describing mathematical models of the process of migration of chemical elements in the subsoil.

MASSEX program is based on three models of migration: a micro-dispersion model, a heterogeneous-block model with a concentrated capacity and a heterogeneous-block model of unlimited capacity.

The process of mass transfer in a uniform unsaturated porous subsoil medium can most adequately be described with the aid of a micro-dispersion model. In the case of a single-layer subsoil, this model will be represented by this system of a differential equation and boundary conditions:

$$n = \frac{\partial C(x,t)}{\partial t} = D \frac{\partial^2 C(x,t)}{\partial x^2} - V \frac{\partial C(x,t)}{\partial x} \quad (1)$$

initial condition :
$C(x,0) = C^0$-background concentration
boundary conditions :
$C(0,t) = C_0$ - initial concentration at point $x = 0$
$C(\infty,t) = C^0$ - concentration at point $x = \infty$
where n is effective porosity, D - coefficient of diffusion, v - rate of seepage.

Solution looks like:

$$\bar{C}(x,t) = \frac{C(x,t) - C^0}{C_0 - C^0} = \frac{1}{2} erfc \left(\frac{nx - Vt}{2\sqrt{Dnt}} \right)$$

where
$$erfc(x) = \frac{2}{\sqrt{\pi}} \int_x^\infty e^{-x^2} dx$$

The use of this solution makes it possible to calculate concentration of a pollutant at any point of the subsoil layer at any moment of time.

To obtain the migration parameters of the mass transfer process in the subsoil, they use breakthrough curves obtained in the tests involving samples of the subsoil of an undisturbed structure under dynamic conditions (Fig. 2). A breakthrough the relationship $C(x, t) = C(t)$ for a sample whose length is designated L. Migration parameters n and D are determined from the obtained relationship and chosen model of migration.

5 SOLUTION OF PREDICTION PROBLEMS

Next stage of studies is solution of prediction problems. A forecast of a pollutant migration in a specific subsoil time-wise enables us to assess it as a natural geochemical barrier. Such an assessment can be carried out in different options
1. For a subsoil of known thickness and known composition of pollutant elements contained in the wastes, it is possible to calculate the distribution of all studied elements depth-wise at any desired moment of time (Fig. 3). In this manner, a most dangerous pollutant (or a group of pollutants) can be discovered and on that basis a decision can be made regarding the dumping of wastes or the need of their pre-treatment.

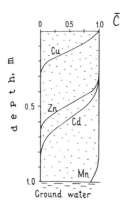

Fig. 3 Nature of pollutants spread in subsoil layer over 5.9 years

2. For a subsoil layer of known thickness and known composition of pollutants it is possible to calculate maximum period of time for tailings dump operation, during which not a single pollutant will be able to reach the aquifer (Fig. 4). The period of safe operation of a tailing dump can be calculated from this formula

(a micro-dispersion model):

$$T = \frac{n\left(2\xi^2 \cdot D + m \cdot V - 2\xi\sqrt{\xi^2 \cdot D + D \cdot V \cdot x}\right)}{V^2} \qquad (2)$$

m - thickness of the subsoil layer in the aeration zone, c - maximum permissible concentration of a pollutant at the lower boundary of the aeration zone.

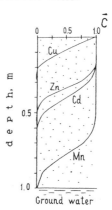

Fig. 4 Nature of pollutants spread in subsoil layer over prognosis time 4.5 years

3. When the design period of a tailing dump operation and chemical composition of the wastes are known, it is possible to calculate the depth of pollutants penetration into the subsoil layer of a particular composition (Fig. 5). This makes it possible to evaluate the need for a man-made screen (e.g. of clay) and determine its thickness by this formula (a micro-dispersion model):

$$m = \frac{1}{n}\left(V \cdot T + 2\xi \sqrt{D \cdot n \cdot T}\right) \qquad (3)$$

T - desired length of time of a tailings dump operation.

Under natural conditions, the aeration zone is frequently made up of several layers of the subsoil, that differ by grain size and structure. Each layer may be characterized by migration parameters organic to it. The existing developments in t of simulation of pollutants migration in multi-layer subsoils have made it possible to suggest the following approach to the solution of prediction problems for multi-layer media (Fig. 6).

The system of equations describing the process of mass

relative concentration $\bar{C}$

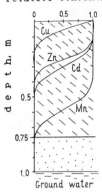

Fig. 5 Forecast of pollutants spread in subsoil layer at the depth of 0.75 m over 3.2 years

transfer for each subsoil layer is supplemented, as a layer boundary condition, by a relationship between the concentration of a chemical element and time. At the boundary between the first and second layers, this relationship will take this form:

$$\bar{C}(R,t)=\frac{1}{2}\left\{erfc\left(\frac{n_1R-V_1t}{2\sqrt{D_1n_1t}}\right)+exp\left(\frac{V_1}{D_1}\right)Rerfc\left(\frac{n_1R+V_1t}{2\sqrt{D_1n_1t}}\right)\right\}$$

(4)

where R-border between 2 layers, n v D -parameters of the 1st layer.

The form of this function will be getting more and more complicated as the number of layers increases.

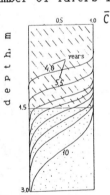

Fig. 6 Nature of Mn spread in two-layer subsoil for different periods of time

The aforesaid approach is used in the MASSEX program, which has made it possible to expand substantially the category of prediction problems under consideration.

6 CONCLUSIONS

1. The study of absorptive capacity of clay subsoils in respect of heavy metals has indicated that the value of sorption capacity N is dependent on their chemico-mineral composition and dispersion. Subsoils of varying composition may have sorption capacity N for the same metal in a wide range of values. For example, N for copper for studied subsoils ranges from 2 to 130 mg/g. The same subsoil is capable of adsorbing metals in this series: Pb > Cu > Zn = Cd > Co > Ni > Mn in the amount from 200 t 6 mg/g for loess carbonaceous subsoils, and from 15 - 20 to 1 - 2 g/g for carbonate-free clay subsoils.

Analysis of the findings enables us to conclude that in some cases, the subsoil layer between the foundation of a tailings dump and aquifer surface may serve a reliable screen, blocking migration of pollutants to ground water.

2. The role of macro-disturbances in the structure of aeration zone subsoil (burrows left by rodents, plant root residues) for the type of pollutants transport to the aquifer has been established.

A method has been developed to eliminate seepage non-uniformity of the subsoils in the aeration zone: a gel-forming solution makes it possible not only to reduce sharply subsoil permeability, but to augment its sorption capacity.

3. Mathematical models are used for the forecasting of chemical elements concentrations in the subsoil layer. Migration parameters also make it possible to determine maximum permissible duration (T) of operating a particular subsoil layer, i.e. the length of time during which a pollutant will be contained within the aeration zone. A method of calculation has been proposed to determine T.

MASSEX program has been developed to establish the migration parameters and solve problems of forecasting transport of pollutants within a multi-layer subsoil.

REFERENCES:

Sergeev V. I., Shimko T. G. et. al. 1993 Procedure of investigating a sub-soil layer as a geochemical barrier for heavy metals. Proc. Geo confine 93:115-121. Rotterdam: Bal kema.

The principle of drawing a map of ground water protection at dumping sites intended for toxic industrial wastes

Principes de la cartographie de la protection des eaux souterraines dans les sites de décharge de déchets industriels toxiques

T.G. Shimko & N.A. Svitoch
Laboratory for Geological Environment Protection, Department of Geology, Moscow State University, Russia

ABSTRACT: The article deals with making special large-scale maps, when choosing suitable sites for dumping industrial wastes that contain heavy metals. Filtration and sorption properties of the soil layers of the aeration zone and technogenic load are taken into account.

RESUME: L'article est consacré à l'élaboration des cartes géologiques en grande échelle pour l'emplacement des lieux de stockage de déchets contenants des métaux lourds. Pour cela on tient compte des propriétés de filtration, la capacité de rétention des sols de la zone d'aération et la charge technogène.

1 INTRODUCTION

In the former USSR and elsewhere it still continues open-air dumping not only of solid wastes but also effluents with toxic elements. The size of dumping areas ranges from 0,5 to 20 or 30 sq.km, the quantity of dumps varying from 3 or 5 thou.cu.m to 200 thou.cu.m per day. It determines the map scale that ranges from 1:2000 to 1:10000.

The dumping sites for solid industrial wastes and effluents containing harmful substances constitute major sources of aquifer pollution in such areas.

Protection of ground waters from pollution is determined by the screening capacity of the soil thickness overlaying the aquafer. That screening effect depends on the thickness, structure and composition of the constituent lithological matter.

The proposed approach to assessment of aquifer protection in waste dump areas consists in the following.

The investigations are confined by the zone of aeration, assuming that no pollutant should reach the aquifer.

The result of the investigations is a map of aquifer protection in the particular area intended for dumping wastes. The map highlights areas that perform as natural screens throughout the entire period of dump operation, and other areas that will function as screens at different times. A map like this shall be the basis for making decisions as to what should be done to protect the aquifers, viz. making all kinds of artificial screens, e.g. of clay or other materials.

The principal approach and factors that are taken into account in mapping aquifer protection are shown in Fig.1.

Ground water protection depends on natural screening capacity of the subsoils in the aeration zone and technogenic load (volums of wastes, concentration of toxic elements in their liquid part and its chemical composition).

Natural screening capacity of the subsoil layer is determined by two factors: filtration protection and absorption capacity of the soils in respect of pollutants present in the wastes.

Such an aproach has preconditioned the need for additional research compared with common investigations in engineering geology which is carried out at the second stage of mapping.

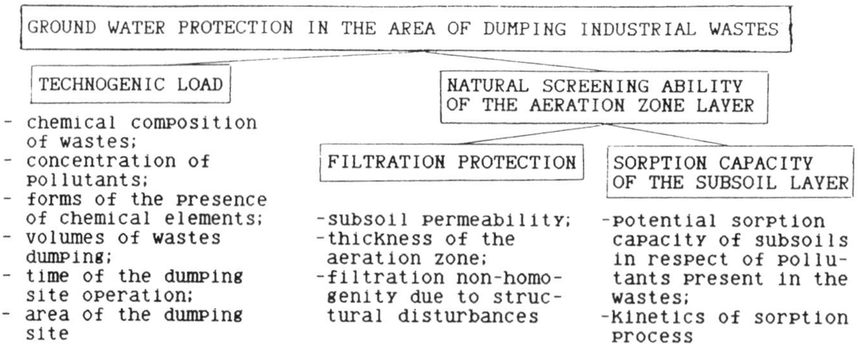

```
┌─────────────────────────────────────────────────────────────────┐
│ GROUND WATER PROTECTION IN THE AREA OF DUMPING INDUSTRIAL WASTES │
└─────────────────────────────────────────────────────────────────┘
```

```
┌──────────────────┐              ┌──────────────────────────┐
│ TECHNOGENIC LOAD │              │ NATURAL SCREENING ABILITY│
└──────────────────┘              │ OF THE AERATION ZONE LAYER│
                                  └──────────────────────────┘
```

- chemical composition
  of wastes;
- concentration of
  pollutants;
- forms of the presence
  of chemical elements;
- volumes of wastes
  dumping;
- time of the dumping
  site operation;
- area of the dumping
  site

```
┌──────────────────────────┐  ┌──────────────────────────┐
│ FILTRATION PROTECTION    │  │ SORPTION CAPACITY        │
└──────────────────────────┘  │ OF THE SUBSOIL LAYER     │
                              └──────────────────────────┘
```

-subsoil permeability;
-thickness of the
 aeration zone;
-filtration non-homo-
 genity due to struc-
 tural disturbances

-potential sorption
 capacity of subsoils
 in respect of pollu-
 tants present in the
 wastes;
-kinetics of sorption
 process

Fig. 1. Factors determining screening capacity of the subsoil layer
in the foundation of a waste disposal site

2 TWO STAGES OF MAPPING

Two levels of mapping can be
alloted. The first is choosing of
a suitable site for dumping; the
second is investigation of ground
water protection at the particular
place.

2.1 Making a map of typological zonning

The work begins from the study of
fundamental geological materials
that characterize the studied area.
Structural-tectonical, facial and
lithological conditions are
analysed. Special attention is
given to the reflection of signs
capable of determining the
filtration protection: spesified
are tectonic disturbance zones,
facia types, lithological
composition of soils in the
aeration zone, thickness of such
soils. At the first stage the
traditional engineering-geological
approach is used. The stage is
finished by making a map where
quazi-homogeneous (by lithological
structure) sections are specified.
An example of such maps can be seen
in Fig. 2a and 3a. On the basis of
them a tentative decision about
waste dumping place location is
accepted.

2.2 The second stage of mapping

Befor final drawing the map of
aquifer protectin special
investigations are made at every
isolated area.

2.2.1 Study of filtration heterogeneity

It can be seen in Fig. 1 a reference
to filtration heterogeneity. Which
means identification of the role of
macro - disturbances in the permiability of the aeration zone (due to
residues of the root system of
plants, burrows left by
rodents, etc.). Why do we stress
this? It is known that clayey soil
has an absorption capacity, but in
the case of filtration along this
disturbances of a liquid containing
toxic elements, the surface of the
contact with the soil and duration
of the contact do not make it
possible for the soil absorption
properties to manifast themselves.
Taking into account this fact is
very important as sometimes discharge through macro-disturbances
can be up to 90% of the whole
discharge. This is typical especialy
for loess grounds. The method of
differential permeability determination is given in the article
"Assessment of the screening
capacity of subsoils in the foundation..." (Sergeev, Petrova, et. al)
in this Proceedings.

2.2.2 Study of sorption properties of the aeration zone subsoils

The second pre-requisite for
making a special map of aquifer
protection at a waste dump is a
study of the soil absorption
capacity regarding chemical
elements contained in the liquid
phase of the wastes.
This study is carried out in two

2552

ways: in batch experiments under static conditions and in percolation tests under dynamic conditions using soil samples of undisturbed structure. Sorption capacities of all types of soils available in the area of a projected waste dump with respect to all hazardous elements present in the liquid phase of the wastes are determined.

As a result of these investigations the parameters of the migration process are obtained, on the values of which it is possible to calculate the time it takes a particular chemical element to reach the water table in each lithological structure of the aeration zone.

A calculation of this nature is conducted with regard to the technogenic load, i.e. the quantity and concentration of dumped elements. The time of reaching the water table is determined for all present elements, whereas maximum allowed time of dump operation is calculated on the basis of the worst sorbed element. The calculation is done with the aid of math modelling (Sergeev, Shimko et. al 1993).

It must be emphasized here that migration parameters are determined taking into consideration the effect of complex formation on the sorption ability of toxic elements.

3 EXAMPLES OF MAPPING GROUND WATER PROTECTION

3.1 Gold-mining plant

The first example is an area of waste dump of gold-mining plant in the South Urals.

The dump bed features the presence of clay soils of Mezozoic mantle occupying around 70% of the area. Their thickness reaches 10-11 metres. The clays are underlain by loamy sands 5 to 80 m thick. The thickness of the aeration zone varies from 3 to 15 m. At the initial stage, soil typization was done to locate quazi-homogeneous areas by lithological composition, structure and filtration properties of the soils in the aeration zone.

Fig. 2a shows this typological diagram. Each of the specified types of area features a characteristic lithological section. Two lithologically-similar zones

can be seen in the map: 1 and Π with different thickness and structure of the aeration zone.

The first zone is characterized by prevalent presence of clays and loams, their filtration coefficient <0.001 m/day. Thickness-wise, 4 sub-zones can be isolated: 1-A - thickness of loams over 8m; 1-B - thickness of loams 6 to 8m; 1-C - thickness of loams 4 to 6m; 1-D - thickness of loams 2 to 4m.

The second zone is characterized by development on the surface of dusty and heavy loamy sands with permeability of 0.01 m/day. Thickness-wise (viz. by depth to the water table), in the aeration zone there are three sub-zones: over 15m thick; 10 to 15 m thick; and 5 to 10 m thick.

Each sub-zone may be sub-divided into sections by type of aeration zone construction: a - in the top part of the section, there is a thin layer (upto 2m) of slightly-permeable loams, underlain by loamy sands; b - entire aeration zone is made up of loamy sands, and c - aeration zone displays presence of fissured rocks.

Table 1 illustrates the types of lithological columns for some areas.

Following this tentative typization for each specified types of the area, we estimated the time it will take the pulp's liquid phase to reach the water table, using the well-known formulae, taking care of the filtration coefficient, soil thickness and hydrolic head. In one of the areas, the time we obtained was 10 years, meaning that even at this tentative stage that area could be regarded as a protected one, requiring no further investigations, providing the time of dump operation did not exeed 10 years.

As far as the other areas are concerned, The Table gives the time of reaching the water table, e.g., 870, 400, etc., days. For these sections a sorption capacity of soils in relation to chemical elements contaied in the wastes should be determined and a forcast of heavy metals transport have to be done.

The procedure of migration parameters determination is described in the article "Assessment of the screening capacity of the subsoil layer in the foundation of sites ..." (Sergeev, Petrova et.al.) of these

Table 1. Estimation of the zones as a screen

Section index, it's area, sq. m	Column,[*] thickness of layers m	Depth till water table m	Filtr. coeff. K m/day	Time of reaching the water table by liq. wastes	Sorp. capacity of subsoils metals, g/cu. m			Protection time, years		
					Cu	Zn	Ni	Cu	Zn	Ni
I-A 3200	8-9	8	0.001	870 days	200	40	8	135	98	141
I-C 224000	4-6 10-26	3-5	0.001 0.015	405 days	200	40	8	72	49	70
II-B-a 59200	0-2 20	10-15	0.001 0.012	– 145 days	200 150	40 35	8 5	– 126	– 102	– 105
II-B-c 24000	0.3	10-15	0.03	0.1 days	150	35	5	4	3	3

*Comment: 1 – clay; 2 – loamy sand; 3 – fractured rock

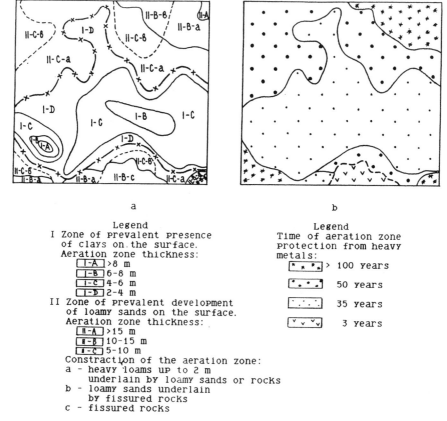

a   b

Legend
I Zone of prevalent presence
of clays on the surface.
Aeration zone thickness:
  I-A >8 m
  I-B 6-8 m
  I-C 4-6 m
  I-D 2-4 m
II Zone of prevalent development
of loamy sands on the surface.
Aeration zone thickness:
  II-A >15 m
  II-B 10-15 m
  II-C 5-10 m
Constraction of the aeration zone:
a – heavy loams up to 2 m
    underlain by loamy sands or rocks
b – loamy sands underlain
    by fissured rocks
c – fissured rocks

Legend
Time of aeration zone
protection from heavy
metals:
  > 100 years
  50 years
  35 years
  3 years

Fig. 2 Maps for the estimation of ground water protection in the foundation of waste disposal site: a – map of typological zonning; b – aquifer protection map

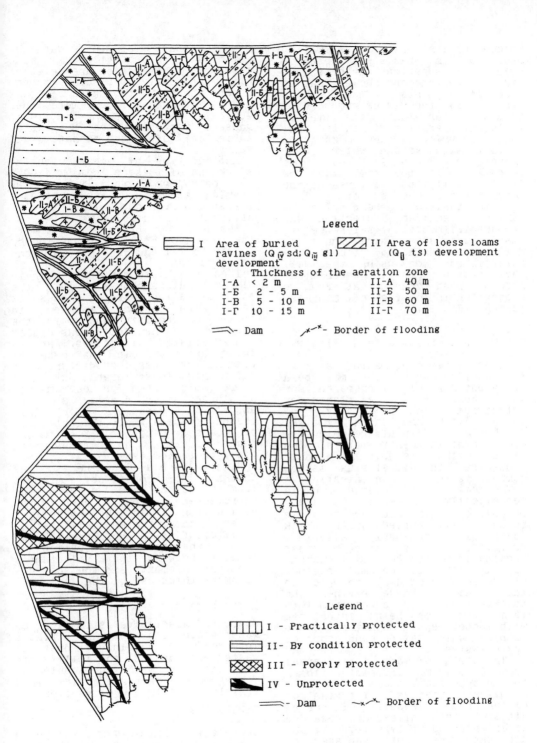

Legend

	I Area of buried		II Area of loess loams

I Area of buried ravines ($Q_{\overline{IV}}$ sd; $Q_{\overline{III}}$ gl) development   II Area of loess loams ($Q_{\overline{II}}$ ts) development

Thickness of the aeration zone

I-A	< 2 m	II-A	40 m
I-Б	2 - 5 m	II-Б	50 m
I-B	5 - 10 m	II-B	60 m
I-Г	10 - 15 m	II-Г	70 m

⟞⟝- Dam   -×-×- Border of flooding

Legend

I - Practically protected

II- By condition protected

III - Poorly protected

IV - Unprotected

⟞⟝- Dam   -×-×- Border of flooding

Fig. 3 Maps of typological zonning (a) and ground water protection (b) at Almalyк, Uzbekistan

Proceedings. A forecast estimation made on the basis of obtained figures of sorption capacity and migration parameters showed that in case of elements concentrations typical for the wastes produced by the particular plant, the time of pollutant migration within the soil thickness prior to the moment of contact with ground waters will amount, e.g., to 98, 49, 105 years. This is the time when the first of pollutants will reach the water table at the section.

On the basis of these results, a map of aquifer protection was prepared with special reference to that dump area where zones of different lengths of time were established for using the soil layer as a natural geochemical screen, and zones that require protection in the form of man-made screens (Fig. 2b).

3.2 Waste disposal site at Almalyk

Another example of making a map for aquifer protection may be a different map (Fig. 3) prepared for waste dump of several plants for extraction and processing of copper, zink and lead at Almalyk, Uzbekistan. Effluents reaching the dump contain heavy metals like copper, zink, lead, molibdenum, manganese and nickel plus macro-components in high concentrations.

The dump is located in a geologically complex area. Paleozoic, Cainozoic and Mezozoic rocks are all buried under thick layer of Quartenary alluvial-proluvial deposits, thier thickness ranging from 1 to 80m. Of fundamental importance is the omni-presence of ravines aged $Q_{iii}gl$ and $Q_{iv}sd$. In the areas of these ravines the thickness of loams ranges from 1 to 15 m, the loams being interlaced with loose loess-like materials and pebbles, which may be conductors of toxic wastes.

The depth of water table ranges from 25 to 90 m, increasing closer to the submontane area.

In the tentative soil typisation of the area, the soils of buried sites of river valleys are regarded as an independent type (type 1) and are subdivided into sub-areas of varying thickness of loess deposits.

The IInd type includes areas featuring loess loams aged $Q_{ii}ts$, 40 and 70m thick.

On the basis of these data the map of aquifer protection is made (Fig. 3b). The section are selected with absolutely protected ground water, where the thickness of loess and clayey subsoils is more then 40m. In these areas, the liquid phase of wastes reaches the water table only after the passage of 25 years.

Another type of areas with the thickness of loesses from 5 to 40 m is only protected from a few chemical elements. These serve as a screen for various pollutants over different periods of time.

Besides, two types of areas with the thickness of loess-type loams, from 0 to 2 m and from 2 to 5 m can be isolated.

Then we should consider the role of macrodisturbances in the anti-filtration properties of soils in the aeration zone. As we said above, macrodisturbances in the structure of the soil, especially characteristic of loesses, are as thick as 5 m. In the case of eliminating these macrodisturbances a 5 m thick loess layer constitutes a reliable screen for heavy metals as these soils display a high absorption capacity. For this reason, zones where the thickness of loesses range from 2 to 5m are regarded as an independent poorly-protected type.

Finally, areas where the thickness of loess soils ranges from 0 to 2m are grouped as unprotected since even the elimination of macro-heterogeneity, given the present filtration coefficient of the soil, will not ensure the required duration of contact between the pollutant's chemical elements and the soil due to small thickness of the soil.

4 CONCLUSIONS

Suggested quantative estimation of water protection at the places intended for dumping industrial wastes provides a possibility to fill the map with any contents. The examples presented in this article are only one of versions.

REFERENCES

Sergeev V. I., Shimko T. G., Kuleshova M. L. & Skvaletsky M. E. 1993. Procedure of investigating a sub-soil layer as a geochemical barrier for heavy metals. Proc. Ceoconfine 93: 115-121. Rotterdam: Balkema.

Underground nuclear waste repositories in deep geologic formations of hard crystalline rocks

Stockages souterrains profonds dans des roches cristallines résistantes

N. Barton, T.L. By & S.A. Yufin
Norwegian Geotechnical Institute, Oslo, Norway

N.N. Melnikov, V.P. Konukhin & A.A. Kozyrev
Kola Mining Institute, Apatity, Russia

ABSTRACT Engineering-geological conditions of the Kola Peninsula in NW Russia are considered from the point of site selection for creation of the regional underground nuclear waste repository in deep geologic formations of hard crystalline rocks. Exploration and research setup is being formed on the basis of previous experience in design and construction of large underground structures in similar geologic media.

RESUME Sont envisagées les conditions géotechniques du presqu' île de Kola au point de vue du choix des sites destinés à l'exécution des stockages souterrains des déchets radio-actifs dans les formations des roches cristallines dures. Les recherches et les études sont basées sur l'expérience d'études et de construction de grands ouvrages souterrains dans les milieux géologiques analogues.

The Murmansk and Arkhangelsk regions of North-West Russia have accumulated large amounts of radioactive waste produced from the Kola nuclear power plant, from the nuclear Northern fleet of the Russian Navy (submarines, aircraft carriers and cruisers), and from nuclear ice-breakers and light carriers of the Murmansk Merchant Marine.

All this waste is contained in temporary surface storage facilities lacking even primitive safety structures. Due to the shortage of even temporary storage facilities, nuclear waste from the Navy is sometimes contained on the ships, and the Navy continues to dump liquid LLW into the sea.

Evaluating this enormous amount of RW one should consider the waste from the Leningradskaya nuclear power plant (to the North of St.Petersburg) which also lacks storage facilities and be aware of a rather special problem of disposal of the decommissioned reactors and other ship structures. The only reasonable solution to this environmental problem is the design and construction of a system of underground disposal facilities in deep geological formations in hard crystalline rocks of the Kola peninsula region.

To ensure the long-term safe performance of a potential underground nuclear waste repository a number of technical aspects of the candidate site, such as geology, geohydrology, rock mechanics, tectonics, seismicity, volcanology and constructability should be investigated. The focus of these investigations will be to determine the characteristics and nature of the physical processses acting at the site, and the expected impact of these physical processes on the ability of the potential repository to permanently isolate radioactive waste.

The ultimate goal of the site characterization process is to determine whether or not a site under consideration is suitable for development as a repository and to evaluate possible permeability changes which may be imposed by the long-term operation of the repository.

Special importance is placed on predicting the effect of physical processes in the disturbed zone around the repository on the performance of the engineered barrier system, which may include layers of specially selected natural geological materials or mixtures.

Kola peninsula relates to regions represented by the early Pre-Caembrian continental earth's crust. The period of it's formation is tremendously long - 1700-3590 millions of years. Geological history of the Kola peninsula has two major stages:

1) - progeosynclinal and geosynclinal;
2) - platformal.

At the first stage plastic upper part of the litosphere was crumpled in folds mainly of north-west strike. After stabilization the consolidated earth's crust reacted to developing stresses by formation of faults and ruptures, mostly of north-west and north-east directions, which divided the territory of the region in blocks of different

Fig.1 Satellite view of the Kola peninsula

shapes and dimensions (Gorbunov et al, 1981). The general picture of the region is clear from the satellite photo on Fig. 1.

Structure of the earth's crust of the Kola peninsula and forming rocks bear traces of impetuous geodynamic processes. Regardless of the geological antiquity and rigid consolidation of the region's crust, modern geodynamic activity here is in evidence. It is manifested by earthquakes, by modern vertical and horizontal movements of earth's crust, by presence of separate abnormally overstressed blocks.

Modern territory of the Kola peninsula is formed by the crystalline rocks, partly covered by thin layers of loose soil. Relief of the surface is stupilated by the complex system of tectonic blocks, raised to different levels. The map of the modern tectonics of the Kola peninsula is presented on Fig. 2.

For the last 10000 years of the geologic development of the Kola peninsula, as of the eastern part of Fennoscandia, the following characteristics of movements are typical:
1) generally arched nature of lifting;

2) more complex character of lifting with block movements and displacements along discontinuities;
3) maximal lifting is referred to the central-glacial region in the north of the Botnic bay;
4) slow down of lifting after deglacialization.

Weak earthquakes are recorded mostly in the zones of active movements. The depth of some of these earthquakes by indirect measurements is not more than 10-15 km. There is tendency of localization of the earthquakes' epicenters in particular zones. In and around Kola peninsula two zones of linear order of the earthquakes' centers-along the Murmansk shore from the river Mezen to the Porsanger-fjord in Norway in the west and along the axis of the Kandalaksha bay to the east to the estuary of the Northern Dvina river and to the west along through the Kovdozero depression, and local concentration in the region of the Khibiny intrusive rock mass (Yakovlev,1980,Panasenko,1980).

In the zones of the Murmansk shore and Kandalaksha significant number of earthquakes' centers were registered. Some of them were of magnitude 5 or a little higher. Distribution of the earthquakes' centers in the

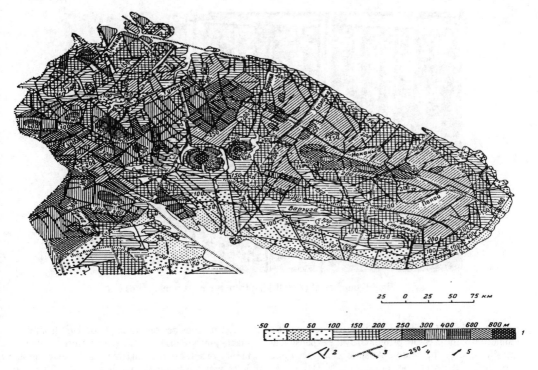

Fig.2 Map of the modern tectonics of the Kola peninsula
1-scale of movement intensities; 2-faults, renewed or formed at the neotectonic stage; 3-deep faults; 4-isolines (isobases) of lifting; 5-directions of the blocks' warps.

region is shown on fig.3.

Decoding of the earthquakes distribution mechanism on the territory of the Baltic shield shows that direction of the regional compression changes from the direction to north-west-east on the territory of southern Sweden, Norway, Estonia to the direction northeast-south-west on the territory of Kola peninsula. Those directions of the maximal compression on the

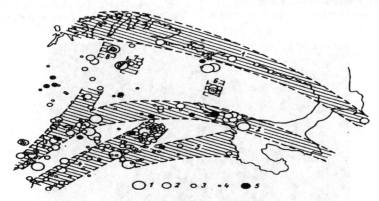

Fig.3 Seismogenic zones of the eastern of the Baltic shield.
1 - 4 - epicenters of the recorded earthquakes
1- M=5;   2- 5<M<4;   3 - 4<M<3;   4- 4<M<3;
5-epicenters, evaluated by indirect measurements

2559

Fig.4 Examples of core disking from the Kola mining sites

Fennoscandia and Kola peninsula territories may be connected to the pressure provoked by the expansion in the rift zone of the Middle-Atlantic mountain ridge extending to the Arctic ocean.

High level of the natural stresses in the rock masses, considered for future siting of the RW repository may be accounted for in the design of underground structures and tunnelling technology. Experience gained from in-situ stress measurements indicates that the technical means for rock mass stress monitoring exist.

The methods used in different countries and institutions are in most cases the same, and measuring equipment as well as the computer techniques for interpretation of the results are gradually being improved. The question is how to use the measured values of stress in practice.

Let us consider one example derived from mining practice in the Kola Peninsula region. Horizontal stresses in rock of some of the Kola mines are as much as 40-50 MPa, with the deepest mining levels (400-500m) reaching values of 50-70 MPa. Intensive core disking as shown on fig.4 is normally encountered.

Kozyrev, 1993, shows that in relation to the rock strength (σ), three stress levels should be distinguished:

$$\sigma_r \leq 0.3 \ \sigma_c \qquad (1)$$
$$\sigma_r > 0.8 \ \sigma_c \qquad (2)$$
$$0.3 \ \sigma_c \leq \sigma_r \leq 0.8 \ \sigma_c \qquad (3)$$

When the stress level σ_r follows equation (1) mining openings are stable. The situation expressed by equation (2) is characterized by intensive spalling and high

Fig.5 Adit after rock burst

2560

Fig.6 Borehole stress release slots

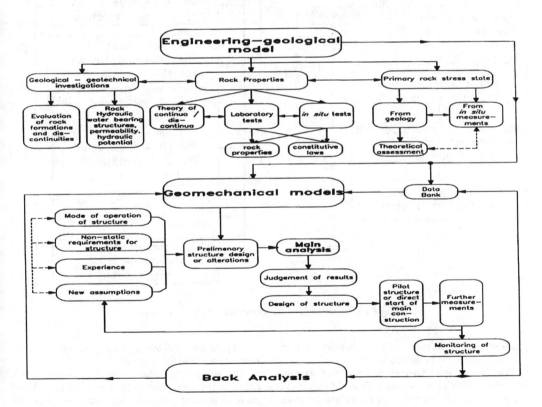

Fig.7 Flow-chart for the design of structures in rock.

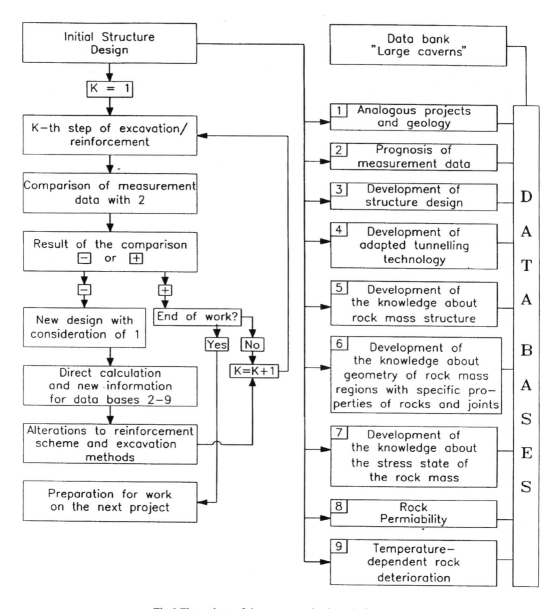

Fig.8 Flow-chart of the computer back analysis system.

probability of rock bursts. The probability of rock failure in case (3) is evaluated as 70%.

The photograph shown as fig.5 is one of the mine adits after the rock burst occured. To avoid dangerous, dynamic effects of rock fracturing and failure due to the high stress level, stress release slots of continuous rows of boreholes were arranged in the arches of adits as shown on fig.6.

Failure of the pillars between the boreholes and slot closure releases some of the accumulated energy, and safe operational conditions in the given region of the mine are restored for some time.

The safety of the mining operations as a whole is controlled by a computerized system which records and analyses acoustic emissions and microseismic events. Most of predictions of possible rock bursts are precise, both in place and time. Shut-downs in the operations in these regions of the mine ensure the safety of the miners. The same information provides a basis for the planning of prevention measures.

N.Barton, S.A.Yufin et al. Underground nuclear waste

2562

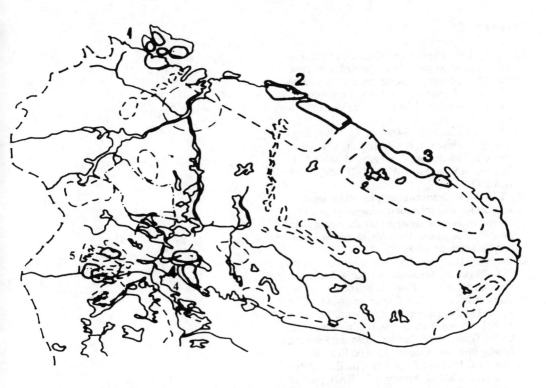

Fig.9 Potential sites for the underground RW disposal facilities on on the Kola peninsula.
1- Sredniy- Rybachiy peninsula (Middle-Fishermen pen.) 2- Dalnie Zelentsy;
3- Guba Ivanovskaya; vil. Gremikha; 4- Kandalaksha and Kolvitsk tundra; 5 - Alakurtti.

Moderately high stresses may be considered as a favourable condition for the siting of the underground RW repository due to rock joints closure. In the research and design of the underground repository on the Kola Peninsula experience gained in the rock characterization using Q-system (Barton, 1993), cross-hole seismic and radar tomography (By, T.L., 1987), finite and distinct element numerical modelling for recent rock engineering projects such as the 60m span Olympic Ice Hockey Cavern in Gjovic, Norway, transportation and hydraulic tunnels, hard-ore mines, Sellafield RW repository in England, etc, (Barton, Yufin & Swoboda, 1994, Yufin, 1993, Yufin et al., 1993), provide the necessary background for the site selection and preliminary design of the geologic repository in the hard crystalline rock of the Kola region in NW Russia.

The whole set of procedures involved in the design and construction of underground structures of a repository may be summarized in the form of the flow chart in Fig.7. Evaluation of the engeneering-geological and consequent geomechanical models are considered to be important part of activity.

Back analysis techniques (Yufin et al., 1993) should ensure reliability of engineering decisions developed during design and construction of the structures. Schematic flow-chart of the computer finite element method-based back analysis system, which is now being developed for implementation at the design and construction stages for underground openings of repositories in Kola and other regions is shown on fig.8.

As candidate sites for regional underground RW repository rock masses of quartz sandstones, granites and gabbro with jointing less than 4 linear km per km square, water contents less than 1%, homogenious, resistant to lixiviation and to radiation doses up to 100 MRoentgen, compressive strength of 200-350 MPa were selected. Those sites are shown on fig.9.

Responsibility of the Project calls for joint effort of Russian and Norwegian scientists for safe design of the underground nuclear waste repository in deep geologic formations of hard crystalline rocks of the NW region of Russia.

REFERENCES

Barton, N., 1993. Physical and discrete element models of excavation and failure in jointed rock. Proc. ISRM Symposium: "Assessment and Prevention of Failure Phenimena in Rock Engineering". Pasamehmetoglu et al. (eds), A.A.Balkema: pp.35-46.

Barton, N., Yufin, S.A., Swoboda, G., 1994. Engineering Decisions in Rock Mechanics Based on Numerical Modelling. Proc. 8th IACMAG Conference, Morgantown, WV. 22-28 May 1994. A.A.Balkema.

By, T.L., 1987. Geotomography for rock mass characterization and prediction of seepage problems for two main road tunnels under the city of Oslo. Proc. Int. Congress on Rock Mechanics. Montreal. A.A.Balkema: pp.37-39.

Gorbunov, G.I., Belkov, I.V., Makievsky, S.I., et al., 1981, Mineral Resources of Kola Peninsula, Leningrad: Nauka, 272pp. (in Russian).

Kozyrev, A.A., 1993. The differentiation of tectonic stresses in the upper part of the earth's crust in order to govern dynamic effects of rock pressure. Dr.Sc. Dissertation, Russian Acad. of Sciences, Mining Institute, Apatity, 402p. (in Russian).

Panasenko, G.D., 1980, Seismicity of the Eastern Part of the Baltic Shield. Apatity, KSC RAN, pp.7-24 (in Russian)

Yakovlev, V.M., 1982, Modern Movements of the Earth's Crust in the Zone of Contact of Khibiny Rock Mass. Geophysical and Dynamic Investigations in the North-East of Baltic Shield. Apatity, KSC RAN, pp.88-95 (in Russian).

Yufin, S.A., 1993. Stability Assessment for Underground Structures in Rock - Recent Trends and Developments. General Report to the ISRM Symposium "Safety and Environmental Issues in Rock Engineering". A.A.Balkema

Yufin, S.A., Postolskaya, O.K., Rechitsky, V.I., 1993. Stability of Rock Caverns as Viewed from the Back Analysis Data. Proc. ISRM Symposium "safety and Environmental Issues in Rock Engineering". A.A.Balkema: pp.751-758.

Geological criterion for toxic waste disposal site selection

Critères géologiques pour la sélection des sites pour stockage de déchets toxiques

M. Komatina & S. Komatina
Geozavod & Geophysical Institute, Beograd, Yugoslavia

ABSTRACT: Determination of rayons convenient for toxic wastes disposal is very actual and complex task. In the paper, approach based on geological and hydrogeological factors, exerting great influence on several others (climatic, hydrological, pedological, etc.) is presented. Geological criterion for choice of toxic wastes disposal potential rayons is analysed and applied to the territory of Serbia (Yugoslavia).

RESUME: Pour des decharges de déchets toxiques, on applique des normes strictes, non seulement pour le choix des sites, mais aussi pour leur realisation. Il faut combiner une multitude de points de vue: geologie, geomorphologie, hydrogeologie, hydrologie, etc. Cet article presente une approche reposant sur les facteurs geologiques, hydrogeologiques, qui exercent une grande influence sur certains autres facteurs (climatique, hydrologique, pedologique). On analyse le critere geologique pour le choix de sites potentiels pour dechets toxiques dans des geosynclinaux, et son application a la Serbie (Yougoslavie).

Toxic wastes disposals are contained of substances endangering human survival, so strict standards not only for locations determining, but also for their arranging, are applied. Successful solving is possible only by multidisciplinar approach, with evaluation of numerous requirements for interesting area (geological, geomorphological, hydrogeological, geochemical, hydrological, meteorological, pedological, etc.). In the paper, approach based on geological and hydrogeological factors, exerting great influence on several others (climatic, hydrological, pedological, etc.) is discussed. Geological criterion for selection of toxic wastes disposal potential rayons is analysed and applied to the territory of Serbia.

SELECTION OF POTENTIAL RAYONS

For successful solving of the various environmental protection problems, complex information on the investigated area geological structure are necessary. Within geological environment, the whole life and human activities are developed; on geological potentials, the whole economy is based; significant number of cases regarding to environment pollution is connected to geological environment disturbing (mining and civil engineering activity, groundwater pollution and overexploitation, exogenetic geological processes reocurring, etc.).

First of all, **geotectonic criteria**, consisted of all an interesting zone geological elements (geological column and space distribution of distinguished lithostratigraphic units, are analyzed. By geotectonic units separating, areas with specific geological composition and texture (in other words, with appropriate neotectonic, seismic and geomorphologycal features) are outlined. In connection with seismic zoning, it is possible to make a zone suitable for toxic wastes disposal narrower.

For further potential zones determining, the territory **hydrogeological zoning** is used. The terrain hydrogeological conditions are very important for final decision in selection of toxic waste landfills. A lot of problems appearing in connection with surface - and groundwater pollution could be avoided by following hydrogeological elements analysis and exploration: 1. rocks hydrogeological features, 2. hanging-wall and the first important water-bearing horizon depth, lithology and filtration characteristics, 3. hard rocks water-repellent, 4. rocks porosity and absorption features, 5. depth to groundwater level and their regime, 6. hydraulic connection between ground- and surface water, 7. groundwater chemical characteristics and quality. Within hydrogeological zoning, rocks categorization on the basis of porosity type and permeability degree is made. According to the classification (dividing rocks into four groups: high-permeable, permeable, low permeable and non-permeable), the terrain water-saturation is also determined, where zones without groundwater are the most interesting for toxic wastes disposing. On the other hand, high-permeable limestones, subjected to direct contamination from the terrain surface, as water-saturated alluvial deposits, are out of the question.

Environmental and groundwater contamination process could be significantly reduced by the zone of aeration (hanging wall) presented as the thick non-permeable layer. During complex geological and hydrogeological exploration, compulsory following alternative locations choice, the protective layer has to be especially analysed.

On **hydrological map**, water resources (rivers, lakes, surface reservoirs) space distribution and areas subjected to innundation (alluvial plains, predominantly) is presented.

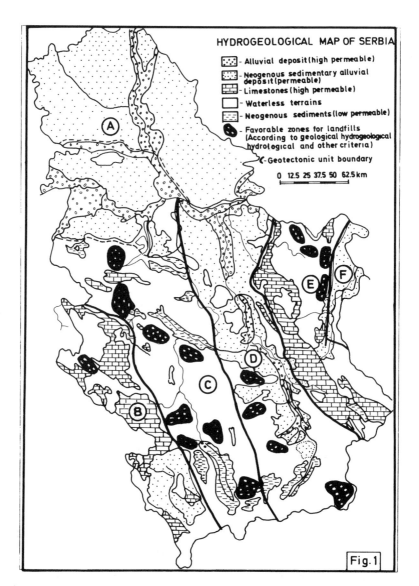

HYDROGEOLOGICAL MAP OF SERBIA

- Alluvial deposit (high permeable)
- Neogenous sedimentary alluvial deposit (permeable)
- Limestones (high permeable)
- Waterless terrains
- Neogenous sediments (low permeable)
- Favorable zones for landfills (According to geological hydrogeological hydrological and other criteria)
- Geotectonic unit boundary

0 12.5 25 37.5 50 62.5 km

Fig.1

Surface reservoirs protection is of great importance, and their watersheds are classified as unsuitable zones for toxic wastes disposal (fig.1).

By cited maps, geological and other important natural factors for investigated area are presented. On the basis of these data, the **territory synthetic zoning** with areas suitable for toxic wastes placement could be performed. In order to state the separated rayons precisely, **geographic map** is used, with settlements, industrial and mining rayons location, on which the landfills distinguishing depends certainly. Not until suitable zone determine on the basis of geological factors, it is possible to apply morphological, meteorological, urbanist, technological and other criteria. Finally, **polluters map**

(hence, industrial and public objects, mines with flotations, oil refineries, land under cultivation, etc.), with disposal types and quantity values, serves for distinguished potential zones ranking according to priority.

DISPOSAL SITES SELECTION FOR TERRITORY OF SERBIA

The territory of Serbia is characterised by very heterogenous lithological composition and complex texture. Within the territory, several geotectonic units with unique structure and special geomorphological and hydrogeological features could be distinguished (A-F zones in fig.1). For example, in

Dinnarides and Carpato-balkanian arc, (B and E), the main water-bearing horizons are presented by karstified limestones; in Vardar zone (C): non-permeable to low-permeable rocks, and in new depressions A, D and F (Pannonian basin, Dakian basin, lake basins within Serbo-macedonian massif) - by alluvial deposits and sands and clays neogenous complex.

In hydrological map of Serbia, geotectonic and hydrogeological regions coinciding is noticeable (fig.1). In Dinnarides (B) and Carpatho-balkanian arc (E), the same areas are characterised by karstified mesozoic limestones and non-permeable rocks of different age, while groundwater discharge is possible exclusively through resurgences. Vardar zone (C) is distinguished by significant groundwater absence, as paleozoic terrains in the southern part of the Serbo-macedonian massif (D). The largest zone of the latter is formed by tertiary sediments and the rivers Velika Morava and Juzna Morava alluvial deposits, while in mio-pliocenous sands and coarse-grained alluvial horizons, huge groundwater resources are accumulated. The same is valid for Pannonian basin (A).

The territory of Serbia synthetic zoning points out that the most suitable conditions for toxic wastes landfills creating is within Vardar zone (even eight potential localities). Seventeen potential rayons is separated, with possible two-three locations.

REFERENCES

Komatina, M.1990. Hydrogeological Investigations, III, Beograd, p.373-391 (in Serbian).

Komatina, M. 1994. Hydrogeological Investigations, IV, Geozavod, Beograd (in preparation) (in Serbian).

Komatina, S., 1994. Geophysical Methods Application in Groundwater Protection Against Pollution; Environmental Geology; p. 53-60, Springer International.

Reuter, F., 1980. Ingenieur geologie; VEB Deutshcher Verlag fur Groundstoffindustrie, Leipzig, p. 511-514.

The role of Engineering Geology in the landfill siting problem – The Greek experience: Recommendations on site assessment aspects

Le rôle de la Géologie de l'Ingénieur dans le choix des sites d'emplacement des décharges de déchets: L'expérience Grecque

Ioannis V. Diamantis & Stylianos G. Skias
Democritus University of Thrace, Civil Engineering Department, Geotechnical Division, Xanthi, Greece

ABSTRACT: Modern landfill is an extremely specialized use of land and expertise interdisciplinary approach is needed for avoiding contamination of ground and water associated to it.

Engineering Geology's principles and methods contribute decisively in achieving an optimum site selection and establishing appropriate design and rehabilitation requirements.

A review at existing selection criteria systems is presented and evaluating methodologies for application in the siting process are proposed.

Two specific cases, refering to landfill siting approach in Greece are presented and faced problems are explained. Certain recommendations on geological suitability, site assessment and presentation of results are, finally presented for optional use.

RÉSUMÉ: Les lieux de décharge de résidus urbains représentent une utilisation très specialisée de la terre qui implique une approche scientifique adequate de façon à éviter la polution aussi bien de la terre que des eaux souterraines ou superficielles. La géologie de l'ingénieur avec ses principes de base et ses méthodes contribue de façon décisive au choix du site approprié et à l' accomplissement des buts du projet.

Les systèmes de critères existants et les méthodes appliquées sont presentés et commentés. Deux cas révélateurs de la façon dont on procéde en Grèce lors de la designation d'un site pour la création d'une décharge, les problèmes qui en surgissent et les solutions proposées sont aussi analytiquement expliqués. Suivant certaines recommandations qui se réferent à l' évaluation de l' environnement géologique et à la fin, en matière de conclusion sont exposés les résultats de cette recherche.

1. INTRODUCTION

Modern waste disposal management moves fast towards a world wide implementation of the strong basic demand to: AVOID, REDUCE and RECYCLE waste. Nevertheless, as far as municipal wastes are concerned, Sanitary Landfilling is indispensable within any sound waste disposal policy.

The prime objective in siting, designing and operating a Landfill should always be the greatest protection of the natural environment from any kind of (negative) impacts associated with waste disposal process during and after the end of the site's life span.

The above basic goal will be safeguarded, to the optimum level, only by introducing quantitative Standards concerning the acceptance of a candidate site in relation to predetermined inadmissible levels of pollutant emissions. Thus, preventing to the highest possible degree the HYDROSPHERE and the BIOSPHERE from being contaminated.

The first step in achieving, eventually, the environmentally optimum solution for our disposal needs, represents the selection of the most suitable site within a given (officially provided) area, among existing candidate sites. This "first step" is the Siting process. Thus, in Landfill projects, design and construction represent the second stage after siting.

Engineering geology has to play a major role in the Siting process through introducing criteria and methods for analyzing and evaluating the geologic environment, as well as, proposing measures for safeguarding and monitoring environmental protection in and around the selected site.

2. TRADITIONAL LANDFILL SITING

Land with little or no economic value was for up to resent years considered good for nothing but waste disposal. Since these marginal land – sites possessed no or very limited natural protection widesprad contamination was subsequently affecting the surrounding and broader area. Especially as far as surface and ground water bodies are concerned. Therefore if the specific characteristics which have lead to the rejection of marginal areas for other uses were combined with other existing (but undetected) natural features of the local environment then this combination would facilitate several environmental damages. Common examples of marginal sites are: stream beds with seasonal flow, karstterrains, wetlands, abandoned, worked out, quarried lands (limestone, sand, gravel etc.) and lands of steep topographic drainage, called ravine lands. It is estimated that until recently, more than 5.000 uncontrolled or semi controlled disposal sites there were existed in Greece.

These marginal lands were often unused and unwanted without human intervention and generally forgotten in the planning process but are attractive to wildlife. It is ironic that such sites are thereby protected now by receiving designation as habitat for threatened and endangered species under the RAMSAR convention.

However marginal land is not the only type of land which is not good enough for landfill. In fact it should be understood that many types of land which are good for a wide variety of uses are Not Capable of providing the kind of environmental protection needed for landfill.

Modern sanitary landfill is an extremely specialized use for land and the selection of a site -siting process- should be conducted with the same attention to detail as the design.

3. MODERN LANDFILL'S BACKGROUND AND SITING FEATURES

Since the very early days of "town planning movement" the basic concept of land use problems has been to identify the BEST and HIGHEST use for every piece of land of an area under planning consideration.

The pioneer planners sought to identify the distinguishing fingerprint of a site of real estate which marked it for a special use.

Ian McHarg in his book "Planning with Nature" showed how this might be achieved by mapping the various aspects of large metropolitan areas on acetate sheets and overlying them to sieve-out the salient features which point the way to future plans.

When applied to the selection of a site for a Landfill however, the process is relatively straight forward. Here the point is to find the areas which development has overlooked and choose the best candidate for use as a landfill under consideration. The Landfill siting problem has been around for a long time. We know that there are many siting systems engaged in the Landfill siting process.

3.1 Criteria selection systems – the "Drastic"

A criteria selection system called DRASTIC has been developed by the U.S. Environmental Protection Agency (EPA) and the National Water Well Association (EPA/NWWA 1985) for evaluating ground water pollution potential using geologic and hydrogeologic settings. This system compares areas by assigning ratings and weights to seven parameters that affect g.w. contamination.

Briefly these seven parameters are:
D – Depth to Water Table
R – Recharge (net infiltration)
A – Aquifer Media
S – Soil Media (surface soils)
T – Topography (Slope)
I – Impact of the Unsaturated Zone Media
C – Conductivity (Hydraulic) of the Aquifer

These factors have been arranged to form the acronym DRASTIC for ease of reference. Each DRASTIC factor has been evaluated with respect to the other to determine the relative importance of each factor and has been assigned a relative weight ranging from 1 to 5 . The most significant factors have weights of 5; the least significant a weight of 1.

In turn, each of these ranges is assigned a rating which varies between 1 and 10. The system allows the user to determine a numerical value for any setting by using an additive model. The equation for determining the DRASTIC Index is:

$$D_R \, D_W + R_R \, R_W + \ldots C_R \, C_W = \text{Pollution Potential}$$

Where the subscripts R and W indicate the rating and the weight respectively as identified for each quantity in the drastic Index.

The best score is 26 and the worst score is 226. This method is applied when there exist maps available for each of the chosen criteria.

Thus there are many and one can create his own selection criteria - system depending on the availability of existing various relevant environmental parameters both as far as the quantity and the quality are concerned.

These systems are basically comparative tools which are used to compare one site to another. However the system can also be used to compare a given site with idealized scores in order to judge comparatively whether or not a site has a merit.

3.2 Siting systems using exclusionary criteria

There exist siting systems where the criteria fell into two distinct categories. The first set consists of exclusionary criteria and failure to meet any one of these resulted in exclusion (of the site) from further consideration. The second category consists of criteria which can be overcome by proper geotechnical engineering and by other means.

The exclusion criteria refer to:
- Specific distances from surficial ground water bodies
- Existence, in the area, of flood plains, wetlands.
- Karst development in the site.
- Existence of Natural processes -natural hazards- in the area which could lead to great problems i.e. seismicity, erosion, drainage, subsidence, slides.

3.3 Proposed methodology for the use of selection criteria in the siting process

After reviewing different existing siting systems and list of criteria related natural environment from the point of view of engineering geology we propose the following procedure in applying criteria for the selection (of the site) process.

1. Use first a set of criteria which, given the features of the local environment, are so important that they preclude landfill no matter what mitigation is considered. These criteria should be mapped as exclusion zones and are summarized under the headline.

Natural Features

- Wetlands
- Flood Plains
- Surface Waters
- Groundwater Resources

- Fault Zones
- Seismic Impact Zones
- Karst areas
- Unstable Areas
- Expansive Soils
- Subsidence Zones
- Threatened and Endangered Species Habitat
- Scenic Areas

A schematic representation of exclusionary criteria mapping process is shown in figure 1.

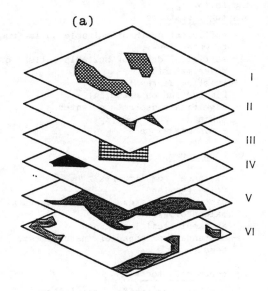

(a)

I
II
III
IV
V
VI

Thematic maps representing each one of the exclusionary criteria

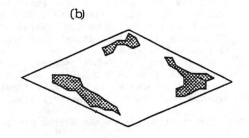

(b)

Composite map showing the remaining areas. All the exclusion zones have been eliminated.

Fig. 1 A schematic representation of exclusionary mapping process

2. After sieving out any exclusion zones, a second set of criteria should be applied on the remaining parts of the area under consideration.

These criteria (which are problematic to a certain degree) must be considered by application of a weighting process which clarifies their level of importance compared to one another, similar to the DRASTIC method presented previously in this work.

Depending on the local environment features a rather broad list for selecting criteria belonging to the second set can be the following:

a. Natural features

- sufficient depth of suitable soils from g. water level.
- sufficient depth of suitable soils for cover material
- Existing depressions
- Limited run on potential
- Significant depth to g. water
  - Minimal well density
  - Minimal slope

b. Land - use

- distance from Airports
- Prime farmland
- distance from areas of architectural, paleontological,
- archaeological importance
- distance from areas of natural scenic beauty

c. Economic factors

- High way and traffic restrictions
- Distance from centroid of waste generation

Sum- Land's ownership and values steps are:

1st step Use of exclusionary criteria

2st step Use of non exclusionary criteria utilizing the method of assigning weights and rates like the DRASTIC method, to get the numerical value of "Pollution Potential" (total score) for each candidate site

3st step In case of equal scores, then apply the Cost - benefit analysis method to determine to most suitable through cost - benefit considerations.

3.4 Important points in the siting process

- It should be noticed here that some of the above criteria of (b) and (c) lists can become, in special circumstances (i.e. short distance) exclusionary criteria.
- Another point of importance is that the very purpose of the siting process is

to make the best use of the land resources locally available.

- Also, within certain constraints, the criteria may be adjusted, consistent with changes in landfill design to increase the number of sites to be considered .
- The first objective when starting the siting process is that there exist: A real to site the Landfill. That means that a "zero landfill site option" is not realistic. Thus, if the result of the siting process is only one, rather "problematic" environmentally, site, we have to evaluate the possibility to overcome the environmental impacts to an accepted degree by the best means of geotechnical engineering.

If the geotechical measures are extremely costly, or the risk of avoiding an unacceptable contamination of the area is very high, then we have to consider extending or the differentiating the geographical area of investigation (through state or municipal consultations) and start a new siting process.

4. GREECE IN TREND OF MODERN LANDFILLING PROCESS

In Greece, there has been, roughly, a decade since we have started exercising geotechnical work concerning Landfills. Under the existing legislation (Law N° 1650/86) the official approval of a site for sanitary landfill construction is based on the results of a proper "Environmental Impact Assessment Study" E. I. A. S. Accordingly the state authority grants permission for the execution of the project under a specific for each case, set of "environmental rerms" which will provide the required under the law, environmental protection of the local natural environment.

The probably most important part of the E. I. A. S. concerns the assessment of the geologic environment.

The Legislation gives only a general outline of the needed assessment and provides the basic principles and goals for its execution. Thus there commonly exist a great lot of disputes and disagreements between consultants and state officials, concerning the contents and the results of EIA reports.

The Greek National Associations of concerned Geologists and Engineers, have started to deal with the problems associated to EIA studies and the siting and construction of Sanitary Landfills. But, they haven't yet reached to an agreement. In the following we present our proposal which comprises, as we believe, a "break through" plan for the rationalization of the Sanitary Landfill siting process (S.L.S.P.).

This proposal document is to be submited to the National Association of Engineering Geology which will forward it to the Ministry of the Environment.

4.1 Basic points of our proposal - document

a. Prior to any Landfill Siting process there must be officially approved (state authority), the geographical area within which the siting process and the E.I.A. will take place.
b. There must be a Law provision for the obligatory participation of specialized engineering geologists in all stages of Landfill: siting, design, construction and monitoring processes.
c. There be must adapted national standards regarding the exclusionary criteria and methods for assigning weights and rates in the evaluation process of geologic and environmental factors, as well as the geotechnical testing requirements.
d. The National standards must cover to some extend the monitoring and quality assurance procedures.
e. All state territory must be covered by engineering geological maps in a scale of 1: 50.000 where all areas protected by the RAMSAR or other international conventions and National regulations as being Ecologically sensitive or archaeologically important should be especially mapped and assigned.
f. Public participation should be involved fully in the siting process.

5. ASPECTS OF THE SITING PROCESS FOR THE SANITARY LANDFILLS (S.L) OF XANTHI AND ALEXANDROPOLIS CITIES

The S.L.s of XANTHI and ALEXANDROPOLIS cities in Thrace, (Fig. 2), represent two of the first cases in Greece where it was planned and executed sound and appropriate research work regarding the geological hydrogeological and geotechnical parameters and the selection of evaluating criteria for site assessment. The existing situation and the geological seting were different in each case.

5.1 The case of Xanthi

After long lasting consultations and discussions among all the municipalities and communities of XANTHI'S prefecture, an

agreement was reached to create one large enough S.L. for the whole region. The site's area should be about 50.000 m^2. A committee was established for supervising

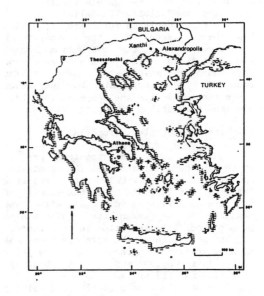

Fig. 2 Map of Greece Politions of XANTHI and ALEXANDROPOLIS cities.

the whole project and the research work was commissioned to our University environmental research team.

The greater investigation area, of several Km^2, covers the plain part of the prefecture's region and is formed by geologically recent soil formations to a depth of about 200 m. We decided to apply two sets of criteria for site selection. The first set was the exclusionary criteria which were the following (8) eight:
- wetlands, flood plains, river deltas, areas of exceptional natural beauty, areas of arcaeological importance, municipal water supply wells, surface waters and Threatened and Endangered Species Habitat.

The above criteria were mapped as exclusion zones on the initial area. Then we selected a second set of evaluating criteria to be applied on the remaining area parts, using a weighting process similar to DRASTIC method presented in section 3.1.

These evaluation criteria, given the specific features of the natural environment and other local features, were the following (7) seven:
- Depth to water table, Permeability of surface soil formations, topography (slope

angles), availability of cover material, distance from centroid of waste generation, absence of groundwater conditions which make monitoring difficult, distance from villages.

Two sites were assigned a similar score regarding the Pollution Potential.

The final decision for site selection has been taken by the supervising committee(to whom the evaluation report was submitted after the end of Environmental assessment and siting process) above economic considerations concerning basically the compensation fees and land's ownership status.

5.2 The case of Alexandropolis

In this case, the research project which was commissioned by the municipal administration to our University's research team, was the Environmental assessment of an already existing disposal site and measures to be taken for safeguarding proper landfill's operation in the future.

The selection of the site has been made by municipality's engineers ten years prior of our intervention and it has been in operation for (9) years.

Thus no selection criteria and siting process method was applied. The first stage of investigation was consisted of analysing the geological setting and the hydrogeological regime of the site and the broader surrounding area. The second stage consisted of the evaluation of Environmental Impacts due to disposal operations and recommendations on design and constructional problems concerning adapation of the site to its new role as a sanitary Landfill site.

After elaborating all data and research findings we reached to the conclusion that due to the specific geological and hydrogeological characteristics of the area, the negative impacts on the local environment were of rather minor importance, but this situation could be worsten due to accumulative processes and some undetected paths of lecheate flow.

Extensive consultations were taking place, for a six months period, among municipal officials and our team, prior to an agreement on the appropriate measures to be taken. Also a cost - benefit analysis was executed by an expert team. The decided measures where the following:
a. Redesigning the whole disposal operation and constructing a proper sanitary Landfill in a site adjacent to the existing one.
b. Constructing a clayey basal lining system as a geotechnical barrier to eliminate the leachate penetration possibility through subsoil to a 300 m away existing stream with seasonal flow. Allthough the existing permeabilities in soil and rock formations are extremely low.
c. Removing all existing quantities of waste and landfilling then in to the new constructed site.
d. Installing a monitoring network for the leachate's flow inspection and the prevention of any contamination which might occur.

The "moral myth" which arises from this case is that: "preventing is far less expensive, by all means, than restoring". In environmental cases the cleaning - up of ground or water bodies from existing contamination is very complex, extremely lengthy and money consuming process. And what is more: Sometimes pollution damage can create rather irrevisible effects.

6. RECOMMENDATIONS ON ENGINEERING GEOLOGICAL SITE ASSESSMENT FOR LANDFILL SITING

The siting and design concepts for Landfills greatly depends on the structure and behavior of the geologic materials. Thus, for proposed new sites a detailed geological, hydrogeological and geotechnical investigation is therefor essential. The type and size of the site investigation depends on the following factors.
- Topography and structure of the area .
- Type and behavior of the waste.
- Geological setting and surficial and ground water regime.

In order assess the suitability Landfill site it is essential to have accurate knowledge of the distribution of groundwater flow paths and barries (aquifers and aquicludes), their hydraulic properties , the deformation behavior of the subsoil and the potential for improving the sealing effect of the subsoil. Consideration must also be given to the need for setting up adequate controls and undertaking subsequent remedial works if appropriate. To assess and evaluate the behavior of the subsoil as a foundation for a Landfill site (base area or perimeter boundaries) it is essential to have knowledge of the local general geological setting, including the following principal aspects:
- characteristics of the morphology;
- structure, extent and geplogical age of the outcropping strata;
- tectonic structures;
- deeper subsoil if it comprises cavities or soluble rocks;
- aquifers and groundwater flow;

- risk of earthquakes and other natural hazards.

6.1 Composition and distribution of superficial soilformations

In order to assess the subsoil for Landfill site it is necessary to know:
- composition, physical and chemical properties and sequence of strata;
- lateral and vertical continuity and distribution of the strata (facies changes);
- porosity;
- permeability (to water and leachate);
- resistance to erosion and washing away of fine particles;
- stress deformation behavior.

6.2 Structure and sequence of soil rock strata

Due to regional geological factors and morphological characteristics, superficial deposits are often relatively thin and therefore the underlying soild strata may have to be included in the survey. Here the following factors need to be considered:
- type of rock, mineralogical composition and stratigraphy;
- state of weathering and weathering resistance;
- solubility in water and leachate or other aggressive solutions;
- type and position of geological boundaries;
- extent, degree of separation and widths of individual joints;
- tectonic and petrographical anisotropies in the rock mass;
- karstification and risk of subsidence;
- deformation behavior of the rock mass;
- permeability to water, leachate, gases and other aggressive solutions (hydrocarbons, etc).

6.3 Determination of hydrological data

Landfill sites must be prevented from having unacceptable impacts on groundwater, surface water, and particularly water abstraction sources. Comprehensive knowledge of the groundwater regime is therefore required, including the following detailed information:
- groundwater regime, direction of flow, gradient and rate of flow, including longterm and seasonal fluctuations;
- permeability (horizontal and vertical) or transmissivity of the outcropping strata, with maximum and minimum values;
- distribution, thickness and depth of aquifers, aquicludes and aquitards,

including the locations of any springs;
- groundwater levels, indicating hydraulic gradients and effective flow velocity in the individual strata components if appropriate;
- groundwater chemistry, including determination of naturally occurring aggressive substances and groundwater quality;
- groundwater protection zones;
- groundwater abstraction and its effects;
- groundwater abstraction rights.
- influence of short – term or long – term lowering of the water table, restoration and extraction or augmentation of groundwater in the future;
- influence of nearby open waters and their relationship with the groundwater system;
- situation in respect to receiving streams, influence of flooding and tides, if appropriate;
- effective rainfall, surface runoff, percolation rate, evaporation and groundwater recharge.

6.4 Presentation of results

Presentation of the investigation results in the form of graphs or diagrams is recommended, particularly where there is a large volume of data. The following may be considered:
- site plans indicating:
. location of boreholes, trial pits etc;
. geological and groundwater level/contour plots;
. groundwater flow direction and effective flow velocity;
. groundwater abstraction (including water resource catchment areas and water protection areas);
. surface water and other hydrological features;
. geochemical zones for groundwater and soil/rock.
- geological sections (indicating borehole records used);
- spatial profile (geological overlays and block models);
- representation of the groundwater system (rainfall distribution, fluctuations in groundwater level, flood and tidal influence).
    In addition to the presentation required by the standards, laboratory test data should be set out in summary tables.

6.5 Engineering geological report

The results of the site investigation should be subject to an analysis and evaluation, taking account all the particular design stages and specific requirements of the general safety plan. This assessment should be set out in an Engineering Geological report.

This report must address the following aspects as a minimum:
- description and representation of the geological structure;
- presence and suitability of natural low permeability strata (thickness, depth, horizontal continuity, permeability, adsorption capacity);
- groundwater regime and permeabilities within the erea to be landfilled and its environs; a groundwater model may be appropriate;
- stability of natural or artificial slopes;
- bearing capacity and deformation behavior of the subsoil;
- faults, possible subsidence, risk of collapse, earthquake risk and other hazard situations;
- overall evaluation of the subsoil as a natural barrier for the site;
- notes on geotechnical measures required to improve the properties of the subsoil as a natural safety barrier.

7. EPILOGUE

Although there is a strong demand to avoid, and to recycle waste, landfills will be indispensable within any sound waste disposal concept. It is an important task of environmental geotechnics to establish principles in the siting design and construction of landfills, in particular with respect to long term safety. Further, geotechnics are strongly involved remedial works for abandoned landfills and contaminated land due to former municipal and industrial activities.

It is an important challenge for engineering geologists to contribute within their profession to the protection of air, soil and water. New pollution and contaminations have to be avoided, as well as existing contaminations in the soil have to be cleaned up. For these new and very complex problems appropriate solutions should be prepared by the geotechnical community, especially by Engineering geology and soil mechanics societies.

In this respect, we believe that Medeterranian countries should take an initiative towards encouraging their national experts, in the relevant fields of Environmental Sciences, to compile, upon agreed principles, through exchanging their experiences and "know - how's"; "REGULATIONS", "STANDARDS" and "CODES OF PRACTICES" for all aspects of WASTE DISPOSAL processes. we strongly believe that this is a realistic and urgently needed Environmental policy upon which all Medeterranian countries should agree to work with devotion and enthusiasm.

REFERENCES

Aivaliotis, V., Diamantis, J., Skias, S. 1992 - '93. Environmental Impact assessment (E.I.A.) for the sanitary landfills of XANTHI and ALEXANDROPOLIS cities. XANTHI: Democritus University Intern. Research Project.

Noble George, P.E. Siting Landfills and other LULUs, 1992. Lancaster, Pennsyl. U.S.A: Technomic P.Co.

German Geotechnical Society, 1991. Geotechnics of Landfills and contaminated Land "G.L.C". Berlin: Ernst and Sohn.

Geological environment and transport roads: Evaluation of technogenous impact (Leningrad region)

Environnement géologique et transports routiers: Évaluation de l'impact technogénique (région de Leningrad)

Irina Chesnokova

Institute of the Lithosphere Russian Academy of Sciences, Moscow, Russia

ABSTRACT: A set of problems related to road construction and exploitation as well as to their impact on the environment appears to be complex and diverse. Only geological-engineering and geocryological aspects are under consideration in the work presented. Economic assessment of the damage was made using the Leningrad region as an example.

RÉSUMÉ: De nombreux problèmes dues à la construction routière, a l'exploitation des artères de transport et leur influence sur l'environnement sont complexes et variés. L'article présent ne traite que les aspects géologiques, géokryologiques et ceux du génie. Sur l'exemple des autoroutes de la région de Léningrad une estimation géoécologique a été effectué.

1 INTRODUCTION

During the past years geological engineers showed a great interest in ecological problems. First of all it could be implied to urbanized territories, where the situation existed requires urgent measures to be taken.

An essential part of ecological problems is related to the transport. You can hardly find any other engineering structure which would so intimately interact with the environment as the main roads do. We can say that the transport means became one of main sources which pollute the atmosphere with exhaust gas, adverse dust particles, etc. Road constructors dig and transfer a great amount of soil and rock disturbing the natural state of ground masses, and disregarding the fact that changes in geological environment practically always entail irreversible changes in ecological balance. Every year the roads capture new agrarian areas good for agriculture or other use.

The geological surroundings are one of the components of environment. According to Academician E.M.Sergeev's (1989) definition "any rocks or soils composing the upper part of the Earth's crust are considered as multicomponent systems being affected by engineering and economic activity of a man that results in changes in na-tural and geological processes as well as in advent of new anthropogenic engineering geological processes which change the engineering and geological conditions of a certain territory".

Changes in geological surroundings depend mainly on road categories which define an offset strip, angles of grades, radius of curves, and other characteristics affecting the scope of influence of roads on the geological surroundings. The quality of construction, the keeping of the project requirements, and environmental protection are also important.

Despite the significance and extent of the problems which arose with road construction until now many of them find no representation in scientific and applied developments. Investigations performed by the author are pioneer with regard to evaluation of the economic damage appearing at highways of the Leningrad region as a result of an adverse effect of engineering geological and geocryological processes. Intense chamges in geological surrounding can be observed in the first years of a highway exploitation dying down gradually or becoming stable within 10-15 years.

The construction of highways produces the noticeable changes in geological surroundings in the north of the European part of Russia. The Leningrad region where the author has conducted investigations, is an advanced industrial area, and that is why the road network is highdensed. All the roads can be divided into four types: state, union, republican, and local. Classification of roads by categories includes the complex evaluation which characterizes the type of road cover (asphalt and concrete, gravel, and ground), the width of the roadbed and of the offset strip. The categories range from the highest (the first category) to the lowest (the fourth category).

Besides, the roads can be classified by age:

1. Roads constructed before the revolution (these were reconstructed in the 60s).

2. Roads constructed before the II-nd World War (usually these were reconstructed in the 60s-70s as well).

3. Roads constructed after the War (mainly short portions of roads, and strategic roads).

The effect of transport over environment (including the deological surroundings) is estimated differentially by features comprising: intensity of traffic, its structure, capacity of passage, width of driving passage, width of sanitary and protecting zones, character of building.

The problem of the highway impact over geological surroundings and a reverse effect of the changed environment over highways is very timely. More than one third of all the roads in the Leningrad region are subjected to heaving. The length of roads under heaving is more than 200 km (it amounts to 5% of all the roads of the Leningrad region). One kilometer of a road costs 1,950 thou. rubles annually, and in some years more than 2.5 mln rubles. Fig.1 shows the territory of the Leningrad region with roads subjected to heaving, a special sign marks the 'sick' areas which were studied by the author.

2 CHARACTERISTICS OF THE REGION

The physical-geographical and geological conditions of the Leningrad region comprise: releif, geological structure, hydrogeological features, vegetation and soils, as well as climate.

The present topography of the region formed due to accumulation and denudation activity of the Quaternary glaciations. Such postglacial exogenic processes as erosion, abrasion, karst phenomena, swamping, slope denudation, etc. developed on the background of insignificant neotectonic epeirogenetic movements. The Leningrad region is situated within two large provinces (the northern part is a province

of denudation and glacial relief, while the rest part is of glacial relief). The till, outwash, lake-glacial and bog plains are more expressed among ten main morphogenetic forms and types of the topography.

Geologically the Leningrad region is located within the conjunction of two geological structures – the Baltic crystalline shield and the Russian plate. The Quaternary sediments cover continuously the whole territory. The region experienced sea transgressions and regressions. The main geological complexes include: the Archean and Proterozoic rocks, glacial, fluvioglacial, Recent marine, Recent lacustrine and lacustrine-alluvial, alluvial, Recent peat deposits.

The Leningrad region is located within the Leningrad artesian basin, and only the northern part of the Karelian istmuth belongs to the Baltic hydrogeological massif. The region is characterized by a seasonal, mainly spring and fall alimentation, being a zone of excessive humidification, where areas with different extent of draining (high, average, low), as well as with a quick change of drainage conditions can be distinguished. Water occurring closer to the surface is of great importance for the evaluation of the technogenic impact of roads over the geological surroundings as well as the reverse effect of the changed environment over the roads. When evaluating there were considered the depth intervals as follows: less than 0.7 m; 0.7-1.5 m; 1.5-3.0 m; more than 3.0 m. The water from the last interval practically exerts no effect.

The Leningrad region is situated within the southern taiga subzone, and only in the norteast – in the transition taiga subzone. In the whole, the region is characterized by spruce forests confined to watersheds with loams and partially loamy sands. Vast areas are mainly occupied by high moors.

Soils of the Region are diverse, podzolic and bog soils prevailing.

The climate of the region is transitional from continental to marine, and is characterized by high precipitation, a moderately warm summer and by a long moderately cold winter with unstable weather. The climate changes considerably from west to east and to a lesser extent from north to south. In the west the climate is milder and warmer.

A detailed landscape-climatic map was compiled as a result of subdivision of the region made by G.Lokshin and I.Chesnokova (1992) on all characteristics. The map demonstrates the microareas of common characteristics, being a base for typification of the geological surroundings and evaluation of the territory by road heaving.

3 GEOECONOMIC EVALUATION OF THE IMPACT EXERTED IN ROADS

An attempt to evaluate the damage resulted from the adverse geocryological process (the heaving process was under consideration) was initiated for the Leningrad region with a well developed road network.

The economic damage is revealed as a cost estimate of negative changes of the geological surroundings caused by both a direct influence of technogenic factors (sources of disturbances) upon the geological surroundings and a reverse effect of the changed environment over the technogenic, social, and natural components of the systems.

In existing literature all the damages are distinguished by: a form of expression (natural damage, intentative indexes, coast (monetary)); a field of expression (economic, social, ecologic damages); a character of expression (direct and indirect damage); a territory of expression (inner and outer); prevention effect (prevented and unprevented); prediction effect (predicted and unpredicted) (Koff 1985). Both the actual and potential (predictable) damages have been evaluated.

The actual damage is evaluated as additional capital and operational expenditures which have been already realized on the studied territory to eliminate the damage. By potential damage we mean predictable loss. A valuational approach which takes account of the probability for hazardous processes to occur is of great importance when estimating the damage. According to this approach the damage can be subdivided into four categories: (1) a catastrophic (extraordinary) damage threatening habitants or a techno-social sphere; (2) a critical damage violating the nature of environment or the operational standard of systems; (3) a warning damage corresponding to the operational standard of systems and unadjastable with system malfunction; (4) a damage which can be adjusted in the course of system operation.

The valuational approach can be expressed by the function $f_m = f (f_1, f_2, f_3, f_4)$, where every category of the damage takes on a value f from zero to extreme f_m being an additional evaluation of expenditures related to prevention or compensation of the damage of a certain category.

A notion of normal (standard) damage was introduced for an approximate evaluation of the damage caused by thechnogenic chan-

ges of the geological surrounding within the territory under consideration.

For the Leningrad region, the author together with Dr. G.Koff conducted the expert estimation of the damage-producing role of various geological-engineering processes. The estimates were performed by packs of statistical programs (Koff 1987).

The expert estimation showed that judging by the damage-producing role and by the potential damage, the heaving on the roads corresponds to the second group of priority processes, giving way to a negative effect of mining and geological processes only. Date on expenditures for the compensation of heaving aftereffects for all highways of the Leningrad region were collected and processed. Table 1 demonstrated these data.

Table 1. Expenditures for road exploitation (Leningrad region).

Highway	Category	Age, yrs	Length of areas under heaving, km	Expenditures on liquidation of heaving (thou roubles/km[*])	
				average	maximum
Yaroslavl'-Vologda-Novaya Ladoga	3-4	40	2.1	1.55	2.06
Dymi-Bor-Kolbeki-Bochevo	4	70	17.8	0.14	0.15
Leningrad-Kiev-Odessa	1-3	100	1.4	2.00	2.80
Leningrad-Tallinn	1	200	0.3	6.00	9.30
Leningrad-Murmansk	2-3	260	7.6	0.98	1.06
Lodeinoe Pole-Tikhvin-Budogoch	2-3	250	17.5	0.15	0.22

[*]Values on expenditures for liquidation of

heaving (thou.roubles) were obtained for
the period preceeding the reorganization
in Russia during the past 1.5 year. Therefore, at present the data on expenditures
are in disagreement with reality. The auth-
or beleives that a coefficient 1000 will
make the expenditures more realistic (for
instance, 1055 r/km should be replaced
with 1,055 thou. roubles per 1 km of a
road).

By analyzing the expenditures correlated
with different characteristics (geological, geomorphological, geocryological) forming the intensity of the heaving the empirical dependences were obtained.
   Using these dependences together with
the map of environment typification by heaving, one can predict the expenditures for
compensation of the economic damage. When
analyzing the expenditures for compensation of the heaving aftereffects we used
data on roads of the same class, same profile, with a similar cover, and similar
depreciation term. It gave us the possibility to assess the changes in natural factors affecting the intensity of heaving,
as well as the expenditures for its compensation.
   For the approximate estimation of an unexpected damage (D) we used the formula:

$$D = l_{i,j} \ D_n^S \ \delta \gamma$$

where $l_{i,j}$ - is a length of the road portion of the i class and j type; D_n^S - a
normal (standard) damage; δ - empiric coefficient depending upon the confiness of
the road to a zone with different ground
wetness (for the Leningrad region ranges
from 4 to 1; γ - empirical coefficient
characterizing the belonging of the road
to the territory with different ground
water level (for the Leningrad region
ranges from 6 to 1).
   In conclusion we should point out that
application of the empirical formula is
difficult with the analysis of expenditures for liquidation of heaving aftereffects on roads with different overall length
of areas under heaving. The more the overall length of areas ender heaving the
less the expenditures for heaving liquidation, as in this case one-time expenditures on the delivery of ground for partial
fill, on explotation of road equipment,
and on other needs can be used for some
'sick' areas. If such work is done only on
a single road area than the similar ocasi-
onal expenditures increase greatly.
   Application of the empirical dependences
found makes it possible to predict roughly

the intensity of damage for different regions, and to plan appropriate expenditures more efficiently.

REFERENCES

Koff, G.L. & I.V.Demidova 1985. On estimation of economic damage caused by technogenic impact on the geological surroundings. Comprehensive estimation and prediction of technogenic changes in the geological surroundings, p.93-101. Moscow,
   Nauka.
Koff, G.L. & A.I. Kartashov 1987. Application of the expert estimation method for
   regional geological and engineering studies "Recent problems of engineering geology of urban territories and agglomera-
   tions", p.257-259. Moscow, Nauka.
Lokshin,G.P. & I.V.Chesnokova 1992. Highways and geological surroundings (Assessment of technogenic impact). Moscow, Nauka.
Efficient utilization and environmental
   protection of cities. In E.M.Sergeev
   (ed.), 1989, p 43. Moscow Nauka.

AREAS OF HEAVING IN ROADS OF THE LENINGRAD REGION

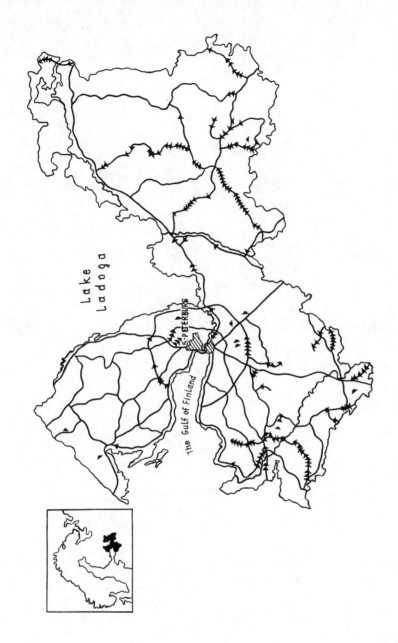

Lake Ladoga

S.-PETERBURG

The Gulf of Finland

Environment water action around dam site in Wan-zhe mountainous area of China

Effets de l'action de l'eau sur l'environnement d'un site de barrage dans les montagnes de Wan-zhe en Chine

Peng Hanxing, Shi Xijing & Liu Jiangang
Hohai University, Nanjing, People's Republic of China

ABSTRACT: Many high dams had been successfully built in 1950's or at the begining of 1960's in Wan-zhe mountainous area of China. Most of them have been working normally for more than 30 years. After the reservoirs impounding, water environment around dam-site varied appearently and geochemical interaction between water and rock took place greatly. As a result, some new environmental engineering geological problems around the dam-sits were produced, such as the acidification and alkalization of groundwater quality, elutes out from drainage bores, softening of the sandwich, slope instability, corrosion of grout curtain and concrete, innormal uplift pressure and so on. These are important reasons for ill and dangerous dams.

Sommaire: Il y a beaucoup de barrages construits à la période 1950-1970 sur les fleuves montagneux des provinces Wan-zhe. La plupart des barrages ont travaillé plus de 30 ans, en causant bien de problèmes hydro-géologiques à l'aspect d'environnement: acidification et alkalization des eaux du sous-sol, instabilité de fondation des barrages, corrosion du rideau d'injection, préssion de fond anormale etc.

1 CHARACTERICSTICS OF THE ENVIRONMENT AROUND DAM SITE

A number of high dams were successfully built on Wan-zhe mountainous area during 1950's or at the begining of 1960's, most of which have been working normally for over 30 years. To a varying degree, some problems relative to the environmental engineering geology around dam sites still exist, such as mechanico-chemical erosion, softening even argillitation of sandwiches, uplift pressure abnormality (i.e. measured value being beyond the designed one) and corrosion on curtain-grouting and concrete body, and others. Case studies showed that the problems above-mentioned result from the action of the environment water around the dam site. After the dam was built, the man-made lake would be developed and as a result, the great changes of the water environment around the dam site would take place. Consequently, some new problems of environmental engineering geology occurred. Hence, some cases were presented so as to study the action of the environ-

ment water around the dam site.

There were four dams, namely Xinanjiang, Chencun, Meishan and Mozitan, situated in latitude between N29°~32°, belonging to humid-subtropical zone. There the annual average temperature was between 15℃ and 17℃ and the annual average precipitation over 1500mm. In addition, the cover of various plants was quite good. The strate around and under the different dam foundation comprised sedimentary, magmatic and metamorphic rocks, respectively, as shown in Table1. Except for Mozitan reservoir, the capacity of others was beyond $2.2 \times 10^9 m^3$ and the height of these dams ranged from 76m to 105m. Under the environment with humid climate, the amount of silt-sand carried by the stream was less and so the thickness of the sediments accumulated on the bottom of reservoirs was smaller than expected. For instance, the sediment of Xinanjiang reservoir, particularly in front of the dam, was 5m or so in thickness during the period of over 30 years.

Table 1. statistics of some large dams

Name	Height (m)	Capacity (10^8m^3)	Types	strata	Time of Construction Completed
Xi-nan-jiang	105	178	Concrete gravity dam with wide-joint	shaly sandstone	1961
Chen-cun	76.3	28.25	concrete gravity-arch dam	shaly sandstone	1978
Meis-han	88.24	22.75	concrete arch dam	granite	1956
Mozi-tan	82.0	3.37	Concrete buttress dam	gneiss	1958

2 CHARACTERISTICS OF THE ENVIRON-MENT WATER AROUND THE DAM-SITE

The environment water around a dam-site usually include the reservoir water, specially in front of the dam, and the ground water around and under the dam foundation.

2.1 *Character of the reservoir water in front of the dam*

The reservoir water in front of the dam was a relatively static water body and the high water level with long period might lead to expanding of various discontinuities in rock masses. The fluctuation of the reservoir water level, particularly the effect from floods, greatly promotes the softening even argillization of sandwiches and the deterioration of the curtain-grouting. Case studies showed that the problems produced were relative to the flood effect, to a varying degree.

Compared with the water quality of the stream before impounding, the reservoir water quality in front of the dam changed significantly, as shown in Table2. The reservoir water at and near the surface was neutral or weak-alkaline one in quality which was similar to that of the stream before impounding, whereas the water near the bottom of the reservoir tended to be weak-acidic water with the temperature of 10℃ or so. The data showed that the content of HCO_3^-, Ca^{2+} and SO_4^{2-} ions increased slightly with the increase water depth, and on the other hand, the content of corrosive CO_2 in the water of the reservoir bottom was quite rich, much higher than that of the water at the surface. Meanwhile, H_2S occured in the water near the botton, for example, the content of H_2S measured in the bottom of Xinanjiang reservoir reached 10mg/L. Obviously, the relatively rich content of H_2S was one of main characteristics of the environment with reduction. Therefore, the water at and near the bottom of the reservoir was of the reduction environment with low temperature, high pressure lack oxygen and weak acidity.

Table 2. Characteristics of reservoir water quality in front of dams

Reservoir	Sampling	Temperature ℃	pH	Corrosive CO_2 (mg/L)	HCO_3^- (mg/L)	SO_4^{2-} (mg/L)	Ca^{2+} (mg/L)	Notice
Chencun	surface	20.0	7.4	2.2	51.87	10.0	10.46	average value meausured in 1992
	30m a·s·l	13.5	6.7	13.2	48.82	12.0	11.90	
	bottom	10.1	6.4	8.2	54.92	26.0	11.54	
Meishan	surface	23.5	7.68	2.16	33.32	5.96	6.71	average value meausured in 1991
	35m a·s·l	13.0	6.34	17.20	41.62	5.67	8.40	
	45m a·s·l	12.0	6.34	15.40	43.28	11.50	8.82	
Xinanjiang	surface	19.9	8	0.9	45.14	3.34	13.6	average value meausured from 1975 to 1983
	70m a·s·l	12.5	7.6	3.9	50.02	3.34	14.0	
	50m a·s·l	9.8	7.3	6.0	54.29	4.30	16.0	
	bottom	/	6.7	9.7	61.0	5.30	17.1	
	before impounding	/	7.6	/	48.06	4.95	14.2	

2.2 *Character of the ground water in the banks around the dam-site*

After the reservoir impounding, the behaviour of ground water flow in the banks varied greatly. The major character was the increase of both hydraulic gradient and flow rate. Taking Xinanjiang dam site as an example. The hydraulic gradient at the right bank was in the range of 1.05 and 1.12 nearly one time as large as that before impounding, shown in Fig. 1. Around the dam site, the change of topography such as dragging of the slope, reform of the platform and others during construction, would be usually suitable for infiltration from precipitation. Again, Fig. 1 demonstrates that the construction of 142m a • s • 1 platform would lead to the increase of the flow rate around the area.

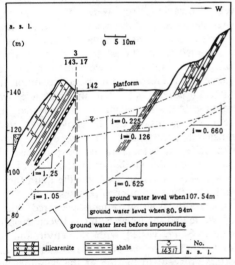

Fig. 1 Cross section of hydrogeology in the right bank of Xinanjiang dam

2.3 *Character of the groundwater under the dam foundation*

As the anti-seepage steps were taken such as curtain-grouting and the drainage system improved under the dam foundation, the ground water flow in the bed domain, tended to be slower and the amount of flow rate decreased with years. In addition, the ground water level behind the curtain grouting was lower and the hydraulic gradient was smaller as a role. However, the ground water near the bank but still under the dam foundation was quite different in that it had relatively larger hydraulic gradient and bigger flow rate. For instance, the amount of flow rate of the dam sections near the bank in Chencun dam has 53.65% of the total flow rate whereas the drainage rate from the boreholes in the bed domain was only 8.27%. Similarly the flow rate of dam sections 2-3 near the right bank had 47.7% of the total flow rate and its variation was quite small in Xinanjiang dam.

Character of the ground water quality under the dam foundation. Usually, the ground water was of HCO_3-Ca in quality and its pH value was between 7-8 before the dam was built. However, the ground water within the domain of the dam foundation after the dam built tended to be of alkali even strong alkali or occasionally of weak acidline, as shown in Table 3. The alkalicization of the ground water quality occured generally behind the curtain-grouting. For example, the pH value of ground water in this particular area was between 9.6-10.1 under Meishan dam and up to 12.6 (dam section 24) under Xinanjiang dam. However, as the increase of distance from the curtain grouting, the pH value of the groundwater under the dam foundation tended to be lower gradually, similar to that of general ground water. Under Chencun dam, the boreholes for drainage were penetrated through the concrete foundation up to over 10m in thickness. As a result, the pH value of the ground water with in the drainage gallery relativelly far from the curtain grouting was up to 11 even 12 as well. It may be concluded that the existence of both curtain grouting and concrete had a great bearing on the variation of ground water quality. Hence, some unusual types in quality such as OH-Ca • Na and CO_3 • OH-Na occurred which were rather rare in nature.

Some of the ground water under the dam foundation was locally of weak-acidity in quality. The ground water with pH=6.5 occured in the boreholes with high water levels, which were located behind the curtain grouting under Chencun dam. The ground water with pH<7.0 slightly bigger than that of the water near the bottom of Meishan reservoir occurred in the dam section near the right bank, and contained

Table 3. Characteristics of ground water quality under some dam foundations

Reservoir	No.	pH	(mg/L)								Location
			Na^+ $+K^+$	Ca^{2+}	HCO_3^-	SO_4^{2-}	CO_3^{2-}	CO_2 (physical)	OH^-		
Xinanjiang	3	6.1	4.7	5.5	27.24	5.5	0	45.55	0		bank
	12	9.4	5.3	16.8	46.03	4.6	13.5	0	5.5		bed
	24	11.6	122.9	50.8	0	8.2	108.1	0	149.2		bank
Meishan	4	10.1	76.0	4.0	29.1	10.7	35.5	0	/		bank
	9	6.4	6.8	8.7	47.3	5.6	0	19.1	/		bed
	13	6.7	19.9	30.8	15.7	2.7	0	41.1	/		bank
Chencun	17	6.8	14.7	80.8	37.2	5.5	0	12.3	0		/
	16*	12.1	124.7	193.0	0	27.2	30.3	0	229.3		bed

Notice, * —borehole designed for measuring uplift pressure, the remainder for drainage.

special dissolved gas, i.e. H_2S. The weak-acidification of the ground water under dam sections 2-5 was more obvious in Xinanjiang dam, the pH value was mostly smaller than 6.4, lower than that of the water at the bottom of the reservoir.

The gas H_2S occurred occasionally in the ground water under the dam foundation, but its content was relatively lower. For instance, the content of this kind of dissloved gas was usually lower than 1mg/L in Xinanjiang dam and mostly from the drainage boreholes, located in or near the bed. Moreover, the number of the boreholes containing H_2S tended to decrease towards the downstream. The relatively lower content of H_2S in the ground water under the dam foundation was of the intergrading between oxidization and reduction as regards the geochemistry environment. This kind of special gas was mostly result from the water at the bottom of the reservoir.

It was demonstrated by a beam of strong light that the ground water from drainage boreholes contained colloidal grains. Therefore, the ground water under the dam foundation was a very complicated solution with some colloid. The one of major characteristics was that the groundwater from the drainage boreholes, fissures in rock masses and occasionally from joints between different dam sections contained the substances with colloidal state, namely eluates. According to statistics, there were 128 drainage boreholes, occupied by 30.7%, from which eluates occurred under Xinanjiang dam foundation. Additionally, there were 18 outcrops of fissures from which eluates occurred as well. However, there were only 36 drainage boreholes, occupied by 15.9% from which those occurred under Chencun dam foundation. It was related to bordholes' top with high elevation. As a result, these were no working for drainage. Therefore, the basic condition of eluates occurred was that the borehole was draining. Site investigation showed that the groundwater inside the boreholes was very transparent, and after flowing out, eluates with different colours would be produced. It could be concluded that the variation of the ground water environment was the main factor leading to the formation of eluates.

Generally speaking, eluates were red, black and white and the intergrading between them in colour. Some varied with the environment. For example, the colour of the eluate near the bottom adjacent to the top of the borehole become black but its composition remained, if red eluate was thicker in accumulation. It was shown as sample 10 in Table4. On the other hand, the black eluate become red one if exposed to air for longer time. As far as the compositions of these eluates were concerned, red eluate was mainly of Fe_2O_3 and its content up to 60~80%. Black eluate was mainly of Fe_2O_3 or MnO, and white one of CaO which might form $CaCO_3$ after depositing. It was shown with X-ray diffraction that the eluates with red and black colours were noncrystal, except for a little $CaCO_3$.

Table 4. Compositions of eluates under dam foundations

Sampling		Colour	(%)					PH
			SiO_2	Fe_2O_3	Al_2O_3	CaO	MnO	
Xinanjiang	No. 10 (top)	red	7. 16	66. 47	0. 90	3. 43	0. 66	/
	No. 10 (bottom)	black	8. 14	54. 63	1. 45	5. 67	0. 41	/
	from fissure	black	8. 21	13. 73	2. 56	5. 60	19. 87	
	No. 3	red	29. 89	30. 74	5. 78	9. 92	0. 20	6. 06
	No. 2	red	2. 41	78. 41	0. 11	0. 29	0. 17	5. 5
Chencu	No. 17	red	13. 36	43. 70	4. 54	9. 05	/	6. 8
	No. 8	black	7. 54	6. 96	1. 87	8. 83	43. 09	8. 5
Meishan	No. 4	red-yellow	17. 49	60. 80	0. 69	4. 00	3. 17	/
Mozitan	joint	red	4. 24	71. 08	1. 83	3. 03	/	/

3 ACTION OF ENEIRONMENT WATER AROUND THE DAM SITE

After impounding, the weak acidification of the water towards the bottom of the reservoir was result from richening of organic substances at the bottom. Under the circumstances of lack-oxygen, the following reaction would take place.

$$CH_2O + H_2O \xrightarrow{\text{bacteria}} CO_2 + 4H^+$$

The process of decomposing of organic substances would release ones with reduced state, suck as H_2S and CH_4. The weak acidity of the water at the bottom of the reservoir was rich because of lower temperature and higher pressure there. This kind of water was the main recharge source of the ground water under the dam foundation. The variation of both environment and water quality resulted in the active interaction between liquid and solid phases (including the curtain grouting and concrete).

3. 1 *The acidification or alkalization of the ground water under the dam foundation*

The alkalization was caused by the hydrolysis of the curtain grouting and concrete through the ground water and the reservoir water. The chemical reaction could be described as follows.

$$CaO + H_2O = Ca(OH)_2$$
$$2CaCO_3 + 2H_2O = Ca(OH)_2 + Ca(HCO_3)_2$$
$$Ca(OH)_2 = Ca^{2+} + 2OH^-$$

Usually, this process was quite slow. However, the ground water with alkalization even strong alkaligation in quality might occur behind the curtain-grouting, as the time elapsed. This kind of water quality was shown in Table 3. The water could be called dead water if its pH value was over 12, such as that in dam section 24 of Xinanjiang dam. Grenerally speaking, the slower the flow under the dam foundation, the larger its pH value. Therefore, the existence of the ground water with strong alkalization could be considered as one of evidences of the fact that the seepage was slow and the anti-seepage effect of the curtain grouting was quite good.

The weak-acidification of the ground water under the dam foundation was related to not only reservoir water but also rocks concerned. For instance, the ground water with pH value lower than that of the water at the reservoir bottom within the dam section near the right bank was relative closely to the rock, i. e. carbonaceous shale. Within this particular domain, the ground water behaviour with big hydraulic gradient and high laterial seepage force promoted greatly the oxidization, decomposition of the organic substances in the a carbonaceous shale.

$$CH_2O + O_2 \xrightarrow{\text{bacteria}} CO_2 + H_2O$$
$$CO_2 + H_2O \longrightarrow H^+ + HCO_3^-$$

The content of HCO_3^- ion in this kind of weak acid ground water was relatively rich, only occurring in where the carbonaceous shale

2587

was distributed. The weak acidification of the ground water intensified the corrosion to the curtain grouting of Xinanjiang dam. For this reason, the effect of strengthening grouting with cement for many times was not so ideal within dam sections 2-3

However, the weak- acidification of the ground water under Chencun and Meishan dam foundation was result from recharge of the water at the reservoir bottom. The evidence was relatively large seepage rate from the boreholes and with H_2S in the ground water. It may be concluded that the existence of weak acid ground water behind the curtain grouting could be thought, as one of evidences of the relatively rapid flow and the curtain grouting being corroded locally if the influence of rocks was not considered

3. 2 *The mechanism of the formation of eluate, under the dam foundation*

The change of the water environment was the main factor resulting in the formation of various eluates. As mentioned before, the water of the reservoir bottom was of weak acidity with the reduced environment and the ground water from the most of drainage boreholes contained H_2S. In addition, it also contained Fe ion with different valences (Fe^{2+} & Fe^{3+}), the content of which was nearly equal. All these demonstated that the environment of the ground water under the dam foundation particularly in the bed domain was the weak reduced environment or the intergrading between the reduced and the oxidized one. Under the environment with weak reduction, the water containing certain content of corrosive CO_2 would make it easily for the element with variable valence in rocks to be transferred in the form of the low valence ion or of the colloid grains. As the pressure tended to be lower and the temperature higher, or the environment alterred from the reduction to oxidization one, the element with low valence such as Fe^{2+}, would be oxidized easily. As a result, the element with high valence occurred, such as Fe^{3+}. Through the action of colloform, eluates with red or brown-yellow colour occurred at the opening of flow out of the drainage borehole. The process could be described as:

$$4Fe(HCO_3)_2 + 2H_2O + \frac{1}{2}O_2 = 4Fe(OH)_3 + 8CO_2 \uparrow$$

$$2Fe(OH)_3 \rightarrow Fe_2O_3 \cdot 3H_2O$$

As the red eluate increased in thickness, the environment near the bottom was transferred into the reduced one again. Consequently, red eluate, $Fe(OH)_3$, was reduced into the colloform, i. e. $FeO \cdot nH_2O$, in which the chemical composition unchanged basically, as shown in sample No. 10 of Table 4.

The element Mn with low valence was transported easily under the reduced environment. However, it would be oxidized and deposited in the form of Mn with high valence under the condition of alkalization. For example,

$$2Mn(OH)_2 + O_2 = 2MnO(OH)_2 \downarrow$$

Moreover, the colloform, i. e. $MnO_2 \cdot nH_2O$, might be formed as a result of $MnO(OH)_2$ oxidized further. therefore, the eluate with black colour occurred mostly within the water environment with alkaization, as shown in Table 4.

Generally speaking, the chemical compositions of these eluates were Fe, Mn largely and the remainder were Si, Al. The elements Fe and Mn were from the water of the reservoir and rocks. The content of these two elements in the groundwater was bigger than that of the water at the reservoir bottom, under Xinanjiang and Chencun dam foundation. Hence, the dissolution of the sandstone of Fe-Mn bearing was one of the important sources of the eluates. However, its influence on rock strength was not obvious because of the lower amount of eluates and the lower content of Fe and Mn element in the rock. The content of compounds SiO_2 and Al_2O_3 in some of red eluate was relatively high and over 30% under the dam sections near the right bank of Xinanjiang dam, as shown in Table 4. It was shown that some of SiO_2 and Al_2O_3 in the shale were transferred and the shale layer itself was softened even argillited locally. As a result, a few drainage boreholes which penetrated through the shale layer collapsed along their wall and the sediment within these boreholes was up to 3m even 4m in thickness.

3. 3 *Action of seepage force*

The seepage force produced by the flood effect of the reservoir would have an important influence on the curtain grouting. The normal designed water level was 119m in Chencun reservoir. Since the dam was built, the steps of strengthening engineering to the curtain grouting were taken because the geological condition was very complicated. Four rows of the curtain grouting formed and the anti-seepage effect was received significantly. Nevertheless, the amount of seepage from a few drainage boreholes increased gradually with year since the higher water level (118.69m) in the reservoir occurred firstly in 1983. The heigest one was 118.81m in 1991. Correspondingly, the amount of drainage was up to the maximum value, in the same year. For instance, the total amount of drainage out of borehole 10-4 was increased from 41.46m^3 in 1984 to 503.7m^3 in 1991. Meanwhile, the water level in this borehole was 5m as higher as that in adjacent one. Consequently, the column of higher water level behind the curtain grouting formed in the domain generally with lower ground water level and flat hydraulic gradient. The measured uplift value nearby was beyond the designed one as well. All these illustrated that the local deterioration of the curtain grouting under Chencun dam foundation existed by the action of higher water level of the reservoir.

The hydraulic gradient and seepage rate were bigger under the dam section near the bank. As a result, the action from the lateral seepage force become the one of main factors influencing on the stability of the dam. For instance, within the dam section near the right bank of Xinanjiang dam, some micro-fissures were found to be filled with the cement stone or cement films by identifying the core from the borehloes for strengthening to the curtain grouting. It could be drawn that ground water might be forced to enter some micro-fissures because of the rich recharge source and bigger hydraulic gradient. Consequently, the interaction between water and rocks was intensified, leading to the weak acidification of water in quality and the softening even local argillization of shale layers with various thicknesses. Correspondingly the measured uplift value locally was beyond the designed one, to which the steps were being taken. Taking Meishan reservoir as another example. As the reservoir water level was up to the maximum value since the dam built, leakage happened suddenly under the dam section near the right bank and the leakage rate measured was 70L/s within the area from arch 14 to 16. Meanwhile, joints were found to occur on the dam body itself. According to investigation, they were produced and developed by both the high pressure of the reservoir water and the action of laterial seepage force. In addition, the shearing strength was dropped rapidly as regards some weathered fissures filled with clay after saturation. Consequently, the rock masses in the area concerned were found to slide slightly along the discontinuity. After the steps taken, this water power station tended to be operated normally again.

4 CONCLUSIONS

4.1   The action of the environment water around the dam site can be indentified by hydrogeochemistry (acidification or alkalization of water, softening even argillizationof sandwiches and chemical erosion) and hydrodynamics. Both were promoted with each other.

4.2   The formation of various eluates resulted from the variation in the water environment around the dam site. They were from either the water at the reservoir bottom, or the chemical erosion on the rocks, or the argillization of samdwiches with regard to material sources.

4.3   The bigger hydraulic gradient in the area of the dam sections near the bank was the main factor controlling the stability of the dam sections concerned, particularly under suitable recharge conditions, such as the establishment of the platform and pools by man-made and the existence of seepage around the dam.

REFERENCES

Blyth, F. G. H., et al, (1984). A geology for engineers Edward Arnold

Peng, H. X., et al, (1991). Hydrogeochemical action in the process of sandwich argillization. J. of Hohai university, No. 1

Peng, H. X., et al, (1994). Characteristics and action of environmental water around the dam site in Xinanjiang water power station. J. of Hydraulic Engineering, No. 2

Shi, X. J., et al, (1993). Study on formative mechanism of exudates from the dam foundation of Xinanjiang water power station. J. of Hohai University, No. 2

Étude des vibrations engendrées lors du battage de différents types de pieux sur un même site

Analysis of vibrations generated by different pile driving techniques at the same site

C. Horrent, D. Jongmans & D. Demanet
LGIH, Université de Liège, Belgique

RÉSUMÉ: Les activités du génie civil génèrent des vibrations pouvant s'avérer nuisibles pour l'environnement. Dans le cadre de la prédiction de vibrations engendrées lors du battage de pieux (spécialement en zone urbaine), une campagne de mesures très détaillée a été réalisée sur un site expérimental en vue de comparer les différentes techniques de mise en place. L'analyse des données dans les domaine temporels et fréquentiels a permis de mettre en évidence les paramètres (distance, profondeur, source) influençant les caractéristiques des signaux enregistrés.

ABSTRACT: Construction operations produce ground vibrations which may have detrimental effects on the environment. In order to improve basic knowledge and prediction ability about piling-induced vibrations, a measurement experiment has been performed on the same site and four different pile-driving techniques have been compared. Signal analysis in both time and frequency fields has shown the influence of the main parameters (distance, depth, source) on ground vibration characteristics.

1 INTRODUCTION

En site urbain, le battage de pieux est susceptible de créer des nuisances plus ou moins importantes aux personnes ou aux habitations. Suite à l'augmentation de la sensibilité aux perturbations de l'environnement, les entreprises de génie civil sont confrontées de manière accrue au problème des vibrations qu'elles génèrent. L'importance du sujet a été reflétée ces dernières années par le nombre croissant de publications concernant l'analyse des vibrations du sol engendrées par le battage de pieux. Parmi d'autres, citons les travaux de Ciesielski *et al* (1980), Massarsch et Broms (1991), Selby (1991) et Hanazato et Kishida (1992). En vue de déterminer l'influence de la source vibratoire et d'améliorer la compréhension physique des phénomènes, une campagne expérimentale comprenant le battage de différents types de pieux et l'enregistrement des ondes générées a été menée en un même

site. La structure géologique de ce dernier a été préalablement reconnue par prospection géophysique (essais simique-réfraction et inversion des ondes de surface) et par essais de pénétration statique (CPT). Le profil de sol standard, comprenant 3 couches superficielles, est présenté au tableau 1.

Tableau 1. Structure géologique du site d'essais et caractéristiques dynamiques des différents terrains.

Terrain	Profondeur (m)	Vp (m/s)	Vs(m/s)
Limon	0,0 à 4,0	310	180
Sable	4,0 à 7,0	400	260
Sable	7,0 à 39,0	510	380

Quatre procédés de battage ont été mis en oeuvre sur ce site: des pieux métalliques battus en pied (PI), des pieux préfabriqués en béton battus en tête (PII), des pieux métalliques battus en tête (PIII) et des pieux métalliques battus en pied de type Franki (PIV). Les vibrations générées ont été mesurées selon les trois composantes du mouvement (verticale, radiale et transverse) par des séismomètres triaxiaux placés à 2,5m, 10m, 25m et 50m du fût du pieu tandis que 12 capteurs verticaux étaient disposés entre 5 et 60 mètres avec un espacement de 5 m (voir figure 1). Des enregistrements ont été effectués à chaque enfoncement de 0,5 mètres du pieu jusqu'à une profondeur de 9 mètres correspondant au refus. Quatre pieux de chaque type ont été battus dans des conditions normales de chantier.

Dans les études de vibration, le mouvement du sol est souvent caractérisé par l'amplitude maximale de la vitesse d'oscillation Vmax. Les valeurs de Vmax vont être analysées afin de déterminer l'influence du type de source, de la distance et de la profondeur d'enfoncement de la pointe du pieu sur le niveau des vibrations générées. La répartition de l'énergie selon la fréquence sera également étudiée par calcul de spectres de Fourier.

2 ANALYSE DES VIBRATIONS MAXIMALES VMAX

En première approche, les signaux mesurés ont d'abord été caractérisés par leur amplitude maximale Vmax. Le traitement de ces valeurs a permis de montrer que les mouvements selon les composantes verticales et radiales étaient généralement prédominants. Ces observations sont en accord avec la radiation théorique d'un point force vertical dans un milieu stratifié. Les valeurs non nulles obtenues selon la composante transversale indiquent cependant le comportement non-axisymétrique de la source et/ou la présence de phénomènes de diffraction dans les couches superficielles.

Les valeurs de Vmax mesurées selon la composante verticale et radiale lors du battage d'un pieu préfabriqué sont respectivement reprises aux figures 2 et 3 en fonction de la distance D et de la profondeur de la base du pieu. Sur chaque figure, les variations de Vmax sont présentées sous la forme d'un histogramme bidimensionnel et d'un diagramme bilogarithmique Vmax = f(D) pour différentes profondeurs. L'examen de ces figures montre que, suivant la composante verticale du mouvement, la vitesse d'oscillation maximale décroît relativement régulièrement avec la distance. Dans le diagramme bilogarithmique, les courbes d'atténuation Vmax = f(D) pour les différentes profondeurs se présentent sous la forme de droites et semblent donc proches d'une loi théorique de la forme Vmax = a D^b. L'ajustement d'une loi de ce type à l'ensemble des mesures (tous types de pieux confondus) a conduit à la relation V= 38 D$^{-0.99}$ avec un coefficient de corrélation de 0.84 (figure 4). Le coefficient b est proche d'une valeur de -1 caractérisant l'amortissement d'une onde de volume dans un milieu homogène. Comme il sera montré plus loin, les vibrations maximales correspondent généralement à des ondes de surface et une valeur pour b de l'ordre de -1 provient sans doute de la combinaison de l'amortissement géométrique des ondes de surface (D$^{-0.5}$) et de l'atténuation intrinsèque résultant de l'anélasticité des sols. Les calculs de régression pour chaque type de pieu conduisent aux droites également présentées à la figure 4. La comparaison entre ces courbes

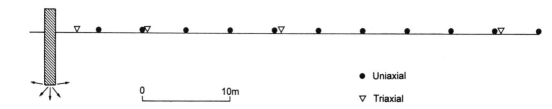

Fig. 1 Dispositif de mesures utilisé.

d'atténuation montre que le battage de pieux tubés battus en pied (PI) génèrent nettement moins de vibrations que les autres systèmes de battage.

L'examen des données pour la composante radiale (figure 3) indique une évolution V(D) relativement semblable pour la plupart des profondeurs. La courbe d'atténuation correspondant à une profondeur de 9,5 m présente cependant une anomalie importante à 10 m de distance par rapport à une loi théorique $Vmax = a\ D^b$. Ce phénomène d'augmentation locale du mouvement à distance moyenne a déjà été observé par Selby (1991) lors d'une étude de battage de pieu. Il sera ultérieurement étudié lors de l'examen de la forme des signaux et du calcul des spectres de Fourier.

A faible distance du pieu (2,5 m), les vibrations maximales sont généralement mesurées entre 4 et 5 mètres de profondeur.

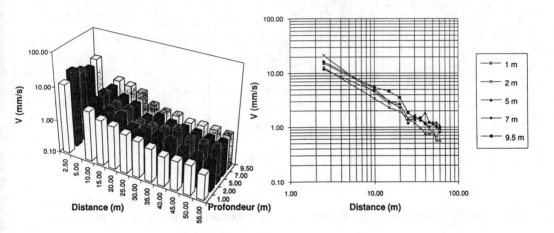

Fig. 2   Evolution de l'amplitude maximale (Vmax) en fonction de la profondeur et de la distance pour un pieu préfabriqué. Composante verticale du mouvement.

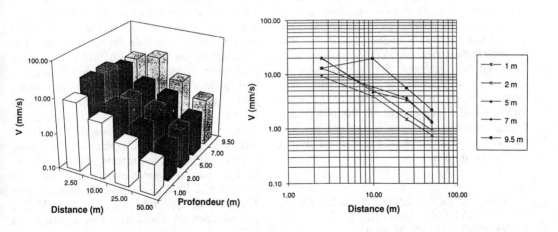

Fig. 3   Evolution de l'amplitude maximale (Vmax) en fonction de la profondeur et de la distance pour un pieu préfabriqué. Composante radiale du mouvement.

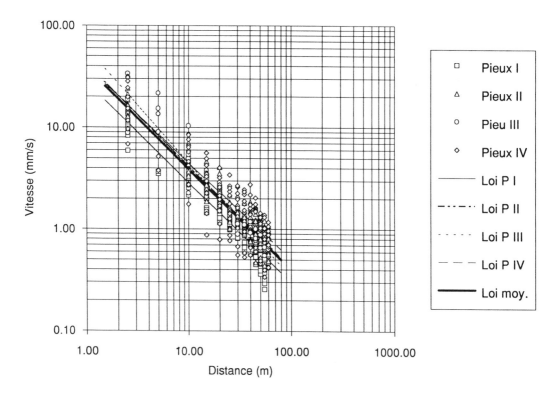

Fig. 4 Loi d'atténuation $V = a\,D^b$ obtenue avec l'ensemble des données de vibrations verticales.

tandis qu'à plus grande distance, elles tendent à augmenter avec la profondeur (figures 2 et 3). A proximité du pieu, le niveau de vibrations résulte de l'influence des deux phénomènes antagonistes: l'augmentation de l'énergie de battage avec la profondeur et l'amortissement géométrique des ondes causée par l'augmentation verticale source-récepteur. Ces deux effets conduisent à un maximum de vibration à une profondeur intermédiaire. Par contre, à plus grande distance source-récepteur, l'influence de la profondeur sur l'amortissement géométrique des ondes diminue rapidement et le niveau des vibrations croit généralement avec la profondeur du pieu.

3 COMPARAISONS ENTRE LES NIVEAUX DE VIBRATIONS GENEREES PAR LES DIFFERENTS TYPES DE PIEUX

En plus de la distance et la profondeur, l'amplitude des vibrations générées dépend également du procédé de battage et du type de pieu utilisé. Rappelons que les quatre types de pieu (pieux métalliques battus en pied (PI), pieux préfabriqués en béton battus en tête (PII), pieux métalliques battus en tête (PIII) et pieux métalliques battus en pied de type Franki (PIV)) ont été mis en place de façon standard. Les vibrations engendrées par les différents types de pieu ont été comparées à différentes distances et profondeur. La figure 5 reprend, pour tous les pieux étudiés, les niveaux de vibration mesurés selon les trois composantes à 2,5 et 10 mètres de distance lorsque la base du pieu était située à 5 mètres de profondeur. L'examen de ces figures illustre l'influence importante du procédé de battage sur les mouvements du sol. En particulier, à 10 m de distance, la mise en place d'un pieu de type tubé battu en tête (PIII) a engendré des vibrations trois fois supérieures à celles mesurées lors du battage des pieux tubés battus en pied (PI).

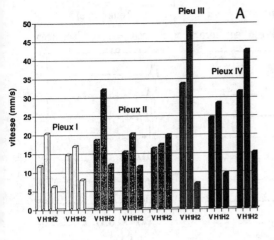

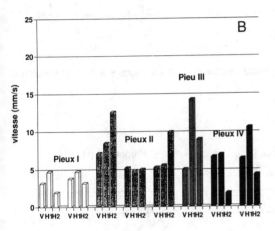

Fig. 5 Vibrations maximales mesurées à 2,5 m (A) et 10 m (B) de distance pour une profondeur de 5 m.

L'examen de l'ensemble des données permet de tirer les conclusions générales suivantes :

• à faible profondeur (<1m), les vibrations maximales sont générées par le battage des pieux PII et PI.
• à profondeur intermédiaire (2-5m), les vibrations les plus faibles correspondent au battage des pieux PI.
• entre 5 et 9 mètres, on a la relation suivante entre les amplitudes des différents pieux :
PI < PII ≤ PIV ≤ PIII

A proximité immédiate de la source, les relations présentées ci-dessus souffrent néanmoins de nombreuses exceptions en raison des variations importantes d'énergie et de l'influence probable d'effets locaux de radiation de la source (champ proche).

4 ETUDE DE LA FORME DES SIGNAUX ET DES SPECTRES DE FOURIER

Les résultats obtenus jusqu'à présent ne se basent que sur l'étude de l'amplitude maximale des vibrations Vmax. Or, l'effet d'un signal sismique sur les structures dépend également d'autres facteurs comme le contenu fréquentiel et la durée du mouvement. Une étude de ces aspects nécessite l'examen de tout le signal et le calcul de transformées de Fourier. La figure 6 reprend les enregistrements (composante verticale) obtenus à 1, 5 et 9 mètres de profondeur pour un pieu tubé battu en pied. L'observation des signaux montre que les amplitudes maximales correspondent généralement à des ondes de surface (de type Rayleigh), sauf à proximité immédiate de la source ou lorsque la base du pieu est proche de 9 mètres. La forme des ondes et la répartition de l'énergie en fonction du temps varient cependant fortement en fonction de la profondeur de la source.

L'examen des spectres de Fourier correspondants (figure 7) montre que le contenu fréquentiel des vibrations, généralement compris entre quelques Hz et 80 Hz, varie en fonction de la distance et de la profondeur. Lorsque la distance augmente, on observe généralement un appauvrissement des hautes fréquences, résultant de l'atténuation intrinsèque des sols dans lequel les ondes se propagent. Parallèlement l'enfoncement du pieu se traduit également par une modification de la forme des spectres de Fourier et de la position des maxima en fonction des couches rencontrées à la pointe du pieu.

En particulier les signaux enregistrés pour une profondeur du pieu de 5 m sont caractérisés par de relativement basses fréquences (de l'ordre de 10 Hz) tandis que les vibrations générées à 1 m de profondeur sont relativement pauvres en basses fréquences.

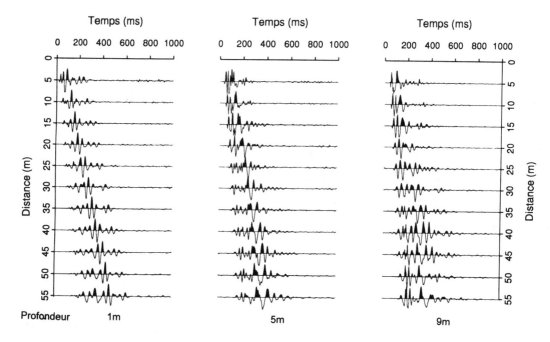

Fig. 6 signaux normalisés mesurés selon la composante verticale lors du battage à 1, 5 et 9 mètres de profondeur d'un pieu tubé battu en pied.

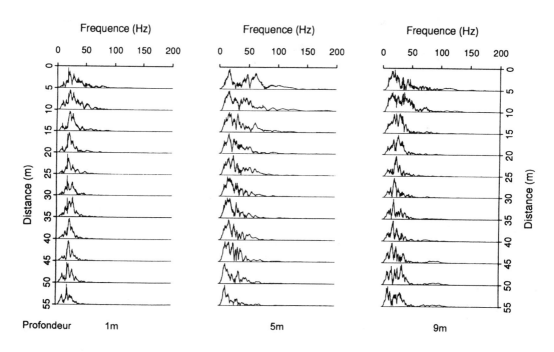

Fig. 7 Spectres de Fourier normalisés correspondant aux signaux de la figure 6.

Les caractéristiques des signaux sont fort semblables selon la composante radiale du mouvement. Une augmentation anormale des vibrations, déjà mentionnée lors de l'étude des vibrations maximales (figure 3), est cependant observée selon cette direction à 10 m de distance lorsque le pieu a pratiquement atteint le refus (9 m de profondeur). La figure 8 reprend, en fonction de la profondeur, les signaux et spectres de Fourier correspondants obtenus à 10 m de distance selon la composante radiale lors du battage d'un pieu de type Franki.

A partir de 8,5 m de profondeur, on y observe l'augmentation importante de l'amplitude du signal associée à l'élargissement des spectres vers les hautes fréquences. A proximité du refus, une grande partie de l'énergie transmise au pieu ne contribue plus à son enfoncement mais à générer des vibrations très impulsionnelles à relativement hautes fréquences. Ce pulse très simple et énergétique

n'apparaît cependant qu'à 10 m de distance (figure 9) et semble résulter de l'interférence constructive entre deux ondes de nature différente (de volume et de surface ?) parfaitement ou partiellement dissociées aux profondeurs inférieures (voir figure 8). Cette hypothèse devra être cependant vérifiée par modélisation numérique.

5 CONCLUSIONS

L'analyse de séismogrammes enregistrés sur un site expérimental lors du battage de quatre types de pieux différents a permis de mettre en évidence l'influence des différents paramètres conditionnant l'amplitude des vibrations observées. Vu le caractère relativement superficiel de la source séismique, les vibrations maximales correspondent principalement à des ondes de surface (de type Rayleigh) et

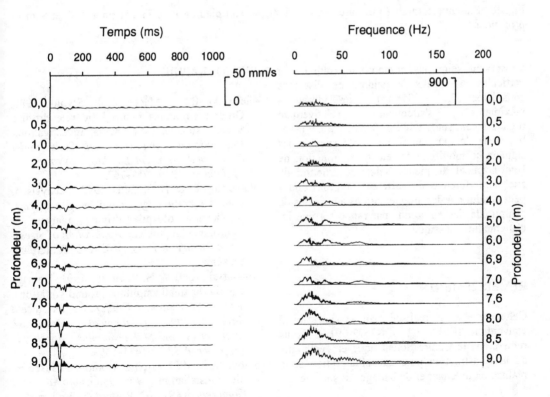

Fig. 8   Signaux et spectres non normalisés mesurés à 10 m de distance selon la composante radiale (pieu de type Franki).

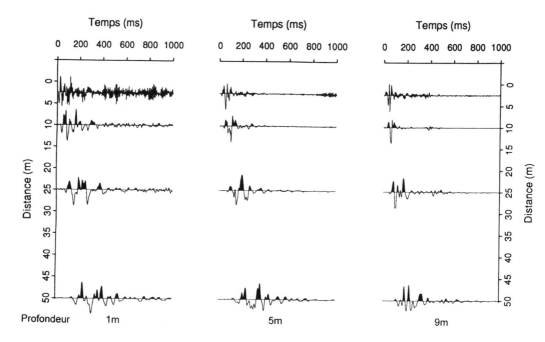

Fig. 9   Signaux normalisés mesurés lors du battage d'un pieu de type Franki pour 1, 5 et 9 m de profondeur.

s'observent selon les composantes radiales et verticales. Si dans la gamme de distances considérée (2,5 à 55 m) , les vibrations maximales s'atténuent relativement régulièrement selon une loi $V_{max} = a\ D^b$ (avec b proche de -1) , la forme et le spectre des signaux se modifient fortement en fonction de l'enfoncement du pieu. Selon la structure du site, l'interférence de différents types d'ondes peut conduire à des augmentations anormales de l'amplitude des vibrations par rapport à une loi d'atténuation classique.

REMERCIEMENTS

Cette étude a été réalisée dans le cadre de la convention IRSIA CI 1/4-7672/091. Nous remercions le Centre Scientifique et Technique de la Construction pour sa collaboration et la réalisation du chantier de battage sur son site.

BIBLIOGRAPHIE

Ciesielski, R., E. Maciag & E. Stypula 1980. Ground vibrations induced by pile driving. Some results of experimental investigations. *International Symposium on Soils under Cyclic and Transient Loading / Swansea 7 - 11 January 1980* : 757-762.

Hanazato, T. & H. Kishida 1992. Analysis of ground vibrations generated by pile driving. Application of pile driving analysis to environmental problem. *Application of Stress-Wave Theory to Piles* : 105-110. Rotterdam: Balkema.

Massarsch, K. R. & B. B. Broms 1991. Damage criteria for small amplitude ground vibrations. *Second International Conference on Recent Advances in geotechnical Earthquake Engineering and Soil Dynamics, March 11-15, paper n° 11.5* : 1451-1459.

Selby, A. R. 1991. Ground vibrations caused by pile installation. *4th International DFI Conference*: 497-502. Rotterdam: Balkema.

7th International IAEG Congress / 7ème Congrès International de AIGI, © 1994 Balkema, Rotterdam, ISBN 90 5410 503 8

An investigation into the effects of vibro replacement stone column construction on the surrounding buildings

Recherche sur les effets de la construction des pieux de gravier par vibro-substitution sur les bâtiments voisins

Wu Yaozhu
Comprehensive Investigation and Surveying Institute of Machinery and Electronics Industry Ministry, Xian, People's Republic of China

ABSTRACT: The detrimental influences on surrounding buildings during construction of vibro replacement stone column were studied in this paper. Through investigating influencing factors, it is found that the main influence is caused by differential settlement due to vibration of subsoil since vibration of vibroflat, and if on collapsible loess site, collapsible settlement will occur under construction as well. Furthermore, the differential settlement due to vibration results from inhomogenety of subsoil, unsymmetry of dead load of structure as well as unsymmetry of dynamic load produced by vibroflot. Based on the above analyses, this paper presents a method of calculating settlement due to vibration and at festing method necessary for required parameters in the evaluation of settlement due to vibration. This method is not only simple in both of its test and calculation, but also practieal at evidently less expense. By analyzing a practical example, it is considered that settlement due to vibration calculated by equivalent static method, in which dynamic stresses are estimated, agrees with practical situation.

1. The problem Raised

Vibro replacement stone column has been widely applied in the treatment of foundation soil containing loose sand, soft soil, etc. After such treatment, the density of loose sand and the anti — liquefaction capacity as well as the bearing capacity of the composite foundation soil based on gravel piles in soft soil have been greatly increased. Therefore, rich experience has been accumulated in this aspect. However, in the course of vibroflotation construction, vibration produces, to various extent, influences on the surrounding buildings, especially on the surrounding buildings based on natural foundations. What factors have something to do with the extent of such influence? How to appraise such influence? These are the problems which designers and constructors care about. The present paper inquires into the factors of influence and the methods of appraisal of vibro replacement stone column construction on the surrounding buildings in loess satuation zones.

2. Characteristics of Vibration of a Vibroflot

As an engineering focus of vibration, the vibroflot has its own features: 1) It is situated inside the foundation soil, producing direct effects on the foundation soil, effects which are greater than those produced by the vibration focus on the surface of the soil; 2) The frequency of vibration is fixed and the frequency is higher (24 H_z). It may be regarded as steady—state forced vibration, and when the electricity of the vibroflot is cut or when a stone column is completed, it becomes free vibration, decreasing with the duration of time. When the frequency at a certain time approaches the fixed frequency of the surrounding buildings, it makes the vibration of those buildings and the foundation soil increase apparently, but the period is rather short; 3) It changes constantly in spatial positions: at the time of the vibroflotation of holes, it goes from downward, and goes upward when the compaction of the materials filled is underway. Besides, the plane relative positions between the vibroflot and the surrounding buildings change constantly; 4) The period of vibration is long with a 7—meter high stone column taking around 20 minutes from filling the materials to its completion, and, generally speaking, the interval between the successive construction of two columns is short.

3. Factors of the Influence Produced by the Construction of Vibro Replacement Stone Columns on the Surrounding Buildings

What with the vibroflotation construction (vibration sluicing method) with the characteristics of vibrationas well as the effects of water, its influence on the surrounding buildings are determined by the following three factors:

3.1 The Factor of Vibration

The vibration of the construction of vibro replacement stone columns is the leading factor of the influence on the surrounding buildings, with all the other problems brought about by the vibration. The intensity or weakness of the vibration is determined not only by the intensity of the force of vibration of the vibroflot, but also by such factors as the distance from the vibrlflot, features of the foundation soil, the lasting period of the vibration, as well as the length of intervals. The greater the power of the vibroflot is, the nearer the distance from the vibroflot becomes, the longer the vibration lasts, and the shorter the intervals become, the more harm will be done to the surrounding buildid-

ngs. The vibration of the vibroflot spreads through the foundation soil around it, and, as a result, causing the vibration of the surrounding buildings, which in turn acting on the foundation soil. In another word, the existence of the surrounding buildings increases the vibration of the foundation soil.

3.2 The Factor of the Features of the Foundation Soil

Under the circumstance when the vibration focus is certain, the nature of the foundation soil is the decisive factor determining the effects on the buildings. Thevibration may make the satuation loose sand produce liqufaction as well as the settlement due to vibraton of the soft soil. Since the liquefaction and plastic flowing zone is limited — — usually less than 3 meters, large —scale lipuefaction and plastic flowing will not happen under the foundation of the surrounding buildings. The result of the effects of the vibration is to make the foundation soil produce settlement due to vibration and deformation to various extent, or the continuous seepage of water into the founation soil under the surrounding buildings reduces the increasing strength of water content of the foundation soil. The dynamic experiments and the analysis of tests conducted by scholars both at home and abroad demonstrate: 1)Loose sandy soil is likely to subside under vibration; 2)Saturated soft soil (including saturated loess soil) is apt to subside under vibration; 3) The relation between the dynamic effective stress of soil σ_d and residual strain ε_p bears the features which are displayed in figure 1: when the dynamic effective stress σ_d applied to the foundation soil is smaller than σ_{dc}, the residual strain of soil ε_p will be very small or zero. At this moment the foundation soil and the buildings above it are absolutely safe, and the settlement due to vibration capacity of the foundation soil will be zero. When the dynamic effective stress σ_d of the foundatin soil is between σ_{dc} and σ_{dc}'', there will be some settlements due to vibration and deformation, though the foundation soil might not suffer damage. If the settlement due to

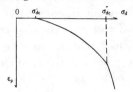

σ'_{dc} :Dynamic loading in failure

σ''_{dc} :Critical dynamic loading in residual strain epual zero

Figure 1 . $\sigma_d \sim \epsilon_p$ curve (with the same frequency and number of vibration)

vibration capacity or the settlement due to vibration difference exceeds the allowable deformation value of the buildings, there will be eracks in the buildings, which might incline. When the dynamic effective stress σ_d of the foundation soil is greater than σ''_{dc}, there will be damage to the foundation soil which might lose its steadiness, severe damage might happen to the buildings; 4) When the dynamic effective stress σ_d of the foundation soil is between σ'_{dc} and σ''_{dc}, the deformation value of settlement due to vibration capacity of soil —S and the repeated dynamic times N have the features as displayed in Figure 2 that is, in the course of constru,

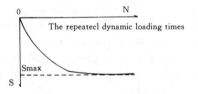

Figure 2 . The relation between the deformation extent S of the settlements of vibration of soil and the number of dynamic loading repeatedly applied

ction the construction of the first few stone columns, more settlements due to vibration and deformation will happen to the foundation soil. Along with the construction (the increase of the dynamic times N which are applied), the settlement due to vibration and deformation value produced by the construction of every stone column to the foundation soil dwindles. When N exceeds a certain value, settlement due to vibration and deformation value will be zero. At this moment, the total value of vibration subsidence and deformation approaches a fixed value S; 5)Loess is very sensitive to water, and along with the increase of moisture content, deformation will obviously increase (moisture subsidence) and the strength will decrease.

3. 3 *The Factor of the Characteristics of the Surrounding Buildings*

If the surrounding buildings are solid with good structure, the influence produced by vibration will be small. The foundation types exert important influence on the value of settlement due to vibration. The bases of piles havethe best capacity of anti—settlement due to vibration, whereas, the strip foundations, independent foundations and raft foundations are comparatively poor. If the bearing capacity practically taken by the foundation soil under the buildings is on the high side, which exceeds or approaches the allowable bearing capacity of the foundation soil, the value of settlement due to vibration is comparatively great. The buil—dings which are in the state of fast settlement and deformation after the completion of the construction are most likely to virate and subside. The longer the years are after the completion of the construction, the settlement tends to be steady, and the smaller will be the value of settlement due to vibration.

To sum up, the influence exerted by vibroflot while vibrating on the surrounding buildings is mainly the settlement due to vibration (in the loess zone of moisture, settlement due to moisture may also happen). However this kind of settlement caused by vibration is different from the subsidence caused by earthquake, because this is determined by the differ-

this kind of settlement caused by vibration is different from the subsidence caused by earthquake, because this is determined by the different natures of the engineering vibration focus and the earthpuake rosus with the vibration of vibroflot being dots with small capacity and on a small scale. However, it is comparative nearer to the buildings with great unsymmetry. As a result, the vibration of the vibroflot is more likely to cause unequal settlement due to vibration. Generally speaking, there are three reasons for the unequal settlements: 1) The layers of earth are differential; 2) The loads of structural objects are unsymmetrical; 3) The dynamic loads are unsymmetrical with heavy dynamic loads near the vibration focus and light dynamic loads far from the vibration focus. The indoor experiments and studies show that under the vibration and in the triaxial shear tests, the residual deformation of test specimen always develop gradually in the direction of the side with the maximum value of scope, displaying obvious unsymmetry. Besides, water demonstrates directions in its seepage with the area near the vibroflotation point effected first, whereas the areas far from the vibroflotation point might not be influenced or the effect is rather late or weak. Thus, the development of unequal settlements due to vibration will be aggra vated. Therefore, the effects of vibro replacement stone column construction on the surrounding buildings are in essence the unequal settlements caused by vibration produced by the foundations under the surrounding buildings.

4. The calculation of Settlements Due To Vibration and the Determination of Safe Distances

The safe distances with the effects of vibro replacement stone column construction on the surrounding builing have something to do with many factors. From the above — mentioned

analysis, it may be regarded that differentiations of settlments carsed by bibration (or inclinations) are the major controlled indexes. That is to say, different distances between vibro replacement stone columns and the surrounding buildings produce different differentiations of settlements caused by vibration. When the differentiations of settlements caused by vibration produced by the foundations of the surrounding buildings within certain distances are within the allowable scopes of the buildings, it is safe to carry out construction within these distances. Therefore, the determination of safe distances means the determination of the differentiations of the settlements caused by vibration of the foundations of such buildings as well as the judgment whether they exceed the allowable deformations of the buildings. A method of calculation to test and analyse comparatively simple settlements caused by vibration is put forth in this essay for this purpose. The main procedures and content of the method are as follows:

4.1 *Dynamic Tests*

The key to the calculation of the settlements caused by vibration lies in the determination of the relation between the residual strain ε_p and the dynamic stress σ_d. Since there are great differentiations in such relations according to the nature of the foundation soil, they have to be determined through dynamic triaxial tests by taking soil samples on the spot. The method of tests is as follows:

Consolidate the soil sample on the dynamic triaxial apparatus in the state of the practical stress on the spot. When the consolidation is steady, apply dynamic loads, and then test the elastic strain and residual strain soas to get the curve of test of the relation between the residual strain ε_p and the dynamic stress σ_d as well as the curve of test of the relation between the dynamic stress σ_d and the dynamic elastic strain ε_d. The following test parameters and the choices of the ways of loading have to be determined in the tests:

1) The determination of the stresses of consolidation of soil samples; the consolidation should be conducted according to the state of the practical stresses on the spot in the depths of the centres of soil where the samples are taken.

2) The determination of the frequencies of dynamic loads and the time of vibration; the frequencies of dynamic loads are taken according to the frequencies of vibration of vibroflots. The time of vibration is the time for equivalent effects for each vibroflot to complete the vibration to the foundation soil. The method of determination is to take the range of depth as the approximate length of the vibroflot as well as the time of vibration and its duration of the vibroflot. For example, the length of the vibroflot is 2.15 meters, and every period of vibration for density after filling is 0.35 minute, the time for vibration and its duration is 20 seconds. Therefore, the time for equivalent effect is :$t=(2.15\times20)/0.35=122$ seconds. Thus 122 seconds may be taken as the time of effect for dynamic load at every level.

3) The amount of dynamic stress and the ways of loading; the dynamic loading on the soil sample had better be applied from the minimum value, thus the position of σ_{dc} on the Figure 2 may be found. When the dynamic loading frequency is 24Hz, and the time of vibration is 120 seconds, in the case of the saturated loess soil (Q3) in Xi'an area, when the dynamic load exceeds 35 to 50 kPa, the soil sample is destroyed. Therefore, the dynamic loads above the sixth grade should be chosen between 0 to σ''_{dc}. One and the same soil sample can be used in the test, and the residual strain and elastic strain with dynamic loads at every grade will be determined in the way in which the loads are applied from small to great amounts (there should be intervals between each two grades). Or several soil samples may be taken from the foundation soil at the same layer, and only one grade of dynamic load is applied to each soil sample, thus the residual strain and elastic

strain of each soil sample at every grade of dynamic loads are acquired. With these two ways of loading, different results of tests will be got for the foundation soil with the same dynamic nature as is shown the author of this essay in Figure 3. suggest that the method of

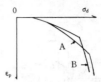

A: one grade of dynamic load is applied to one soil sample
B: loading grade by grade

Figure 3. The characteristics of the curve of the relation between the residual strain and the dynamic stress of the soil in different ways of loading

loading grade by grade be adopted . As a result, a large number of soil samples will be saved, and the effects on the dimensions of settlements caused by vibration of the foundation soil exerted by repeated loading (the construction of multiple stone columns) are also considered.

4.2 *The Determination of the Value of Extent of Dynamic Load σ_d of Various Layers of Soil at Different Distances from the Vibration Focus*

It may be determined approximately with the following two methods:

1) The static method: according to this method, the average contact pressure q between vibroflot and the foundation soil is regarded as static pressure on the foundation soil, and then turn 90 of the condition in the theory of the calculation of Boussinesq's linear elastic half — space superimposed stress, so as to determine the superimposed stresses in different distances from the vibroflot, and take the value of this stress as the extent value of dynamic load σ_d which is exerted on this spot.

2) The method of dynastic elastic half — space theory: this method is to regard vibroflot

as vibration of the round foundation with e-
quivalent effects on the surface of the space (as
shown in Figure 4). First of all, determine the
horizontal extent of vibration and the speed of
vibration at the spot with the horizontal distance of z from the center O of the vibroflot,
and then work out the shear strain γ_d of this

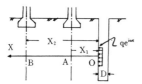

Figure 4. The calculation of the scope of vibration

spot according to the speed V of the distortional wave of the foundation soil, so as to determine the dynamic elastic axixl strain ε_d, and
then determine the dynamic σ_d of this spot according to the test curve of ε_d and dynamic
stress σ_d the course and formula are as follows:
① The horizontal amplitude of vibration Ax
(m) at any spot from the centre x of the vibroflot in vertical angle is calculated with the
following formula:

$$A_x = \frac{qr_0^2(1-2\mu_d)}{4G_d(1-\mu_d)x} \quad (1)$$

In the formula, q — — the average contact
pressure value of extent (kPa).

$q = Qo - - - - - - - - - - - - - - - -$
$- - - /(DL)$, Qo is the value of extent (kN)
of the general force of vibration of the vibroflot;

D stands for the diameter of the vibroflot
and L for its length;

r_0 — — is the radius of the round foundation with equivalent effects (m);

$r_0 = 0.55 \sqrt{LD}$;

G_d — — dynamic midodulus of elasticity in
shear (kPa) of the foundation soil, which may
be determined according to the speed of the
distortional wave;

μ_d — — dynamic Poisson's ratio.

The above calculated extent of vibration

which changes with distances tallies with the
results of practical tests of a certain project as
shown in figure 5.

②The determination of dynamic shear strain

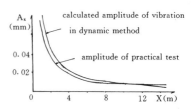

Figure 5. Comparison between the calculated
scope of vibration and the practical
tested scope of vibration

$$\gamma_d = \frac{V}{V_s} \quad (2)$$

In the formula, V stands for the speed of
vibration (m/s) at the spot x of calculation, V
$= \omega Ax$, ω is the circular frequency (Hz) of the
vibroflot;

V_s — — the speed of the distortional wave
of the foundation soil (m/s).

③The dtermination of axial dynamic and elastic strain

$$\varepsilon_d = \frac{\gamma_d}{1+\mu_d} \quad (3)$$

④The dtermination of σ_d according to ε_d curve
combination of test data from each layer of the
foundation soil will be carried out according to
the results of the tests of dynamic and elastic
strain ε_d and dynamic stress σ_d, then the combined curve equation of ε_d to σ_d at every layer of
the foundation soil will be acquired, and finally
σ_d will be determined with the formula of relation.

4.3 *The estimate of the extent (or differentiations)of the settlements caused by vibration.*

After the estimate of σ_d according the above —
mentioned procedures, and then determine the
residual strains ε_p, at different time σ_d, and the
extent of the settlements caused by vibration S

(m) is worked out with the following formula:

$$S=\sum_{i=1}^{N}(\sum_{j=1}^{M}\varepsilon_{p_{ij}}h_j) \qquad (4)$$

In the formula, N stands for the number of vibro replacement stone columns in the construction;

M stands for the number of layers of the foundation soil with in the range of the effects of vibro replacement stone columns at the spot of the calculation;

$\varepsilon_{p_{ij}}$ stands for the residual strain produced by i stone column on j soil layer;

h_j stands for the thickness of j earth layer (m).

Since the above—mentioned calculation is rather complicated, a PC—1500 computer program has been complied. The calculation of settlements caused by vibration of over a dozen spots with dozens of stone columns takes only dozens of minutes. When the extents of the settlements caused by vibration at the spots with different foundations are worked out according to the above—mentioned method, the differentiations of the settlements due to vibration between various spots may be determined. In the analysis of the settlements due to vibration with the above method, the tests of the relation between dynamic stress σ_d and dynamic residual strain ε_p of the foundation soil is very important and essential. In the tests, the dynamic stresses applied to the soil samples cannot be too great, and it had better be $\sigma_d < 50 \mathrm{kPa}$.

5. Analysis of examples

5.1 Features of Buildings and the Construction of Vibro Replacement Stone Column

The example is about a site situated northwest of the New Town Square of Xi'an. The building near the construction site was completed, handed over and put into use in 1980, a civilian residential building of six storeys with mixed structure of frames and bricks as well as raft foundation. The average pressure on the plane raft foundation was 119 kPa. In early March, 1990, on the north side of the building the construction of vibro replacement stone column was conducted, beginning from the south end of the side near the surrounding building. The order of the construction was a paralled row with the surrounding building. After its completion the second row will be carried out, thus from south to north row after row. In the course of the forst three rows, the enclosing wall between them collapsed and there appeared crevices in the walls of the near by building. Most of the crevices appeared beneath the windows of the vertical walls in vertical directions which slightly inclined. The shortest distance between the side of the vibro replacement stone column and the line on the north side of raft foundation under the near by building was 1.5 meters, with the shortest distance from the stone column on the north side was 2.3 meters. ZCQ — 30 type of vibroflot was used in the construction, and the fill was supplied continuously into the bore holes, with the vibroflot raised upward continuously. It took 10 to 20 minutes from filling to the completion for each and every stone column with the compact electric current less than 50 to 60 A.

5.2 The Nature of Physical Mechanics of the Structure of the Foundation and Soil of the Construction Site

The nature of physical mechanics of the structure of the foundation and soil of the construction site is displayed in Table 1. It can be seen from the nature of the earth, the void ratio of loess $Q_3(2)$ is big, its liquidity index high. The bearing capacity of the earth in soft or plastic states is low, and that of loess$Q_3(3)$ is also low. These two soil layers are the main bearing ones and are apt to suffer from settlements due to vibration under the effects of vibration.

5. 3 *The Test of the Dynamic Features of Soil*

The test was conducted on the dynamic triaxial apparatus of electromagnetic oscillator of DSD type, and the consolidation was conducted according to the ratio of 1. 8 to 2. 0 of consolidation pressure. When steady, loading was applied with the frequence of 24 Hz, and the time of vibration at every grade of loading was 120 seconds, and then the dynamic and elastic strains as well as the dynamic residual strain were determined. The results of the data of the tests after curve combination are shown in Table 2.

5. 4 *The Calculation of the Extent of the Settlements Due to Vibration*

The extents of the settlements due to vibration of the centres of stone columns No. 1 to 8 of the foundations B to B of the surrounding buildings calculated with the method mentioned above in this essay are shown in Table 3. It can be seen from the results of the calculation that the estimated results of the general extents of the settlements due to vibration S differ greatly. According to the practical on — the — spot analysis, the estimated results worked out with the static method conform more to reality, where as the estimated results worked out with the dynamic method obviously do not conform to reality. It is discovered from calculation that in the estimate of σ_d with the latter, it decreases comparatively slowly along with the increase of distances from the vibration focus as is shown Figure 6 there fore, through the accumulation of multiple stone columns, the values of the settlements due to vibration become greater. Owing to the lack of experience in this aspect, it is difficult to decide which method is correct. However, it can been seen from this example that if σ_d is estimated with the static method so

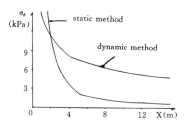

Figure 6. Comparison between the two methods in the estimates of dynamic stresses

as to calculate the extents of the settlements due to vibration, they tally with what actually happened (cracks in the walls of the surrounding buildings) on the spot.

6. Postscript

In this essay, various factors exerting influences on the surrounding buildings by the construction with vibro replacement stone columns have been roughly analysed, thus acquiring the conclusion that the settlements due to vibration produced by the foundation soil of the surrounding buildings, especially the unequal settlements are the important factors effecting the safe use of the surrounding buildings. Then a simple and convenient method of estimate of the extents of settlements is put forth, so as to solve the problems concerned in the design and construction of vibro replacement stone columns. Due to my limited level, criticisms are welcome.

Table 1. Table of dynamic mechanics of soil

name of soil	depth (m)	N. M. c (%)	D. S. T (%)	V. R e	O. M MPa	L. I I_L	C. P. T (kPa)	S. P. T $N_{63.5}$	D. S S_t	B. C (kPa)
loess $Q_2(2)$	1. 3	30. 0	83	0. 983	4. 83	0. 92	1038	2. 8	11. 3	100
loess $Q_3(3)$	5. 6	28. 8	93	0. 843	5. 73	0. 87	1458	3. 7	11. 0	140
loess $Q_3(4)$	6. 7	26. 9	99	0. 740	18. 30	0. 60	2002		6. 0	170
ancient soil $Q_3(5)$	9. 1	26. 9	93	0. 822	5. 80	0. 73	2002	5. 0		170
ancient soil $Q_3(6)$	10. 7	23. 6	93	0. 691	11. 00	0. 39	3010	8. 7		230
loess $Q_2(7)$	15. 1	24. 5	99	0. 652	8. 20	0. 62	3010	9. 8		230
loess $Q_2(8)$		21. 7	97	0. 602	15. 90	0. 24	3941	14. 5		300

Table 2. Table of dynamic nature of soil

name of soil	S—W. V V_s(m/s)	D. M. E. S G_d(kPa)	D. P. R μ_d	L. E. P. C K_o	σ_d(kPa)~ε_d(%)	ε_p(%)~σ_d(kPa)
loess $Q_3(2)$	125	28080	0. 42	0. 595	$\sigma_d=127.44\varepsilon_d^{0.439}$	$\varepsilon_p=0.0698\sigma_d$
loess $Q_3(3)$	150	42590	0. 45	0. 613	$\sigma_d=65.45\varepsilon_d^{0.227}$	$\varepsilon_p=0.0698\sigma_d$
loess $Q_3(4)$	175	59310	0. 45	0. 725	$\sigma_d=84.64\varepsilon_d^{0.313}$	$\varepsilon_p=0.0652\sigma_d$
loess $Q_3(5)$	175	59310	0. 45	0. 725	$\sigma_d=84.64\varepsilon_d^{0.313}$	$\varepsilon_p=0.0652\sigma_d$

Table 3. Table of the calculation of the extent of the settlements due to vibration from No. 1 to No. 8

estimate method	NO. of column	1	2	3	4	5	6	7	8
	distance(m)	3. 2	1. 9	1. 4	1. 4	2. 0	3. 4	15. 8	
static method	S(m)	29. 3	11. 7	8. 0	6. 2	4. 9	3. 7	2. 5	0. 7
	D. V. S(cm)	17. 6	3. 7	1. 8	1. 3	1. 2	1. 2	1. 9	
dynamic method	S(m)	129. 1	114. 9	109. 2	105. 5	102. 2	98. 5	93. 2	78. 5
	D. V. S(cm)	14. 2	5. 7	3. 7	3. 3	3. 7	5. 3	14. 7	

REFERENCES

F. E. Richart, Jr. R. D. Woods J. R. Hall, Jr. 1970. Vibrations of soils and foundations Englewood cliffs, New Jersey.

Shi Zhaoji, Yu Shousong, Weng Lunian. 1988. 11. Seismic settement evaluation for Tang gu New port area. Chinese journal of civil engineering.

Yan Renjue. 1981. 3. Introdution of half—space theory of dynamic foundation. China construction industry press.

Highway route selection through the interaction between the environmental impact assessment and the design

Choix du tracé des autoroutes par interaction entre l'étude d'impact sur l'environnement et le projet

M.L.Galves

Instituto de Pesquisas Tecnológicas, São Paulo, Brazil

ABSTRACT: This paper presents a critical review of the main environmental impact assessments of highways prepared in the state of São Paulo, Brazil, in the last five years. This review, with emphasis on physical components of the environment, concentrated on whether the design and the environmental impact assessment interact. Based on the gathered information, an approach to select the route for a highway is proposed so that the impacts on the environment can be reduced.

RÉSUMÉ: Cet article fait le bilan des études d'impact routières les plus importantes élaborées à l'état de São Paulo, Brésil, au cours des cinq dernières années. Ce bilan, axé sur des composantes physiques de l'environnement, a eu pour but vérifier s'il y a et comment se fait l'intégration entre le projet et l'étude d'impact. A partir des informations obtenues, il est proposée une demarche pour le choix du tracé d'une autoroute de façon à ce que les impacts sur l'environnement soient réduits.

1 INTRODUCTION

Environmental impact studies of civil engineering projects have been carried out frequently in Brasil since 1986. This was caused by legal requirement for the preparation of prior environmental impact assessments (EIAs) and environmental impact reports for the construction, operation or extention of activities which modify the environment.

Highways are included among the civil works for which environmental studies are required; in this way for the construction of such work one must consider, besides the technical and economic aspects, the impacts on the environment.

The legislation meets the growing concerns of the society as to the quality of the environment, defining responsibilities and basic criteria for environmental impact assessments and reports. However, the importance of highway transportation for the economic growth of our country cannot be ignored, which can be translated into the need for the construction of new highways in the near future.

Due to the fact that Brazilian environmental legislation is relatively recent, there is a gap between those professionals in charge of the design and the team responsible for the environmental impact studies.

The main environmental impact assessments of highways prepared in the state of São Paulo, Brazil, in the last five years have been analysed in this paper. This analysis concentrated on whether the design and the environmental impact assessment interact. It should be pointed out that in spite of emphasizing its physical components, the environment is defined herein by the dynamic interaction of physical, biological, socio-economic and cultural components, as it is shown in Figure 1 (adapted from Instituto de Pesquisas Tecnológicas 1992).

2 INTERACTION EIA - DESIGN: PRESENT SITUATION

Traditionally, during the engineering studies of a proposed highway, different alignments are considered. The methods utilized for route selection are generally based on geometric characteristics, traffic performance and economic aspects of the studied alignments.

According to Brazilian legislation, the environmental impact report, the document available to public consultation which summarizes EIA results and conclusions, will recommend the highway route to be considered to construction. The EIA and the corresponding environmental impact report should therefore supply technical inputs for the decision as to the most favorable alignment.

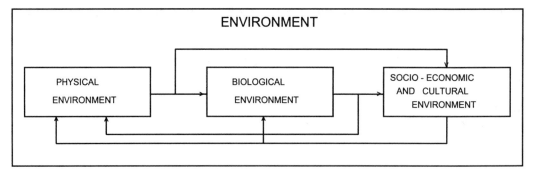

Figure 1. Environment defined through energy and matter flow.

If, on one hand, the engineering studies analyse the technical and economic attributes of the proposed highway, on the other hand, the EIA takes into account the environmental attributes. However, they both have to converge to one single result, that is the choice of the highway location. This implies that the engineering and the environmental studies should interact as much as possible.

The analysis of the main EIAs of highways presented in the state of São Paulo enabled to recognize some characteristics that are common to all of them in a higher or lower degree. From the different phases of an EIA, such as diagnosis of the environment, impact assessment, impact mitigation and monitoring program, the first two ones are emphasized in this analysis.

The first point to be considered is related to the project characterization. The inputs for this phase are provided by the engineering studies. The detailed description of the project is of great importance to the EIA for the assessment of the impacts is conditioned by the location and the technological characteristics of the highway facility. Unfortunately this aspect has been neglected in the majority of the EIAs and the project is many times poorly characterized.

The diagnosis of the environment, specially on what physical components are concerned, is based on the description of isolated elements, such as soils, rocks and relief. Frequently the description is quite exhaustive but it lacks an objective relation with the project.

The impact assessment phase reflects the consequences of an incomplete project characterization and a diagnosis of the environment of little interest for application in EIA; besides, the methodologies of impact assessment generally adopt arbitrary scales for impact valuation. The concept of impact is not made clear in many cases: alterations in natural processes of the environment, consequences of these alterations and even techniques of construction are used as synonyms of impact.

The conclusions drawn from what has been observed are that there is little interaction of the EIA with the design and that EIAs use to be too generic to allow the route selection to be properly made.

3 ENVIRONMENTAL PROCESSES AND ROUTE SELECTION

The interaction between the environment impact assessment and the design should be initiated in the phase of diagnosis of the environment. Once some preliminary alignments have been considered within limits set by the points of origin and destination of the proposed highway, the approach suggested for the diagnosis of the environment is based on the identification of the environmental processes within the area of influence of the project.

This approach, developed by the Instituto de Pesquisas Tecnológicas (1992), takes into account the dynamics of the environment through its processes. In the case of the physical components of the environment, such as soils, rocks and water, the processes of interest are: erosion, landslides, draining of surface water, flooding, subsidence, among others.

The knowledge about the existing environmental processes should be incorporated into the engineering studies, for the definition of possible alignments (Figure 2). Each proposed route can then be characterized in terms of its technological processes, that is, the various methods and techniques developed during the construction and operation of the highway; examples of technological processes are: cut and fill, retaining structures, drainage.

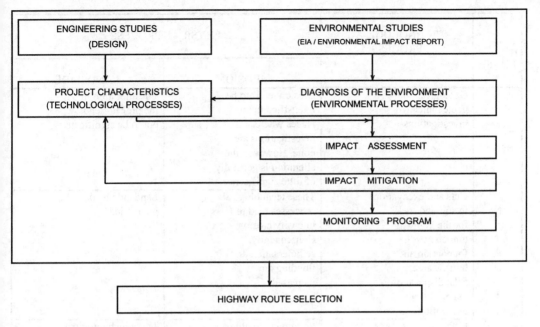

Figure 2. Interaction between the EIA and the design

The impact assessment phase consists in analyzing how the environmental processes can be altered by means of technological processes; in this paper impacts are defined as significant alterations in environmental processes. The following aspects should be considered when determining the significance of environmental alterations: the importance of the environmental attribute in question to project decision makers; the distribution of change in time and space; the magnitude of change; and the reliability with which change has been predicted or measured (Duinker & Beanlands 1986).

The analysis of potential alterations can be performed by cross-checking between the environmental and the technological processes; Figure 3 shows an example of this matrix arrangement, emphasizing physical processes of the environment. This is of course a generic example that should be detailed for the alignments and environmental characteristics in each particular case.

The results of the impact assessment, together with those of the engineering studies, enable the alternative routes to be compared in terms of their technical characteristics, environmental impacts and costs. The costs are not only those of construction and maintenance, but also those associated with impact mitigation measures. From this comparison, the highway route should then be selected.

4 CONCLUSION

This paper focuses on the necessary interaction between the design and the environmental impact assessment so that the highway route selection can be achieved in an objective and comprehensive manner. The approach presented herein is based on the technological processes of the project, the environmental project, the environmental processes and the assessment of impacts through the cross-checking of these two kinds of processes. Although developed to highways, this approach can be applied to any engineering project.

REFERENCES

Duinker, P.N. & G.E. Beanlands 1986. The significance of environmental impacts: an exploration of the concept. *Environmental Management* 10: 1-10.
Instituto de Pesquisas Tecnológicas 1992. *Alterações no meio físico decorrentes de obras de engenharia*. São Paulo: IPT. Publicação 1972.

PHASE	TECHNOLOGICAL PROCESS	PHYSICAL PROCESS	
		EROSION	LANDSLIDE
C O N S T R U C T I O N	Geological and geotechnical investigations	Erosion tends to be intensified in the places where investigations are done. However, the alteration is generally of little significance.	The alteration in the process tends not to be significant.
	Cuts and excavations in soil and rock for the highway, tunnels, bridge foundation and borrow areas; fills; disposal of material from excavation	These technological processes tend to intensify erosion. Consequently, sedimentation and flooding are increased.	Landslide tends to be intensified.
	Retaining structures; drainage and surface protection of cuts and fills	Retaining structures tend not to alter erosion. Drainage and surface protection control or eliminate erosion.	These technological processes control or eliminate landslides.
	Pavement	Pavement controls or eliminates erosion in the road surface.	The alteration in the process tends not to be significant.
O P E R A T I O N	Traffic	The alteration in the process tends not to be significant.	The alteration in the process tends not to be significant.
	Maintenance of: retaining structures, tunnels, bridges, drainage, surface protection of cuts and fills, pavement.	Maintenance guarantees the conditions for erosion control.	Maintenance guarantees the conditions for landslide control.

Figure 3. Alterations of physical processes by technological processes
(adapted from Instituto de Pesquisas Tecnológicas 1992).

Ecological problems related to the interaction between the geological medium and hydraulic structures

Problèmes écologiques de l'interaction du milieu géologique et des ouvrages hydrauliques

A.A. Kagan
The Hydroproject Institute, St. Petersburg, Russia

N.F. Krivonogova
The B.E. Vedeneev VNIIG, St. Petersburg, Russia

ABSTRACT: Hydraulic structures exert combined physical, physicochemical, chemical and thermal effects on the geological medium. The analysis of the effects enables various models of the geological medium to be developed.

RESUME: Les ouvrages hydrauliques exercent sur le milieu géologique une influence à la fois physique, physico-chimique, chimique et thermique. L'analyse de ces influences permet de mettre en place des modèles de la protection du milieu géologique.

Hydraulic structures can have combined physical, physicochemical, chemical and thermal effects on the geological medium. The kind, character and extent of the effects are mainly determined by features specific to the geological medium and by designs and operating conditions of hydraulic structures.

The physical effect due to the dead weight of a structure and that of water gives rise to the increase in static and dynamic loads and to disturbances in underground water conditions. These changes may terminate in considerable reduction or complete loss of stabilities of structure foundations and slopes, hillsides, in seepage deformations, dangerous rise of ground water table, reservoir bank transformation, river channel scours downstream from dams. As a result in zones affected by hydraulic structures landslides and piping processes develop, bogs and karst holes form more intensively.

By way of example we refer to the structures incorporated into the St. Petersburg flood barrier. The foundation of the structures is composed of weak glacial-lacustrine clays (of mean moisture content 44%, dry density 1.1 t/m^3, the flowage index 1.58). When sand was stockpiled on the dyke and the design load was exceeded, the clays were squeezed out up to the height of 4 m and the dyke slope failed.

Before the construction of the Plyavinskaya hydro power plant on the Daugava river underground water from the Upper Devonian aquifer discharged within the reservoir area. Once the construction was completed the same reservoir foundation turned to a zone of water supply and at a distance of 200 to 500 m downstream from the dam numerous gryphones are now observed at lowered levels of water in the tailrace (Pestovsky & Razumov 1972).

Water load of a reservoir filled not infrequently leads to the development of seepage deformations

For one example, the Almaatinskaya hydro power plant structures are erected on deposits of eluvial and colluvial large-fragmental soils of highly non-uniform grain size. On reservoir filling more and more springs of constantly increasing discharge appeared on the tailrace floor year-to-year. At every reservoir drawdown funnel sinks were found in the dam abutments, they were plugged up but reappeared at new places after the next rise of a reservoir water level.

Transformation of reservoir banks and a rise of ground water table are most obvious ecological consequences of the influence of a reservoir on the geological medium. They lead to bog formation at vast territories, deterioration of soil qualiti-es, development of slope processes, and primarily of landslides. Areas subject to the above processes may vary within a very wide range from 3 to 6% of the water-surface reservoir area as far as the bank transformation is concerned (the Gorkovskaya, Saratovskaya, Bratskaya, Ust-Ilimskaya hydro power plants) and up to 100% where the ground water table rise is involved (the Novosibirskaya hydro power plant) depending on the geological, hydrogeological, hydrological and climatic conditions of a region as well as on the dimensions of a reservoir, its filling and drawdown schedules etc. (Parabuchev 1990).

The diversion tunnel of the Tereblya-Rikskaya hydro power plant in the Transcarpathians cuts a slope composed of alternating Cretaceous-Paleogene sandstone and clay shale covered by deluvium and eluvium formations. Originally the ground water table at the end stretch of the tunnel was at the depth of 40 to 50 m. Within a few years after reservoir filling the underground water table rose and the strength of soils decreased due to water leakages. Landslides in covering beds have led to the development of a ring fracture in the tunnel lining. Excessive surface water was diverted and so further development of deformations was prevented (Molokov 1985).

The physicochemical effect shows up as soil soaking and swelling and later as considerably deteriorated properties of soils, all resulting in the reduction of slope stability and intensified transformation of reservoir banks.

Engineering geological examination of soils in the region of the Kuibyshev hydro power plant reservoir planned to be bridged in the city of Ulyanovsk revealed that Lower Cretaceous hard clays included occasional sand lenses and interlayers. The reservoir bank slope at the site of a bridge was stable and showed the tendency to a further increase in the stability due to flattening on account of deluvium washing off.

The situation changed abruptly with the reservoir filled by 18 to 20 m when water reached the same elevations as they were in the Volga river of the Upper Pleistocene, i.e. in the epoch of active landslides formation. Water-saturated clays acquired flowable plastic consistency and their swelling pressures increased, on average, from 0.6 to 1.2 MPa. Over 30 years of reservoir service numerous slides of floating mats and most serious displacements of soil blocks occurred due to the deterioration of clay qualiti-es. The banks have retreated by 100 to 140 m. Since the revival of ancient landslides and formation of new ones are predicted for the bank zone of 500 m width the system of bank-protecting structures and slide-fighting measures are developed, the location of the bridge is changed (Ziangarov et al. 1993).

The reservoir basin of the Tsymlyanskaya hydro power plant is composed of Tertiary deposits, mostly of clay soils, the slopes covered with deluvium having evidence of landslides at individual areas. Active landslides were recorded immediately after filling the reservoir. In two years additional deluvium slides appeared and in the successive two years large landslides formed in lower middles of bank slopes not only in deluvium but also in Tertiary deposits. After a lapse of two years more landslides were seen over the whole height of the slopes, they even reached the dividing ridge (Klyueva 1966).

The water flow rejected downstream from the dam spillway can scour a river bed of very strong dolerite and the like having the ultimate compression strength of

204 MPa (the Vilyuiskaya-I hydro power plant) or of crystalline schist with the ultimate compression strength of over 140 MPa (the Sayano-Shushenskaya hydro power plant), the scour holes of 5 and 10 m depth being formed, respectively.

The chemical effect of reservoir water involves melting, leaching, salting and desalting of soils and underground water. It is essential that any changes in the chemical composition of soils, surface and underground water are of importance from the ecological point of view. As a result, in soluble soils karst holes can increase in number as it was the case on reservoir banks and in the tailrace of the Kamskaya hydro power plant. In loess deposits subsidence phenomena become more conspicuous.

Characteristic features related to the geology and hydrogeology of the Charvakskaya dam site included karsting of markedly dislocated carbonaceous soils of the Early Carbonaferous period and the availability of deep thermal ground water of high mineralization. They have motivated changes in hydrochemical and thermal conditions of the aquifer. In a non-controlled river channel the underground water discharged upstream from a dam to be constructed. With the reservoir filled up to low elevations the underground water continued to discharge upstream from the dam and bottom layers of reservoir water started to be salted. At the reservoir filled up to higher elevations surface water mixed with underground water, seepage paths, water temperatures and mineralization being altered (underground water discharges downstream the dam). Under natural conditions the underground water mineralization was 1 g/l, later with the reservoir filled it reduced and is now equal to 240 to 500 mg/l depending upon the reservoir water levels, temperatures of water in sources are being lowered by 1.0 to 1.5°C per year (Leonov et al. 1983).

In several years of the service of the Charvakskaya hydro power plant reservoir water pollution with nitrites and nitrates was noted. The reason was traced to discharges of raw sewages of health-resort and recreation bases constructed, those of the settlement and the ambient air pollution of the chemical plant. In 1991 the nitrates concentration was as high as 0.2 mg/l. The figure exceeds the maximum permissible concentration, so the drinking water supply is impossible here without the proper sewage treatment.

At the design stage of the Katunskaya hydro power plant in the Katun river basin including the Kurai mercury zone the concerns were voiced that the mercury would enter the river runoff from the soils flooded. However according to the results of extensive investigations only 0.01 g/year of mercury would enter the reservoir. It is considerably less than the quantity of mercury (3.64 g/year) entering the Katun river at the present time. No adverse impact of this mercury on the nature and man was assumed and proved (Kayakin et al. 1993).

The problem of the thermal effect of hydraulic structures and water on the geological medium is of particular importance for regions with the abundance of permafrost. In these regions it manifests itself either in lowering of temperatures and freezing or in temperature rise and thawing of soils in the zones affected by hydraulic structures.

Such temperature variations together with other effects discussed above induce an active formation of thermal karst holes, thermal erosion, thermal abrasion, solifluction and other slope processes as well as alterations in hydrodynamic and thermal conditions of underground water, its chemical composition.

An example exists as the Khantaiskaya hydro power plant whose reservoir banks are composed of permanently frozen ice-rich silt.

Thermal effect and wave action of water on the geological medium are on their own responsible for thawing of soils and the thermokarst subsidences in reservoir bottom and banks, active thermal abrasive transformation of the banks. As a result the volume of the reservoir has become increased to such an extent that even after 40 years of service the reservoir still remains to be filled up to the design elevation. In addition the changes in microclimate of the zone affected by the hydro power plant which show up as higher temperatures and humidity of ambient air, precipitations of varied quantities and regimes have also led to the intensive development of thermal karst and thermal erosion processes, solifluction, cryogenic heaving in the band 1 km wide around a reservoir (Kronik & Onikienko 1980).

At the site of the Adychanskaya hydro power plant under design the forest and all the other vegetation were removed, which fact promoted the intensified thawing of ice-rich soils on one of the reservoir bank slopes. The process was accelerated due to the local fire occurred in the upper portion of the slope. As a result, the rapid solifluction happened and a soliflucting stream formed an erosion ditch of the depth of 1.5 m on the slope and adjacent terrace (Kagan & Krivonogova 1991).

Thermokarst holes and funnels in the land abutments and also slope damages are typical of the medium- and low-head embankment dams constructed on ice-rich soil foundations. Plugging of thermokarst holes and funnels is sufficiently effective only when additional freezing is provided or other seepage-control measures are taken.

It is possible to minimize adverse effects of the interaction between hydraulic structures and the geological medium with the help of the model of the geological medium protection. The model is developed by the following consecutive stages:

a) estimation of a kind(s) and extent of the influence which will be exerted by a structure upon the geological medium;

b) forecast of dimensions of an area which will be under the influence mentioned above;

c) prediction of changes in the geological medium, of their rates and consequences;

d) assessment of economical and social consequences of changes in the geological medium;

e) development of measures intended to minimize detrimental effects of changes in the geological medium;

f) assessment of social and economical consequences of accomplishing the above steps.

As a preliminary the kind and character of the interaction between structures and the geological medium is determined from the analysis of the behaviour of all the structures incorporated into a hydro power plant. For instance, when a low-head embankment is constructed on highly fissured rock most attention is to be concentrated on the physical action of reservoir water on the embankment foundation and land abutments.

To predict dimensions of a geological medium zone affected by a structure use should be made of both computation methods (e.g. in order to determine the thickness of a soil layer under compression, the seepage water contour etc.) and the analysis of the geological medium behaviour. In particular the foundation of the St.Petersburg flood barrier mentioned above is a weak clay deposit 28 m thick underlain with very dense glacial loam (dry density of 2.04 g/m^3). Inasmuch as the pressure due to the dead weight of the barrier is 0.12 MPa it is evident that the depth of the lower boundary of the geological medium zone affected by the dyke and that of the upper boundary of loam coincide.

A geological engineer can make a major contribution into the construction of a model of the geological medium protection when various protecting measures are being developed.

It is important to remember that at the stage of selecting a site for a hydro power plant adverse effects of hydraulic structure on the geological medium timely forseen may force the site favourable from the geomorphological or other viewpoit to be rejected. For example, the site of the Mokskaya hydro power plant was transferred from the most narrow to a wider section of the Vitim canyon as the motion of thick stone streams was predicted in the former case.

When the site is selected water

levels of a reservoir and power parameters of a plant can be varied.

In the design of the Amguemskaya hydro power plant on the Amguema river the normal water level of the reservoir was so assigned that the ground water risen could not reach the zone of ice-rich soils and endanger the settlement and arable land in the case of these silt soils being thawed. With the same object in view protecting dykes were designed for the least favourable areas of the reservoir banks.

When designing the Beloporozhskaya hydro power plant on the Kem' river the normal water level of the reservoir was lowered by 1 m to overcome similar difficulties.

A great variety of construction measures may have extensive application on a hydro power plant site selected to ensure safety of principal structures and to protect the geological medium from deterioration.

By way of example let us mention the Farkhadskaya hydro power plant where the drainage is organized so that it regulates both hydrochemical conditions of underground water and related deformations in loess foundation of the structures at the water front. In addition salting of arable land is decreased during the operation of irrigation systems (Rodevich 1988).

Landslides in the left-bank abutment of the Dzora dam are found to be induced by creep strains developing in pelitic tuffs at the base of the bank slope and in the clay layer of complex genesis running under the river-channel section of the dam. It is precisely these landslides that are responsible for discontinuities formed in the water front structures. The displacement of overlaying andesite-dacite blocks occurs along the clay layers mentioned. The landslides are aided by disturbances in seepage conditions of dam foundation and earthquakes which happen here every now and then. In particular the earthquake of magnitude of 7 was registered at the Dzora dam site during the Spitak earthquake with epicentre at a distance of 200 m away.

To provide the normal behaviour of the Dzora hydro power plant structures a specific amortisseur was designed with rates and extent of landslides motion taken into account. It was placed between the dam and the left-hand bank. At present the service life of the amortisseur is over, so a new measure is suggested. Sliding soil masses of the left-hand bank are projected to be strengthened with a diaphragm cut off wall. A deep ditch for the wall will cut the left bank off and be filled with the material capable of absorbing displacements of the upper portion of sliding soil masses, their lower portion being stabilized by itself.

The results of creating a model can be presented as a series of maps. The first of them depicts changes forseen for the geological medium as induced by the construction of a hydro power plant. Besides it gives money losses due to each of probable changes. The next map shows measures aimed at the protection of the geological medium. The third map relates each measure cost to the total cost of the construction.

Such modelling enables the feasibility of a hydro power plant to be estimated together with the economical and social aspects of its construction and operation. The model is instrumental in developing economically and ecologically well-proved designs.

REFERENCES

Kagan,A.A. & N.F.Krivonogova 1991. Forecasts of transformation of reservoir banks in regions of permafrost. Gidrotekhnicheskoe Stroitelstvo. 4:11-14.
Kayakin,V.V., A.V.Mulina, Dmitrieva I.L. 1993. Lessons from the expert appraisal of the design of the Katun hydro-electric power station. Gidrotekhnicheskoe Stroitelstvo. 10:12-14.
Klyueva,V.A. 1966. Dynamics of the Tsymlyansk hydro power plant reservoir banks. In Sb. Works of the Tsymlyansk Hydromet. Observatory. Vyp.3.

Kronik,Ya.I. & T.S.Onikienko 1980.
The influence of thermal abrasion of banks of the Khantaiskaya hydro power plant reservoir on the increase in its
volume. Mater. konf. i soveshch.
po gidrotekhnike. p.14-17.
Leningrad.

Leonov,M.P., N.V.Mokhova, L.I.Svi-
telskaya. 1983. Variations in
the thermal and hydrochemical
regime of drainage water at the
Charvak hydro power plant. Izvestia VNIIG. 165:17-22.

Molokov,L.A. 1985. Engineering geological processes. 206 pp. Nedra.

Parabuchev,I.A. 1990. Hydraulic
engineering and geological medium. Sb. nauch. tr. Gidroproek-
ta. 143:15-17.

Pestovsky,K.N. & V.K.Razumov 1972.
The Plyavinskaya dam on the Da-
ugava river. In "Geology of
Dams". VI:101-107. Moscow.
Energia.

Rodevich,I.V. 1988. Hydrodynamic
investigations of salined loess
foundations of hydraulic structures aimed at chemical piping
control. Izvestia VNIIG.
209:40-48.

Ziangarov,R.S., A.L.Ragozin,
B.A.Snezhkin et al. 1993. Combined engineering and geological
estimation of the sliding slope
in the city of Ulyanovsk. Geo-
ekologiya. 1:89-93.

Interaction entre les routes et l'environnement souterrain (Mandat de l'office fédéral des routes)

Interaction between roads and the underground environment

S. de Coulon, A. Parriaux, M. Bensimon, J. Tarradellas & J.-C. Vedy
École Polytechnique Fédérale de Lausanne, Suisse

RÉSUME: L'objectif de cette recherche est de déterminer s'il est possible de mettre en évidence dans le milieu souterrain des substances spécifiquement liées à la route. La vulnérabilité du milieu souterrain en fonction du contexte géologique est étudiée en vue de déterminer des mesures constructives adaptées à l'infiltration d'eau de ruissellement typiquement routières.

ABSTRACT: The objective of this research is to determine whether it is possible to put in evidence in the underground environment specific substances in relation with roads. The vulnerability of the underground environment according to the geological context is studied in view of determining constructive measures adapted to road run-off infiltration.

1. OBJECTIFS DE LA RECHERCHE

Cette étude est basée sur la caractérisation chimique des divers types d'eaux de ruissellement routier ainsi que sur les processus de transfert dans le sous-sol, qui servent à l'évaluation des risques sur la qualité des eaux (vulnérabilité des nappes souterraines).

La connaissance du devenir de certaines substances telles que les métaux lourds notamment lors de leur transfert dans le milieu souterrain, est de première importance. Il existe une tendance générale à considérer que la présence à l'état de traces de produits toxiques dans l'eau est uniquement due à des causes anthropiques. Nos recherches montrent que plusieurs d'entre eux sont issus du milieu naturel, en particulier les roches dans lesquelles les eaux circulent. Il est complexe de déterminer l'origine des substances rencontrées dans le milieu souterrain, et particulièrement le lien privilégié de certaines de ces substances avec les routes.

Ce projet contribue à résoudre les problèmes suivants:

1) Origine des substances trouvées dans l'environnement souterrain: existe-il des traceurs typiquement routiers?
Dans le cortège de substances anthropiques toxiques ou non toxiques présentes dans le milieu souterrain, notre but est de déterminer s'il est possible de mettre en évidence des substances spécifiquement liées à la route. Utilité pratique dans la détermination objective de l'impact des ouvrages d'infiltration sur ce milieu.

2) Définition d'une notion de vulnérabilité du milieu souterrain en fonction des conditions naturelles:
La variabilité spatiale contenant divers types de sous-sol doit absolument être prise en compte lorsque l'on veut établir une synthèse du risque. Définir une notion de vulnérabilité, différenciée en fonction des conditions naturelles du milieu souterrain, par rapport à la réinfiltration d'eaux de ruissellement typiquement routières à une utilité directe dans la conception de mesures constructives adaptées aux contextes géologiques rencontrés dans la pratique.

3) Implications constructives dans la gestion des eaux de chaussées: La gestion des eaux de route implique la réalisation d'ouvrages qui doivent tenir compte du type de route et de la vulnérabilité du sous-sol. Des recommandations seront établies selon des configurations typiques en Suisse.

Cette recherche est menée par le laboratoire de géologie de l'EPFL (GEOLEP), prof. A. Parriaux, en collaboration avec:
- Groupe de recherche en écotoxicologie, EPFL, prof. J. Tarradelas.
- Laboratoire de pédologie, EPFL, prof. J.-Cl. Védy.

Démarche globale effectuée:

Définition d'un réseau d'observation:
L'accent a été mis sur une étude expérimentale d'un nombre réduit de sites choisis dans divers contextes géologiques représentatifs. Le choix du réseau d'observation a donc été guidé dans un premier temps par les différentes lithologies à prendre en compte: plaine alluviales, fluvioglaciaire, molasse, karst calcaires, karst évaporitiques et massifs cristallins.
Pour une même situation géologique, les eaux prélevées en amont et en aval de la route sont analysées de façon à avoir des comparaisons de données "input/output" avec ou sans influence de la route, ceci pour différents types de tronçons routiers. Des cas où l'eau de chaussée transite à travers un sol ont aussi été intégrés.
Le réseau d'observation de base est constitué par une quarantaine de points d'eau, répartis en Suisse occidentale et centrale.

Le réseau est composé de plusieurs types de sites:
- Le réseau de base: Constitué d'une trentaine de point de prélèvement d'eau (source, puits, piézomètres) répartis selon les différentes géologie à tester, il a été choisi de façon à permettre les comparaisons d'aquifères de même géologie, avec et sans influence anthropogène.
- B.A.C. (Bassin Amortisseurs de Crue): Points de prélèvements particuliers d'ouvrages de traitement et d'infiltration des eaux de ruissellement routières. Quatres de ces ouvrages ont été échantillonnés dans cette première phase de l'étude. L'un d'eux, situé à la hauteur de Vernayaz (VS), a pu être équipé grâce au Service des Routes Nationales valaisans de six piézomètres battus placé selon un profil transversal à l'étang d'infiltration. Ce site particulier permet un suivi précis des variations de concentrations dans le sous-sol aux abords de sites d'infiltration.
- Stations sols: Il est important de ne pas négliger le rôle du sol au sens pédologique lors de transferts à travers les talus routiers. Dans ce but, en collaboration avec le Laboratoire de pédologie de l'EPFL, quatre stations pédologiques ont été définies dans deux sols types, choisis pour leur grande représentativité régionale. Ces sols ont été échantillonnés à la tarière dans un premier temps. Les quatres stations seront équipées début 1994.
- Scories: L'impact de l'emploi de scories d'incinération de déchets ménagers (mâchefer) est testé aux abords de deux sites particuliers dans les cantons de Genève et du Valais. Dans les deux cas les scories ont été employées comme remblai de talus lors de la construction d'autoroute.

2. DÉFINITION DE L'INPUT; RECHERCHE DE TRACEURS ROUTIERS MINÉRAUX ET ORGANIQUES:

Une phase importante de l'approche de l'impact des routes sur l'environnement souterrain est de définir l'input routier. Nous nous sommes attachés dans un premier temps à décrire ce qui provient des voitures, de la route ou de son environnement direct et qui, par usure, écoulement, etc., va parvenir sur la chaussée pour être lessivé lors des pluies.
Dans cette première phase d'analyses, nous nous sommes concentrés sur la définition de l'apport en éléments traces dans le milieux souterrain liés aux eaux de ruissellement routiers. Des échantillons d'huile neuves ou de vidange ainsi que des échantillons de divers types d'essence brute ont été analysés pour permettre une meilleure compréhension de l'origine des éléments traces dans les eaux de ruissellements. Un suivi de l'impact du salage routier est également en cours.

fig. 1: définition schématique de l'input.

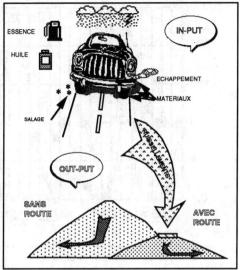

2.1. Recherche de traceurs minéraux

Méthodologie d'analyse:

Les analyses des traces élémentaires sont effectuées au laboratoire de Géologie (GEOLEP) où nous disposons depuis peu d'un spectromètre de masse haute résolution: High Resolution Inductively Coupled Plasma Mass Spectrometer (HR-ICP-MS).
L'ICP-MS est le mariage de 2 systèmes performants: la source d'ionisation à plasma (ICP) et le pouvoir incomparable de détection du spectromètre de masse (MS) (Gray & Date, 1983; Douglas, 1983). L'ICP-MS utilisé actuellement est le modèle Plasma Trace HR-ICP-MS de Fisons. Grâce à cet appareil, l'analyse des composés est poussée très avant pour acquérir une caractérisation extrêmement complète des éléments traces et par là garantir une identification sûre de l'origine d'un ou d'une série d'éléments.

Les échantillons sont prélevés dans des récipients en polyéthylène neufs. L'échantillon après filtration si nécessaire est acidifié avec de l'acide nitrique Suprapur concentré (65%), de façon à ce que son pH soit voisin de 2. Un pH bas inhibe l'adsorption des ions métalliques sur la surface du récipient et prévient la formation de précipités de métaux en traces, ou la coprécipitation des métaux en traces avec d'autres constituants majeurs (Taylor, 1989)
A partir des spectres de masse de chaque échantillon, on peut déduire la présence des traces et majeurs suivants: Li, B, Na, Mg, Al, K, Ca, Sc, V, Cr, Fe, Mn, Ni, Co, Cu, Zn, Br, Rb, Sr, I, Ba, La, Pb, Bi et U.

a) Salage routier:
Le B.A.C. de Mardièra en Valais nous a servi grâce à l'aide du service des routes Nationales à Sion, de site test pour le suivi du salage routier. Durant 5 mois des échantillons de ruissellement routier, d'eau de la partie étang du B.A.C. ainsi que des prélèvements effectué dans un réseau d'un dizaine de piézomètres implantés aux abords du B.A.C. ont été récoltés. Le dépouillement des analyses de chlorures ainsi que des statistiques de salage routier sur le tronçon considéré (mises gracieusement à notre disposition par le centre d'Entretien des Routes d'Indivi (VS)) est en cours.
Depuis octobre 93, un suivi du salage routier au col de Pierre-Pethuys est par ailleurs en cours en collaboration avec la commune de Tavannes.

b) Huiles et essences: .
Un échantillonnage de différents types d'essence (super, super +, sans plomb, diesel) a été effectué. Des échantillons d'huile neuve et d'huile de vidange ont aussi été récoltés.
Les éléments solubles sont extrait par acidification à l'HNO3 ~2%. La phase aqueuse est ensuite analysée. On obtient de cette manière les concentrations d'éléments solubles présents dans l'échantillon.

Essences: Les analyses d'essences nous sont essentiellement utiles comme référence "input" lors de l'étude des analyses du réseau d'observation. On y observe une nette prédominance du Pb et du Br dans la super. Les autres concentrations restent inférieures aux objectifs en matière de qualité des eaux pour les eaux courantes et les retenues.

Huiles: La comparaison entre une huile neuve et un échantillon de vidange du même type d'huile montre une nette augmentation de la concentration en éléments tels que le bore, le zinc, le brome, le molubdène, le barium et le plomb dans les huiles de vidange.

Différents types de mécanismes peuvent être envisagés pour expliquer cette augmentation. Nous en citons quelques unes, sans prétendre être exhaustifs:
- Minéralisation des molécules d'additifs des huiles ou des benzines.
- L'usure des pièces du moteur constitue un apport sans nul doute important en éléments métalliques.
- Apport d'éléments de l'essence dans les huiles dans les cylindres.

c) Eaux de ruissellement:
Les eaux que nous appelons ci-dessous eaux de ruissellement "brutes" sont des eaux de ruissellement routier ayant eux un minimum d'interaction avec le milieu naturel avant le prélèvement de l'échantillon. Ces échantillons d'eau brute ont été prélevés directement en bordure de routes cantonales ou nationales, à la sortie des conduites d'évacuation des eaux de ruissellements. La majorité des échantillons d'eau de ruissellement viennent de Bassins Amortisseurs de Crues du canton de Vaud et du Valais.

La comparaison des résultats d'analyses de micro-traces minérales sur des eaux de ruissellement et sur une source sans influence anthropogène est présentée ci-dessous à titre d'exemple.(figure 2):
Le référentiel "milieu naturel" sans influences anthropogènes est représenté par la source de la Sarve (SAR), vidange de pied de versant droite de la Plaine du Rhône. Référence aquifère type de roches karstiques carbonatées.
Les eaux de ruissellements routiers peuvent être considérée comme des eaux usées. Les concentrations élémentaires obtenues lors des analyses ont donc été comparées de façon indicatives aux objectifs en matière de qualité des eaux pour les eaux courantes et les retenues.

Figure 2:

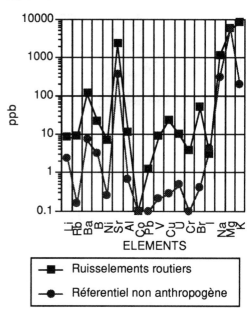

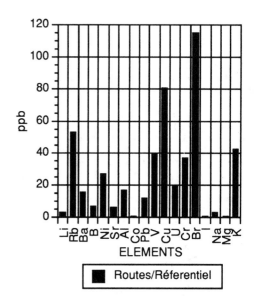

2622

Bien que cette première approche de la définition de l'input ne permette pas encore de définir de façon nette des traceurs typiquement routiers, elle amène néanmoins les observations suivantes:

- Les concentrations élémentaires de Hg, Cu et Cd dans les eaux de ruissellements routiers sont nettement supérieures aux valeurs définies comme objectifs en matière de qualité des eaux pour les eaux courantes et les retenues. Les valeurs limites d'exigence pour le déversement dans les eaux sont même atteinte dans le cas du Hg, ainsi que dans le cas du cuivre pour un échantillon (BAC, déshuileur Gd St Bernard).

- L'ordre de grandeur des concentrations d'éléments tels que Ni, Ba et U dans les ruissellements autoroutiers est 10 à 20 X supérieur aux référentiels sans influences anthropogènes. Dans le cas du Cu, Br, Rb, l'ordre de grandeur des concentrations élémentaires est respectivement de 80, 115 et 55 X supérieure aux référentiels sans influence anthropogène.

2.2. Recherche de traceurs organiques

Démarche

Dans le cadre de la recherche de traceurs organiques, le choix s'est porté sur des groupes de composés présents de façon ubiquiste dans les effluents de routes. Une recherche bibliographique et notre propre expérience nous ont permis de sélectionner un certain nombre de familles de micropolluants. Ce dosage ciblé des échantillons, permet de réduire le temps d'investigation lors de la mise au point des méthodes analytiques.

Durant l'année de recherche 1992-1993, les analyses ont été limitées à des échantillons d'eau issus ou non de canalisation de route, mais appartenant au réseau d'observation.

Les résultats des analyses sont considérés selon deux aspects: l'un quantitatif et l'autre qualitatif. Le premier permet d'évaluer le degré de contamination de l'eau. Tandis que le deuxième sert à identifier les sources en comparant les profils (fingerprint) de contamination.

La contamination des sites est apportée de manière relativement diffuse. Pour des raisons de simplification ce sont les profils des eaux de ruissellement qui sont pris en compte pour la comparaison des profils de contamination.

Parmi les problèmes liés aux analyses des molécules organiques présentes dans les effluents de route, celui des particules en suspension est de loin le plus difficile à résoudre. En effet 90 à 95% de la contamination apportée par les molécules organiques est liée aux particules. Ceci engendre une difficulté dans l'interprétation de la contamination. Il a été décidé de tamiser les eaux à 20 µm, puis l'échantillon est analysé brut.

Définition de l'entrée des polluants organiques (input)

Description de l'échantillon

Les eaux analysées au cours de l'année 1992-1993 ont été divisées en deux types:

1) Les eaux chargées en particules (mélanges):
Eaux de canalisation de route, de déshuileurs, de bassin d'infiltration ainsi que de parking.

2) Les eaux sans particules (solutions):
Eaux de sources

Description des polluants pris en compte pendant l'année 1992-1993:
- Polychlorobiphenyles (PCB)
- Hydrocarbures aliphatiques C12-C44
- Hydrocarbures aromatiques polycycliques (PAH)
- Esters phosphoriques

Les polychlorobiphenyles (PCB).:

C'est une famille de produits donc la structure moléculaire est composée d'une molécule biphenylée substituée par différents chlores. Ces composés ont été largement utilisés comme fluides isolants pour les transformateurs, comme retardant de flamme dans les plastiques, comme fluide hydraulique, comme lubrifiant, etc...

Malgré de fortes restrictions concernant leur utilisation une grande quantité de ces produits se trouve encore en circulation. C'est pourquoi, les investigations concernant ces polluants continueront dans cette étude, notamment dans tous les fluides utilisés dans le fonctionnement des véhicules automobiles.

Les hydrocarbures aliphatiques C12-C44:
C'est la fraction obtenue par cracking du pétrole qui compose les carburants diesel. Lors d'une contamination de site par des hydrocarbures, se sont les composés qui sont dosés en priorité .

Les hydrocarbures aromatiques polycycliques (PAH).:
Certains composés de la famille des PAH sont produits volontairement. Mais la plupart du temps leur source est le résultat d'une combustion incomplète d'hydrocarbures. C'est pourquoi on les retrouve dans les gaz d'échappement et plus particulièrement dans ceux des moteurs diesel. Ce sont des produits mutagènes.

Les esters phosphoriques.:
Ce sont des produits qui sont largement utilisés dans l'industrie: comme retardateurs de flamme dans les résines phenoliques et oxyde de phenylène qui entrent dans la fabrication de composés électriques, d'éléments pour les automobiles et dans les acetates de cellulose qui composent les films photographiques. Ils ont aussi des applications comme fluides hydrauliques et lubrifiants.

Les polluants suivants sont retenus pour la suite du travail:
-Hydrocarbures aromatiques polycycliques
-Polychlorobyphenyles

Le dosage des esters phosphoriques est abandonné. Les résultats n'ayant pas mis en évidence une présence systématique de ces produits.

Les résultats pour les pièzomètres et le bassin de Mardiéra en Valais, ne nous ayant pas permis de mettre clairement en évidence les différences anthropiques et naturelles entre les hydrocarbures aliphatiques, le dosage de ces produits est également abandonné.

Le dosage des hydrocarbures aromatiques polycycliques a permis de mettre en évidence une prédominance de certains composés (fluoranthène, pyrène, benzo(a)pyrène et benzo(e)pyrène). Pour quelques échantillons types issus de la route, le rapport des quantités de ces produits est similaire à ceux trouvés dans la littérature.
A l'heure actuelle nous ne disposons pas encore assez d'analyse pour généraliser ce qui a été mis en évidence dans quelques échantillons.
De plus nous devons prendre en compte la contamination diffuse apportée par les eaux de pluie. Pour cela il est prévu de doser les retombées atmosphériques en un endroit choisi près d'un bassin d'infiltration.

Les résultats de l'analyse des PCB ont montré la présence de ces composés dans certains pièzomètres. Etant donné que l'installation de ces pièzomètres remonte à plusieurs années et qu'à cette époque, certains lubrifiants de machine pouvaient contenir de grandes quantités de PCB, le dosage de ces composés a été momentanément interrompu, le temps de déterminer, si à l'heure actuelle, la route est source de PCB.

Les polluants suivants sont également potentiellement intéressants pour la suite du travail.
-Hydrocarbures aromatiques polycycliques nitrosés
-Hydrocarbures monoaromatiques (série BTEX: benzène, toluène, ethylbenzène xylène)
-Phenol (phenols substitués)

Les hydrocarbures aromatiques polycycliques nitrosés:
Cette famille de composés est émise de la même façon que les PAH. Ils sont produits en nettement moins grande quantité que leurs homologues PAH, mais ils sont beaucoup plus dangereux du point de vue toxicologique.

Les hydrocarbures monoaromatiques (série BTEX):
Se sont des composants de l'essence des moteurs à combustion (sans plomb, super, normal) obtenus par cracking du pétrole. Leur présence dans les eaux résulte de fuites des réservoirs des véhicules automobiles, ainsi que d'imbrûlés. Comme se sont aussi des solvants très utilisés dans l'industrie, leur présence dans l'environnement s'est généralisée .

Phenol & phenols substitués.:
Certains congénères phénoliques sont retrouvés dans les produits pétrolier. De plus, étant donné leur large utilisation dans l'industrie, d'autres composants des véhicules sont suspectés d'en contenir.

3. COMPORTEMENT PÉDOLOGIQUE

Dans ce projet, il a été décidé de porter une attention spéciale à l'étude du transfert des eaux de ruissellement des routes au travers du compartiment insaturé, en se focalisant sur le contrôle exercé par le sol sur la composition de ces eaux.

Le comportement des éléments dans le sol dépend d'un grand nombre de processus physico-chimiques qui contrôlent la répartition de l'élément entre la phase solide, et la phase liquide qui peut percoler plus profondément dans le sous-sol et qu'on qualifie souvent de phase mobile. La répartition entre ces deux phases dépend du mode d'arrivée des éléments au sol, sous forme dissoute ou sous forme de particules de tailles diverses, mais également de toute une série de réactions physico-chimiques telles que précipitation, adsorption (ou complexation de surface sur les colloïdes minéraux et organiques), diffusion dans les minéraux cristallins, absorption par des organismes vivants, en particulier par des bactéries, etc.

Investigations

Pour étudier le comportement des traceurs élémentaires choisis, on utilisera plusieurs approches:

1) comparer les profils de concentration des éléments dans les sites choisis, de la surface du sol jusqu'au voisinage de la roche-mère;

2) extraire la phase liquide du sol par un réactif choisi ou par centrifugation des échantillons de sol;

3) prélever les eaux de percolation à l'intérieur du sol en place par des lysimètres sans tension (dans ce cas des gouttières en polypropylène enfoncées dans le sol à une profondeur déterminée), de façon à obtenir un bilan qualitatif des traceurs retenus.

Choix des sites

Deux types de stations ont été retenues, l'une sur calcaire Kimmeridgien et l'autre sur molasse Burdigalienne. Pour chacune des deux stations, on a choisi un site au voisinage d'une route et un site de référence loin de toute voie de circulation.

Les sols ont été prélevés par carottage (tarière Humax); toutes les carottes ont été découpées en tronçons correspondant à des couches de sols de 10 cm; on n'a donc pas tenu compte de la stratification naturelle des sols en divers horizons, mais cette procédure a été retenue afin de pouvoir comparer plus facilement des sols de nature et de profondeur très différentes.

Stations	Calcaire Kimmeridgien		
Influence de la route	avec	sans	sans
Sites	Jura route	Jura forêt	Jura prairie
Couvert végétal	Prairie	Forêt	Prairie
Types de sol	Rendzine brunifiée humifère rendosol RP 92	Brun calcique calcisol RP 92	Rendzine brunifiée humifère rendosol RP 92
Profondeur du sol [cm]	≈ 30	≈ 30	≈ 30

Stations	Molasse Burdigalienne		
Influence de la route	avec	avec	sans
Sites	Lucens	Belmont	Lutry
Couvert végétal	Prairie	Forêt	Forêt
Types de sol	Brun calcique calcisol RP 92	?	Brun acide alocrisol RP 92
Profondeur du sol [cm]	> 200	> 200	> 200

Après séchage et tamisage à 2 mm des échantillons sur un tamis en nylon, on en a mesuré le pH (H_2O et KCl 1:2.5) et dans le cas de la station "molasse", la capacité d'échange cationique au pH du sol, la granulométrie à 60-70 cm de profondeur et à la profondeur maximum de prélèvement. Chaque échantillon a été extrait par HNO$_3$ 2M à chaud selon les directives OSOL afin de pouvoir déterminer les profils de concentration des indicateurs élémentaires.

Les premières analyses de la fraction facilement solubilisable des éléments minéraux ont été faites sur des solutions obtenues par extraction du sol par une solution NaNO$_3$ 0.1 M (selon les directives OSOL).

Résultats obtenus

La comparaison des premiers résultats obtenus (profils de pH, de concentration de Ba, Mn et Zn, et composition de quelques extraits de sol par NaNO$_3$ 0.1 M.) permet les observations suivantes:
On n'a observé aucun signe d'hydromorphie dans les sols analysés; le milieu est toujours oxydant, le pe peut donc être considéré comme invariant et le principal paramètre régulant le comportement chimique des traceurs est donc le pH. Pour le même substrat minéral, les sols sous forêt sont plus acides que leur équivalents sous pâturage; les analyses granulométriques des sols sur molasse Burdigalienne donnent des résultats différents: les sables sont plus abondants à "Lutry" qu'à "Lucens": un drainage moins efficace dans le sol "Lucens" peut expliquer les différences que l'on observe entre les deux sols sur molasse (différence de pH, de niveau de décarbonatation et de saturation du complexe absorbant).
Les profils de concentration des trois éléments traceurs déjà dosés ne montrent pas de différences nettes en fonction de la distance séparant le site de prélèvement de la route. Dans la solution du sol, du plomb n'a été détecté qu'en surface du sol brun acide de "Lutry".

Développements futurs

Une fois que les profils de concentration des indicateurs élémentaires auront été mesurés dans les sols choisis, on pourra extraire la solution du sol par centrifugation des échantillons les plus caractéristiques. On pourra ainsi suivre l'évolution de la concentration des éléments minéraux dans les eaux capillaires de la surface du sol jusqu'au substrat rocheux.
Dans ces solutions extraites par centrifugation, on pourra déterminer l'importance des transferts particulaires: la séparation sur filtre cellulosique 0.45 μm (norme internationale) nous permettra d'obtenir d'une part la fraction dissoute des indicateurs élémentaires, et d'autre part la fraction particulaire, c'est-à-dire la proportion des éléments retenue sur le filtre, déterminée après minéralisation du filtre.
Il est également prévu de récolter des eaux de percolation par prélèvement in situ en équipant les profils de lysimètres sans tension (gouttières en polypropylène).

4. COMPORTEMENT DU SOUS-SOL:

Les analyses sur les eaux de chaussées ont montré que l'input à l'entrée dans le sous-sol est composé de deux fraction très différenciées :

a) les suspensions colloïdales

Ce sont des colloïdes divers, retenus sur les filtres en laboratoire, qui ont tendance à fixer sur les particules une grande quantité de substances organiques et minérales, dont des métaux lourds. Cette fraction est retenue en partie dans les déshuileurs. Le reste aboutit à l'entrée du sol ou du sous-sol. Dans le cas des sous-sols granulaires fins, ces colloïdes seront très rapidement fixés dans le squelette. Dans les sous-sols granulaires grossiers et dans les roches karstiques et fissurées, ils pourront en revanche migrer, dans les cas les plus défavorables jusqu'aux exutoires. Ils seront retrouvés à l'état colloïdal dans l'output.
L'analyse de cette fraction s'est limitée actuellement aux tests méthodologiques. Les résultats quantitatifs seront obtenus dans la deuxième phase, surtout dans le cadre de l'analyse détaillée des output.

b) les solutions vraies

Ce sont les filtrats de l'input recueillis dans la filtration à 1μ en laboratoire. Cette fraction présente une capacité de migration dans le sous-sol qui est nettement plus généralisée que pour les colloïdes. Dans le sous-sol, elle ne reste pas inerte pour autant. Tous les phénomènes d'interaction géochimique avec le squelette vont modifier cette solution originelle avant l'exutoire. Cette fraction a été analysée dans l'input.

Premières analyses des traces minérales sur l'output

L'été 1993 a été consacré à du travail de terrain, permettant l'échantillonnage du réseau d'observation. Les premiers échantillons ont été analysés à l'ICPMS -HR au cours de l'automne pour une caractérisation des traces minérales uniquement. Les nouveaux échantillonnages comprendront également des analyses organiques.

Les analyses des eaux de l'output ont été réalisées sur les eaux brutes aux exutoires, sans filtration. Ce choix a été motivé par la prise en compte d'éventuelles atteintes à la potabilité des eaux. En effet, la plupart des ressources en eau souterraines sont consommées sans traitement. C'est donc d'abord le bilan global des substances ingérables par le consommateur qui doit être considéré, indépendamment de la forme sous laquelle se présentent ces substances. Dans des recherches plus spécifiques, la spéciation des substances devra être déterminée pour préciser les éventuelles atteintes.

Une première comparaison entre les output de systèmes d'écoulement avec et sans route fait ressortir le point suivant. Les éléments présents dans les solutions de l'input et dans les lixiviats acides d'hydrocarbures (B, Ni, Cu, Br, Mo, Cr, V et Ba) sont présents dans les output d'eau souterraine dans des concentrations très variables selon l'élément considéré et le site. Dans certains cas, les concentrations semblent plus élevées que dans les analogues naturels. C'est le cas par exemple avec le Brome, le Rubidium et le Baryum dans l'exemple présenté à la figure 3.

Figure 3.: Comparaisons entre trois aquifère karstiques carbonatés, avec et sans influence anthropogène.

-fig. 3a: Concentrations des traces élémentaires en µg/l.

-fig. 3b: Comparaison avec influence anthropogène/sans influence anthropogène.

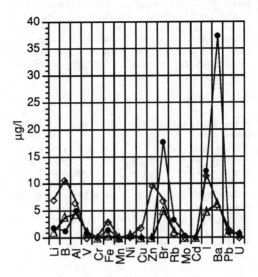

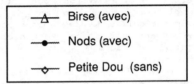

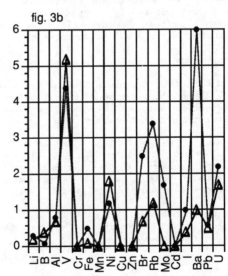

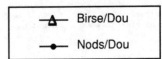

Nous ne sommes toutefois pas encore en mesure, à ce stade des analyses, de fixer pour tous les éléments et tous les sites l'origine anthropogène ou naturelle. De plus, dans les marquages anthropogènes, la spécificité de la route devra encore être précisée et assurée.

5. CONCLUSION:

Les résultats acquis au cours de cette première étape montrent que la motivation du projet est confirmée. En effet, les premières analyses permettent une bonne caractérisation des composantes. Le réseau d'observation donne une image représentative pour la Suisse.
La phase de développement des méthodologies d'analyses est maintenant à terme pour la plupart des substances organiques choisies comme traceurs. Les analyses minérales effectuées sur le réseau d'observation pourront être complétées dans le courant de l'année 1994 par celles des traceurs organiques, permettant une image plus globale de l'impact routier sur le milieu souterrain.
Finalement le suivi de ces traceurs au moyens des stations pédologiques affinera encore la compréhension du devenir de ces substances lors de leur transfert vers le milieu souterrain.

Bibliographie:

Attéia O., 1992, Rôle du sol dans le transfert des éléments traces en solution- application à l'étude de quelques écosystèmes d'altitude, Thèse EPFL n° 1031, Lausanne Suisse.

de Coulon, S. , Parriaux, A. , Bensiomon, M., Tarradelas, J. , Vedy , J.-C. (1993). Interaction entre les routes et l'environnement souterrain. Actes colloque int. environnement et géotechnique, Paris 1993, pp. 163-171.

Hamilton, R., Harrison, R.(1991) Highway pollution. Studies in environmental sciences 44. Elsevier

Hampton, C. V.; Pierson, W. R.; Schuetzle, D. & Harvey, T. M. ; (1983). "Hydrocarbon Gases Emitted from Vehicles on the Road. 2. Determination of Emission Rates from Diesel and Spark-Ignition Vehicles." Environmental Science & Technology, 17(12), 699-708.

Keller C., 1991, Etude du cycle biogéochimique du cuivre et du cadmium dans deux écosystèmes forestiers, Thèse EPFL n° 916, Lausanne Suisse.

Muschak, W. ; (1989). "Strassenoberflächenwasser-eine diffuse Quelle de Gewässerbelastung." Vom Wasser, 72, 267-282.

Office fédéral des routes . Forschungsauftrag 57/90, 1992: Bodenverschmutzung durch den Strassen- und Schienenverkehr in der Schweiz.

Office fédéral de la protection de l'environnement, 1987, Directives pour le prélèvement d'échantillons de sols et l'analyse de substances polluantes.

Parriaux, A. & Dubois, J.-D. (1990) Groundwater typology before and after the introduction of the ICP-MS. CHIMIA 44, pp. 243-.245.

Parriaux, A. & Lutz, T. (1988) Contribution to an informatized concept to the chemical analysis of groundwater series. Proc. 21st IAH Congress, Guilin (China).

Parriaux, A., Gabus, J.-H. & Bensimon, M. (1991) Characterization of natural waters using trace element analysis obtained in plasma source mass spectrometer. J. trace and microprobe techniques, 9(2&3),pp. 81-93.

Rogge, W. F.; Hildemann, L. M.; Mazurek, M. A.; Cass, G. R. & Simoneit, B. R. T. ; (1993). "Source of Fine Organic Aerosol. 3. Road Dust, Tire Debris and Organomettallic Brake Lining Dust: Roads as Sources and Sinks." Environmental Science & Technology, 27, 1892-1904.

Simoneit, B. R. T. ; (1986). "Characterization of Organic Constituents in Aerosols in Relation to Their Origin and Transport: A Review." International Journal of Environmental Analytical Chemistry, 23, 207-237.

Transportation research record 1017 (1985). Surface drainage and highway runoff pollutants.

Williams, R.; Sparacino, C.; Petersen, B.; Bumgarner, J.; Jungers, R. H. & Lewtas, J. ; (1986). "Comparative Characterization of Organic Emissions from Diesel Particles, Coke Oven Mains, Roofing Tar Vapors and Cigarette Smoke Condensate." International Journal of Environmental Analytical Chemistry, 26, 27-49.

Deformation and seepage and geothermal fields of a reservoir-induced earthquake

Champs de déformation, de percolation et géothermique d'un séisme induit par un réservoir

Chunshan Jin
Department of Civil Engineering, Dalian University of Technology, People's Republic of China

ABSTRACT: A reservoir-induced earthquake is another form of a structural earthquake induced by sypracrustal dilatancy measure's equilibrium conditions being ruptured when a reservoir water-injection occurs. Dilatancy measure is original earthquake strtum, which is formed by natural defomation field in reservoir area; Both water-hammer pressurization caused by seepage flow field and geothermal pressurization caused by additional geothermal field are force sources of induced earthquakes; The combined efforts of them form a potentical earthquake focus.

RESUME: Le séisme induit par le réservoir est une autre forme de séisme tectonique provoqué par la destruction des conditions d'équilibre dans le système de la couche dilatante de la surface de l écorce terrestre à cause de l'injection d'eau. Le système de la couche dilatante formé dans le champ de déformation naturelle de la zone du réservoir est la couche où se produit le séisme; et la pression du coup de bélier, faite par le champ d'infiltration additionnel, ainsi que la pression géothermique accomplie par le champ géothermique constituent les deux sources d'energie qui pourraient induire le séisme. L'action conjointe de ces trois champs forme une zone de l'hypocentre latente.

1 INTRODUCTION

A reservoir-induced earthquake (RIE) is a whole process that both the natural-geological environment and engineering projection act counter to natural setting. Its occurence is selective and it is another structural earthquake induced by natural equilibrium conditions of crust supracrustal which are ruptured by man's engineering activities. It gives great influence to engineering stability and it is not a certain result of impoundment in nexervoir, but reservoir water-injecter can induce the earthquake disturbance.

The reservoir water-injecter causes a new adjustment of natural deformation field, seepage field and geothermal field, then forms a earthquake-pregnant position, which will convert a non-earthquake-pregnant state into an earthquake-pregnant state. The earthquake-pregnant position of RIE is an outcome of union effect the three fields which perfectly reveal the principle of massive and compromise. It's an important angular on how to know the formation of RIE.

2 THE GENETIC CHARACTERS OF RIE

Deep research on genetic characters of RIE is an effective way to explore its genetic mechanism, because genetic characters are presage of genetic mechanism, which can expound that the mechanism of characters are reliable, at the same time, the characters are to expound and magnificate these genetic mechanism, which may trace the possiblity of the genetic enviornment from the genetic characters.

The genetic characters of RIE reveal collectively:

1. It is distributed in upwarping tensional tectonic province, active normal-slip fault or near the strike-slip fault. There is little or even no RIE occured in reversed fault or a reservoir near the flat fault. In geothermal anomalous area and the local region in the aseimic-weak and aseimic area, the earthquake volume is small.

2. Magnitude (in normal, $2-4M$), Shallfocuc ($3-5Km$), high-intensily, short-predominate period, foreshock closey and swarm earthquake. RIE is a high b—value process under the lower tectonic stress state, and it is also a nonlimited stress earthquake of the energical advanced release.

3. The basic conelitions inducing earthquake are the water permeability of the deep mass and the hydraulic connection. The earthquake activities take place at the same time that or after the reservoir is impounded.

These charaelers and data are model theory and test in themselves, because these approaches and models got from data are full of vitalily. They also declare that it is not a touch-and-go source environment before impoundment, but a pregnant formation process. It's a new adjustment caused by reservoir water-injecter about natural deformation, seepage field and geothermal field and forms a potential source zone.

3   DEFORMATION FIELD OF RIE

3.1 Natural deformation field of RIE

Natural deformation field refers to the inherent character of geologic body, and it mainly relates to the ground stress state and causative medium in condition of pregnant earlhquake. The stress increass and migration of underground fluid are two big factors leading to RIE, namely, RIE is the result of crustal dilatancy and dilatancy fluid migration. Earthquake-origine reservoir must possess the superstructure setting which has special reaction on water of reservoir-crustal dilatancy. This is the process of earlhquake formation in the rock mechanic field. Special minute structure of causative medium formation is dilatancy structure which refers to dilatancy fissure of rockmass and the formation and development of indusion area.

The dilatant earth deformation mainly reflects land uplift and effect of boundary (earth tilting). The said area is the local uplift region in neotectonic active area. The function of dilatancy effect that inducts such fluid earthquake is desisive. If the tectonic stress doesn't convert, rockmass possibly occur dilatancy structure in the early-stage of destructure. Dilatancy area is corresponding to tensional upwaping area (Fig. 1)

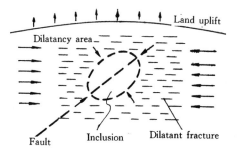

Fig. 1   Dalitant pattern of shallow strata for earth crust

Dilatant enviornment of rockmsaa decides whether the pregnant process of RIE will exist or not. The existence of dilatancy area certainly leads to the formation of inclusion area. RIE is chiefly distributed in dilatant measure of shallow crustal. Dilatant measure, among which inclusion area is potential source zone, is composed of caustive bed and earthquake-pregnant area.

Field studies display that dilatant effect affected by tectonic stress in major region is not regionality distributed, but it is contined to shallow section in fault zone and nearby position of which the rock strength is low. Dilatancy generally occurs in the depth of less than 10km and most sitable scope of depth is about 3 to 5km, which is correspondent to the local character of RIE and focal depth. When Volume stracn $\varepsilon_V < 0$, rockmass occur dilatantion and its discriminant is that

$$(1+\varepsilon_x)(1+\varepsilon_y)(1+\varepsilon_z)-1<0$$

ε_x, ε_y and ε_z in the formaule are respeclively the linear strain in the direction of x, y and z.

The stress state of rockmass dilatant is initial stage of fracture. It is clastic volume spreading of brittle rockmass appearing before unsteady-state extends in fracture stage. The stress state is equal to yield strength of rock. According to dilatant theory and model experiment, when the stress reaches 0.33 to 0.67 fold of disruption stress, rockmass occurs dilatancy; when the stress level approach or just a little overtake the proportional limit of rock, the clustered fissures in dilatant area form the inclusion area. Dilatant stress is the stress conditions of forming RIE. Dilatancy indicates the fracture feature of rockmass in non-limiting stress state. Therefore, lower stress state of RIE formation can be summarized by dilatant stress state. It is necessary to point out that inclusion area is generally in unsaturated state (similar-vacum, negative pressure) because of dilatant effect, therefore it possesses intensely water-absorbing effect. The upper limit of the power releasing in advance which is induced by reservoir water action can only be this value. In general, in the position where the void-fluid pressure is equal to lithostatic pressure, if the strength of supracrustal stratum in tectonic province estimates about 1×10^8Pa, the value of dilatant stress is about 300 to 600 bar which is close to the value of crustal stress 400 ± 200 bar in lower stress state presented by M. B. G sovski.

Therefore, the stress state in the dilatant measure is consitent with the lower stress state of RIE. The b-Value in magnitude-frequency retation is as well as a mirror of the character of crust tectonic activity and stress state. The high b-Val-

ue process of RIE displays the high fissuring of medium, how strength, the heterogeneity of earthquake focus are and low stress level. And the phenomenon that the b—Value after earthquake is lower than that before earthquake just indicates the releasing of energy in advance.

Therefore, the recent stress state which induce RIE is the condition of lower stress and the stress function stage is yield strength which is corresponding with the dilatant mechanism, then dilatant stress make dilatancy measure enter the earthquake pregnant stage.

3. 2 Additional deformation field of RIE.

Additional deformation field of RIE is an adjustment domain which locates in the stress adjustment environment when natunal deformation field is counteracted by mankind engineering activity— impoudment in reservoir. The water load which is formed by impoudment in reservoir is not enough to adjust primitive stress conditions to induct earthquake. Earthquake activities don't inevitably relate to the water load, not to the highest water depth in reservoir either. The additional stress field which was produced by water load and weight of dam is hard to constitute the leading factor that inducts earthquake.

In brief, natural deformation field stands up decide function and adelitional deformation field can be neglected, therefore, the dilatancy mechanism in natural deformation field takes leadling function. Natural deformation field is reatic stable so it is in reatic static state. It only presents earthquake—pregnant medium—the superdeep seepage flow passage and inclusion area in dilatancy measure—in lower stress stale.

4   SEEPAGE FLOW FIELD OF RIE

The hydrodynamics mechanism of seepage flow field is the study of relation between earthquake fracture and fluid activity. RIE is an effect of crustal dilatancy and dilatancy — fluid migration. Void — fluid is an essential condition of inducting earthquake and an exposition on how to form anomalous void — fluid pressere. RIE take place at the same time that and after the reservoirs are impounded, and the lower stress state and high — b value process of the natural deformation field, therefore, the water mass has huge effect.

4. 1 Natunal seepage flow field of RIE

Before impoindment, the present conditions of seepage flow field is natural seepage flow field in the dilatant measure which is an independent hydrogeological element. The transmissibilily of rock of shallow crustal has such a general trend that is reducing with the increasing of the depth. When the depth is over 4 — 5Km, the rock void is compressed, closure, and it's difficult for water to seep into the deep-seated. But in dilatant area, the abnormal phenomenon appeared because the dilatant mechanism have formed a permeable placement which is over — deep and high porosity. The earthquake — pregnant stress level of intense permeable medium is much lower than that of nonpermeable medium; In other words, the earthquake — origine opportunity of the intense permeable medium is much more than that of the nonpermeable medium. The natural seepage flow field possesses his own recharge, runoff and discharge system, in which mainly under the natural condition, the water mass void — free flow forms a normal void — fluid pressure which is difficult. Therefore, the natural seepage flow field is static normal void — fluid pressure which is difficult to induct earthquake because the level doesn't worth mentioning.

4. 2 Additional seepage flow field of RIE

This is an analyze for evolutionary trend of natural seepage flow field after the reservoir is injected.

The over — deep seepage flow placement, formed by dilatant mechanism, with anormal high porosity is a concentration seepage flow zone, and is a closed hydrogelogic element. The reservoir water is its recharge area. It is thought in the dilatancy and water — hammer patter that the crux of RIE lies in the formation of underground new water system. The underground new water system, which is formed in shallow crustal dilatancy measure after reservoir injecting is composed, is composed of reservoir water, dilatancy passage, and stress inclusion catchment. It's a local underground hydrodynamic circulation system with over open and under closed. This system has an protruding magnification effect for void — fluid pressure, that means the tuo processes which conducted alternately between the reservoir water injection and absorbing reservoir water in inclusion area (Fig. 2), the former process is only a under hydrostatic pressure produced by void — free flow, the latter one is that when the dilatant velocity is faster than the velocity of reservoir water injection, it forms a unsaturated state of dilatant rockmass, which makes the inclosion area is similar — vacuum state. When it is connected with the reservoir water, it can be absorbed quickly in the reser-

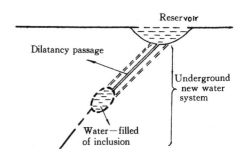

Fig. 2 Pressurization system of dilatancy structure

voir water to form a high velocily fluid flow and produce hydrodynamic pressure. When the inclosion area is full, the fluid flow is restraited, it will produce a water — hammer pressurization and form an immense water — driving pressure. When the value of abnormal high void — free pressure approach to the overlying load pressure, the local rockmass of the earthquake focus area find itself in floating state, which induce earthquake $\lambda = Pa/Sz$, just as $\lambda < 0.464$, it's under void — fluid pressure, just as $\lambda < 0.464$, it's under void — fluid pressure, just as $\lambda \approx 1$, $Pa = Sz$, the overlying rockmass is made to lose balance, finding itself in floating state.

Therefore, the earthquake takes place in inclusion area in dilatant measure. In inclusion area, which is not only the most active factor of the undrground new water system but a hydraulic gradint mega position caused by concentration secpage, gives a striking magnification play to void — fluid pressure. The engineering effect is shown not only on high percolation rate, but water — hammar pressurization possibly caused by moment large capacity injection. So, the natural seepage flow field and formation field are transformed intensely by the underground new water system, which concertrate the shearing stress.

The underground new water system is an additional seepage flow field. Its Pa is only related to the formation of its system velocity water flow and the restricted condition fluid flow, but not with the storage capacity and dam height. So, whether the additional seepage flow field is stable or not is up to the hydraulic head functioned on earth court in seepage flow field, hydrauliec gradient and formed water — hammer pressurization. According to an analogy between RIE and deep — well — injection earthquake, the value of the injection earthquake's critical void — fluid pressure is $250 - 460$ bar, that is when Pa reaches to 3/4 confining pressure, it comes close to $300 - 500$ bar. Therefore, Pa is the focus of force of induced earthquake, and RIE is fluid earthquake. Pa, a quickly applied load formed by water — hammer effect in underground new water system, displays a spasmodic of dynamic process, but Pa has a structural signification for the formation of RIE.

In brief, the normal state of natural seepage fluid in the dilatancy measure system is static, which is under void — fluid pressure state. But the additional seepage flow field in dynamic static, which transform intensely the natural deformation field and seepage flow field, form a potential seismics focus, so it possibly induces earthquake.

5 GEOTHERMAL FIELD OF RIE

5.1 Natural geothermal field of RIE

The distribution of RIE active area is generally corresponding with that of geothermal anomalous area. Quite a few earthquake — origine reservoirs have dense therme groups especially four of them ($M > 6$) occured. 50 percents of foruteen earthquake — origine reservoirs in China have thermes which are well — developmed and all of them are endogenous type. This proves that heat source in earthquake focus area geothermal anomal, which is a magnifier of RIE formation mechanism, is an important basis for analysing destability of RIE focus enviornment and inspecting the reliablity of forming genetic hypothesis. It is universally acknowledged that geothermal anomaly is one of discrimenant mark of RIE. It was indicated definitely in the Eirst International Conference about inducting earthquake(1976): "Reservoir area locates in residual heat of earth crust and the mergin of stable block". Zhangzhuo yuan pointed out in 1981, "High geothermal flow is a general condition of the examples of RIE". H. N. Kaluhuh and other people indicated (1982), One of three conditions that form RIE is geothermal anomaly…". It was in the severn discriminate marks of RIE which was proposed in Planning & Engineering Institute of Hydraulic and Hydroelectric in China in the fist 1980's, " There are themes in reservoir area." Lizu wu indicated in 1984, "Earthquake occurred in the area where the residual heat of earth crust is pretty high"; Ding Yuanzhang indicated in 1989, " RIE generally locates in distritutive region of themes and geothermal anomalous area". All the person who have talked about the discriminant mark of RIE mentioned the existence of geothermal anomaly which is well—developed without exception. This reveals that the geothermal flow value of earthquake — origine reservoir is high and ther is geothermal anomalous body in reservoir bottom. The existence of therme groups also show the development of under ground passage in earthquake—origine reservoir area and the intense comparation of water exchange circulation. The fact that the tempratue of thermes is not high explain that the depth of circular passage is only 2 to 3 Km. The geothermal anomalous mark of RIE is obviouser than that of natural earthquake.

It's necessary to point out that geothermal anomaly of reservoir area not form during the earthquake, not before or after earthquake either, neverthcless, the geothermal anomalous area, which has the geotherrnal history, has formed befor earthquake originates. The characters of geothermal anomaly are: heat of shallow focus is anomalous and earthquake focus area is generally geothermal anomalous body with 5Km depth; geothermal anomaly has no connection with regional deep—tecto—nically activity and its sphere is limited; there is not particular heat source (for instance, volcano) or radioactive heat source; the geothermal history forms before the reservoir is built. The above characters that the occurence of earthquake only has genetic relationship to geothermal system of shallow structure.

According to the above exposition, we have to assume that geothermic mechanism is dilatancy — hydrothermal effect, which is an abnormal geothermic produced for the formation of shallow crustal dilatancy measure. Because rock dilatancy destroys the strenching stress among rock molecule and the heat diffusion which the volume change by dilatancy fracture causes, partially geothermal gradient is caused to increese and the underground water circulating in it is warmed. Although the heat effect has no obvious influence to heat state inside the earth, the local characters are enough to increase in temp of rockmass and form geothermal anomalous area.

D. Craw and some other people indicated " The heat energy caused by the increasing of geothermal gradient which crustal quick upwarping brings about heats the rainwater circulating in it. K. N. G. Fuller's (1977) study shows that the increasing of unit surface enengy is because of a lot of heat loss when the extended velocity of crack is high. The end of cracks may even thermoluminescence and plastic zone will extend. Therefore, geothermal anomalous area intimately correlates to the structural setting of strata dilatancy and dilatancy measure is just geothermal formation of surface in active structure area. When the dilatancy fissure in the dilatancy measure forms inclusion area, heat energy inclines to form high—temprature anomalous body. This process from appearance of dilatancy crack to formation of inclusion body and further connection is a persistant exothermal process. In the dilatancy measure produced in the certain depth (within 15Km); On condition that tectomic stress field is relatively slable, dilatancy effect presents impulsive state; faster dilatancy velocity is more heat releases; and the dilatancy heat forming in a moment is replenished constantly. Therefore, geothermal field the result of releasing energy by strata dilatancy.

In addition, there are violent creep area in some parts of active fault and creep cause volume change of stratum and temprature increasing of terrestrial heat. According to P. Bregman's study, 85 to 100 percent of energy of plastic deformation convert into thermal energy.

In brief, stratum constantly releases heat and changes into thermorock because of dilatancy and creep so that shallow buried geothermal field is formed by water — rock interaction. Natural geothermal field, which is stable and in static state, is steady—state thermal anomaly.

5. 2 Additional geothermal field of RIE
Additional geothermal field nefers to the thermal stress caused by exchange circulation between thermorack or hydrothermal water and reservoir water injection or seepage which reach the deep in-

clusion area; in addition, hydrotherma may be pueumalotysised and driven to a tectonic stress concentration placement, then be hindred and e-menge gathering of thermal energy which induces earthquake. Under the condition of no void volume change, when temperature raise 1℃, void—water pressure increase 10 bar; under the same condition of no viid—water pressure charge, void—water produce pneumatolysis afterward, volume increase and produce pneumatolytiz expasion which makes void — fluid pressure increase. Therefore, additional geothermal field is a dynamic system hydrothermal pressurization action which is an important factor inducing earthquake. It can be said that RIE is thermal induced earthquake.

6 CONCLUSION

Genetics of tectonic type of RIE are comparatively complicted and its occurence is selective. We can correctly undefstana and predict RIE by means of mastering natural deformation field, additional geothermal field and additional seepage flow field and studying synthetically.

The analysis of natural deformation field displays that dilatancy measure is earthquake — origine stratum of RIE and natural deformation is in relatively state; the analysis of seepage flow field displays that water — hammer pressurization formed by additional secpage flow field — Pa, makes deformation field and seepage flow field produce new deformation trend. Geothrmal field is additional field of natural deformation field and additional seepage flow field. And geothermal pressurization of additional geothermal field is an important factor of inducing earthquake union effect of deformation field, seepage flow field and geothermal field in reservoir area causes reservoir area converted into earthquake — pregnant state from nonearthquake — pregnant state, which induces the earthquake.

REFERENCFS

Jin Chunshan. 1986. A dilatancy and waler — hammer patterm of reservoir—induced seism. Proceeding 5th International Congress International A ssociation of Engineering Geology, Volume 4: 1359—1365

Jin Chunshan. 1990. Reservoir — induced earthquake and deep — well — injection earthquake. Proceeding Sixth International Congress International Association of Engineering Geology. Volume 4: 2757—2762

Cui Zhengquan. 1992. An Introduction on systems Engineering Geology.

Engineering geology in the technogene

Géologie de l'ingénieur et technogène

G. Ter-Stepanian

Laboratory of Geomechanics IGES, Armenian Academy of Science, Yerevan, Armenia

ABSTRACT: The great changes in the environment during the last milleniums and especially during the current century caused by human activity resulted in the transition from the Quaternary to a new geological period - the Technogene. The attack on Nature at present surpasses all reasonable limits. We must protect not Nature in general but exactly that one to which we are adapted. We must act together with other two international geotechnical societies for soil and rock mechanics and demand not only the revision of building codes and make them more stringent but also to introduce environmental codes for limiting the arbitrariness reigning presently in the use of Nature. Since the pollution of the earth's surface, water and air does not recognize frontiers, the conservation of life of higher mammals, Man included, on Earth should not depend on the goodwill and possibilities of individual governments but should be realized on an international basis, even under compulsion. Therefore an organization standing above nations should be formed, and corresponding rights be secured.

RÉSUMÉ: Les grands changements environementaux pendant les derniers millénaires et en particulier lors de notre siècle excités par l'activité humaine ont eu pour effet la transition de la période quaternaire à une nouvelle période géologique - la technogène. L'attaque sur la nature a présentement dépassé toutes les limites raisonables. Il ne nous faut pas protéger la nature en général, mais exactement celle auquelle nous sommes adaptés.Pour cela il nous faut collaborer avec aux autres deux sociétés internationales géotechniques de méchanique des sols et des roches, et de demander non seulement de reviser les codes de la construction et de les faire plus rigoureuses, mais aussi d'introduire les codes environementaux pour limiter l'arbitraire qui règne actuellement dans le traitement de la nature. Étant donné que la pollution de la surface terrestre, de l'eau et de l'air ne reconnaît aucune de frontières, la conservation de la vie des hautes mammifères, l'homme y compris, ne doit pas dépendre de la bonne volonté ou de possibilités des gouvernements individuels, mais doit se réaliser sur une base internationale, même par contrainte. À cause de cela une organisation supranationale doit être formée et les droits correspondents doivent être assurés.

1 THE BEGINNING OF THE TECHNOGENE

The Technogene is a new geological period we live in. The Technogene has started with the Holocene under influence of a new previously unknown geological agent -

human activity. Human beings appeared on the Earth several million years ago, long before the Holocene. However the influence on Nature of ancient men, Paleolithic man included, did not differ in principle from the influence of other living beings such as plants or animals – they extracted from the environment no more substances than was necessary for their existence. Man maintained his existence by hunting and food gathering, and led the same predatory mode of life as his surrounding. This very long time interval covering much of anthropogenesis was characterized economically by food appropriation. The Paleolithic age was long, – more than 2 million years and progress was slow. The human activity of that time fell into a common group: "Activity of organisms".

In the Holocene, Man went over to agriculture and cattle-breeding i.e. to food production (the Neolithic revolution). Man became capable of extracting from his environment much more substances than was necessary for his existence. This caused a population explosion, the origin of the family and private property, accumulation of wealth and the transition from matriarchy to patriarchy. The Neolithic age was very short and dynamic; it lasted only several thousand years.

On the geological time scale the transition was extremely rapid and the changes turned out to be substantial. This makes it necessary to separate human activity from the beginning of the Holocene by virtue of new important distinguishing features from the common group "activity of organisms" and consider it as an independent geological agent. Man is able to affect, reproduce, imitate, prevent or impede many exogenous, a number of endogeneous and biological processes, and even some extraterrestrial ones as well (Ter-Stepanian 1988).

2 THREAT TO THE EXISTENCE OF HIGHER MAMMALS

The attack on Nature at present surpasses all reasonable limits. The environmental situation is getting worse. The unlimited high birth rate and overpopulation in many countries, accelerated urbanization, pollution of the environment by industrial wastes and domestic garbage, accelerated mining of natural resources, deforestation, desertification, destruction or oppression of many species of living beings, and the menace of a thermonuclear World War III create a real danger for the preservation of Nature and the development of mankind. Man is, as other higher mammals, strongly specialized to certain environmental conditions and cannot fit quickly to new ones which are being produced as a result of his technical activity.

If mankind will non act reasonably, the result will be extinction of higher mammals, Man included, and the Earth will be transformed into a contaminated world where only insects and scorpions may flourish, if anything. We must protect not Nature in general but exactly that one to which we are adapted.

3 MUCH DEPENDS ON THE ENGINEERING GEOLOGY

The situation in our profession is rather unfavourable. Realization of wrong, short sighted projects is playing a great part in the disturbance of Nature's brittle equilibrium. Among them may be mentioned construction of great water reservoirs in lowland river valleys leading to flooding of fertile bottomlands of valleys and to landslides on vast territories; deflation of sands as a result of destruction of thin vegetable layer by development of oases, causing sand storms; land subsidence as a result of oil, gas and water extraction causing formation of shallow lakes and swamps; drainage of high moors leading to dust storms, etc. The responbility in many cases is on engineering geology, which failed to forecast events and to eliminate the danger.

The changes which previously needed

geological time intervals are now progressing during one human lifetime; they take place everywhere right before our eyes. Blagodat mountain in the Middle Urals which was in the nineteen twenties 316 m above the neighbourhood is now excavated completely and a 800 m deep opencast mine is being exploited in its place. Knob-and-basin topography with up to 100-180 m heights was formed during many millions of years of development of the Middle-Russian upland. As a result of the working of the Mikhailovsky mining integrated plant 100 m deep opencast mines and 80 m high spoil heaps were originated during several years.

During the last decades a great part of the Amu Darya river flow in Middle Asia was diverted from the Aral sea and directed for irrigation of the Kara Kum desert; the canal was unlined and the desert began to swamp; the Aral sea being deprived of the Amu Darya's run-off is now being transformed into a technogenous desert. Was it realy impossible to predict such results?

Often such situations are caused by neglecting the engineering geological requirements and an absence of long-term prognoses. The general schemes of large-scale national projects are worked out by experts in other branches - economics, energetics, transport, irrigation, mining, but not engineering geology. The principal decisions are giving due consideration to the nearest advantages and benefits of the scheme but they fail, being based on poor understanding of laws of engineering geological development, and without prognoses of environmental changes in the remote future.

Engineering geologists are drawn often to secondary jobs; after the main decisions are taken, the scheme is accepted and funds allocated, the engineering geologists are charged with separate tasks such as to evaluate seepage losses from water reservoirs or canals, to measure rock pressure on a tunnel lining, and similar necessary but not very important problems.

Representatives of disciplines having taken principal decisions should be blamed not only for the wrong schemes but also for not having suspected their helplessness.

It follows from the above that in all cases when environmental changes are awaited the engineering geologist must be among those who has to forecast the results of technogenous impacts on Nature. He must be able to make such prognoses.

Forecasting results of short sighted projects, an engineering geologist will be impeding the realization of seemingly profitable proposals. To be able to do it, the engineering geologist must be a competent expert and brave man.

4 THE ROLE OF ENGINEERING GEOLOGY IN THE ENVIRONMENTAL PROTECTION

Our science being on the boundary between geology and engineering has a mission to study, carry out and control different problems in using the upper layers of the Earth's crust for building, transport, industrial, mining, urban, agronomic and other kinds of human activity. Owing to enormous technical possibilities and great volume of construction all over the world, it is doubtless that those countries which will use engineering geology reasonably for mitigation detriment to Nature will develop and flourish in the next century while those where engineering geology is neglected will degrade and be transformed into contaminated deserts. Therefore the attention to engineering geology is now of prime importance.

The above-said about the exceptionally important role of engineering geology is not a slip of the tongue dictated by hypertrophied ideas about the importance of one's profession. The engineering geology and hydrogeology are now indeed the most important fields of geological science, and errors here are incomparable in their consequences compared to other fields of geology.

If an engineering geologist would provoke a landslide by his wrong recommendations in calm conditions or the bursting of a dam causing flooding in a valley with many victims and material losses it would be a great tragedy. If a hydrogeologist would cause radioactive or chemical contamination of an artesian basin or bacterial contamination of a mineral water basin the mistake would be fatal and irreparable.

Understanding of the responsibilities of engineering geologists in our rapidly changing world puts on them special obligations. The geologist should be not only a pathologist making post-mortem examinations for understanding the causes of disasters, or a surgeon eliminating the emergency conditions, but first of all an internist who foresees events and prevents the initiation of undesirable changes in the environment. He must not only give correct diagnoses but make prognoses of events.

5 NEED OF RESOLUTE ACTIONS

The summary result of poorly controlled use of land are pollution of Earth's surface, its bowels, water, and air, radioactive and chemical contaminations, permafrost degradation, landslides, floods, deforestation, soil erosion, desertification, swamping of deserts, etc. The continuation of such an abnormal situation is intolerable.

We must act resolutely together with international sister-societies for soil and rock mechanics and demand not only to revise building codes and make them more strict but to introduce environmental codes for limiting the arbitrariness reigning now in the use of Nature, based on law of private property on land and its bowels. This law is not to be denied but also not to be abused.

The same refers to the sacred principle of state sovereignity. Since hazards and their consequences do not recognize frontiers, conservation of life on Earth should not depend on the goodwill or possibilities of individual governments but rather must be performed on an international level, even under compulsion. With that end in view a special environmental organization standing above nations should be created within the UN as United Nations Environmental and Mankind Development Organization.

The unique distinguishing feature of the Technogene - the ability of living beings to affect purposefully the course of geological and biological processes should be used to make the Earth a prosperous and reasonable planet. For the first time, Geology is not as much concerned with the past as with the future.

REFERENCE

Ter-Stepanian, G. 1988. Beginning of the Technogene. Bull. Int. Ass. Eng. Geol., Paris, v.30, pp. 133-142.

La pollution dans le milieu naturel de l'Amazonie équatorienne causée par l'exploitation des hydrocarbures

Environmental pollution caused by the exploitation of oil fields in the Amazonic basin of Ecuador

L. E. Torres
Politécnica Nacional, Equateur

Abstract
The exploitation of the oil fields in the amazonic bassin of Ecuador have caused an important damage in the phisical, biological and social economical medium of this region whose closest area is so fragile because of the biodiversity that is typical in the Amazonas.

Résumé
L'exploitation des champs pétroliers du bassin amazonien de l'Equateur ont causé une grave détérioration du milieu physique, biologique et socio-économique de la région, dont l'entourage est très fragile à cause principalement de la biodiversité typique de l'Amazonie.

1. Description du milieu

1.1. Milieu physique

Les aspects physiques analisés se rapportent surtout à la géologie, la climatologie, la morphologie, l'hydrologie, les risques géodynamiques, l'utilisation du sol, et la végétation.

L'objet de cette analyse est de connaître le milieu physique où l'exploitation pétrolière se développe et les rapports entre les deux.

1.1.1. Climat, météorologie et hydrologie

La région amazonienne équatorienne (RAE) a des caractéristiques climatologiques tropicales, influencées par la présence de la forêt humide comme source d'eau évaporée, avec un régime de pluies permanentes pendant presque toute l'année.

Précipitations annuelles	2500 mm	à

Tiputini jusqu'à 6500 mm au Reventador

Température moyenne	26 °C
Humidité relative	88%
Lumière naturelle	1440 h/an
nuvosité	6/8

Nombre de jours avec pluviosité

supérieure à 1 mm250
supérieure à 10 mm80
Précipitation maximale en 24h 160 mm

Vitesse moyenne des vents 1m/s

1.1.2. Géologie

Les conditions géologiques éxistantes jouent un rôle important dans l'analyse des paramètres qui agissent sur le milieu. Plusieurs phénomènes géologiques, volcaniques, tectoniques et en général tous ceux d'origine géodynamique sont au début de la déstabilisation des pentes, des écroulements, des déplacements, des effondrements, de l'érosion, etc,; provoqués par l'activité de l'homme et la construction d'oeuvres d'infrastructure.

La région amazonique équatorienne (RAE) où se situent les champs pétroliers, se trouve dans une unité téctonique régionale appelée Plataforma del Alto Amazonas (Plateforme du Haut Amazonas), qui s'étend de la frontière est de l'Equateur avec le Pérou, en passant par Lago Agrio, jusqu'au sobrecorrimiento AYAGAMA (nord – sud), à 65 km à l'ouest de Lago Agrio.

Cette plateforme qui se trouve à la limite nororientale du pays, s'approfondie vers l'occident en constituant un pétrin sedimentaire, dont l'axe se trouve près du champ pétrolier de Lago Agrio. Dans cette région on ne trouve que des rochers sédimentaires plio-quaternaires, coupées dans les dépressions des fleuves par des alluvions actuels, la plupart provenant du fleuve Aguarico.

La granulométrie des matériaux déposés sur cette zone augmente vers l'ouest, avec la suivante distribution relative par rapport à l'aire: argile - gravier 58%, argile 18%, conglomérats 16% limolites   8%.

Cette plateforme est divisée en sous-zones, en vertue de la présence d'autre sobrecorrimiento orienté en sens nord - sud, dans la zone appelée LOROYACU qui a élevé les conglomérats paleo-éocéniques de la formation Tiyuyacu.

D'autres alignements NE des fleuves Conejo et Agua Blanca Chica, sont interprétés comme sobrecorrimientos mineurs, tandis que le changement de direction de la cordillère de Lumbaqui est interprété comme une faille de direction dextrale. Cependant, cette zone est considérée celle de plus grande stabilité tectonique - dynamique.

Vers l'ouest cette plateforme du haut Amazonas a comme limite le soulevement Napo, une structure de sous-bassement jurasique, élevée en différentes périodes dont les trois principales sont: le premier soulèvement du paléocène, la période des réglages tectoniques du pliocène et l'emplacement du volcan Reventador au quaternaire.

1.1.3. Géomorphologie

Dans la région amazonienne équatorienne (RAE) on trouve surtout des planices en deux niveaux avec des petites collines et des secteurs de relief tabulair, ainsi que la présence de terrasses fluviales et quelques cônes de piedemont. Il a souvent des zones innondables et des marais à cause des ruptures des digues sur les lits majeures des courants d'eau pendant la période d'hiver, ce qui est à l'origine de la "varzea" ou forêt des plaines de débordement des principaux fleuves. Les légères pentes évitent les écroulements, de facon que l'élément plus perturbateur de la zone est d'origine antropique.

1.1.4. Aptitudes d'utilisation du sol

La situation actuelle de déforestation de la région amazonienne équatorienne obéit à un processus historique en rapport avec l'occupation du sol de cette zone, celle qui a augmentée considérablement dès le début de l'activité pétrolière, puisque les milliers de kilomètres de chemins et de tuyaux qui envahissent la région ont attiré les colonisateurs sans que ce phénomène soit controlé.

Chaque unité géomorphologique possède des aptitudes qui dépendent de ses propres caractéristiques: pente, type de matériel, âge, consolidation, géometrie, drainage, etc.

Pour orienter l'utilization du sol dans la RAE on a préparé une carte d'aptitudes d'utilisation du sol (échelle 1:250000) où on a déterminé les unités d'aptitude naturelle suivantes:

Forêt de protection	B
Forêt (exploitation et culture) et pâturage	F
Agriculture et élevage	A
Conservation	C
Cultures de saison	T

Les zones basses où on recomende une Forêt de protection sont surtout des zones avec de grandes réserves de faune et flore et aucune autre utilization est possible. Parmis ces zones on trouve les parques nationnaux de Cuyabeno et Yasuní, ainsi que les réserves de Vie Sauvage de Lagartococha et la réserve biologique de Limoncocha. On doit noté cependant que ces régions sont actuellement des zones de recherche ou/et exploitation pétrolière.

Les zones hautes présentent des problèmes à cause des pentes, de la destruction du sol, du niveau d'humidité et de leurs bas potenciels. Elles ne supportent pas l'activité agricole ni l'élevage ni l'exploitation forestière.

Dans l'unité des cultures de saison , les tranches qui correspondent aux terrasses fluviales et quelques lits des fleuves, dans les zones qui possèdent un bon drainage, peuvent être utilisées pour des cultures périodiques, mais il faut tenir compte au climat et le débit d'eau. Il s'agit généralement de zones fertiles, sauf quand les éléments nourrissants ont été complétement détruits.

Pour conclure on peut dire que:

- La plupart de la RAE possède des sols dont les restrictions sont assez importantes, à cause de leur composition, drainage et potenciel, donc, leur exploitation doit être limitée et contrôlée.

– Une grande partie de cette zone doit être protégée ou conservée à cause de la fragilité de l'écosystème (B et C).

– Seulement un pourcentage peut être utilisé de façon continuelle pour l'agriculture et l'élevage (A),tandis que des petits secteurs admettent des cultures périodiques (T).

– La partie qui reste doit être l'objet d'études détaillés pour préciser quelle est l'idéal de ferme d'exploitation agricole et forestière qui ne sera pas nuisible et qui sera en même temps économiquement rentable. D'autre part on doit déterminer le type d'exploitation forestière factible ainsi que les espèces plus adéquates et les techniques qui vont être développées.

1.2. Milieu écologique et biologique

1.2.1. Introduction

Tous les écosystèmes ont des rapports très importants entre eux à travers les différents éléments biotiques et abiotiques dont ils sont composés. Ces relations forment un équilibre de flux d'énergie ‾ travers les différents maillons des chaînes "trofiques". A partir des organismes producteurs (ceux qui fabriquent leurs propres aliments à base d'éléments nourrissants et lumière) on arrive aux organismes herbivores, carnivores et finalement aux décomposeurs qui ferment le cycle.

Ces interrelations peuvent être perturbées par l'intervention d'éléments étrangers aux écosystèmes qui affectent un composant quelconque, en modifiant l'équilibre établi. Ces perturbations peuvent être causées directement ou indirectement par l'activité humaine.

Les études écologiques développés dans ce document incluent les principaux éléments biotiques des écosystèmes qui composent la partie nord de la RAE.

1.2.2. Zones de vie

Dans la RAE on trouve fondamentalement une zone de vie appelée Forêt Humide Tropicale (bh–T).

Il existe deux types de forêt dans cette zone de vie; celle qui est innondée la plupart du temps (Igapó y Varzea) avec des espèces comme CRTON TESSMANNII, SCHEELEA BRACHYELADA, MAURITA FLEXOUSA, BROWNEA ARIZA et PTEROCARPUS SP. et celle de terre (non innondée) avec les suivantes espèces: IRIARTEA DELTOIDEA, PHYTHELEPHAS MACROCARPA, BROWNEA ARIZA, NEALCHORNEA VARUPENSIS, PENTAGONIA SPATHICALYX et LEONIA CRASSA.

Dans le modèle pluvieux forestier, forêt humide tropicale, on a réalisé des repérages pour la flore et la faune dans la zone du puits # 3 du champ Mariann, qui se trouve à 76° 22' ouest et 06° 08' sud, à 73 km à l'ouest de Lago Agrio.

On a placé 33 points de repérage pour l'analyse physique et chimique, ainsi que pour la description hydrologique de l'eau.

1.2.3. Flore

La région néotropicale est la plus riche en espèces dans le monde entier, on pense à 90 000 pour le nombre d'espèces de plantes supérieures dans les néotropiques, tandis qu'on estime que le nombre en Afrique tropicale est de 30 000 et 35 000 en Australie tropicale.

Au néotropique continental, les forêts des terres basses sèches possèdent généralement 50 espèces par 0.1 hec., les forêts humides ont de 100 à 150 espèces, les forêts très humides 200 et les forêts pluviales 250 (Gentry, 1976).

Plusieurs études réalisées en Equateur montrent qu'il existe une grande diversité d'espèces végétales, ce qui est un fait très important en tenant compte qu'il ne s'agit pas d'un pays avec un grand territoire. Les dernières études faites dans l'amazonie équatorienne, sur la voie Hollín – Loreto, communauté Challvayacu, ont démontré l'existance de près de 150 espèces (Gentry, 1989) ; à Jatun – Sacha et Misahuallí dans l'analyse de transects en 0.1 hec. on a pu trouver plus de 260 espèces (Gentry, 1988); à Dureno Lago Agrio, en 0.4 hac. 138 espèces ont été identifiées. On doit aussi noter un fait important: avant d'étudier la forêt de Jatun – Sacha, les forêts tropicales de Chocó en Colombie et d'Iquitos au Pérou étaient considérées celles ayant la plus grande diversité d'espèces au monde (260 espèces), aujourd'hui l'Equateur pourrait être au même niveau ou les dépasser.

Plusieurs travaux d'éthnobotanique réalisés par de nombreux auteurs démontrent que les indigènes de ces zones de l'amazonie utilisent une grande quantité de plantes sylvestres, ce qui pourrait devenir important dans l'économie des forêts tropicales, à travers des études biologiques. Les Quichuas de fleuve Napo utilisent 212 espèces, les Quichuas des fleuves Arajuno et Huambino 120 espèces médicinales, les Quichuas des flancs du volcan Sumaco 173, les Sionas et les Secoyas 224, les Huaorani 120, les Cofanes de Dureno 292, les Aschuar 130 et les Shuar de Bomboiza 200 (Cerón, 1990).

Une revision de l'herbier de l'Université Catholique de l'Equateur nous a fait connaître 600 espèces utiles dans l'amazonie équatorienne. Les quantités qu'on vient de citer doivent être considérées basses, des études plus approfondies sont nécessaire car le nombre de plantes utiles d'un groupe ethnique quelconque en Amérique Latine peut aller de 120 à 650 espèces et chaque groupe ethnique connaît 250 espèces utiles sans les superposer à celles d'un autre groupe (Toledo, 1986).

L'activité pétrolière affecte directe et indirectement la forêt tropicale:
- Directement , à cause de la coupe des forêts pour la construction des installations pétrolières, l'aplanissement du terrain pour les plateformes des puîts, voies d'accès, campements.
- Indirectement, à cause de l'action des colonisateurs qui utilisent les voies pétrolières pour entrer dans la forêt et grandissent les zones ravagées. Ces impacts sont peut être les plus importants dans la RAE.

Résultats au champ Mariann
Ce lieu est représentatif de la zone étudiée.

Diversité
A Mariann, la diversité de la forêt primaire de sol ferme est de 180 espèces, avec une espèce dominante: IRIARTEA DELTOIDEA; dans la forêt innondée il y a plus de 100 espèces dont CROTON TESSMANNII est l'espèce dominante, malgré l' importante présence de MAURITIA FLEXOUSA qui n'occupe qu'une troisième place dans cette forêt.

Composition végétale
Pour les forêts de sol ferme, les familles plus importantes sont les ARECACEAE, avec des espèces comme IRIARTEA DELTOIDEA qui représente 5.8% en 0.1 hac., PHYTHELEPHAS MACROCARPA 3%, les CAESALPINIACEAE avec

BROWNEA ARIZA 3%, les EUPHORBIACEAE avec NEALCHORNEA YAPURENSIS 1.8%, les RUBIACEAE avec PENTAGONA SPATHICALIX 1.8% et les VIOLACEAE avec LEONIA CRASSA 1.8%.

Pour les forêts innondées les familles plus importantes sont les EUPHORBIACEAE avec l'espèce CROTON TESSMANNII 7.5%, ARECACEAE avec les espèces SCHEELEA BRACHYELADA 6.5% et MAURITIA FLEXOUSA 6.1%, CAESALPINIACEAE avec BROWNEA ARIZA 5.8% et les PAPILIONACEAE avec PTEROCARPUS SP 4.4%.

Physionomie végétale
Dans la forêt primaire non innondée on a pû observer que la couche supérieure de la forêt est occupé par les différentes espèces de palmiers, les plus importantes dans cette région. Les niveaux moyen et inférieur sont occupés par d'autres espèces y inclus les palmiers.

Dans la forêt primaire innondée (Varzea), la couche supérieure est formé par des palmiers, même si l'espèce dominante n'appartient pas à ce groupe. Le niveau moyen est dominé par les palmiers en association avec d'autres espèces, tandis que dans le niveau inférieur il y en a très peu.

1.2.4. Faune

Les forêts tropicales se caractérisent par leur énorme diversité de faune, dûe aux conditions écologiques très diversifiés et qui possèdent des avantages pour le développements d'espèces vertébrées et invertébrées. L'Equateur présente une variété d'écosystèmes fragiles où habitent un grand nombre d'espèces; par exemple dans le bassin du río Napo on peut trouver jusqu'à 500 espèces de poisson, ce qui représente le plus grand nombre d'espèces pour un bassin dans le monde (D. Stewart, R.Barriga et M. Ibarra, 1987); le nombre d'espèces d'oiseaux dans la forêt tropicale orientale arrive à 700 (Thomas et Butter, 1975), 220 espèces ont été identifiées parmis les espèces d'oiseaux les plus communnes dans la zone du Cuyabeno (Paz y Mi±o, 1989) , cependant cette liste inclut aujourd'hui 400 espèces et au Yasuní plus de 500 espèces peuvent être trouvées.

Dans cette même zone, au Yasuní, on a pu observer 83 espèces de mamifères (Albuja, Gallo, Cerón, Mena, 1988), et seulement dans la région de Sumaco on a rencontré 98 espèces (P. Mena, 1990).

Le travail réalisé reflette la détérioration

qu'on souffert deux groupes de vertébrés (oiseaux et mamifères), à cause des activités pétrolières dans les forêts tropicales et les effets directs et indirects de cette pratique. On montre la diminution de la diversité faunistique et les altération dans sa distribution et abondance. On a utilisé l'Indice de Diversité comme base de ce travail.

Pour l'analyse de la diversité on a du créer un inventaire des vertébrés, c'est-à-dire, déterminer le plus grand nombre d'espèces pendant le travail de champ, dans les forêts primaire et secondaire, dans les zones d'influence des activités pétrolières.

Dans la RAE le lieu de repérage élu a été le champs MARIANN, où on a eu les résultats suivants:

Nombre total d'espèces d'oiseaux déterminées: 84, avec 16 ordres, 33 familles et 68 genres. Les groupes plus abondants sont les passeriformes, les pisciformes et les psitaciformes. Parmis les 84 espèces déterminées 75 se trouvent dans la forêt primaire avec 16 ordres, 32 familles et 62 genres. Dans la forêt secondaire on a déterminé 36 espèces qui correspondent à 9 ordres, 18 familles et 32 genres.

Nombre total d'espèces de mamifères déterminés dans la forêt primaire: 40, avec 9 ordres, 18 familles et 36 genres. Les groupes plus abondants sont les rongeurs, les quiroptères et les primates. Dans la forêt secondaire on a trouvé seulement 16 espèces qui correspondent à 5 ordres, 9 familles et 15 genres; les plus abondants sont les rongeurs et les quiroptères.

En comparant la diversité d'oiseaux de la forêt primaire avec celle de la forêt secondaire, on se rend compte que celle-là a diminué en chaque groupe taxonomique plus de 43% et le nombre d'espèces perdues arrive à 52%.

Dans le cas des mamifères, le plus important pourcentage perdu appartient au nombre d'espèces et de familles (60% et 50% respectivement), le pourcentage de genres perdus arrive à 58% et celui des ordres à 44%.

L'activité pétrolière liée au processus de colonization atteint la diversité des espèces d'oiseaux et mamifères. Dans tous les lieux de repérage on a pu constater une perte considérable de la diversité de la faune de la forêt secondaire.

La coupe et la brulûre des forêts amazoniennes commencées par les entreprises pétrolières et continuées par les colonizateurs et les entreprises qui exploitent du bois atteignent les différents habitats des espèces. La diversité perdue ne peut pas être récupérée.

1.2.5. Hydrobiologie

L'analyse des structures des communautés bentiques et leur réaction aux perturbations de diférents types, donnent les bases pour établir la qualité de l'eau des corps naturels.

Les macroinvertébrés qui habitent les lacs, les fleuves et autres corps d'eau, sont les indicateurs biologiques plus utilisés pour déterminer la qualité de l'eau. Ces organismes peuvent être facilement identifiés à cause de leur grandeur et de leur mobilité restrincte, ce qui les limite à des zones particulières. S'ils subissent aucun changement, la composition de la communnauté bentique change aussi.

On a voulu déterminer la faune bentique supérieure à 0.5 mm trouvée au fond des principaux corps d'eau qui traversent les différents champs d'exploitation pétrolière, pour établir une Ligne Base et déterminer le degré de polution de l'eau. On a élu 33 points de repérage dans les champs pétroliers.

Les macrophites aquatiques jouent un rôle important dans les écosystèmes aquatiques puisqu'elles fournissent une grande quantité de matière organique et d'énegie aux consommateurs et en même temps elles servent comme réfuge pour une grande variété de macro et micro organismes, on a évalué aussi la présence de plantes aquatiques et leur importance dans l'écosystème des fleuves élus pour le repérage, de même que l'identification des espèces.

On a trouvé des macrophites aquatiques seulement dans 6 courants.

Neuston: les organismes du neuston habitent dans la limite supérieure des corps d'eau et ils se trouvent généralement dans les eaux dormantes des fleuves et dans les ravins: ils ne résistent pas les courants de vitesse. On les trouve souvent dans les lacs et les marais. Pour cette étude on a considéré aussi très importan l'analyse des invertébrés du neuston.

Des 33 points de repérage élus, seulement 25

possédaient ces organismes.

Résultats: d'après l'analyse des Indices de Diversité, Indice d'Abondance, Indice Biotique, présence de macrophites, organismes associés et Neuston, enregistrés tout au long de cette étude, comme base pour mesurer la qualité de l'eau, on s'appercoit que dans l'amazonie équatorienne il y a une baisse de la qualité de l'eau à cause de la polution dûe à la présence d'hydrocarbures, matière organique provenant des campements et des nouveaux villages, pesticides utilisés par les entreprises agro-industrielles et par un haut niveau de déforéstation qui produit l'érosion et apporte des sédiments vers tous les fleuves.

La communnauté bentique était représentée par 93 espèces, dont 75 correspondent aux insèctes, cést-à-dire 78.2%. Parmis les 18 espèces qui restent les oligochoeta représente 16.9%, les hirudineos 2.2%, molusques 1.8%, aphasmidia 0.2%, crustacea 0.12%, turbelaria et porifera 0.8%, aracnida 0.04%.

L'Indice de Diversité de Margalef a déterminé que 9 points étaient dans le rang des eaux très pérturbées, 24 dans le rang des eaux moyennement perturbées avec un indice entre 1 et 3.

L'Indice Biotique a déterminé que 18 points de repérage se trouvent au dessous de la limite (5), ce qui exprime une condition d'eaux perturbées (limite de qualité biotique de l'eau).

Avec l'Indice d'Abondance on a établi l'existence d'une relation directe avec l'Indice Biotique. Il faut dire que si nous avons un Indice d'Abondance élevé on aura aussi un Indice Biotique élevé.

Comme groupes bioindicateurs on a trouvé des plécoptères sur deux points de repérages, les tricoptères en 12 points et les éphéméroptères en 16 points.

1.2.6. Qualité physique et chimique de l'eau

Dans ce chapitre on présente et analyse les résultats des paramètres physiques et chimiques déterminés par les échantillons d'eau. Le repérage a été dirigé à l'identification de la qualité des courants superficiels dans la zone d'influence de l'activité pétrolière.

Les paramètres physiques et chimiques des eaux superficielles sont utilisés comme point de repère en cas de versements dans les champs de production, dans les lignes conductrices (oléoductes et poliductes) et les refineries.

On a choisi 35 points de repérage pour réaliser l'analyse et on a tenu en compte les paramètres suivants:

Température: tous les courants répondent aux critères établis.

p.H.: les valeurs générales du p.H. sont relativement bas comparés avec les niveaux de qualité (entre 6 et 9).

Oxygène dissous: 75% des courants sont dans la limite établie pour la qualité de l'eau qui peut être consommée par l'homme (6 mg /l d'oxygène), 14 % est légèrement au dessous de cette limite (5.7 mg/l d'oxygène), tandis que 11% possèdent des concentrations assez basses.

Troubles: 93% des courants sont dans la limite de qualité (100 NTU), seulement 7% possèdent des indices non permissibles.

Bases: par rapport à la classification des types de bases (hydroxydes, carbonates et bicarbonates) on peut dire que les bicarbonates sont à l'origine des milieux basiques dans tous les courants.

"Fenoles": 93% des courants analysés présentent des valeurs au dessous de 0.002 mg/l et seulement 6.2% est au dessus de cette valeur.

Solides dissous: On peut dire que tous les courants sont dans les limites conventionnelles (1000 mg/l).

Huiles: La norme établie que l'huile doit être absente de la superficie de l'eau. Le repérage a registré 4 courants avec une pélicule visible.

Clorures: Seulement deux courants ont présentés des valeurs de clorure hautes par rapport aux autres courants.

1.3 Milieu socio-économique

La population de Napo et Sucumbios, provinces nororientales de l'Equateur, est de 200 000 habitants, dont 100 000 habitent les zones pétrolières où la structure démographique est principalement composée par des jeunes hommes.

La population permanente de la RAE s'occupe

surtout des activités agricoles, tandis que la population flottante s'occupe des activités technique et administratives de l'industrie du pétrole.

Dans la région nororientale du pays, près des champs pétroliers se trouvent trois zones naturelles protégées: Cuyabeno, Yasuní et Limoncocha.

Dans cette région habitent ancestralement plusieurs peuples indigènes tels que les Cofanes, Quichuas, Sionas, Secoyas, etc., qui sont les plus affectés par l'activité pétrolière, puisqu'ils ont perdu une grande partie des territoires qui leurs appartenaient traditionellement et leurs possibilités de survivre culturellement, de même leurs pratiques traditionelles de chasse et de peche ont été affectées, à cause de la déforestation et de la polution de l'eau et du sol.

Le processus de colonisation dans cette zone pendant les années 70, la création de chemins liés àux activités d'exploitation du pétrole représentent la destruction d'un million d'hectares de forêts primaires.

Les milieux plus sensibles, avec leur diversité en flore et faune, ainsi que les peuples indigènes ont été gravement atteints par les activités d'exploitation d'hydrocarbures, l'action des colonisateurs et l'exploitation du bois, dûs à la présence des entreprises pétrolières dans la RAE. La carte d'aptitude d'utilisation agricole du sol démontre que la région amazonienne intervenue possède des espaces différents et très complexes. Les terrains aptes pour les activités agricoles ainsi que les forêts fragiles ont été intervenues par l'activité pétrolière et après, par une colonisation sans controle.

2. Description de l'activité pétrolière

2.1. Généralités

Tous les champs pétroliers en production actuellement se situent dans la région amazonienne au nororient du pays. Dans cette zone se développent les activités de recherche et d'exploitation. L'influence de ces activités atteint surtout deux provinces orientales: Napo et Sucumbios.

Perforation de recherche: cette étape est principalement liée à la perforation de puits d'épreuve et elle commence avec le transport de l'équipement et des machines dans la zone pour les travaux d'aménagement du territoire.

Deux piscines sont construites dans chaque location, avec un volume de 2000 m3 chacune pour retenir la boue et l'eau dégagées pendant la perforation.

Dans ces piscines la boue et l'eau sont traitées et reciclées postérieurement.

La boue produite par la perforation est composée généralement par eau et argiles, avec d'autres substances comme la "barita," "carboximetil", cellulose, soude et "lignosulfonatos". La méthode de perforation plus utilisée est rotatoire.

Exploitation: après l'identification des zones contenant des hydrocarbures, et quand les réserves de pétrole trouvées sont suffisament importantes pour atteindre de bons résultats commerciaux, les perforation nécessaires sont faites pour pouvoir extraire le pétrole. Si le puits est sec, il est clôturé avec du béton, sinon il est équipé pour pouvoir controlé son flux et sa pression. En accord avec le potentiel du yaciment, on construit l'infrastructure nécessaire pour la récolte du pétrole pompé: chemins, lignes de transfer, emplacement des puits, installations de production.

Il existe, actuellement, à peu près 600 puits perforés dans 26 champs pétroliers distribués tout au lon de la RAE.

Production: le flux provenant d'un puits est un mélange d'hydrocarbures, liquides, gaz, eau et quelques solides. La principale fonction du processus de production est celle de séparer les éléments de ce flux et de retenir ceux qui ont un intérêt commercial, de telle facon que les hydrocarbures liquides et gazeux puissent être commecialiser.

Dans une station de production le flux provenant des puits est recu et soumi à une séparation, lavé et stabilisé. Ces stations peuvent stocker le pétrole et le redistribué dans un oléoducte secondaire. Le projet inclus 32 station de production, presque une par champs.

Ces champs ont produit 1 700 millions de bariles de pétrole depuis 1972 jusqu'en décembre 1993, ce qui veut dire une moyenne de 225 000 bariles par jour. (1 baril = 158.98 l)

Transport et stockage du pétrole: le transport du pétrole dans le continent se réalise à travers des conduits, principalement à travers l'Oléoducte Transéquatorien et les lignes de transfer réunies à chaque puits dans les stations de production, ce qui forment un réseau de transport assez complexe.

L'oléoducte transéquatorien (SOTE) a 503 km de long, avec 26' de diamètre dans la partie plus longue de son trajet, et une capacité installée pour transporter 300 000 bariles par jour. L'oléoducte part de Lago Agrio (RAE) et arrive à Balao (Côte du Pacifique) et il transporte le pétrole vers les raffineries et les centres d'exportation. Il existe cinc stations de dilatation et quatre stations pour réduire la pression.

Activités complémentaires: L'activité pétrolière a besoin pour son développement, en plus des activités techniques, une structure complexe très importante pour atteindre les objectifs de production proposés. Il est donc nécessaire dans les étapes de recherche et d'exploitation la construction de voies d'accès et de chemins, campements provisionnels et permanents pour permettre le séjour des personnes chargées de l'opération et l'entretien des stations, des raffineries, machines et équipement, et un réseau de communication.

3. Diagnose

3.1. Introduction

Dans ce chapitre on essaira de faire une évaluation du problème probablement causé ou généré par les activités liées aux hydrocarbures. Il s'agit d'identifier les effets et les impacts sur le milieu physique, naturel, socio-économique et culturel. de la zone d'influence de ces activités. Les analyses et les résultats comprennent les aspects suivants:

Diagnose des milieux: la Ligne Base du Milieu, developpée dans la description du Milieu Existant présente l'état actuel des composants du milieu où s'inscrit le projet. On décrit alors le problème identifié face à l'état de conservation, qualité et fragilité de ces composant, résumé à travers la Matrice d'Interaction du Milieu.

3.2. Diagnose par secteur

Après le développement des diagnoses physique, biologique, écologique et de l'eau, et socio-économique par secteurs, à

travers les analyses et les recherches, on est arrivé à obtenir certaines références de la detérioration du milieu, ensuite décrit.

Les impacts sur le milieu physique provoqués par le activités liées à l'exploitation des hydrocarbures sont:
- Déplacement et compactation de la couche organique du sol.
- Mouvements de terre qui cachent des zones d'intérêt botanique
- Procès erosives erreversibles, qui provoquent des changements du relief et modifications du drenage.
- Polution des eaux avec déchets solides et liquides, huiles, combustibles et composés chimiques.
- Intervention sur les zones de protection naturelle: parques nationnaux de Cuyabeno, Yasuní et réserve biologique de Limoncocha.
- Polution de l'air à cause du bruit, des vibrations, gaz, vapeurs et particules.
- Epanchement de pétrole causés par fautes techniques ou phénomènes géodynamiques.
- Polution avec eau de formation (eau salobre)

Les impacts sur le milieu biologique et écologique sont les suivantes:
- Un des impacts plus important est lié à la présence de zones naturelles protégées proches ou dans les mêmes champs pétroliers.
- La production de poluants solides en suspension, dissous et sédimentables, de matière organique, combustibles et huile qui affectent la qualité biologique des eaux naturelles et donc des organismes vivants: bentiques, neuston, poisson, anphibiens et flore aquatique des fleuves, ravins, lacs.
- Les effets sur les organismes aquatiques sont principalement la diminution de l'oxygène et de la quantité de lumière, la rupture de la tension superficielle de l'eau ce qui cause la dispersion, éloignement et mort des microorganismes.
- Les organismes aquatiques supérieures (poisson, anphibiens) sont aussi affectés par la chasse et la peche de colonisateurs et des travailleurs de l'industrie pétrolière.
- L'habitat de plusieurs espèces est dérrangé par la présence et l'opération des machines, par la construction des campements, le bruit et la présence de l'homme.
La construction de voies, de chemins et de l'infrastruture nécessaire ont détruit la végétation terrestre et la forêt naturelle, et ont causé la perte d'un volume de biomasse considérable.
- Ces altérations de l'ecosystème provoquent un problème encore plus grave, la perte progressive de la diversité

biologique, source et matière première du développement génétique et de la médecine futurs.

Les impacts sur le milieu socio-économique sont les suivants:
- Intervention à travers les concessions pétrolières dans les zones naturelles protégées et dans les zones de réserve des indigènes.
- les activités pétrolières et la construction des oeuvres d'infrastructure limitent l'utilisation des ressources naturels parmis les communnauté natives autonomes. principalement l'eau et la terre atteints par la polution et le mouvement des terres
- la qualité de vie des etnies amazoniennes est détériorée, puisque la polution affecte leur santé, ainsi que la limitation dans les sources d'aliments.
- La colonisation intensive augmente la déforestation et les nouvelles utilisation de la terre, elle crée aussi des conflits sociaux entre les ethnies locales
- La demande de main d'oeuvre augmente avec les effets qu'on vient de citer.
- Le développement du réseau de voies et chemin est à l'origine d'une polarisation espaciale qui augmente la densité de la population et la colonisation. Les conflits économiques et sociaux s'accentuent et se diversifient.

Geology and the high-speed railway project in the Northern Apennines (Italy)

Géologie et projet du chemin de fer à grande vitesse dans les Apennines du Nord (Italie)

A. Bucchi
Istituto di Infrastrutture viarie e Geotecnica, Facoltà di Ingegneria, Bologna, Italy

G.C. Carloni
Istituto di Topografia, Geodesia e Geofisica mineraria, Facoltà di Ingegneria, Bologna, Italy

A. Orlandi
Istituto di Trasporti, Facoltà di Ingegneria, Bologna, Italy

Abstract

High-speed railroads in Italy (the 'Alta velocità' project) are planned in order to stimulate a harmonic and well-balanced development of all rail infrastructures, eventually providing efficiency to the whole transportation system supporting national economy.

High-speed rail can be viewed as the most rational use of energy supply, the effective response to the modern European demand for environmental protection even in overcrowded areas.

The new 'Direttissima' route from Florence to Rome (262 Km) was begun in 1970 and completed in 1990, the maximum speed of trains currently being 250 Km/h. The planned Bologna-Florence route crosses the northern Apennines in a very complex structural and geomorphological environment which gives rise to many problems for engineering geologists.

The present paper, after having described the general geological aspects of the area, focuses attention on the relationships between the planned railroad route and slope stability. Following on this analysis, an alternative route is proposed which crosses the Romagna Apennines east of the route currently planned. This alternative route is slightly longer than the one currently planned but, thanks to more stable geological structures and less landslide-prone rock formations, shorter tunnels at lower costs could be constructed, thus making for overall cost effectiveness and shorter construction times. More importantly, this alternative route would entail a lesser impact on the highly populated metropolitan area of Bologna, human health being considered by the authors as the essential aspect in evaluating a project affecting a densely populated area and for ensuring proper environmental protection.

Résumé

La réalisation de lignes de chemin de fer à grande vitesse permet un développement harmonieux et équilibré des infrastructures à même de mieux répondre à la domande dans tous les secteurs dans lesquels le système ferroviaire est appelé à jouer un rôle efficace dans l'économie nationale. En outre, il constitue le moyen le plus approprié pour une utilisation rationelle des ressources énergétiques, tout en satisfaisant les exigences européennes touchant la protection de l'environnement, y compris dans les zones à forte densité de population.

La construction de la ligne express Rome-Florence (262 km) commencée en 1970, a été terminée en 1990, avec des trains atteignant une vitesse maximale de 250 km/h. Elle sera prolongée jusqu'à Milan et Naples, par l'ajout de 515 km. Le tronçon Bologne-Florence constitue, en ce qui concerne le tracé, la partie la plus difficile car il traverse les Apennins, zone très tourmentée au double point de vue géologique et géomorphologique. C'est sur ce tronçon que se concentrent les considération de géologie appliquée à une oeuvre de génie civile grandieuse et complexe.

Dans cette étude, après une présentation géologique générale, sont abordés les problèmes de la stabilité des versants et du choix du tracé. A titre alternatif, on a localisé un tracé situé plus à est par rapport aux tracés proposés par l'avant-projet; ce tracé, s'il présente une longueur légèrment supérieure, comporte des coûts plus réduits et des délais de réalisation plus courts. En effet, il traverse des zones géologiquement plus stables, et nécessite par conséquent de tunnels plus courts, ce qui représente une réduction non selement des coûts, mais également de l'impact sur le tissu très dense de la zone urbaine de Bologne. Les auteurs estiment en effet que l'impact sur l'environnement n'a pas de prix, surtout pour la sauvegarde, non seulement du milieu physique, mais surtout de la santé de l'homme, quand sont concernées des zones fortement urbanisées.

1. INTRODUCTION

The construction of the so-called Rome-Florence 'Direttissima' railroad (262 Km), begun in 1970 and completed in 1990 and designed for trains travelling at a maximum speed of 250 Km/h, will be extended to Milan and Naples with the construction of a further 515 Kms (Figure 1).

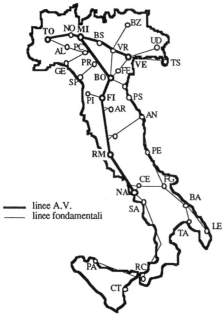

Fig. 1 - Projected high speed lines compared to the existing network in Italy

The completion of this line is part of the more general high-speed development and overall national rail network expansion project entailing the construction of new lines capable, unlike in France, of catering for both high-speed and ordinary passenger and freight trains, so as to constitute a fully integrated rail transportation system. In addition to railroad construction, the project also entails acquisition of a fleet of trains capable of travelling at a maximum speed of 300 Km/h as well as the construction of new maintenance installations, the estimated number of passengers being 20 million a year. It should be observed that

the choice of a mixed or integrated system is particularly suitable for the physical characteristics and settlement patterns of the Italian territory. In fact, not only is it characterized by physical features (the Apennines) with rather limited spaces available for the construction of travel routes, which are sometimes extremely narrow corridors already traversed by roads, motorways and railroads, but also by numerous small urban settlements.

The decision to adopt an integrated system also depends on the general trend in railroad development currently undertaken in other European countries where high-speed lines are already operating, under construction or currently being planned in order to adapt existing lines to higher speed standards. It should however be noted that, unlike in the Italian situation, none of these projects is faced with the problem of having to traverse a mountain chain such as the Apennines (Figure 2).

The northern Apennines, in fact, have always constituted a considerable problem in the construction of roads, railroads and motorways given the complex geological and structural characteristics of the territory. It should also be borne in mind that the Bologna-Florence connection, which is part of the more general link-up between the Po river valley area and Tuscany, has always played, and still plays, a major role in the context of the development and improvement of communications between Central Europe and Southern Italy. It has long been considered as one of the major bottlenecks adversely affecting both commercial and tourist traffic. This is perhaps due to the fact that the Emilian-Bolognese Apennines have in the past always been considered to be the only viable link-up route. In fact, even in the course of the last ten years, when a project for the construction of an alternative route to the Autosole, whether solely for freight haulage or as a variant proper, was being

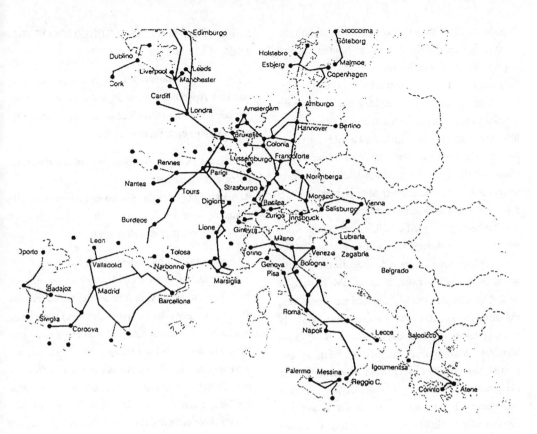

Fig. 2 - European high speed rail network

considered, attention was focused once again on the Reno river valley of the Bolognese Apennines, and in particular on that of its tributary, the Setta, as the only possible route at least on the northern side. Notwithstanding the considerable economic development in this area, for example between the towns of Casalecchio and Sasso Marconi and beyond, the time has come, at least as far as the choice of the route for the high-speed railroad is concerned, to consider other alternatives so as to relieve this area of excessive traffic.

After having analysed possible alternative routes still through the Bolognese Apennines, attention has finally been given to areas located further south-west. Bearing, however, in mind the particular geological characteristics of the northern Apennines, the authors suggest that a more south-westerly route be taken into consideration than those so far identified. Starting from the town of Firenzuola (a starting point on the Tuscan side convenient for our discussion) and following a route reaching across to the northern side and then down along the geologically stable Santerno river valley, this route is particularly viable also from a civil engineering point of view (tunnels, viaducts, slopes and bends). Moreover, the low number of densely populated areas, the type of valleys and their morphology, all contribute to reducing environmental impact and to make this route also viable from an economic point of view. This route seems all the more ideal if one also bears in mind the considerable difficulties associated with more north-westerly routes. In fact, apart from the less favourable geomorphological conditions of the territory, the existence of great engineering works built for the Direttissima railroad before Second World War and, later, for the Autosole motorway, is also to be taken into account.

2. GEOLOGICAL CHARACTERISTICS OF THE AREA

From a general geological point of view the northern Apennines between Bologna and Rimini can be divided into three distinct areas. From north-west to south-east, these include:

A. the area between the Reno and Sillaro rivers, or Bolognese Apennines;
B. the area between the Sillaro and Savio rivers, or Romagna Apennines;
C. the Marecchia river valley area.

Leaving aside a more detailed description of this latter area, as it is outside the scope of our discussion, to begin with it should be noted that the three above areas have a single element in common, that is they are characterized by Miocene-Pleistocene formations constituting the so-called 'Pedeappennino'. These formations are present in the hill belt which runs almost parallel to the Via Emilia in a monoclinal orientation dipping towards the Po river valley plain shell. This belt is a direct continuation of the similar belt bordering the Apennines between Piacenza and Bologna. The belt will not be directly affected by the High-Speed project since in the corresponding stretches of the valleys concerned the route shall run only along the valley bottoms.

A detailed description needs to be given of areas A and B (Figure 3) which shall be partially or totally involved in the project depending on the route selected.

2.1. The Bolognese Apennines

This area is characterized by a wide expanse of the heterogeneous Chaotic Complex (scaly clays and 'Ligurides' as defined by the authors) incorporating floating blocks, in some cases very extended, of

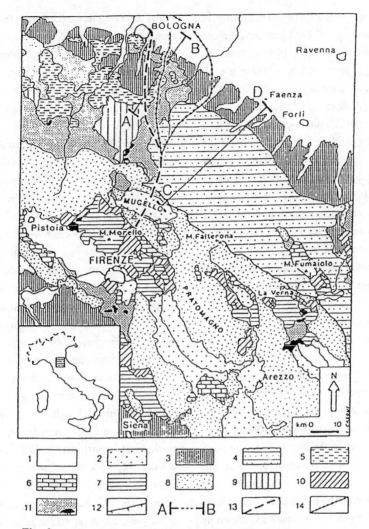

Fig. 3 - Schematic geological map of the Northern Apennine section
concerned in high speed rail planning

1 = Depositi alluvionali recenti e terrazzati *(Quaternario)*. **2** = Sedimenti fluvio-lacustri Villafranchiani *(Pliocene - Quaternario)*. **3** = Argille, sabbie e conglomerati marini *(Pliocene - Pleistocene)*; F.ne Gessoso - solfifera *(Miocene sup.)*. **4** = F.ne Marnoso arenacea *(Miocene inf. - med.)*. **5** = Areniti di Loiano e Bismantova, marne del Termina *(Eocene sup. - Miocene med.)*. **6** = Argille e calcari del Complesso di Canetolo *(Eocene Miocene inf.)*. **7** = Flysch calcareo - marnoso di M. Morello *(Paleocene - Eocene med.)*. **8** = Flysch arenacei toscani, Scisti varicolori e Scaglia toscana *(Cretaceo sup. Miocene)*. **9** = Flysch calcarei ed arenacei liguri *(Cretaceo sup. - Paleocene)*. **10** = Flysch arenaceo della Pietraforte *(Cretaceo sup. - Eocene med.)*. **11** = Complesso caotico eterogeneo, unita' liguri indifferenziate: prev. argille e calcari, con of ioliti *(Trias sup. ? - Cretaceo)*. **12** = Linea di ricoprimento. **13** = Tracciato previsto dalle FF.SS. **14** = Proposte alternative. A-B, C-D = Tracce delle sezioni geologiche.

lithoid formations which differ considerably in terms of facies and age. These formations include marls, limestones, ophiolites, flysch, turbiditic limestones, clays, arenites, etc. . These are all allochthonous and parautochthonous materials making up the gravitational plastic nappe which have translated from a south-west to a north-east direction reaching the Po river valley edge of the Apennines. The extent of this translation, which continued from the Miocene to the beginning of the Quaternary, measures about 10 to 15 kms on the front. This enormous allochthonous mass of the 'Chaotic Complex', together with the floating blocks, was associated, in its translation both along its front and sides, with veritable underwater landslides invading the surrounding areas for several kilometers and upon which normal sedimentation continued.

The neoautochthonous series rests on the allochthonous mass and is characterized by two large outcrops; one is represented by the internal or 'intra-apennine' syncline which from the Monte delle Formiche, along the Livergnano-Monte Badolo-Monte Adone-Sasso Marconi- Mongardino axis, continues as far as the Samoggia river valley and beyond, while the other is represented by the already mentioned 'pedemontana' monocline. The former is characterized by outcrops of sand, gravel and shore and delta molasses which are coeval to the sand and clay of the monocline belt dipping under the plain (Figure 4).

2.2. The Romagna Apennines

The heterogeneous chaotic complex which characterizes the Bologna Apennines is practically absent from this area except for small remnants in special orientations. A largely widespread autochthonous series comprising the 'Romagna marly-sandy' formation is however present and amply described in the literature. This consists of a sequence of re-sedimented clastic sediments which throughout the Miocene have filled up a graben stretching lengthwise along a northwest-southeast direction, from the Sillaro river valley to the Marecchia river valley. The graben is bordered by the Tuscan-Romagna ridge to the west and by the more recent 'Pedeapennines" belt to the north- east.

The lithology of this formation is characterized by alternating sequences of arenites, siltstones and marls to which are sometimes associated conglomerates, calcarenites and marly limestones. The arenites and siltstones have practically the same composition with an abundant phyllosilicate carbonate cement matrix and frequent carbon frustules. The marls contain calcium carbonate in varying quantities, and generally an amount of sand and silt greater than 50%. Conglomerates and calcarenites are present only locally.

From a tectonic point of view, the marly-sandy formation has an overal thickness of about 8000 metres, 5300 of which outcropping, and features a subparallel fold and fault pattern with a generally northwest-southeast orientation and a great horizontal continuity. The synclines are considerably wide and asymmetric, with the south-west side more inclined and sometimes very narrow and the north-east one gentler and often quite wide. The interbedded anticlines are on the other hand extremely narrow andreversed,bending towards north-east with the reverse side and axis characterized by a large compression fault (Figure 4).

The south-west edge of the formation is cut by a large thrust of the Tuscan complex on the Romagna autochthonous one. The formation has only been slightly affected by the presence of gravitational slide shells of the Liguride type in the Bolognese Apennines and Marecchia river valley. These are clearly a tectonic nappe covering the marly-sandy

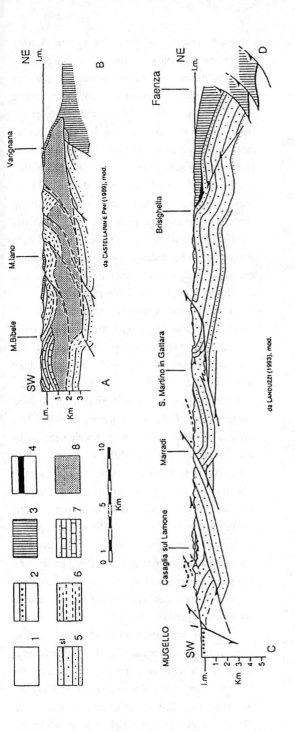

Fig. 4 - Schematic geological cross-sections

1 = Depositi alluvionali recenti e terrazzati (*Quaternario*) 2 = Sedimenti fluvio-lacustri Villafranchiani (*Pliocene - Quaternario*).3 = Argille, sabbie e conglomerati marini (*Pliocene - Pleistocene*) 4 = F.ne Gessoso - solfifera (*Miocene sup.*). 5 = F.ne Marnoso - arenacea (*Miocene inf. - med.*); sl = orizzonte caotico di franamento sottomarino (*Miocene med.*). 6 = Areniti di Loiano e Bismantova, marne del Termina (*Eocene sup. - Miocene med*). 7 = Flysch arenaceo di Monghidoro e facies calcarea di M. Venere (*Cretaceo sup. - Paleocene*) 8 = Complesso caotico eterogeneo, unità liguri indifferenziate: prev. argille e calcari (*Trias sup.? - Cretaceo*).

formation as has been confirmed by the deep hole drilled by AGIP.

3. SLOPE STABILITY AND CHOICE OF THE ROUTE

For the above given reasons, a comparison of the two areas described in relation to slope stability and proness to landslide in general, shows that the Romagna area is clearly more suitable for the purpose than the Bolognese one.

Following the preliminary project and the environmental impact study conducted by the Italian State Railways in April 1992, two corridors (east and west) of about 10 Kms in width have been identified within which four possible routes have subsequently been traced. An environmental analysis was then conducted along each route for a width of about 5 Kms based on six thematic aspects, that is according to geomorphology, hydrology, landscape, historical-settlement patterns, natural features and protected areas.

On the basis of this analysis, the east corridor route was finally chosen as it was seen that the territory offered "least intrinsic resistance to being crossed by a linear transportation structure".

The general layout of the route linking up Bologna and Florence from S.Ruffillo (Bologna) to Via Salviati (Florence) as contemplated in the preliminary project (Figure 5) entails engineering works along open stretches for about 11.928 Kms, of which 4.649 of 18 bridges and viaducts of varying lengths, and for the greater part tunnels for about 65.967 Kms. It should be noted that, despite a less steep morphology which could suggest the presence of sufficiently favourable geological structures along this route, major problems of stability both along the open stretches and underground would in fact be encountered.

Furthermore, even though current tunnel excavation technology would permit to overcome the difficulties encountered during the pre-war construction of the Bologna-Florence Direttissima line, the construction of viaducts and bridges raises major problems. In fact, the large number of foundations which would have to be laid for these structures would entail confronting complex problems and considerable risks due to both the lithological heterogeneity and the poor carrying capacity of the Liguride units. Even though initially rigid enough, the latter could in fact, as has already happened, subsequently turn out to be exotic masses completely embedded in the clays and therefore scarsely or completely unreliable.

In conclusion, therefore, having taken care to avoid the geologically risky areas, especially those subject to landslide, crossing of the more difficult areas will require the adoption of the best structural and technical solutions, which are the most appropriate also from a geological point of view, while bearing in mind landscape and false economy aspects as well. The actions which need to be undertaken include the careful rearrangement of the cut slopes by restoring the original vegetation or by planting new one, and the construction of effective defence works against leaching, run-off and infiltrating waters by perfectly governing both surface and underground drainage. By avoiding erosion and appropriately channeling the streams, maintenance costs and work could be reduced to a minimum, thus also eliminating delays or even worse interruptions in the traffic.

A brief mention should also be made of hydrographic aspects associated with the Savena, Zena, Idice and Santerno river valley basins on the Po river valley side and with those of the Sieve and Mugnone river valley basins on the Tuscan side. These rivers are torrential with considerable seasonable variations. In the Mugello area in

TRATTO (da progr. km a progr. km)	SVILUPPO (M)	TIPO DI OPERE	
3+720 - 5+292	Scoperto L = 1572 m	Viadotto Savena (4+310-5+060)	L = 750 m
5+292 - 15+778	Galleria L = 10486 m	Galleria Pianoro	
15+778 - 16+555	Scoperto L = 777 m	Viadotto Laurinziano	L = 180 m
		Viadotto Corcione	L = 60 m
16+555 - 19+585	Galleria L = 3030 m	Galleria Sadurano	
19+585 - 19+845	Scoperto L = 260 m	Viadotto Zena	L = 120 m
19+845 - 21+089	Galleria L = 1244 m	Galleria Prato Donne	
21+089 - 21+153	Scoperto L = 64 m	Scatolare	
21+153 - 21+292	Galleria L = 139 m	Galleria Fornacetta	
21+292 - 21+631	Scoperto L = 339 m	Viadotto Campagna	L = 180 m
21+631 - 28+788	Galleria L = 7157 m	Galleria di Monte Bibele	
28+788 - 29+771	Scoperto L = 983 m	Ponte	L = 30 m
		Viadotto Idice S.S.E.	L = 750 m
29+771 - 40+232	Galleria L = 10461 m	Galleria Raticosa	
40+232 - 40+392	Scoperto L = 160 m	Viadotto Diaterna I	L = 120 m
40+392 - 40+902	Galleria L = 510 m	Galleria Le Piagnole	
40+902 - 40+957	Scoperto L = 55 m	Ponte	L = 20 m
40+957 - 43+212	Galleria L = 2255 m	Galleria Scheggianico	
43+212 - 43+338	Scoperto L = 126 m	Ponte Brentana	L = 30 m
43+338 - 43+637	Galleria L = 299 m	Galleria S.Pellegrino I	
43+637 - 43+690	Scoperto L = 53 m	Scatolare	
43+690 - 43+900	Galleria L = 210 m	Galleria S.Pellegrino II	
43+900 - 44+271	Scoperto L = 371 m	Viadotto Santerno	L = 285 m
44+271 - 44+328	Galleria L = 57 m	Galleria artificiale	
44+328 - 44+430	Scoperto L = 102 m	Ponte	L = 30 m
44+430 - 57+654	Galleria L = 13224 m	Galleria Firenzuola	
57+654 - 61+293	Scoperto L = 3553 m	Ponte Bagnone I	L = 12 m
		Ponte Bagnone II	L = 12 m
61+293 - 62+384	Galleria L = 1176 m	Galleria Art. Cappuccini	
62+384 - 64+573	Scoperto L = 2189 m	Ponte Bagnone III	L = 30 m
		Viadotto Sieve	L = 1440 m
64+573 - 65+058	Galleria L = 485 m	Galleria La Ruzza	
65+058 - 65+328	Scoperto L = 270 m	Viadotto Faltona I	L = 210 m
65+328 - 65+493	Galleria L = 165 m	Galleria Faltona I	
65+493 - 66+030	Scoperto L = 537 m	Viadotto Faltona II	L = 390 m
66+030 - 66+673	Galleria L = 64 m	Galleria Faltona II	
66+673 - 67+109	Scoperto L = 436 m	Posto di Comunicazione di Borgo S. Lorenzo	
67+109 - 78+380	Galleria L = 11271 m	Galleria Monte Senario	
78+380 - 78+460	Scoperto L = 80 m	S.S.E. di Caldine	
78+460 - 81+615	Galleria L = 3155 m	Galleria Monte Rinaldi	

In definitiva si ha, per il progetto in definitivo:

- sviluppo totale: m. 77.895
- linea su corpo ferroviario allo scoperto: 11.928 m.
- linea in galleria: m. 65.967 con 18 gallerie
- linea in viadotto: m. 4.649 con 18 ponti o viadotti

Fig. 5 - Schematic review of the F.S. preliminary planning (April 1992), about the Bologna-Florence section of the high speed rail

particular, they are associated with the presence of wells and springs. For these reasons, these basins will have to be carefully analysed and studied in relation to the possible impacts of the project on the hydrological environment. The project, in fact, could interfere with the wells and springs along the routes considered, thus leading to an impoverishment or pollution of the more or less deep water tables which supply them.

4. DESIGN PARAMETERS

It is clear, therefore, that the Bologna-Florence stretch of the Milan-Naples high-speed railroad is the most complex as it crosses a geologically disturbed area. The geometric design parameters depend mainly on the desired speed to be kept along the line. The problem associated with determining this speed is essentially an economic one. In fact, even though a higher speed may be considered as providing a better service, it may require higher construction costs. It should also be borne in mind that increasing the speed beyond certain limits does not lead to any appreciable advantages while infrastructure costs both for construction and running and maintenance increase in a more than linear fashion. Moreover, it should be remembered that such an important project does not only entail the solution of technical problems but also of management and organizational ones as well. On the basis of these considerations, the ideal maximum speed between Bologna and Florence can be identified as being around 250 to 270 Km/h. At such speeds, the minimum radius of the bends would have to be equal to 4,000 metres, taking into account a non-compensated acceleration of 0.6 m/sec2.

As regards the maximum gradient, this must be determined by considering both speed and tunnel length. In fact, both speed and tunnel length

decrease as the gradient increases. In this specific case, by maintaining the crossing altitude unchanged, and by increasing the gradient to a maximum of 15%, the route which can thus be traced is more flexible and therefore it can be better adapted to the geomorphological configuration of the territory. For high-speed lines, the maximum gradient is generally given as between 12% and 20%, this latter value being the one adopted in France.

In designing the profiles, it is very important to determine the crossing altitude especially in relation to the problems caused to traffic by snow falls and ice formation. It is therefore advisable that this altitude be as low as possible and that route requirements be optimized according to morphological features. In the Apennines area in question, it is thought that the crossing altitude should not be, and in fact, cannot be below 500 m.

It is also important that the tunnels not be too long in order to reduce infrastructure costs and, above all, construction times as these directly affect the putting into operation of the line. For these reasons, it is felt that maximum tunnel length should not be more than about 10 Kms.

5. ROUTE CONSTRAINTS AND ENVIRONMENTAL IMPACT

The constraints imposed on a linear transportation infrastructure such as a high-speed railroad are many and conflicting. The route must be determined on the basis of a critical evaluation of the various constraints in order to arrive at a solution capable of optimizing the requirements of both the line and the territory. This means that both static and environmental characteristics must be fully evaluated. Once the design parameters of the geometric characteristics have been defined, an analysis must immediately be made for the purpose of introducing the planimetric and altimetric connections and for coordinating the planimetry with the profile. The behaviour of the vector in use is then simulated by means of a computer in order to evaluate the travelling conditions and the stresses affecting the passenger so as to optimize travelling comfort at such high speeds.

Generally speaking, the topographic characteristics of the Apennines railroad stretch examined within the wide corridor between Florence and Bologna are not significantly different from one route to another so that the same service level can be obtained following any of these routes. On the other hand, the geomorphological characteristics of these routes differ considerably along this stretch whether west or east of the Firenzuola- Bologna axis on the northern side. As already mentioned, these differences considerably affect relevant slope stability and consequently the static characteristics of the works making up the infrastructure. Notwithstanding the fact that present-day civil engineering technology is such as to sufficiently guarantee the stability of such works, the relevant costs are considerably higher on scarcely stable terrains. These costs are greater not only with respect to the works as such but also to the time required for their completion. Moreover, maintenance costs also increase as a consequence of the inevitable reduction in speed which such conditions entail, but which ideally should be avoided in any high-speed railroad.

Finally, with regards the environmental impact of such a line, this aspect should not only be seen as representing a constraint affecting the decision as to the route to be chosen but it should also be taken into consideration in relation to the social costs involved. In fact, these could be such as to offset the benefits to the community deriving from the infrastructure and which ultimately determine its general validity and functionality, the quality of living being adversely affected not only by pollution but also by limitations in mobility due to the presence of physical obstacles such as an overly invasive transportation line.

Having said this, it should however be noted that the possibility of significantly reducing or completely eliminating environmental impact in densely populated urban areas is rather slim. In fact, for similar type of vectors, the reduction of noise pollution by means of special barriers is always possible, but only within certain limits, while the vibrations transmitted via the structures and the ground inevitably increase as speed increases. The cancer risk due to the harmful electromagnetic fields generated by the power lines and contact points must also be borne in mind. The space required for the vector to pass through cannot obviously be reduced by more than so much.

Ultimately and ideally, therefore, minimization of environmental impact is achieved by locating the impacting infrastructures as far away as possible from areas which are more susceptible to being adversely affected, and in particular from densely populated areas. Proper location not only means reducing the polluting potential of the project, but also mitigating possible constraints. In conclusion, therefore, it must be said that the densely populated areas along the southern routes which directly converge on Bologna are environmentally incompatible with the planned high-speed railroad.

6. POSSIBLE ROUTES

It should be borne in mind that whatever route is ultimately chosen to link up Florence and Bologna, the project will entail the construction of numerous tunnels. Moreover, even if the chosen route were not to traverse areas featuring marly-sandy formations, the open stretches would in any case be crossing particularly critical areas from the point of view of slope stability. In addition to the poor quality of the geotechnical characteristics, which are closely related to the various typologies involved, especially in the areas featuring outcrops of the "Chaotic Complex" (defined as "scaly clays" by the authors), the project will also have to take into account the particular geomorphological and geological aspects of the territory concerned. It will therefore be necessary to contemplate from the very start special works for slope consolidation and stability. In fact, it is foreseeable that the general engineering works required by the project will be such as to further aggravate disrupting phenomena, especially where there are slopes which are alreadyunstable or in particularly precarious conditions. The effects of the project on the parts of the territory, even if of a moderate extent, subject toenvironmental modifications due to the construction of bridges, viaducts, embankments and ditches and to the setting up of work sites, cannot be overlooked.

Having therefore determined that from every point of view the construction of a high-speed railroad for the Florence-Bologna link-up on the northern side cannot be made to run through the Reno river valley to the west and the Savena river valley to the east of Bologna as these areas are densely populated and already feature major transportation infrastructures and the environmental impact would be extremely acute and hardly tolerable, it is thus necessary to identify a more geologically suitable area located further east in the Romagna Apennines.

We shall now examine in greater detail the main features of these valleys.

The Reno river valley, although very wide, is densely settled especially in the stretch between Sasso Marconi and Casalecchio, which is already traversed by the Autostrada del Sole (to be doubled), the Porrettana national road 64 (also to be widened) and the Bologna- Pistoia railroad.

Moreover, at the exit of the river valley, towards the plain, urban conditions along the Via Emilia axis are such as to cause considerable difficulties for a possible link-up to the Bologna-Milan railroad.

As far as the Savena river valley further east is concerned, not only is it rather narrow but it is also densely settled up to and beyond Pianoro. Moreover, it is already traversed by the current Bologna-Florence 'Direttissima' railroad as well as by the Futa national road 65 and by the so-called "fondovalle Savena" local access road, which carries considerable local traffic. Despite the importance of this latter road, it has been practically impossible to solve the problems for linking it up to the Bologna ring road, and this in itself is indicative of the difficulties which the construction of yet a further transportation line would entail. In conclusion, therefore, it is clear that neither the Reno nor the Savena river valleys can be exploited for the crossing of the high-speed railroad without causing enormous damage to the environment.

Two routes, A and B, have been studied further east which, from the Apennines ridge, descend towards the plain. These routes are located along the Idice and Santerno river valleys, respectively, and can be directly linked up to the already existing Ancona-Bologna railroad between Imola and Bologna (Figure 6). A river valley closer to Bologna, that of the Sillaro River, has not been taken into account owing to its geomorphological conditions which are the most unstable of the entire region and which would therefore involve enormous difficulties for the implementation of the project. Let us now examine each of the two above-mentioned routes in more detail.

6.1. Route A

This route runs through the Idice river valley. Although scarcely populated, this valley is less

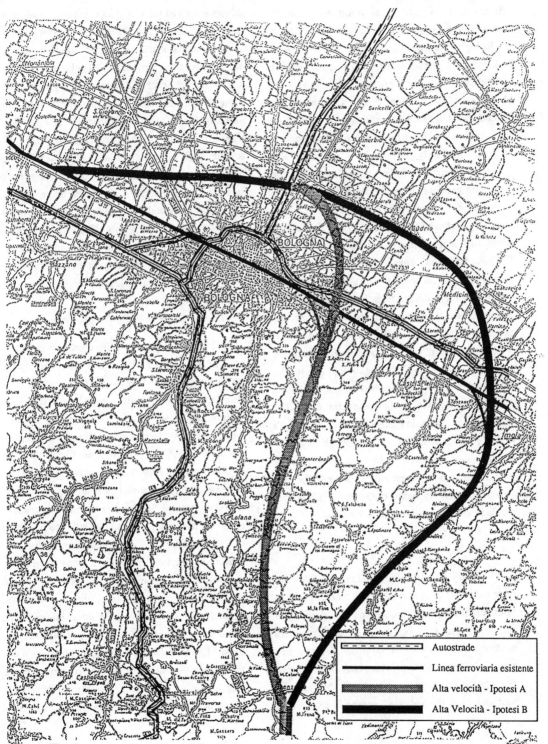

Fig. 6 - Alternative tracking hypothesis from the northern side of the Apennine belt to the town of Bologna

suitable from a geological point of view as compared to the other. The route branches off from the Bologna-Ancona line at about 10 Kms from Bologna, at Mirandola di Ozzano. It then heads directly south towards Castel de' Britti, Mercatale, Monterenzio, S.Benedetto del Querceto and Campeggio. Up to this latter town, no tunnels of considerable length are contemplated. The ones to be built are only intended to optimize the planoaltimetric configuration of the line, while the maximum gradient along this stretch is 1.5%. The crossing tunnel, at an altitude of about 460m, would start a short distance after Campeggio. The tunnel, about 10.7 Km in length, would pass under the Raticosa Pass and come out close to Firenzuola. After Firenzuola, on the Tuscan side, the line would continue through another tunnel, of about 7.9 km in length, located at an altitude of about 440m, passing under the Giogo Pass and coming out at Borgo S.Lorenzo. Upon having crossed the Mugnone river, the line could then be linked up to the 'Faentina' close to Caldine. Having passed below Fiesole, the line would then reach the Campo di Marte railway station in Florence. The overall length of this route is 92.7 Km, 11.1 of which along the Bologna- Ancona line. This latter stretch, of course, would have to be appropriately improved. The new stretch between Mirandola di Ozzano and Florence would cover the remaining 81.6 Kms.

6.2. Route B

As already stated, the route along the Santerno river valley is ideal for engineering works thanks to both its scarce population and its geological stability. These conditions are essential for reducing environmental impact to a minimum. Overall, this route is longer than the previous one. Nevertheless, the stretch to be newly built is shorter and, moreover, has the added advantage of requiring a shorter length of tunnels. As the previous one, this route branches off from the Bologna-Ancona line at Toscanella, at about 27 Kms from Bologna, with a Y- junction the construction of which would be much easier as compared to that of the junction for route A. It then runs close to Dozza Imolese, entering the Santerno river valley at Fabbrica. Running close to Borgo Tossignano and Castel del Rio, it reaches Firenzuola with a gradient which never exceeds 1.5% and without requiring long tunnels. At Firenzuola, it links up with the previously described route, and then continues on to Florence. Its overall length is 105.5 Kms of which 26.2 Kms along the already existing Bologna-Ancona line and 79.3 Kms along the new stretch to be built between Toscanella and Florence.

6.3. Link up to the Bologna-Ancona line

The crossing of the city of Bologna by a new line, and moreover by a high-speed one, with trains arriving to and departing from the Bologna Central Railway station, would cause a considerable environmental impact. In particular, as the line would run along the existing one in the stretch from Bologna S.Ruffillo to Bologna Centrale, the adjoining settled areas would be adversely affected. Vice versa, a transapenninic line linking up with the Bologna- Ancona line by means of a Y-junction or of an external ring line, could be made to run about 6-7 Kms north of the city through an agricultural area, approximately following a route close to the so-called "Trasversale di pianura" (plain transversal line). This external ring line, which would then continue along the high-speed railroad line to Milan, could be linked up at Anzola via a Y junction to a branch line from Bologna Centrale. Thus, the trains from Florence could use either the Bologna-Ancona line, which would have to be appropriately developed to cater for this extra traffic and linked up directly to Bologna Centrale or to the the Bologna S. Donato station via the existing ring line, or, alternatively, run along an outer ring line which could eventually be built at a later stage if the existing inner network were to become saturated.

This solution would afford great system flexibility as it would allow the trains to travel both directly into Bologna and outside the city, thus causing less disturbance to the suburban areas. A further advantage of this solution is that it could use the existing Bologna- Ancona line without the latter having to be substantially modified. In fact, between Castel S.Pietro and Anzola, the line has already been quadrupled and it can easily be expanded in the remaining stretches without excessive environmental impact. An exchange station could be built on the outer ring line at the intersection with the Bologna-Venice line in Castelmaggiore, which could be linked to Bologna Centrale by means of a shuttle train service.

7. CONCLUSIONS

As has been seen, there are several possible routes for a high-speed railroad linking up Bologna and Florence. Despite the fact that the routes thus far considered by the National Railways which run directly into Bologna are shorter than those running further east, the overall costs are practically the same. In fact, the overall length of the line cannot be the only factor taken into consideration when determining construction costs. In addition to tunnels, the construction of which may cost up to two-three times more than for the same length of open stretch, the costs for geomorphological or hydrological- related works, especially required for mountain area reclamation, should also be borne in mind. Moreover, tunnels inevitably prolong construction time, thus delaying the coming into service of the line. An additional problem with tunnels is the disposal of the excavated material, which also raises costs. The routes along the corridor of the current Direttissima Bologna-Florence line run through heterogeneous geological areas characterized by complex and highly unstable formations. They also require the construction of very long tunnels; for example the current S.Benedetto Val di Sambro-Vernio tunnel, is 18,150m long. Even if shorter than the ones suggested by us as alternatives, therefore, the construction costs of lines along these routes would be higher and the time required for completion of the works longer. Moreover, these lines would require greater maintenance.

In conclusion, it is worth underscoring the fact that no price can be set on environmental impact. In fact, it can never be simply reduced to a problem, for example, of demolishing existing buildings, but it must above all be considered in terms of its direct effects on human health and well-being. This impact, it should be noted, is clearly greater whenever densely populated areas are involved in any project such as the one described here.

References

AUTORI VARI: *"L'alta velocità ferroviaria nel territorio bolognese"* Seminario dell'Istituto di Trasporti. Quaderni dell'Istituto Bologna 1993.

Atti dfel I° Convegno Internazionale su *"Le ferrovie nei trasporti degli anni 2000."* Bologna 1989.

CARLONI G.C. 1987: *"Le autostrade , la geologia e l'ambiente fisico"* I° Congresso Internazionale di Geoidrologia su "L'antropizzazione e la degradazione dell'ambiente fisico". Firenze

CASTELLARIN A. et alii 1985: *"Analisi strutturale del fronte appenninco padano"*. Giorn.Geol. sez.3, vol.47/1-2 Bologna.

Atti del Convegno su *"Le soluzioni tecniche possibili per l'attraversamento autostradale dell'Appennino Tosco-emiliano"* Modena 1987

Atti del Convegno dei Lincei su *"La funzione della Geologia nelle opere di pubblico interesse"* Roma 1961.

Bilan de la sédimentation dans une retenue eutrophisée, quinze ans après sa création

Balance of sedimentation in a reservoir that became eutrophic fifteen years after its construction

A. Jigorel
Laboratoire de Minéralogie et Géotechnique, INSA de Rennes, France

J. P. Morin
Service Départemental de l'Agriculture et de l'Environnement, Conseil Général des Côtes d'Armor, St. Brieuc, France

RESUME: Une étude des sédiments de la retenue sur le Gouet (Côtes d'Armor, France) a été réalisée à l'occasion d'une vidange partielle, 15 ans après sa création. Le réservoir subit à la fois une érosion des pentes latérales et un envasement important du fond. La sédimentation résulte des apports des berges, des apports des rivières et des fortes productions phytoplanctoniques internes favorisées par l'état d'eutrophisation de la retenue.

Les sables grossiers apportés par les rivières, lors des plus fortes crues, se déposent dans des aires limitées à leur exutoire et à la bordure des chenaux. Les vases qui ont une origine mixte détritique et biogène, recouvrent assez uniformément le fond plat du lac. Elles sont caractérisées par un taux élevé de matière organique (8 à 17 %) par l'abondance des frustules siliceuses de diatomées et une texture d'autant plus fine que la fraction biogène est importante. Le taux élevé de sédimentation (3 cm an $^{-1}$) est amplifié par la morphologie du réservoir et la faible concentration des vases (250 à 500 g l $^{-1}$).

ABSTRACT: The sediments of the Gouet reservoir (Côtes d'Armor, France) were studied during a partial emptying of the lake, 15 years after it was created. The lateral slopes are clearly eroded while the bottom is covered with a thick layer of muddy deposits. Sedimentation comes from shore erosion, clastic inputs of the rivers and a high endogenous phytoplankton production enhanced by the eutrophication of the lake.

Coarse sand brought during major floods, is laid near the river mouths and along the sublacustrine channels. Muddy deposits which have a clastic and biogenic origin cover evenly the bottom of the reservoir. They have a high rate of organic matter (8 to 17 %) as well as many diatom frustules, the grain - size distribution of these deposits being finer where biogenic fraction is more important. The very high sedimentation rate (3 cm year $^{-1}$) is increased by the morphology of the reservoir and by the low concentration of the muddy deposits (250 to 500 g l $^{-1}$).

1 INTRODUCTION

L'importance de la sédimentation dans un lac dépend de facteurs, climatique, géologique et anthropique. L'accroissement des taux de sédimentation observé dans le massif armoricain, pendant les dernières décennies, a généralement été attribué à une recrudescence de l'érosion, favorisée par la restructuration des terres agricoles et l'évolution des pratiques culturales. Aux apports de l'érosion des sols, il faut ajouter, dans les lacs eutrophisés, la sédimentation biogène induite par les fortes productions internes.

La vidange partielle d'un lac artificiel, alimenté par des rivières, nous a permis de préciser l'importance et l'origine de son envasement.

2 CARACTERISTIQUES DE LA RETENUE

La retenue sur le Gouet est située au centre-nord du département des Côtes d'Armor, France (fig. 1). Elle a été crée en 1978, essentiellement pour assurer l'alimentation en eau potable de la région de St Brieuc. Comme toutes les vallées côtières de la région, celle du Gouet est étroite et encaissée et son verrouillage par un barrage de 40 m de haut a donné un réservoir dont la capacité maximale est de 7,9 millions de m^3. Le plan d'eau présente une forme allongée (longueur de 6 km). et sa profondeur s'accroit régulièrement de l'amont vers l'aval pour atteindre 37 m au niveau du barrage.

Les berges sont constituées de roches métamorphiques granitisées et de granodiorite, recouvertes par des formations meubles d'altération qui ont localement des épaisseurs de plusieurs mètres. Les pentes sont dans l'ensemble très élevées, voisines de 30 à 40 % lorsque le substratum rocheux est subaffleurant, mais plus douces (15 à 25 %) dans les secteurs où les formations d'altération sont bien développées. Le fond de la retenue est assez plat car l'ancien lit majeur de la rivière a été remblayé pendant l'holocène par des alluvions sablo-limoneuses.

Le réservoir est alimenté, en queue de retenue, par le Gouet, tributaire principal et la Maudouve. Quelques petits ruisseaux à écoulement intermittent se jettent directement dans la retenue. Le bassin versant qui a une superficie totale de 195 km^2 au niveau du barrage est constitué pour l'essentiel de terrains granitiques.

3 LES METHODES D'ETUDE

Les épaisseurs de vase sur le fond ont été mesurées en deux étapes. Une première série de mesures (95 points) a été effectuée à partir d'un bateau, au fur et à mesure de la vidange, en utilisant une perche graduée et une sonde.

Après la vidange, des mesures directes (100 points) ont été réalisées selon les profils transversaux, dans tous les secteurs où les vases en voie de dessèchement, avaient une portance suffisante pour permettre un déplacement à pied. Vingt-cinq carottages ont été effectués pour valider les mesures directes et pour déterminer la concentration des vases. Les analyses des sédiments ont été faites sur des échantillons prélevés à l'aide d'une drague manuelle et sur des carottages.

L'analyse granulométrique a été réalisée, après destruction de la matière organique avec H_2O_2, par sédimentométrie à la pipette de Robinson et par tamisage. La nature des sédiments a été précisée par un examen au microscope électronique à balayage (MEB)et un examen au microscope optique de la fraction sableuse. La teneur en matière organique a été déterminée selon la méthode de la perte au feu à 550 °C.

4 RESULTATS

La retenue est soumise à un double phénomène, une érosion des berges et un envasement important du fond de la cuvette.

Le périmètre rapproché est totalement boisé et donc bien protégé de l'érosion. Par contre les berges sont soumises au battement des vagues sur une hauteur de 5 à 6 m qui correspond aux variations saisonnières de la cote du plan d'eau. Les arènes granitiques qui recouvrent superficiellement le substratum rocheux ont été érodées, voire totalement décapées. Le déchaussement de souches sur une hauteur de 30 cm témoigne de l'importance du

phénomène. La fraction sableuse résiduelle des arènes remaniées constitue sur les pentes inférieures à 40 %, tantôt une couche continue de 10 à 20 cm d'épaisseur qui présente des rides en surface, tantôt des petits cordons, parallèles aux isobathes du réservoir.

L'envasement concerne tout le fond relativement plat de la retenue, la partie inférieure des pentes latérales (pentes inférieures à 10 %) et l'exutoire des petits ruisseaux. Les sédiments présentent un faciès sableux et un faciès de couleur sombre, à texture très fine de type vase. Les sables sont rencontrés en queue de retenue jusqu'à la confluence Gouet - Maudouve et à l'exutoire des petits ruisseaux. Ils se présentent selon leurs zones de dépot, sous forme de bancs, de microlits interstratifiés dans les vases ou de cônes.

4.1 *Epaisseur des sédiments*

Les épaisseurs maximales voisines ou supérieures à 1 m ont été mesurées à l'exutoire des rivières et dans le chenal correspondant à l'ancien lit mineur du Gouet. Celui-ci était totalement comblé avant vidange, sur toute la longueur comprise entre l'aval de la confluence Gouet - Maudouve et l'amont du Pont - Noir. Pendant la vidange, la rivière a remis en suspension tous les sédiments accumulés dans son ancien lit et les a transporté dans la partie aval de la retenue non vidangée où ils ont décanté.

Les épaisseurs mesurées sur le fond plat sont généralement comprises entre 30 et 60 cm et varient surtout en fonction de la morphologie originelle du fond de vallée. Les vases qui sont très fluides, ont en effet tendance à combler toutes les dépressions. Les épaisseurs les plus fortes ont été enregistrées sur le bord convexe des méandres et dans une bande d'une trentaine de mètres située dans l'axe central du lac. L'importance des dépots augmente du bas des pentes latérales vers le chenal central. Les épaisseurs les plus faibles ont été mesurées, dans un vaste secteur situé sur la rive gauche en queue de retenue, et de part et d'autre du Pont - Noir.

La concentration des vases sous eau est comprise entre 250 et 500 g l^{-1}. Elle croit légèrement avec la profondeur du fait de la compaction, mais varie surtout en fonction de la teneur en sable.

Les teneurs en eau des vases (exprimées par rapport au poids sec) étaient comprises entre 130 et 200 % pour les prélèvements effectués après vidange. Les sédiments alluviaux sablo-limoneux sous jacents aux vases lacustres avaient des teneurs en eau beaucoup plus faibles, inférieures à 35 %.

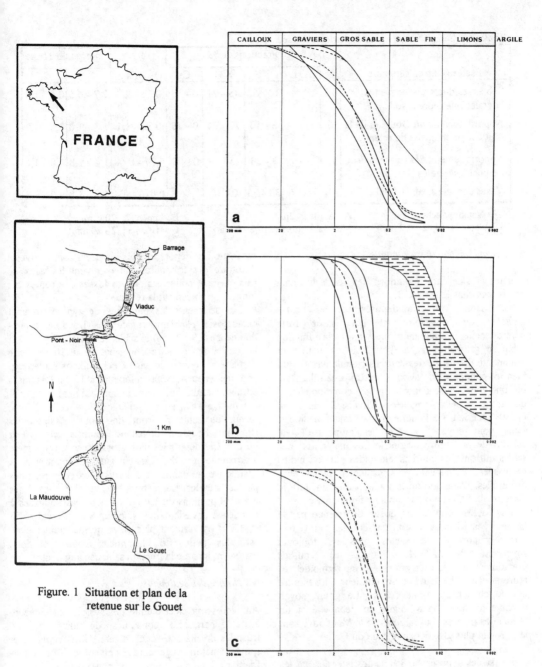

Figure. 1 Situation et plan de la
retenue sur le Gouet

Figure.2 Courbes granulométriques cumulatives des sédiments
a - Sables sur les pentes latérales
b - Sables apportés par le Gouet et la Maudouve et vases
c - Sables apportés par les petits tributaires

2669

Tableau 1. Caractéristiques granulométriques des sédiments *

Faciès sédimentologique	N	Fractions en %				Q_2 en mm	Indices de Trask	
		A+L	S F	S G	Gr		S_o	S_k
Sables résiduels sur les pentes latérales inférieure à 40 %	15	-	10 - 55	35 - 77	0 - 55	0,16 - 2,3	1,7 - 4,5	0,5 - 1,3
Apports sableux du Gouet et de la Maudouve en queue de retenue	6	-	3 - 10	77 - 94	0 - 18	0,38 - 1,1	1,3 - 1,8	1,0 - 1;2
Apports sableux des ruisseaux sur les pentes latérales	10	-	3 - 23	55 - 93	0 - 23	0,27 - 0,68	1,4 - 2,8	1,0 - 1,6
Vases déposées sur le fond	40	25 - 70	30 - 75	0 - 10	-	0,01 - 0,04	1,5 - 3,8	0,2 - 0,8

*N : Nombre d'échantillons   A : Argile (< 2µm)   L : Limon (2 - 20 µm)
SF : Sable fin (0,02 - 0,2 mm)   SG : Sable grossier (0,2 - 2 mm)   Gr : Graviers (2 - 20 mm)

4.2 Granulométrie des sédiments

Les principales caractéristiques des sédiments sont présentées dans le tableau 1.

Les sables résiduels déposés sur les pentes latérales du réservoir ont une assez forte hétérogénéité granulométrique, mais la fraction sableuse grossière est le plus souvent nettement dominante. La forme des courbes cumulatives traduit bien les conditions de dépôt (fig. 2a). Les sables qui ont une courbe à tendance rectiligne correspondent à des arènes à peine remaniées. L'indice So de Trask (1930) supérieur à 3 indique un classement médiocre. Leur grain moyen Q_2 varie beaucoup et dépend surtout de la texture originale des matériaux. Les sables qui ont des courbes sigmoïdes plus ou moins redressées proviennent d'un remaniement plus intense des arènes. L'ablation de la fraction fine améliore le classement, So est alors toujours inférieur à 2,5.

Les sables apportés en queue de retenue par le Gouet et la Maudouve sont caractérisés par la très forte proportion de sables grossiers toujours supérieure à 77 %. Toutes les courbes granulométriques présentent une forme sigmoïde très redressée (fig. 2b). L'indice So inférieur à 1,8 montre que le classement est excellent. Le grain moyen diminue lorsque l'on s'éloigne du débouché et du chenal des rivières. Les apports de la Maudouve sont légèrement plus grossiers que ceux du Gouet.

Les sables déposés sur les pentes latérales par les ruisseaux qui se jettent directement dans le lac ont une fraction sablo-graveleuse supérieure à 77 %. Les courbes granulométriques ont une forme sigmoïde qui présente une concavité vers le bas, lorsque le sable comprend une fraction graveleuse (fig. 2c). Le classement est très bon à normal et l'indice So est d'autant plus élevé que la pente latérale du réservoir est forte. Les dépôts les plus grossiers sont trouvés à l'exutoire des tributaires qui reçoivent les eaux de ruissellement collectées par les fossés des routes qui longent ou traversent la retenue.

Les sédiments les plus fins de type vase sont caractérisés par l'importance des fractions argilo-limoneuse (25 à 70 %) et sableuse fine (30 à 75 %) et la quasi absence de fraction sableuse grossière. Cette dernière a été trouvée dans 8 échantillons seulement, et à une teneur toujours inférieure à 10 %. Les vases à fraction argilo-limoneuse dominante ont une courbe cumulative d'allure rectiligne tandis que celles à forte teneur en sable fin, ont des courbes sigmoïdes (fig. 2b). Le grain moyen est compris entre 10 et 40 µm. Les dépôts les plus grossiers sont rencontrés à proximité immédiate des chenaux du Gouet et de la Maudouve, au débouché des ruisseaux et au pied des pentes latérales. Les dépôts les plus fins sont situés dans les portions du lac qui ne sont pas soumises à l'influence des tributaires. L'indice So varie entre 1,5 et 3,8 et est d'autant plus élevé que le grain moyen est plus petit. Le classement s'améliore donc nettement lorsque la fraction sableuse augmente.

4.3 Nature des sédiments

Au microscope électronique à balayage, les sédiments bruts apparaissent constitués de minéraux, de frustules de diatomées, de divers débris organiques (phytoplancton, végétaux, zooplancton ..) et d'une fraction fine qui forme un ciment (fig. 3). La proportion de ces différents constituants varie en fonction de la granulométrie. Les dépôts les plus fins contiennent une forte proportion de diatomées. Ce caractère biogène se vérifie également dans la fraction fine, par la présence de matière organique et d'abondantes particules à l'aspect phylliteux, qui proviennent de la désagrégation de frustules

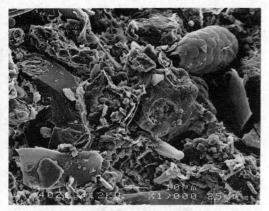

Figure.3 Vue au MEB du sédiment brut

Figure 4 Vue au MEB de la fraction fine

siliceuses (Fig. 4). En outre, nous avons souvent observé une chlorophycée, *Staurastrum gracile*, dans un parfait état de conservation.

L'examen aux rayons X a montré que la fraction fine est pauvre en minéraux argileux. Elle comprend de la kaolinite, de l'argile micacée et divers édifices interstratifiés. C'est le cortège habituellement rencontré dans des sols et formations d'altération du quaternaire récent de la région (Esteoule et al. , 1971).

La fraction sableuse comprend du quartz nettement dominant, des feldspaths, des micas et quelques minéraux secondaires. La composition minéralogique est constante et reflète la nature granitique du bassin versant.

Les teneurs en matière organique des vases sont très élevées, comprises entre 8 % et 17 %. D'une manière générale, les teneurs augmentent lorsque le grain moyen diminue. Les valeurs les plus fortes, supérieures à 15% ont été mesurées au débouché du Gouet et dans la partie centrale du lac.

5 DICUSSION

L'étude des dépots de la retenue sur le Gouet, effectuée à l'occasion de sa vidange partielle, confirme les résultats obtenus lors d'un suivi en continu à l'aide de pièges (Jigorel & Bertru 1993). L'envasement résulte des apports des tributaires, de l'érosion des berges et des fortes productions phytoplanctoniques internes. Les sédiments présentent deux faciès bien distincts, des sables grossiers propres et des dépôts plus fins de type vase dont l'origine est à la fois détritique et biogène. La répartition de ces deux faciès résulte de paramètres hydrologiques, de la morphologie et du mode d'exploitation du plan d'eau.

5.1 *La sédimentation détritique*

La courbe des débits moyens journaliers du Gouet montre sur un cycle pluriannuel, des variations très importantes (fig. 5); Aux brèves périodes de crues, succèdent de longues périodes d'étiages pendant lesquelles le débit est très faible souvent inférieur à $1 \text{ m}^3 \text{ s}^{-1}$. Les plus fortes crues dont le débit moyen journalier est supérieur à $10 \text{ m}^3 \text{ s}^{-1}$ sont peu fréquentes (2 pendant les 5 ans de la période de référence), mais apportent des quantités importantes de sédiments comme le montrent les mesures de matière en suspension à l'entrée de la retenue et les résultats obtenus dans les pièges.

Les estimations faites pour l'année 1990 indiquent que les apports de la crue de Février ont représenté environ 80% des apports annuels. Ils comportent des sables grossiers transportés par saltation (Passega 1969) et qui se déposent sous forme de bancs de part et d'autre des chenaux des deux tributaires principaux. Les sables fins et la fraction argilo-limoneuse transportés dans une suspension granoclassée, décantent progressivement dans une aire beaucoup plus vaste et se trouvent ainsi associés aux sédiments biogènes. Les microlits de sables interstratifiés dans les vases à l'amont de la confluence Gouet - Maudouve marquent les périodes d'apports détritiques les plus importants. Les répartitions longitudinale et transversale des apports détritiques, présentent en queue de retenue, une analogie avec celles mises en évidence par Hakanson (1976).

Les petits ruisseaux qui se jettent directement dans le lac déposent également des sables grossiers dans

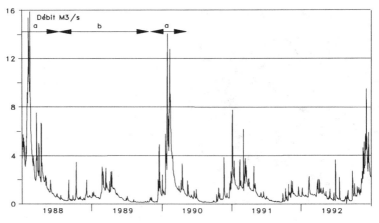

Figure. 5 Débit moyen journalier du Gouet
a - sédimentation détritique b - sédimentation biogène dominante

des cônes qui présentent une structure de "microdelta" (Reineck & Singh, 1973), lorsque la pente latérale est forte (fig. 6). Les lits sableux supérieurs subhorizontaux recoupent les lits sous-jacents à stratification oblique (pente de 40%) soulignée par des petits graviers. La superposition de deux ou plusieurs cônes à l'exutoire des ruisseaux résulte des variations de côte du plan d'eau. Les apports les plus importants ont lieu pendant les crues hivernales, alors que le lac est maintenu à sa côte minimale.

Figure. 6 Cône sableux à structure de "microdelta"

L'érosion des berges est continue, mais connait deux périodes plus intenses, la première pendant la phase de remplissage maximal du lac au printemps , la seconde pendant la vidange partielle avant les crues hivernales. Elle se traduit par le dépot de sable grossier sur les pentes latérales et par la dispersion sur le fond de la fraction fine issue du remaniement des arènes.

5.2 Les sédiments fins détritiques et biogènes (vases)

L'origine biogène des dépôts est démontrée par les fortes teneurs en matière organique, qui ne peuvent s'expliquer par des apports allochtones (Wetzel 1974), et par l'abondance des débris phytoplanctoniques, notamment de diatomées. Dès sa mise en eau, le réservoir a subi un processus d'eutrophisation qui s'est manifesté par de fortes proliférations algales en période estivale. Pendant leur transit vers le fond, les algues sénescentes subissent une dégradation bactérienne qui pourrait affecter selon Jewell & Mc Carthy (1971),   80 % de la biomasse phytoplanctonique. Dans le cas présent, la production du lac excède largement sa capacité de dégradation qui se trouve limitée par la désoxygénation chronique de l'hypolimnion. Les traitements répétés au sulfate de cuivre effectués pour inhiber le développement algal contribuent à accroître l'apport de matière organique, pendant des périodes où règnent des conditions anoxiques peu favorables à la minéralisation.

La sédimentation biogène varie donc chaque année en relation avec l'importance des productions phytoplanctoniques et le taux de dégradation de la matière organique. Le suivi effectué avec les pièges montre qu'elle est maximale de Juillet à Octobre alors que le débit des rivières est minimal (Jigorel & Bertru 1993). De longues périodes de sédimentation mixte détritique et biogène succèdent donc aux courtes périodes d'apports détritiques des crues hivernales. Cette double origine des vases se vérifie dans leurs caractéristiques granulométriques.

La figure. 7 montre que tous les points représentatifs des échantillons analysés, sont alignés sur une droite bissextrice de l'angle S. La texture varie donc en fonction de l'importance relative de la fraction sableuse fine, c'est - à - dire des apports détritiques. Ceux-ci sont maxima dans l'aire d'influence des rivières et au pied des pentes latérales (fig. 8).

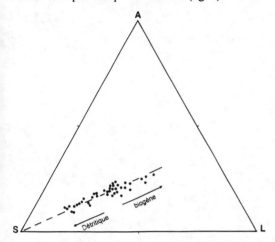

Figure. 7 Texture des vases   A: Argile (< 2 µm)
L : Limon(2 - 20 µm)
S : Sable (> 20 µm)

La succession dans le temps de phases d'apports exclusivement détritiques, puis de sédimentation mixte détritique fine et biogène se traduit par la superposition de faciès sédimentaires bien différenciés à l'exutoire des rivières. Les sables grossiers apportés par les crues et disposés en bancs sont recouverts par les vases organiques déposées en été alors que le lac est à sa côte maximale. La sédimentation biogène estivale est suffisamment importante pour ne pas être totalement effacée par les remises en suspension, lors des crues hivernales suivantes. De ce fait, de nombreux bancs sableux se retrouvent interstratifiés dans les vases.

La répartition relativement uniforme des sédiments sur le fond s'accorde bien avec l'origine biogène. Le phytoplancton prolifère dans la couche euphotique qui concerne toute la surface du lac (80 ha) tandis que l'accumulation des dépots se fait seulement sur la partie plane du fond et sur les pentes inférieures à 10 % (43 ha). La concentration des vases sur une surface réduite, favorisée par les pentes latérales, accroit le taux de sédimentation d'un facteur 1,9.

Les plus faibles épaisseurs de vases ont été mesurées dans la zone située de part et d'autre du Pont - Noir où débouchent plusieurs ruisseaux et des fossés routiers qui peuvent avoir des débits importants lors de pluies violentes. Ceux-ci

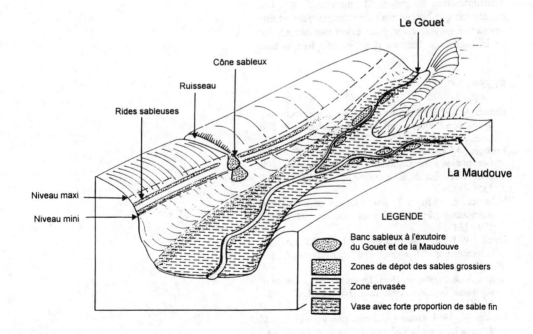

Figure. 8 Schéma montrant la répartition des faciès sédimentaires

2673

provoquent une remise en suspension des sédiments sur le fond, facilitée par la faible largeur du lac. Ce phénomène localisé se traduit par une texture plus sableuse et une épaisseur moindre des sédiments.

6 CONCLUSION

Le taux de sédimentation atteind 5 cm par an dans les zones les plus envasées et est voisin de 3 cm par an si l'on considère la surface au fond de la retenue. Il résulte pour une part essentielle des proliférations phytoplanctoniques dues aux teneurs excessives en nutriments (phosphore et azote) des eaux. La diminution de l'envasement ne pourra être obtenue que par une amélioration de l'état trophique du lac. Les mesures à prendre concernent tout d'abord la réduction importante des apports de nutriments, puis dans un second temps le curage des vases accumulées sur le fond. Celles-ci contiennent un stock énorme de phosphore (la teneur en phosphore total est de 0,25 % du poids sec des sédiments) facilement bio-disponible du fait des conditions réductrices qui règnent à l'interface eau-sédiment.

Les résultats de l'étude montrent que les prévisions d'envasement des réservoirs d'eau doivent non seulement tenir compte des phénomènes d'érosion, mais également de la sédimentation biogène dont l'importance varie avec le degré d'eutrophisation du milieu. Il faut prévoir dès leur conception que bon nombre de retenues devront être dévasées à moyen terme, pour éviter une dégradation de la qualité de l'eau et pour rétablir leur volume utile.

REFERENCES

Bietlot, A. 1940. Méthodes d'analyses granulométriques, application à quelques sables éocènes belges *Ann. Soc. Géol. Belg.* 44 : 80 - 174

Esteoule, J. Esteoule - Choux, J. Perret, P. 1971. Constitution minéralogique et origine des limons de la côte Nord- Est de la Bretagne. *C.R.A.S Paris* 273: 1355 - 1358

Hakanson, L. 1976. A bottom sediment trap for recent sedimentary deposits. *Limnol. Ocean,* 21, n°1 : 170 - 174

Jewell, W.J. & Mc Carthy, P.L. 1971. Aerobic décomposition of algae. *Env. Sci. Technol.* 5 : 1023 - 1031

Jigorel, A. & Bertru, G. 1993. Endogenic development of sediments in a eutrophic lake. *Hydrobiologia* 268 : 45 - 55

Passega, R. 1964. Grain - size representation by C - M patterns as a geological tool *J. Sediment Petrol.* 34 : 830 - 847

Reineck, H. E. & Singh, I. B. 1973. Depositional sedimentary environments. *Springer Verlag Berlin.*

Wetzel, R. G. 1974. Allochthonous organic carbon of a marl lake. *Arch. Hydrobiol.* 73: 31 - 56

Engineering geology and environment

La géologie de l'ingénieur et l'environnement

P.T.Cruz

Escola Politécnica, University of São Paulo, Brazil

ABSTRACT: The paper deals with predictions, calls the attention to the inerent difficulties in making predictions, in a world that is continually in movement. Shows how properlly done predictions, have failed in underdeveloped countries, due to political and economical problems. The importance of environment is detached.

RÉSUMÉ: Cet article traite des prévisions. Tire l'attention sûr des difficultés de faire des prévisions, dans un monde qui est en changement continue. Montre comme des prévisions convenablement faites, n'ont pas réussi dans les pays sousdéveloppés, à cause des problèmes politiques et économiques. L'importance d'environnement est detaché.

1. INTRODUCTION

Man inhabits the Earth for millions of years, but the Planet will survive man.

Within the last 100 years, that is but a fraction of second in geological time, the Earth has seem an uncontrolled process of devastation of its surface, where man is born, lives, procreates and die.

The devastation of the natural habitat of man, means his own destruction. The Planet, however, will survive.

Geology as a Science Mater, has been able to describe this devastation process, working in geological time (millions of years).

Engineering geology has denounced and admonished for the consequences of devastation to the survival of man, considering engineering times (decades).

Within this end of century and in the next to come it will be necessary to work in record times, may be seconds, to stop this process of devastation of the Earth surface, not only explaining the phenomena and the geological processes, but searching with the Engineering, radical, efective and immediate solutions.

To the times of crisis and obscuratism, like the plague, the mediaeval time, the war, the breakdown of the finantial system, a new era of social reorganization and change of values has followed.

Today we face a crisis, and a world crisis with no precedents. And it is within this crisis, and from this crisis, that it is necessary to grab, to uproot new values and a new way of live.

And because the Geologist understand the Earth, and because the Earth is threatened, we expect from them an urgent and objetive contribution.

2. PREDICTIONS

In the sixties a group of americans and venezuelans spent long months inland Venezuela collecting information on population, agriculture, clima, formal and informal economy, topography, geology, hidrology, religion, family organization, political views and political parties, and pratically all aspects of life inland Venezuela. All this information was properly organized and computorized.

An artificial flood, associated with the failure of a dam in a mountaneous region was created, in order to predict the consequences.

The system produced daily papers written by the computer, including losses of lives and properties, the extension at the flooded areas, agricultural losses, and also reports of people opinions and feeling, articles written by local politicians, workers, farmers, engineers.

The project, develloped at MIT, was supported by the "Aliança para o Progresso", and had the purpose of making predictions, using Informatica resources.

In the sixties the Cold War was a world affair and differents efforts, in differents countries, with differents methods, and differents finantial supports were made to Predict what could happen if a small country, or a big country all of sudden decided to change his political system, and or if a natural disaster would occur.

Within the last 40 years the world has changed so drastically, that it seems, that in spite of the very sophisticated resources to make predictions, Man has never knew so little of what is coming, and how his children will face the next century.

The so considered endless resources of energy, water, forests, mineral, and raw materials, used without any consideration of economy, by the consumption society, are giving signs of fatique or failure.

It seems that no one was allert in 1960, that in 1987 the year consumption, per capita, of "steel" could reach more than 400 Kg in the USA, West Germany, Japan and Sovietic Union; that the average cost of the World Propaganda per capita, in 1990, would aproach 50 US dolars; that the americans in every year dispose of 180 millions of razor blades, plastic plates and cups, and enough paper to wrap 6 sandwichs for the world population, and enough aluminium tins to build 6 thousand DC-10; that the japanese throw away 30 millions of photografic cameras per year, and that the english discart 2,5 millions of children diapers.

The consumption economical society that has made the USA what it is today, and that was extended to Europe and Japan after World War II, is close to bankruptcy.

Today 8 millions americans and 4 millions germans are unemployed.

The world division in rich and poor nations, developed and underdeveloped contries, nations that rule their own economy, and nations that depend on the economy of others, have some basic differences, that must be consider when predictions are made.

The statistics show that 3 million brazilians workers are out of job. If we consider the population of Brazil, in comparison with the population of the Unites States and Germany, the percentage of unemployment looks the same. What is not mencionated in the statistics is that there are another 50 to 60 million brazilians that receive less than US$ 65,00 per month, and that they are unable to provide home, food, education and health for themselves and their families. If we add this figure to the 3 millions workers, we will have almost 40% of unemployment in Brazil.

One of the consequences of this fact is that the governamental tax in home applyances, construction materials and food are much higher in Brazil than in the developed countries.

A recent earthquake in Los Angeles, made 40 or 50 victims, and left 30 or more billions dollars of material prejudices. Another earthquake of the same magnitude in Sumatra, left 134 death and 1.000 wounded, half in serious conditions. There is no notice of properties losses, but in no case they would add to much more than 1 billion dollars.

Another problems reffers to the political regimes that prevail in different countries.

Out of my 60 years of life. I lived under dictatorship from 1934 to 1945, and from 1964 to 1985, 53% of the time. We are living today under a democratic regime, but no one can tell what is coming within the next 60 years. The USA has lived under a democratic regime for more than 200 years. The same can't be said about European Countries.

Predictions are a complex matter.

In the beginning of the automation process in industry, there was a general fear of massive unemployment. The hope was that the workers required to build and develop "robots", would surmount the losses in jobs of unskill people, and that the salaries would raise.

And that relly happened during a certain period, but ins't the actual situation.

The Brazilian Volkswagen has produced in 1993, more cars than in his best years of the past, but only a few new job oportunities were offered.

Another astonishing figure is the fact that more than 50% of the jobs offered today are in business that were unexistent, 40 or 50 years ago.

The very rapid development in tecnology and the so called post-tecnological society, will face this situation in a much shorter period of time.

Predictions in spite of all the very sophisticaded programs that can be used, are in one side necessary, but in the other side subject to failures. A few examples will ilustrate this affirmative:

An hydroeletrical plant, was built in Angola, with the prediction that energy would be supplied for many years, even if the country would show the same rate of growth. An internal war has changed the situation drastically.

The Assuan dam, built in Egypt to provide eletricity, had a negative impact on the agricultural production, along the historic fertile margins of the Nile. Very long and expensive canals were open to compensate for those losses, trying to conquer new agricultural land, but the results were poor.

A program in China to provide food for her fantastic population, include plantations everywhere, even in steep slopes, never planted before. A generalized erosion developed, and after some years the total production of grains was reduced.

The Trans Amazonic road, built in Brazil, during the military regime, included a pretentious agricultural program. The intention was to provided areas of cultivation, for the people of northeast Brazil, submited to long periods of dryness. The forest in a few years reconquered its space, and the agrovilles were rapidly destroyed.

Many african countries are facing internal wars, for wich armament is supplied by the "rich and poor nations of the world". The U.N. has interfere in these wars, suppling food, but also armament to the peace troops. If the money spent in arms and destruction was used to supply food, education and health to these countries they would be much better off.

Now what is wrong in these projects? Lack of tecnology? Inadequate projects? Construtive problems? Doesn't seem so. In all case the very modern tecnology and machinery were employed and the cost of these enterpraises were always high.

Are predictions wrong? It looks so.

Engineering prediction are made out of statistical records of natural, economical, and social data. But very rarely they consider either political instability, and or the overall impact of the enterpraise.

I want now to consider two different models or criteria for predictions:

Wiener, in the sixties, when at MIT introduced a new branch of science: The Cibernetics. By that time almost no one could understand what he was talking about. One very curious contribution of Wiener, was a criteria to identify geniality. In a simple graph he proposed to put in the abcissas the amount of available knowledge of a specific field, and in the ordinates the corresponding cronological time (Fig 1).

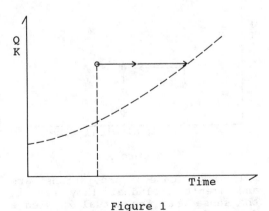

Figure 1

A genius would be identified by his position in the graph. The genius would be always some years or some decades ahead of his time. Leonardo da Vinci, Julio Verner, Beethoven, Jesus Christ, Van Gogh, Terzaghi, Einstein, and of course Wiener, could be easily identified.

A Second example reffers to the very actual Plate Theory, or what is also called the New Global Tectony.

Hasui (1992 - in Geologia Estrutural Aplicada - ABGE-IPT) summarizes these new conceptions:

"Important contributions were made related to the expansion of the oceans, and the derive of continents. Works on the transforming zones of great structures in the global sphere, continental margins, global sismology, magnetic anomalies, paleo magnetism, geo-termal flow, magnetism, metamorphism, orogenesis, etc.

"This volume of data, quickly delineate the litosphere plates, their limits, velocities of displacement, geometric and dinamic characteristics and the mezozoic-cenozoic evolution, consubstantiating the existing data as well as clarifying and unifying the interpretation of general features of the ocean and continents. The Theory of Plates was borning, also called "New Global Tectony".

Geology as well as Antropology, and sometimes History seems to work in the opposite direction. That is, they use the present knowledge, to return to the past, and explain the present. See Figure 2.

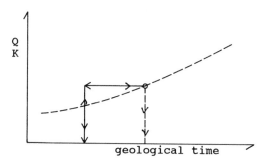

Figure 2

In both cases, predictions are and aren't included. They are in the sense that the actual knowledge for genius, allow them an extrapolation to the future, and very important qualitative steps in science. The Plate Theory can be ex-

tensively used to predict earthquakes, large structural massive movements, and continental displacements.

3. ENVIRONMENT

Environment is now an overall cry in the world, and a unparallel amount of money has been spent in trips, talks, papers, books, courses, academic work, congresses, symposiums, workshops, conferences, and in all branchs of science and overall knowledge, "new experts are being produced". We have environmental geology, environmental engineering, environmental medicin, environmental pharmacy, environmental economy, environmental turism, environmental industry, environmental preservatives, environmental political parties, environmental societies, groups, armies, environmental religion, and so forth.

What we have above all are Environmental Predictions, the majority of them disastrous, catastrophics, alarming, terribles, conusing, as for example the Introduction to this paper.

The proplem is that nothing is wrong with those predictions. They are far more correct than other type of predictions related to economy growth or internal markets; unemployment in industry, inflation, oil production, or cost of wars.

The problem is that environmental conditions aren't properly considered in large enterprises that are mainly concerned with specific itens, like the plantation of soya beam, hydroeletric power, space-war weapons, or TV programs.

The genious could foresee the future, in his own, and many times ermetic brain; the Plate Theory work with geological times, that may be inadequate for men's life.

What I want to emphasize is that either in engineering projects or applied geological issues, men cannot anymore be enclosed in his own sphere of knowledge, but any predictions that are made have to absorb many other aspects of knowledge, and its environmental.

But, in the other side environmentalists can't disconsider the working tools of practical and applied sciences.

Predictions are in today's world, milestones for the human wellfare in the future.

Environmental sciences are leading the way in this aspects, but their cries and lamentations alone are not enough.

The present economical order of the world, may subsist for some time, but clear signs of deterioration are visible and if these signs are not properly consider in any predictions, whatever we do, has a good chance of having the same destiny of our irresponsable consumption society.

REFERENCES

Cruz, P.T. 1993. Geologia de Engenharia e Meio Ambiente. 7º Congr. Bras. Geol. Eng. Vol.1.

Hasui, Y. 1992. Geologia Estrutural Aplicada. ABGE-IPT

Candotti, E. 1992. Reflexões e Refrações de uma Eco. Revista Estudos Avançados. Nr.15. Vol.6 - USP

Duming, A. 1991. State of World.Worldwatch Institute, Washington D.C. (also in Technology Review - MIT)

Daily papers and Magazines.

Geotechnogenic structures as a new approach to the working out of the solutions on the foundations arrangement

Structures géotechnogéniques, une nouvelle approche des solutions des systèmes de fondation

B. N. Melnikov & I. V. Cherdantsev

Institute of Geology & Geochemistry, Ekaterinburg, Russia

Annotation:Here it is described methodology of elaboration of the technical solutions of the underground parts of buildings arrangement, as the systems of changing the structure of the initial ground massives into the nature-technogenic (geotechnogenic) structures, capable to apprehend and distribute all the additional outer influences as distinct from the traditional foundations systems. It is formulated the main idea and principles of structures creation means of their estimation and the ways of their regularity. It is noted the examples of effective and rational types of structures, substantial economical expediency of their application is shown.

Résumé:La metodologie d'elaboration des resolutions techniques de l'arrangement des parties souterraines des batiments comme des systemes de changement de structure des massifs terrains initials en structures naturelles technogeniques (geotechnogeniques) qui sont apte a percevoir et repartir toutes les influences complementaires exterieures a la difference de systemes traditionels de batiments est montrée. L'idée essentielle et les principes de formation des structures, les moyens de leur evaluation est formule ici. Il y a les examples des types effectues et rationnels des structures; la rationalite essentielle economique de leur application est montrée aussi.

When working out technical solutions of the underground parts of buildings arrangement, it appears a necessity of considering multiparameter and multilevel interactions of technogenic objects and geological medium. It defined the necessity of seeking new approaches to their elaboration. One of such approach is the creation of geotechnogenic structures. The working out of their technical basis is carrying out jointly with acad. V.I.Osipov and is basing on the considering of acad. V.I. Vernadsky ideas about noosphere.

The main idea of creation geotechnogenic structures, including qualitively different objects - geological and technogenic - is the maximum realization of the effect of mutuel work of all structural elements, taking into consideration their heterogenousness, multilevel scheme of composing and multifactor character of development.

For its realization it is formulated the following main principles of the technical solutions elaboration on the arrangement of geotechnogenic structures and their optimization.

1. All technogenic influences upon the nature massif - mechanic, heating, elec-trical, the substantial content changing, and as well the requirements on the arrangement of the underground rooms, commu-nications running and so on have been including in a common system, which is considered as the additional one to the system of the nature interactions.

2.Geotechnogenic structure for the whole period of its elaboration, arrange-ment and exploitation has been regarding as the single whole- the system. Any mea-sures on the engineering preparation of territories, ground foundations, the ar-rangement and erection of building con-structions are determined as the measures on the structure regulation of the ground thickness.

3. With the usage of the worked out by us classification of rational types of structures from the point of view of per-ceivement and distribution of outside influences in natural massif there have been placing such structures or their fragments,and measures on their completion have been working out.

4.Regulation of the massif's structure includes the removal of the redundant ele-ments, the change of grounds characteris-tics in some volumes of space, creation of the additional elements.

5. The solutions on the geotechnogenic structure arrangement include measures on protection of their natural and technogenic elements.

In such setting geothechnogenic structures seem to be rather complex systems. Special works are devoted to the rendering of researching methods of these systems. Here we shall stop only at some moments of these methods.

Out of many peculiarities characterising heterogeneity of object we shall mark that fact that the massif's areas differing by density, strength and other properties at one and the same influences change differently. As far as these areas form the common system, natural influence upon each other takes place. Thus the changing of properties of the separate volume of the massif displays itself by two effects – the effect, connected with the marked changes of characteristics and the effect of mutual influence of this volume and the space surrounding it. The intensity of the first effect displaying is more considerable than of the second one. However it's action manifests itself only in the limits of the volume of properties changing. The second effect is less considerable on intensity, as it is secondary.However its volume of displaying is sufficiently more than of the first one. The task of changing of the massif's structure is in the searching of the most rational combination of the marked effects.

It is clear that when the massif's homogeneity is increasing, the displaying of the second effect is decreasing. It determines the irrationality of the regular properties' changing of the whole massif, its fastening and makes it expedient the putting in order its heterogeneity character,that is the regulation of the massid's structure. Thus the sufficient change of the massif's properties can be reached by the local strengthening of the ground, that is by the creation of rigid inclusions, which volume comprises 1/10 - 1/25 of the common massif's volume. At the maximum counting of the massif's multievelness and creation of different volume and form inclusions the relative volume of inclusions one can decrease up to 1/30-1/200, and in case of frame-alveolate alveolate structure creation the same effect is obtained at the relation volume of inclusions 1/200 -1/500.

At the model presenting of heterogeneity it is expediently to consider the

massif's space as the system of matrix and inclusions. According to correlation of rigidities the inclusions can be both of bigger and smaller rigidity than the matrix is.

Around every inclusion it is distinguished an influence zone, in which the interaction of matrix and inclusions is realizing. In the influence zone it is displaying the stabilizing role of inclusion to the matrix. The size of the influence zones is determined by the matrix properties, the size and the form of inclusions. When making the form of inclusion more complex the relative volume of the influence zone is increasing at the expense of rizing the inclusion surface and decreasing at the expense of overlapping of the influence zones from its different parts. At the practical elaborations these dependences determine the ways of searching effective forms of inclusions.

The degree of stabilizing influence is determined by the relation of the deformation inclusion modulus a nd the matrix – by relative inclusion modulus. One can make out the active interval of these moduli changing in the limits of which their influence upon the appropriate moduli of the whole system is sufficient. The establishing of this interval has a great practical meaning. So make artifical inclusions is expedient with such moduli, which quantitative meanings do not exced the given interval. The expenses connected with reaching of higher moduli are needless, as they are practically don't exercise any influence upon the properties of the whole massif.

At the presence of many inclusions , including differing on content, type, form and other features, it is expediently to consider the effect of joint work of all the massif's elements (of one structural level) with the help of structural nets presented by three kinds of space : the matrix, the inclusions and the influence zones.

Out of all the variety of questions connected with the research of the multilevelness of massives, as the examples we shall mark their peculiarities defining correlations of different levels. Let's imagine the following hierarchical row of geotechnogenic structures:fields /n-I=n1/; massives /n-2=n2/; blocks of massives /n-3=n3/; nodes /n-4=n4/; details, blocks of rocks /n-5=n5/; aggregates /n-6=n6/; mineral parts, monomineral grains /n-7=n7/. Bearing in mind the difficulty of interaction in massives the position of many structural formations can be defined on some variants. For example, the parts of geotechnogenic massif - n2-I, n3+I ; the nodes - n2-I, n2-2, n3; combined blocks - n2-1, n3+1; macroaggregates - n5-I, n6; microaggregates - n5-1, n5-2, n5-3, n6 and so on.

Let's note some examples of matrix and inclusions interactions in the multi-level structures and the ways of their using in practical elaborations.

The surface, lower of which the additional tensions connected with the outside loadings are less of some value can be taken as a roof of rigid element. Stabilizing influence of late upon the overlying thickness is expedient to use in practical elaborations. How to do this ? If to carry out the underground part of building and its foundation as the horizontal bearing element, then it may be considered as the inclusion with its own zone of influence. Thus we have two inclusions - the upper one (horizontal bearing element) and the lower one (underlying thickness of the ground) between which it is located the intermediate layer of natural ground, suffering the stabilizing influence of the pointed inclusions.This scheme corresponds to rather effective three-layered construction widely used in technique. With the help of the marked scheme the properties of the underlying thickness of the ground are being used as much as possible. Changing the properties of the intermediate layer so that in a fuller way to use the stabilizing influence of the underlying thickness and horizontal bearing element one can come up to minimum the expenses on its arrangement.

Here is the second example. In mining and building mechanics it is long ago known and widely used the vaulty (archy) effect. In the ground massif with the elements of different modulusness vaulty effect manifests itself in case when the structure character provides the resting of arch on the more rigid elements, and the correlation between the thickness of the overlapping strong layer and the distance between the elements of resting on provide the vault formation.

Both pointed structures are three-layered constructions and the vault systems have been applicated by in many building objects.

The given examples are refered to the structures of the second order - to the massives / n-2=n3 /. Let's consider the examples of the third order structures - geotechnogenic blocks /n-3 =n3/.

We have marked before that the difference of the deformation moduli of the contacting bodies should not exceed some limit. Depending on the type of structural scheme of massif and the form of separate bodies this limit changes in the interval of 5-100. So it is expedient to arrange inclusions and elementary alveoles of spectral type, when strong elements are framing by less strong ones. For example, on the central axis of element is located metallic or reinforced concrete bar. Concrete block is placed around it, in the edge parts of which it is situated soil-concrete, then it goes the coating of their compact ground, which gradually comes into the body of incompact ground. As the examplex one can give the injectional piles of calcination, piles of underpressure, etc.

When working out of economizing decisions a great importance has the finding of the effective form of inclusions, among which may be marked linear, laminar, cross-like, three-laminar and others. It is important to bear in mind, that the form of inclusions and other bodies of the ground massives is one of their structural scheme regulating, so it is impossible to consider the bodies'form in loosing touch with the spacial interposition, correlation of deformation ability moduli of contacting bodies, sizes of influence zones, their superposition upon each other All these features it is expedient to take into consideration at the example of the elementary cells of structure, which combination determine the common spatial massif's scheme.

In practical elaborations it is important to take into account the variety of interactions of the massif's elements with each other, multifactorness of their manifestation. For example in the ground of the crust weathering over the location areas of bedrocks ridges physical-mechanical properties ofthe grounds are usually higher that at the areas of the nest pockets. At the same time the ridges are the areas of deformations spreading, and the nest pockets are the areas of the counter streams. So in the nest pockets the influence zones of the massive rocks are rather considerable and the arrangements on the massif's increasing by means of its structure regulation are rather effective.

The packing of the ground by impact loads can promote to the destruction of structural bounds and discompaction of the ground in the surrounding zone. It is especially seen in territories of the massive grounds spreading at some depth, which play the role of the reflecting screen for the shock waves. In some cases as such ones screen can serve as the level of ground water.The use of reinforced concrete or metallic elements in the structure of the massif can considerably change the scheme of distribution and the intensity of stray currents, that is necessary to take into account when building electrolysis departments or near electric transport routes.

Each factor of interaction of the massif's elements it is expediently to distinguish separately as functions, which we determine conditionally as functions of interaction. This system also includes a great number of purpose functions, which are determined by the building object. Taking into account this system are being fixed the decisions on the regulation of the natural grounds structure at the ar-

rangement of geotechnogenic massif.

We have marked only some peculiarities of geotechnogenic structures as complex multiparameter systems. They have been considered at the empirical level. For analytical research of such systems it is elaborated a conception of structural space.

Separately taken or the interacting massives of rocks and technogenic objects are considered by us under the common term "structural space". Each object or phenomenon appears before us as a certain volume of space which the studying object is taking or in which the considering phenomenon takes place. The display of structural space peculiarities as the objects or the phenomens are called of its realizations.

Among realizations of structural space there have been making out structural realizations and realizations of exchange. The features of structural realization determine those peculiarities of the studying objects, which at the given moment

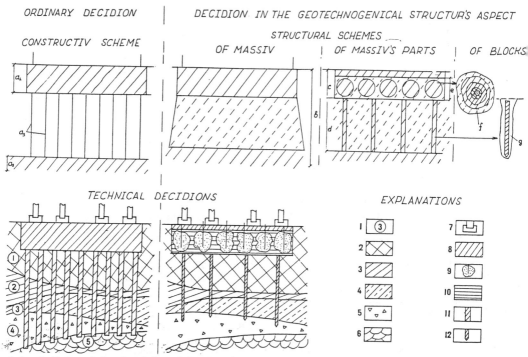

a1-hard disk, a2-hard slab, a3-hard pivots, b-three-layered structure, c-two-layered structure, d-pilar-cell structure, e-frame-cell structure, f-isometrical structure, of spectral type; g-lining structure of spectral type; 1-number of ground, 2- technogenical grounds, 3-deluvial loam, 4-alluvial loam, 5-eluvial loam, 6-rocky ground, 7-ferro-concrete structures, 8-ferro-concrete, 9-injectioning body, 10-solid ground, 11-drilling-stuffing's pile, 12-ferro-concrete pile.

of time are not connected with their changes. The degree of order of these realizations in the long run is presented by combination of fragments of four types of structures: of chaos, the structures : of chaos, the structures of spiral (turbulent) type of structural nets, of closed structures.

The signs of realizations of exchange characterize the peculiarities of realizations changing at t -> 0, where t is time. It's distinguished the following types of the degree of order of exchange realization (distribution effects) : bundle distribution, distribution effects of the ring, cylindric and globe yokes, etc.

The working out of practical decisions on basements'arrangement is enclosed in the finding of rational combination of the degree of order types of structure and exchange realizations. On the base of such combinations it is woorked out and used in the practice of construction some economic systems of the basements: metastructural basement, carcass and alveolate structures, three-layered systems, the structures of spectral type and others. Example of elaboration of the technical solutions as form of the geotechnogenic structure showed on the figure.

Geotechnogenic structures are widely enough applied in the practice of construction, with their usage it is erected more than 120 objects. The application of geotechnogenic massives at present time allows to obtain the decreasing of the labour expenses for 25-30%, energetical and material expenses - for 18-45%. When working out the technological aspects, and as well at the elaborating of special machines and mechanisms it is provided the lowering of the power and material expenses in 2-2.5 times.

Effect of human activities on the abrasion and sliding progress at waterwork Nechranice

Effet des activités humaines sur l'érosion et les glissements de terrains sur les rives du barrage Nechranice

Tamara Spanilá
Institute of Rock Structure and Mechanics, Academy of Sciences of the Czech Republic, Czech Republic

ABSTRACT: The construction of water reservoirs causes a sudden and profound alteration of hydrological, hydrogeological, geomorphological and engineering and geological conditions which triggers the dynamic developments especially in the banks area and its neigbourhood. Dams as man - controlled structures are still exposed to important effects of various natural, in particular, hydrometeorological factors. Therefore any waterwork should be conceived as complex natural and technical systems and any neglection of this fact could result in large or considerable damages. The attention is paid to the development and significance of typical exogenous processes which take place in loamy, sandy and clayey sediments on the banks of Nechranice Water Reservoir.

RÉSUMÉ: La construction des barrages mène vers les changements profondes et inattendus des conditiones hydrologiques, hydrogéologiques, géomorphologiques et ingénieur - géologiques qui provoque les processus dynamiques dans la zone litorale et dans leur entourage. Les barrages sont des chantiers dirigés par l'homme, mais quand même ils sont influencés par les facteurs naturels et avant tout hydrométéorologiques. Les ouvrages hydrauliques, on doit les comprendre comme les systèmes naturele compliqués et techniques. Négliger ce fait pourrait mener vers des dommages énormes. Cette oeuvre prete l'attention au développement et à l'importance des processus typiques exogènes qui se déroulent dans les sédiments terreux, sableux et argileux sur la rive du barrage "Nechranice".

INTRODUCTION

Large waterworks, especially dams, represents serious interferences in the natural environment thus contradicting as a rule, to the principles of its protection.

Nevertheless, whilst waterworks are man-controlled structures, they can not be taken out of the natural environment. They are still exposed to important effects of various natural factors, the significant role among them being played by those hydrometeorological by their nature. Therefore, as the subject of research, utilization and management they should be covered both by natural and technical disciplines, including the projection of the potential future ecological destabilisation of the territory concerned.

If a waterwork is conceived from the point of view of simultaneous operations of human and natural factors, it can be utilized more rationally, whilst any neglection of this reality could result in large and/or considerable damages.

In general, the effects of waterworks on the natural environment could be classified as favourable and unfavourable, direct and indirect. This work concerns the direct and indirect effects exerted by the Nechranice Water Reservoir ("NWR") on the commencement of land-sliding movements within its banks zone and devastating processes of the surrounding area caused by unreasonable human activities.

NECHRANICE WATER RESERVOIR

NWR was built with the primary goal to provide industrial and power generation facilities with supply water and, in the second place, population with drinking water. The dam is also used for direct power generation and recreational as well as sport activities. Another important task of NWR consists in protecting territories under the dam from floods, in improving the quality of water under the dam, etc. NWR as the largest earth dam belongs also to the biggest water reservoirs as to surface area and the volume of dammed water in the Czech

Republic (see Picture 1).

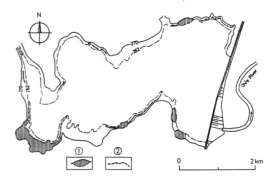

Fig.1 Inundation zone of the Nechranice reservoir, with different deformations on the banks. 1 - landslides, 2 - abrasion.

On the top water-level of 273.05 m above sea the surface area totals almost 14 km², the volume of dammed water 288 million m³ and the bank line ca 20 km. The dam, put into operation in 1968, is located in the north-western part of the country. The dam on the Ohře River was built under very unfavourable geological conditions without any technical defects, however, certain secondary effects appeared in a scale not predicted by the results of the geological survey.

The inundation area of the dam, through which an important tectonic fracture runs, consists mainly of the tertiary rocks, to a lesser degree, by crystallinicum rocks and, sporadically, by Permian and Cretaceous sediments. The oldest rocks are represented by gneisses, mostly heavily affected by caolinic weathering. The relicts of Cretaceous sediments are represented by conglomerates and by block, caolinic and glauconic sandstones.

The tertiary rocks are mostly rocks of vulcanic and sedimentary series. Vulcanogenic rocks can be found in the inundancy area in a narrow zone only where agglomeratic tuffs, tuffits and tuffit clays prevail. The sedimentary rocks are represented in the larger part of the inundancy area. To the surface mostly the sediments of an overlaying series of silty clays through argillates come out, whilst in a narrow zone along the main tectonic fracture they are substituted by a lignite overlaying series.

The inundancy area and its wider surroundings provides by its geological structure and physical and mechanical characteristics of tertiary rocks favourable conditions for the occurrence of landslides. Their commencement and development there is connected, in particular, with the erosive activities of the Ohře River. Another factor favourable for landsliding consists in the presence of two rock aggregates with different physical and mechanical characteristics placed one on the other, e.i., quite permeable pebble gravels and gravel sands and almost unpermeable clays of overlaying or, as the case may be, interlaying series. Thus, the dam and the larger part of the inundancy area are located in the area with engineering and geological conditions extremely unfavourable for the stability of the bank zone.

The fluctuation of water levels, and its amplitude especially, result from the purposes for which NWR was built. The average amplitude of water-level fluctuations for the whole period of operations ranges in spring between 0.61 and 14.82 m, in winter between 0.74 and 14.22 m, in summer the fluctuations are small and ranges around 2 m. The yearly amplitude of water-level fluctuations on NWR for the whole period of operations ranges between 1.56 and 19.84 m, thus involving up to 45 per cent of total dammed water.

The water level almost permanently fluctuates up and down with a speed ranging from some centimeters to some dozens of centimeters or even meters per day.

On the basis of engineering and geological survey the bank zones threatened by landsliding and abrasion were identified. Some dangerous zones, critical for the body of the earth dam were fortified in advance. Other parts of banks were permanently monitored.

Long-term observations of the banks deformations enable to correct the assumptions on which the original projections have been based, to forestall catastrophic developments and to evaluate major factors causing deformations. On the basis of regular measuring the fortification works on new landslides could be started in time.

The banks of NWR are exposed to landslides, abrasion and accumulation. As a matter of fact, only landslides and abrasion are important from the point of view of banks deformations. The scope of individual types of banks deformations is shown on Picture 1. Landsliding movements take place, first of all, on the banks consisting of miocene sediments and as only exceptionally, on the banks consisting of Permian and Cretaceous rocks.

One of a few factors on which stability of banks is conditioned is abrasion. Active abrasion in the upper part of the bank contributes to its stabilisation, however, if it takes place at the foot of the abrasion cliff with low original stability or directly in the underwater abrasion platform. Large range of water levels (yearly, seasonal, daily) have identified a sufficiently wide bank zone exposed to the permanent effect of water-level fluctuations. The recession of the banks caused by abrasion is measured regularly (Picture 2). Almost one half of the ters long, retreated in some sectors by up to 100 meters. The abrasion is most active on north-west oriented banks which are the most exposed to the strongest and most frequent winds. A decisive period

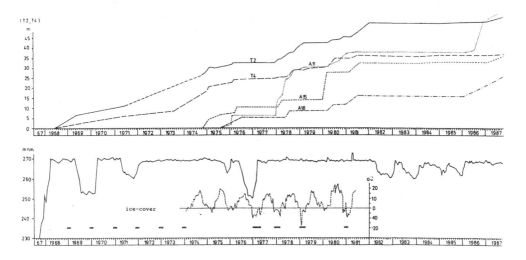

Fig.2 Effect of water level fluctuation and climate factors on the abrazion progress at the waterwork Nechranice.

banks line length suffers from exogenetic processes. A zone where abrasion and local landslides take place, which had represented before 1989 a bank line of about 2 000 meters for the development of abrasion is winter, if the dam is not ice-bound. The most interesting landslide occurred in the tectonically unstable area on the left

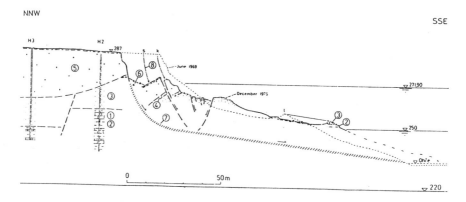

Fig.3 Geological profile of landslide (by Rybář)
Permian formation:1.sandstones, 2.alternation of claystones and sandstones; Cretaceous: 3.kaolinic and clayey sandstones, 4.glauconitical sandstones; Tertiary: 5.bazalts, tuffs; Quaternary: 6.debris, 7.sliding plane, 8.components of points s, k, shifting

bank of the water reservoir. The slope was impaired after the first considerable lowering of the water level. The water level moved at a maximum speed of 2 meters per month, whilst the total decline amounted to almost 20 meters. The slid-down slope consists in its upper par of basalts and basalt tuffs. These volcanits overlay the remnants of Cretaceous glauconitic sandstones, with additions of clays in the base. Permian sandstones, claystones and silts. The probable depth of sliding surface is about 40 meters (Picture 3). The engineering and geological survey of this landslides has shown that the danger of major extension of deformations over the upper edge of the slope does not exist, if a certain

of operation of the dam which is based on the projected goals of NWR utilization.

However, an unreasonable permission for building a large recreational cottage campus was issued covering an area endangered by landsliding movements after the dam was put into operation. The building works and the cottages built have activated the landslides. At present, some of the structures and plots of land, located on the surface of the active landsliding zone, have been already destructed or imminently endangered (see Picture 4). This is the result of indirect influence of the water reservoir on the deformation of banks and devastation of a large surrounding area.

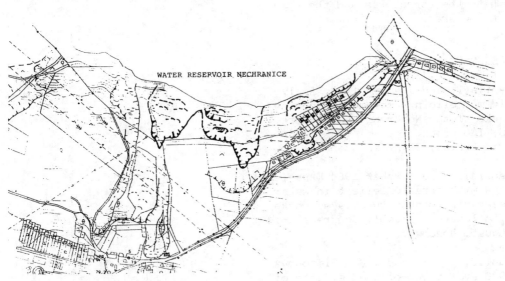

WATER RESERVOIR NECHRANICE

Fig.4 Situation of sliding area

limit of lowering the water level and, in particular, a certain pace of discharging water from the reservoir is not exceeded.

The described types of banks deformations have been caused by the way

CONCLUSION

When building a waterwork it is necessary to focuse on its incorporation into the natural environment, whilst paying the exclusive atten-

2691

tion to a one-purpose structure
from the point of view of the lo-

west investment expenses spent on
the building of the waterwork and
an insensitive approach to its operation would not be reasonable.

The results of monitoring the banks
deformations of dams clearly show
that the consequences of such process are more serious than those
normally projected within the framework of solutions proposed by water managers.

Unreasonable human, either direct
or indirect interferences can lead
to large deformations of banks resulting in large damages incurred
both to the structures and the surrounding area.

REFERENCES

Spanilá,T. and J.Rybář 1983. Analysis of sliding movements on the
banks of the Nechranice waterwork,
MS, IGG Czech. Ac.
Sci.,Prague, 33pp.(in Czech)

Spanilá,T. and J.Rybář 1984 Experience with abrasion effect in water
rezervoir banks in clayey rocks.
27th Internat. Geol. Congress,4-14
August, Moscow.

Spanilá, T. and G. Simeonova
1989.Evalution of the effects of
technogenic factors on the conversion of water reservoirs banks,
Vodní hospodářství 39, 3, 78-82 (in
Czech).

Techno-environmental system: 'Structure-geological medium'
Main subject for investigations of the engineering-geological science

Système technique de l'environnement: 'Milieu géologique-structure'
Thème majeur de recherche de la géologie de l'ingénieur

Igor A. Parabuchev
'Hydroproject' Institute, Moscow, Russia

ABSTRACT: The paper deals with the function of the engineering-geological science in solving the problem on optimization of establishment and functioning of complicated techno-environmental system on the basis of the analysis of the experience gained in designing, construction and performance of major hydroelectric projects in the former Soviet Union and Russia from the point of view of safety of scheduled structures and need to minimize the potential damage to the nature (environment). The most effective ways and methods are substantiated to solve this burning problem.

RÉSUMÉ: Dans le présent rapport, sur la base de l'expérience acquise dans le domaine d'études de construction et d'exploitation de grandes ouvrages hydro-électriques dans l'ancien URSS et en Russie, on examine le rôle de la génie géologique dans la solution du problème d'optimisation de la création et du fonctionnement des systèmes technogènes tant au point de vue de la sécurité des ouvrages futures que dans le but de réduire au minimum le dommage probable à l'environment. Sont justifiés les moyens et les méthodes les plus efficaces de la solution de ce probleme actuel.

1 INTRODUCTION

The paper, presented for your kind attention, has, first of all, to be considered as an effect to find the ways to solve the ever going continued contradictiones between the real problems for construction of modern engineering structures, combining undeniably design safety, the high economic effectiveness and facilitation of the minimum damage to the environment on one side and on the other - obviously out of time traditional methods of their solution. The thing is, that it is impossible to create the best from the feasibility and social-ecological points of view such complicated techno-environmental system, which, for example, is taking shape in the process of construction of hydroelectric projects by impruving separate elements of the unified technological procedure: investigations - designing - construction - performance. Their interaction and interrelation are so closely associated, that the optimisation, for example, only the engineering-geological investigations themselves, being the first link of this chain, although may bring in certain positive effect, but not shurely shall impose the

desired effect to the other links and may give birth, in the begining, to the best design and then - to the improved structure. That's why, in our idea, the only promising way to solve the problem - is to consider it in association with all interacting elements of the created bu us, when designing, construction and performance of the techno-environmental system, making it, on the basis of estimation of the origin and scale of these processes, which may develop as a result of such interaction, adopted on the basis of the said forecast of design concepts and immediate chance for their adjustment during the construction. Certainly such an approach is far and away from the limits of conventional engineering-geological tasks and requires the reconsideration of the place and the role of an engineering-geologist in building up and functioning of techno-environmental systems. For sure it should involve the change of main objects to be investigated or in the respective methods of their examination. The possibilities to find the best solution to this problem are governed by a number of issues, the consideration of which is the main goal of this report. We"ll try to tackle upon them thoroughly in application to the engineering-geological investigations for the aims of hudroelectric projects construction.

1 FIRST ISSUE

The first, out of the considered issues, which is the most important, may be read in the following way:

During the execution of engineering-geological investigations to substantiate the construction of hydroelectric projects the main task of them is to obtain the scientifically supported estimated information on the qualitive and quantitive characteristics of the future techno-environmental system, which come to exist as a result of interaction of the designed structures with geological medium in the process of their construction and performance.

The main idea of this issue is to change the widely spread point of view for the geological medium as a passive element of engineering domestic activities of human beings, which inevitably breaking the established equilibrium with interaction of natural elements, shall bring in this or that state of degradation of the medium frought with all negative ecological consequences. But in life, as proved by the experiance gained, the geological medium does not cease to exist, it is intensively associates with other effects, trying to regain the lost equilibrium, but with involvement of the newly built man made element. The new system comes to exist as a result of the said: geological medium - structure, the origin of interaction of elements of which, their qualitive performances and estimated development with time should be the goal of the investigations undertaken by an engineer-geologist. Having anderstood the problem this way the conventional investigations of the engineering-geological conditions of the project area, become only the first, although a very essentive phase of engineering-geological investigations, on the basis of which, with the necessity to have an appropriate proffesional knowledge of structural features of the designed structure, the methods of its construction and performances, the following main problems should be raised and solved:

1. To establing the extent and specific features of the area of interaction of the scheduled structure with the geological medium, tracing the origin, scale and trend of development within the boundaries of the main techno-environmental processes, imposed by the project.
2. To establish the sequence in development of main processes, connected with alteration of origin of techno-environmental effects at various phases of construction and after commissioning of a structure, to detect the most infavourable of them from the view point of safety of construction work and stability of the structures as vell as from the point of view their ecological consequences.
3. To elaborate the monitoring system of techno-environmental processes, developing in the area of interactions of constructed and running structures, including the observations either over their qualitive and quautitive changes of the state and properties of geological medium, or over the processes governed by the said changes, taking place in the structures themselves.
4. To elaborate the recommendations to be included into the contents of the design of the special preventive and protective arrangements, aimed at warning, location, and if necessery, rapid removal of consequences of development of unfavourable thecno-environmental processes either during construction of the structure or during its performance.

3 SECOND ISSUE

The second issue logically coming out from the first. It reflects the principal methodological approach to solution of the problem, from the considered point of view, of optimization of engineering-geological investigations with the use of the main criteria, of the scientific substantiation and practical support of the correctness of the methods used, the results of long-term field observationes over the dynamics of processes in the newly born techno-environmental system. It may be identified in the following manner:

The metodological background of effective solution of the problem on optimization of engineering-geological investigationes required for the hydroelectric projects should be the results of monitoring processes over the interaction of the constructed and running structures with the geological medium, as the most appropriate criteria of assessment of correctness of engineering-geological estimations for functioning of the newly born techno-environmental system and, being the unified safety instrument, to implement the "feedback" principal. .

At present the observatines over the processes of interaction of the hydroelectric prijects at certain heads, being constructed and already running, with geological medium, except for some rare cases, are not made systematically and the utilisation of their results to establish the trends of development of the said interaction and updating of the scientific forecasting is not satisfactory at all. Even more depressing picture is faced in arrangement of observationes over the techno-environmental prosesses, taking place within the boundaries of areas of Russia, where reservoirs exist. The designing-research organisations, designed the hydroelectric projects with reservoirs, having undertaken the supervision over the construction and given the forecasts for the origin, extent and potential consequences due to development of techno-environmental processes, practically, are beyomd the control of the correctness of the respective design solutions and the quality of their support by the results of investigations, meaning without the real adequate feedback to update the creative development of it. It leaps to the eye, that such an approach has to be changed. That"s why in the effort to find the solution to the said problem a versatile role of an engineer-geologist is essential in arrangement and functioning of the monitiring system for the process of hydraulic structures with the most effective means of control over their dynamics and utilization of results of monitoring for the comparison assessment of the estimated and actual characteristics of the trend and extent of the said processes, establishment of the reasons of traced discrepancies and to elaborate, in case of necessity, quick arrangements on warning or elimination of unfavourable consequences either for geological medium or for the structures themselves. Meanwile it should be kept in mind, that the monitiring problem here has two separate, although closely interconnected aspects.

The first is mainly related to the issue of interaction of the head structures being constructed and running with the geological medium from the point of view of warsening of construction properties of the latter. We"ll name it "geotechnical aspect".

The second one is connected with ecological consequences of the large scale techno-environmental effects on the geological medium of already constructed hydroelectric projects, first of all their reservoirs, often damaging it. We"ll name it "ecological aspect".

Taking into consideration the both aspects the particular tasks of monitoring of techno-environmental processes with many common issues in ahaping and functioning of the system, may have their own specific features. Thus, from stand poin of "geotechnical" aspect, one may say actually about the same torgets and purposes, which are facing, already sufficiently well theoretically developed and implemented into the engineering-geological practice, "litomonitoring geological medium". The difference is, that we include the complex of field researches of the structures themselves into the system of monitoring processes of interaction of hydroelectric structures wirh geological medium, without which the problem of effective control over the established by us techno-environmental system may not be solved completely.

The ecological aspect of the considered monitoring is mainly connected with the wide range of effects on the environment (including geological) of the reservoirs operating. It tackles upon the different fields of engineering-domestic activities, links the impact on the living conditions of many human beings and, due to this, it turned out to be the burning social-political issue.

4 THIRD ISSUE

The third issue is closely connected with the first one and it reflects the urgent necessity to redistribute the scope and extent of comprehensiveness ofdesigning-investigation phases in favour of those performed earlier, without which it is simply impossible to obtain the appropriate substantiated assessments of the origin of changes of the state and properties of geological medium under the impact of construction and operation loads and, respectively to adopt the best design solutions. It is proposed to formulate it in the following way:

"Since the essential design solutions, governing the type, design, arrangement of the project main structures, the conditions of its "matching" the geological medium and meeting the requirements of ecological protection should be adopted at the phase of the Feasibility Report (technical-economic substantiation), this pbase of designing-investigation activities, in particular, is the most essential from the view point of reliability of the engineering-geological assessments of dynamics of the structures with geological medium and, respectively, requires the substantial increase, in comparison with other phases, of the scope of comprehensive investigations and thorough appraisal".

As well established durung a long period of time the basis for determination of the contents and quantities of engineering-geological investigations to substantiate the hydroelectric projects in the former Soviet Union, at different phases of desiquing-investigation activities, was the instructions and regulatory papers issued by the high authorized organizations, which were elaborated on the basis of accumulative, mainly during the previous years and the first after war five-year plans, experience gained in construction of hydroelectric plants in various regions of the former Soviet Union. This experience, accounting for the specific features of geology and engineering-geological conditions of the valleys of the Volga, Dnieper and other plain rivers, proved the authorily of gradual increment of the quantities of main types of investigation activities, if required, updating the extent of thoroughness of investigations from the predesign elaborations to as-built drawings.

At the same time, even then, at certain projects there were found the negative consequences of insufficient coverage of the first phases of elaborations of the designs (Feasibility Studies) resulting in a number of essential engineering-geological factors, causing the serious complications during construction. Among the said project we may recollect: Buchtarma, Onda, Ult-Kamenogorsk, Gorkiy HPP and a number of others.

The situation became even more critical when mastering of new types of hydroelectric projects, first of all, the high head dams on rock foundations often with underground power houses, requiring to solve the main new engineering-geological problems and, respectively, the utilisation of new methods of investigations. It is quite natural, that the underestimation of these or those governing factors, at the same time, "matching" of the future structures with the geological medium and the insufficiently supported assessments of the origin and the extent of development of unfavourable techno-environmental processes, imposed hy the projects, caused by the vivid insufficiency of the basic data at the phase of the Feasibility Study, often led to substantial complications during the process of construction and operation of structures to the extent of emergencies. The similar situation came out when cnstructing the Krasnoyarsk, Miatla, Charvak, Kolima, Djiunvali, Mingechaur HPP etc.

5 FOURTH ISSUE

The fourth issue governs the foundamentals for selection of the most effective practices and methods for taking the solutions with the use of a set of the already available means of the practical part of the problem on optimization of engineering-geological investigations at the places of designing and construction of hydro-electric projects. This formulation is proposed in the following way: acconting for the complications and unique techno-environmental systems created during the construction of hydroelectric projects, the most promising trend to improve the effectiveness and the quality of engineering-geological investigations is the wide application of the principals of "active designing" with the adequate possibilities of adjustment of rated parameters of the state and properties of geological medium and, naturally, the adopted design solutions on the geological supervision activities over the construc-

tion work, engineering- geological logs of excavations monitoring of the techno-environmental processes an various methods of deformation modelling.

The idea of "active designing", based on the above considered issues, is the baby of the new approach fo assessment of the role of an engineer-geologist i designing-investigation process implemented in life an its involvement in the control over the development c the process of interacted structures with geologica medium. The basis of substantiation of the given prin cipal is the fact, that the wide varition and complexity o rock masses, serving the basis or the medium of hy dropower structures, limit the possibilities of detectio of all distinquishing features of their formation an properties. The effort to obtain the comprehensiv information for these distinquishing features by mean of simple increase of the scope of investigations, as rule, turned out to be a failure and only gave birth t unneeded confidence in perfection of our knowledge not reflecting the real situation. That's why the problen on optimization of investigations requires a search fo such methodological effects, which may allow to get th maximum effect with the most rational labou consamption. This effect is the implementation of cer tain sequence and aim in execution of designing-inves tigation activities at all phases of designing under th exact conditions of the hand in hand cooperation of o engineer-geologist and a designer.

At the predesigning phase, on the basis of th available data usually of a regional character, on th geological structure, hydrogeological conditions, char acter of geophysical fields and other elements of th natural conditions with the use of informative-explor atory systems the selection of analoques is made and th solution of problems, allowing to obtain the preliminar estimation of all engineering-geological factors, whic to this or that extent, may govern the conditions of th construction and operation of structures. The origir and extent of this phenomenon have to be established Simaltenously with similar modelling, accounting fo the data, obtained on the basis of it, there are preliminar design elaborations aimed at selection of the mos rational under the given conditions of the structures and their arrangements are made as well as the comparative estimation of the extent of impact of the main natural factors on the selection of the adopted design solutions At the same time the boundary of the "sensibility" of the designed structures to the changes of characteristic of geological medium. Thus, the three main complexes of the problems are solved:

1. Detection of the engineering-geological issues, the solutions of which, govern the principal approach to

the selection of design solutions: to determine the origin and extent of impact of the foundation properties on the parameters of structures and their performances; determination of the necessary level and adequacy of accurancy of the studies of the engineering-geological characteristics.

2. Estimation of the origin, trend and the extent of possible deviations from the values of the parameters of structures and the properties of the real foundation from its model, which may be constructed with the use of these or those methods on the basis of certain scheduled investigationes.

3. Selection of the most rational, applied to the certain natural conditiones, types, designs and arrangements of structures; elaboration of the flexible design solutions, accounting for the possibility of varioticus, within certain boundaries, of some characteristics of the state and properties of the foundation, allowing to make the due adjastment in the process of construction.

From the commencement of construction work such adjustments of the design solutions are implemented by the site resident designers on the basis of the information obtained from the geological supervision team resident at the project site and the engineering-geological logs of excavations made, the analysis of the information, obtained by the field observations when monitoring the techno-environmental processes as well as the data of the informative modelling.

These are the fundamentals, which, for sure, reflect only the essential methodological approaches to the problem of optimization of the engineering-geological investigations to substantiate the construction of hydroelectric projects under the present conditions and the most, in my idea, effective ways and methods of its solution.

Mechanism and prediction of thick soil mass deformation due to mining subsidence

Mécanisme et prédiction de la déformation d'un massif épais de sol causée par la subsidence minière

Sui Wanghua, Di Qiansheng, Shen Wen & He Xilin
China University of Mining and Technology, Xuzhou, Jiangsu, People's Republic of China

Rao Xibao & Zhu Chaofeng
Yangtze River Scientific Research Institute, Wuhan, People's Republic of China

ABSTRACT: The measurements on mining-induced subsidence in some coal mines covered by thick soil layers indicate that the surface ground movement and deformation show certain special phenomena in China in recent years. Based on measurement data, centrifuge tests and numerical simulation, the universal mechanism causing the particularity of subsidence, i. e. the interaction of subsidence induced deformation and pore water pressure has been presented. The coupled numerical model of Biot's theory and Cambridge model is introduced into the subsidence induced deformation calculation of soil mass. As an example of the prediction method suggested in this paper, the soil mass deformation due to slice mining of thick coal seam at a mine in Yanzhou mining area, China, was predicted and the result has been confirmed.

RESUME: Depuis quelques années, les mesures en surface de la subsidence due aux exploitations dans des mines de charbon couvertes par des couches de sols (en Chine) ont révélé des phénomènes spéciaux. A partir des donées mesurées in situ, des tests centrifuges et de simulation numérique, nous avons analysé le mécanisme qui cause les particularités de subsidence, c'est-à-dire celui de l'interaction entre la déformation due à la subsidence et la pression de pore. Le modèle numérique couplé de la théorie de Biot et de modèle de Cambridge est introduit pour calculer la déformation du massif de sol due à la subsidence. Comme un exemple d'application de la méthode de prédiction, proposée dans cet article, la déformation du massif de sol due à l'exploitation d'une couche épaisse dans une mine au Bassin charbonnier de Yanzhou (en Chine) est prédite, et les résultats ont été confirmés.

1 INTRODUCTION

The surface subsidence due to coal mining is one of the exogenic geologic hazards developing in coalmines. Not only does it damage farmland, buildings and surface water bodies at the present time, but also the environmental problems caused by the geomorphologic changes and geologic process during subsidence will influence the coming generations. There are more and more contradictions between mining and environmental protection in the coal-producing areas of East China, North China and Northwest China, which are generally located in populous and flourishing cities and nearby regions. In order to correctly predict the subsidence hazards induced by mining and take comprehensive measures to handling the relationship between resources exploitation and environmental protection, it is necessary to study the mechanism, regulation and calculation method of subsidence.

The measurements on subsidence in some coal mines covered by thick soil layers indicate that the ground surface movement and deformation usually show some special phenomena in China in recent years. For example, (1) subsidence factor is on the high side or even larger than 1. 0, i. e. the maximum subsidence can be near to or larger than the

shear height; (2) the range of subsidence can expand to a large scale; (3) horizontal movements can be larger than vertical movements in the boundary of subsidence basin; (4) angle of maximum subsidence is near to 90°; (5) subsidence is violent and concentrating during active period; (6) subsidence can last for longer time ; etc (Li 1987, Ma 1989, Di 1991, Sui 1992).

The systematic study on the mechanism of those phenomena has not been made and the reasonable explaination has not been obtained at home and abroad so far. We take it for granted that the mechanism of the coupled action of soil mass deformation and pore water pressure during subsidence should not be ignored (Di 1991 and Sui 1992) in the study on this problem. As pointed out by Buckpincki (1984), the mechanism research is very important to the establishment or selection of prediction methods, we must understand the movement of the all strata from coal seam to groud surface in order to study surface subsidence, because surface is one part of the moving strata and ground surface subsidence is a comprehensive result of the subsidence of each strata. Only when the engineering geological properties and the internal deformation regulations of soil and rock strata have been understood, can the problem of subsidence perdiction be thoroughly solved.

In this paper, first of all, the deformation mechanism of soil mass during subsidence is studied by the analyses of measured data and physical and numerical simulation. Centrifuge model is employed for physical simulation. Then, the results of soil mass deformation mechanism during subsidence gained on the model tests is used for the prototype, and according to it, a numerical model can be established and used to study the deformation characteristics of various soil mass structures. Based on the above-mentioned study, a quatative and quantitative method, premining engineering geological prospecting and predicting method has been presented and used to the prediction of soil mass deformation in the first mining district of a coal mine in Yanzhou mining area, China.

2 CENTRIFUGE MODEL TEST OF SOIL MASS DEFORMATION MECHANISM DUE TO MINING SUBSIDENCE

The application of centrifuge modelling to geotechnical engineering has the history of about sixty years. During the last two decades in which the technique has became increasingly developed, a hundred of centrifuges in various size have been constructed in former USSR, USA, Japan and Europe. There have been about 10 large scale geotechnical centrifuges to operate in China since the first one was completed in Yangtze River Scientific Research Institute in December 1983. The certrifuge model test provides an effective tool for the study of soil mass deformation mechanism , the establishment of prediction method and the explaination of the characteristics of surface movement and deformation in the mining areas covered by thick soil mass.

2.1 Scale factors

The dimensionless products of mining subsidence induced soil mass deformation model and its prototype deduced from Buckingham's law of dimension analysis are listed in table 1. It is seen that in the common model (i. e. 1g model) the most factors can hardly fit the demand of the law of similarity, while in the centrifugal model (ng model) the most factors can fit that except grain size and time scaling. In our models, fine sands and clays were used for centrifuge test, the ratio between partical size D and dimension of mined-out area L, i. e. D/L is less than 1:50, so the scale effects can be ignored. The time scaling in centrifuge model test depends upon the main physical phenomenon we considered, e. g. , it is $1/n^2$ for consolidation, if ng is the artificial gravity during centrifugation, 1.0 for creep and $1/n$ for soil movement. The consolidation and movement are the main problems in our case, so the time scaling varies between $1/n^2$ and $1/n$ in different phases of subsidence, which should be determined by the comparison between the test results and the measured data.

Table 1. The dimensionless products of soil mass deformation model and prototype

variable	symbol	prototype (1g)	common model (1g)	centrifuge (ng)	model scale factor
1. void ratio	e	e	e	e	1
2. degree of saturation	S_r	S_r	S_r	S_r	1
3. angle of internal friction	φ	φ	φ	φ	1
4. density of liquid	ρ_l	ρ_l	ρ_l/ρ	ρ_l/ρ.	1
5. Poisson's ratio	μ	μ	μ	μ	1
6. grain size	d	d/l	$d/l/n^*$	$d/l/n^*$	1
7. coefficient of viscosity	η	$\eta/\rho_l d\sqrt{gl}$*	$\eta/\rho_l d\sqrt{gl/n}$	$\eta/\rho_l d\sqrt{ngl/n}$	1
8. cohesion	c	$c/\rho gl$	$c/\rho gl/n^*$	$c/\rho ngl/n$	1
9. permeability	k	$k\eta/d^2\rho_l lg$	$k\eta/(d/n)^2\rho_l lg$	$k\eta/(d/n)^2\rho_l lng$	n
10. average stress	σ_m	$\sigma_m/\rho gl$	$\sigma_m/\rho gl/n^*$	$\sigma_m/\rho ngl/n$	1
11. pore pressure	u	$u/\rho gl$	$u/\rho gl/n^*$	$u/\rho ngl/n$	1
12. modulus of deformation	E	$E/\rho gl$	$E/\rho gl/n^*$	$E/\rho ngl/n$	1
13. time					
laminar flow	t_1	$t_1 k/l$	$t_1 k/ln^*$	$t_1 nk/n/l^*$	$1/n^2$
soil mass movement	t_2				$1/n$
creep	t_3				1

* Not similar

2.2 Test apparatus and measuring technique

The tests were carried out on the centrifuge of Yangtze River Scientific Research Institute. It has an effective radius of 3m, a maximum acceleration of 300g, a maximum mass of 1000kg and a designed load of 1800g- kN. The inner space of swing basket is 0.7×0.7×0.82m. The relative error of stress in the model is approximately ±3% (Wang 1988). The mining of coal seam performed inflight by draining away a liquid used inside the excavation to replace the coal. The markers embedded on the top and side of the model and LVDTs were used for measuring the displacements of different points of the model. Six pore water pressure micro transducers were adopted and IMP-microcomputer aquisition system was employed for recording very large amounts of pore pressure data at the different time.

2.3 Scheme of model and test procedure

The general geological and mining conditions in coal mines performing the particularity of subsidence covered by thick soil mass are as follows:

(1) The thickness of soil mass is near to or larger than that of overburden . The latter generally constitutes 20-50% of toal thickness of overburden strata.

(2) Overburden rocks include sandstone, sandy claystone, claystone and etc. The structure of rock mass in weathered zone beneath the soil mass was damaged and weathered fissures developed. The fissures generally filled by clay minerals and calcareous components. The weathered strata has been cemented and consolidated to a certain degree under gravity pressure of overlying thick soil mass in a long period of geological history.

(3) Thick soil mass consisted of interbedded sediments of clayey and sandy layer in horizontal layers can be divided into several aquifers and aquifuges.

(4) Longwall mining method along the strike is generally adopted. The shear height varies between 1. 8 and 3. 5m and average rate of advance between 20 and 40m per month. It is assumed that supercritical mining was reached in the model test, i. e. the dimension of mined-out area is 1. 4 times of the depth of excavation. The width of coal pillars are larger than the depth. On the basis of the abovementioned conditions, and in consideration of capacity of centrifuge and measuring technique, a model scale of 200 (200g gravity) was chosen. The cross section and dimension of a model is shown in fig. 1.

The following steps have been taken in centrifuge tests. The centrifuge was accelerated to run and operated under200g, untill measured displacements of LVDTs and pore water pressures of transducers became roughly stable, i. e. the process of consolidation ended basically. After the machine was stopped, the model was taken out the basket for measurement of displacements of markers. The model was installed in the basket again, then the

centrifuge was started and operated under 200g. When deformation and pore water pressure reached the stable value before the stop last time, the electrothermic switch was closed for 1 minute and liquid was drained out to simulate the seam mining. The displacements and pore water pressures were recorded untill they reached relative stable value. Then the machine was stopped and the displacements of subsidence markers were measured.

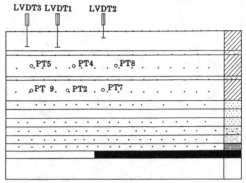

PTi - pore pressure transducers
• - subsidence markers

	model (cm)	prototype (cm)
size of mined-out area	42	84
thickness of coal seam	1. 6	3. 2
depth of excavation	22. 5	55
thickness of soil layers	16	32

(containing three aquifers and two aquifuges)

Fig. 1 Cross-section of the model

2.4 Variation of pore water pressures in soil mass during subsiderce

The variation of observed pore water pressures during subsidence, together with premining consolidation process of a period of 41 minutes is illustrated in fig. 2.

The pore pressures measured with transducers PT2, 9, 4 and 5 installed in the soil mass above the mined-out area dropped for a temporary period of 15 ~20 seconds (model time) after the start of excavation. Then, as the developing of deformation, the pore pressures at those locations rose up and excess pore water pressure appeared. Accompanied by the trend of increment of pore water pressure, the dissipation of excess pore pressure occurred in different locations. The variation characteristics of observed pore pressure are listed in table 2. It is clear that the pore pressures change considerably on PT7,8,2, 4 and 5. The reconsolidation of soil mass in those positions occurred due to the dissipation of excess pore water pressures in the meantime of soil mass movement. During the initial phase of mining, the soil layers above the goaf performed a little

2701

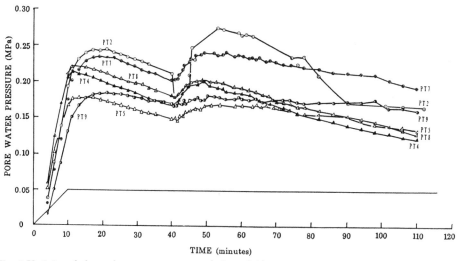

Fig. 2 Variation of observed pore water pressures during subsidence

expansive deformation (which was soon replaced by compressive deformation) associated with the dissipation of negative excess pore pressure. During the later phase of subsidence, the soil layers above the pillar also performed some expansive deformation adding to the former compressive deformation.

Table 2 Variation of pore water pressures in soil mass during subssidence

transducer	excess proe water pressure $\triangle u$ (kPa)	$\dfrac{\triangle u}{\sigma'_{z0}}$ (%)
PT9	13. 4	5. 3
PT2	50. 0	19. 7
PT5	21. 9	15. 0
PT4	32. 0	22. 0
PT7	40. 3	15. 8
PT8	22. 8	15. 6

* σ'_{z0}: effective geostatic stress of the measured point

2. 5 Displacement and deformation analysis of soil mass during subsidence

According to LVDTs and subsidence markers, the displacement and deformation of soil mass during and after mining were determined. Fig. 3 shows isogram of settlement of the model. As can be seen, the dip of isolines to the pillar in the upper soil layer above the goaf indicates the uniform tensile effect; the dip of them to the goaf in the middle soil layer indicates the compressive effect which can

be by 2-4 % of the thickness of the layer and the dip to the pillar in the lower soil layer above goaf shows the vertical extension. Isolines above the pillar indicate the vertical compression.

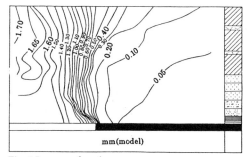

mm(model)

Fig. 3 Isogram of settlement

The time- dependent process of settlement is shown in fig. 4. OA stands for the premining consolidation of soil mass. After the start of the excavation, the speed of settlements is different from each other in the different locations. It has the maximum value above the center of the mined- out area and the minimum above the pillar. The course of settlement can be obviously divided into three segments: AB, BC and CD. The displacement of segement AB is composed of the movement of soil mass and the reconsolidation associated with the dissipation of excess pore pressure, the proportion of movement is a major cause of displacement. The displacement of CD is composed of the reconsolidation and secondary consolidation of soil mass. BC is the transitional stage of AB and CD.

Measured subsidence factors of different models range between 1. 05 and 1. 11 and horizontal displacement factors between 0. 33 and 0. 42.

2. 6 Preliminary summary

The centrifuge test indicated that the coupled actions existed between the soil mass deformation and the pore water pressure in the process of mining subsidence. In the case of no- drainage of sand layers in model, the internal deformation of soil mass is the comprehensive results of the reconsolidation deformation associated with the dissipation of the excess pore pressure and the expansive deformation in the lower layers above the goaf in initial phase and in the upper layers above the pillar in the later phase during mining subsidence. The reconsolidation of soil mass is the immediate cause for that the subsidence factor larger than 1. 0 and larger horizontal displacement. It is thus evident that the results of centrifuge tests can give us a relative satisfied explaination about the characteristics and mechanism of soil mass subsidence.

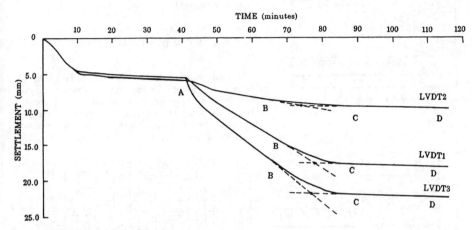

Fig. 4 Time-dependent settlement

3 NUMERICAL SIMULATION OF DEFORMATION AND PORE WATER PRESSURE OF THICK SOIL MASS DURING MINING SUBSIDENCE

In order to analyze the distribution of pore water pressures and deformation of soil mass, the coupled numerical analyses of Biot's theory and Cambridge model (Shen 1989) were attempted for mono-layered and multi-layered soil mass.

3. 1 Mono-layered soil mass

It is assumed that the land surface is free draining surface, and nodes in the surface of baserock and water flowing fractured zone are also free draining ones. At the end of excavation the distribution of excess pore pressure is given in fig. 5. It can be seen that the positions that pore water pressures increase are located in the soil layer above goaf. The maximum increment occurs in the central and upper layers. The positions pore water pressure decrease are located in the soil layers above pillar. The distribution of pore water pressures along the cross-section above the center of goaf seems analogous to that of one-dimensional consolidation. The boundary of increment and decrement of pore-water pressure is roughly located above the pillar line.

3. 2 Multi-layered soil mass

If two sand layers are intercalated in the soil mass which have the same dimensions as shown in fig. 5, the distribution of pore water pressure is shown in fig. 6. As the existence of sand layers, the distribution of excess pore water pressure induced by mining varies and the maximum value and hydraulic gradient increase obviously. The maximum value of pore pressures in the same layer along horizontal direction appeared in the soil layer above the center of goaf also, and the range of incremet of pore water pressures expanded into the soil layers above pillar. In the multi-layered soil mass, the dissipation of ex-ecss pore pressure has completed in a period of 670 days after mining end because of the existence of sand layers, while in the mono-layered soil mass, 1500 days after mining end, residual pore water pressure was also up to 75% of that at immediate end of mining. So the intercalated sand layers in thick soil mass play a significant role in the accumulation and dissipation of excess pore water pressure during mining subsidence. Thus, the deformation of soil mass intercalating sand layers is different from that of no sand layers. For instance, in the mono -

layered layer the subsidence factor is 0.67 at the end of excavation and reaches 0.71,1500 days later (fig. 7(a)); while in the multi-layered layers containing two sand layers the subsidence factor is 1.0 at the end of excavation and reaches 1.11,1500 days later (fig. 7(b)). Therefore, the interbedded structure of the permeable and low compressible sand and the compressible clay is an important condition causing the surface settlement larger than shear height in a not long time after mining.

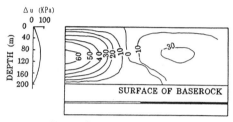

Fig. 5 Distribution of pore-water pressures in mono-layered soil mass

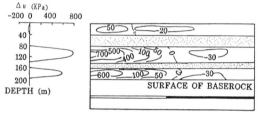

Fig. 6 Distribution of pore-water pressures in multi-layered soil mass

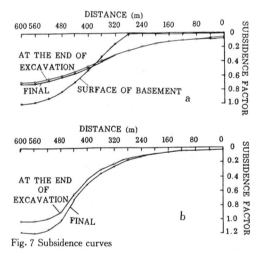

Fig. 7 Subsidence curves

4 PREDICTION OF MOVEMENT AND DEFORMATION OF THICK SOIL MASS DUE TO MINING SUBSIDENCE
- a case study of a coal mine in Yanzhou mining area

The engineering geological prospecting and hydrogeological investigations were conducted at the first mining district. Then the engineering geological and hydrogeological characteristics of soil and rock masses related to coal mining were analyzed in detail. A vast amount of physical and mechanical indices of overburden were gained by laboratory tests. After the fractured height of overburden had been predicted and the size of water proofing coal/rock pillars determined, the coupled action model of pore water-pressure and deformation was used to predict the deformation of soil mass. The movement and deformation and their time-dependent variation process were gained.

4.1 Engineering geologic conditions of the mining district

The coal seams excavated in the studied coal mine belong to Shanxi Formation of Permian System which is a series of coal-bearing clastic rocks of transition facies and continental facies and composed of greyey white to dark grey sandstone, siltstone, silty claystone, claystone and coal seams.

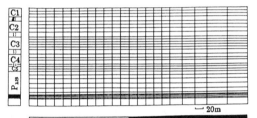

Fig. 8 Calculation model

The coal seam No. 3 is the predominant seam which is stable all the mining district. It ranges from 8.7 to 9.3m thick and 9.0m on the average, dipping with an angle of between 8° and 15°. The immediate roof of the seam is silty clay rock, main roof fine sandstone or siltstone, floor clay rock, siltstone or limestone. The coal measure strata are overlain by 148.21 to 189.10m thick soil mass composing mixed impermeable clay layers with water-bearing sands. The clay layer in the central part of the column is stably distributed in the mining district. It can cut off the direct hydraulic connection between the upper and lower aquifers, so the water level in the upper aquifers will not be influenced by the mining. In the bottom of the thick soil mass there exists a clay layer ranging from 3 to 5m thick, which can cut off the direct hydraulic connection of the lower aquifer and weathered zone of baserock.

High pressure oedometer tests show that the majority of clay layers in thick soil mass are low compressive and partialy medium compressive. It can be judged from this that the reconsolidation of soil layers during subsidence will not be large.

4.2 Prediction of soil mass movement and deformation due to mining

The deformation and displacement of a strike cross-section was calculated. The calculation model is shown in fig. 8. The soil layers at that cross-section are divided into 9 layers containing 5 clay layers (c1-c5) and 4 sand layers (s1-s4). The calculation parameters are listed in table 3.

Table 3. Parameters for prediction of deformation of thick soil mass

layer	buoyant unit weight γ' (kN/m^3)	lateral pressure factor k_0	vertical coefficient of permeability k_y (10^{-7}cm/s)	ratio of horizontal and vertical coefficient of permeability k_y/k_x	shear modulus G_0	compression index C_c	expansion index C_e	experimental parameter M
C_1	10.40	0.43	24.00	2.0	0.52	0.159	0.019	0.084
C_2	10.59	0.53	23.57	2.0	0.33	0.174	0.022	0.57
C_3	10.01	0.43	0.046	2.0	0.46	0.088	0.024	0.38
C_4	10.59	0.43	18.70	2.0	0.46	0.171	0.019	0.30
C_5	10.40	0.43	14.40	2.0	0.46	0.123	0.050	0.58
S_1	10.40	0.33	249200	1.0	0.52	0.159	0.019	0.17
S_2	9.61	0.33	249200	1.0	0.60	0.183	0.033	0.47
S_3	9.71	0.33	249200	1.0	0.60	0.183	0.033	0.57
S_4	9.42	0.33	249200	1.0	0.60	0.183	0.033	0.35

The distributions of excess pore water pressure in layers c1- c5 after the excavation of first slice are different from each other because their properties and relative locations are different. For example, the coefficient of permeability of layer c5 is much larger than that of c3, so the excess pore-water pressure in c5 dissipates much faster than that in c3 does. Deformations of different sand layers and clay layers during subsidence are different. As can be seen from table 4, the compression of clay layers is the main course of additional settlement of surface, while the compression of sand layers is much little.

Table 4. Compression deformation in a period of 1 year after the excavation of first slice

layer	consolidation settlement a year (cm)	relative compressibility (cm/m)
C1	3.38	1.62
S1	0.05	0.062
C2	1.61	0.59
S2	-0.311*	-0.28*
C3	0.254	0.085
S3	0.009	0.01
C4	0.238	0.10
S4	0.009	0.018
C5	0.041	0.033

* Swelling

The subsidence curves are shown in fig. 9. The settlement due to reconsolidation in the soil layers above the center of goaf is 8.08cm at the end of excavation and reaches to 13.6cm one year after the completion of excavation. The subsidence factors are 0.74 at the end of excavation and 0.764 one year later, respectively.

After the completion of excavation of the first slice, the second, third and fourth slices are excavated every second year. The predicted variation of the maximum pore pressure in layer c3 is shown in fig. 10. The excess pore pressures caused by different slices are interacted and the reconsolidation are associated with the dissipation of the excess pore pressures. The subsidence factor at the time of one year after excavation of the second and third slice are 0.75 and 0.83, respectively. The subsidence factor at the completion of the fourth slice mining is 0.96 and it increases slowly along with the dissipation of excess pore- water pressure in layer c3. The calculation result has been partially proved by the preliminary measuring data in situ.

5 CONCLUSIONS

The coupled action between subsidence-induced deformation of thick soil mass containing aquifers and pore water pressure has been presented and proved. The coupled numerical model of Biot's consolidation theory and Cambridge model has been introduced into the deformation calculation and prediction of soil mass in the case of slice mining of thick coal seam. A case study at a coal mine in Yanzhou mining area, China, showed that prediction results of subsidece were in agreement with in-situ measuring data.

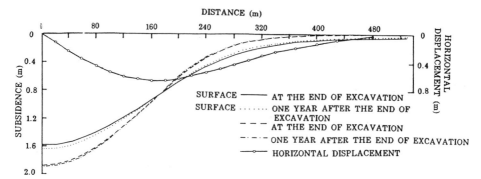

Fig. 9 Subsidence curves due to first slice mining

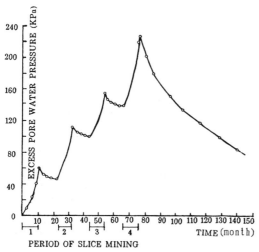

Fig. 10 The accumulation and dissipation of pore water pressure induced by slice mining

REFERENCES

Di Qiansheng and Huang Shanmin 1991. The study of water-soil couple mechanism due to minig subsidence, *Proceedings of the 2nd international symposium on mining technology and science*, China University of Mining and Technology.

Li Jinzhu and Liu Tianquan, 1987. A preliminary study on the surface movement calculation due to the drainage of aquifers, *Coal Science and Technology*, 1987. No 1 (in Chinese).

Ma Weimin and et al 1989. Study on the mechanism and parameters of subsidence due to mining under thick aquifers in East China, *Reports of Coal Science Foundation*, China University of Mining and Technology, (in Chinese).

Shen Zhujiang 1989. Evalution of consolidation deformation model of soft soil foundation, *Journal of civil Engineering*, Vol. 17, No 2.

Sui Wanghua and et al 1992. An analysis of deformation mechanism of soil mass in mining-induced subsidence and experimental study of relevant compressibilty parameters, *Journal of CUMT*, Vol. 21, supplement, 1992.

Wang X. 1988. Studies on the design of a large scale centrifuge for geotechnical and structural testing, *Centrifuge 88*, A. A. Balkema/Rotterdam: 23- 28.

Some legal controls for environmental damage to engineering structures in Nigeria

Contrôle réglementaire des dégats causés par l'environnement sur les structures en Nigéria

B.O. Ezenabor
Department of Geology, University of Benin, Nigeria

ABSTRACT: Alteration in the physico-chemical setting of the environment by human activity or natural factors may affect the engineering stability of structures, for example by changing the ground water levels or soil/water chemistry which may reduce the load bearing capacity of the soil. Such activities as mining, quarrying and excavation operations may induce landslides or subsidence resulting in both damage to the environment and to engineering structures. In Nigeria there is a legal liability for such activities if they lead to environmental degradation and deterioration of engineering structures. For example pollution of water resources through mining or other activities may be the subject of court action if it results in changes in soil chemistry which affect the engineering stability of structures. However, damage to structures resulting from de-watering of mines, quarries and excavations or normal ground water exploitation are not actionable, even when extensive subsidence results. This paper discusses the legal protection to the environment provided by the English Common Law and its application to the Nigerian situation, including local statutory provisions and case laws.

RÉSUMÉ: La modification de l'arrangement physo-chimique du milieu effectuée par l'activité humaine ou les facteurs naturelles peuvent influer sur la stabilité des structures. Par exemple, c'est en modifiant les niveaux piézométriques de la nappe d'eau souterraine ou la chimie du sol/d'eau qui pourrait réduire la charge utile du terrain. Des opérations telles que les exploitations de mines, de carrièrres et l'excavation peuvent produire les éboulements ou l'effondrement consécutif aux dégâts du milieu et aux structures. Au Nigéria, il y a une responsabilité légale pour telles opérations si elles provoquent une dégradation et une détérioration du milieu des structures. Par exemple, la pollution des ressources en eau par l'exploitation de mines ou d'autres opérations pourraient être l'object d'action tribunale si elle résulte aux modifications de la chimie du sol qui influe sur la stabilité des structures. Cependant, des dégâts causés aux structures due au dénoyage des mines, des carrièrres et des excavations ou l'exploitation normale d'eau souterraine ne sont pas poursuivis, même s'il y arrive un effondrement étendu. Cette étude discute la protection légale du milieu du droit civil d'Angleterre et son application à la situation nigérienne, y inclus des provisions statutaires locales et des précédents jurisprudentiels.

1 INTRODUCTION

In 1988 Decree No 58 of the Federal Military Government of Nigeria established the Federal Environmental Protection Agency, charged with the responsibility for the protection and development of the environment. Amongst other duties, the Agency was to

a) develop research on the chemical, physical and biological effects of various activities on the environment;

b) establish environmental criteria, guidelines, specifications or standards for the protection of the nation's air, land and waters from degradation;

c) establish such procedures as necessary to minimise damage to the environment from industrial or agricultural activities and to control concentrations of substances in the air which are likely to result in damage or deterioration of property.

2 ABSTRACTION OF SUBTERRANEAN WATER

In the British case of Popplewell v Hodkinson the court held that "an owner of land has no right at Common Law to the support of subterranean water". This principle was applied in the British case of Langbrook Properties Ltd v Surrey County where the plaintiff company claimed damages against the defendants and alleged that by pumping out an excavation in the vicinity of the plaintiffs' land, the defendants had abstracted water percolating beneath the plaintiffs' land, thereby causing subsidence and settlement of the buildings on that land. The Court held that the plaintiffs had no cause of action as the defendants were entitled to abstract the water under their land, percolating in undefined channels, to whatever extent they choose, not withstanding any injury their action might have caused to the plaintiffs. The Court observed that any such damage was "damage without legal injury" for which there was no cause of action. In the case of Bradford Corporation v Pickles, the court held that the above rule applied even where such abstraction of subterranean water was prompted by an improper, selfish or malicious motive.

3 ABSTRACTION OF MINERALS

For the above principle to apply, the abstracted material has to be water; hence where minerals such as clay, silt and brine were withdrawn, the rule ceased to be applicable.

In the British case of Jordeson v Sutton Southcoates and Drypool Gas Company, the defendants, in the course of excavation in their land, penetrated an underground stratum of "running silt" or quicksand which also extended under the plaintiff's land. In draining the excavation the defendants withdrew a large quantity of the running silt from under the plaintiff's land and thus caused a subsidence of the surface and structural damage to his houses. The British Court of Appeal observed that the plaintiff's land was supported by a bed of wet sand or running silt, not by a stratum of water and held the defendants liable in nuisance at Common Law. A similar decision was given in the case of Lotus Ltd v British Soda Co Ltd where serious damage had been caused to buildings on a factory site belonging to the plaintiff due to the subsidence of the land as a result of pumping on nearby land. The pumping extracted "wild brine" - ground water saturated by the dissolution of rock salt. As the brine was removed, more water flowed under the plaintiff's land and dissolved further

salt. In an action for damages caused to the plaintiff's buildings by withdrawal of support due to the pumping of brine, the court held that the surface land had a right to be supported by a subjacent strata of minerals and that accordingly the plaintiff was entitled to damages.

4 REMOVAL OF SOIL SUPPORT

At Common Law a landowner has the absolute right to the adjacent and subjacent support of his land in its natural state (Clerk and Lindsel, 1975). This right is not an easement but a natural incident of his ownership. In the British case of Humphries v Brogden the occupier of the land surface brought an action against the miner of the subjacent minerals for not leaving sufficient pillars and supports during the working of the minerals that the surface gave way. The Court held that the owner of the surface was entitled to support from the subjacent strata.

It appears there is no natural right of support for a building by the foundation of another building. In the British case of Peyton v The Mayor and Community of London, the defendant pulled down his house whose foundation lent support to the foundation of the adjoining plaintiff's house. As a consequence, the plaintiff's house was impaired and in part collapsed. In a court action against the defendant for not giving notice of the pulling down of the house, the court held that the plaintiff had no cause of action and was bound to protect his house by shoring.

Right of support for buildings may be acquired, as an easement by implied grant where the house has stood for more than twenty years. In the case of Hide v Thornborough the plaintiff and the defendant were the owners of adjoining lands and the plaintiff's house had for more than twenty years been supported by the adjoining land of the defendant. The defendant dug a foundation for some proposed buildings so near the plaintiff's house that it collapsed. The Court held that as the plaintiff's house had been supported for such a long time, to the knowledge of both parties, the plaintiff had a right to such support as an easement and was therefore entitled to damages.

A similar judgement was made in the case of Dalton v Angus where two adjoining dwelling houses, each built independently by its owner but each on the extremity of the owner's land, benefited from lateral support from the soil on which the other rested. The two houses co-existed in this situation for more than twenty years but subsequently one of the houses was converted into a coach factory; the internal walls being removed and girders inserted into a stack of brickwork in such a way as to throw a greater lateral pressure on the soil under the adjoining house. More than twenty years after the structural conversion the owners of the adjoining house employed a contractor to pull down their house and carry out excavations. Deprived of the lateral support of the adjacent soil, the plaintiffs' stack sank and fell, bringing with it most of the factory. The House of Lords held that the plaintiffs had acquired a right of support for their factory by the twenty years enjoyment of such support following the structural conversion and were therefore entitled to damages.

In the case of Hooper v Rogers the defendant used a bulldozer to deepen a track which cut across a steep slope around the plaintiff's house, thereby interfering with its natural angle of repose and expose it to the process of soil erosion which would eventually deprive the footings of the plaintiff's house of support and cause it to collapse. The court ordered the defendant to reinstate the natural angle of repose of the slope. Similarly, in the case of Redland Bricks Ltd v Morris, farmland of several acres sloped down towards and adjoined land from which a brick company excavated earth and clay. As a result of this excavation, which began some 20m away from the plaintiff's boundary, landslide was induced which progressively damaged the plaintiff's land and engineering structures. The court awarded damages and restrained the defendants from further excavations.

5 THE RULE OF STRICT LIABILITY

The legal principle of strict liability (Rylands v Fletcher) was restated in the case of the National Telephone Co v Baker:

"If the owner of land uses it for any purpose which from its character may be called non-natural/extraordinary user, such as for example the introduction on to the land of something which on the natural condition of the land is not upon it, he does so at his peril, and is liable if sensible damage results to his neighbours' land, or if the latter's legitimate enjoyment of his land is thereby materially curtailed".

In the case of Hoare v McAlpine, in preparing a site for a large building in the heart of a city the defendants drove a very large number of piles into the soil, thereby setting up such a heavy vibration as to cause serious structural damage to a building belonging to the plaintiffs. The

court followed the rule in Rylands v Fletcher, as proposed in the case of the National Telephone Co v Baker and held the defendants liable.

In the Nigerian case of J P Investment Nig Ltd v Foundation Construction Ltd the plaintiffs claimed damages for withdrawal of support to their building by driving piles on the adjacent plot. The defendants were restrained from continuing the pile driving operations unless adequate steps were taken to prevent further withdrawal of support from the plaintiffs' building.

6 DISCUSSION

Both sections 77 of the Nigerian Mineral Act (1958) and 35(2) of the Quarries decree (1969) make provision for the payment of adequate compensation for any disturbance of the surface rights of land owners/occupiers and for any damage done to the surface of the land upon which mineral prospecting or mining is being or has been carried out. Compensation may also be awarded to the owner of buildings and engineering structures damaged in the course of mining, quarrying or excavation operations.

The above provisions appear to conform with the Common Law principle on withdrawal of support and damage to engineering structures due to the abstraction of minerals and other excavation operations. A major discrepancy exists, however, between the provisions of Common Law and Nigerian Statute in relation to injury suffered due to abstraction of subterranean water. By section 49(1) of the Mineral Act (1958) it is forbidden for any person to make any such alterations in the water supply of any lands as may prejudicially affect the water supply enjoyed by any other person or lands in Nigeria. By this provision, therefore, abstraction of subterranean water which prejudicially affects the water supply of neighbouring lands is actionable in Nigeria. This is contrary to Common Law doctrine as established in the case of Bradford Corporation v Pickles. There appears to be no specific local regulation in Nigeria in relation to environmental damage to engineering structures resulting from ground water abstraction, however, either in the course of de-watering of mines, quarries, and excavations or during normal ground water exploitation. At Common Law there is no liability, as established in the British case of Langbrook Properties Ltd v Surrey County Council. This Common Law rule should be regarded as applicable to Nigeria as there was no express or implied intention to the contrary by the Statute. This view was held in the case of R v Morris by

Justice Byles where he stated that:

"It is sound rule to construe a Statute in conformity with the Common Law rather than against it, except where or so far as the Statute is plainly intended to alter the course of the Common Law."

Whatever might have been the intention of the Common Law in establishing this rule, the author considers it likely to bring hardship to the population, especially in Nigeria where heavy ground water exploitation has been implicated in the occurrence of widespread land subsidence. (Oteze, 1983).

7 CONCLUSION

The law appears to have made good provision for ensuring that environmental activities in form of construction works, mining, quarrying and excavation operations do not threaten the stability of engineering structures. However, the Common Law position on damage caused by ground water abstraction needs to be reviewed in the interest of the populace who might suffer from subsidence resulting from such activity.

ACKNOWLEDGEMENT

This article is part of an on-going research project on the application of legal principles to Earth Sciences in Nigeria.

REFERENCES

Clerk, J.F., Lindsell, H.H.B. 1975: on Torts No. 3, Common Law Library 14th ed., Sweet & Maxwell, London.

Ezenabor, B.O. 1990: "Some legal aspects of water resources activities in Nigeria", Proceedings, The International Hydrological Programme (UNESCO) 1st National Biennial Hydrology Symposium 26th - 28th November, Maiduguri, Nigeria.

Ezenabor, B.O. 1991: "Riparian Rights and Water Resources Development in Nigeria". Water Resources Journal published by Natural Resources Development Co. Ltd., Lagos. Vol. 1, No. 2, pp 4 - 9.

Oteze, G.E. 1983: "Groundwater levels and ground movements". In Ola, A. (ed.) Tropical Soils of Nigeria in Engineering Practice, A.A. Balkema, Box 1975 3000 B.R. Rotterdam, Netherlands; 172 - 195.

CASES/CITED

Bradford Corporation v Pickles (1895) A.C. 587.

Dalton v Angus (1881) 6 A.C. 740.
Hide v Thornborough 2 CAR & K. 249.
Hoare & Co. v McAlpine (1923) 1 Ch. 167.
Hooper v Rogers (1974) W.L.R. 329.
Humphries v Brogden 12 Q.B. 739; 116 E.R. 1048.
Jordeson v Sutton Southcoates & Drypool (1899) 2 Ch. 217.
J.P. Investment Nig. Ltd. v Foundation Construction Co. Ltd. (1971) N.M.L.R. 121.
Langbrook Properties Ltd. v Surrey County (1968) 1 W.L.R. 161.
Lotus Ltd. v British Soda Co. Ltd. & Another (1972) 1 Ch. 123.
National Telephone Co. v Baker (1893) 2 Ch. 186.
Peyton and Others v The Mayor & Commonality of London 9 B & C 723; 109 E.R. 269.
Popplewell v Hodkinson (1869) L.R. 4 Ex. 248.
R.V. Morris (1867) 1 CCR 90 at 95.
Redland Bricks Ltd. v Morris & Another (1970) A.C. 653.
Rylands v Fletcher (1867) L.R. 3 H.L. 330.

STATUTES/CITED

Federal Environmental Protection Agency Decree No. 58 1988.
Mineral Act Cap. 121 1958.
Quarries Decree No. 24 1969.

Study on the characteristics of slope and groundsurface deformation caused by underground excavation and open pit mining

Étude des caractéristiques des pentes et de la déformation de la surface du sol causées par les excavations minières souterraines et à ciel ouvert

Nie Lei & Wang Lianjun
Engineering Department, Changchun Earth Science University, Jilin, People's Republic of China

ABSTRACT: Openpit mining and underground excavating often lead to openpit wall, groundsurface and nearby natural slope deformation at the same time, which effects the stability of openpit wall, groundsurface buildings and slope. To prevent effectivily groungsurface deformation hazards the main reason of deformation must be determined. In this paper the characteristics of deformation caused by openpit stripping and underground excavating are studied through in-situ measurement, finite element analysis and physical similar material simulation test. By these studies, the different deformation characteristics of these two kinds of mining methods are given. It is shown that these two kinds of mining methods have different deformation velocity, and different displacement vector characteristics and subsidence curves have different shapes not only at the perpendicular direction of openpit wall but also at the parallel direction.

RESUME: L'exploitation à ciel ouvert et souterrain souvent conduisent à la déformation de muraille de carrière à ciel ouvert, surface du sol et de pente naturelle près en meme temps. La déformation influence la stabilité de muraille de carrière à ciel ouvert, des constructions à la surface du sol et de pente. Pour empecher efficacement la catastrophe causée par la déformation. de surface du sol, le facteur principal de déformation doit être déterminèe. Dans ce papier les caracteristiquses de la déformation provoquée par l'explloitation en carrière et souterrain sontétudiees à l'aide de mesure sur place; l'analyse d'unite limitée et du test analogique par matière analoque physique. Par ces études les caracteristiques de déformation différente dans les deux méthodes d exploitation sont présentées. Cette étude démontre que ces deux méthodes de l'exploitation ont la vitessede déformation différente; la caractère du déplacement vecteur différent et les formes des courbes d'affaissement différentes nor seulement dans la direction perpendiculaire mais aussi dans la direction parallèle du muraille de carrière à ciel ouvert.

1 THE CHARACTERISTICS OF DEFORMATION ALONG SLOPE TREND

Openpit mining often forms a huge openpit along the run of the ore vein and the wall trend of openpit will be as same as the run of the ore vein. So along slope trend often have similar engineering geological condition except when some big faults exist. On another aspect, because of the confine of openpiting method and transport condition, openpiting is taken place on a serial horizontal levels. So along slope trend direction in a long time the change of piting depth and slope angle are almost uniform. In openpit mining, because the boundary condition and engineering geological condition are uniform along slope trend direction, the slope and ground deformation caused by openpiting are homogeneons along slope trend direction.

On the other hand, for undergronnd excavating, even though engineering geological conditions are uniform along slope trend direction, but the slope and surface deformation is controled by underground excavating method, the scale of excavating area, the filled formation and filled ratio ect. which cause the characteristics of deformation along slope trend direction are not uniform. These factors will be greatly effect the deformation quatity of slope and groundsurface and the deformation characteristics on slope and groundsurface will be related closely to these factors.

This kind of change regularity of the deformation curve charactoristics on the direction of slope trend caused by openpiting or underground excavating is observed in some mines. For example, on the north wall of Fushun West Openpit Coal Mine, the defferent deformation charactoristics on slope and groundsurface are very notable. The deformation amounts at the different part of the slope trend direction are strongly different which show the different active degree of openpiting and underground excavating. The in-situ measurement deformation curve of the upper part of Fushun West Openpit north wall shows in Figure 1. This subsidence measurement line which is parallel to the north wall on the Xinpin road has been measured for more than 30 years. In figure 1 from local mine coordinate of W25 to E75 there is a coal pillar used to prevent water. It is the boundary between Shenbu Mine region and Shanli mine region. From E75 to E700 there is the region of number 510 underground excavation belong to Shengli Mine. It is a band mining region, that is excavating a band of 28 m width then set 38 m coal pillar and shale is used to fill the excavated area. The filling proportion is above 85%. The region of W25 to E700 is nearly 800 m long, because there are lots of coal pillars which sustain the roof of upper strata, the groundsurface subsidence amount caused by underground excavation is very small not only at now when underground excavating have been stopped for a long time but also at the time when excavating is taking place. The subsidence of the upper rock strata of excavation area is commonly less than 30 cm, The in-situ measurement subsidence on the groundsurface is only several centimeters then subsidence trends to stop. The subsidence of slope and groudsurface in this region is mainly caused by openpit mining. The subsidence amount is commonly less than 50 cm in the region from W25 to E700 in 33 years from 1959 to 1992, especially in the year from 1980 to 1991 the subsidence rate is stable which shows that the effect of openpiting to slope and groundsurface deformation is much less than that of underground excavation.

In the region between E700 and E1400, there are the underground excavating regions of Shengli Mine. Even though the excavating regions are filled by oil shale, but the exist of huge excavation area has caused a large number of slope and groud-surface subsidence from 1950 year to now as figure 1 shown. In the region between W25 to E700 the groundsurface subsidence is mainly caused by openpiting, from 1950 to now in 33 years, the subsidence amount caused by openpiting is only 7% about to that caused by underground excavating. From above we can see that the underground excavating has much more strong action to strata than that of openpiting and caused much more

deformation.

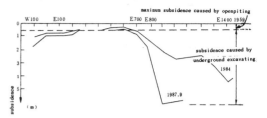

Fig. 1 The subsidence curve of groundsurface

It also can be seen from figure 1 that the subsidence is also larger in the west of coordinate W25 than that between W25 and E700 region which shows the effection of Shenbu Mine is still exist although it has been stoppped for nearly 35 years.

2 THE CHARACTERISTICS OF DEFORMATION IN THE PERPENDICULAR DIRECTION OF SLOPE

Besides the notable difference of deformation characteristics on parallel slope direction (along slope trend direction) caused by openpiting and underground excavating, in the perpendicular direction the effect of these two kinds of mining methods are different not only on the effect scope and the deformation curve shape, but also to the diretion of up and down of mine.

On the aspect of effect scope, the deformation caused by underground excavating is mainly located on the slope upper part and groundsurface in the demountain direction, the maximum subsidence close to slope upper part and deviate from underground excavating area. Underground excavating don't caused the deformation of middle, the lower part of this wall and another side wall. When mining is exsisted at the bottow of openpit, different with the underground excavating, it not only cause the demountain direcron slope and groundsurface deformation, but also caused the other side wall and groundsurface deformation. When the groundsurface boundary of openpit is limitted, the deformation effection of slope and groundsurface on the mine vein upper part direction is mainly the effect of openpit deepening, but on the demountain direction besides the effect of openpit deepening, the slope and groundsurface also will suffer the effect of slope steepening. Although openpit stripping causes the slope and groundsurface deformation not only on demountain direction but also on the upmountain direction, but the effect on downhill direction is stronger and the effect on uphill direction is weaker. When the stripping is on the middle or upper part of wall, it mainly effects the upper wall near stripping. On the stripping area it will cause a little elestic rebounded deformation. It has been observed by similar marterial model test and by numerical analysis. Because of the limit of miming condition it is not easy to measure out in-situ measurement in openpit.

The typical deformation curve of underground excavating shows in Figure 2. The subsidence curve shows as a basin shape, the subsidence centre remove to the downhill sides, the steeper the ore vein is, the larger the removement amount is. When the ore vein dip angle reachs a certain value which is determined by rockmass properties, the thickness of overlay strata, ect. the whole subsidence basin removed to demountain side. On both sides of subsidence basin horizontal displacement is towards subsidence basin centre.

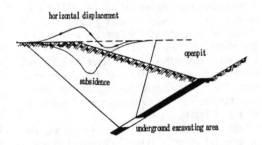

Fig. 2 The typical deformation curve of underground excavating

The deformation caused by stripping have not above typical curve charactaristic, have not subsidence basin exist, and horizontal displacement is not towards any point. From above we can see there are much more different of deformation characteristics between stripping and underground excavting. Through in-situ displacement measurement on a mine with stripping and underground excavatting, the deformation curve can be given, and the main deformation reason will be distinguished, that is the study on the characteristics of deformation curve is one of the most important foundation to distinguish the reason of slope and groundsurface deformation. A example is the study on the slope and groundsurface deformation of Fushun West Openpit. In the deformation serious area of Fushun West Openpit north wall (the west of W25 and E700), on the slope perpendicular direction the deformation curve is similar to Figure 2. It shows that the effect of underground excavating is very seriously at these large defomation wall part.

3 THE DIFFERENT CHARACTERISTICS OF DISPLACEMENT VECTOR

The characteristics of displacement vector caused by stripping and underground mining are different notablely. For most openpit and undergroud excavating mine the ore vein often is dip strata, so the mining to dip strata is mainly discussed about here.

From a huge number of mine in-sitn measurement data, numerical analysis results and model tests it is determined that there are some difference on the characteristics of displacement vector at the upper part of openpit and groundsurface between stripping and underground excavating. The displacement vector caused by strippng and underground excavating are shown in figur 3. The vectors in figure 3a are caused by underground excavating and the vectors in figure 3b are caused by stripping. From figure 3a it can be seen that at groundsurface the displacement caused by underground excavating is much larger than that caused by stripping. The groundsurface displacement vector caused by underground mining is mainly the horizontal displacement vector and the perpendicular displacement vector is smaller. The characteristics is the same not only at the groundsurface but also at the depth of strata. The ratio of perpendicular displacement to horizontal displacement is between 0.4 to 0.9. The farther the distance to underground mining area is, the smaller the proportion is, and the more gently the displacemen vector is. The farther the distance is, the smaller the displacement value is. On the other side, the groundsurface displacement value caused by openpit stripping is much less than that caused by underground excavating. The main displacement vector is perpendicular displacement vector, and the horizontal displacement vector is smaller. When deformation is caused by openpit stripping the proportions of perpendicular vector to horizotal vector are between 1.4 to 4.0. The farther the distance to openpit is, the larger the proportion is at this kind of condition.

In a certain scope of groundsurface the displacement value caused by stripping is stable, no notabl decrease and in a depth range the displacement vector value and direction are similar to that at the groundsurface.

So there are notable different in the direction and at the value of displacement vectors caused by stripping and underground excavating. The vector direction caused by underground excavating point openpit. On the other hand, the vector direction caused by openpit stripping don't point to openpit, but point to underground excavation area or more

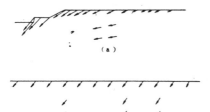

Fig. 3 The displacement vector caused by stripping and underground excavating

deeper area.

When underground excavating is take place in dip strata, the typical displacement vectors which can be got by the displacement curve are shown in figure 4. It must be pointed out that the displacement vector direction is effected by a lot of factors such as the depth of underground excavating area, the scope of mining, the strata dip angle ect. For example, to the effect of strata dip angle, in a horizontal strata the horizontal displacement vectors point to the underground mining area centre from all around. But to a gently dip strata the horizontal displacement vector deviate toward the downhill direction. The deeper the mining area is, the more deviated amount take place. To a steep strata all horizontal displacement vectors point to the uphill direction of mining area. All above show the huge effect of strata steep degree on the direction of displacement vector. Other factors, such as the depth of mining, mining method, mining area scope, strata control, the exist of fault ect. also have great effect on the direction of displacement vector.

Fig. 4 The displacement vectors of a dip strata

4 THE EFFECT OF THE CHANGE OF TECTONIC STRESS TO GROUNDSURFACE DEFORMATION

In different filling method area of underground excavation, the change of tectonic stress will yield the different deformation forms of slope and groundsurface. When tectonic stress decreases, the displacements of a openpit elestic-plastic finite

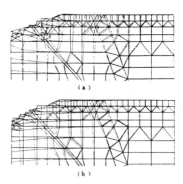

Fig. 5 The deformation caused by tectonic stress decrease

Figure 5a shows the slope and groundsurface deformation when the tectonic stress decrease, there underground excavation area is filled densely by shale or there are coal pillars exist. It shows that tectonic stress decrease will yield the groundsurface that above the densely filled excavation area produce a little deformation which direction deviate from the underground excavation area. The main deformation is the horizontal displacement, the perpendicular displacement is very much little.

Figure 5b shows the slope and groundsurface deformation at an unfilled underground excavation area when tectonic stress decreases. In this condition the deformation charactaristics of slope and groundsurface are wholly as same as the deformation charactoristics of excavating underground area. At the upper part of slope and groundsurface the main deformation is horizontal and the perpendicular displacement is smaller. The total displacement vectors point towards openpit and subsidence curve shows a basin shape, the horizontal displacement has a typical charactoristics of dip strata suffered underground excavating. It shows that the decrease of tectonic stress in an unfilled underground excavating area has same effection as excavating underground mine area, the tectonic stress decrease will cause old underground excavating area reactivation. The decrease of tectonic stress can be caused by fault activity or some other facters.

5 THE GROUND DEFORMATION RATE EFFECTED BY STRIPPING AND UNDERGROUND EXCAVATING

Because the boundary condition on openpit and

underground mining area is notablly different, the shapes of displacement rate curve and total deformation curve is different. The groundsurface deformation rate and total deformation ~ time relationship curve of stripping and underground excaving show in Figure 6. The curve ① is the groundsurface subsidence rate curve caused by underground excaving, curve ② is the total subsidence ~ time relationship curve caused by underground excaving, curve③is the groundsurface subsidence rate curve caused by openpit stripping, curve ④ is the total subsidence ~ time relationship curve caused by openpit stripping.

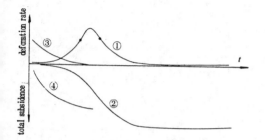

Fig. 6 Deformation rate and total deformation~time relationship curve

From Figure 6 we can get that the deformation curve appears maximum value when some times have been passed after undergrownd excavating takeing place, the curve from concave turn to convex shape, after maximum value deformation point the deformation rate slowly decrease then the deformation rate curve from convex turn to cancave, at last tends to stable. On the other hand the groundsurface deformation rate caused by openpit stripping reach the maximum at the same time of stripping, then the deformation rate decrease slowly, the shape of whole deformation rate curve is concave, and the deformation time is much shorter than that of underground excavating. Total subsidence curve form a turn point near the time when deformation is reach maximum value, at the beginning of underground excavating total subsidence curve is convex, at last it turn to concave. The total subsidence curve of openpit stripping has not turn point and whole curve is concave. Because the displacement rate and amount curve caused by different mining methods are different, in a real project the study on deformation main factors through the study on the deformation rate and amount curve characteristics got by in-situ deformation measurement, main deformation effection factor can be got.

6 CONCLUSIONS

By studing on the effection of openpit stripping and underground excavating using deformation in-situ measurement of slope and groundsurface, numerical analysis, physical similar material model test following results are given:

1. Because of the difference of mining methods, filled forms and mining depth etc. along strata trend the deformation of slope and groundsurface appear notable different amount, the deformation in in-situ measurement is a important basis for determination the deformation main reason.

2. In perpendicular direction, underground excavating form typical subsidence basin which stripping do not form, and horizontal displacements are notable different each other.

3. At the groundsurface where is a litlle far from mining area, the horizotal displacement vector caused by underground excavating is larger than perpendicular displacement vector, the total displacement vector point to openpit. But in openpit stripping the perpendicular displacement vector is larger than horizontal displacement vector, and total vector point to the diretion of strata demountain.

4. When tectonic stress decrease, different deformation types are yielded at the different filled form of underground excavating area. The deformation form at unfilled mining area is as same as the deformation form of mining same time, it shows that the decrease of tectonic stress has much stonger action to unfilled mining area than that to filled mining area.

5. There are notable different on groundsurface deformation rate and total deformation ~ time relationship curves between openpit stripping and underground excavating, it is a important foundation to distinguish deformation main reason.

In project by study on the factors discussed above, combine with engineering geological condition analysis, we can determine the main factor that caused ground deformation and take corresponding prevention policy which will be the important mathod to solve the ground deformation problem caused by mining.

REFERENCE

China mining university, 1981, The groundsurface deformation of coal mine strata, The publisher of coal industry.
Coal ministry of P. R. China, 1986, the rules of the coal pilar set under buildings, water, railway. The publisher of coal industry.

Repetitive examination of land damages in mining areas

Enregistrement continu des dégats en surface dans une région minière

J. Pininska
Faculty of Geology, Warsaw University, Poland

ABSTRACT: The exploitation of industrial minerals causes permanent conflicts of various extent between economy and ecology on local, regional, country or global scale. This conflict results in progressive degradation of building terrains. In order to obtain retrospective data on degradation of the Walbrzych City caused by coal mining, the old and new aerial photos, taken in 1958 and 1988 respectively have been processed together with the newest field data acquired in 1992.

RÉSUMÉ: La gestion industrielle des matières premières minérales provoque des conflits entre l'économie et l'écologie a plusieurs échelles: locale, régionale, nationale et globale. On a fait une tentative d'utiliser les anciennes photographies aériennes afin d'obtenir un inventaire des terrains de la ville de Walbrzych affectés surtout par l'exploitation souterraine de charbon dans la période de 1958-1988 a 1992.

1 INTRODUCTION

Exploitation of minerals and processing of raw materials causes ecological conflicts in urbanized areas. The greater is the industrial development the more urbanized becomes the area. Such a trend, known in economy as the "Thunen's rings", results of natural human tendency of shortening access roads between the work place, settlement and market place.

The Walbrzych City is sited in South-Western Poland, (Fig.1) in the Sudetes mountainous country, in morphologically and geologically complex environment.

Selected as the test polygon it is a typical mining agglomeration, increasingly plagued with the all eco-eco (ecological-economic) conflicts (Pininska 1990). The most of the old settlements were located, as engineering geological surveys show, in poor subgrade conditions, which further deteriorated in a result of underground mining and development of processing plants. Further deterioration of subgrade quality is expected after the closing down of worked-out mines followed up by cease of pumping-out of mine waters.

According to the annual inventory of building terrains for 1989, of 85 km^2 of total area of township, good for siting (of good engineering-geological condition) terrains cover 31.5%, terrains sufficient

Fig.1. The Walbrzych city; G, P, S, W – subsidence areas; 1 - dump, 2 - pool, 3 quarry, 4 - area of low risk, 5 - area of fair risk, 6 - presented in Fig. 3.

for siting cover 32.8%, and the 35.7% are insufficient for siting (Fig.2).

a. Existing
1. ━━━━━━━━━ 31.5% (good)   ¦ for
2. ━━━━━━━━━ 32.8% (sufficient) ¦siting
3. ━━━━━━━━ 29.1% (not sufficient)¦
4. ━━ 6.6% (dumps, tills, e.t.c.)
b. Predicted for 1989
1. ━━━━━━ = = = = 81.6% (undegraded)
2. ━━━━ 14.0 % (low risk)¦ (degraded)
3. ━━ 3.4% (fair & high risk) ¦ (degraded)
c. Predicted for 2010
1. ━━━━━ = = = = 74.0% (undegraded)
2. ━━━━ 15.0% (low risk)   ¦ (degraded)
3. ━━ 11.0% (fair & high risk)¦ (degraded)

Fig. 2. Share of various engineering-geological siting condition in Walbrzych.

The engineering-geological aspect of conflicts lyes mainly in devaluation of building terrains within the existing or planned urban areas, in direct vicinity of mines and processing plants (Dragowski et al 1983, Gallagher 1978, Goszcz 1993), the main factors being:
- continuous and discontinuous surface deformations in a result of rock massif deformations;
- damages of surface drainage system and waterlogging;
- diminishing of geotechnical properties of soil;
- increased infiltration, leaching and sinkholing;
- dumping of waste and forming of settling basins;
- changes in chemistry of soil caused by direct disposal of wastes, sewages and brine;
- excavations, pitting and quarrying;
- uncontrolled soil devastation - pollution by oils, erosion by heavy traffic, non cultivated tills, e.t.c.
Limitation of ecological conflicts in urbanized areas and a proper development planning depends on proper assessment of trends of degradation factors and requires projected data on degradation/time ratio. For accomplishing that requirement the two sets of available, standard aerial photographs, the most recent (1984) and the oldest (1958) were subjected to stereoscopic comparable examination of the state of advancement of terrain damages (Fig.3).
The results of photographs interpretation were verified in the course of geological field traverses, and mapped together with the complex data on mining activity - geological documentation, development plans, expected extent of subsidence, repetition levelling profiles, etc.

2 STATE OF THE ART OF GEOLOGICAL-
-ENGINEERING CONDITION

Deterioration of geological-engineering condition within the administration boundaries of the Walbrzych City is under strong impact of natural processes and all classes of engineering activity, what in synoptic form was presented altogether with geological background on a detail topographic map (Pininska 1993), part of which is presented in Fig.4.
The geological-engineering synoptic map gives in abbreviated form an integrated information on:
- Engineering-geological conditions: geology, morphology, hydrology, surficial geological processes (landslides, erosion).
- Land degradation caused by human activity: mining (subsidence, subsiding troughs, fissuring, settling basins, spoil heaps, spoil heaps), open pit excavation, irrigation and drainage canal systems, siting of constructions, dumping of wastes etc, communication systems (railways, access roads, slack cable-ways, power and other open wire lines), made grounds and uncontrolled soil abrasion by cross-country heavy traffic, and a total disorder of abandoned objects, earthworks and constructions around industrial plants;
- Mining facilities: mining block extents, quarries, health's resort mineral waters protection zone, mine shafts;
- Prognoses: based on mine operational forecasts (the prognostic extent of surface deformations for selected subsequent years and on retrospective analysis of recorded data from mine reports, surveyor's reports and municipal archives and partially, from detail cartographic materials, and reconstructed data based mainly on aerial photos and partially on cartographic materials.
The most significant components of terrain damage in the Walbrzych City are continuous (troughs) and disjunctive (fissures) terrain deformations concentrated in southern part of the City, slake dumpings and settling basins, uncontrolled terrain devastations in vicinity of industrial plants, waterlogging and flooding upon local depressions.
Subsidence, covering 18% of the Walbrzych City, strongly developed above exploitation blocks of the Thorez, Victoria and the Walbrzych coal mines at four sites (see Fig. 1): in the Sobiecin quarter (S), in the Podgorze quarter (P), near the Railway Central Station (W) and in the Glinik Nowy quarter (G). These areas tend to merge each other into single vast subsidence area within the downtown limits, what agrees well with the mining damage

2718

a.

b.

Fig. 3. Photomosaics of Walbrzych south: a - 1958, b - 1983, (after Pininska 1993).

forecast for the 2010. Moreover, at sites S, P and W are many disjunctive deformations, sinkholes and fissures.

Spoil dumps and settling basins, covering >6% of the area, are located in close proximity to compact settlements. The total acreage of rather small quarries is about 0.2% of total area, but it is bothersome due to their scatter distribution over building terrains.

3 RETROSPECTIVE ANALYSIS

The knowledge of contemporary state of degradation - caused by a series of past events has been gained by means of retrospective methods. The most useful appeared comparative interpretation of two sets of aerial photos taken in 1958 and in 1984, which covered almost a three-decade span of time. The degradation trends as fo■

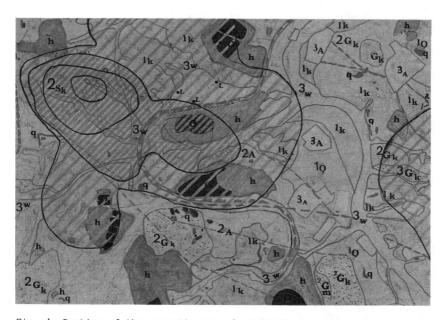

Fig. 4. Portion of the synoptic map of the Walbrzych city; the Sobiecin quarter; original scale 1:10 000.
Engineering-geological conditions:
1 - favorable: subsoil of good geotechnical properties, ground water lever over 2 m deep, gradient less than 12%: 1_Q - Quaternary deposits: fluvial-glacial sand and gravel, 1_k - Carboniferous sandstone, conglomerate and some mudstone forming detritus and clay-soils;
2 - sufficient: - subsoil of fair geotechnical properties (s): 2_{sk} - Carboniferous deposits, weathered sandstone, conglomerate and mudstone with coal seams, mostly forming clayey soils, 2A - native soil in part disturbed by human activity; - gradients (G) 12%..30%: 2_{Gk} - Carboniferous deposits, clayey soil and debris cover on native rocks, 2_{Gm} - Carboniferous-Permian igneous rocks; - ground water level (w) 1 .. 2 m deep: 2_{wk} Carboniferous deposits, clayey soils and debris cover on sandstone, conglomerate and mudstone;
3 - insufficient: gradient (G) over 30%: 3_{Gk} - Carboniferous rocks, undifferentiated; ground water level (w) 0 .. 1 m deep: 3_w - Quaternary mud, older substratum deposits undifferentiated, 3_A - geodynamic and human-activated processes: uncontrolled fills, road crossings, railways, etc.;
Impact of mining activity: underground coal mining subsidence risk:

- low,   - fair,   L.- high risk subsidence-or discontinuous deformation (sinkholes, thresholds and fissures),

q - quarry,   h - dump,   - pool,   - fault.

the most affected quarters of the Walbrzych City are presented in Fig.5. In 1958 were already slight impacts of the Victoria mine noticeable on plough grounds over the Sobiecin subsidence trough.

There are seen small dip-and-strike steps developed due to activated downdip interbed slides, which contour south and west boundaries of the trough. Already existed there waterlogged local depressions, settling basins, slake dumpings and acres of railway embankments.

In 1984 lithologic steps bordering the subsidence area became more distinct, and new ones are seen within the area. On both sides of the local road leading to compact settlement area, developed several wet depressions and pools of total acreage of 3 km^2. Areas covered by slake dumpings and settling basins increased in size and uncontrolled terrain devastation significantly contributed abandoned plant facilities in direct contact with township buildings. Several dozen kilometers of abandoned bypass railway remained as a string of escarpments, embankments, ruined bridges, culverts, e.t.c., at periphery of the area, while many new railway embankments were constructed inside the area.

The real extent of subsidence trough at Sobiecin as recorded on both 1958 and 1984

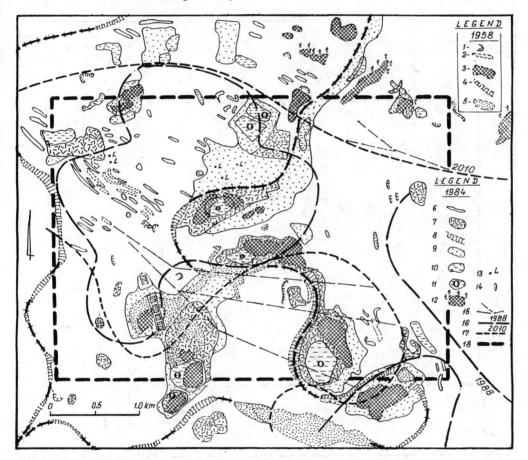

Fig. 5. Terrain degradation in the 1958 through 1984 period, aerial photos interpretation;
1958: 1 - quarry, 2 - threshold, 3 - dump, 4 - railway embankment, 5 - uncontrolled devastation;
1984: 6 - threshold, 7 - dump, 8 - railway embankment, 9 - uncontrolled devastation, 10 - overfloodded depression, 11 - pool, 12 - recultivated dump, 13 - sinkhole, 14 - quarry;
Other units: 15 - faults, 16 - prognosed extent of subsidence (for 1988), 17 - prognosed extent of subsidence (for 2010), 18 - area shown in Fig. 4.

sets of aerial photos is marked with the
series of elongated, blur steps trending
NW-SE. The steps are almost parallel to
swarms of faults at the SW margin of that
trough. One can assume that this steps are
extended over the Podgorze subsidence
trough, what needs confirmation based on
additional studies, as the terrain
separating both areas is obscured with
constructions, slake dumpings and railway
embankments. However, the railway
embankment is being strongly expanded
there, indirectly proving a progress in
subsidence. In northern part of Sobiecin
are present typical disjunctive
deformations developed upon abandoned mine

shafts and other old shallow workings.
Similar observations apply to other areas
of subsidence at the Walbrzych City. The
extent of mine deformations at the Podgorze
quarter was also greater in 1984 than
forecasted for 1988.

Towards the south west terrain
deformations extend upon urbanized areas
beyond the Niepodleglosci street, proving
that already in 1984 this subsidence trough
if fact merged with the subsidence trough
developed upon the Walbrzych mine. Northern
extent of deformations reaches the Three
Roses hamlet i.e. over 1 km farther than
that prognosed for later period. There are
present many slumps, sinkholes and

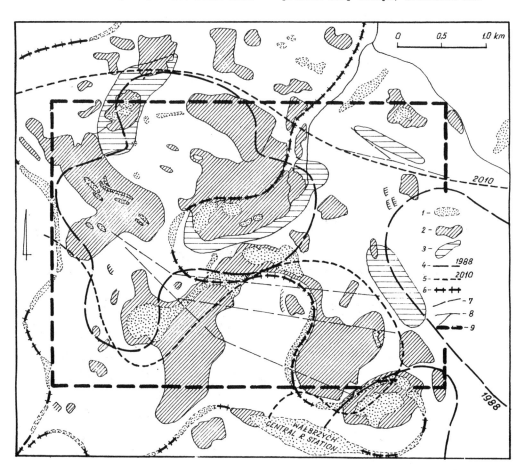

Fig. 6. Extent of damaged land, according to the repetitive examination, valid for the
following periods:
1 - till 1958, 2 - 1958 - 1990, 3 - the recentmost;
Extent of areas to be subjected to damages according to forecast for:
4 - 1988, 5 - 2010;
Other symbols: 6 - railway, 7 - road, 8 - fault, 9 - area shown in Fig. 4.

strike/dip steps, witnessing an old setting of that zone. Noted irregularities in extents of subsidence zones are resultants of geological structural trends and a shape of mine workings.

In the terrain balance of the Walbrzych City, as much as 25.1% of the township administration area was degraded, out of which 33.3% were terrains good and sufficient for building purposes. It is worth to underline, that at places the forecasted extent of mining deformations for the end of 1988 is lesser than that recorded on aerial photos taken in 1984 - e.g. for the Sobiecin quarter.

4 COMMENTS

According to the forecasts of development for 2010, the degradation figures increase to 50% of total terrains, out of which 46% will include terrains good and sufficient for building. Results of repetitive verification showed, that in 1984 already 40% of the area was degraded by subsidence, and that the areas covered with the waste-rock heaps and settling pools increased two-fold since 1958, penetrating into densely urbanized areas. This trend was magnified by another trend - an expansion of new settlements towards and among the existing pools and heaps. Thus the ecological-economy conflict deepened in accordance with the Thunen's law. It is dated back to the XIV century in Walbrzych, where people used to locate their living quarters as close to their work places as possible even in the most insufficient living conditions (in narrow gorges, at landslide prone slopes, in waterlogged areas). And that insufficient conditions deteriorated further on due to mining damages in 1988 to 1992 (Fig.6).

It has been also discovered, using the retrospective analysis, that many refuse dumpings, railway embankments and plant facilities were sited at terrains of good engineering-geological categories. And opposite - many uncontrolled dump-ups and man-made grounds were located upon abandoned quarries, masking them totally. The degradation of building terrains, therefore, must not be determined statically. In determining future development of complex factors, both the existing state of the terrain as well as its dynamic processes and actual trends must be considered. The retrospective analysis, on the other hand, verifies a correctness of prognoses and provide base information on possible anomalies (or defects) of geological substratum: e.g. an existence of multistage man-made grounds of poor quality with their blurring effects, a discovery of now buried sinkholes or pits, quarries, etc., old slope failures, landslides and ponds, and even abandoned underground constructions.

5 REFERENCES

Dragowski, A., Kaczynski, R. & Pininska J. 1983. The Influence of an Underground Vein Ore Mine on Engineering Geological Conditions of Urbanized Areas. Bull. I.A.E.G. No.28. Paris.

Gallagher, C.P., Henshaw A.C., Money M.S. & Tarling D.H. 1978. The Location of Abandoned Mine-shafts in Rural and Urban Environments. Bull.IAEG No.18. 179-195. Paris.

Goszcz, A. 1993. Influence of coal mining in Upper Silesia Coal Basin on environment. Przegl. Geol.7:473-481.

Pininska, J. 1990. Data Base for Land Assessment in Mining Regions (in Polish). Proc.IX-th National Conf. SSMFE. 433-439.Krakow.

Pininska, J. 1993. Application of retrospective method in assessment of degradation caused by coal mining in urbanized areas. Proc.

Stability and environmental rehabilitation of the bulky landfill in coal quarrying areas

Stabilité et réhabilitation de l'environnement dans des zones de mines de charbon á ciel ouvert

Florica Stroia, Petre Bomboe & Daniel Scradeanu
Bucharest University, Romania

ABSTRACT: Anthropical store systems the waste dumps are available thanks to the need of sterile store result from coal exploitation in open pits. They are particular sedimentary structures due to primarily by the nature and diversity of the sterile cover from various open coal pits conferred here and by the second hand through the chaotically manner of his laying down. These elementsin addition with the position and the topographical place and appearance of any specific Hydrogeological factors difficult to check up may generate into sites of the waste dumps earth slidings whose dynamic and structure can be solved out through quantitative and qualitative methods.

RÉSUMÉ: Dans la zone de SO de la Roumanie le terrils, résultant de l'exploitation de charbon en carrières représentent un monades permanent de l'environnement. L'hétérogénéité lithologique, les conditions hydogéologiques et la dynamique actuelle de glissements sont étudiées pour le zonage du risque de dégradation de l'environnement et également pour la prévision de ces types de dépôts.

1 INTRODUCTION

The waste dump Stiucani lies in the basin afferent to the valley with the same name and it is one of the sterile store coming from the open pit mine Rosiuta, coal basin Motru.

The material was deposited down up to criteria required by the storage technology and not according to a concerned to assure the stability of rock mass.

Through systematically tests,followed by a statistical processing it was settled the proportion of sand to clay. Thus, the granulometry of the stored material is; clay 82% and sand in 18%.

The investigation means used were basically the topometric maps existing in 1985 and 1992, the profiles with topometric marks placed for watching the change of the place and a geophysical investigation meted, namely appearance resistivity.

an attentively description of the waste dump morphology, it may be detach the zones with a important degree of instability and it was made a prognostication regarding the behaviour of the rock mass.

2 THE WASTE DUMP MORPHOLOGY

For being able to estimate quantitatively the size of the waste dump we used the existent topometric maps on Stiucani valley before beginning the store and for the resulted surface after a period in which it was carried out the waste dump.(fig.1,2).

The store surface Stiucani had by 1992 level 1980m length and about 1.035 skm area.

As it result from the map with equal lines of down laying uniform material thickness, in the centre zone the stores reached et 50 to 70m maximum thicknesses. The line of maximum thickness is overlapping on minimum share line from the pre-existent valley(fig.3).

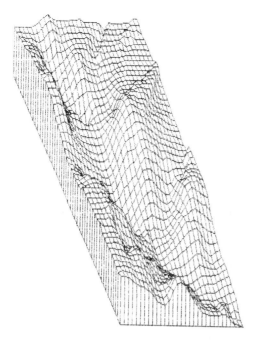

Fig.1.Stiucani Valley

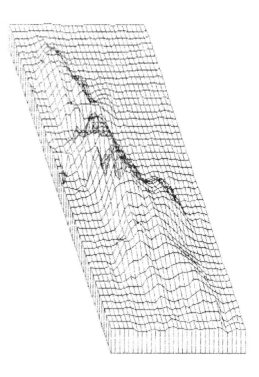

Fig.3.The Bulky Landfill..

A specific morphological phenomena is the "hanging lakes". The waste dump has around it a number of eight such water eyes, part of witch are previous to the stores and other seven water mirror were growing up on the sterile in 1993 summer.

Also appear gravitation detached cracks in meter size. Sometimes the appearance such as openings on quasi-horizontal surfaces.

3 WATCHING OF THE CHANGES THROUGH TOPOMETRIC MARKS

In the S-E part of the waste dump there are three profiles fitting out with movable marks for change of places registration; they were under observation in February-June 1992 period.

We carried out the displaces measurements are x,y and z directions. After stereographic projection principle on make the project of displace vector in horizontal plan.

In figure number 4 is reproduced the projection of displacement vectors in flat level,the stereographic

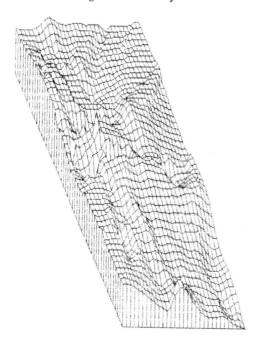

Fig.2.Stiucani Waste Dump

In 21.02-27.03.1992 the marks do not suffer displacements, it was a period of displacement stagnation.

One of the next period in that the waste dump was active happened between 08.05-02.06.1992. Following an analysis after the same principles such as the previous result: the displacement in vertical direction are mainly down side; the general direction of the movement is changing; the average displacement speed was 5-10cm a month.

5 INSTABILITY PHENOMENA

For reasons concerning the technology of the stratification of steril it wasn't possibly to respect the projects to ensure the best stability conditions.

Therefore the waste dump is formed through laying downs difficult to be checked up and the instability phenomena don't respect a general valid law for the whole stored steril body but by various reasons appearing local movements which may involved even large rock masses.

At least three are the reasons which caused significant local displacements: gravitationally detaching, earth slides thanks to overtake stress solicitation for the hill side growth up by steril storing,loading up the previous under consolidated manmade sediments over the available limit.

For exempla we enclose by a few concrete situations (fig.4;fig.5).

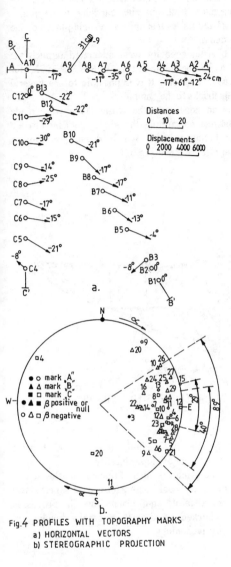

Fig.4 PROFILES WITH TOPOGRAPHY MARKS
 a) HORIZONTAL VECTORS
 b) STEREOGRAPHIC PROJECTION

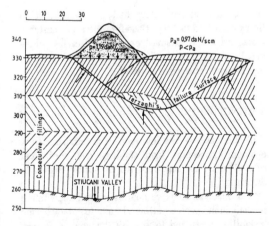

Fig.5 THE PROBLEM OF OVERBOARD BEARING CAPACITY

projection and the statistic remaking in shape of a histogram for 19.02-21.02.1992 period. From the analysis, the vector projection result the following remarks:on small distances the displace vector change sometimes spectacular in direction and size; in vertical direction numerous marks were getting up; the general direction of the movement is in gravitation way; the average displacement speed 5-10cm on three days.

In the course of time,the stores are growing up through successively steril laying down made from layer which are less than 20m in thickness thanks to the technology of earthfilling. The mechanical and physical properties are improved through consolidation. In this way in areas with maximum thickness of the deposits (70m) on establish four diminution step for cheering resistance, from depth to surface. Thus the base layer has the following shear parameters: angle of internal friction = 21 and cohesion c=45 kN/sm and at the some time for the surface layer the characteristics are : =10 and c=15 kN/sm.

On such lithological succession follows a new steril lay down which transfer to the earth a bearing about 1.2 daN/scm. By computing the bearing capacity the allow pressure is maximum 1.4 daN/scm and can take a minimum of 0.6 daN/scm. The result is that the loading exceeds the bearing capacity, that explains driving back forces that appear in some places on the waste dump.

The bearing capacity was calculated through the work technique indicated by Vesic in 1975, for shallow foundations. It was possible to use several variants thanks to the computer.

In varied zone from the waste dump were made calculations concerning the slope stability through Nilmar-Jambu method, which doesn't required a circular slide surface (fig.6).

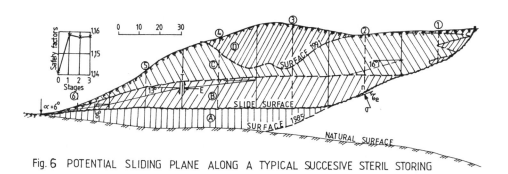

Fig. 6 POTENTIAL SLIDING PLANE ALONG A TYPICAL SUCCESIVE STERIL STORING

The coefficients of safety had values ranging in limits from 1.15 to 1.93. Therefore there are unstable areas, usually with small extension. But the slides may be propagated step by step and it is possible to involve great masses of rocks.

5 CONCLUSIONS

With this case we are involved in an environmental problem concerning land stability and ecological rehabilitation on an area which suffer manmade damages.

It was made an attentive qualitative and quantitative investigation which allowed us to detach the instability areas and to specify the causes. It was necessary a large number of data prelucration technics : to clear up the waste dump morphology; to remake the data furnished by topographic marks; the interpretation of profiles made through geophysical resistivity survey method; estimate of bearing capacity; estimate of slope stability.

The conclusion is that Stiucani waste dump, belonging to the coal basin Motru is in a permanent search for a stabile equilibrium state. the processes can by checked up through a careful heading of earthfilling technology.

REFERENCES

Chen, Hangey. 1990..A care study of the morphology on the unstable slopes ,.6th International IAEG Congress ,Rotterdam,Balkema

Winterkorn,H.F.&Y.F.Hsai 1975.Foundation engineeringhandbook,Van Nostrand Reinhold Company,New York.

Zamfirescu,F & R.Comsa & L.Matei 1985.Rocile argiloase inpractica inginereasca,Bucuresti

Geoecological effect of mining on the shore of Lake Baikal

Effet géoécologique de l'activité minière dans le littoral du lac Baikal

N. I. Demianovich

Institute of the Earth's Crust, Siberian Branch of Russian Academy of Sciences, Irkutsk, Russia

ABSTRACT: Recent geoecological conditions of the Lake Baikal shore claim a specific regime of its nature management. The impact of mining as a technogenic factor is discussed on the example of a Sliudyanka mining site. Phlogopite and marble can be regarded as clean or positive raw materials in terms of ecology. Although there is considerable regional need of cement raw material, the prospects for broader development of mining are to be found at alternative sites outside the Baikal catchment.

RÉSUMÉ: Dans ce rapport sont décrites les conditions géoécologiques contemporaines de la partie littorale du lac de Baïkal qui déterminent le régime tout à fait particulier de son utilisation. Comme le fait technogène on y présente l'estimation de l'activité de l'exploitation minière. Sur l'exemple du chantier minier de Sludianka on a fait l'analyse de l'influence de l'exploitation du marbre et de la phlogopite et ces derniers se sont avérés non-dangereux au point de vue écologique. Vu le besoin élevé en ciment dans la région, on peut supposer que l'élargissement de sa production est associé aux gisements équivalents, situés hors du bassin du lac.

1. INTRODUCTION

Being a unique geological phenomenon, Lake Baikal draws attention and concern as an object of nature preservation that requires strict limitations on industrial activities within its drainage area. As a natural catchment, Lake Baikal accumulates polluted products of technogenesis from vast territories. Industrial development might be extremely deleterious for its shoreline. According to zoning of the drainage territory by the technogenic impact on the waters of Lake Baikal (Demianovich & Pisarsky 1993), the shore is regarded as a zone of direct influence since self-purification of waters is impossible due to short filtration ways.

Ecological problems including protection of the quality of Lake Baikal waters as the environment of unique ecological systems require establishment of a zone of rigorous nature management. For reduction and eventual lowering of the technogenic impact on the lake, it is necessary to assess possible contributions of different industrial activities into the total man-induced effect on the waters of Lake Baikal and to elaborate an adequate strategy for management of the geological environment of the lake basin (Pinneker 1991).

Being represented merely by a Sliudyanka mining site, mining puts rather a low contribution into development of the area studied. Since most of the geological, engineering, hydrogeological and ecological characteristics of this site are typical of other deposits in this territory, conditions and consequences of exploitation of the Sliudyanka mining site are analysed to assess the prospects for broader development of rich mineral resources in the Pribaikalian region.

2. GEOECOLOGICAL ESTIMATE OF ENGINEERING-GEOLOGICAL CONDITIONS

The Sliudaynka mining site is located on the southern shore of Lake Baikal at the junction of a chain of narrow proluvial-lacustrine basins and the mountainous framework. These morphostructures comprise a province of formation and transfer of technogenic products, their sources being represented by the mining complex in the mountainous part and other industries in the piedmont plain.

Two geoecological regions are recognized in the area studied according to the directions of natural and technogenic material transfer. Figure 1 shows a map of A-1 - the Sliudaynka and A-2 - the Kultuk regions. Since by a number of factors which comprise the geoecological potential of the territory, these regions are considered

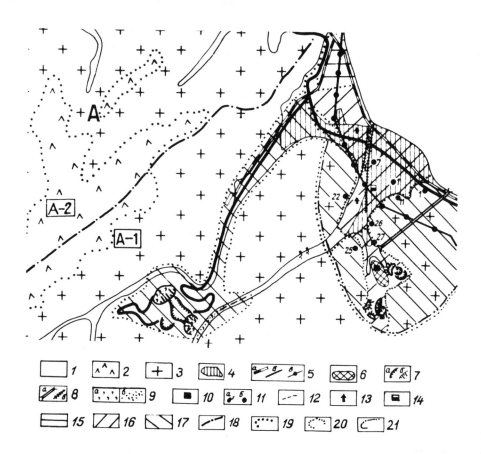

Figure 1. Schematic map of engineering geoecological conditions of the Sliudyanka mining site. 1-3 - engineering geological settings: 1 - loose, mostly gravel-cobble sediments, 2 - the Cenozoic effusive basalts of high permeability, 3 - strongly cemented rocks of metamorphic and magmatic formations of the Archean and Proterozoic. Sources of technogenic impact: 4 - town, 5 - railway (a), road (b), electric line (b); 6 - mining field; 7 - quarry (a), dump (b); 8 - drainage tunnel (a), anti-mudflow dam (b). Results of technogenic impact: 9 - technogenic earthwork (a) and inwash (b), 10 - land subsidence above underground workings, 11 - lowering of spring discharge (a) and of ground water level (B) resulting from drainage tunnel, 12 - areas of surface runoff loss, 13 - areas of critical elevation of the ground water level, 14 - deformation of buildings. Scale of changes in geological medium: 15 - high, 16 - middle, 17 - low. Boundaries: 18 - geoecological regions (A-1 - Sliudyanka region, A-2 - Kultuk region), 19 - areas of different engineering geological setting, 20 - zones of different changes in geological medium, 21 - influence zone of the Baikal drainage tunnel.

the most vulnerable parts of the shore, they are regarded as the first-order protection zone (Regulations... 1987). In this paper we discuss environmental problems of the Sliudaynka region as mining is developed there.

Geoecological peculiarities of the Sliudaynka region are predetermined by geological evolution of the area. The first factor is uprise of the Baikal domal uplift, the south-western flank of which hosts the deposits mined. The second factor is formation of the Lake Baikal depression, as the lake basin accumulates the surface and ground waters and technogenic streams. With respect to these factors and according to the degree of

susceptibility of the geological medium to natural and technogenic impact, three types of engineering geological setting are distinguished in the territory studied.

The first type of engineering geological setting is typical of slopes of the mountainous framework. It is characterized by development of strongly cemented rocks of metamorphic and magmatic formations overlaid by a thin loose sedimental cover.

The crystalline rocks differ in petrographic composition, their strength values varying in a wide range from 46 to 293 MPa. Thus these rocks are highly resistant to loading. Fractures, fault and shear zones weaken the rocks. The strength of granite-gneisses is lowered by anisotropy of their properties.

Fissuring in the range from 4 to 8% is irregular both in space and depth. In carbonate rocks, it is enhanced by karst processes. The strength of gneisses and crystalline schists is considerably lowered in fault zones wherein the rocks are crushed to fragments 0.5 - 0.8 m in diameter. The weathering waste cover of considerable thickness (up to 20 m) overlies the zones of rock breakage. Strong physical weathering and erosion intensify and further develop these features. Thus, the rocks of the mountainous framework of the basin provide the solid phase of mudflows.

Karst processes play an ambiguous role in the regional engineering geological conditions. On the one hand, karst-vein waters occur in carbonate rocks below the erosional incises, that makes underground working even more complicated. On the other hand, high filtration ability of rocks subjected to karstification favours deep drainage of the mined carbonate massives. Therefore the quarries as sources of disturbance of the geological environment are isolated features.

Tectonic fracturing of rocks is the most important factor in terms of geological ecology. First, it predetermines the presence of fissure, fissure-karst and vein waters (Pisarsky 1968). Secondly, fractures and faults which may penetrate down to 70-100 m depths allow active interaction of ground and surface waters and thus heighten the danger of their mutual pollution.

The geoecological potential of the topography as a territorial resource is moderate. Owing to highly dissected topography this type of the engineering geological setting is characterized by scarce areas which are convenient for development. However, the local topography facilitates open-cast mining in exposures due to deep drainage and hence low water content in rocks. Geodynamic instability of the relief is manifested by landfalls and landslides at the slopes of valleys which provide solid material for mud-rock floods. The Sliudaynka belongs to the rivers of high mudflow activity in the South Pribaikalie.

The second type of engineering geological setting is characteristic of flattened divides. The basalts covering metamorphic rocks are remarkable for their strength of 76-288 MPa. The basalt body is fissured and porous through its whole thickness. Under weathering the basalts produce coarse fragments and argillo-arenaceous material. According to the rock strength and the geodynamic conditions, the discussed type of environment is resistant to technogenic impact, however its zone lies outside the area of development.

The third type of engineering geological setting is characteristic of the piedmont plain where alluvial-proluvial, alluvial-lacustrine sediments are predominant. The thickness of these loose clastic, mostly incoherent rocks varies from 50 m near the Sliudaynka river outlet to 90 m at the settlement of Kultuk. Incompressible and slightly compressible rocks are dominant. Shallow ground waters, swamping and mud-rock floods cause complications in development of this zone.

However, almost the whole space of this zone has been already developed. A number of industrial and urban buildings and constructions are situated in the hazardous zone of the piedmont plain wherein mud-rock flows generated in the mountainous framing may be discharged. Occurence of mudflows becomes even more probable under conditions of high seismicity and heavy rainfalls.

With respect to preservation of Lake Baikal, the geoecological potential of the territory is predetermined by the capability of the geological environment to produce and transfer natural and technogenic contaminants. The petrogenic composition of the mined crystalline rocks excludes likelihood of occurence of deleterious agents above limited amounts. High capability of technogenic contaminants transfer is controlled, first, by openness of the ground water to warming effect of the lake water in the absence of a permafrost waterproof strata, secondly, by increased permeability of quarried crystalline rocks and loose rocks and by high velocity of water currents which favour close interaction of the surface, ground and lake waters while self-purification is impossible. These geoecological factors suggest high vulnerability of the Lake Baikal shores to technogenic impacts on the water component of the geological environment.

3. THE SLIUDYANKA MINING SITE AS A TECHNOGENIC FACTOR

Peculiarities of exploitation of the Sliudyanka mining site are predetermined by the regional topography. Man's activities are mainly confined to the piedmont plain, its spacial resources being almost exhausted. The mountainous framework is suitable only for mining. As a technogenic factor, the Sliudyanka mining site is represented by underground and open-cast working.

Formation of the Sliudyanka industrial complex started in the early 1900s with underground working of a phlogopite field. For almost 20 years the deposit has been closed down. Nonetheless, mine-workings branching down to 150 m and the Baikal drainage gallery still function as a constant source of disturbance of the underground hydrosphere. The disturbed area approximates 21 square kilometers. Abandoned quarries and small-volume dumps are additional local sources of technogenic impact on the environment.

Mining is carried out by open-cast working of marble at three localities. The "Pereval" quarry is distinguished by its large mining area, the maximum height of its rock face is about 430 m. The technogenic effect is imposed by quarries and dumps with slopes of 33 and 38 as well as by hole firing and processing of the raw material. Within the limits of the mining site, the major amount of goods traffic is done by motor transport. Outside transportation is carried out by railway. Only at the "Pereval" quarry a cable way is used to deliver the raw material to the processing complex.

As a technogenic factor, the open-cast working has the following peculiarities which are of importance in terms of ecology.

1. In highly drained rocks the disturbed zone is represented by isolated spots which correspond to the localities which have suffered technogenic effects.

2. The areas subjected to technogenic processes are of almost no utility for any other activities except mining.

3. The material mined does not contain contaminating ingredients and its processing does not involve chemical treatment.

Thus the impact of mining discussed below is only typical of mineral raw material which can be regarded as clean or positive in terms of ecology. However, the long-term experience of the Sliudyanka mining site exploitation provides representative data for the analysis of consequences caused by mining. Trends of environmental changes have been revealed with respect to topography (a mountainous framework or a piedmont plain), mining techniques (underground or open methods), types of raw materials being mined and the lifetime of technogenic relief forms.

4. IMPACT OF MINING ON GEOLOGICAL ECOLOGY

The most typical changes in geological conditions within the limits of the Sliudaynka mining site are concerned with changes in the state of stress of the rocks in quarry walls, in the vicinity of mines, underneath the dumps and after the phlogopite mine-drainage. Due to high strength of the material mined, such changes do not induce considerable deformation. On slopes of quarries the most fractured rocks suffer rockfalls, landslides and screes which are sometimes initiated by earthquakes or explosions. Land subsidence occurs above the shallow underground workings.

Under conditions of high precipitation, increasing permeability of quarried rocks is another nuisible effect. As erosion removes much of the upper regolith, fractures and open-joint fissures become exposed. The infiltrational waters wash the fracture infill out and develop fissure-karst caverns. Erosion and weathering can affect greater depths. In crystalline rocks, this effect is slight, though karst process may be enhanced. In terms of ecology, it is of importance that atmospheric, ground and surface waters interact more closely in quarry fields. The consequence manifests itself by the sharp elevation of the ground water level which was heightened to produce water gushes in the bore-holes located in the valley of the Sliudyanka river after a 60 m/sec mud flow had been consumed by a quarry (Solonenko 1961).

Contaminating agents washed away by rain waters can be transferred by the above described ways. In case of the Sliudyanka mining site, the scale of such pollution is quite moderate whereas mining of raw materials with noxious ingredients would be detrimental to the lake ecology.

Waste rock piling causes considerable environmental changes. Small-volume dumps cover the valleys and their slopes around the sites where underground workings come to the surface. Open-cast mining also causes degradation of the natural landscape around quarries and dumps. Dumps are the most injurious when placed on mudflow-active river banks. Lumps and boulders are transported to bottoms of streams. The lower parts of large-volume dumps may be washed out by spring floods. Landwaste is easily subjected to erosional and suffosional processes, landslides and landslips. These processes cause contamination of surface waters by suspended solids during heavy rainfall periods when loose waste material is transported to river beds. In addition, crushed material is stockpiled at localities where attempts are made to improve the topography since the areas suitable for construction are scarce.

The mine-drainage of a phlogopite field has produced the worst effect on the environment. The Baikal drainage tunnel caused dewatering of the rocks in the area of 21 sq km. The ground water slope and velocity changed in the direction to the tunnel (Pinneker et al. 1979). Such changes have not considerably affected the high-strength rocks with deep surface of ground waters, though critical elevation of the ground water level took place in the areas of loose rocks. Due to formation of an influence zone around the drainage tunnel, the Sliudyanka river surface runoff is lost in the town of Sliudyanka. Therefore water pumping is complicated there, deformation of buildings is often noted.

The injurious effect of mining on the piedmont plain is evidenced by a shallow (up to 3-4 km) quarry of the early 1900s which provided material for the permanent railway. The shallow ground water level of this technogenic depresssion is anomalous relative to that of the lake shore. In consequence, the residential area regularly suffers from critical elevation of the ground water level.

Dust around quarries and dumps is another negative factor in terms of ecology. According to studies of chemical composition of snow carried out by the Sosnovgeologia enterprise, the daily technogenic solid phase approximates 600 kg/sq km in the vicinity of quarries, the values tailing away in the direction to the lake.

The ecological potential of the area studied is lowered by degradation of the natural landscape and inevitable forest extermination on large areas. For instance, the "Pereval" quarry has destroyed 150 hectares though only 74 hectares are occupied by open-cast mining. However, after mining is brought to a close, natural afforestation will take place. As typical for disturbed areas, dark conifers will be substituted by broad-leaved forest. Such afforestation has already started at some dumps around the abandoned underground workings.

Contamination of ground waters on the territory of the town of Sliudyanka immediately affects the waters of Lake Baikal. According to regime observations carried out by the Irkutsk geological survey, bacterial and chemical pollution of ground waters is registered down to 50 m, contamination from mining being slight. Metal-working industry servicing the railway and manufacturing industries is the major contamination source in the town.

Thus, low-scale technogenic changes occur in the zone of open-cast mining (see Fig. 1) due to moderate negative effects. Middle-scale technogenic changes are revealed in the limits of the underground mining field, in the lower part of the Sliudyanka river valley and in the urban area, since natural relationships between components of the geological medium have been disturbed. High-scale technogenic changes are typical of the low-lying lake shores. However, they result from a rise in the lake level as Lake Baikal is the major water body in the system of water reservoirs on the Angara river.

5. CONCLUSION

Mining at the southern shores of Lake Baikal plays a minor role in the total balance of the lake water contamination. The reason is that the mineral material being mined is free of noxious ingredients and its processing does not involve any chemical treatment. The ecological impacts of mining are concerned with degradation of the natural landscape, contamination of the air and soils by dust, increase in mud-flow activity of rivers. Owing to openness of ground waters to pollution and their close interaction with surface and lake waters, mining and processing of ecologically dangerous material is impermissible in the first-order protection zone of the lake. Prospects of broader development of cement production in the Sliudaynka geoecological region are fairly moderate. In terms of ecology, it is reasonable to develop alternative deposits located outside the catchment area of Lake Baikal.

REFERENCES

Demianovich, N.I. & Pisarsky, B.I. 1993. *Technogenic geological systems of hydrotechnical constructions, natural and artificial water reservoirs. Lake Baikal basin. In Problems of geological medium protection (on the example of East Siberia).* Novosibirsk: Nauka, 117-123 (in Russian).

Pinneker E.V. 1991. *Some problems of geological medium protection and studying.* Engineering geology 3: 3-8 (in Russian).

Pinneker, E.V., Pisarsky, B.I. & Blokhin, Yu.I. *Preliminary results of the International symposium on radiometric equipment and methods of studies of ground water in karst areas. In Ground runoff on the territory of Siberia and methods of its studies.* Novosibirsk: Nauka, 109-118 (in Russian).

Pisarsky, B.I. 1968. *Hydrogeology of mineral deposits. In Hydrogeology of Pribaikalie. Moscow: Nauka, 148-155 (in Russian). Regulations of protection of waters and natural resources in Lake Baikal basin (1st edition) 1987.* Moscow: Soyuzgidrovodkhoz, 48 p. (in Russian).

Solonenko, V.P. 1961. *Mud-rock floods at Lake Baikal.* Priroda 5: 61-64 (in Russian).

Geomechanical aspects of control over the geological medium in mineral resources mining and underground construction

Aspects géomécaniques du contrôle sur l'environnement géologique dans l'exploitation des ressources minières et la construction souterraine

N. N. Melnikov, A. A. Kozyrev & V. I. Panin
Mining Institute, Kola Science Centre, Russian Academy of Sciences, Apatity, Russia

ABSTRACT: The Kola Peninsula is taken as an example here to show the relationship between large volumes of mined out and transported rock masses and mining-induced geodynamics and seismicity of the region. The ways to reduce mining-induced effects on geological medium are shown. The system and results of geomechanic monitoring carried out in the Khibiny and Lovozero massifs are presented.

Due to the general worsening of the ecological situation during last years, the interest increases to the problems of control and prediction of mining-induced changes in a geological medium in the regions of large mining complexes (Zaitsev 1986). Besides, the ecological effect of the present mining industry is not confined only by the change of the surface relief being the result of waste dumps and milling wastes storage. The world experience witnesses that large-scale mining essentially violates the regime of a geological medium, which creates the additional ecological tension in mining regions. This is proved by powerful mining-induced tectonic rock bursts registered in the Utah coal mines, USA, in the Verra salt mines, Germany, in the Tashtagol deposit, the Rudny Altai, in the North- and South-Ural boxite mines, in the khibiny and Lovozero mines, the Kola Peninsula, Russia. These powerful seismic events result in destruction of surface and underground structures, mining becomes very dangerous to be carried out, material costs needed to avoid the consequences of these seismic events become substantial.

By their energy of (10^9-10^{12}) J and the force of action exerted on the environment, such powerful rock bursts may be qualified as mining-induced earthquakes the probability of occurrence of whose becomes greater with the increase in scale and depth of mining. It is illustrated by the research results obtained in apatite mines (Kozyrev 1993). The correlation between the volumes of the rock masses mined-out and some characteristics of geodynamic regime of the Khibiny massif has been established. The analogous regularities have been registered

in hydrotechnical engineering in impounding water reservoirs, in oil production regions in pumping-out oil and gas from deep holes.

By the existing notions, the mining-induced earthquakes differ from natural ones in the mode of seismic energy release due to the anthropogenic factors action. They are characterized by a great number of foreshocks, i.e. antecedent shocks or mining-induced rock bursts, and slow decay of aftershocks, as well as an increased magnitude because of the site of their origin being located not very deep. Therefore, for these mining-induced earthquakes to be initiated, needed is lower energy; the plot of occurrence of seismic events is steeper.

The main conditions favourable for the mining-induced earthquake formation are the presence of hard brittle rocks characterized by high tectonic stresses in the mining region, large areas and volumes of mining, as well as technological explosions playing the role of a trigger mechanism in a shock-like motion of large blocks or avalanche-like fracturing.

To eliminate or to substantially decrease the affect of mining on the environment or underground structures is possible if three main conditions are observed, namely:

1. The reliable information on the geomechanical state of the geological medium in the regions of creation of any mining enterprise or any large important object construction should be available at all the stages of site engineering and geological surveying and development of the appropriate area.

2. Working out and application of organizing and technological measures ensuring the most appropriate conformity of mining

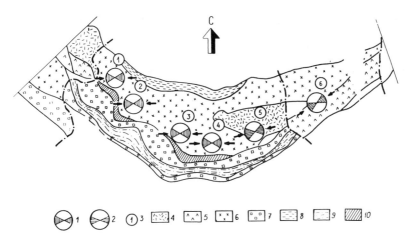

Fig. 1. The directions of tectonic stresses in the exploited and perspective deposits of the Khibiny massif:
  1 – by instrumental determinations;
  2 – forecasted – in deposits:
1 – the Kukisvumchorr, 2 – the Yuksporr, 3 – the Apatite Circus deposit, 4 – the Rasvumchorr Plateau, 5 – the Koashva, 6 – the Oleny ruchey
  Enclosing rocks: 4 – nepheline syenite, 5 – massive micaceous rischorrite, 6 – rischorrite, 7 – ijolite-urtite, 8 – gneissic rischorrite, 9 – trachitoide ijolite, 10 - ore body.

working designs and the elements of mining with geomechanic and mining-technical conditions of the geological medium.

3. Geological and geophysical medium monitoring in the area of geomechanic space near the mines and underground structures.

The main characteristics of geomechanic state of rock massifs are the parameters of rock properties and state of stress of rocks in situ. A special procedure is worked out including both the well known ways and methods of stress determinations in situ and the new original ones. Of most interest among them is the evaluation technique of the stress field parameters by the geological prospecting drilling data (Melnikov 1991).The information about state of stress of rocks in the area studied is obtained without any additional expenditures. This technique has been widely tested in additional prospecting of deep horizons being mined and in prospecting new deposits. Fig. 1 shows the results of determination of tectonic stresses directions both in mined and in perspective deposits in the Khibiny massif.

Using the information about state of stress of rocks in the massif and about their physical properties, the strategy of mining is planned, which comes mainly to decrease the stress concentrations in the zones of active mining or to decrease the ability of near-contour rock masses to accumulate the critical levels of magnitudes

of potential energy of elastic strain (Melnikov 1993).

But even in most optimal mining it is impossible to completely avoid the affect of mining on the environment. Mining out and handling the substantial volumes of rock masses change the mode of strain in the surrounding geological medium, the technological explosions act like triggers producing catastrophic dynamic phenomena, which is confirmed by the appropriate practice. Therefore, working out the ways to predict and to prevent such phenomena is one of the most ecological problems and the solution of it depends to a greater extent on geomechanical monitoring of geophysical medium of the mining region. Besides, according to (Radionov 1986) the volume under control may be determined from the expression:

$$V \geqslant (0.4 \cdot 10^8) L_0^3$$

where L – is the dimension of a newly formed parting resulted from a dynamic phenomenon.

In the Khibiny apatite mines the mining-induced tectonic rock bursts result in L = 100 m, which corresponds to energy of a 7 or 8 energetic class of earthquakes (E = $10^7 \div 10^8$ J), therefore, the area of about 40x35x30 km cub. should be observed, i.e., practically, the whole area of the Khibiny massif to the depth of about 30 km. It is supposed that the extent of the site of origin for maximum earthquake will not exceed

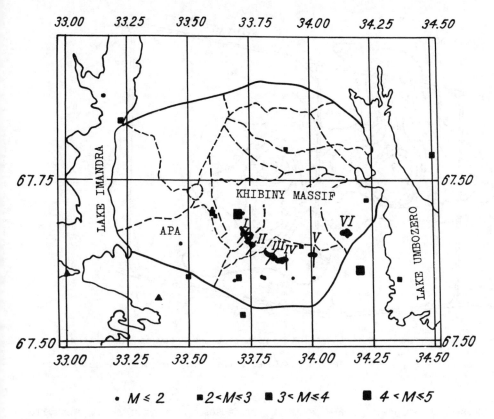

Fig. 2. The position of mines (I–VI) in the Khibiny Massif together with fault structures and earthquakes. Mines I, II and III are underground, whereas IV, V and VI are open-pit mines. The location of the Apatity seismic station (APA) is also shown.

L_{max}=0.8 km, which corresponds to a K = 12.5 energetic class event, i.e. the earthquakes of M 5.1–5.2. The area of preparation of such an event, i.e. of manifestation of a number of possible portents of the event, equals about 10 000 km .

The most informative seismological portents of the earthquakes are the parameters of the seismic regime and strain processes at levels of a different scale. Basing on it, worked out and realized is the programme of works carried out in the Kola geodynamic testing ground including :

1. Stress and strain control;
2. Regional seismological observations;
3. Seismoacoustic emission registration within the shaft fields and horizons;
4. Joint acoustic and electromagnetic emission registration in the local areas of mining workings
5. Physical and mathematical modelling of strain and failure processes in rocks and of stress distribution in the structures of different types and systems of mining

workings.

Stress determination is carried out in mining workings by stress relief and ultrasonic methods, as well as by the parameters of the prospecting holes failure. Special stations are equipped with high precision apparatus for strain control and periodic high class levelling is carried out in underground and surface testing grounds. Some results are presented in (Kozyrev 1988) and (Kozyrev 1993).

Seismological observations are made in the Apatity seismological station which is reckoned among the members of a single seismological network of Russia and North-West Europe (Kremenetskaya 1992). The seismicity of the Khibiny massif is shown in Fig. 2 where the earthquake centres are grouping in the area of mining works, predominantly near the Kirovsky, Yuksporsky and Central mines, which is explained by maximum volume of the rock masses mined out and replaced just in this area. The second group of the earthquakes gravitates towards the south boundary

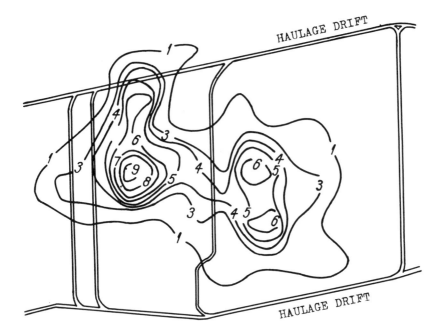

Fig. 3. A map of seismic activity in one of high-stressed areas of the Kirovsky mine (as of the state of mining works in 1985).
The figures on isolines are the average values of the relative energy magnitudes of seismic events.

of the Khibiny massif where located are the tailing dumps of concentrating mills and of thermoelectric power plant, ore storages as well as large industrial and civil objects, i.e. to the area where the mining-induced effect on the surface is rather great.

To localize the dynamic events within the shaft fields and horizons of the operating mines, the automated system to carry out the control over rock seismicity in a massif is worked out, its frequency band is to 80 Hz at the seismic energy level of (10^2-10^6) J (Melnikov 1987). Fig. 3 shows the map of the seismic activity in one of high-stressed areas near the Kirovsky mine of the Apatit industrial association. Using such maps it is possible to determine the areas of dangerous stress concentrations on the plans of mining works and topredict the point of rock bursts as well as to evaluate the efficiency of antishock preventive measures.

The state of some mining workings is controlled by the data of acoustic and electromagnetic emission registered in the frequency band from 20 to 100 Hz.

To interpret the field research data and to determine new ways to evaluate the state of rocks in a massif, some laboratory experiments are carried out under the conditions of high pressure to study strain and failure

processes occurring in rocks. For instance, the test results obtained by the nepheline syenite samples have shown the actual possibility to evaluate the state of stress of rocks by the emission of gases of hydrocarbon group and by rhodon (Nivin 1993).

The systematic observations made in the geodynamic testing ground allows to determine the portents of catastrophic events and their threshold values, and to work out the proper prediction models.

In conclusion, it is necessary to pay attention to the international aspect in the solution of this problem. A lot of efforts is paid in many countries lately to work out the methods to predict the earthquakes, including those of mining-induced nature. But such recent catastrophic earthquakes which took place in Armenia, Georgia, USA, India, Japan testify that the solution of this problem has not been found yet. It appears, that the international co-operation should be closer to solve this problem. The most realistic and effective form of such a co--operation could be joint researches in geo dynamic testing grounds where the latest equipment and apparatus could be tested, informative portents of the seismic events of different classes determined, the software to process the geophysical data available perfected, and the appropriate prediction methods worked out and tested.

Thus, the attention should be paid to the problems of geological medium control at all the stages of the deposit mining or in any underground structure construction. Besides, the information base for the necessary organization or technological decisions to be realized, should be supplied by the proper monitoring systems.

REFERENCES

Kozyrev, A.A., Lovchikov, A.V., Osika, V.I. and Popov, E.I. 1988. Monitoring of present cristal movements in the Lovozero massiv by high-precision methods. Journal of Geodynamics, 10, p.263-274.

Kozyrev, A.A., Panin, V.I. 1993. Effect of the area and mine-induced seismicity manifestation. Safety and Enviromental Issues in Rock Engineering. Rotterdam, Balkema, p.841-845.

Kremenetskaya, E.O. & Trjapitsin, V.M. 1992. Induced seismicity in Khibiny massif (Kola Peninsula). Semiannuale Technical Summary. NORSAR Scientific Report, N1, p.125-127.

Melnikov, N.N., Raspopov, O.M., Yerukhimov, A. Kh., Kuzmin, I.A. 1987. A new instrument of investigations in mining geophysics. Vestnik AN SSSR, N 5, p.6-15.

Melnikov, N.N., Kozyrev, A.A., Panin, V.I. & Gorbunov Y.G. 1991. The prediction of stress state in situ at the stage of prospecting drillihg. Pr. 7 Int. Congress on Rock Mechanics. Aachen, p.1725-1727.

Melnikov, N.N., Kozyrev, A.A., Panin, V.I. 1993. Geomechanical controlof mining in high-stressed rock mass. Assesment and prevention of falure phenomena in rock engineering. Rotterdam, Balkema, p.699-701.

Nivin, N.I., Belov, A.N. & Petrov, A.N. 1993. On practicable application of gas phase as an indicator of rock tectono-physcal state in construction of underground storages. Geoconfine 93. Rotterdam, Balkema, p.93-98.

Radionov, B.N., Sizov, I.A., Tsvetkov, V.M. 1986. The principles of geomechanics, Moscow, Nedra, 1986, 301pp.

Zaitsev, A.S. & Ustinova, Z.G. 1986. Control over the state of geological medium in the regions of large mining complexes activity. Prospecting and control over the entrails of the earth. N 3, p.33-36.

Geotechnical and geochemical properties of some clays occurring in Ilorin, Nigeria and the environmental implications of their mode of exploitation

Propriétés géotechniques et géochimiques de certaines boues à Ilorin en Nigéria et les implications environnementales de leur mode d'exploitation

O. Ogunsanwo & U. Agbasi
Department of Geology, University of Ilorin, Nigeria

E. Mands
Institut für Angewandte Geowissenschaften, Giessen, Germany

ABSTRACT: In-situ derived clays occur in several parts of Ilorin, Nigeria with the largest deposit occurring in the Okelele ward. These clays have been utilised since time immemorial locally for pottery. Investigations show that they are very good materials for this purpose as well as for some engineering construction works. The clays are obtained by means of shallow hand dug pits by several individuals working in isolation. As a result, the area of operation is littered with several of such pits.

The health hazards which are likely to be caused by the stagnant water in the pits are temporary and may be adequately taken care of. However attempts to fill the pits are grossly inadequate. This has created an uneven profile within the terrain with attending geotechnical implications especially in foundation works.

RÉSUMÉ: Les boues provenant de l'interieure de la terre se forment en differents points à Ilorin, an Nigeria avec le plus grand depôt se formant dans l'arrondissesment de Okelele. Ces boues ont été utilisées localement depuis le temps immemoriaux pour la poterie. Les investigations montrent qu'elles sont de très bons materiaux pour ce besoin (poterie) aussi bien que pour certaines travaux de construction. Les boues sont obtenues par le moyen des creusements manuels des puits par plusieur personnes travaillant en groupe isolées. De ce fait, la region d'operation est jonchée de plusienurs puits de ce genre.

Les santés hasardeuses dont l'origine de situe dans les eaux stagnantes de ces puits, sont temporaires et peuvent être adequatement contrôlées. Ce pendant, les tentatives de remplir les puits sont inadequates. Ceci a avec une situation peu confortable dans ce terrain avec les implications géotechniques specialement pour les travaux de fondation.

1 INTRODUCTION

The present world wide recession has caused diversification of activities: new avenues are being sought while the existing ones are being reviewed. Cottage industries utilising clays for pottery have been quite successful. However, the recent situation which calls for the optimum necessitates critical appraisals of the clays so as to ensure their most economic and fullest utilisation.

The clays which occur in the Okelele ward of Ilorin, Nigeria have been utilized based essentially on the experiences of the workers as no scientific data on the properties of the clays have ever been reported. The first part of this study delves into the geotechnical and geochemical properties of these clays with a view to finding the best uses into which the clays can be put.

The clays are exploited by all sundry - everyone working in isolation and this has led to a more or less haphazard mode of quarrying via pitting. The clay horizon is rather thin. As a result, the hand-dug pits are soon abandoned and new ones dug. The area of operation is thus littered with such pits which are filled to the brim with rain water during the rainy season. During the dry season some attempts which totally lack geotechnical expertise are made to fill up the pits. The second part of this study considers the health hazards posed by the stagnant water in the pits. It also highlights the geotechnical problems that will arise as a result of the ordinary way of filling of the pits.

1.1 Geology and derived soil

Nigeria may be divided geologically into two: the area underlain by the crystalline rocks (the basement complex) and the area underlain by sedimentary rocks - each area constituting roughly 50%. The city of Ilorin and its environs are underlain by the rocks of the basement complex which are pre-cambrian in age.

Clays are classified geologically on the basis of origin into two groups: (i) primary or residual clays and (ii) transprted or sedimentary clays. The primary or residual clays are those formed in-situ The clays may be products of intensive chemical weathering, hydrothermal or pneumatolytic alteration of different rock types and the characteristics of the resulting clay will obviously depend on the physico-chemical factors dominating the environment of formation, climate and the parent rock mineral assemblage (Pettijohn, 1975). The clays used in this study are believed to have been formed in-situ by the weathering of a granite gneiss which is the rock type in the sampling locality. The sampling locality, Okelele, is situated north west of Ilorin township and slightly southwest of the Sobi Hills. (Ilorin lies approximately on longitude 4°35'E and latitude 8°30'N). Samples were taken from the hand dug pits - the soil profiles and the sampling horizons are shown in later text.

2 METHODS OF TESTING

The classification, Proctor compaction (standard and modified) and the unconsolidated-undrained (U-U) shear box tests were carried out following the guidelines of BS 1377 (1975). The shear box was employed because it is much more adaptable than the triaxial compression equipment. It has also been found that the soil parameters ϕ (angle of internal friction) and (cohesion) obtained by the shear box method are about as reliable as the triaxial values (Bowles, 1981). The shear tests were performed on the standard proctor compacted samples.

The mineralogy of the soil was investigated via an X-ray Diffractometer (XRD) and the chemistry via an X-ray Flouresccence (XRF)

3 RESULTS

The clays contain appreciable amounts of sand and silt (Fig. 1), their fractional percentages being gravel = 0 - 2; sand 21 - 36, silt = 14 - 16 and clay 48 - 63. However, the clays are sieved through 1.2 - 1.4mm diameter sieves before use in the pottery industry locally. The fractional

percentages of the "processed soil" are thus virtually the same as those of the natural soil but without gravel.

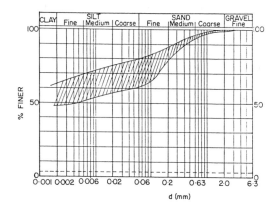

Fig. 1 Grain size distribution

The bulk densities of the clays range from 1.99 to 2.00 t/m^3 at a natural moisture content of 22%. Specific gravity = 2.66. The values of the Atterberg consistency limits obtained were constant viz: liquid limit (w_l) = 58%, plastic limit (w_p) = 23%. Index of plasticity (Ip) = 35%. The clays thus plot in the CH zone of the plasticity chart implying that they are inorganic clays of high plasticity. The linear shrinkage values vary from between 6% at 105° and 9% at 900°, loss on ignition = 7%.

The maximum dry density (MDD) of the clays increased appreciably (1.72 to 1.89 t/m^3) when the compaction energy was increased from that of standard proctor (0.60 MNm/m^3) to that of modified proctor (2.69 MNm/m^3). The optimum moisture content decreased from 13 to 12 (Fig. 2). The plot of shear stress against normal stress obtained for the soil shows that it possesses a c value of 28KPa and a ϕ value of zero.

Mineralogical analyses show that this soil is a kaolinitic clay that contains 40 to 50% native quartz, 40 to 45% kaolinite. The kaolinite is a smectite mixed layer and probably contains minor amounts of chlorite and or vermiculite that is in a decomposing stage. Results of the chemical analyses are shown on Table 1.

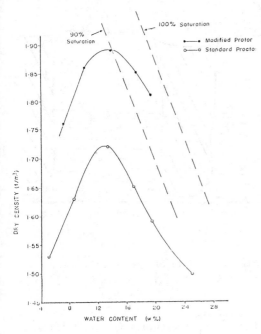

Fig. 2 Compaction curves

Table 1. Results of the chemical analyses of the Okelele clay

SiO_2	56.36%
Al_2O_3	22.53%
Fe_2O_3	4.88%
MgO	0.47%
CaO	2.93%
Na_2O	1.45%
K_2O	0.56%
TiO_2	0.95%
Cu	24ppm

4 DISCUSSIONS

4.1 Geotechnical

The liquid limit (58%) obtained for the clays shows that the soil does not contain appreciable amounts of swelling clays. This then means that the mixed smectite layer contains more chlorite than Montmorillonite. The low linear shrinkage obtained (6% at 105°) is also in support. According to the Engineering Use Chart of Wagner (1957), soils that plot in the CH zone of the plasticity chart should possess poor shearing strength when compacted.

The results obtained for this soil (c = 28KPa, $\phi = 0°$) conforms with the statement. Wagner (1957) also stated that this type of soil should have a very low permiability when compacted. It may then be useful as core of earth dams. However as the clay occurs as a thin layer with a small areal extent making it a small deposit, it is necessary to find other avenues where it can be more economically utilised.

4.2 Mineralogical and chemical

From the mineralogical and chemical analyses, this soil will find application as Refractory bricks (Parker, 1967), brick clays (Murray, 1960) and ceramics (Singer and Sonja, 1971). These of course will be in addition to use in the already successful pottery industries.

5   ENVIRONMENTAL CONSIDERATIONS

Health: The pits from which the clays have been excavated are filled with rain water during the rainy season (Fig. 3). The wa-

Fig. 3 The pits filled with rain water

ter filled pits naturally serve as breeding grounds for mosquitoes and other undesirable animals. Also, the water supply in this area is far from being adequate. There is therefore the possibility of the inhabitants of this area of resorting to this murky water.

2743

Engineering geological: Attempts are made during the dry season to level out the excavated area (Fig. 4). This process fills

Fig. 4   The filled-up pits

up the pits but causes a general depression of the area. This fiiling up of the pits with some of the overburden soil is done loosely - it is not accompanied by any form of compaction. Samples from the fills taken at the end of the rainy season/the onset of the dry season have natural densities ranging from 1.52 to 1.57 t/m^3. (The Okelele clays have an average natural density of $2.00 tm^3$).

The profile thus obtained is as shown in Fig. 5 and this can pose serious problems for engineering construction works. A soil investigation procedure that misses out the filled portions of the terrains and uses only the results obtained for the Okelele clays will most likely run into difficulties. This is because the clays, based on texture and density considerations, are bound to possess geotechnical properties which will be quite different from those of the fill material.

6   CONCLUSION

The Okelele clays are quite suitable for certain engineering construction works. The clays will find application in the manufacture of ceramics and refractory bricks. In fact they have been successfully used for pottery since time immemori-

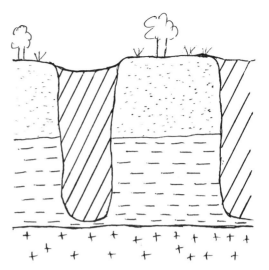

Fig. 5   Profile of the terrain after the filling-up

LEGEND

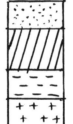

Overburden ($\gamma = 1,66$ t/m^3)

Fill material (derived from overburden) ($\gamma = 1.54$ t/m^3)

Okelele clay ($\gamma = 2.00$ t/m^3)

Weathered bedrock

al.

However, the mode of exploitation of the clays poses certain environmental problems. It is advisable that the individuals who dig the numerous pits work together as a team in a systematic manner. In that way, there will no longer be any pits and the filling of the terrain will result in a uniform profile.

REFERENCES

B.S. 1377. 1975. Methods of Testing Soils for Civil Engineering Purposes. British Standards Institution.
Bowles, J.E. 1981. Engineering Properties of Soils and their Measurement. McGraw-Hill, Tokyo.
Murray, H.H. 1960. Clay Industrial Minerals and Rocks, 3rd Ed. An. Inst. of Mining, Metall. and Pet. Engrs., N.Y.
Parker, E.R. 1967. Materials Data Book for Engineers and Scientists. McGraw-Hill,

N.Y.

Pettijohn, E.J. 1975. Sedimentary Petro-
    logy. Harper and Row, N.Y.
Singer, F. & Sonja, S.S. 1971. Industrial
    Ceramics. Chapman and Hall, London.
Wagner, A.A. 1957. The use of unified
    soils classification system by Bureau
    of Reclamation. Proc. 4th ICSMFE,
    London, Vol 1: 125

Acknowledgements: The author is grateful
to the authorities of the University of
Ilorin, Nigeria for partly financing this
work through its Senate Research Grant.

Impacts caused by mining subsidence in the Germunde coal mine (NW of Portugal)

Impacts sur l'environnement dus à la subsidence minière dans la mine de charbon de Germunde (NW du Portugal)

A. Fernandes Gaspar, C. Mendonça Arrais, J. Pinto Barriga & P. Bravo Silva
Empresa Carbonífera do Douro, Castelo de Paiva, Portugal

C. Dinis da Gama
Instituto Superior Técnico, Lisbon, Portugal

ABSTRACT: The exploitation of coal using a variant of the sublevel caving method, causes the fissuring of the adjacent rock mass, inducing subsidence at the surface.

The limits of the subsidence cone are essentially determined by discontinuities, geological contacts and faults, resistance characteristics of the rock mass and depth of exploitation.

The methodology used to quantify the damages at the surface was based on the individual classification of the buildings, which consisted in assigning a punctuation involving the type of construction, state of maintenance and proximity of fissures.

The buildings were grouped in classes with different hazardous indices, which allowed to delineate a strategy to minimize the socio-economic and environmental problems related with the degradation of the subsidence zone.

RESUMÉ: L'exploitation du charbon par l'abatage du toit - méthod type "sub-level caving"- produit une fissuration progressive du macif rocheux jusqu'en surface, avec l'affaissement du terrain - c'est la subsidence minière. Le domaine du cone de subsidence est controlé par plusieurs paramètres, dont les plus importants sont les discontinuités, les contacts géologiques et les failles, les caracteristiques géomecaniques des roches et la profondeur de l'exploitation.

La methodologie apliquée pour evaluer les dommages induits en surface a été basée sur un certain classement des bâtiments, en attribuant des poids (exprimés en chiffres) au type de construction, à l'état de conservation, à la proximité de grosses failles ou fractures et à la vitesse de subsidence dans l'endroit.

Avec ce critère-ci il serait possible de classer les bâtiments dans des groupes probabilistiques de tendance à la degradation et, par conséquence, d'établire une strategie pour amoindrire les problèmes décurrents socio-économiques et environnementales.

1 INTRODUCTION

Although the exploitation of coal in Germunde began in the early years of this century, mining was only developed from 1945/46, and until 1966 the upper levels in slope exploitation were mined, and only at that date began the first level, with access by shaft.

Around 1966 the cut and fill method, was subsituted for a variant of the sublevel caving. In terms of surface consequences, there are two distinct periods: until 1966, and from then until the present days.

During the first period the surface damage was not significant, apart from the places where exploitation reached the surface.

After 1966, as the excavations have not been refilled, the collapsing of the rock mass, with the swelling and later consolidation of the material, began to be the main responsible for the fracturing in the neighbourhood of the stopes and to its propagation to the surface, affecting a much larger area.

2 GEOLOGICAL SETTING

The Germunde coal mine belongs to the Douro Coalfield, which forms a narrow strip of continental Carboniferous formations striking NW-SE. In the Germunde area, the footwall of the Carboniferous formation is formed by Upper Precambrian to Lower Cambrian shales and greywackes.

The hanging wall, on the northeastern side of the trough, is formed by Ordovician slates and quartzites, integrating the southwestern limb of the Valongo anticline.

The contact between the Carboniferous and the Ordovician formations is a thrust fault filled with an important amount of clay which in part controls the infiltration of water into the mine.

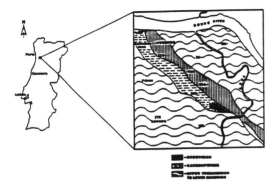

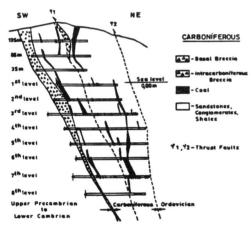

Fig. 1 Geological map (A.) and cross section (B.) of the Germunde coal mine

The structure of the Douro coalfield is formed by a complex system of subvertical layers. The most common lithologies are sandstones, conglomerates, schists and coal. The coal seams are branched in a sigmoidal irregular shape, which makes the mining extremely difficult. Douro river runs along the northern border of Germunde mine, defining a limit to exploitation and confering particular importance to the hidrogeological characteristics of the formations.

3 SURFACE DEGRADATION AS A RESULT OF SUBSIDENCE

The consequences of mining subsidence, concerning surficial degradation, causes

the worst social impact, involving high costs in its resolution.

In the case of Germunde mine, besides controling the phenomena in the afected area, it is sought to determine the maximum extension of the degraded area, after the end of exploitation.

The main damages caused by subsidence at the surface are observed in:
- agricultural fields,
- roads,
- the water distribution,
- the electrical distribution system,
- buildings,
- waterheads and wells.

The damages in agricultural fields are mainly those caused by a reduction of flow in waterheads, as a consequence of lowering the water table.

The topographic changes resulting from the ground settlement have little influence in the agricultural production itself, making sometimes difficult the accessibilit of machinery.

The damages in the secondary roads and pathways which cross the degraded area, are corrected by continuous levelling with waste material from the treatment plant.

The damages in the water distribution system are worsened by the fact that they were not adapted to suffer displacements. To solve this problem the pipes were placed on the surface and junctions with appropriate supports have been applied to confer some flexibility.

Subsidence also affects the electrical distribution systems, mainly causing tilting of posts, sometimes dangerously approaching the lines to the buildings and trees.

It was necessary to change the crossing of the river by a high tension line, because the post on the south margin was placed in an area highly affected by differential settlements.

The buildings are one of the main concerns, because considerable sums of money are involved in their repairing and compensation. They are structures of rural type, very badly adapted to the absorption of stresses related to differential settlements.

Most of their foundations are built in stone, mainly schists and quartzites from the Ordovician complex, and sandstones and schists from the Carboniferous.

Subsidence lowers dramatically the phreatic level, which causes serious problems in water impoundings, most of times making them completely dry.

The small section tunels excavated to find water, sometimes with more than a hundred meters long, offer priviliged conditions to study the subsidence phenomena.

It is clear that the slippery planes which show clear evidence of movement, are coincident with bedding, faulting, or jointing planes, which show the importance of discontinuities in the subsidence process.

4 TOPOGRAPHIC CONTROL OF THE SUBSIDENCE

The evolution of subsidence in Germunde is controlled by topographic leveling of marks in alligned sections (Fig. 2), which reflect the surficial movements of the ground, and its relationship with the position of mining works, providing important information to the study of its future progression.

Initially the control was made by altimetric survey considering the observation marks with fixed horizontal coordinates, which means that only vertical displacements were measured.

Subsidence process involves tilting of blocks, which implies the levelling of horizontal coordinates, this was done from November 1990, and brought out several difficult problems from the practical point of view due to the large amount of topographic marks.

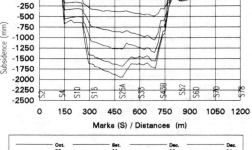

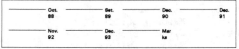

Fig. 2 Subsidence profile (S) transversal to the direction of Carboniferous

The distribution of measuring points should be as regular as possible, and so the square net would be recommended, although in the field that was impossible to achieve once the surface is not plane and it is covered with vegetation, the reason why marks were placed mainly along pathways and roads.

Irregular profiles introduced errors in the extrapolation methods, used to obtain curves of isovelocity and subsidence contours which affect the reability of results.

The use of vertical transversal and longitudinal sections is very important to the study of subsidence, and the accuracy in its determination can be seriously affected by this extrapolation.

5 OTHER TYPES OF SUBSIDENCE CONTROL

Besides the topographic sections, which are used to control the subsidence, it is necessary to control critical zones, and the behaviour of underground water.

Individual marks have also been placed in structures that by their importance are object of special attention like water reservoirs and electrical posts.

To characterize the behaviour of underground water which circulates in the rock mass, a set of piezometers in 6 vertical boreholes 50 m deep was installed.

Differential movements of the fractures are controled by triorthogonal joint meters.

6 DELIMITATION OF THE SUBSIDENCE ZONE

The surface area affected by subsidence has increased, as exploitation of coal reaches deeper level in the mine.

The sequence of rock mass fracturing, which can be roughly established by the complaints about damages in various buildings, shows clearly that after a new fracture is installed, other fractures develop inside that block.

In the NE of the Carboniferous, in the Ordovician and Lower Cambrian formations, the marks where no subsidence is detected are aligned with the last fracture detected in water mines, and define the limit of the present subsidence are (Fig. 3).

It is very important to determine the evolution of this boundary until the end of exploitation, and after the closing of the mine, once it is the only way to do a correct evaluation of the number of buildings that will be affected by subsidence.

The exploitation of coal has slightly decreased yearly since 1990, with a strong reduction in the previsible coal production for 1994. It is curious to detect that in these years the zone of subsidence has not expanded.

It has been observed in the stopes that the area disturbed by exploitation, forms a cone with an angle of no more than 30º with the vertical, which is similar to the dipping of the contact between the Lower Cambrian and the Carboniferous.

As these angles are almost coincident, and there is a clear boundary in what geomechanical properties of both units are concerned, it is not expected that the

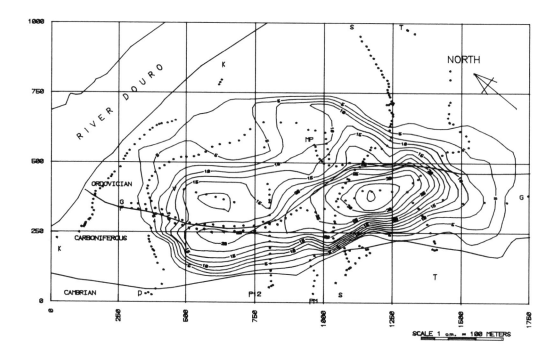

Fig. 3 Isovelocity subsidence curves (mm/month) between Nov./90 and Dec./93

Lower Cambrian will be affected by subsidence.

Is has been impossible to make a clear correspondance between each fracture and the depth of exploitation, and to determine the interval of time between exploitation of a certain level and the development of a fracture at the surface, which makes extremely difficult to predict the final subsidence area.

7 QUANTIFICATION OF DAMAGES

To evaluate house damage, a quantitative method based on the description of each building and the factors directly related to surface subsidence was developed.

Each house was evaluated, in an individual classification sheet according to three characteristics: type of construction, present state of maintenance and proximity of fractures. It is also mentioned its location, orientation, geology of the terrain, and velocity of subsidence, elements that often help to understand the origin of fractures.

It's pointed out the date, namely of the first complaint, which is roughly coincident with the beginning of movement, and the cost of all repairs performed so far.

Each group was subdivided in 4 smaller subgroups, with classifications of 10, 7,

4 and 1, reflecting the degree of danger in the house.

In the type of construction were considered the following subclasses

Table 1

Subclass	Description	Punctuation
A1	Wood made building	10
B1	Masonry building without cement or structure	7
C1	Masonry building without structure	4
D1	Masonry building with structure	1

In the classification sheet, for each subclass, it is recorded the building material.

The subclasses considered to group buildings according to their present situation of maintenance, are displayed in Table 2.

As proximity of fractures is concerned the adopted criteria to the different subclasses is the one presented in Table 3.

The sum of partial punctuations defines a class which intends to project future evolution of each building.

They were defined 4 different groups according to final punctuation in Table 4.

Table 2

Subclass	Description	Punctuation
A2	Highly fractured building with differential settlements	10
B2	Fractured building without important damages	7
C2	Lightly fractured buildings without structural damages	4
D2	No fractured buildings	1

Table 3

Subclass	Description	Punctuation
A3	Building placed on fracture, in its neigbourhood (< 15m), or in an area with high subsidence velocity (< 20mm/month)	10
B3	Building near fracture (15-30m), or in a zone with subsidence velocities 10 to 20mm/month	7
C3	Built on the Carboniferous or on the Ordovician with subsidence velocity (< 10mm/month)	4
D3	Built outside Carboniferous and of the Ordovician, or with subsidence velocity (< 5mm/month)	1

Table 4

Subclass	Description	Punctuation
1	Downfall will not occur. Eventual damage will have no importance.	25-30
2	Probable long term ruin	20-24
3	Improbable ruin	15-19
4	No ruin will occur	3-14

After classifying all the buildings in the subsidence area, it was possible to group them according to their hazardous degree as showed in Table 5.

This is an important base to analyse the problem, from the economical point of view, once it will be necessary to build new houses, and to do repair work on others. It will be convenient to remember that the subsidence will continue after the closing of the mine.

It is possible to detect some buildings in which frequent repair work was accomplished until a certain date, after which there wasn't any complaint, reflecting a reduction of differential settlement.

This type of procedure enabled the distinction of the buildings whose degradation is not caused by mining subsidence.

Table 5

Class	Nº of buildings	%
1	14	13
2	25	23
3	24	22
4	45	42

8 CONCLUSIONS

This investigation allowed to circunscribe the superficial subsidence zone above Germunde coal mine, and through the used methodology of damaged analysis, it was achieved a classification for buildings, according to different degrees of risk.

On this basis it was possible to delineate a convenient strategy for compensation and relodging.

ACKNOWLEDGEMENTS

The authors wish to express their gratitude to the European Community of Steel and Coal for their sponsorship in the research projects number 7220/AD/761 and 7220/AF/001 which provided the basis for the present article.

REFERENCES

Brady, B. H. G.; Brown, E.T., 1985, Rock mechanics for underground mining. George Allen & Unwin, London, 527 pp.

Dutra, J. I. G., 1993, Análise da subsidência induzida pela mineração em maciços rochosos muito fracturados. PhD thesis, Instituto Superior Técnico, Lisboa.

Empresa Carbonífera do Douro, S.A., 1992, Modelagem computacional da subsidência mineira em jazigos de carvão muito inclinados, C.E.E. final report DG XXVII, convention nº 7220/AD/761.

Hoek, E., 1974, Progressive caving induced by mining and inclined orebody. Transactions Institution of Mining and Metallurgy, (Sect. A: Min. Industry), Vol. 83, pp. A133-A139.

Kratzsch, H., 1983, Mining Subsidence Engineering. Translated by Fleming, R. F. S., Springer-Verlag, Berlin

Heidelberg, NY, 543 p.

National Coal Board, 1975, Subsidence engineers handbook, 2nd (rev)edn., London, National Coal Board Mining Dept.

PENG, S. S., 1986, Coal mine ground control. 2ª Ed., Jonh Wiley & Sons, NY, 486 p.

SILVA, P. B.; Chaminé, H. I., 1993. A subsidência mineira na mina de carvão de Germunde - contribuição da geologia de superfície. Geologos. 1:1-10, Porto.

Les effets érosifs et environnementaux de l'exploitation de l'or dans la forêt amazonienne: Le cas du sud-est Péruvien

Erosion and environmental effects of the gold exploitation in the Amazonian jungle in the southeast of Peru

Susana Kalafatovich C.
INCADI, Cusco, Perú

Raúl Carreño C.
École Polytechnique Fédérale de Lausanne, Suisse

RESUME: L'exploitation de l'or dans le piedmont subandin et la forêt amazonienne commence à montrer une série d'effets pernicieux sur l'environnement. Les processus incontrôlables d'érosion, la déforestation, les changements du régime hydrologique, la pollution chimique sont, parmi d'autres, les effets les plus frappants et dangereux pour ces écosystèmes extrêmement fragiles.

ABASTRACT: The gold explotation in the subandean region and in the amazonien jungle, begins to show a series of pernicious effects on the environnement. Erosion process, deforestation, hydrological regime changes, chemical pollution are the main and most dangerous process that affect these extremely fragile ecosystem

1 INTRODUCTION

Les vallées et le piedmont subandins du sud-est péruvien (départements de Cusco, Puno et Madre de Dios) ont subi une véritable fièvre de l'or ces dernières années. Bien que l'extraction de l'or alluvial signifie une source de survie pour une très importante partie de la population, son impact sur l'environnement est aussi très important, par cause notamment de l'érosion-sédimentation, du déboisement et de la pollution chimique.

Compte tenu du type des gisements, des techniques d'exploitation et de la situation sociale du pays, les effets futurs et accumulatifs de cet activité sont encore plus imprévisibles, même non maîtrisables, et les retombées dans un cadre environnemental plus large, touchent l'échelle planétaire, car ils font partie de la détérioration générale du milieu amazonien.

2 LE CADRE GÉOLOGIQUE

Les différentes unités morpho-structurelles contenant de l'or forment une transversale plus ou moins parallèle à la direction générale de la cordillère andine dans cette zone (SE-NW). Ce piedmont fait part du bassin amazonien, et s'avère l'un des écosystèmes les plus fragiles de la planète de par leurs pentes raides, leurs sols minces, leur haute pluviosité, etc.

Sur les versants orientaux de la chaîne andine affleurent notamment les roches métamorphiques du Paléozoïque, en spécial les schistes.

Les filons-mères qui alimentent les gisements aurifères alluviaux du piedmont et morainiques des montagnes, se trouvent justement dans les roches siluro-dévoniennes de la formation Ananea.

Sur la limite entre le piedmont subandin et la plaine amazonienne, Bonnemaison et al. (1985) ont identifié les unités géologiques suivantes, du Néogène au Quaternaire, et à caractère sédimentaire, toujours en discordance angulaire sur les unités paléozoïques et meso-cenozoïques:

1. le bassin de Quincemil, d'origine probablement palustre, où l'on trouve la formation Huajiumbre (couches de grès limoneux à contenu organique avec d'horizons minces de gravier et de nodules à pyrite) et la formation Cancao, d'environ 250 m de puissance, formée de graviers assez hétérométriques; les deux formations étant séparées par une discordance érosionnelle et affectées par une néo-tectonique de compression N-S.

2. La formation Mazuko, formée notamment de grès, de limons et de conglomérats, avec de restes organiques et de paléosols. Cette formation est déformée par un plissement N100° à N110° (probablement d'âge oligo-miocène), et par quelques failles inverses E-W.

De différentes séries de terrasses échelonnées sont présentes dans toutes les vallées de la région.

Morphologiquement on a affaire à une série de vallées parallèles, plus ou moins perpendiculaires à la chaîne andine, qui forment un piedmont étendu, dans la transition à la plaine amazonienne.

3 LA NATURE DES GISEMENTS

Dans tous les cas, il s'agit de gisements secondaires d'origine alluviale qui, de son côté, proviennent de l'érosion des gisements détritiques d'origine glaciaire et fluvio-glaciaire situés à plus de 4000 msm, dans les départements andins de Cusco et Puno (cordillères de Carabaya, du Vilcanota-Ausangate et de Camanti).

L'or se trouve disséminé dans les sédiments du Miocène au Quaternaire récent (Bonnemaison et al. 1985) sur environ 15'000 km^2. La puissance des couches minéralisées est variable.

Fig. 1
Situation de la region aurifère subandine du sud-est péruvien.

Les filons primaires appartiennent aux formations paléozoïques situées dans l'haute montagne, où les veines de quartz aurifère traversent les ardoises siluriennes.

L'histoire des gisements est complexe et montre une succession de cycles d'érosion-accumulation, encore non bien définies, due à un processus géomorphologique lent, à maintes rémobilisations des sédiments, à une hydrographie favorable dans une région à faible pente. Le régime méandriforme actuel des fleuves en est l'évidence ainsi que les dépôts aurifères situés dans les terrasses, les plages, les lits des rivières et dans les anciens bassins fluviaux, maintenant couverts, au milieu de la forêt.

Les teneurs en or sont assez aléatoires, variant de 5 à 10 mg/m^3; parfois on y trouve des concentrations de 40-50 mg/m^3 dans la formation Mazuko.

L'étude faite par Bonnemaison et al. montre que aucun des échantillons avait moins de 2 mg/m^3.

Si on compare ces teneurs avec ceux du gisement morainique de San Antonio de Poto, dans le département de Puno (0,100 à 0,500 mg/m3, d'après Kihien 1985), et qui alimente les alluvions du piedmont andin, on voit que le facteur de dilution est de 10, pour une distance de moins de 200 kilomètres.

4 MÉTHODES ET TYPES D'EXPLOITATION

Une grande partie des problèmes attachés à l'exploitation de l'or dérive des techniques minières rudimentaires. Sauf quelques entreprises bien implantées dans la zone et qui exploitent l'or avec de moyens mécaniques modernes (tracteurs et dragues), toute l'extraction se fait de façon artisanale, avec un minimum d'outils et de ressources.

Les types d'exploitation obéissent aux conditions saisonnières (disponibilité d'eau) dans les trois endroits où l'on fait les travaux: "en monte", "en cumbre" et "en río" (Mosqueira 1992).

Le premier type, "en monte" (dans la région on appel *monte* à la forêt en soi), se fait au delà du lit majeur des rivières, dans les zones boisées, sur les anciens dépôts alluviaux. Normalement on y trouve de teneurs en or un peu plus hauts que dans les plages, à cause de l'enrichissement dû au remaniement successif des matériaux, mais par contre, il est nécessaire d'enlever plus de matériau stérile, et toute la couverture biologique.

Le second type d'exploitation *"en cumbre"* (en sommet) concerne les parties élevées des terrasses alluviales et d'autres dépôts sédimentaires dans les bois, même les plus anciens. Là aussi les mouvements de matériaux sont très importants, ainsi que les besoins en eau pour la concentration du minerai. La disparition de la végétation y est incontournable.

Le dernier type, *"en río"* (en rivière), se fait soit dans les plages, dans les zones d'anastomose et de méandre, soit dans le lit des fleuves, par dragage.

La couche stérile atteint normalement plus de trois mètres, et il faut l'enlever totalement et la jeter dans les rivières pour qu'elle ne gêne pas les travaux. Cette couche est stérile du point de vue minière, mais très féconde biologiquement.

L'extraction de l'or comporte un travail appelé "lavado" (lavage), destiné à éliminer les matériaux les plus grossiers et à concentrer, par gravité, les sédiments lourds, qui sont amalgamés, afin de récuperer l'or.

La récupération de l'or se fait par le moyen du "refogado" , dans un brûloir artisanal, où n'on récupère qu'une partie du mercure utilisé, le reste partant volatilisé dans l'atmosphère.

Un autre type de exploitation à petite échelle, pour profiter des particules plus fines et des colloïdes, utilise un morceau de laine de mouton imbibé de mercure, laissé dans les ruisseaux. Cette façon de faire est plus dangereuse, car elle comporte une perte encore majeure de mercure, parfois dans les ruisseaux qui alimentent en eau les foyers.

Les entreprises plus performantes utilisent de dragues, pour extraire les sédiments du lit des rivières ou des bords de plages.

Le mouvement de masses est facile, compte tenu de la nature peu cohésive des matériaux.Dans le cas du sol végétal, la tâche est encore moins gênante, car on pratique l'incendie du bois et de la couche d'humus, avant d'atteindre les formations sédimentaires.

5 LES EFFETS NÉGATIFS DE L'EXPLOITATION AURIFÈRE

Le cercle de cause à effet créé par la fièvre d'or subandine montre un impact dont la dimension est surprenante. L'éventail des phénomènes concomitants à cette exploitation irrationnelle est varié, desquels ont peut citer:

5.1 La déforestation

La première constatation qui ressort d'une observation des exploitations minières est la déforestation, laquelle doit être complète pour permettre d'exploiter l'or sédimentaire. Cela est fort remarquable dans un environnement tropical.

Il n'y a pas une suivie du rythme de déforestation due non seulement à l'extraction de l'or, mais de toutes les activités de colonisation de la forêt subandine-amazonienne, mais dans les photos satéllitaires les zones déboisées se font de plus en plus évidentes au milieu de la forêt.

En ce qui concerne notre cas, la surface totale des concessions minières aurifères atteint plus ou moins les 2,2 millions d'hectares, seulement dans le département de Madre de Dios, plus de 250'000 hectares dans le département de Cusco et un chiffre similaire pour le département de Puno. C'est cette surface-là donc qui pourrait éventuellement être déboisée à long terme à cause de l'or

Les terrains soumis à l'exploitation sont entièrement perdus, car les sédiments fins sont évacués vers les cours d'eau et les autres plus grossiers (comme les cailloutis, les blocs et les graviers) restent empilés, sans que la végétation puisse plus y reprendre.

Une déforestation secondaire, dont le rythme de l'érosion dérivée est moindre, résulte de l'habilitation de terrains pour une agriculture de subsistance et de l'ouverture de chemins d'accès aux exploitations.

L'activité forestière, que parfois sert de complément à l'activité minière, est une source additionnelle du déboisement .

5.2 L'érosion -sédimentation

Celles-ci sont l'effet immédiat de la déforestation et de l'activité minière, tant à cause du remaniement des masses impliquées pour préparer le chantier de la future carrière, que par les précipitations intenses qui tombent dans la région, rangée parmi les plus humides du monde.

La végétation ne pouvant plus reprendre sur les sols dépourvus de matière végétale et de sédiments fins, le ruissellement érosif en amont devient généralisé.

Effet de l'érosion accélérée: les falaises des anciennes terrasses, au bord de quelques rivières, commencent à subir une série de glissements de terrain, qui entraînent une turbidité additionnelle.

Par contre, là où la sinuosité des fleuves est haute, à la limite de la forêt tropicale ((La Riva 1992) la sédimentation dérivée de la surcharge solide, pose un autre type de problème, par la surélévation du lit majeur et les inondations concomitantes.

5.3 L'augmentation des débits solides

Comme on l'a déjà dit , la technique minière et la typologie des gisements obligent de laver les sédiments plus fins pour récupérer l'or. Les eaux usagées contenant un débit solide très élevé, doivent être déversées dans les rivières pour continuer l'exploitation.

Dans le peu des cas où les mines ont de réservoirs de sédimentation, la capacité de celles-ci est restreinte et, finalement, les eaux de relave contaminées doivent être rejetées.

Les retombées de cette pratique sont néfastes: la turbidité presque permanente des eaux a commencé à réduire la masse biologique des rivières et des lacs, et à augmenter les inondations, selon les témoignages des gens. Le sols inondés gagnent en imperméabilité à cause de la précipitation des argiles.

Bien qu'il n'y a pas de données quantitatives, l'impact qualitatif y est évident et le régime hydrologique montre une altération dangereuse, en accentuant la sinuosité et les changements soudains du lit des rivières, par la migration des méandres.

Cela touche aussi les "cochas", ou lacs en arc, riches en faune et flore particulières, car elles reçoivent plus de l'apport sédimentaire et se comblent plus vite.

Pour avoir une idée des volumes de sédiments libérés par l'extraction d'or, on montre la composition granulométrique des gisements de montagne, qui alimentent les gisements alluviaux subandins (Kihien 1985):

- blocs. 5%
- cailloutis 4%
- gravier 23%
- sable 28%
- limon et argile 40%

En faisant un calcul élémentaire, selon les données de production et de teneurs métalliques, on peut trouver que les volumes de sédiments déversés dans les rivières est énorme: environ 400 millions de mètres cubes par année sur une aire concentrée de moins de 5'000 km^2. A ce chiffre il faut ajouter les volumes dus à l'érosion naturelle et aux mouvements de masses, ainsi que l'érosion grandissante arrivant des versants andins.

5.4 L'exploitation partielle des ressources.

La technique rudimentaire d'exploitation fait que n'on puisse exploiter ni de l'or colloïdale, ni des particules fines ni de l'or incrusté dans les cailloutis. Le pourcentage des pertes est fort important: entre 5% et 30%, selon quelques estimations (Carreño 1990).

La présence d'autres minérales lourds, tels que la monacite, la wolframite, etc. a été annoncé, mais soit par méconnaissance, soit par d'autres problèmes techniques ces matériaux son négligés.

Un traitement métallurgiquement convenable à partir de la récupération des boues post-lavage, seraient une manière tout à fait faisable et rentable tant de améliorer la production que d'assurer un contrôle écologique des déchets miniers, ainsi que pour récupérer les sédiments fins et réaménager les terrains dénudés.

En gros, seulement environ 30% de l'or sédimentaire est exploité par les méthodes actuelles d'extraction et métallurgie artisanale.

L'exploitation future des couches minéralisées inférieures, avec de moyens plus performants est envisageable, ce qui signifierai d'autres remaniemennts encore plus importants de matériaux.

5.5 La pollution chimique

L' amalgamation est le moyen métallurgique le plus simple pour les mineurs de la région. La séparation de l'or se fait, comme on a dit, par le moyen d'un brûloir.

Même s'il n'y a pas une étude approfondie à ce sujet-là, on a pu estimer indirectement une perte de mercure d'environ 5 à 10% par cycle d'amalgamation. Ce mercure-là, sublimé, part dans la nature ou, pire encore, est inhalé par les gens concernés.

Nous ne disposons pas d'une information valable pour calculer le taux de pollution par mercure dans la zone, mais dans un autre site -Nazca- où l'exploitation artisanale d'or est aussi pratiqué dans des conditions semblables, les boues résiduelles ont 600 ppm en mercure (la limite tolérable étant de 0,1 ppm).

A Nazca, les points critiques atteignent 12 mg/m3 de mercure (120 fois le taux tolérable de 0,1 mg/m3) (Cruz 1992). Les proportions doivent être presque les mêmes dans les exploitations subandines et amazoniennes, ce qui donne un point de référence du problème qui nous occupe.

On envisage d'introduire la cyanuration pour la récupération de l'or colloïdale dans les boues métallurgiques. Ça est déjà fait ailleurs, ayant comme résultat une pollution par cyanure qui atteint un taux de concentration de 100 fois plus haut que la moyenne tolérable = 0,1 ppm (Cruz 1992).

Seulement pour les vallées de Caichive et Mazuko (moins du 10% de la surface totale en exploitation de la région), on a calculé que 419 kilogrammes de mercure par année sont jetés dans les rivières et la forêt (Mosqueira 1992). Si on élargit ce calcul pour toute la région aurifère subandine, on aurai un total de plus de cinq tonne de mercure qui partent vers la nature chaque année.

Malgré l'obligation légale d'acheter le mercure avec une autorisation préalable auprès du Banco Minero (Banque minière du Pérou), le contrebande de cette matière est la pratique la plus répandue. Tout contrôle de pollution devient donc impossible.

CONCLUSIONS

1 L'extraction de l'or alluvial dans le piedmont subandin et la plaine amazonienne du Sud-est péruvien provoque d'importants processus d'érosion, de déboisement, et de pollution chimique.
2 Les causes du problème sont dus notamment à la typologie des gisements secondaires (à faible teneur en or), aux méthodes d'exploitation artisanales et au manque de contrôle sur les concessions minières.
3 Un contrôle et une surveillance étendus des carrières étant impossibles, seul un changement du type des exploitations pourrait permettre de restituer la forêt originale et d'arrêter la détérioration écologique de ce milieu fragile.

Un plan d'exploitation combiné pourrait être une sollution, avec de l'extraction de l'or par étapes d'abord, et ensuite, une restitution des bois ou le réaménagement des terrains pour qu'ils soient destinés à l'agriculture. L'exemple des mines d'étain alluvial dans les vallées de Malaisie ou de quelques mines d'aluminium du Brésil dans la forêt amazonienne, qui furent aménagés après l'extraction des minérais, en serait un bon antécendent.

RÉFÉRENCES

Bonnemaison, M.; Fornari, A.; Galloso, G. & Grandin, G. 1985. Evolución geomorfológica y placeres de oro en los Andes surorientales del Perú.*Bull. de la Sociedad Geológica del Perú*, vol. 75: 14-32. Lima: SGP.

Carreño, R. 1990. *Oro y erosión* . Rapport INCADI. Cusco.

Cruz, R. 1992. Control ambiental en minería aurífera informal de Nazca. En *Trabajos t écnicos II Symposium nacional de minería aurífera*: 315-319. Cerro de Pasco: UNDAC.

Kihien, A. 1985. Geología y génesis del yacimiento aurífero de San Antonio de Poto. *Bulletin Sociedad Geológica del Perú* vol. 14:17/26. Lima:SGP.

La Riva, J. 1992. Análisis de sinuosidad de un sector del río Madre de Dios y su relación con los depósitos de placeres auríferos. En *Trabajos técnicos II Symposium nacional de minería aurífera*: 59-61. Cerro de Pasco: UNDAC

Mosqueira, G. 1992. *La economía del oro en Madre de Dios*. Cusco: CBC.

Geoenvironmental changes from quarry activities in Bulgaria

Impacts sur l'environnement des carrières en Bulgarie

K.Todorov & D.Karastanev

Geotechnical Laboratory, Bulgarian Academy of Sciences, Sofia, Bulgaria

ABSTRACT: The most unfavourable effects on geoenvironment from quarry activities in Bulgaria during the last decades have been described and analyzed. Appropriate measures are proposed to restrict the detrimental changes in geoenvironment and to recover the damaged by intensive quarry works terrains.

RÉSUMÉ: On décrit les plus importants changements défaborables en environment géologique, qui sont provoqués par l'exploitation des carrières différantes en Bulgarie pendant les dernières années. On propose des conditions convenables pour limitation des changements nuisibles et pour rétablissement des terrains géologiques detruits par exploitation intensive des carrières.

INTRODUCTION

The intensive quarry activities during the last decades substantially changed the outlook of vast territories, damaged great areas of arable land and cause a constant pollution of environment. The unfavourable alterations of geoenvironment and ground hydrosphere often provoke or activate destructive geologic processes and phenomena which are characteristic for Bulgaria - landslides, rock falls, plain and linear erosion, soil collapsibility and liquefaction, etc. Considerable economic losses are caused by these harmful events. Quarry activities affect various types of rocks and most of the uncohesive and cohesive soils.

The protection of environment and the rescuing of the cultivable land, which is the most valuable national wealth, imposed the performance of the present geological engineering assessment of the changes caused by the different types of quarry activities, and the choice of the most suitable and effective measures for maximum restriction and elimination of the detrimental impact.

EFFECTS FROM GRAVEL PITS

More than 30 million m^3 of aggregates (sand, gravel and mixture of river gravel and sand) are being produced annually for constructional purposes. About 70 % of this quantity are extracted from alluvial sediments in the middle- and downstream valleys of big rivers and their affluents. There were 178 gravel pits in the river valleys in 1989. The total output of aggregates during this year from the operating 144 gravel pits is 21.2 million m^3. Most of the gravel pits are situated in the western and middle part of the country. There are no big rivers with thick alluvial sediments in East Bulgaria. Only several gravel pits along the Kamchia river valley are working in this region.

The damaged areas from gravel pits operation amount to more than 56 million m^2. The greatest areas of destroyed land are along the big river valleys (Table 1).

Table 1. Damaged areas of arable land by gravel pit operation along the big river valleys in Bulgaria (data for 1989)

River	Number of gravel pits	Annual output, thousand m^3 per year	Damaged areas, thousand m^2
Danube	7	4570	8258
Iskar	21	7212	7423
Maritsa	22	2461	5724
Strouma	8	686	4515
Toundzha	20	1598	3009

Sand and gravel are obtained not only from river beds and flood terrace plains but also from high river terraces, from talus-prolluvium mountain foot cones and sandy Tertiary sediments. In 1989 12.6 million m^3 of aggregates were produced in 103 quarries of this type situated on an area of 13.7 million m^2. Sand and gravel quarries outside the river beds are mainly in South Bulgaria.

The scraping up of large quantities of aggregates changed in a short-term period the established centuries ago river beds. The deepening of the rivers during the gravel pits operation in the river beds activates bottom and side erosion. The hydraulic breaking of river banks causes landslides, destroys stabilization and capture facilities, endangers bridge stability and safety. Similar cases have been established along the river valleys of most of the big rivers in Bulgaria.

The vast quantities of sand and river gravel and sand mixture scraped up from the Danube (4.5 million m^3 annually) are the reason for the activation of a multitude of landslides along the southern bank and for changing of the river fairway.

The gravel pit operation often leads to lowering of the ground water level and to the necessity of deepening of pumped wells. Along a left affluent of the Maritsa river for instance, the pumped wells were made from 30-40 to 150 m deep. The water supply of some settlements in the Plovdiv region suffered complications and was even ceased since pump stations and dug wells ran dry. The lowering of ground water level leads to drying up of the surface soil layer and disturbing the water regime in the aeration zone. This is the reason for plant withering and cracking of surface clays and of the structures and facilities built on them.

The gravel pit operation causes serious damage to agriculture since it destroys the most fertile land in the terrace plains. The recovery of the damaged land is expensive and takes a long period of time. About 20 million m^2 have been recultivated till 1990.

Gravel pits are a source of environmental pollution. Abandoned gravel pits are turned to large marsh-land. Various domestic and industrial wastes are stored in them without any preliminary preparation thus polluting the soil and ground water.

The production of aggregates causes economic troubles too. Nowadays aggregates are used in Bulgaria at higher rates than these for their natural reproduction. The materials formed in the course of millennia are quickly exhausted and in a lot of regions of the country they are deficient. Aggregates production is characterized by high expenditure of labour and energy and on the other hand their insufficiency hampers building industry.

The increased output of sand and gravel by means of crushing of suitable rock material and the substitution of natural aggregate in road construction with stabilized soil or hard industrial wastes will restrict the scraping up of uncohesive soils from river valleys. This is the best way to solve the above mentioned ecological and economic problems. Recently lagooned ashes from thermal electric plants (Todorov et al. 1976) were successfully used for embankment and road base construction. The recultivation of already abandoned gravel pits is also indispensable.

EFFECTS FROM QUARRIES IN ROCK MASSIFS

More than 600 quarries for various construction materials have been built in rock massifs. Only in 1989 the output is 40 million m^3 of building materials from 458 quarries (Table 2).

Table 2. Number of quarries in rock massifs and corresponding output (only for 1989)

Type of rocks	Number of quarries	Annual output, thousand m^3 per year
Intrusive (granites, syenites)	28	2550
Effusive (andesites, rhyolites)	49	5910
Sedimentary	294	25044
Metamorphic - non-schistous (marbles)	48	3674
- schistous (gneisses, amphibolites)	39	3150

The quarries in sediment rocks (limestones, dolomites, sandstones, marls) predominate and are encountered throughout the country. The quarries in other rock types are situated mainly in mountain massifs in South Bulgaria.

The development of most quarries in rock massifs is not in conformity with the modern ecological requirements. This type of quarry activities damaged more than 35 million m^2 arable land. Their operation disturbs the natural state and stability of the rock massifs with all unfavourable consequences following from this. The natural relief is changed and the landscape is deformed for a long time. The operation of quarries in rock massifs leads very often to deforestation of vast territories which contributes to stronger plane erosion. The rock material blasting brings to turbidity and pollution of surface water and to dust loading of the neighbouring regions. Strong explosions change sometimes spring flowrate.

The damaged rock massifs are recovered by vertical and horizontal planning, anti erosion measures and biological recultivation (spreading of a soil layer and aforestation). It is to be regretted that because of insufficient financial funds there is a very small number of recovered rock terrains.

In single cases the empty spaces in the rock massifs formed by the anthropogenetic activities may be used for various economic purposes. It was established that the abandoned limestone quarries near Rousse are suitable for storerooms.

With the view of diminishing the harmful consequences from rock massifs quarries the application of ecological technologies for their operation as well as the full and effective recovery of the damaged terrains are intended. The restriction of the output of rock building materials by more wide use of suitable substitutes will also aid to the solution of the problem. This will stop the enlargement of the existing quarries and the development of new ones.

EFFECT FROM OTHER QUARRY ACTIVITIES

The open-cast coal and ore mining, the non-metal minerals and building materials production are other activities when great quantities of soil are scraped up and redeposited (Table 3). The total volume of the scraped up and redeposited soil mass only from the open-cast lignite coal production in the Maritsa

Table 3. Quantities of redeposited soils in 1988

Origin	Quantity, thousand m^3
Rock and soil masses from open-cast mines	268726
Wastes from non-ferrous ore mining and ore dressing	17570
Wastes from ferrous ore mining and ore dressing	327
Wastes from coal mining	686
Wastes from building materials production	253

coal basin exceeds 2 billion m^3, while the damaged areas are more than 130 million m^2. It is expected that these areas will reach 325 million m^2 at the final stage of exploitation of the basin (Dimitrov 1989). The damaged areas by the copper ore mining in the Sredna Gora Mountains are already 127 million m^2 and by the kaolin and bentonite production - 11.7 million m^2. The above mentioned activities affect not only arable land but also towns and villages, roads, railways and communication systems.

The removal of soils from their natural state and bedding conditions leads to changes in their density and moisture content and affects unfavourably their geotechnical properties. The frequent presence of over saturated clays with plastic consistency sharply decreases the slope stability of the formed soil embankments.

Great areas are damaged by the storage of the vast quantities of waste materials from the output and mining of coal, ores and non-metal mineral resources (Table 3). The deposited wastes are uncompacted and unconsolidated state and possess low bearing capacity (Todorov 1987). Their properties are heterogeneous and vary at small distances in horizontal and vertical direction.

The removed and redeposited wastes are concentrated in the industrial regions of the country (the Maritsa coal basin, the Devnya valley, the Sofia kettle, the Pernik mine basin, the West Sredna Gora Mountains). It is considered that till now this quarry activity has destroyed and damaged more than 400 million m^2 of land, suitable for agricultural and construction purposes. Most probably the amount of badly affected areas is much higher since quarry activities for building materials production are developed in the vicinity of each settlement and they are difficult for planning, control and statistics.

The removed soils and deposited wastes are regular pollutants of air, water and soils. The formed technogenic landscape is ugly and is harmful for nature and tourism. The use of this land for building sites without expensive preliminary works is almost impossible.

The open-cast coal and ore production activates erosion processes so that landslides occur often in such regions. The observed landslides in the Maritsa coal basin are 500-700 m wide, 300-500 m long, up to 100 m deep and amount to tens of millions m^3 (Frangov 1990). The rate of movement during the active stage reaches several meters per minute and the total displacement towards the dug out space is up to 100 m.

The removed soil and deposited wastes change the environment not only by their presence but also by the influence that they exert on the natural sediments under and around them.

In the case when weak uncompacted and over saturated soils lay under the redeposited masses, a considerable hazard exists for various deformations and failures. For example, several years ago plastic soil foundation masses were pushed out and moved aside. The total volume of the moved masses reached 500 thousand m^3 and resulted in the destruction and damage of buildings and engineering structures.

The natural relief is also changed by the scraping up and redeposition of soils as well as by the waste storage. The anthropogenetic alterations in the relief lead to fading or increasing of important physico-mechanical processes. The anthropogenetic relief changes very quickly because of further compacting or erosion. This hampers the prediction of hazardous geologic processes or phenomena, activated by relief variations.

The regions affected to a greater degree by this quarry activity suffer from disturbed water and temperature balance between the atmosphere and the lithosphere. Sometimes changes in the micro climate are observed (increased precipitation, higher annual average temperature, etc.).

For a small and densely populated country like Bulgaria, the most substantial harm from this anthropogenetic activity is the damage of arable land, which amounts at present to 11 million hectares. For this reason priority should be given to biological re-cultivation. This will contribute to the recovery of the greater part of damaged land so that it can be used by agriculture again. Another possibility to solve the problems concerning the oversaturated soils and deposited wastes is their effective application as building bases (after preliminary improvement), raw materials for the industry containing valuable components or cheap building materials. The performed investigations in this direction are rather encouraging. Good effect will also be achieved by the elimination or restriction of a series of unprofitable and harmful for ecology enterprises.

REFERENCES

Dimitrov, T. 1989. Ore mining and environment - state of the art, problems and goals. *Minno delo.* No 3: 3-5 (in Bulgarian).

Frangov, G. 1990. Main regularities in the distribution of technogenic landslides in Bulgaria. *Review of the Bulg. Geological Society.* vol. L1, part 3: 79-85 (in Bulgarian).

Todorov, K. 1987. Strength and deformation properties of technogenic soils in Bulgaria. *In: Proc. of the 1st Conf. on Mechanics*, Praha, vol.6: 178-181.

Todorov, K., D.Evstatiev & P.Slavov 1976. Investigating lagooned ashes utilization in road construction. *In: Proc. of 5th Int. Congress IAEG*, Buenos Aires: 1693-1696.

Environmental problems associated with mining activity in hilly terrain

Problèmes de l'environnement associés à l'activité minière dans une région montagneuse

T.N.Singh & M.Goyal
Department of Mining Engineering, Institute of Technology, Banaras Hindu University, Varanasi, India

J.Singh
Department of Civil Engineering, Institute of Technology, Banaras Hindu University, Varanasi, India

ABSTRACT: Exploitation of mineral resources from the hilly terrain is somewhat different from the planner areas. Mining activity in hilly terrain, which is highly eco-fragile, results in even more intense environmental problems in the form of landslides, deforestation, change in isostancy of the region, geohydrological problem, health hazards, ecosystem changes, water pollution, subsidence and solid waste disposal. Indian hilly terrain are prone to frequent seismic activity which further aggrivates the problem.

The growing awareness and concern among people to ensure that any developmental activity is eco-friendly or atleast has minimum impact on environment and ecology and mining is no exception.

This paper deals with the environmental problem caused due to mining activity in hilly terrain and some suitable control measures have been suggested so as to sustain developmental activities. The socio-economic impact is also dealt herein for feasible and effective environmentally acceptable control measures.

1 INTRODUCTION

Accelerated industrialisation in any country requires sustained inputs of raw materials which are chiefly obtained from mineral deposits. Mineral development is important not only economically but also for ensuring social welfare. This demands not merely inputs of mineral resources but other equally crucial natural resources like clean air and fresh water, without which survival of life is impossible (Goyal et al.,1994; Maudgal and Kakkar,1993; Singh et al.,1993).

The need for the day is to maintain a balance among the natural resources such as air, water and land while taking up exploitation of natural resources. With the general awakening among the masses, environmental protection and mitigation has become an important issue.

To strike a balance between nature and development's impact on the surrounding environment, an environmental protection element needs to be introduced at the conceptual stage of the mining project itself. This is the modern concept of sustainable development (Bandyopadhyay and Siva,1985; Mohan,1988).

2 HIGH ECO-FRAGILITY OF HILLY TERRAINS

Hilly terrains have highly fragile and sensitive ecosystem. These areas are characterised by following features:

a. Weak and young geological formations,
b. Unique property of wilderness,
c. Intrensically low resiliance i.e., poor rate of recovery towards normalcy following disturbance, due to harsh climate, low nutrients, slow recovery rate and steep slopes,
d. High species richness and bio-diversity
e. Susceptible to species loss, especially rare and extinct ones,
f. Aquifiers and water recharge zones of mountaion springs,
g. Active geological faults and seismic hazards.

A small disturbance to environment in such areas may bring a noticeable change in existing ecosystem (Maudgal & Kakkar, 1993).

3 ILL EFFECTS OF MINING IN HILLY AREAS

Mining in hilly areas causes various ill effects some of which are described below:

Fig. 1. Land degradation due to mining in hilly terrain

3.1 *Visual intrusion and land degradation*

Degradation of scenic beauty is caused by mining and waste disposal. The damage done shows in clear contrast to lush green vegetation cover all around (Bose,1993; Chaulya et al.,1992; Mathur,1985). Fig. 1 shows a typical example of land degradation due to mining in hilly terrain.

3.2 *Soil and rock removal*

Removal of overburden is necessary for assess to mineral body. This has to be transported and dumped at a new location. But only restricted space is normally available in hilly terrain and such dumped material is washed off by rain finding its way into natural drainage channels (Dube,1993).

3.3 *Uprooting of vegetation and deforestation*

Surface mining can not be done without deforestation. The vegetation like grasses, shrubs and trees are removed to clear the land for the mining and allied operations. This vegetation cover is essential for protection and generation of soil cover. It gives various forest products for the use of the society and also acts as water retaining agent. Areas robbed off vegetation are exposed to erosion.

3.4 *Blasting vibrations*

Blasting operation is very difficult in hilly terrain because of non availability of free faces and most of the explosive energy is wasted in form of fly rocks, dust, noise and ground vibration (Singh,1987). It is a repeatedly occuring exercise contributing to weakening of the hill mass as a whole and giving rise to land slides (wiss and Lineham,1968).

Vibrations have psychological effect on the nearby population and the fauna of the area causing complaints and hostality among the local residents and migration of birds and wild life to other areas (Chaulya and Singh,1993).

3.5 *Land slides and destablisation of hill mass*

The hilly region generally has a past history of intense tectonic upheavels like faulting and folding resulting in the presence of abnormal insitu and residuals stresses. Highly undulating surface results in large variation in vertical stress due to gravitational loading of overlying rocks. Any sudden change in the geometry of the hill mass may cause stress concentrations in and around. The rocks themselves may have been considerably weakened and problems of squeezing and swelling grounds coupled with heavy rains destabilises the rockmass. Steep slopes comprising of fragile rock formations cause slides which pose problems both for safety and stablisation of mines (Chaulya et al.,1993; Sawarynski,1981). The blasting shock may trigger land slide in weak and factured rockmass (Bose,1993; Dube,1993). Proneness of the area to earthquakes and seismic events further enhances the chances of potential failures.

3.6 Noise pollution

Noise is generated due to blasting, operation of plant and machinery and movement of trucks and dumpers. It is a nuisance as it decreases efficiency of workers and causes discomfort to the people in the vicinity. It creates loss of hearing leading even to permanent deafness (Down and Stocks,1978; Grange,1977).

3.7 Change in surface and ground water regime

Diversion of streams is often inevitable for carrying out mining activities, thus effecting the water regime and the down stream users. Flora and fauna are also influenced resulting in ecological imbalance. Mining changes the topography of the area and hance the natural drainage pattern is disturbed. Mine excavations usually have high water in-flux from rainfall or due to ground water flow. This unwanted water has to be pumped out. It can be contaminated by particulates, oil and grease, unburnt explosives, ore chemicals etc. The mine waste and scree gets discharged into the natural drainage channels. This causes overspilling and choking up of the channels, diturbing the natural flow of water. Even flash floods may occur. Use of surface water for mining requirements upstream may reduce the flow further downstream. Artificial lowering of ground water table and depletion of aquifers causes changes in hydrodynamic condition of rivers and underground recharge basin and there is reduction in volume of subsurface discharge to rivers. Percolation of fines reduce effective transmissibility of aquifers and their water yielding properties. Adverse hydrochemical alteration occurs due to chemical pollution from mining wastes (Muthreja, 1993; Sinha, 1990). Water present in the voids causes some liquification phenomena due to daily blasting and ultimately it results into landslides.

3.8 Air pollution

All the mining operations generate noxious gases and/or dust thus contributing to air pollution. Blasting operation generates huge quantities of gases and dust. Operation of diesel engines of various machines also give out noxious fumes. These gases are harmful for health (Agrawal,1980; Bagchi et al.,1990). Dust is generated at transfer points, crushers, near dumping yards, etc. and due to movement of vehicles. It affects human health and also reduces visibility, increases wear of machine parts, discolouration of buildings, contamination of soils, vegetation and water (Bates,1971; Brandt & Rhoades,1972; James,1970; Muthreja,1993; Strahler & Strahler,1973).

3.9 Mine waste management

Huge quantitites of soil and rock waste is generated due to mining posing an acute problem for its disposal. In hilly terrain, sufficient space is never available for waste disposal. Also, the waste stored near the drainage channels or depressions may slide into them causing pollution. Fig. 2 shows a dump in a valley obstructing the natural run of surface water.

3.10 Migration of fauna

Deforestation, change in water drainage pattern, disturbance of habitat, noise and vibration as well as the human interference leads to the emigration of birds and wildlife from the vicinity of the mining area (Dube,1993; Mondal,1993).

3.11 Change in isostancy of the region

Cutting down of trees, changes in topography, hydrological changes, air pollution, etc. effect the temperature and rainfall in the region thus causing drastic climatic changes. Most of the young hilly terrains are seismically active and these activities are continuously going on within the earth crust. Even a small change caused by the mining activity may make the path for disturbance in the form of earthquake or volcanism.

3.12 Socio-economic impact

There is significant impact due to minig in complex natural situation seriously affecting the living and the non living. The most complicated is the shifting and

Fig. 2. Waste dumped in valley obstructing the water flow

resettlement of the population from the lease area. There is problem of disruption of social fabric of oustees and their unemployment. There is rise in criminality, vandalism, prostitution, loss of morals and other forms of social evils. Increase in mortality rate and lowering of life expectancy is there because of accidents and health problems due to pollution. Prices of essential commodities rise and problems due to urbanisation like slums and traffic congestion come up. Exploitation of simple folk, loss of cultural identity, increased labour problem, domination by external enterprenuers and steep hike in land and building costs are some of the other socio-economic impacts worth mentioning. Growth in variety of employment opportunities , improved communication, modernisation of infrastructure, improvement in lifestyle and better standard of living are some of the positive impacts which could not be ignored (SIngh et al.,1993; Soni,1993).

4 PREVENTIVE AND CONTROL MEASURES

To control the environmental detoriation, exploitation can not be stopped. Sustainable development can be achieved only by giving priority to expoitation of natural resources and at the same time taking preventive and control measures side by side.

With the help of the present day science and technology, it is possible to achieve the coexistance of mining and the society around it in an environmental friendly manner (Dube,1993).

The various mitigative techniques available for achieving the objective of eco-friendly mining in hilly terrain can be broadly classified according to different stages of mining operations, as follows:

4.1 *Planning and development stage*

This stage includes measures like Adoption of scientific mining processes, doing away with outdated transport system and mismanagement of wastes, selection of feasibly best possible ecofriendly mining method involving disturbance of bare minimum area and preference for undergroung mining as an alternative to opencast mining wherever feasible (Bose,1993; Dube,1993; Mohnot and Prasad, 1993).

While planning a mine, the sequence prescribed should be in decending order from hill top.

An environmental Management Plan (EMP) must be prepared at the planning stage itself and vegetation should be undertaken on large scale to compensate for future deforestation and to create green belts concealing ugly scars.

4.2 *Extraction of mineral*

Major contribution to environmental pollution comes from this stage and therefore it need special attention. Wet drilling, use of sharp bits and dust collectors must be adopted for drilling blast holes (Vines,1975). Blasting should be done with provisions of pre-splitting, use of delay detonators and muffling for preventing fly rocks (Dube,1993; Singh et al.,1993).

Non conventional drilling, blasting and muck removal methods suited to fragile ecosystem in hilly areas should be prefered. Use of semi manual, semi mechanised rock breaking process like ripping and mechanical breakers and silent chemical explosives may be adopted (Mohnot and Prasad,1993).

4.3 *Transportation and handling of excavated material/ore beneficiation (if needed)*

Noise level can be brought down by using efficient and noiseless machines, regular maintenance and by creation of tree barriers.

Semi mechanised or manual means should be resorted to as they cause lesser pollution in comparision to the higher degree of mechanisation. Alternative to road transport like conveyors, aerial ropeways, glory hole method, shaft and adit system, etc proves to be less pollution causing and also more economical in hilly areas (Mohnot and Prasad,1993). Transportation in covered or totally enclosed vehicles results in reduction of dust pollution (Singh et al., 1993).

4.4 *Management of rejects*

Removal of overburden contributes to large volumes of solid wastes which requires site for dumping as well as further treatment for stabilisation. The top soil cover should be removed and stocked for reuse in growing vegetation after reclamation of mined out area (Singh et al.,1993). The waste dumps should be confined by embankments and stabilised by benching into smaller slopes, plugging of gullies through check dams and plantation and mulchng on slopes (Bose,1993).

Water from the mining area should be reused in a closed circuit system, preferably with zero discharge into natural drainage system. Preplanning is essential for minimising contamination. Diversion detches, settling tanks, oil traps and sumps can arrest the speed and large portion of contamination (Mondal,1993). Erosion can be prevented by regrading, compaction, diversion and revegetation methods. Installation of under drains in dumps enhances their stability along with reducing contamination. Isolation of wastes from water and sealing off of abandoned mines will give favourable results (Muthreja,1993). Chemical treatment may also be resorted to (Mishra,1985).

Recycling of mineral wastes can also be thought of (Goodman and Chadwick,1973; Goyal et al.,1994; Mohan,1988; Mukherjee & Halan,1993).

4.5 *Reclamation of mined out area*

Reclamation should be done considering the profitability of the final outcome. The mining activity can be done with an objective to curve out an area which may have a ready use. The reclamed land may be devoted to developing colonies, fields, ponds and tanks, gardens, tourist resorts and air strips. Various excavation activities undertaken in the form of canals, road cutting, bunds, terracing etc. for creating useful land in hilly area can be integrated with mining process, thus achieving the dual objectives (Sharma,1987).

5 CONCLUSIONS

Following conclusions could be drawn from the above discussion:

5.1 Preplanning for pollution control due to mining activity in future should be efficient as well as feasible.

5.2 Effective monitoring of different environmental parameters is necessary to facilitate timely mitigative action

5.3 Lease size for mining should be sufficiently large so that scientific exploitation is feasible.

5.4 Only deposits of very high value and of stratagic importance should be allowed for exploitation in order to meet national needs.

5.5 Special provisions for environmental protection should be made in the mining and environmental laws and regulations for carrying out mining in hilly areas.

5.6 Innovation is always necessary to improve slope stability, recontouring and for faster revegetation. Revegetation will improve the binding of slope material and hence the stability.

6 REFERENCES

Agrawal,M.C. 1980. Environmental pollution due to surface mining -Some controls and remedies. Ind. Min. and Engg. J. Vol.19(102), pp 13-18.

Bagchi, A. and Gupta, R.N. 1990. Surface blasting and its impact on environment. Environmental management of mining operations. Ashish Pub. House, N. Delhi. pp 262-279.

Bandyopadhyay, J. and Shiva, V. 1985. The conflict over limestone quarring in Doon valley, Dehradun, India. Environmental conservation. Vol 12(2), pp 131-139.

Bates, D.V. 1971.The air environment & our health. CIM bull. Vol. 64(712),pp 56-59.

Bose, A.N. 1993. Restoration of mined out areas with reference to Himalayan region, proc. Course on environmentally viable mining tech. for fragile ground cond., Delhi, pp 43-50.

Brandt, C.J. and Rhoades, R.W. 1972. Effect of limestone dust accumulation on composition of forest community. Air Poll. Vol. 3(3), pp 217-225.

Chaulya, S.K. and Singh, T.N. 1993. Environmental effects of blasting in limestone quarry, Ind. J. Cement Rev., Vol 7(7), pp 21-33.

Chaulya,.K, Singh, T.N. and Sharma, Y.C. 1992. Environmental pollution and its prevention in limestone quarries, Ind. J. Cement Rev. Vol 7(9), pp 9-16.

Chaulya, S.K., Singh, T.N. and Singh, J. 1993. Effect of mine waste disposal on Environment and its protection. EUROCK 1993, Portugal. Balkema Ed., Netherland, pp 283-291.

Down,C.G. and Stocks, J. 1978. The environmental impact of large stone quarries and Openpit non-ferrous metal mines in Britain.H.M.S.O., London.

Dube, A.K. 1993. Some aspects of environment friendly mining technology, proc. Course on environmentally viable mining tech. for fragile ground cond., Delhi, pp 134-166.

Goodman, G.T. and Chadwick, M.J. 1978. Environmental management of mineral wastes. Sijhoff & Noordhoff Publ., The Netherlands.

Goyal, M., Kumar, P., Dastidar, A.G. and Singh, T.N. 1994. Impact of waste disposal on environment with particular reference to lead-zinc mines, Proc. Int. Conf. on Beneficiation, Nagpur,India.

Grange, G.H. 1977. Environmental control in the mining industry, J. S. Afr. Inst. Min and Met.Vol 78(2), pp 19-23.

James, N.C. 1970. The effect of strip mining on a natural system -A water quality study of Piednont lake, Ohio, Canada.

Mathur, V.N.S.1985. Environmental problems in mine & mineral processing - Present and future, Proc. Recent trends in Min. Industry, Varanasi. pp. 373-386.

Maudgal, S. and Kakkar, M. 1993. Mining operations in ecologically sensitive regions, proc. Course on environmentally viable mining tech. for fragile ground cond., Delhi,pp 19-20.

Mishra, G.B. 1985. Mine environment and ventilation. pp 75-151.

Mohan, I. 1988. Environmental awareness and urban development. Ashish Publ. House, New Delhi, India.

Mohnot, J.K. & Prasad, V.V.R. 1993. Alternatives to blasting and environmental considerations for fragile ground conditions of Himalayas, proc. course on environmentally viable mining tech. for fragile gd. cond., Delhi, pp 222-235.

Mondal, S.C. 1993. A case study of limestone mining at Lambidhar Project, proc. Course on environmentally viable mining tech. for fragile ground cond., Delhi, pp 117-128.

Mukherjee, B. and Halan, M. 1993. Say yes to refuse. Business Today, India, June 7-21, pp 90-96.

Muthreja, I.L. 1993. Measures to control air pollution and water pollution in eco logically fragile areas, proc. Course on environmentally viable mining tech. for fragile ground cond., Delhi, pp 106-116.

Sawarynski, J.W. and Conn, K. 1981. Refuse pile design consideration. Min. Engng., 33(12), pp 1724-1728.

Sharma, Y.M.L. 1987. Land reclamation in mining operations - A view point. Nat. W.shop on Envtal. Management of min. operations in India, Varanasi. pp 23-26.

Singh, D.P. 1987. Environmental effects of blasting. Nat. W.shop on Envtal. Management of min. operations in India, Varanasi.pp 95-108.

Singh, T.N., Goyal, M. and Singh, J. 1993. Environmental pollution due to mining and its control, Ind. Cement Ind. Desk book, Commercial Pub., Bombay, pp 82-93.

Sinha, S.S. 1990. Impact of Mining on water regime. Environmental management of mining operations. Ashish Publ.House, N. Delhi. pp 357-367.

Soni, A.K. 1993. Socio-economic conditions of the inhabitants of mineral belts of Himalaya : an analysis, proc. Course on environmentally viable mining tech. for fragile ground cond., Delhi,pp 202-203.

Strahler, A.N. & Strahler, A.H. 1973. Environmental geoscience. Hamilton Publ.Co.

Vines, D.H.1975. The additive spray method of dust control, quarry management and products. Vol 2(7), pp 179-184.

Wiss,J.F. & Lineham, P.W. 1968. Control of vibration and blast noise from surface coal mining. USBM OFR. pp 103-179.

Environmental impacts of the Achimota Quarry, Accra, Ghana

Les impacts écologiques de la Carrière d'Achimota, Accra, Ghana

Kwami Edwin Noble Tsidzi
Geological Engineering Department, Institute of Mining & Mineral Engineering, University of Science & Technology, Kumasi, Ghana

ABSTRACT: The Achimota Quarry which is located in the northwestern part of Accra , the capital city of Ghana, has since the late 1960s been producing variegated rocks mainly for the purpose of decorating buildings and fence walls, and occasionally as crushed aggregates for concrete works within the city and its environs. However, the quarry has in recent times been posing some serious environmental problems which need to be addressed immediately in order to forestall disastrous consequences. Following a comprehensive field study, supported by some laboratory analyses, the environmental problems associated with operations in the Achimota Quarry have been investigated in order to provide a sound basis for a possible amelioration. It is recommended that the Ministry of Environment, Mines and Energy in collaboration with the Accra Metropolitan Assembly (AMA) and should draw up effective educational and monitoring programmes so as to ensure that operations in the Achimota Quarry are carried out in an environmentally-friendly manner.

RÉSUMÉ: La Carrière d'Achimota est située au Nord-Est d'Accra, la capitale du Ghana . Cette carrière a depuis la fin des années 1960 produit des pierres aux couleurs variées principalement pour intention de la décoration des bâtiments et des murs de clôture et parfois comme des granulats broyés pour des travaux du beton dans la ville d'Accra et ses alentours. Toutefois dans ces jours recents cette carrière a posé des problèmes écologiques sérieux pui doivent être abordés immédiatement pour prévenir des conséquences désastreux. Suite à une étude compréhensive sur le terrain, soutenue par quelque analyses laboratoires, les problèmes écologiques associés avec les fonctionnements à la Carrière d'Achimota ont été examinés, afin de fournir le fondement solide pour une amélioration possible. Il est conseillé que le Ministere de l'Environnement, Mines et Energie en collaboration avec Accra Metropolitan Assembly (AMA) doivent faire des programmes d'éducation et de surveillance effectifs afins d'assurer que les fonctionnements à la Carrière d'Achimota soient faits dans une façon écologique amicable.

1 INTRODUCTION

A recent demographic upsurge in the Accra-Tema metropolis has made it highly demanding for individuals or estate developers to put up an increasing number of residential buildings and for government to provide more office buildings, schools and hospitals to cater for the growing population. All these undertakings require the use of substantial quantities of natural building materials such as facing stones and concrete aggregates, the procurement of which has particularly led to an indiscriminate proliferation of stone quarrying activities within the metropolis.

The Achimota Quarry which was established in the late 1960s primarily for the production of variegated decorative stones is one of such quarries where even the production of concrete aggregates is gaining a wider dimension during recent times. This quarry, therefore, has now become a major stone producing centre in the Accra-Tema metropolis where a sizeable proportion of the energetic 'rural-urban drift' youth usually earn their living. The net result of these sporadic quarrying activities is that the environment is constantly being endangered.

The paper discusses the various aspects of the environmental impact of operations at the Achimota Quarry and offers proposals for a possible amelioration.

2 LOCATION

The Achimota Quarry, which is a cluster of quarries, is located in the north-western part of Accra at the foothills of the Akwapim Mountain Range (Figure 1). It trends approximately north-south, extending for about 1.5km.

This cluster of quarries is bounded on the west by the Accra-Kumasi railway, which is about 100m north of the Accra-Kumasi motor road, and on the east by the Achimota Schools.

The area falls within the dry equatorial climatic zone with a coastal schrub and grassland type of vegetation which experiences a mean annual rainfall of about 1000mm and a relative humidity of 65% - 90% (Dickson & Benneh, 1970).

3 GEOLOGICAL CHARACTERISTICS

The quarry area is underlain by rocks of the Precambrian Togo Series which strike nearly north - south with low dips of about 20° towards the east (Kesse, 1985). These rocks have been regionally metamorphosed to the greenschist facies and include mainly phyllites, phyllonites and quartzites with the latter rocks being sometimes weakly to moderately foliated. Quartz

veins are prominent in the Togo Series and they tend to be parallel to the foliation.

The bottom part of the quarry consists of a grey, glassy, massive quartzite which is overlain by a grey, medium grained, weakly foliated quartzite. The foliated quartzite is, in turn, overlain by phyllonite and then phyllite. These latter rocks are persistently light brown in colour, having undergone an intense chemical weathering under the prevailing humid conditions and thus resulting in the formation of about 1.5m capping of dark brown residual soil which is partly lateritic. It should be noted that the degree of weathering does not consistently vary vertically from grade I at the quarry floor to grade VI at the top but dependent on the competence of the lithology as dictated by stratigraphy, composition and fabric. For that matter, it is not uncommon to find that some portions of the quarry have weathering grade II material being sandwiched between grade III materials.

Petrographic studies conducted on fresh massive quartzite and completely weathered phyllite samples from the quarry, using optical microscopy and X-ray diffractometry respectively, revealed that the quartzite is characterised by about 60% of strained quartz while the phyllitic saprolitic soil has illite/montmorillonite as the principal clay minerals of modal composition of about 75%.

The main discontinuities observed in the quarry include faults, joints and foliation, some of these discontinuities being occasionally healed by quartz veins.

At one particular location in the quarry were encountered two small faults with dip/dip direction values of 72/308° and 50/110°. These two faults occurred within a distance of about 10 m, forming a zone of strongly brecciated and highly weathered material.

The rocks are well jointed with the most prominent joint having an average dip/dip direction of 75/320°. The joints are medium spaced and generally tight.

The foliation generally has a north - south strike direction with a dip of 30° - 40° to the east.

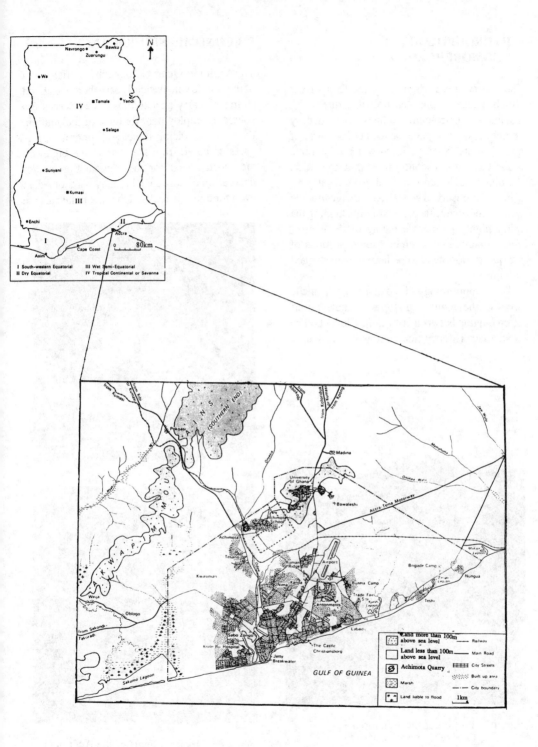

Figure 1 Location map of Accra showing the Achimota Quarry

4 HYDROGEOLOGY & GEOMORPHOLOGY

The quarry area is drained by the Odaw river which is heavily polluted probably due to the discharge of petroleum products from a nearby vehicle assembly plant or to the discharge of quarry waste. Such polluted water may move along the many discontinuities that abound in the quarry area into the local groundwater regime. Also observed was water seepage in some parts of the quarry face.   Notwithstanding the fairly highly permeable nature of the ground within the quarry precincts, some portions of the quarry floor tend to be largely waterlogged (Figure 2).

The geomorphology of the quarry site is made up of gentle undulating slopes with elevations often ranging between 30m and 140m O.D. The quarry face is fairly steep but generally stable.

Figure 2 Portion of the Achimota Quarry showing water-logging (foreground) and undercutting of high tension electric pylon (background)

5 ECONOMIC SIGNIFICANCE

The Achimota Quarry is of much benefit to most of the people who basically earn their livelihood from the quarrying activities since it serves as a source of employment for most of the youth in the area. The facing stones and concrete aggregates procured from the quarry (see Figure 3) are mostly used for the construction and/or decoration of buildings in many parts of Accra even though it is realised that the materials are

Figure 3 Rock products of the Achimota Quarry; Decorative stones in top picture and chippings in bottom picture

not always wholly suitable for use in concrete works. The relatively fresh massive quartz aggregates may not produce good quality concrete on account of the presence of strained quartz. Also, the highly weathered quartzite and phyllite do not acquire the requisite strength for the purpose since the aggregate crushing and porosity tests conducted on these rock materials in accordance with the British Standards specifications (Anon. 1975) gave mean values of 43% and 2.7% respectively.

Interviews conducted in the quarry area indicated that even though the quarry workers tend to be fully aware of the fact that they sometimes produce low quality materials, which are relatively much easier to win, the prospective buyers invariably prefer these materials to the relatively more expensive high quality types produced in other quarries. It is evident, therefore, that quality is being sacrificed for cost, which is very unfortunate indeed.

6 ENVIRONMENTAL HAZARDS

With regard to environmental aspects, four key issues are considered in this discussion relating to the Achimota Quarry. These include potential earthquake damage, health hazards associated with dumping of refuse in the disused parts of the quarry and ponding of water, undermining the stability of the high tension electric pylons and trees as well as the occupational hazards associated with unauthorised blasting in this quarry.

6.1 Potential earthquake damage

Despite the fact that Ghana is not located within any of the world's major seismic belts, southern Ghana has generally been experiencing some significant seismic activities since 1615 when a major earthquake was reported to have damaged the fortress of Sao Jorge at Elmina, west of Cape Coast in the Central Region.

In the city of Accra and its environs, in particular, major seismic events had been experienced in 1862, 1906 and 1939 in addition to the numerous minor shocks which, nevertheless, are of much engineering significance. Most of these seismic events have their epicentres located either along the Akwapim Fault, about 20km west of Accra, or offshore of Accra along the Coastal Boundary Fault. It is worth noting that the Achimota Quarry is located within the Akwapim Fault Zone.

The 1939 Accra Earthquake, undoubtedly, is the best documented local major seismic event in Ghana. According to Junner et al. (1941), this earthquake with a focal depth of 13km and an epicentre located approximately 40km offshore to the southwest of Accra, shook most of West Africa on June 22, 1939 at about 19:20 hours (GMT) and produced shocks of intensity VII and IX on the ten-point Modified Mercalli (Indian) Scale. Within the city of Accra, the event had an estimated magnitude of 6.5 on the Richter Scale with the main shock having a duration of about 30 seconds. A detailed study undertaken by Junner et al. (1941), relating the extent of structural damage caused by the earthquake in the city of Accra to the ground conditions, largely revealed that the effects were appreciable at the contacts of grounds of dissimilar geotechnical properties such as soft waterlogged alluvium and competent rock. Apart from some structural damage experienced in central Accra, which was at the time a developed area, there were also other significant effects of the 1939 earthquake such as the occurrence of extensive liquefaction within the alluvial and lagoonal areas, rockfalls in the Weija/McCarthy Hill area and in the coastal cliffs, as well as the formation of fissures in the more competent ground. Sixteen people were reported killed by this earthquake and about 130 injured; nearly 1500 houses had to be demolished and over 600 needed to be rehabilitated before re-occupancy.

The quarry site is quite near to an active seismic zone. The Akwapim thrust fault which has brought the older Dahomeyan rocks into a thrust contact with the relatively younger Togo series rocks runs northeast - southwest through a point southeast of the quarry. Another thrust fault bringing the Togo Series rocks into contact with the relatively younger Voltaian System rocks passes northwest of the quarry and trends nearly east - west. These two thrust faults intercept the Romanche transcurrent fault about 50 km southwest of the quarry, thus giving rise to active seismicity of the area. Hence, the Achimota Quarry site can experience relatively severe seismic damage in the event of a major earthquake. The necessary precautionary measures, therefore, need to be taken.

6.2 Waste disposal hazards

The use of the quarry site by the Accra Metropolitan Assembly (AMA) for the disposal of human waste and domestic refuse imposes a lot of health hazards to the people working at the quarry and also to the inhabitants within the vicinity of the quarry.

The non-engineered manner in which the AMA personnel dispose off the waste leaves much to be desired. A polluted stinking atmosphere is often created which can lead to an epidermic in the area.

6.3 Mosquito-infestation of ponds

Most parts of the quarry become waterlogged during the rainy season. This water-logging, apart from rendering quarrying operations difficult, also serves as breeding grounds for mosquitoes which are noted for the transmission of malaria and other related diseases.

6.4 Undermining the stability of high tension electric pylons and trees

Of major environmental concern is the fact that the quarry workers, out of ignorance or neglect, are largely not mindful of the existence of the high tension electric pylons that carry the nation's hydroelectric power from Akosombo/Kpong to the northern parts of the country. The rocks are, therefore, quarried rather dangerously close to these structures thereby undermining their stability. Figure 4 shows a classic case of this situation. The fall of the high tension pole is a potential danger to human life or property both within the quarry area and elsewhere.

Also, the stability of some huge silk cotton trees is impaired since most of their supports have been removed through quarrying (see Figure 5).

Figure 4   An undercut high tension electric pylon at the Achimota Quarry

6.5 Occupational hazards

The rock materials are sometimes won in the quarry by unauthorised blasting instead of the conventional hammer-and-chisel technique.

It must be emphasised that, apart from the occupational hazards associated with the use of explosives especially by untrained personnel, the blasting itself could cause such environmental hazards as ground vibrations, flyrocks and dust pollution. There have been frequent complaints from residents that the continuing blasting of stone in the area poses a threat to their electrical equipment and also causes cracking damage to their buildings.

Figure 5 An undercut silk cotton tree at
the Achimota Quarry

7 CONCLUDING REMARKS

It is observed from this study that the siting of
the Achimota Quarry has served a very significant purpose in the area by offering a
source of employment for the youth and also
providing raw materials for building and
decorating houses within the Accra -Tema
metropolis.

These positive impacts notwithstanding,
operations in the quarry are largely endangering the physical and social environment.
There is a serious threat to the safety of the
people within the quarry area.

However, with such good quarrying practices as provision of adequate drainage, avoid-
ance of winning materials so close to the high
tension electric pylons and large trees, judi-
cious usage of explosives and routine engineering testing of quarried materials (quality
control) as well as improvement in AMA's
technique of waste disposal at the disused

parts of the quarry, it is envisaged that quarrying activities in the Achimota Quarry could
be continued into the future with minimal
environmental disturbance.

To achieve these, it is recommended that
the Ministry of Environment, Mines and
Energy in collaboration with the Accra Metropolitan Assembly (AMA) should draw up
effective educational and monitoring programmes for operations in the Achimota
Quarry.

ACKNOWLEDGEMENT

The author is grateful to workers at the
Achimota Quarry for providing information
relevant to this paper.

REFERENCES

Anon. 1975. British Standard (BS) 812:
Methods for sampling and   testing of
mineral aggregates, sands and fillers.
London: British Standards Institution.
Dickson, K.B. & G. Benneh 1970. A new
geography of Ghana. London:
Longmann.
Kesse, G.O. 1985. The mineral and rock
resources of Ghana.Rotterdam:Balkema.
Junner, N.R., D.H. Bates, E. Tillotson & C.S.
Deakin 1941. The Accra earthquake of
22nd June 1939. Bull. No. 13, Geological
Survey of the Gold Coast,Accra, Ghana.

Hydrogeological characteristics of fossil valley in the Sahel region

Caractéristiques hydrogéologiques des vallées fossiles de la région du Sahel

Kenji Nakao
Taisei Corporation Technology Research Center, Yokohama, Japan

Motoo Fujita
Myu Giken Consultants, Tokyo, Japan

ABSTRACT : Some fossil valleys distributing in the Sahel region cut the basement rocks downward and are filled with sand and gravel which are most suitable container for the storage of groundwater. Hydrogeological characteristics of fossil valley are described in relation with construction of subsurface dam. Characteristics of the water resources and of the proposed afforeststion project in this region are also summerized in view point of utilization of groundwater.

RESUME : Les vallées fossiles réparties dans la région du Sahel coupent le socle vers le bas,et sont remplies de sable-gravier,le conteneurle mieux adapté au stockage des eaux de souterrain. Les caracteristiques hydrogeologiques de la vallée fossile sont décretes en relation avec la construction du barrage de sous-surface. Cette étude résume quelques-unes des caracteristiques des ressources en eau et du project de reboisement proposé pour cette regeon du point de vue de l'utilisation de l'eau souterrain.

1. INTRODUCTION

One of the most serious social problems resulting from desertification is a decline in agricultural productivity due to the environmental disruption and subsequent disruption of local economies.

Probably one of the most effective measures for preventing desertification and such economic calamity from occurring is to develop far-reaching afforestation policies, several of which have already been proposed and adopted. In this study, we tested the applicability of one such plan to the Sahel region of Africa. Undertaken by a consortium of companies, the project involves an groundwater storage system which utilizes undeveloped water resources and solar energy in an effort to prevent desertification. It seeks to achieve the following objectives as one the countermeasures against the desertification :

1) To develop undeveloped water resources, especially shallow groundwater, as circulatory water resources.

2) To design subsurface dam systems to temporarily store groundwater for later use in irrigation.

3) To develop water pumping and supply systems powered by solar-generated electricity.

4) To utilize irrigation for large-scale afforestation projects and associated social development.

The results of these studies have been reflected in measures proposed to prevent desertification in the Sahel region. In this paper, the results of studies related to water resource development in Mali and Niger, which occupy large part of the Sahel, will be summarized in the following section.

2. DESCRIPTION OF WATER RESOURCES

As we can see in Figure 1, the Sahel which is probably a transitional zone of steppe being converted into desert borders of the southern Sahara, stretching some 4000km from east to west and 300km from north to south. This region is said to be losing about

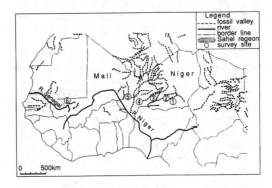

Figure 1. Distribution of fossil valley

ten kilometers annually to the southern advance of the Sahara. If this encroachment were to continue, the entire Sahel region could be expected to become part of the Sahara desert within some ten years.

Figure 2 shows the southward advance of declining annual rainfall precipitation which has been observed in recent years. Here we can see that the precipitation curve shows about 350-600mm of annual rainfall for what is believed to be the northern limit of arable land, an area which comprises the Sahel. We may consider the southern advance of this belt to be an indication of an expanding desert.

There are several types of water resources in the Sahel. The main ones are described as follows :

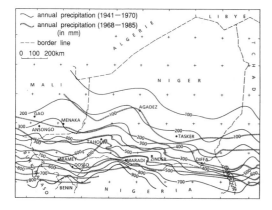

Figure 2. Declining annual rainfall precipitation

2.1 Rivers and streams

During periods of drought in this region, the only major bodies of flowing water are the Niger and Senegal rivers ; therefore, development of flowing water is only feasible along these two waterways. Regions within the Sahel which are experiencing especially serious effects from the advancing desert are mostly those which cannot use these river resources.

There are intermittent streams or "wadi" within the Sahel which flow during the rainy season but disappear during the dry season ; therefore, systems which directly utilize surface water have not been developed in these areas. Building traditional dams on such wadis could probably be able to supply water even during the dry season, but there is the problem of salinization which would probably occur as a long term result of the high annual rate of evaporation. Furthermore, most of the large wadis have become "fossil valleys" whose sites are consisted of thick riverbed sediments, making it impractical to build dams there.

2.2 Standing water bodies (lakes and ponds)

Within the nations of the Sahel, there are standing bodies of water. For example, in Niger there are roughly 200 lakes and 3400ha of ponds. There are still water bodies which have not yet been exploited and are potential sites for future development. However, it is well-known that water levels here are highly susceptible to extreme fluctuations caused by climatic changes. A typical example is Lake Tabalak Kehehe in central Niger, which for the last 40 years has been sporadically drying up during dry periods and reverting back to a lake during the rainy season. Although it became a "permanent" lake in 1953 as a result of unusually long rains, it dried up several years later, only to become "permanent" again within the last 5 years or so.

Such standing water bodies in the Sahel are extremely unstable and have little potential as dependable, long-term water resources. Furthermore, they, like the man-made aboveground dams, are vulnerable to evaporation-induced salinization.

2.3 Groundwater of the wadis

In both Mali and Niger, there are large wadis such as Dallol, Goulbi, etc., which mostly have their origins in "fossil valleys". "Fossil valleys" are the remnants of large rivers and streams which formed during past geological eras (roughly 8000 years ago). They are presently hidden underneath the desert sands, buried by ancient riverbed sediments. Therefore, flowing water in the drainage basin is collected here, and there are many hydrogeological conditions where groundwater can easily infiltrate.

Since the groundwater of such wadis is recharged annually by precipitation, it has been gaining notice in recent years as a potential circulatory water resource. Previous studies on such relatively small wadis as Goulbi Maradl in Niger have estimated potential groundwater recharge at 20 - 40 million tons annually. However, even though there is ample groundwater in wadis during the rainy season, ground water levels become markedly low at the end of the dry season. Moreover, there is a considerable decline in groundwater during drought.

2.4 Shallow groundwater in the plateau

As we can seen in Figure 3, the sandstone layers within the Continental Terminal, a group of typical Tertiary sedimentary formation, can become important groundwater resources as aquifers in the plateau. These formations are widely distributed, making them an important source of water for the widely-scattered population of this region and therefore an important item for water resource development.

In recent years, however, numerous wells using

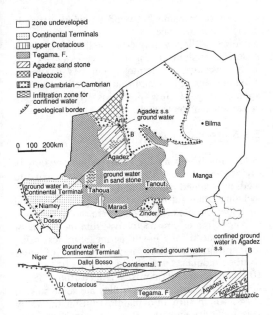

zone undeveloped
Continental Terminals
upper Cretaceous
Tegama. F.
Agadez sand stone
Paleozoic
Pre Cambrian~Cambrian
infiltration zone for confined water
geological border

0 100 200km

Figure 3. Geology and water resources in Niger

these aquifers have been going dry during the dry season as a result of advancing desertification. Previous studies have estimated potential groundwater recharge to this aquifer at about 13.0 thousand tons/km^2/year. Moreover, there are problems associated with concentrated pumping of water from one location. If more water is pumped out than recharged, groundwater levels decline, adversely affecting surrounding wells.

2.5 Deep groundwater

In both Niger and Mali, as we can see in Figure 3, there are numerous aquifers containing deep groundwater which have been supplying water for deep wells. However, there are several problems associated with exploiting this type of resource, indicating that we cannot expect it too much to meet future water demands.

The so-called "fossil water", or deep groundwater, has taken thousands of years to accumulate within aquifers. Therefore, if much of it is used for irrigation in some region, resupply will not catch up, which may in turn lead to declining water levels in aquifers and greater difficulty in pumping water. Based on these considerations, the potential groundwater recharge from the Continental Terminal, an aquifer of deep groundwater in Niger, has been estimated "zero".

Furthermore, since it took thousands of years for his water to collect, the permeation of salt from the

sorrounding formations would generally raise salt concentrations in the groundwater. Therefore, the use of this type of groundwater could accelerate the salinization of the soil.

Table 1 considers various perspectives to evaluate the suitability of using this type of water resource to irrigate the Sahel. As we can see from the table, a system of sub-surface dams for storing and pumping water has been determined to be the most feasible for exploiting water resources in those regions which are currently suffering from the onslaught of desertification.

Table 1. Classification of water resources in the Sahel region

WATER RESOURCES	SURFACE WATER			SUBSURFACE WATER		
	river		lake or pond	shallow ground water		deep ground water
SYSTEM	dam + pipe line (open channel)	pump + pipe line (open channel)	pump + pipe line (open channel)	subsurface dam + borehole pump	shallow well + pump	deep well + pump
objectives	river	river	lake or pond	fossil valley	wadi or fossil valley	deep aquifer
available quantity	large	large	small~medium	medium	small	small
seasonal quantity variation	small~medium	small~medium	large	small~medium	medium~large	none
influence on environment	large	small	medium	small	small~medium	small~medium
quality of water	good	good	good~poor	fairly good	fairly good	good~poor
REMARKS	·big project ·high construction cost ·evaporation loss ·long work period	·limitation on available area	·water quality control ·limitation on available area	·recharge method ·high construction cost	·counter-measure in dry season ·small water quantity	·surface subsidence after long term pumping ·drying up ·small water quantity
EVALUATION	FAIRLY GOOD	GOOD	POOR	FAIRLY GOOD	POOR	POOR

In order to make this kind of system a reality, we must have a satisfactory understanding of the hydro-geological properties of "fossil valleys". Therefore, in Mali and Niger, we have conducted both field investigations and laboratory tests such as borings, electric sounding and pumping test. Selected test sites are shown by numbers in Figure 1. The following section will use these test results to describe the hydrogeological characteristics of fossil valleys.

3. GEOLOGICAL AND HYDROGEOLOGICAL CHARACTERISTICS OF THE FOSSIL VALLEYS

3.1 Distribution

The large-scale fossil valleys in Sahel region, most of which stretches over Mali and Niger as shown in Figure 1, has been recognized since several decades ago. The accurate location and scale can be reconfirmed through a photogeological analysis of the picture from Landsat. Figure 4 show some results of the analysis. These results indicate the extent, width,

2777

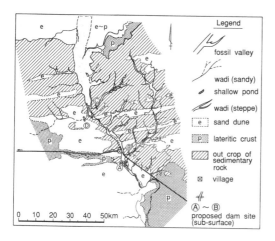

Figure 4. An example of analysed Land-Sat image (Mali ⑤)

drainage basin, and surrounding vegetation of the valley ; however, information on surrounding geological structure is insufficient.

3.2 Topography and geology

As mentioned before, most of the fossil valleys are "wadis", which flow during the rainy season but disappear and turn into complete sandy places during the dry season. Photo 1, displays typical configurations of fossil valleys. Owing to their origin, fossil valleys have various sectional forms and channels which have washed out the basement rocks. Table 2 shows configurations of some fossil valleys in Niger and Mali confirmed through field investigations and other exploration.

In general, the forms of the valleys can be The described as : both banks are 10m to 30m higher in

the form of embankment ; and the width varies in proportion to the scale.

valley is filled with riverbed sediments or sandy materials, as shown in Figure 5. The basement rock at the investigation site mostly consists of Tertiary siltstone and the Cambrian crystalline schist, apparently with few problems of cutoff wall construction underground.

3.3 Sediments in fossil valleys

Sediments filling up the fossil valley are largely divided into three : aeolian sand ; fluvial fine-to medium-grained sand ; and riverbed sediments including small gravel. Typical geologic columns and geological profile of each valley, are shown in Figure 6, 7 and Figure 8, respectively. From the results of sample analysis and in-situ permeability test using test pitting, coefficient of permeability of these riverbed sediments was obtained as 1×10^{-2} to 1×10^{-3} cm/sec, and its porosity ranged from 35% to 40%. On the surface of the valley are some sunken spots, which turn into puddles during the rainy season but dry up within several tens of days. They are covered with a solid crust-like layer consisting of evaporation residue, whose thickness reaches to 1m in some parts.

3.4 Hydrological properties and water balance

It is generally considered that a hydrological environment, where surface water and underground water collect from the drainage basin of old rivers, is maintained in the fossil valley.

The investigations up to present covers groundwater level, the precipitation in the surrounding areas, and the surface condition and geological structure of the drainage basins. Therefore, the details of hydrogeology must be investigated further.

Photo 1. An example of typical fossil valley

Table 2. Configuration of fossil valleys

COUNTRY		NIGER				MALI	
NAME OF VALLEY		① Goulbin N' Kaba (A)	② Tarka Valley	③ Dallol Mauri (A)	④ Dallol Mauri (B)	⑤ Goulbin N' Kaba (B)	⑥ Serpent Valley
INVESTIGATIONS	Photo analysis	O	O	O	O	O	O
	Field survey	O	O	O	O	O	O
	Electrical sounding	O	O	O	O	O	O
	Boring	O	−	−	O	O	O
	Laboratory test	O	O	O	O	O	O
	Water level observation	O	O	O	O	O	O
CONFIGURATION OF FOSSIL VALLEYS		±1800m 10m~20m 23m	>2000m 20m~30m	±1000m 30m~35m 5m	±1200m 20m~30m 14m	1000m~1800m 20m~30m 20m	±2000m 10m~15m 20m
BASEMENT		silty cohesive clay (Quaternary)	silty cohesive sand stone (Tertiary)	hydrated lime stone (Tertiary)	silty shale (Tertiary)	silty cohesive clay (Quaternary)	green schist
IMPERMEABILITY OF BASEMENT		good ~ poor	good	good	good	good ~ poor	good ~ poor
WATER LEVEL (dry season)		−20m	−10m	−2m	−5m	−9m	−10m
WATER LEVEL DIFFERENCE		3m	2m	2m	6m	9m	8m
ENVIRONMENT		savanna	dune-pasture	savanna-laterite top soil	pasture~laterite top soil	pasture ~ dune	savanna

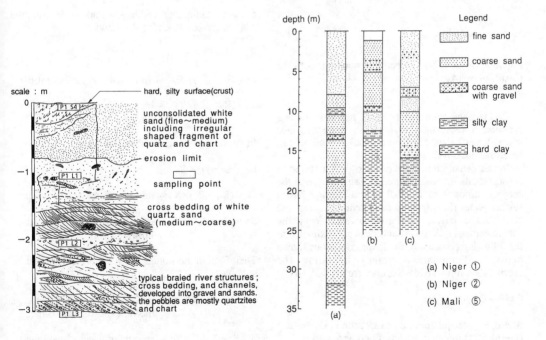

scale : m

P1 S4 — hard, silty surface(crust)

unconsolidated white sand (fine~medium) including irregular shaped fragment of quatz and chart

erosion limit

sampling point

cross bedding of white quartz sand (medium~coarse)

typical braied river structures ; cross bedding, and channels, developed into gravel and sands. the pebbles are mostly quartzites and chart

depth (m)

Legend

fine sand
coarse sand
coarse sand with gravel
silty clay
hard clay

(a) Niger ①
(b) Niger ②
(c) Mali ⑤

Figure 5. River bed sediment of fossil valley (Niger ②)

Figure 6. Geological column of fossil valley

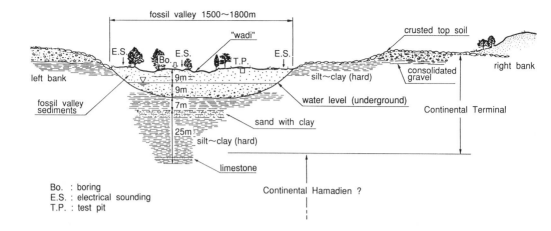

Bo. : boring
E.S. : electrical sounding
T.P. : test pit

Continental Hamadien ?

Figure 7. An example of geological profile of fossil valley (Mali ⑤)

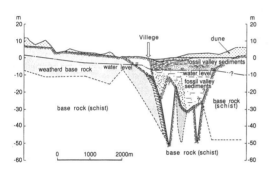

Figure 8. An example of electrical sounding profile of fossil valley (Mali ⑥)

On the assumption that the fossil valleys in Table 2 function as subsurface rivers, which collect surface water, water balance in case the valley is cutoff can be estimated as follows.

In this calculation, the amount of water stored in subsurface dam was estimated as a summation of the seepage amount of rainfall in the storage basin, surface water flowing from the catchment basin into the storage basin, and groundwater from the catchment basin. Water balance was approximated in the following equation. Calculation conditions shown below and numerical values including the precipitation from TAMS Report were applied.

$$R = E + D + S$$

where, R : precipitation ; E : evaporation ; D : surface run-off ; S : amount of underground water stored

When P denotes the area of the plateau, which is considered to be the catchment basin of sub-surface dam, Q denotes the area of the "wadi" as a storage basin, and X and Y denote the rates of surface run-off and underground seepage respectively, the following equation can be obtained :

$$P/Q = C \qquad X_P + Y_P = 1.0$$

$$X_Q + Y_Q = 1.0$$

Surface and underground inflow into "wadi" from the plateau can be calculated as follows :

$$R_P - E_P = D_P + S_P$$

$$D_P = (R_P - E_P) \times X_P \quad \text{— into "wadi" as surface inflow}$$

$$S_P = (R_P - E_P) \times Y_P \quad \text{— into "wadi" as underground inflow}$$

In the storage basin of sub-surface dam,

$$R_Q - E_Q = D_Q + S_Q$$

$$D_Q = (R_Q - E_Q) \times X_Q \text{ — into "wadi" as surface inflow}$$

$$S_Q = (R_Q - E_Q) \times Y_Q \quad \text{— into "wadi" as underground inflow}$$

Therefore, in the storage basin of sub-surface dam as a whole,

$$D = D_Q + D_P \times C \times X_Q \text{— surface inflow from the storage basin of sub-surface dam}$$

$$S = S_Q + S_P \times C \times Y_Q \quad \text{— amount of water stored in the storage basin of sub-surface dam}$$

2780

The field investigation shows that "wadis" are composed of permeable riverbed sediments, while the plateau is composed of a highly permeable laterite top layer and an impermeable bed rock below it. Thus, if X_P and Y_Q are given each 0.95 and 0.80, effective porosity 5% which was obtained as 15% of average porosity from in-situ test and the average thickness of aquifer be 20m, then amount of water stored as in Table 3 was obtained. As all the calculated "maximum potential recharge" is not necessarily utilized, the amount of water storage in Table 3 presents only 10% of it.

Table 3. Result of water balance calculation

Site No.	Catchment (km^2)	Area of storage (km^2)	Thickness of aquifer (m)	Storage volume (m)	Rise of water level (m)	Amount of water storaged *) m^3/year	m^3/$km^2 \cdot$day
No.1	4360	250	20	250×10^6	6.6	83×10^6	900
No.2	3600	112	20	112×10^6	7.6	43×10^6	1040
No.3	450	140	20	140×10^6	0.6	0.4×10^6	8
No.4	2290	68	10	34×10^6	9.1	31×10^6	1250
No.5	4980	310	20	310×10^6	2.7	40×10^6	315
No.6	2500	150	20	150×10^6	10.0	75×10^6	1350

(* 10% of maximum potential storage)

4. STUDIES ON THE FEASIBILITY OF CONSTRUCTION OF SUBSURFACE DAM

4.1 *Amount of groundwater storage*

Water balance based on the investigation on some fossil valleys implies that the effective potential groundwater recharge can reach several millions cubic meters per year, if these valleys extending over so many kilometers are fully utilized.

The average hydraulic gradient of the storage basin given here is assumed to be extremely small due to its configuration. Accordingly, the flow and storage of underground water might depend on a slight change of configuration and a partial underground geological structure. If so, the storage basin must be divided into smaller sections to be investigated in detail. At present, the investigation is under way on assumption that the hydraulic gradient of fossil valleys complies with that of the ordinary valleys, that is, from upstream to downstream. This is one of the subjects which require further investigation.

The infiltration in the drainage basin affects the calculation of the amount of groundwater storage significantly. The values such as 5% for the plateau and 20% for the sand stratum are general values, since the actual measured values have not been available. Depending on the configuration or geological conditions at specified construction sites, we should probably consider the additional measures to increase the value of infiltration. This also requires

further investigation.

4.2 *Groundwater containment and impermeability of basement*

In case of construction of cut-off wall underground, impermeability of the foundation at the construction site and the surrounding areas is one of the important factors, on which the construction cost depends.

As far as we can judge from the geological profile of the investigated fossil valley, the foundation bed is composed of the Tertiary siltstone and the Cambrian crystalline schist, which are highly impermeable. The selection of the construction site might be relatively easy, though further investigation is required. The problem is that the cut-off wall extends according to the scale of the valley, which invites economical difficulties.

As for the aquifer which stores underground water, the effective porosity is crucial. Generally, thickness of sediment of fossil valley reaches around 20m ; however, its geological structure varies. The measured permeability and porosity of some typical fossil valleys proves their high storage capacity. Coefficient of storage was given 5% with a margin for safety, so it can be increased up to 10%.

4.3 *Effects on the surrounding area after construction*

Effects of several million cubic meters groundwater storage per year and pumping on the surrounding areas must be examined carefully. In particular, thorough consideration must be given to the downstream area where people use wells as household water. The rise of water level in the storage basin also affects the surroundings. As the water level rises, the difference in water level with the plateau will reduce, and the drop of water level in wells around "wadi" during the dry season is expected to ease. If the water level rises significantly, surface water will remain longer. This causes the change in vegetation and the problem of salinization due to the increased rate of evaporation. In order to avoid these changes, the crest of the sub-surface dam must be constructed several meters below the ground surface.

5. CONCLUSIONS

Studies have been conducted on the feasibility of using undeveloped shallow groundwater as a water resource in irrigation planning for the Sahel region of Africa. We can conclude from various investigations, such as field studies, that :
1) There are fossil valleys in the Sahel which contain plentiful groundwater during the rainy season.
2) Due to the geological structure of these fossil

valleys, there are places where it would be feasible to build sub-surface dams.

3) Since fossil valleys are buried by riverbed sediments, they have suitable water-storage properties.

4) There are catchment basins where groundwater has collected on such a scale as to ensure the volume of water needed for the project.

From these conclusions, we can see that in this region, the construction of sub-surface dams will most likely be able to provide several million cubic meters of groundwater annually for irrigation. A conceptual diagram underground storage system which has been proposed for water resources development is shown in Figure 9. Although it is believed that there are no technological problems associated with building such dams, there is presently a lack of detailed information regarding specific conditions for designing. In particular, the infiltration rate of surface water into the ground is one of the most important factors needed to be considered when calculating the volume of groundwater recharge, yet it is extremely difficult to measure. In addition, when deciding on a specific location, we should probably consider the construction of artificial structures to increase the value of infiltration, the application of which is considered not to be difficult.

Acknowledgement

We would like to express our sincerest gratitude to the many people who helped to make this report possible, especially to the staff of the Sahel Greenbelt Project Consortium for their assistance in conducting field surveys and compiling the results. We would also like to thank all the government agencies involved, including the Ministries of Hydrology of the Republic of Mali and Niger, for their tremendous contributions.

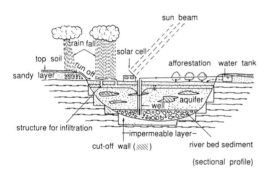

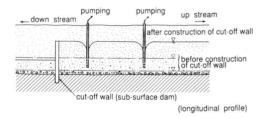

Figure 9. Concept of underground storage system

REFERENCES

1) U.S.AID (1982) : Natural Resources Management Action Program-NIGER.
2) Comité Interafrican D'Etude Hydrauliques (C.I.E.H) (1976) : Notice Explicative de la Carte de Plantification-Des Résources en eau Souterraine de L'Afrique Soudan-Sahelienne.
3) Ministére de Mines et de L'Hydraulique, République du Niger (1978) : Atlas des eaux Souterraines du Niger-Etat des Connaissances.
4) Ministére de L'Hydraulique et de L'Environment, République du Niger (1987) : Actualisation de L'Atlas des eaux Souterraines du Niger-Etude Financee par le Fonds D'Aide et de Cooparation de la République Francaise.
5) TAMS Engénieur (1985) : Le Résources Terrestres au Mali, U.S.AID Grant Agrément, 688-0105.
6) Nakao K.et al (1991) : Sahel Greenbelt Project and Subsurface Water Storage System, International Conference on Climatic Inpacts on the Environment and Society (C.I.E.S).
7) Kenji NAKAO, Yoshiyuki OHTSUKA & KUWAHARA (1993) : Water resources in the Sahel and its Storage System, Journal of the Japan Society of Engineering Geology, 33-6.

Effect of aquifer dewatering and depressurization on the compaction properties of sandstone

L'effet de drainage et dépressurisation des aquifères sur le compactage d'un grès

H.R.Nikraz & M.J.Press
School of Civil Engineering, Curtin University of Technology, Perth, W.A., Australia

ABSTRACT: This paper describes the research carried out to evaluate the compaction associated with dewatering as an aid to surface subsidence prediction in the Collie Coal Basin.
A triaxial test technique has been adapted to determine the compaction characteristics of the sandstone aquifer in the Collie Basin. The technique, which allows the strata stress regime to be reproduced by triaxial loading with zero lateral strain and provides a precise evaluation of lateral stresses and consequently Poisson's ratio under in situ conditions.
The paper contains details of equipment commissioning and test techniques, and the analysis and interpretation of derived data obtained.
The testing and evaluation techniques are general in nature and are applicable to field situations in locations where similar weak sandstones occur

RÉSUMÉ: Cet article décrit un travail de recherche qu'était fait pour évaluer la compacité associé au sèchement comme une aide à la prédiction de l'affaissement de la surface dans le Bassin Houiller de Collie.
Une analyse triaxiale avait été adapté pour déterminer les traits distinctifs de la compacité, du grès aquiffer dans le Bassin Houiller de Collie. La technique permet que le régime de pression soit reproduit par la charge triaxiale avec la pression latérate a point zéro, et en même temps, offre une évaluation précise des pressions laterales, et par conséquent le coefficient de Poisson, suivant les conditions in situ.
Cet article contient les détails des instructions d'equipement et les techniques des essais et, aussi l'analyse et l'interprétation des informations correspondantes.
Les techniques d'essai et d'évaluation sont d'un caractère général et sont applicables aux situations sur le terrain dans les emplacements où se trouvent des grès tendres pareilles.

1 INTRODUCTION

The Collie Basin lies nearly 200 km south-southeast of Perth in Western Australia and is 27 km long by 13 km wide, covering an area of approximately 230 km^2 (see Figure 1). It contains extensive reserves of good steaming coal which is currently being mined by both open cut and underground methods.

The Collie Coalfield has a long history of strata control problems. They manifest themselves in the form of localised poor roof control, surface subsidence, slope instability and mine abandonment (due to a sand-slurry inrush). Major sources of these problems include the very

extensive, weak, saturated, sandstone aquifers. As a result, underground operations have been limited to room and pillar extraction, presently carried out by continuous miners and road-heading machines. Approximately 30-40% recovery by volume is being achieved by this method.

In order to increase the recovery to approximately 70%, the Wongawilli method of short-wall mining has been introduced. Caving of the immediate roof is integral with this method. Extensive aquifer dewatering was carried out to enable this mining method to be applied. The porous and weak nature of the aquifers provides a potential source of subsidence with a significant risk of environmental instability on a large scale. This is particularly critical adjacent to townsites and industrial complexes and therefore requires an enhanced understanding of strata mechanics to enable confident application of engineering design. Controlled strata deformation was required for safe operations and limit surface subsidence.

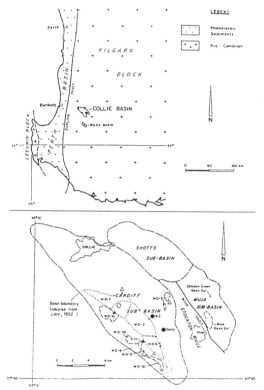

Figure 1 - Collie Coal Basin, location and regional geological setting

The roof dewatering/depressurisation procedure involved a combination of in-mine vertical roof drainage holes and conventional dewatering bores construction from the ground surface above the mining area (see Figure 2). A full account of the dewatering strategy may be found elsewhere (Humphreys and Hebblewhite, 1988 and Dundon et al, 1988).

Of prime concern is the effect of pore pressure reduction upon strata compaction. To simulate those effects it is necessary to perform tests under triaxial conditions of the same order as experienced in situ. The pore pressure effect phenomenon is not a new concept. However, investigation of the effect under triaxial conditions is relatively new.

Although rock bulk compressibility figures are generally larger for high porosities, simple compressibility porosity correlations do not exist. Furthermore, compressibility data reported for poorly-consolidated sandstone differs greatly. This paper describes the equipment and the techniques and procedures used in carrying out deformation characteristics of poorly-consolidated sandstone in Collie Coal Basin.

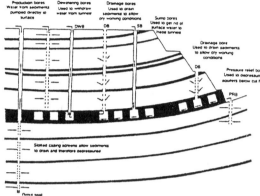

Figure 2 - Schematic diagram of production dewatering drainage sump and pressure relief bores

2 REGIONAL GEOLOGY AND HYDROLOGY

The Collie Basin is comprised of two unequal lobes in part separated by a fault controlled, basement high known as the Stockton Ridge. The basin itself consists of three sub-basins, the Cardiff, Shotts and Muja (see Figure 1).

The Collie Basin sediments are mainly cyclic, high energy fluviatile sandstones with thin gravel and conglomerate lenses. Siltstones and shales occur as overbank, lacustrine or paludal deposits. Coal seams are remarkably uniform in thickness and composition over considerable distances.

The Collie Basin sediments can be described as saturated and weak, and have been altered through weathering or post depositional processes. Lowry (1976) estimated that the coal measures were composed of 65% sandstone, 25% shale and claystone and 5% coal.

The whole Collie Basin can be thought of as an inter-related groundwater system of Permian coal measures bounded by Archaean basement.

Permeable aquifers comprise fine to granular quartzose sandstones with little fines content. Moderately permeable material consists of silty-clayey sandstones. Siltstones represent the low to moderately permeable aquifers whilst mudstone, shale and coal layers form the system aquitards.

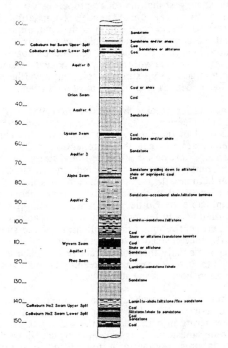

Figure 3 - Generalised hydrostratigraphy

All coal seams in the deep mines are bounded by aquifers. In some locations aquifers are situated directly above or below the seams, however, most areas have aquitard barriers of variable thickness separating the mining seam from neighboring aquifers (see Figure 3).

3 GEOMECHANICAL PROPERTIES OF THE COAL MEASURES

The geology of the Collie Basin can vary within short intervals, both vertically and laterally. There are also marked variations within the major lithologies (sandstones, shales, siltstones, laminites). Each has a wide range of engineering properties, dependent on past and present geological processes. Table 1 below lists typical ranges of compressive strengths, elastic moduli, cohesive strengths and friction-angle for each major lithology of the collie Coal measures. This table highlights the weak and plastic nature of Collie sediments and also illustrates that coal strengths are in the order of 3-4 times greater than non-coal lithologies. In terms of subsidence, the resistance to movement of non-coals is small, and thus there is the possibility that coal seams will deform differentially and lead to bed separations at coal contact.

4 ROCK MOVEMENTS CAUSED BY DE-WATERING IN POORLY CONSOLIDATED SANDSTONE

Land subsidence is caused by a number of mechanisms. Two such mechanisms are the withdrawal of fluid and the collapse of underground openings. This study is concerned with subsidence that results from the withdrawal of fluid.

The deformations resulting from equilibrium disturbance of the aquifer rock due to water pressure decline, are either elastic or non-elastic. Elastic deformations are mostly of a negligible extent with respect to both the involved surface subsidence and the reserve of the stored water, being only of importance in respect to the variation of the rate of flow.

The extent of the non-elastic deformation are due to compaction or migration of the rock material. The former, depends on the geotechnical characteristics of the rock, and on the extent of the pore pressure reduction. The extent of migration, on the other hand, depends on the pressure gradient (the flow velocity). The compaction may cause regional subsidence, while the migration of the rock particles causes local displacement, both phenomena being dependent upon the characteristics of the aquifer rock and the extent of de-watering.

Several techniques are available for predicting subsidence due to fluid withdrawal. They have been classed by Poland (1984) into three broad categories : empirical, semi-theoretical and theoretical. Empirical methods essentially plot past

Table 1 - Typical mechanical properties of Collie Coal Measures

Lithology	UCS (MPa)	Elastic Modulus (MPa)	Cohesive Strength (MPa)	Friction Angle (Deg.)
Sandstone	5.2	300	0.5	32
Siltstone	4.7	600	0.6	25
Laminite	4.7	700	0.7	25
Shale	7.0	1200	0.8	22
Wyvem Coal	19.8	2000	2.0	42

subsidence versus time and extrapolate into the future based on a selected curve fitting technique. However, empirical methods suffer from the lack of well documented examples to establish their validity. Semi-theoretical methods link on-going induced subsidence to some other measurable phenomenon in the field. Theoretical techniques require knowledge of the mechanical rock properties, which are either obtained from laboratory tests on core samples or deduced from field observations. Essentially, however, theoretical techniques use equations derived from fundamental laws of physics, such as mass balance.

Geertsma (1973) has shown in a theoretical analysis that reservoirs deform mainly in the vertical direction and that lateral variations may be discarded if the lateral dimensions of the reservoir are large compared with its thickness. For the one-dimensional compaction approximation, the vertical deformation of a prism of the aquifer material can be computed by :

$$\Delta h = C_m \, h \, dP \qquad (1)$$

Where Δh is the change in the prism height, C_m is the one-dimensional compaction coefficient, h is the prism height, and dP is the change in pore fluid pressure.

A similar approach to that used by Geertsma's (1966, 1973) was adapted by Martin and Serdengecti (1984). Martin and Serdengecti (1984) report that in most cases C_m is the most difficult of the three one-dimensional compaction parameters to determine and they suggest that the best way to obtain values of C_m is to measure it on core samples in the laboratory.

The one-dimensional compaction coefficient 'C_m' of friable sandstones can be measured by different methods : (1) indirect measurement by measuring rock compressibility 'C_b' under hydrostatic load and estimated Poisson's ratio of the rock : (2) direct measurement by equipment which simulates the aquifer boundary condition of zero lateral displacement (such as, Oedometer cell test, or a modified triaxial cell test). Although the triaxial test method is laborious and time consuming, its unique experimental conditions make it essential as they produce aquifer stress quite well. In addition, the triaxial set-up has the advantage that the circumferential pressure needed to prevent lateral stain is measurable. The Poisson's ratio of the rock sample can therefore be determined independently from the ratio of lateral to vertical stresses.

5 LABORATORY-DETERMINED COMPRESSIBILITIES

The cores taken from Collie Basin show marked variations in both porosity and grain correlation.

Medium to high porosities are found in consolidated and semi-consolidated sections. In addition, the nonhomogeneous appearance of the cores, suggest that the rock's properties vary over short distances consequently, compaction is expected to vary considerably with depth, implying that the cores must be sampled systematically at short intervals to obtain a reliable compaction profile. As this involves compaction measurements on a large scale, a simple, rapid, but nevertheless reliable measuring technique must be adapted.

The earlier studies by Grassman (1951); Biot (1941); Geertsma (1957) and Van der knaap (1959) resulted in the theory of pore elasticity. They demonstrated that the compaction behaviour depends only on the effective frame stress, i.e. the difference between external and internal stresses. Furthermore, the results obtained by Nikraz (1991) confirmed that the effective stress theory is applicable to Collie sandstone. Therefore, to stimulate aquifer compaction in a laboratory experiment requires the application of the stress difference instead of the actual stresses. Thus, experimentally the most attractive approach is to load the samples externally, keeping the pore water pressure constant and atmospheric.

Thus, a triaxial technique was adopted to predict the compaction behaviour of strata due to dewatering in particular for the weakly cemented Collie sandstone. This technique, which allows the strata stress regime to be reproduced by triaxial loading with zero lateral strain, also provides a precise evaluation of lateral stresses and consequently Poisson's ratio under in situ stress conditions. The condition of zero lateral strain during triaxial compaction test was achieved by both preventing any volume change in the cell-water system surrounding the specimen and by using the modified piston and top cap (see Figure 4). This piston was of the same diameter as the sample, therefore induced the triaxial stress in the sample, but the deviator stress. Because bulk volume change was detected from pore volume changes, the pores of the specimens had to be completely saturated. A full detail of the equipment design may be found in Nikraz (1991).

The experimental procedure comprised of two stages : (1) the preparatory stage, in which the specimen was brought into an "initial" loading state prior to the test; and (2) the test itself, which further compacted the specimen.

In order to eliminate possible membrane penetration effects during the test and thereby cause errors in test results, the specimens were first loaded hydrostatically to a pressure of 1.25 MPa, the volume change related to this pressure was assumed as a reference point. The axial stress was then measured continuously at a constant rate, until the desired axial stress was achieved. The cell pressure was adjusted simultaneously to prevent

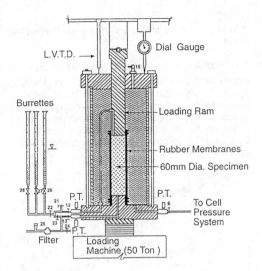

Figure 4 - Arrangement of apparatus for compaction test

any lateral strain. However, the maximum axial stress level was confined within cell pressure limitation (maximum cell pressure limited to 12 MPa).

Therefore to check the zero lateral strain, the following relationship had to be satisfied :-

$$\Delta V = (A\ X)/1000 \qquad (ml) \qquad (2)$$

Where ΔV = volume change (ml); A = cross sectional area of the specimen (mm^2) and X = axial deflection (mm).

To determine the effect of loading history on compaction, the axial stress was released incrementally to approximately 1.5 MPa. Consequently, the confining pressure was adjusted to satisfy equation 2. The loading and unloading were repeated for another two cycles. A total of six tests were made on specimens at strain rate of 2 x 10^{-4}min^{-1}.

6 RESULTS ANALYSIS AND INTERPRETATION

Measurements were made on six core samples taken from four locations in the Collie Coal Basin (see Table 2).

Typical axial stress/uniaxial compaction and lateral stress/uniaxial compaction are shown in Figures 5 and 6 respectively. Similar behaviour was observed in other five specimens.

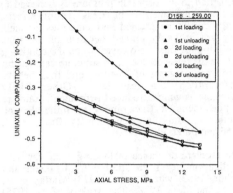

Figure 5 - Typical axial stress-strain compaction relationship with three loading cycles

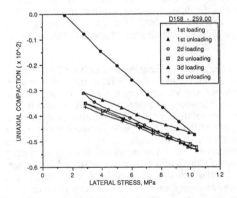

Figure 6 - Typical lateral stress-strain compaction relationship with three loading cycles

Table 2 - Sandstone properties

Sample Number	Location	Depth (m)	UCS (MPa)	Initial Porosity (%)	Permeability (10^{-8} m/s)
1	D156	286.00	3.351	20.50	36.22
2	D157	262.88	5.895	17.65	18.33
3	D157	264.26	5.201	18.05	20.62
4	D157	266.70	4.783	18.78	21.64
5	D158	259.30	3.481	22.09	34.07
6	Western 6	125.00	2.311	23.10	41.311

The three significant features of the stress/uniaxial compaction curves are their non-linearity, hysteresis and irrecoverable compaction on loading. Microstructural changes which produced permanent strain are a likely source of cycling effects. For example, assume that a microstructural element such as an asperity contributes to the elastic response of a rock by separating two grains. If the asperity is crushed subsequently at a high pressure, then later strain curves will be different because of the absence of the asperity. The unloading to atmospheric pressure is believed to have a significant role in stress cycling effects. When cracks are created and asperities crushed, they are probably pinned because of the high confining pressure. However, when the confining pressure is released, the microstructure can deform along new degrees of freedom and thus behave differently when reloaded. Other likely mechanisms which produced permanent strain are displacement of fines and clay minerals and frictional sliding on grain contacts (Brace, et al, 1966 and Batzle et al, 1980).

The problem of choice of loading cycle for field application has been studied by Knutson and Bohor, 1963; van Kesteren 1973; Mattax et al, 1975; and Mess, 1978. The work of Mess (1978) suggests that for fully undisturbed unloaded core material, compressibility values derived in laboratory tests should be lower than in situ values for reservoirs that are not over consolidated. For over consolidated reservoirs they could be either too low or too high for in situ application, depending on the degree of overconsolidation of the reservoir rock.

It is suggested by Knutson and Bohor (1963) that, a reasonable compressibility value may be obtained, by averaging values from the first and subsequent cycle. However, in extensive laboratory and in situ tests on relatively soft rock, Mattax et al (1975), suggested that the first cycle compressibility is the most realistic measure of in situ response to changes in effective pressure that occur during reservoir depletion. It was mentioned, however, that erroneously high values of first cycle compressibility are obtained in laboratory tests on unconsolidated sands because of systematic experimental error (caused by freezing and thawing of the sample, and some grain crushing). It was therefore recommended that about two thirds of the first cycle compressibility be taken as representative of in situ compaction.

The uniaxial compaction curves representing the six samples tested are plotted in Figure 7 for the first loading cycles. The graph shows an almost linear compaction/stress relationship for higher stresses, so that average compaction per unit stress can be calculated for this range. Further, it is noted that the compaction curves are parabolic thus, there is an observed relationship :-

$$\varepsilon_1 \, \alpha \, 1 \sqrt{\sigma_1'} \qquad (3)$$

Where ε_1 = axial strain and σ_1' = axial effective stress.

To demonstrate the observed relationship the axial strains have been replotted against σ_1' as shown in Figure 7. This plot provides strain lines, although it is noted that some points deviate slightly from linearity. By using the linear relationship as shown in Figure 8, the uniaxial compaction coefficient, C_m, may be calculated over the relevant stress interval.

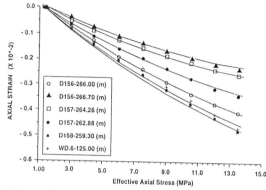

Figure 7 - Relationship between axial strain and effective axial stress for first loading

Consider the simulation of dewatering operations for a typical specimen such as D156-286. Assuming an average overburden density of 2.5 t/m^3, the initial in situ hydrostatic effective stress 7 MPa would increase to 9 MPa to simulate the effects of dewatering. Hence the uniaxial compaction can be calculated by :-

$$C_m = \frac{(\varepsilon_1)_9 - (\varepsilon_1)_7}{9 - 7} \qquad (4)$$

Where $(\varepsilon_2)_7$ and $(\varepsilon_1)_9$ are axial strain at hydrostatic effective stresses of 7 and 9 MPa respectively.

$$C_m = \frac{(0.288 - 0.277) \times 10^{-2}}{2} \qquad (5)$$

The uniaxial compaction coefficient data corresponding to first, second and third loading cycles are plotted as a function of initial porosity in Figure 8. It appears that compaction is greater for the first loading, indicating loading history influences on compaction. However, those correlations serve to assess a reliable average field value of the uniaxial compaction coefficient, which is required for a prediction of field compaction.

In an early study (Nikraz, 1991), the average porosity obtained from 105 samples tested as 20.77 percent of bulk volume. Although variation in porosity between holes was considered to be minor. This, and the near linear relationship between uniaxial compaction and porosity prompted the acceptance of 20.77 percent porosity for the determination of an average value of the uniaxial compaction coefficient.

Based on the first loading cycle, Figure 9 indicates a uniaxial compaction coefficient of 3.124×10^{-4} $(MPa)^{-1}$. The effects of stress relief upon sampling are accommodated within this value. However, the second and third loading cycles exhibit elastic compaction characteristics and provide an average value of uniaxial compaction coefficient for the second and subsequent loading cycles of 1.6409×10^{-4} $(MPa)^{-1}$.

The difference between two values indicates the elastic component of compaction. Considering the strain-hardening and core disturbance arguments one may expect the true compaction to be somewhere in between. In view of the quite small difference between maximum and minimum values, the most practical approach seems to be to take the average as a working value, thus reducing the uncertainty to an acceptable limit. Thus, a mean value of 2.382×10^{-4} $(Mpa)^{-1}$ was used to represent the in situ compaction coefficient.

Applying these results to a 12.5 m thick aquifer above the Collieburn No. 2, with an ultimate reduction in pore water pressure of 2.0 MPa could produce a vertical compaction of :-

$$\Delta h = - C_m \, h \, \Delta P \qquad (6)$$

$$= 2.382 \times 10^{-4} \times 12.5 \times 10^3 \times 2$$

$$= 5.96 \, mm$$

The Poisson's ratio of the specimens tested can be determined independently using the ratio of lateral to vertical stresses. The ratio of lateral to vertical stresses under isotropic conditions suggested by Teeuw (1971) is :-

$$\frac{\sigma_h}{\sigma_v} = \frac{v}{(1 - v)} \, 1/n \qquad (7)$$

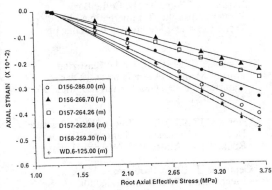

Figure 8 - Relationship between axial strain and root of effective axial stress for first loading.

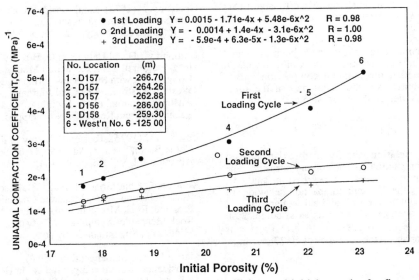

Figure 9 - Relationship between uniaxial compaction coefficient and initial porosity for first, second and third loading

2789

Where v is the Poisson's ratio and n is the exponent in relationship of the uniaxial compaction/axial pressure in Figure 3. The exponent reflects the deformation of the contact points and/or contact areas between grain (Brandt, 1955). According to Hertz's theory (Timoshenko, et al, 1951), for perfect spheres n = 2.3, while for linear elastic media such as non-porous quartz and steel, n = 1 reducing Equation 7 to the well known equation,

$$\frac{\sigma_h}{\sigma_v} = \frac{v}{1-v} \qquad (8)$$

For ideally elastic materials, thus a variation in n reflects a change in grain sphericity at the point of contact between adjoining grains. The values of n for the specimen tested range from 0.869 to 0.982. This range is higher than the value of 0.677 for spheres and indicates flatter contact surfaces.

7 CONCLUSION

Special purpose-designed triaxial testing equipment has been designed, tested and commissioned. A series of uniaxial compaction tests were performed for laboratory determination of compressibilities and in situ behaviour of the Collie sandstone. The following conclusions are drawn.

Whilst recognizing the early stages of development of subsidence prediction; some deformation has been postulated based on laboratory observations. In situ monitoring of strata deformation will be required for verification of the actual deformation mechanisms at work.

It has been observed that the uniaxial compression of Collie sandstone is characterised by significant non-linearity, hysteresis and an irrecoverable strain on unloading.

Uniaxial compaction curves have been presented for the sandstone aquifer in the Collie Basin. It was found that the uniaxial compaction curves were parabolic over the major part of the stress range. This yielded the expression :

$$\varepsilon_1 \propto \sqrt{\sigma_1'} \qquad (9)$$

A good correlation was found to exist between uniaxial compaction coefficient and porosity. The correlation quantified by regression analysis. Considering the different compaction behaviour of the specimens in the first and subsequent loading cycles, an average value for uniaxial compaction coefficient equal to 2.382×10^{-4} $(MPa)^{-1}$ was obtained for an average porosity of 20.77 percent.

REFERENCES

Batzle, M L, Simmonds, G, and Siegfried, R W (1980), Microcrack Closures in Rocks Under Stress : Direct Observation. Jnl Geophys. Res. 85, pp 7072-7090.

Boit, M A (1941), General Theory of Three -dimensional Consolidated. Jnl App. Phys 12, 2, pp 155-162.

Brace, W F, Paulding, B W and Scholz, C, (1966), Dilatancy in the Fracture of Crystalline Rocks, J. Geophys, Res. Vol. 71, No. 16, pp 3939-3958.

Brandt, H (1955), A Study on the Speed of Sound in Porous Granular Media, Jnl App. Phys., March, Vol. 22, p479.

Dundon, P J, Humphreys, D and Hebblewhite, B (1988), Roof Depressurisation for Total Extraction Mining Trials, Collie Basin, Western Australia, 3rd International Mine Water Congress, Melbourne, Australia, pp 743-752.

Grassman, F (1951), "Uber die Elastizitat Poroser Medien", Vierteljahrschrift der Naturforschenden Gesellschaft in Zurich, 96, 1, Sec. 1-51.

Geetsma, J (1957), The Effect of Fluid Pressure Decline on Volumetric Changes of Porous Rock, Trans, AIME, Vol. 210, pp331-430.

Geertsma, J (1966), Problems of Rock Mechanics in Petroleum Production Engineering, proceedings, First Congress of the International Society of Rock Mechanics, Lisbon, Vol. 1, pp 584-594.

Geertsma, J. (1973), Land subsidence above compacting oil and gas reservoirs, Jnl of Petroleum Technology, Vol. 25, pp 734-744.

Humphreys, D and Hebblewhite, B K (1988), Introduction of Total Extraction Mining in Conjunction with Strata Dewatering at Collie, Western Australia. The Coal Journal, Australian Journal of Coal Mining Technology and Research, No. 19, pp45-56.

Lowrt, D C (1976), Tectronic History of the Collie Basin, Western Australia, Geol. Soc. Australia Jnl, Vol. 23, pp 95-104.

Knutson, C F and Bohor, B F (1963), Reservoir Rock Behaviour under Moderate Confining Pressure, Rock Mechanics Proceedings, 5th Rock Mechanics Symp., University of Minnesota, Macmillan, New York.

Martin J C and Serdengecti, S (1984), Subsidence over oil and gas fields. In Holzer, T L (ed), Man-induced Land Subsidence, Rev. Engng Geology VI, The Geological Soc. of America, pp23-34.

Mattax, C C, McKinley, R M and Clothier, A T (1975), Core Analysis of Unconsolidated and Friable Sands, Jnl of Pet. Techn., p.1423.

Mess, K W (1978), On the Interpretation of Core Compaction Behaviour. Surendra, K S (ed), Evaluation and Prediction of Subsidence, publ. by American Society of Civil Engineers, pp76-91.

Nikraz, H R (1991), Laboratory Evaluation of the Geotechnical Design Characteristics of the Sandstone Aquifer in the Collie Basin, PhD Thesis, Curtin University of Technology, p.317.

Poland, J F (1984), Guide to the Study of Land Subsidence due to Ground-water Withdrawal, United National Educational, Scientific and Cultural Organisation, Paris.

Timoshenko, S and Goodier, J N (1951), Theory of Elasticity, McGraw-Hill Book Co Inc., New York

Van der knaap, W (1959), Non-Linear Behaviour of Elastic Porous Media, Trans. AIME, Vol. 216, pp 179-187.

Ven Kesteren, J (2973). Estimate of Compaction Data Representative of the Groningen Field, Verhandelingen Mijnbouwkundig Genootschap, Vol. 28, pp33 -42.

On prevention and control of hazardous gushing water in railway tunnels

Au sujet de la prévention du hasard des venues d'eau dans les tunnels des chemins de fer

Shi Wenhui

Fourth Survey and Design Institute, Ministry of Railways Wuhan, Hubei, People's Republic of China

ABSTRACT : This paper comprehensively describes the commonly encountered hazardous problems of water gushing and water outburst which are extremely hazardous to railway tunnels on the basis of theoretical analysis and practically proved information, puts forward theproduction and development mechanisms, i. e. the key factors to determine the total gushing water are the underground water feeding width, the water head gradient and the percolation coefficient ofsurrounding rocks; the key factors to determine the water outburst are the existence and break through of the discontinuous plane of underground water, and expounds and proves the prediction and forecast methods, assessment and basic law for hydrodynamic balance method, total water gushing and water outburst, infiltration coefficient method, hydrodynamic method, hydrological statistics method, well spring feefing method, surface water flow difference method and static storage method. Principles for control measures are also outlined.

RÉSUMÉ: Le travail décrit les problèmes des infiltrations aléatoires d'eau dans les tunnels de chemin de fer, y compris les gonflements importants, et analyse les facteurs qui conditionnent la quantité d'eau jaillissante: le ravitaillement des nappes le gradient et le niveau phréatique le coefficient de permeabilité des terrains envellopants et l'intersection des fissures du massif. Des méthodes pour la prédicticon et prévention du jaillissement d'eau dans les tunmels sont presentés et analysés. Des principes pour le contrôle de l'eau et pour le soignement de ce phenoméne sont aussi présentés.

INTRODUCTION

During 166 years in the world 102 years in China for the railway tunnel construction, water gushing is the most commonly encountered geological hazard and extremely hazardous. The key point to prevent and control it are accurate prediction and forecast as well as formulation and execution of effective control measures.

Prediction is the task in the survey and design stage. It mainly investigates and studies regional and site hydrogeological conditions, selects the favorable conditions and avoids the unfavourable ones to proposal guiding prevention and control measures. Forecast is the task in construction stage. It mainly studies and perfects forecast data topropose execating measures of engineering treatment.

Railway tunnel gushing water is generally classified into tunnel total gushing water and outburst water. The former refers the total gushing water (including the water outburst afer attenuation stabilizing) that the tunnel flows out stably and steadily. It influences the type and scale of water proof and drainage facility during construction and operation. The latter refers suddenly occurred and extremely hazardous transient gushing water. Prevention and relaxation measures need to be taken to ensure construction safely.

1 PREDICTION AND FORECAST TECHNIQUE OF TUNNEL TOTAL GUSHING WATER

1. 1 Production and development machanism of total gushing water

Tunnel total gushing water indudes static storage, dynamic storage and artificial infiltration of underground water.

During the building process of a new underground drainage passage, i. e. tunnel, key factors to determine tunnel total gushing water are underground water feeding width (in karst area, the location andinfluence of underground watershed need to be considered), water head gradient and the percolation or infiltration coefficient of surrounding rocks.

Due to the gradual release of static storage of water and the large water head gradient at the beginning of the construction stage, total gushing water is relatively large (karst waters may gradually connect each other due to crack filling by the effect of underground water addition and erosion aggravation, so it may be small at first and largeat last). Then it may gradually decrease, tend to approach a relatively stable value corresponding to a dynamic storage, i. e. feeding water isequal to gushing water. This is a common law for tunnel gushing water and a theoretical basis for analyzing and studying gushing water.

1. 2 Prediction and forecast of total gushing water

Prediction methods of total gushing water mainly includes hydrogeological analogy method, hydrodynamic method, hydrodynamic blance method, simplicied hydrodynamic blance (infiltration coefficient) method, hydrologic statistics (dry period water flow modulus) method, well—spring feeding method, etc. Computing equation are shown in Tab 1.

Total gushing water is influenced not only by hydrogeological conditions but also by construction conditions. Becuse hydrogeological conditions are complicated and changeable, for instance, seasonal precipitation, rock mass heterogeneity, lithological characters difference, etc. Will influence the water head of underground water respectively, it is very difficult at the design stage to control the change of the surrounding rock percolation conffient and feeding area (including ground surface feeding area and underground feeding area) width and construction organization (construction sequence, schedule and method). Therefore, at the construction stage, it is necessary to carry out long and medium distances gushing water forecast to supplement and perfect construction measures in time by means of groundsurface water and underground water dynamic monitoring, pilot prospecting (pilot advance, pilot drilling, shallow semic refraction prospectiong sonic prospecting etc.) according to the scheme and requirement predicted at the design stage.

2 PREDICTION AND FORECAST TECHNIQUE OF TUNNEL OUTBRUST WATER

2. 1 Production and development mechanism of outburst water

Discontinuous plane of underground water often exists between water—resting layer (rock formation with muddy and linear fissure rate less than 3%)and water—bearing strata because of the rock formation heterogeneity. When the tunnel construction breaks through water—bearing layer, it will trigger underground water outburst with high pressure in the water—bearing strata. This location is called water outburst point(belt). Outburst water volume may attenuates as time passes generally based on a logarithimc proportion.

2. 2 Prediction and forecast of outburst water point (belt)

The key point for prediction of outburst water point (belt) is to make full investigation and research on lithological characters, structure, fissure system and hydrogeological parameters of tunnel surrounding rocks, in combination with construction arrangement so as to make acorrect analysis and judgement. During counstruction forecast, monitoring shall be strengthened and necessary pilot probing shall be taken to perfect control measures and prevent serious outburst water.

In outburst water point (belt) prediction and forecast, sepcial attention shall be paid to the following locations of possible outburst water and the omen of impending outburst water.

1. Possible outburst water location

Open fault fractured zone hanging wall (active wall), tension fault fractured zone or fault fractrued zone intersection;

Open compression fractured zone with clayey particle content more than5%, muddy rock formation, water—beaing layer is lated by rock formation with linear crack rate less than 3%;

Position from open non—soluble rock to karsted soluble rock contact zone;

Position from open water—resting layer to loose water—bearing layer;

Position of open large—scale surface water seepage point (belt);

Position of passing through karst dissoluted depression and karst horizontal circulating zone.

2. Omen of impending outburst water

During tunnelling, surrounding rock cohesiveness increases and gushing water volume decreases and possibly, there is a water—bearing layer in the front;

Table 1. The computing equation of tunnel total gushing water.

No	Meathod	Equation	
1	Hydrogeological analogy meathod	$Q_T = q_o FSL$	(1)
		$q_0 = Q_0 / F_0 S_0$	(1,1)
		$q_1 = q_o FS$	(1,2)
2	Hydrodynamicl meathod	Obtaining various parameters accerding to hydrogeological exploration and seleting each corresponding equation base on different boundary conditions.	(2)
3	Hydrodynamic balance meathod	$Q_T = Q_c + Q_g$	(3)
		$Q_c = \mu v$	(3,1)
		$Q_g = \omega \Lambda X$	(3,2)
4	Simplicied Hydrodynamic balance (Infiltration coefficient) meathod	$Q_T = \omega \Lambda X$	(4)
5	Hydrologic statistics (dry period water flow modulus) meathod	$Q_T = qeLB = qe\Lambda$	(5)
		$qe = Qe / lb = Qe / a$	(5,1)
		$q_T = qeB$	(5,2)
6	Well—spring feeding meathod	$Q_T = \sum q_i + \sum q_r$	(6)

Note: Q_T Tunnel total gushing water (m³/d)

Q_c Underground water static storage (m³/d)

Q_g Underground water dynamic storage (m³/d)

Q_e Dry period surface water flow within the tunnel influence (m³/d)

Qo Total gushing water of existing adit (tunnel) (m^3/d)

qo Unit area and unit drop depth gushing water of existing adit (tunnel) $(m^3/d \cdot m^2 \cdot m)$

qt Tunnel unit length gushing water $(m^3/d \cdot m)$

q_T Tunnel unite length gushing water $(m^3/d \cdot Km)$

qe Dry period surface water flow modulus within the tunnel influence $(m^3/d \cdot Km^2)$

ql Left side well and spring water flow in dry period of the tunnel underground water feeding area (m^3/d)

qr Right side well and spring water flow in dry period of the tunnel underground water feeding area (m^3/d)

Fo Cross—section area of the existing adit (tunnel） (m^2).

F Cross—section area of the tunnel (m^2).

S Tunnel underground water table drop (m).

So Existing adit (tunnel) underground water table drop (m)

A Tunnel underground water feeding region area (Km^2).

a Surface water catchment area within the tunnel influence (Km^2).

L Tunnel length (Km).

l Surface water trunk length within the tunnel ifluence (Km).

B Tunnel underground water feeding width (Km).

b Surface water catchment area width within the tunnel influence(Km).

μ Rock yield of water , linear fissure rate, linear karst rate (Ky).

V Surrounding rock volume of tunnel excavation (m^3/d).

X Average daily precipitation in tunnel area (m/d).

α Surface water infitration coefficient.

Soft rock formation is broken, desquamation worsened and gushing water on the working face reduced or exhausted;

Gushing water becomes muddy during tunnelling;

Working surface fissure or borings during tunnelling while water or mud or sand spraying occurs.

2.3 Prediction and forecast of outburst water volume

Outburst water volume is predicted and forecast by means of hydrodynamic method, surface water flow difference method; well—spring feeding method and static storage method, etc.

Hydrodynamic method refers that calculation is carried out by selecting corresponding formulus based on different boundary conditions and various parameters obtained according to the hydrogeological prospecting.

Surface water flow difference method refers that tunnel outburst water volume is obtained by measuring the flow difference at the adjacent surface water sections on the same day in different seasons.

Well — spring feeding method is proposed based on the theory that dry period water flow volume of well — spring within the tunnel influence width is corresponding to the underground water feeding volume (dynamic storage) is this area. Therefore it can be used as a prediction method of total gushing water volume (stable flow). Meanwhile considering that the area in which well—spring lies is often controlled by fractuer zone of geological structure, well—spring flow within different tunnel drainage influence width in different seasons may be also used as the outburst water volume prediction method for tunnel passing through this structural fracture zone in the corresponding season.

Static storage method is used in closed and no—feeding water — filling karst section or artificial cave and water—filling body section, with its static storage being the outburst volume for the tunnel passing through these sections.

3 ASSESSMENT OF TUNNEL GUSHING WATER PREDICTION AND FORECAST METHOD

It can be seen from the comparison (Fig1, Tab2 and 3) between construction total gushing water volume actually measured and predicted by various means in the survey and design stage for Nanling Tunnel (shallow seated, passing through mountain and karst area) on Jingguangline.

3.1 Hydrodynamic statistics method (B = 0.5b) and well—spring feeding method (B = 1.0Km) obtain the approximate value with the total gushing water volume in the dry period of the tunnel construction. That is , these methods may be used to predict total gushing water volume in the dry period of the tunnel construction. The total gushing water volume in wet period may be obtained based on the influence width widened volume or an uneven coefficient of λ = 2.5～4.0 (the actual value being 1.96～3.29).

3.2 Considering the feature that total gushing water volume in the initial construction period being larger than that stablized in later period, hydrodynamic balance method (Ky method, percolation coefficient method), well—spring feeding method (B = 2.0Km), hydrodynamic method, all these are rational for Predicting and forecasting total gushing water volume in wet period, with accuracies respectively being 1.168, 1.147, 1.052 and 1.522.

3.3 By the above—mentioned methods, the accuracy of

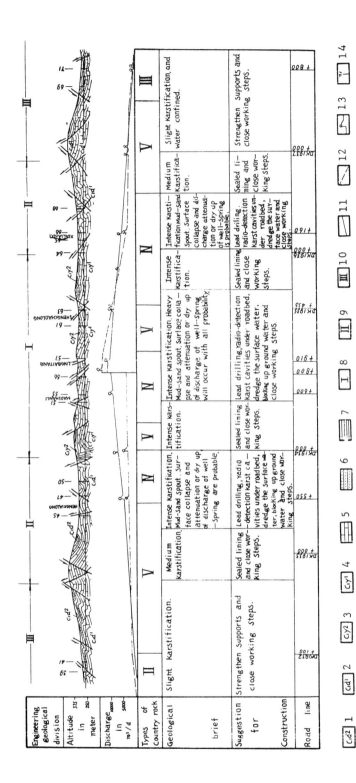

Fig. 1　longitudinal engineering geological section of Nanling tunnel on railway Beijing——Guangzhou

1.Ceshui Stage，Datang Formation. 2.Shidenzi Satang Formation. 3. Upper Shale，Yanguan Formation. 4.Lower limstone Stage，Yanguan Formation. 5. Limstone. 6.Sandstone. 7.Shale. 8.Divsion of rift intense karstification. 9.Division of rift medium karstification. 10. Division of fissure slight karstifca-tion. 11.Predicated hydrogroph curve of total gushing water valumes. 12. Actually—measured valumes of total gushing water valumes. 13. pred-icated and actually—measured position of outburst water point. 14.Drilling and its number.

Table 2. Total gushing water compaarison between the actually—measured value and the predicted value for Nanling Tunnel.

Meathod		0.5b	b	1km	2km	Karstic rate meathod (Ky)	Section calulated	Whole calulated	1km	2km	Hydrodynamic meathod	Values adopted in the design	Dry Period	Wet Period	Wet/Dry
QT (m³/d)	North section	2001.67	4003.34	2778.76	5557.53	9661.54	9523	12480	6184.4	10272.5	5253.4	10859	2325	4550	1.96
	South Section	4071.68	8143.36	4048.45	8096.89	10059.87	9860	13020	3768.8	7490.7	20443.6	11311	3745	12335	3.29
	Total	6073.35	12146.7	6827.21	13654.42	19721.41	19383	25500	9953.2	17763.2	25697	22170	6070	16885	2.78
QT/QB	Dry Period N.Section	0.861	1.722	1.195	2.390	4.156	4.096	5.308	2.660	4.418	2.260	4.671			
	S.Section	1.087	2.174	1.081	2.162	2.686	2.633	3.477	1.006	2.000	5.459	3.020			
	Total	1.001	2.001	1.125	2.249	3.249	3.193	4.201	1.640	2.926	4.233	3.052			
	Wet Perod N.Section	0.440	0.880	0.611	1.221	2.123	2.093	2.743	1.359	2.258	1.155	2.387			
	S/Section	0.330	0.660	0.328	0.655	0.816	0.799	1.056	0.306	0.607	1.657	0.917			
	Total	0.360	0.719	0.404	0.809	1.168	1.147	1.510	0.589	1.052	1.522	1.313			

Table 3. Outburst water comparison between the actually—measured Value and the Predicted Value for Nanling Tunnel.

No.	Place	Kilometerage	Length (m)	Point	condition	Volume (m³/d)	Calulated meathod	RearKs	Point	Volume (m³/d)	Remarks
1	Heng—xia—lung	DK1933+400 ~ DK1933+600	200	DK1933+550	Karstic horicontal circulation Zone	737.5	Well—Spring feeding meathld	Possibly with gushing mud and sand	DK1933+360 ~+570	Smaller	Outburst water volume reduced by improving the scheme
2		DK1933+920 ~ DK1934+020	100	DK1934+020	Cutting throught the water—resting layer	855.2			DK1934+000 ~+051	Extremdy small	
3	Mao—shan—li	DK1934+500 ~ DK1934+690	190	DK1934+600	Karstic horizontal circulation zone	373.3	Hydrodynamic meathod	With gushing mud and Sand			
4	Ling-bai-tang	DK1934+810 ~ DK1934+980	170	DK1934+910		1276	Well—spring feeding meathod	Possibly with gushing mud and sand	DK1934+880 ~+890	1130	
5	Sheng—chau—long	DK1935+335 ~ DK1935+535	200	DK1935+435		2140	Hydrodynamic meathod	With gushing mud and sand	DK1935+450 ~+463	Small	With mudding, the Pilot preinjection grout making the outburst water paint move forward.
									DK1935+915	1000	
6	Xia—lian—xi	DK1936+110 ~ DK1936+210	100	DK1936+160		1567	Well—spring feeding meathod	Possibly with gushing mud and sand	DK1936+173 ~+269	Heavy rain 2400, average 1400, min. 240.	Outrurst water spraying distance 12m, water pressure 400/KPa. Outburst mud and sand 3000m³, When conecting with surface water, Outburst water Volume reaches 11143m³/d.
									DK1936+960	Small	
									DK1937+064	Extremly small	
7	Lino-ji-awan	DK1937+600 ~ DK1937+800	200	DK1937+800	Cutting through the water—resting layer	875	Hydrodynamis meathod		DK1937+800 ~+850	Small	

comparison betweenthe predicted and the actually — measured outburst water point location for Nanling Tunnel reaches 70~80%. Outburst water volume is less than that of the predicted because of the improvement in north section scheme and the pilot pre—injection groutin south section.

3. 4 Prediction and forecast of total gushing water volume shall be made by distingguishing different hydrogeological units, rationally determining parameters and using various methods in each subsecions to analyze, compare and adopt rational or average values as the design values so as to avoid large errors. For instance, in Nanling Tunnel, by using various methods in each subsections to compare and adopt average values as the design values, the forecast total gushing water volume is 1. 313 times as the actual total gushing water volume during construction. The forecast value is more practical if the total gushing water volume of initial stage is larger than the stabler stage at constructing is considered.

4 BASIC LAW OF TUNNEL GUSHING WATER

According to the actual evidence of 65 foreign and domestic tunnel as Aomori — Hakodate, Taiseisvi, Dayaoshan, Nanling, Dabashan, etc, following basic laws for tunnel gushing water can be drawn.

4. 1 Total gushing water volume

1. In general condition, unit tunnel gushing water volume of hypogenic magmatic rock is $300 \sim 1000 m^3/d \cdot Km$, that of fragmental rock is $500 \sim 2000 m^3/d \cdot Km$, and that of karst rock is $500 \sim 5000 m^3/d \cdot Km$. Uneven coefficient of hypogenic magmatic rock is $1.5 \sim 2.0$, that of fragmental rock is $1.5 \sim 2.5$, that of karstic rock in vertical and horizontal zones which qrecircularly alternate is $2.5 \sim 4.0$, that of karstic rock invertical circular zone is $5.0 \sim 10$, or more.

2. Tunnel drainage influence width is related to tunnel buried depth, watertable and water permeability of rock formation. By actual measurement, Naling Tunnel is within $4000 \sim 1600 m$; tunnel in Japan are generally within $4000 \sim 1000 m$; developed fault fracture zone of Paleozoic formation and volcaniclastic rock are within $2000 \sim 4000 m$. Considering the geological conditions in China, it is generally within $500 \sim 1000 m$, it may be within $1000 \sim 2000 m$ in fault zone developed region and karsted tunnel.

3. After the tunnel completion, underground waater will gradually become balanced because of the lining closing. So the total gushing water volume will reduce accordingly. The reduction degree will differ with the closing degree. For instance, after completion, the total gushing water volume of Jappanese tunnel is only $0.125 \sim 0.2$ times of that in the construction period, where as in China, the number is as high as $0.2 \sim 0.8$ or more. No doubt it is related to the complex and changeableness at the hydrogeological conditions (popularly developed karst, etc) in China, but the design principle and some of the construction technologies still need to be discussed.

4. 2 Outburst water volume

1. Under general condition, outburst water volume is about $2 \sim 10 m^3/min$. Under special condition, for example, in a large scale structruefractrue zone or encountering a large scale surface seepage water, etc, it is $100 m^3/min$ or more.

2. Mostly, outburst water volues obtained by hydrodynamic method arestable flow values. So it is on the small side comparing with theinitial outburst water volume. According to the ratio of actuallymeasured initial outburst water volume to the attenuated stableoutburst water volume, Aomori — Hakodote Tunnel is $1.6 \sim 36.7$, DayaoshanTunnel is 10, Nanling Tunnel is 1.7, Guanjiao Tunnel is $16.7 \sim 20$, Daheishan Tunnel is $1.2 \sim 2.4$, etc. So it can be drawn that initial instant max outburst water volume is about $2 \sim 10$ times of the stable outburst water volume. Under special condition, for example, surrounding rock break, surface water intrude, etc, this value reaches $20 \sim 40$ times.

5 CONTROL OF TUNNEL GUSHING WATER HAZARD

According to the true situations of tunnel gushing water and outburst water, drainage blockage or the combination of these shall be adopted respectively. In the case of deep — seated non — karst tunnels with serious gushing water and outburst water, drainage shall be taken as a primary means, including drainage in pilot advance, dewatering by boring, etc. For protect the eclogical environment equilibution, that karst, underwater and shallow — seated tunnel, blockage shall be taken as aprimay means, adopting bi — liquid grouting seal to minimize sudden drop of underground watertable, surface water intrude, avoid ground subsidence and well spring exhaustion. To prevent outburst water hazard, a excavation sequence of water — resting layer (including compression fault) first and water — bearing layer second shall be avoided aspossible as it can be in the construction organization. In — advance water diversion or pilot grouting may be adopted to reduce pressure, dewater, stablize surrounding rock and avoid or abate outburst water hazard.

REFERENCES

Hydrogeological Cadres Program for Advanced Studies by Ministry of Geology and Mineral. 1963. Pit water prediction method. China Industry Publishing House.

Shi Wenhui. 1982. Preliminary exporation on Karstic hydrogelogicalproblems of Nanling Tunnel. Proceedings of Second Karst Symposium of China Society of Geology; P. 145~151. Science Press.

Shi Wenhui. 1982. Discussion on engineering geological problems around Nanling Tunnel in karst rock. Proceeding 4th International congress, IAEG, Vol. IV, p. 153~166, New Delhi.

Simulation de l'effet d'un mur emboué et de fondations profondes sur les écoulements souterrains: Cas de la place St Lambert à Liège

Modelling the impact of a slurry wall and deep foundations on the groundwater levels: Case study of the St Lambert square in Liège

A. Dassargues & A. Monjoie
Laboratoires de Géologie de l'Ingénieur d'Hydrogéologie, et de Prospection Géophysique (LGIH), Université de Liège, Belgique

RESUME: Un modèle numérique utilisant la méthode des éléments finis a été construit pour simuler les écoulements souterrains dans la nappe d'alluvions et de pieds de versant située sous la place St Lambert dans le centre de la ville de Liège en Belgique. Des travaux importants du génie civil sont prévus, comportant l'implantation de murs emboués traversant le sous-sol de la place et des fondations profondes sous certains nouveaux bâtiments. Après calibration du modèle sur la situation piézométrique mesurée avant les travaux, les simulations tendent à déterminer l'impact des travaux projetés sur les hauteurs piézométriques de la nappe. De plus, une situation de chantier avec pompages intensifs, a été simulée permettant d'estimer les rabattements provoqués. En effet, les rabattements autant que les remontées sont à craindre, en raison de la présence d'un important bâtiment historique (le Palais des Princes Evêques) dont les fondations sur pieux en bois risquent d'être sensibles à toute modification importante de la surface libre de la nappe et de sa frange capillaire.

ABSTRACT: A finite element numerical model has been used to simulate the groundwater flow conditions in the alluvial and colluvial aquifers below the St Lambert place in the city centre of Liège. Important Civil Engineering works are foreseen with realization of slurry walls crossing the underground of the whole place and deep foundations under important new buildings. After the calibration procedure of the model, completed on the previous measured piezometric situation, the simulations tend to determine accurately the impact of the Civil Engineering works on the piezometric heads of the aquifer. Moreover, a field-work situation with intensive pumping has been simulated in order to determine the induced drawdowns. Any important drawdown or recovery is to be avoided due to the presence of very old wooden foundations belonging to an historical building (the Bischop-Princes palace). These wooden pile foundations should be very sensitive to any important change of the water table and of its capillary fringe.

1 INTRODUCTION: MODELES NUMERIQUES D'ECOULEMENT SOUTERRAIN

Quelque soit le but final, la meilleure façon d'utiliser au maximum toutes les données disponibles relatives à une nappe aquifère, est de combiner celles-ci aux lois physiques appropriées (exprimées sous forme d'équations) et de réaliser ainsi un modèle mathématique (Dassargues, 1991a). Ces modèles numériques ont pour but de décrire et de quantifier les comportements observés.

En utilisant, dans leur formulation, les équations régissant les processus qu'on veut représenter avec les vraies variables impliquées, on s'assure de la signification physique des simulations et de la fiabilité des résultats. Pour fournir des solutions ayant un sens physique le plus précis possible, il est nécessaire de tenir compte de la complexité des réservoirs aquifères. Les hétérogénéités, variations spatiales, anisotropies des terrains géologiques, sont autant de sources de complications, alourdissant la réalisation des modèles mathématiques en hydrogéologie. La construction d'un modèle mathématique requiert les principales étapes décrites ci-après (Bear & Verruijt, 1987) :

(a) La synthèse des données disponibles doit aboutir à un modèle conceptuel de l'aquifère étudié, à partir duquel seront choisies les dimensions spatiales, les conditions aux limites et la formulation en régime permanent ou transitoire.

(b) La formulation en équations mathématiques des processus physiques est suivie de la recherche de l'expression numérique de celles-ci et de leur implémentation dans le programme.

(c) Le programme est testé quant à sa convergence, sa stabilité et sa précision; des comparaisons avec des solutions analytiques connues, en utilisant de grands intervalles de variation des paramètres, sont recommandées.

(d) La calibration est menée en ajustant la répartition de la valeur des paramètres (et éventuellement des conditions aux frontières) de manière à ce que les résultats calculés soient similaires aux données mesurées pour une même sollicitation du système.

La loi de Darcy, établissant que le flux est inversément proportionnel au gradient hydraulique, s'écrit :

$$q_i = K_{ij} \frac{\partial h}{\partial x_j} i, j = 1, 2, 3 \qquad (1)$$

où K_{ij} est le tenseur des perméabilités du milieu, q_i le vecteur flux, x_j coordonnée généralisée et h la hauteur piézométrique.

Cette loi, écrite sous forme scalaire, consiste en fait en 3 équations avec 4 inconnues : les q_i et la hauteur piézométrique h. La 4ème équation scalaire est fournie par l'expression du principe de conservation de la masse. En combinant la loi de Darcy et l'équation de continuité, on obtient l'expression de l'équation de diffusivité :

$$\frac{\partial}{\partial x_i}\left(K_{ij}\ \frac{\partial h}{\partial x_j}\right) - Q = S_s \frac{\partial h}{\partial t} \qquad (2)$$
$$i, j = 1, 2, 3$$

où S_s est le coefficient d'emmagasinement spécifique, t le temps et Q le terme de flux externe (pompages, réinjections, infiltrations).

Lorsque la nappe étudiée peut se réduire à un problème 2D, l'équation est ramenée à 2 dimensions par intégration sur l'épaisseur de la hauteur saturée. Cette simplification repose sur l'hypothèse que les vitesses d'écoulement horizontales sont très supérieures aux vitesses verticales (hypothèse de Dupuit).

$$S\frac{\partial h}{\partial t} = \frac{\partial}{\partial x}\left(T_x. \frac{\partial h}{\partial x}\right) + \frac{\partial}{\partial y}\left(T_y. \frac{\partial h}{\partial y}\right) - Q \qquad (3)$$

où S est le coefficient d'emmagasinement, Q le flux externe, T la transmissivité et x, y les coordonnées horizontales. Cette équation est uniquement valable pour des flux horizontaux en nappe captive, mais elle est également utilisée comme approximation en nappe libre avec $T = K.e$, e représentant l'épaisseur des terrains saturés de la nappe.

Une nappe aquifère a une extension limitée dans l'espace et, sur ces limites, les échanges d'eau avec l'extérieur sont régis par les conditions aux frontières. Ces conditions sont de trois types :

1°) *Condition de Dirichlet ou de potentiel imposé* :

$$h = \underline{h} \qquad (4)$$

La valeur du potentiel h est alors spécifiée sur la frontière considérée, cette condition est typiquement celle d'un contact nappe-rivière, le potentiel constant imposé étant égal à la cote de la surface libre.

2°) *Condition de Neuman ou de flux imposé* :

$$\frac{\partial h}{\partial n} = \frac{\partial h}{\partial x}e_x + \frac{\partial h}{\partial y}e_y = \phi \qquad (5)$$

où n est la normale extérieure à la frontière considérée, e_x et e_y les cosinus directeurs de cette normale et ϕ le flux imposé à la frontière.

La valeur du gradient de potentiel normal à la frontière est alors imposée. Dans le cas particulier où $\phi = 0$, cette condition exprime, par application de la loi de Darcy, que la composante d'écoulement normale à la frontière est nulle. Les équipotentielles sont donc perpendiculaires à cette frontière et les lignes de flux parallèles.

3°) *Condition de Fourier ou mixte* :

$$h + \lambda \frac{\partial h}{\partial n} \quad \text{imposé} \qquad (6)$$

Ce troisième type de condition permet d'imposer une relation entre le potentiel et le flux, comme cela est le cas lors :
- de la drainance par une frontière séparant la nappe aquifère d'un plan d'eau;
- du suintement à la frontière d'un milieu poreux, au contact de l'atmosphère.

La définition ad hoc de ces conditions aux frontières est indispensable à la résolution correcte du problème de champ posé.

2 DONNEES ET CONDITIONS DU MODELE

Les données géologiques et hydrogéologiques de la zone à étudier ont été tirées des rapports antérieurs des L.G.I.H. de l'Université de Liège sur le sujet (Monjoie & Calembert, 1975). Les données concernant la géométrie des ouvrages et travaux prévus ont été fournies par la Société des Transports En Commun (TEC), le Service de l'Urbanisme de la ville de Liège et l'Association Momentanée des Bureaux d'Ingénieurs-Conseils.

La carte piézométrique mesurée en 1972, avant les travaux importants, est reprise à la figure 1. Les perméabilités de la nappe alluviale dans cette zone peuvent varier entre 1.10^{-3} et 5.10^{-2} m/s selon la granulométrie des alluvions. La figure 2 présente l'implantation du mur emboué ainsi que la localisation des trois piézomètres de mesure situés dans la zone étudiée. On distingue également sur cette figure, le tracé d'un égoût drainant ainsi que la zone correspondant aux fondations profondes de l'îlôt Tivoli. Sur base de ces informations hydrogéologiques et de la géométrie des travaux projetés, un modèle d'écoulement souterrain utilisant la méthode des éléments finis a été élaboré.

Des éléments finis "briques" à 8 noeuds sont utilisés pour la discrétisation. Leur épaisseur reprend systématiquement l'épaisseur du milieu poreux saturé via l'utilisation des transmissivités.

Le problème est ainsi ramené en pratique à une simulation à deux dimensions dans le plan

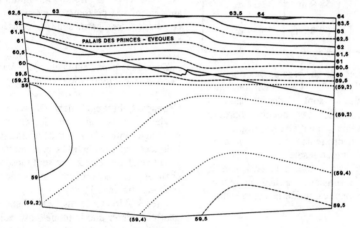

Fig. 1 Carte piézométrique mesurée avant tout travaux.

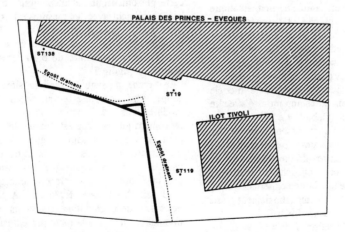

Fig. 2 Plan de situation montrant l'implantation du Palais des Princes-Evêques, de l'Ilôt Tivoli, du mur emboué et de l'égoût drainant.

horizontal par intégration sur l'épaisseur de la hauteur saturée. Ces fondations influencent la perméabilité moyenne équivalente à prendre en compte dans le modèle d'écoulement.

Les mesures piézométriques aux trois piézomètres de référence sont résumées au tableau 1 pour différentes conditions de chantier.

Tableau 1. Mesures piézométriques.

	Avant travaux	Pendant travaux (avec pompage)	Pendant travaux (avec pompage) +mur emboué
ST 138	61.2 m	≈ 56.0 m	57.62 m
ST 19	59.25 m	58.9 m	59.22 m
ST 119	59.38 m	58.85 m	59.37 m

La discrétisation effectuée sur base de ces données est reprise à la figure 3. Les conditions aux frontières latérales sont des hauteurs piézométriques imposées car aucune bordure réellement imperméable ne peut être envisagée dans cet environnement hydrogéologique. La base du modèle est supposée imperméable.

Les infiltrations efficaces dues à la pluviométrie sont négligées et la simulation a lieu en régime permanent pour une nappe considérée comme libre.

3 CALIBRATIONS DU MODELE

Ce n'est que lorsqu'un modèle mathématique représente de façon précise les phénomènes simulés, que son application peut être envisagée pour d'éventuelles prédictions. La procédure appelée "calibration" consiste à minimiser la différence entre mesures et résultats par ajustement des données d'entrée jusqu'à ce que le modèle reproduise les conditions du champ mesuré avec un niveau de précision acceptable. Le plus souvent, les données modifiées sont essentiellement les valeurs et répartitions des perméabilités - transmissivités car les autres données résultent de mesures plus fiables. Cette approche requiert de la part de l'hydrogéologue une bonne expérience et un bon jugement afin d'obtenir finalement une calibration fiable (Dassargues, 1991b).

Dans notre cas, la calibration a débuté en distinguant deux zones de perméabilités différentes :

- Zone de la plaine alluviale   : $K = 8.10^{-3}$ m/s
- Zone du colluvium du versant   : $K = 1.10^{-4}$ m/s.

Au départ, les piézométries calculées lors des premiers passages (essais) sont quelque peu différentes de la piézométrie mesurée (figure 1). On remarque quelques différences notamment dans la zone de transition entre la zone de la plaine alluviale et celle du versant. La calibration a été poursuivie pas à pas. Sans entrer dans les détails, quelques résultats intermédiaires vont être exposés ci-après, afin de montrer le cheminement réalisé.

Au 4ème passage, on peut considérer le modèle comme calibré sur la situation piézométrique mesurée de 1972. La carte piézométrique calculée est reprise à la figure 4, montrant un bon ajustement à la piézométrie mesurée (figure 1). A ce stade, les perméabilités utilisées sont les suivantes (figure 5) :

- Zone de la plaine alluviale : $K = 5.10^{-2}$ m/s
- Zone du colluvium de versant : $K = 1.10^{-4}$ m/s.

Cependant, il est également nécessaire de calibrer le modèle sur la situation du chantier avec les pompages intensifs qui ont rabattu la nappe, de telle façon que les niveaux piézométriques calculés soient similaires à ceux qui sont mesurés pendant le chantier et notamment aux piézomètres de référence (tableau 1).

Après différents essais, au 15ème passage, on peut considérer que le modèle est calibré sur cette deuxième situation, tout en imposant un minimum de changement dans les valeurs de perméabilité. La carte piézométrique calculée (figure 6) montre un bon ajustement avec les mesures. Les valeurs des perméabilités aux éléments 306 et 307 ont été ramenées à 1.10^{-4} m/s et une zone de pied de versant, de perméabilité de 5.10^{-3} m/s a dû être distinguée (figure 7).

Le 16ème passage reprend la simulation de la nappe, sans pompage, et en tenant compte des modifications de perméabilités. La carte piézométrique calculée est proche de celle obtenue pour le 4ème passage (figure 4).Au vu des piézométrie obtenues avec et sans pompage de chantier, on remarque que les pompages ont provoqué des rabattements sous une grande partie du Palais des Princes Evêques. A partir de ces deux situations calibrées (situations avec et sans pompage), les simulations présentées ci-dessous établiront l'influence sur les écoulements souterrains de la construction du mur emboué et des fondations profondes prévues pour l'Ilôt Tivoli.

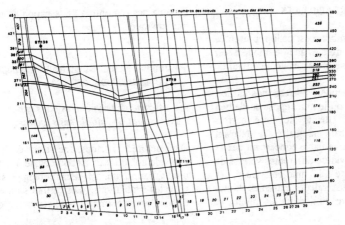

Fig. 3 Discrétisation de la zone en éléments finis. Une géométrie un peu tordue a permis de tenir compte explicitement de la géométrie du mur emboué.

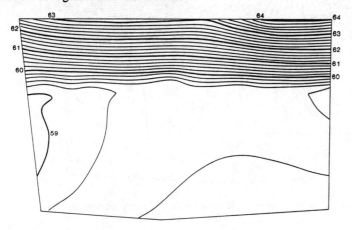

Fig. 4 Résultats calculés au quatrième passage (essai) de la calibration.

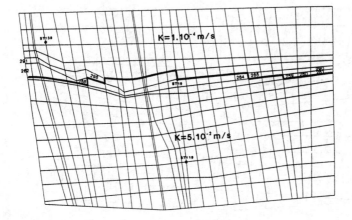

Fig. 5 Valeurs des perméabilités utilisées pour le quatrième passage de la calibration.

4 SIMULATIONS ET RESULTATS

La géométrie du mur emboué est celle représentée à la figure 2. Etant donné que le modèle est monocouche, la "perméabilité équivalente du mur"

à introduire dans les éléments correspondants doit tenir compte de la zone où le mur repose sur le bed rock altéré et où des débits de fuite parfois importants peuvent être enregistrés. N'ayant aucune donnée quantitative fiable concernant ces débits de fuite, on a considéré différentes valeurs de "perméabilité équivalente" du mur emboué. Les résultats obtenus pour les valeurs extrêmes de ces perméabilités permettront de déduire les minima et

maxima des remontées de la nappe en présence du mur emboué.

4.1 Simulation de l'écoulement de la nappe en présence du mur emboué

4.1.1 *Calcul des remontées par rapport à une situation initiale sans pompage*

Dans un premier temps, une valeur de "perméabilité équivalente" du mur emboué égale à 1.10^{-7} m/s et une situation initiale sans pompage sont choisis pour les simulations. Les résultats obtenus (figure 8) montrent notamment une remontée de la nappe

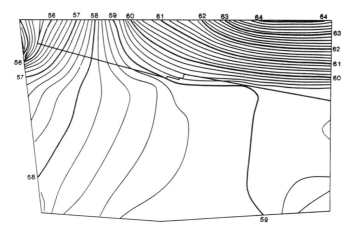

Fig. 6 Carte piézométrique calculée, calibrée sur la situation de chantier avec pompages.

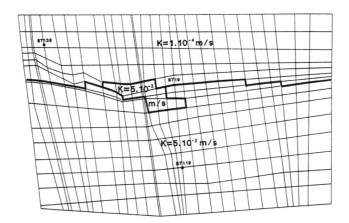

Fig. 7 Valeurs et répartition des perméabilités utilisées pour la situation calibrée.

de 0.89 m au ST138, de 0.09 au ST19 et de 0.08 m au ST119. Avec les mêmes conditions initiales mais une "perméabilité équivalente" du mur emboué égale à 1.10^{-8} m/s, les résultats montrent une remontée de la nappe quasi identique : 0.92 m au ST138, 0.09 m au ST19 et 0.08 m au ST119. On peut en déduire que le fait de diminuer la "perméabilité équivalente" du mur emboué en dessous d'une valeur de 1.10^{-7} m/s ne modifie que très légèrement la remontée de la nappe.

En reprenant ensuite les mêmes conditions initiales et une "perméabilité équivalente" du mur emboué égale à 1.10^{-6} m/s, il est logique que les résultats obtenus montrent une remontée moins importante de la nappe dans la zone de ST138 (près du coin du Palais des Princes Evêques): 0.68 m au ST138, 0.09 m au ST19 et 0.08 m au ST119.

En augmentant encore la "perméabilité équivalente" du mur emboué à 1.10^{-5} m/s, les remontées obtenues se limitent à : 0.20 m au ST138, 0.09 m au ST19 et 0.08 m au ST119.

Rappelons que l'ensemble de ces résultats correspondent au calcul de la remontée de la nappe par rapport à une situation initiale sans pompage. On constate une relative sensibilité de la remontée de la nappe dans la zone du piézomètre ST138 en fonction d'une "perméabilité équivalente" du mur emboué passant de 1.10^{-5} m/s à 1.10^{-6} m/s. Cependant, dans la pratique, il est réaliste de considérer que cette "perméabilité équivalente" devrait être inférieure ou égale à 1.10^{-6} m/s. Les remontées de la nappe sont, dans ce cas, comprise entre 0.7 et 1.0 m près du pignon Ouest du Palais des Princes Evêques (ST138) et inférieures à 0.10 m dans la partie centrale de la zone simulée (ST19 et ST119).

4.1.2 *Calcul des remontées par rapport à une situation avec pompages*

Dans un deuxième temps, une valeur de "perméabilité équivalente" du mur emboué égale à 1.10^{-7} m/s et une situation avec pompages sont choisis. Les résultats obtenus (figure 9) montrent notamment une remontée de la nappe de 0.87 m au ST138, de 0.21 m au ST19 et de 0.17 m au ST119 (par rapport à la situation initiale calibrée correspondante). Avec les mêmes conditions de pompage et une "perméabilité équivalente" du mur emboué égale à 1.10^{-8} m/s, les résultats montrent une remontée de la nappe quasi identique : 0.89 m au ST138, 0.21 m au ST19 et 0.17 m au ST119. Comme dans le cas des simulations sans pompage, on peut en déduire que les remontées de la nappe

sont très peu dépendantes d'une diminution de la "perméabilité équivalente" du mur emboué en dessous de 1.10^{-7} m/s.

En reprenant ensuite les mêmes conditions de pompage et une "perméabilité équivalente" du mur emboué égale à 1.10^{-6} m/s, les résultats montrent une remontée moins importante de la nappe dans la zone de ST138: 0.70 m au ST138, 0.21 m au ST19 et 0.17 m au ST119.

En augmentant encore la "perméabilité équivalente" du mur emboué égale à 1.10^{-5} m/s, les remontées obtenues se limitent à : 0.26 m au ST138, 0.20 m au ST19 et 0.17 m au ST119.

On constate que la diminution de la "perméabilité équivalente" du mur emboué au dessous de 1.10^{-7} m/s n'a que peu d'influence sur les remontées de la nappe.

D'autre part, et c'est plus grave, l'ensemble des résultats montre également que la remontée provoquée par la mise en place du mur est largement inférieure aux rabattements provoqués par les pompages de chantier au droit de la partie Ouest du Palais des Princes Evêques.

4.2 Simulation de l'écoulement de la nappe en présence du mur emboué et des fondations profondes de l'Ilot Tivoli.

La zone des fondations de l'Ilot Tivoli concerne les éléments 78 → 83, 107 → 112, 136 → 141 et 165 → 170 du modèle (figures 2 et 3). Si les fondations sont discontinues, elles ne constituent pas un obstacle complètement imperméable pour les écoulements de la nappe alluviale. Dans le modèle, des perméabilités équivalentes (Schneebeli, 1966) doivent être choisies dans les éléments finis concernés afin de tenir compte de l'imperméabilité totale des "plots" de fondation et de la perméabilité des sédiments alluviaux. Si on considère que 1/100 de la surface occupée par l'Ilot Tivoli est laissée libre pour l'écoulement entre les fondations, on obtient une perméabilité équivalente de l'ordre de :

$$K_{eq} = \frac{1}{100} K_{plaine\ alluviale} + \frac{99}{100} K_{béton} \qquad (7)$$

Dans notre cas, on aurait:

$$K_{éq} = 5.10^{-4} m/s$$

Par le même raisonnement, une perméabilité équivalente de 1.10^{-8} m/s correspondrait à une imperméabilité quasi-totale des fondations de l'Ilot

Tivoli par rapport à la perméabilité des dépôts alluviaux de 5.10^{-2} m/s.

4.2.1 *Calcul des remontées par rapport à une situation initiale **sans** pompage*

Les conditions sont similaires aux passages décrits au paragraphe 4.1.1 avec une perméabilité équivalente du mur emboué à 1.10^{-8} m/s. On y ajoute la modification de la perméabilité dans la zone de l'Ilot Tivoli, prenant $K_{eq} = 1.10^{-8}$ m/s dans cette zone. Les résultats obtenus montrent qu'une remontée additionnelle est quasi inexistante. Les passages avec des perméabilités équivalentes (de la zone Tivoli) supérieure à 1.10^{-8} m/s fournissent évidemment des résultats encore moins marqués.

4.2.2 *Calcul des remontées par rapport à une situation **avec** pompages*

Les conditions sont similaires aux passages décrits au paragraphe 4.1.2 avec une perméabilité équivalente du mur emboué prise à 1.10^{-8} m/s. On y ajoute la modification de la perméabilité dans la zone de l'Ilot Tivoli, avec $K_{eq} = 1.10^{-8}$ m/s dans cette zone. Les résultats montrent une remontée additionnelle nulle près de l'Ilot Tivoli; ailleurs, l'influence des pompages serait même favorisée, expliquant des remontées plus faibles dans certaine zones.
Les résultats des passages avec des perméabilités équivalentes (de la zone Tivoli) supérieures à 1.10^{-8} m/s sont encore moins marqués. Dans ces conditions, aucune remontée due spécifiquement aux fondations de l'Ilot Tivoli n'est constatée au ST19.

4.3. Simulations avec d'autres valeurs de la perméabilité de la nappe alluviale

Les remontées calculées, dans le cas de la situation initiale sans pompage et pour la situation avec pompages, sont très faibles de par la perméabilité très grande des dépôts alluviaux, obtenue par la calibration (K = 5.10^{-2} m/s). Afin de se situer du côté de la sécurité et donc en se fixant des conditions plus défavorables, des simulations ont été réalisées en divisant par 10 et par 100 les valeurs de la perméabilité de la nappe alluviale.

4.3.1 *Calcul des remontées par rapport à une situation initiale **sans** pompage*

Les simulations sont réalisées avec les mêmes paramètres qu'au paragraphe 4.2.1 si ce n'est une perméabilité de la nappe alluviale divisée par 10 (K = 5.10^{-3} m/s) et puis par 100 (K = 5.10^{-4} m/s). Les résultats obtenus (figure 10) montrent que, dans ces dernières conditions notamment, les remontées seraient plus importantes, et surtout dans la zone proche de ST19.

4.3.2 *Calcul des remontées par rapport à une situation **avec** pompages*

Les simulations sont réalisées avec les mêmes paramètres qu' au paragraphe 4.2.2 si ce n'est une perméabilité de la nappe alluviale divisée par 10 (K = 5.10^{-3} m/s) et puis par 100 (K = 5.10^{-4} m/s). Les résultats obtenus (figure 11) montrent que dans ces dernières conditions, les remontées seraient plus importantes, sauf dans la zone proche de ST138 fort influencée par le pompage.

5 CONCLUSIONS

Au total trente-six simulations ont été réalisées, dont 16 pour les besoins de la calibration du modèle, étape nécessaire pour que les résultats du modèle puissent être interprétés avec fiabilité.
De la piézométrie calibrée sur la situation ayant cours avec les pompages de chantier, on peut tirer une première constatation : les pompages de chantier ont induit un rabattement non négligeable (parfois de l'ordre de 5 m) de la nappe sous une grande partie du Palais des Princes Evêques. Les fondations de celui-ci ont donc été, sous la cote 64, désaturées totalement ou partiellement. Les simulations relatives à l'effet du mur emboué et des fondations de l'Ilot Tivoli ont été menées en distinguant :

•une situation initiale avec ou sans pompages;
•des perméabilités équivalentes du mur de 1.10^{-8} m/s à 1.10^{-5} m/s;
•des perméabilités équivalentes des fondations Tivoli de 1.10^{-8} m/s à 1.10^{-5} m/s;
•des perméabilités de la plaine alluviale de 5.10^{-4} à 5.10^{-2} m/s.

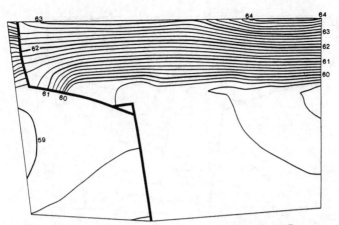

Fig. 8 Piézométrie calculée sans pompage et avec le mur emboué à $K_{éq}= 1.10^{-7}$ m/s.

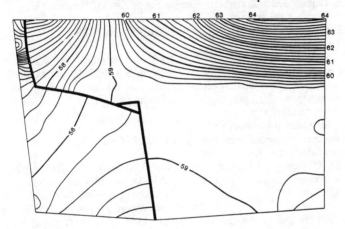

Fig. 9 Piézométrie calculée avec pompages et avec le mur emboué à $K_{éq}= 1.10^{-7}$ m/s.

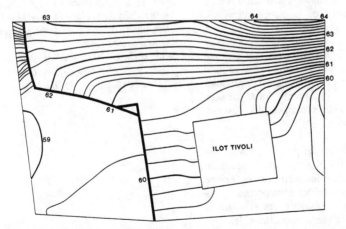

Fig. 10 Carte piézométrique calculée, sans pompage, avec le mur emboué et les fondations Tivoli à $K_{éq}= 1.10^{-8}$ m/s, l'aquifère alluvial avec une perméabilité de K= 5.10^{-4} m/s

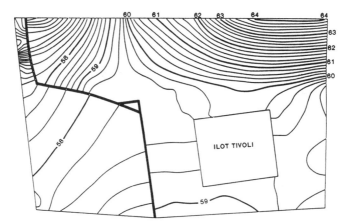

Fig. 11 Carte piézométrique calculée, avec pompage, avec le mur emboué et les fondations Tivoli à $K_{éq} = 1.10^{-8}$ m/s, l'aquifère alluvial avec une perméabilité de K= 5.10^{-4} m/s

Sur base des résultats décrits ci-dessus, les conclusions suivantes peuvent être tirées :

•les remontées de la nappe alluviale induites par l'implantation du mur emboué et les fondations de l'Ilot Tivoli sont peu importantes. Sans pompage, elles pourraient atteindre un maximum d'environ 1.5 m dans les endroits les moins perméables de la nappe. A ces valeurs maxima correspondraient des hauteurs piézométriques d'environ 62.5 m au ST138 et 60.85 m au ST19. Dans ces conditions, il serait souhaitable de rendre drainant l'égoût traversant la place St. Lambert.

•l'effet séparé des fondations peu perméables ou imperméables de l'Ilot Tivoli est négligeable pour les conditions dans lesquelles le calcul a été réalisé (perméabilité importante de la nappe alluviale). Néanmoins, tout ouvrage qui tendrait à "imperméabiliser" de quelconque façon que ce soit la zone située entre l'Ilot Tivoli et le mur emboué doit être proscrit ou tout au moins étudié avec soin. Des calculs supplémentaires simulant l'obstruction de ce passage important indiqueraient inévitablement une nette remontée des niveaux piézométriques en amont.

Bien que ne disposant que de peu de données, les exercices de simulations réalisés dans le cadre de cette étude ont permis indubitablement de mettre en évidence les facteurs réellement importants pour le problème posé. Les résultats des simulations réalisées pour des intervalles importants dans les valeurs des paramètres principaux, permettent cette analyse.

REMERCIEMENTS

Tous nos remerciements vont à la société IBM Belgium S.A. pour le soutien qu'elle nous accorde actuellement au niveau des équipements informatiques.

REFERENCES

Bear, J. & Verruijt, A. 1987. *Modeling groundwater flow and pollution.* Reidel, 414 p.

Dassargues, A. 1991b. *Paramétrisation et simulation des réservoirs souterrains, Couplages et non linéarités.* PhD Thesis in Applied Sciences, University of Liège, unpublished

Monjoie A. & Calembert L. 1975. Ouvrages souterrains de la place St Lambert. Rapport STIL. non publié, Faculté des Sciences Appliquées, Université de Liège.

Schneebeli, G. 1966. *Hydraulique souterraine.* Eyrolles. Collection du centre de recherches et d'essais de Chatou. 357p.

Heavy metal retention by a clay soil – Experimental studies

Rétention des métaux lourds par un sol argileux – Études expérimentaux

B.K.Tan
Universiti Kebangsaan Malaysia, Bangi, Malaysia

R.N.Yong & A.M.O.Mohamed
McGill University, Montreal, Ont., Canada

ABSTRACT: This paper presents the results of experimental investigations into the potential use of a particular natural clay soil as a landfill liner and adsorbent of pollutants. The natural clay soil is characterised by having moderately high specific surface areas of 90-206 m^2/g, low cation exchange capacities (C.E.C) of about 10 - 20 meq/100 g, maximum dry densities of 1.83 - 1.84 Mg/m^3, and low permeabilities of 10^{-9} m/sec for the compacted soil samples. Increase in the densities with corresponding decrease in the permeabilities can be attained by mixing the natural clay soil with other soil admixtures. Heavy metal (Pb^{2+}, Zn^{2+}) retention studies employing the batch - equilibrium method and leaching column tests of the natural clay soil and soil admixtures indicate that the heavy metals retained by the soil can exceed the C.E.C. of the soil, thus implying that other processes or mechanisms are involved besides cation exchange, such as precipitation, etc. While the natural soil alone performs satisfactorily in terms of permeability (low permeability of $\sim 10^{-9}$ m/sec), pH (high pH of 7 - 8), high adsorption or retention of heavy metals, the addition of a particular soil admixture either enhances or decreases this performance, depending on the characteristics and mineralogical compositons of the particular soil admixture.

1 INTRODUCTION

An investigation into the potential use of southern Ontario hydrobiotite-vermiculite soil as a landfill liner and adsorbent of pollutants was conducted recently at the Geotechnical Research Centre (GRC), McGill University, YONG ET AL. (1991). Soil samples were obtained from the Stanleyville hydrobiotite-vermiculite deposit which represents a previously mined deposit located in southern Ontario, Canada. This paper summarises the results of the investigation.

2 CHARACTERISATION

Basic characterisation of the soils include: specific surface area, cation exchange capacity, compaction and permeability test. The "vermiculite" in actual fact consists of phlogophite, serpentine and mixed layer clay minerals which expand and exfoliate on heating. More recent mineralogical analysis has determined that the coarser fraction consists of intergrowths of phlogophite-lizardite, in a finer grained matrix of talc, smectite and aliettite (FARKAS, 1991).

2.1 Specific Surface Area

Specific surface areas of the soil samples were determined using the Ethylene Glycol Monoethyl Ether (EGME) adsorption method, CARTER ET AL. (1965). The values of the specific surface areas range from 90 to 206 m^2/g, i.e. moderately high, indicating perhaps mixtures of mainly illitic/micaceous minerals and others. Surface area of illite $\sim$60-200 m^2/g, while vermiculite has much higher specific surface areas of $\sim$ 400-800 m^2/g (UEHARA & GILLMAN, 1981).

Table I. Permeability Tests (Falling Head)

Sample No.	γ_d(Mg/m³)	w_o(%)	k(m/sec)
TR1A	1.54 (1.84)	25.4 (14.4)	1.2×10^{-9}
No. 10	1.82' (1.83)	16.2 (16.1)	2.3×10^{-9}
No. 10 + kaolinite	1.84 (1.85)	14.2 (15.9)	1.4×10^{-9}
No. 10 + Champlain Sea Clay	1.90 (1.91)	14.3 (14.5)	2.9×10^{-10}

Table 2. Heavy Metal Retention with Leachate (low heavy metal concentrations)

Sample No	Pb^{2+} ppm	meq/100g	%	Zn^{2+} ppm	meq/100g	%	Σ (meq/100g)
ST1A	35.0	0.338	98.87	38.3	1.17	100.00	1.51
ST1B	35.3	0.341	99.72	38.3	1.17	100.00	1.51
ST1C	34.6	0.334	97.74	38.3	1.17	100.00	1.50
TR1A	35.2	0340	99.44	38.3	1.17	100.00	1.51
TR1B	34.6	0.334	97.74	38.3	1.17	100.00	1.50
TR2A	35.3	0.341	99.72	38.1	1.17	99.48	1.51
TR2B	34.7	0.335	98.02	37.7	1.15	98.43	1.49
TR4A	35.3	0.341	99.72	37.8	1.16	98.69	1.50
CF1	33.8	0.326	95.48	37.9	1.16	98.96	1.49
Leachate	35.4	0.342	--	38.3	1.17	--	---

Table 3 Heavy Metal Retention (High Concentrations) (soil No. CFI,CEC - 18.59 meq/100g)

Test No.	Pb^{2+} Influent (ppm)	Effluent (ppm)	Adsorbed (ppm)	% Adsorbed	Zn^{2+} Influent (ppm)	Effluent (ppm)	Adsorbed (ppm)	% Adsorbed	Pb^{2+}/Zn^{2+} or $Pb^{2+}+Zn^{2+}$ Adsorbed (meq/100g)
CIIg	-	-	-	-	1800	1400	400	22.2	12.24
CIIh	-	-	-	-	2400	1650	750	31.3	22.94
CIIi	-	-	-	-	375	175	200	53.3	6.12
CIIIa	500	260	240	48.0	150	93	67	38.0	4.05
CIIIb	1000	630	370	37.0	300	194	106	35.3	6.81
CIIIc	1500	1160	340	22.7	450	350	100	22.2	6.34
CIIId	2000	1050	950	47.5	600	370	230	38.3	16.21
CIIIe	3000	1975	1025	34.2	900	630	270	30.0	16.15
CIIIf	4000	3125	875	21.9	1200	900	300	25.0	17.82
CIIIg	*5000*(4500)	3600	1000	22.2	*1800*(1700)	1280	470	26.9	24.03
CIIIh	*5000*(3400)	3075	325	9.6	*2400*(2350)	1690	660	28.1	23.33
CIIIi	1250	665	585	46.8	375	215	160	42.7	10.64
Co (blank)	-	0	-	-	0.6 1.0	0.78	-	-	-

* Note: Value in bracket indicates actual concentration

Table 3 Heavy Metal Retention (High Concentrations) (soil No. CFI,CEC - 18.59 meq/100g)

Test No.	Pb^{2+} Influent (ppm)	Effluent (ppm)	Adsorbed (ppm)	% Adsorbed	Zn^{2+} Influent (ppm)	Effluent (ppm)	Adsorbed (ppm)	% Adsorbed	Pb^{2+}/Zn^{2+} or $Pb^{2+}+Zn^{2+}$ Adsorbed (meq/100g)
CIa	500	44	456	91.2	-	-	-	-	4.40
CIb	1000	330	670	67.0	-	-	-	-	6.47
CIc	1500	750	750	50.0	-	-	-	-	7.24
CId	2000	885	1115	55.8	-	-	-	-	10.78
CIe	3000	1800	1200	40.0	-	-	-	-	11.58
CIf	4000	2525	1475	36.9	-	-	-	-	14.24
CIg	5000	4050	1050	32.5	-	-	-	-	16.82
CIh	*6000*(5380)	5000	1380	21.6	-	-	-	-	13.32
CIi	1250	830	420	33.6	-	-	-	-	4.05
CIIa	-	-	-	-	150	78	72	48.0	2.20
CIIb	-	-	-	-	300	162	138	46.0	4.22
CIIc	-	-	-	-	450	272	178	39.8	6.46
CIId	-	-	-	-	600	234	366	61.0	11.20
CIIe	-	-	-	-	900	590	310	34.4	9.46
CIIf	-	-	-	-	1200	850	350	29.2	10.71

2.2 Cation Exchange Capacity (C.E.C)

Exchangeable cations (Na+, K+, Mg2+, Ca2+) and the cation exchange capacity of the soil samples were determined using two methods, namely: a) the batch equilibrium test, ASTM D4319 (1984a), and b) the silver-thiourea method, CHHABRA ET AL. (1975). The results show that the C.E.C. values are rather low, i.e. ranging from 7 to 19 meq/100 g soil (batch equilibrium method), or 3 to 14 meq/100 g soil (silver-thiourea method). The low C.E.C. values once again indicate that the soils are mainly illitic/mica in composition. C.E.C. of illite ~ 15-40 meq/100 g soil; C.E.C. of vermiculite ~ 120-200 meq/100 g soil, MORRILL ET AL. (1982).

2.3 Compaction

The Standard Proctor compaction method was used, ASTM, D698 (1984b). The results are as follows: maximum dry density = 1.84 or 1.83 Mg/m3, and optimum moisture content = 14.4 or 16.1% for the respective sample. It was noted that the compacted soil lacks cohesion and disintegrates readily. In addition, compaction tests were also performed on soil samples mixed with soil "additives". This was done to investigate further the effects of adding other soil "additives" on the geotechnical and adsorption characteristics of the "vermiculite"/natural soil. Thus, compaction tests were also conducted for i) natural soil + kaolinite (20% by dry weight), and ii) natural soil + Champlain Sea Clay/Quebec Clay (25% by dry weight). The results are shown in Fig. 1. The results show that the addition of kaolinite or Champlain Sea Clay causes an increase in the maximum dry density obtained, from 1.83 Mg/m^3 (natural soil) to 1.85 Mg/m^3 (natural soil + kaolinite) and 1.91 Mg/m^3 (natural soil + Champlain Sea Clay). There is a corresponding decrease in the optimum moisture content from 16.1% to 15.9% and 14.5% respectively. The increase in maximum dry density with the addition of kaolinite or Champlain Sea Clay is advantageous to the proposed use of the soil as

a clay liner since it would increase the shear strength of the liner and reduce its compressibility and permeability.

2.4 Permeability

The falling head permeability test was conducted. The samples were compacted at or close to the optimum moisture content/maximum dry density except for one which was compacted wet of optimum moisture content. The results are shown in Table 1. The results show that the permeabilities of the natural soil and soil admixtures are low, i.e. ranging from 2.3 x 10-9 to 2.9 x 10-10 m/sec. The addition of kaolinite or Champlain Sea Clay causes a reduction in the permeability. In the case of addition of Champlain Sea Clay, a reduction in one order of magnitude was obtained.

3 BATCH EQUILIBRIUM TESTS

Heavy metal retention by the soil samples were studied by means of the batch equilibrium test. The heavy metals studied were Pb2+ and Zn2+. Two phases of studies were conducted, namely: i) using leachate with low heavy metal concentrations, and ii) using solutions with high heavy metal concentrations.

3.1 Low Heavy Metal Concentrations

The leachate used was produced in an actual landfill site receiving only municipal solid wastes. Since the initial heavy metal concentrations in the leachate were very low or nil, the leachate was spiked with Pb2+ and Zn2+ to slightly increase the Pb2+ and Zn2+ concentrations to 35.4 ppm and 38.3 ppm respectively. The results of the heavy metal retention studies are shown in Table 2. As can be seen from Table 2, the % heavy metals adsorbed were mostly high, i.e. > 95% in all cases. The sum of heavy metals adsorbed was only ~ 1.5 meq/100 g, i.e. << C.E.C. of the soils. Thus, the exchange capacity of the soils has not been exhausted yet, and it can be inferred that for low heavy metal concentrations, adsorption by the soils is virtually complete - i.e.

Table 4. Leaching Test - Dimensions & Compaction
(Compaction by Compression Machine in 3 Layers - Static Compaction)

No.	Diameter (cm)	Height (cm)	Volume (cm³)	Pore Volume (P.V., cm³)	Weight of Compacted Soil (g)	Bulk Density of Compacted Soil (g/cm³)	wopt.* (%)	γ_{dry} (g/cm³)	$\gamma_{d'max}$ (g/cm³)
V1	11.4	12.1	1235.1	398.1	2646.6	2.14	16.1	1.84	1.83
V3	11.4	12.1	1235.1	398.1	--	--	16.1	--	1.83
V5	11.4	12.1	1235.1	398.1	2654.2	2.15	16.1	1.85	1.83
K1	11.4	12.1	1235.1	390.4	2676.1	2.17	15.9	1.87	1.85
K3	11.4	12.1	1235.1	390.4	2666.5	2.16	15.9	1.86	1.85
K5	11.4	11.8	1204.4	380.7	2615.2	2.17	15.9	1.87	1.85
Q1	11.5	9.5	986.8	290.0	2138.1	2.17	14.5	1.90	1.91
Q3	11.4	9.5	969.7	285.0	2125.3	2.19	14.5	1.91	1.91
Q5	11.5	8.7	903.7	265.6	1973.3	2.18	14.5	1.90	1.91

*: w opt. and $\gamma_{d\,max}$ from Standard Proctor Compaction

Table 5. Chemical compositions of the Reconstituted Leachate.

(all ionic concentrations are in ppm).

pH	1.33
Specific Conductivity (mS/cm)	16.833
Na⁺	346
K⁺	164.8
Mg²⁺	43.8
Ca²⁺	95.4
Pb²⁺	1372.2
Zn²⁺	1141.6
Cl⁻	5258.4

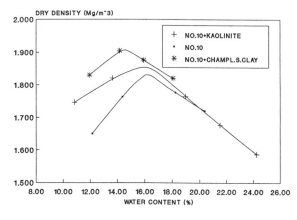

FIG.1 - COMPACTION CURVE (STD.PROCTOR)
SAMPLE NO. 10 , + ADDITIVES

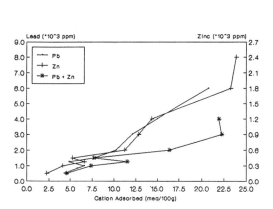

Fig.2 Heavy Metal Retention
Soil# 10 (CEC=14.89 meq/100g)

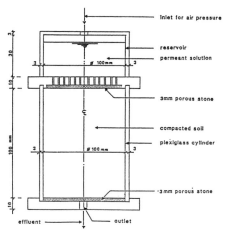

Fig.3. Schematic Representation of Laboratory Leaching Cell
Used in This Study.

2812

total attenuation or retention of the heavy metals.

3.2 **High Heavy Metal Concentrations**

The second phase of heavy metal retention studies involved the use of solutions with high heavy metal concentrations. Three series of tests (1,11, 111) were conducted, with the concentration of Pb2+ and Zn2+ as shown in Table 3. (Note that in 1 and 11, Pb2+ and Zn2+ are applied singly while in series 111 Pb2+ and Zn2+ are applied as composite solution). Only two soils were tested, namely sample No. CF1 (C.E.C. = 18.59 meq/100g), and sample No. 10 (C.E.C. = 14.89 meq/100g). Results of the heavy metal retention studies are shown in Table 3, as well as Figure 2. The results show that: i) retention of Pb^{2+} or Zn^{2+} increases with the initial cation concentration (Series 1 and 11). ii) retention of Pb^{2+} and Zn^{2+} also increases with the initial cation used (Series 111). iii) retention of Pb^{2+} in Test Series 111 (Pb^{2+} + Zn^{2+} composite solutions) is lower than the retention of Pb^{2+} in Test Series 1 (Pb^{2+} only) for the corresponding initial influent concentration. Similarly, retention of Zn^{2+} in Test Series 111 is also lower than the retention of Zn^{2+} in Test Series 11 (Zn^{2+} only). This difference is attributed to competition between the cations for the exchange sites. However, the sum of Pb^{2+} + Zn^{2+} retained in Test Series 111 (composite solutions) exceeds that of Pb2+ or Zn2+ retained for Test Series 1 or 11 (single cation solutions).
iv) it appears that the maximum amount of heavy metals retained can exceed the C.E.C. of the soil (eg. Pb^{2+} or Zn^{2+} for CF1, and Pb^{2+}, Zn^{2+} or Pb^{2+} + Zn^{2+} for soil No. 10). Thus, other processes are involved in the retention of heavy metals besides cation exchange. These other processes, for example, could be precipitation, adsorption by amorphous substances, etc. However, they were not investigated further in this study.
v) the % heavy metal adsorbed is high (~ 90%) only at initial 500 ppm Pb2+ (Test 1a). With the other higher initial concentrations, the

% adsorbed is mostly low (<50%). Thus the "efficiency" of the adsorption mechanism decreases with increasing initial cation concentration although the amount retained on a weight basis increases.

4 **LEACHING COLUMN TESTS**

Leaching column tests were conducted to study the adsorption characteristics of the soils, in particular with respect to the retention and migration of heavy metals through the soil columns. Leaching column tests were conducted on sample No. 10, sample No. 10 + kaolinite, and sample No. 10 + Champlain Sea Clay. The samples are designated as follows:
i. Natural Soil "vermiculite", designated as V,
ii. Natural Soil + Kaolinite (20% by dry weight), K,
iii.Natural Soil + Champlain Sea Clay/Quebec Clay (25% by dry weight), Q.
The soils were compacted at their optimum moisture contents using static compaction (compression machine) into plexiglass cylinders which were later reassembled as leaching cells. A typical leaching cell is shown schematically in Fig. 3. For each soil type mentioned above (V, K, Q), three leaching cells were prepared, for eg. V1, V3, V5, representing leaching with 1 pore volume, 3 pore volumes and 5 pore volumes of leachate. The dimensions and details of the compaction of the soil samples used for the leaching tests are shown in Table. 4.
Leaching was done using Municipal Solid Waste (MSW) leachate spiked with heavy metals (Pb^{2+}, Zn^{2+}) and cations (Na^+, K^+, Mg^{2+}, Ca^{2+}) in the form of chlorides. The pH of the reconstituted leachate was also lowered by adding some concentrated HCl. The chemical compositions of the reconstituted leachate are as shown in Table. 5.
Leaching was carried out under a constant applied air pressure of 12.0 or 15.0 p.s.i., ie, equivalent to a water head of 8.4 or 10.6 m depending on the sample height. This resulted in a similar hydraulic gradient of 87-89 for almost all the samples, except for sample Q5 which has a slightly

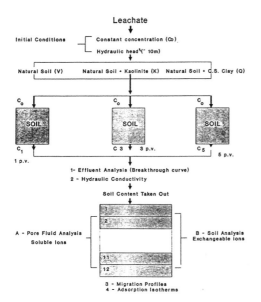

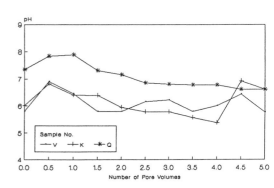

FIG. **6.** EFFLUENT pH
COMPARING SAMPLE V,K,Q (5 P.V.)

Fig**4.** Test Scheme

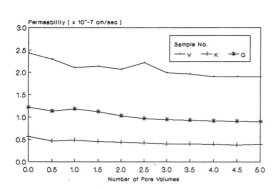

FIG. **5.**PERMEABILITY VARIATIONS
COMPARING SAMPLE V,K,Q

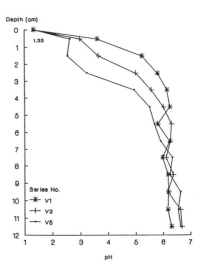

FIG. **7.**pH VARIATIONS
SAMPLE V (NATURAL SOIL)

higher hydraulic gradient of 97. The test scheme adopted is depicted in Fig. 4. During the leaching process, effluents were collected at every 0.5 p.v. (pore volume) and analyzed. At the end of the 1 p.v., 3 p.v. and 5 p.v. series, the soil samples were extruded, cut into 1 cm thick slices, and the soil slices analyzed for pore fluids contents (soluble ions) and

exchangeable cations. The analytical procedures were carried out in accordance with the Geotechnical Research Centre (GRC) Laboratory Manual. Cations and heavy metals were determined using Atomic Absorption Spectrophotometry (AAS). Chloride was determined using titration with $AgNO_3$. The pH and conductivity were determined using a pH meter and the

Electrophoretic mass-transport analyzer respectively.
Results of the leaching column tests are presented below:

4.1 Permeability

The variations in the permeability values during the leaching process are shown in Fig. 5. For the V soils, the permeability values show a slight decrease from 0-1.0 p.v. (V1, V5), then remains almost constant after 1.0 p.v., tapering off at 1.9×10^{-7} cm/sec. The K soils (e.g. K5) show an almost constant permeability tapering off at 0.4×10^{-7} cm/sec. Similarly, the Q soils (eg. Q5) show a slight decrease in permeability at the beginning of leaching, then tapering off to a constant value of 0.9×10^{-7} cm/sec. Thus, leaching only caused a very slight decrease in the permeabilities of the 3 types of soils examined at the initial stages of leaching.

4.2 pH

The pH of the original MSW leachate was 7.06 (neutral). However, for the leaching column tests, the pH of the reconstituted leachate was lowered to 1.33 (highly acidic) by the addition of concentrated HCl. This was done to facilitate the study of the migration of the heavy metals (Pb^{2+}, Zn^{2+}) through the soil columns since it is well known that heavy metals generally would precipitate out of solutions if the solution pH's are high (e.g. Pb^{2+} precipitates at pH > 5, etc.). Figure 6 shows the effluent pH of the samples V, K, Q (5 p.v. series), indicating a general gradual decrease in the pH's of the effluents of the three soil types with increasing permeation by leachate. The figure also clearly shows that the effluent pH's of Q > V > K. This is attributed to the differences in the mineralogical compositions of Q. The Champlain Sea Clay used for Q came from the Matagami site and contains abundant carbonates, up to 30-40%, WANG (1984), QUIGLEY ET AL (1982), LI (1985). The carbonates (calcite, $CaCO_3$, and dolomite, $CaMg (CO_3)_2$) contribute to higher pH on leaching. On the other hand, the K soils contain kaolinite which produces lower pH conditions.
The variations of the pH's of the pore fluids (of the leached soils) with depth of the soil column are shown in Fig. 7 & 8. As can be expected, the curves show a general increase in the pH from the top (upper slices) to the bottom (lower slices) of the soil column. This is clearly seen for the V and K soils. The pH increases from about 3 at the uppermost slice (slice no. 1) to about 6.5 at the lowermost slice (slice no. 12). With increasing permeation by the leachate (p.v. = 1 to 5), the pH of each slice decreases progressively as shown by the V soils. The Q soils show high pH's (pH > 7) throughout the profile. This indicates the high buffering capacity of the Q soils, obviously due in part to the presence of carbonates in the Q soils. Also, comparing soil types, pH of Q > V > K.

4.3 Breakthrough Curves

A breakthrough curve shows the variation of the relative concentration (Ce/Co) of a particular ion with the number of pore volumes of leachate permeated. Ce = concentration of a particular ion in the effluent, Co = original concentration of the ion concerned in the influent leachate. Breakthrough of the ionic species is said to have occurred when Ce/Co = 0.5 (BOWDERS ET AL, 1986). Breakthrough curves for Pb^{2+} & Zn^{2+} and the anion Cl^- have been plotted for the 3 types of soils (V, K, Q) studied. The following sections discuss the results.
i) Lead, Pb^{2+}: The breakthrough curves for Pb^{2+} are shown in Fig. 9. Note that the relative concentration values are all extremely low, ie. Ce/Co are in the order of 10^{-4}. This shows that most of the Pb^{2+} is retained in the soils, i.e. high attenuation of Pb^{2+}. Breakthrough thus did not occur in any of the soils tested. Nevertheless, the breakthrough curves show the increasing amounts of Pb^{2+} passed through the soil columns with increasing number of pore volumes of leachate permeated. The retention capabilities of the three soil types with respect to Pb^{2+} are similar, with all three showing high attenuation of Pb^{2+}.
ii) Zinc, Zn^{2+}: For Zn^{2+}, again

Depth (cm)

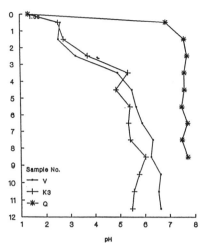

FIG.8. pH VARIATIONS
COMPARING SAMPLE V,K,Q(5 P.V.)
(K=K3)

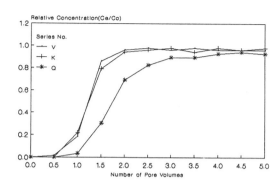

FIG.10.BREAKTHROUGH CURVE
COMPARING SAMPLE V,K,Q - Cl

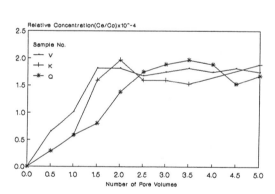

FIG. **9.**BREAKTHROUGH CURVE
COMPARING SAMPLE V,K,Q - Pb

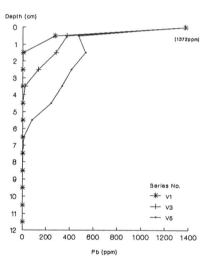

FIG.1**1.** MIGRATION PROFILE
·SAMPLE V (NATURAL SOIL) - Pb

high retention by all three soil types is indicated, Ce/Co ~ 10^{-4} as in the case of Pb^{2+}. The V and K curves are similar, while Q soils show a slightly higher retention of Zn^{2+} compared to V and K.
iii) Chloride, Cl^-: Breakthrough curves for Cl^- are shown in Fig. 10. The chloride ion is considered to be a very mobile and non-interacting anion (conservative contaminant). As such, attenuation of the Cl^- is low as depicted in the figure. The V and K soils show very similar curves, with breakthrough of Cl^- at p.v. ~ 1.2, and Ce/Co ~ 1 at 2 p.v.. The Q soils show a slightly higher attenuation of Cl^- with breakthrough at p.v. ~ 1.7, and Ce/Co ~ 0.9 at 3 p.v. Incidentally, close correspondence or good agreement of results for the 1 p.v., 3 p.v. and 5 p.v.

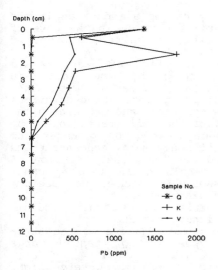

FIG. 12.MIGRATION PROFILE
COMPARING SAMPLE V,K,Q(5 P.V.) - Pb

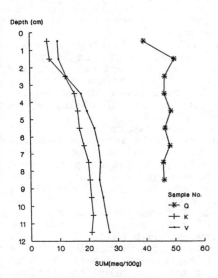

FIG. 13.EXCHANGEABLE CATION
COMPARING SAMPLE V,K,Q(5 P.V.) -SUM E.C.

series in the breakthrough curves of Cl^- (as well as the heavy metals) was obtained.

4.4 **Migration Profiles**

The migration profiles depict how a particular ionic species migrates or moves through the soil column with increasing permeation by the leachate. Migration profiles for the two heavy metals (Pb^{2+}, Zn^{2+}) are discussed:

Heavy Metals (Pb^{2+}, Zn^{2+}): Migration profiles for Pb^{2+} are shown, for eg., in Fig. 11. Note the drastic reductions in the Pb^{2+} values from 1372 ppm (initial Pb^{2+} concentration in leachate) to zero ppm -i.e. high attenuation of Pb^{2+} by the soils. With increasing permeation from 1 p.v. to 5 p.v., the Pb2+ concentrations increase progressively and Pb^{2+} penetrates deeper into the soil column. Comparing soil types, the order of Pb^{2+} retention capability is Q >> V >K, Fig. 12. Note that the Q soils show almost complete attenuation of Pb^{2+} within the first uppermost centimeter of the soil column.
The migration profiles for Zn2+ are very similar to those of Pb^{2+}. As for the case of Pb^{2+}, the migration profiles for Zn^{2+} also show high attenuations of Zn^{2+} by all the three types of soils (V, K, Q) tested, from 1142 ppm (initial Zn^{2+} concentration in leachate) to zero ppm. The Zn^{2+} concentrations increase with increasing number of pore volumes of leachate permeated, and the depth of migrations also increases accordingly. For the V and K soils, total attenuation of Zn^{2+} occurred at depths > 8 cm; for the Q soils, total attenuation occurred within the uppermost 2 cm. The order of retention capability for Zn^{2+} is Q >> V > K.
Comparing Pb^{2+} and Zn^{2+}, Zn^{2+} is more mobile (greater depth of migration of up to 8 cm for Zn^{2+} compared to 6 cm for Pb^{2+}; and slightly higher Zn^{2+} concentrations compared to Pb^{2+}, in spite of the lower initial concentration of Zn^{2+}, i.e. 1142 ppm Zn^{2+} versus 1372 ppm Pb^{2+} in the leachate).

4.5 **Exchangeable Cations**

To better illustrate the adsorption or retention of the cations by the soil columns, plots are made showing the variations of exchangeable (adsorbed) cations with depth. The exchangeable cations were determined using NH4OAC through batch equilibrium tests.
i) Heavy Metals (Pb^{2+}, Zn^{2+}): High adsorption or retention of Pb^{2+} occurred at the upper portions of the soil column, and with increasing permeation more Pb^{2+} is adsorbed and the P^{b2+} also migrates

2817

deeper down into the soil column. Comparing soil types V,K,Q, the similarity between the V and K soils is striking; also the adsorption capacity of Q>V>K. Similar results are obtained for Zn^{2+}, with Zn^{2+} showing greater depth of penetration/migration (more mobile).

ii) Sum of Exchangable Cations: The sum of exchangeable cations (Na^+, k^+, Mg^{2+}, Ca^{2+}) represents the cation exchange capacity (C.E.C.) of the soil. From Fig. 13, it is noted that the C.E.C.'s of the soils are in the order of Q > V > K, ie. with Q ~ 40-50 meq/100 g, V ~ 10-30 meq/100 and K ~ 10-20 meq/100g. These results are to be expected since Q contains Champlain Sea Clay which is known to have a higher C.E.C. of, for eg., around 60 meq/100 (WARITH, 1987), while K contains kaolinite which has lower C.E.C. of around 5-15 meq/100g (YONG AND WARKENTIN, 1975). The"vermiculite" samples have C.E.C. of about 10-20 meq/100 g.

5 CONCLUSIONS

The following concluding remarks can be made:
i) the specific surface areas of the natural "vermiculite" soils range from 90 to 206 m2/g, i.e. moderately high, indicating perhaps mixtures of mainly illitic/mica minerals with other,
ii) the cation exchange capacities of the soils are low, ranging from 7 to 19 meq/100 g (batch equilibrium test), or 3 to 14 meq/100 g (silver-thiourea method), again in the range of illite/mica,
iii) maximum dry density and optimum moisture content of the "vermiculite" soils are 1.83 or 1.84 Mg/m^3 and 14.4 or 16.1% respectively. Addition of kaolinite or Champlain Sea Clay increases the maximum dry density to 1.85 Mg/m^3 and 1.91 Mg/m3 respectively, with a corresponding decrease in the optimum moisture content to 15.9% and 14.5% respectively,
iv) the permeabilities of the natural soil and soil admixtures are low, ie, ranging from 2.3 x 10$^-$9 to 2.9 x 10^{-10} m/sec. The addition of kaolinite or Champlain

Sea Clay causes a reduction in the permeability, which in the case of the latter, is up to one order of magnitude.
v) the retention of heavy metals (Pb^{2+}, Zn^{2+}) by the soils in suspension increases with the initial heavy metal concentrations in the solution, and can exceed the C.E.C. of the soil (retention of up to about 24 meq/100 g under the range of heavy metal concentrations studied). Thus, the retention of heavy metals could involve other processes or mechanisms besides cation exchange,
vi) results of leaching column tests indicate that:
a. the permeabilities of the three soil types (V,K,Q) tested range from 0.4-1.9 x 10^{-7} cm/sec. Leaching caused only a very slight decrease in the permeabilities of the soils at the initial stages of leaching - this attests to the compatibility between the soils and the leachate.
b. the effluent pH's range from 5.4 to 7.9, in spite of the very low (highly acidic) influent leachate pH of 1.3. This indicates the high buffering capacities of the soils, in particular the Q soils, where effluent pH's of Q > V > K,
c. the pore fluids pH's of the columns generally increase from the top to the bottom. For the V and K soils, the pore fluids pH's range from about 3 (at the top) to about 6.5 (at the bottom). For the Q soils, high pH's (pH > 7) are maintained throughout the profile, indicating higher buffering capacity of the Q soils,
d. the breakthough curves indicate that breakthrough (Ce/Co = 0.5) occurred for Cl^- below the maximum 5 pore volumes of leaching conducted. On the other hand, there is high adsorption retention of Pb^{2+} and Zn^{2+}. The Q soils show higher attenuation of the heavy metals and Cl^- compared to the V and K soils,
e. the migration profiles indicate high attenuation of the heavy metals Pb^{2+} and Zn^{2+} by all the three soil types (V, K, Q) tested. The order of retention capability is Q > V > K.
f. the exchangeable cations variations with depth indicate high cation exchange or retentions at

the upper portions of the soil columns, in particular the heavy metals Pb^{2+} and Zn^{2+}. The cation exchange capacities (C.E.C.) of Q > V > K, thus the greater adsorption capacity of Q,

g. the differences in the retention/adsorption/buffering capacities of the Q, V, and K soils are attributed to differences in the mineralogical compositions of the soils, notably the presence of carbonates in the Q soils,

h. while the V soils (natural "vermiculite" soils) alone performs satisfactorily in terms of permeability (low, $\sim 10^{-7}$ cm/sec), pH (high, pH = 7-8 naturally), high adsorption/ retention of cations, especially Pb^{2+} and Zn^{2+}, the addition of Quebec Clay/Champlain Sea Clay (Q soils) enhances or improves its performance while the addition of kaolinite (K soils) reduces its performance in terms of adsorption or retention of cations. The C.E.C. of Q $\sim$ 40-50 meq/100g, V $\sim$ 10-30 meq/100 g and K $\sim$ 10-20 meq/100g.

i. retention of heavy metals is mainly through precipitation (High pH's of > 5) and cations exchange, the latter confined to the upper portions of the soil columns only.

REFERENCES

American Society for Testing and Materials, ASTM, D4319-83 (1984a). Standard Test Method for Distribution Ratios by the Short-Term Batch Method. Annual book of ASTM Standards, Volume 04.08, Soil & Rock; Building Stones, pp. 766-773.

ASTM, D698-78 (1984b). Test Methods or Moisture-Density Relations of Soils and Soil Aggregate Mixtures, using 5.5 lb. (2.49 kg) Rammer & 12-inch (305 mm) Drop. ASTM Annual Book of ASTM Standards, Volume 04-08, Soils & Rock; Building Stones, pp. 201-207.

Bowders, J. J., Daniel, D. E., Broderick, G.P., and Liljestrand, H. (1986). Methods for testing the compatibility of clay liners with landfill leachate. Hazardous and Industrial Solid Waste Testing: Fourth Symposium, ASTM STP 866, J.K, Petros, Jr., W.J. Lacy and R.A. Conway, Eds, American Society of For Testing

and Materials, Philadelphia, 1986, pp. 233-250.

Carter, D., Heilman T. and Gonzalez, J. (1965). Ethylene Glycol Monoethyl Ether for determining surface area of silicate minerals. Soil Science J., March 1965: 356-361.

Chhabra, R., Pleysier, J., and Cremers, A. (1975). The measurement of the cation exchange capacity and exchangeable cations in soil: A new method. Proceedings of the International Clay conference, Applied Publishing Ltd., Illinois, U.S.A., pp. 439-448.

Farkas, A. (1991). Ontario hydrobiotite-vermiculite as a potential landfill liner and adsorbent. Part A: Mineralogical study. Ministry of the Environment Research and Technology Project 485C, Ontario, Canada.

Li, Y.L. (1985). Compositional differences between Norwegian and Canadian clays with similar sensitivities. M. Eng. thesis, Dept. of Civil Engineering & Applied Mechanics, McGill Univ., 180 pp.

Morrill, L., Mahilum, B. and Mohinddin,S. (1982) Organic compounds in soils: sorption, degradation and persistence. Ann Arbour Science Publishers Inc., Ann Arbor, MI, 326 pp. (pg. 38).

Quigley, R.M., Sethi, A.J., Boonsinsuk, P., Sheeran, D.E. and Yong, R.N. (1982). Geologic control on soil composition and properties, Lake Ojibway clay plain, Matagami, Quebec. 35th Canadian Geotechnical Conference, Montreal, Quebec, Canada.

Uehara, G. & Gillman, G. (1981). The mineralogy, chemistry and physics of tropical soils with variable charge clays. Westview Press, Boulder, Colorado, 170 pp (pg. 9).

Wang, B.W. (1984). Compositional Effects of Soil Suction. M.Eng. thesis, Dept. of Civil Engineering & Applied Mechanics, McGill Univ., 205 pp.

Yong, R.N., Mohamed, A.M.O., & Tan, B.K. (1991). Ontario hydrobiotite-vermiculite as a potential landfill liner and adsorbent. Part B: Cation exchange capacity, leaching tests and geotechnical characterizations. Ministry of the Environment Research and Technology Project 485C, Ontario, Canada.

Yong, R.N. and Warkentin, B.P. (1975). Soil Properties and Behaviour. Elsevier, New York, 499 pp.

Alternation of drainage ability on the subsided Quaternary lowland by pumping

Oscillation de la capacité de drainage dans les plaines Quaternaires par pompage

Yushiro Iwao & Mashalah Khamehchiyan
Saga University, Japan

ABSTRACT: Lands affected by fluctuating surface water level are classified as lowlands. In lowlands, though the assessment about the drainage ability is the most important for the municipal drainage planning and river improvement, but it is scarcely obtained because of the confused drainage network and difficulty of experiment. Heavy rainfall in July 2, 1990, caused flooding in Saga plain, a lowland in north of Ariake sea in Kyushu, Japan, which was hazardous by land subsidence. Simple simulation based on the network model with 6 tanks reproduced the actual flooding of July 2, 1990, in Saga plain on personal computer. It showed that the comprehensive ability of drainage can be got by such an actual flooding simulation. It proved that the real drainage ability had been decreased by the land subsidence.

RESUME: Les territoires pris par l'oscillation de la surface aquatique sont classifiés comme les plaines. Bien que l'évaluation de la faculté de drainage dans les plaines soit la plus importante pour la planification, on ne peut la realiser qu'avec de grandes difficultés, en raison de la complexité du réseau de drainage ainsi que des problèmes en cas d'expérimentation. De fortes précipitations du 2 juillet 1990 ont produit des inondations dans la plaine de Saga - au nord de la Baie Ariake sur l'ile de Kjuschu au Japon. Cette plaine est prise par la subsidence. Par une simple simulation d'un modèle de reseau de 6 citernes, on a reproduit l'inondation du 2 juillet 1990 de la plaine de Saga sur l'ordinateur personnel. On a démontré que l'aplitude du drainage peut être evalué à l'aide de simulation d'une vraie inondation. De même on a montré que l'aptitude d'asséchement a été baissée par la subsidence.

1 INTRODUCTION

Lowlands have been defined as lands affected by fluctuating surface water level, e.g. by tides, flood, etc. Usually, lowlands are characterized by small difference of altitude with sea level. In lowland areas in which the resourced ground water is extracted by pumping up, are frequently alternated on the drainage network and its ability. Therefore, the drainage system is very important from point of view of sanitary aspect in these areas.

Saga plain a lowland of less than 5 meters above sea level with an area of more than 400 square kilometer lies in north of Ariake sea, in Kyushu island, Japan. The tidal range of the Ariake sea is about 6 meters, so several dikes are constructed to prevent flooding during heavy rainfall. Reclamation in Saga plain has rapidly developed in past years by gradual construction of dikes step.

The Chikugo, Kase and Rokkaku are the main river systems that flow through east, middle, and west of Saga plain, respectively and enter the Ariake sea in the south.

Torrential rains caused by the seasonal rain front hit the Kyushu island from July 1 to July 2, 1990, and damaged northern and middle parts of Kyushu. In the Saga plain, many artibutaries of the Rokkaku river system and of the Chikugo river system overflowed, resulting in a great disaster involving more than 10 thousand houses submerged in the water in Saga City. The flooded area in and around Saga City is shown in Fig. 1.

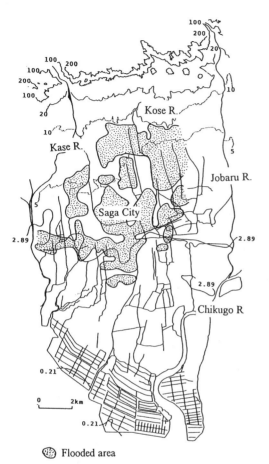

○ Flooded area

Fig. 1 Flooded area in and around Saga City, July 2, 1990

2 HISTORICAL SUBSIDENCE

Since 1957, land subsidence due to the withdrawal of ground water has been noted in the Saga plain. Also during this time, substantial ground water withdrawals were occurring in the Saga area from large capacity of industrial and agricultural water wells with resulting reduced aquifer pressures and associated land subsidence. By 1994, these withdrawal has caused more than 100 centimeters of land subsidence. While restraining of pumpage and utilization of surface water was being introduced through the construction of water conveyance and dam, the

levels of subsidence exceeded 57 and 103 centimeters in the east and west area of Saga plain respectively. Total land subsidence in Saga plain is given in Fig. 2. This subsidence resulted in permanent flooding of central urban area and substantially increased flooding in areas subject to the tidal surges.

In 1980, drainage by big turbine pump has been constructed on the seashore bank to prevent the flooding.

3 SCOPE OF STUDY

The scope of this study includes two major components of the riverine flooding and its analysis in the east area of Saga. Riverine flooding analysis portion of study is an evaluation of ability of drainage system in flooding time by using simulation model. Before detection of land subsidence in Saga plain, flooding was scarcely and weak on the record of old flooding. This study is aimed to evaluate the drainage ability when the ground surface was subsided.

Fortunately, the leveling about flooding and ground level based on the annual record of bench mark has been carried out. By these data, we can simulate and evaluate the flooding and ability of drainage.

4 HEAVY RAIN AND FLOODING

4.1 Rainfall pattern

The rainy season usually is from early June to July in Japan. On the final stage of June, 1990, seasonal rain front went down south to the northern Kyushu island and stayed there till July 3. On July 2, the front caused torrential rains in northern and middle parts of Kyushu. As shown in Fig. 3, Saga city was also hit by a heavy rain in early morning on July 2. The maximum rainfall per 60 minutes and per 24 hours were 72 millimeters respectively. The daily rainfall was 285.5 millimeters and the peak was 164 millimeters in three hours on July 2. The total rain fall from June 29 to July 3 reached 461 millimeters (Saga Meteorological Observatory).

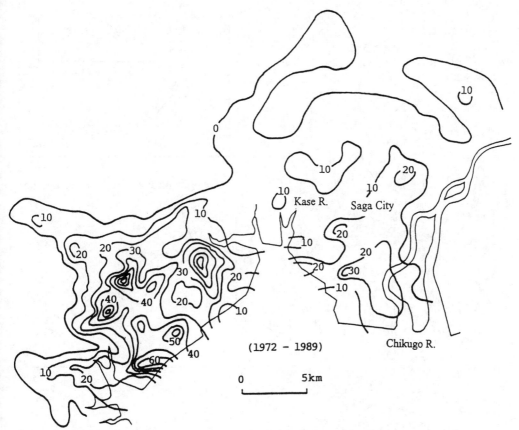

Fig. 2 Total land subsidence in saga plain from 1972 to 1989 (contours are in centimeter)

4.2 Riverine flooding

Flooding in coastal area can be directly correlated with land subsidence. While the land surface is lowered, the sea level and storm surge level remain constant. Thus, each meter of subsidence results in an increased depth of flooding of one meter. In riverine flooding, the relationship between subsidence and increased flooding is not clearly apparent, and the channel capacity and rate of flow are controlling factors, rather than tidal elevation. Channel capacity is a function of the geometry of the channel cross section and the slope of the energy gradient for a given flow (gradient of flow). Land subsidence affect only the gradient of flow. Potack (1991) reported that in riverine flooding, the maximum increase in flooding depth is less than one third of the related

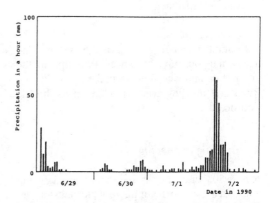

Fig.3 Heavy rainfall in Saga City, 1990

subsidence of the ground. In urban areas, urban activities also contribute to magnitude of flooding, because they cause a delay in reaching

2823

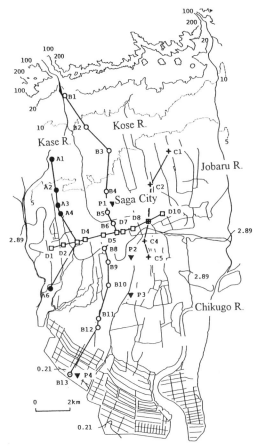

Fig. 4 Distribution of bench marks and leveling routes in studied area

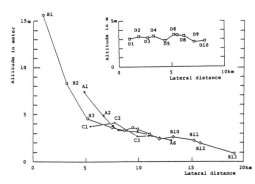

Fig. 5 Topographical profile of selected routes

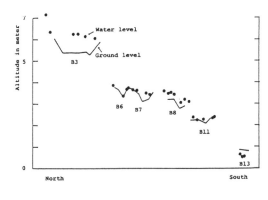

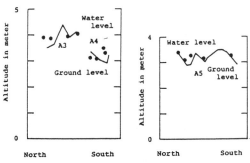

Fig. 6 Ground levlel and flood water table, july 2, 1990

the rainfall water flow to drainage system. The heavy rain on July 2 caused flooding on the urban area and rice field near the sea (Fig. 1). Road and housing were flooded up to the the floorboards.

4.3 Leveling of flooding

About 80 bench marks has been setted in the Saga area for the leveling of land subsidence. They are checked every years. Some of them is shown in Fig. 4. The course of traverses (Fig. 4) were selected to find primary simulation data. The topographical profiles on the north-south and east-west lines is given in Fig. 5. North of studied area is mountain foot composed of

granite, and most of south area is artificial reclaimed land. Northern area is a fan with a gradient of 2.2-3/1000, and topographical slope change in the north of Saga city (nearly between B3 and B4, Fig. 4 & 5)and the area is a reviewed plain with a gradient of 0.25-0.28/1000 by marine regression.

Ground level and flood water level on the north - south direction (selected path in simulation model) are given in Fig. 6. The surface of flooding is not horizontal as a sea. Partially flooding level was high and low with the difference of about 20 centimeters on the course of 500 meters. The depth of flooding was shallow at the rice field and suburb area. Greater flooding detected at the northern part is characterized by break down of Kose river bank.

5 ANALYSIS OF FLOODING

5.1 Modeling

The flooding occurred on the area of Fig. 1 was simulated by the tank model shown in Fig. 7. Six tanks and drain pipes and connection pipes resemble the flood field and drainage river system in the north south direction. The bottom height of every tank is equal to the level of ground.

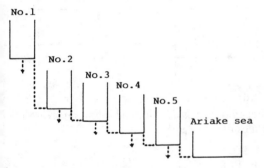

Fig. 7 Configuration of tank model

When the tank i and tank j are connected, the conductivity between them is shown as a_{ij}. The volume of water flow from tank i to tank j in the time of dt is:

$$x = a_{ij}.(u_i - u_j)dt \qquad (1)$$

where: u_i and u_j water level of tank i and tank j
$\sum_j$ is the summed conductivity to tank i from all of the tanks. When the tank i and tank j are not connected, it is definited as $a_{ij}=0$, and also there is the following derivation (Togawa, 1977).

$$\sum_j a_{ij} = -a_{ii} \qquad (2)$$

and:

$$\sum_j a_{ij}.(u_i - u_j)$$
$$= u_i . \sum_j a_{ij} - \sum_j a_{ij}.u_j$$
$$= -a_{ii}.u_i - \sum_{j \neq i} a_{ij}.u_j$$
$$= -\sum_{j=1}^{n} a_{ij}.u_j \qquad (3)$$

The water level in tank i after a very short time dt is:

$$u_i(t+dt)$$
$$= u_i(t) + \sum_{j=1}^{n} a_{ij}.u_j(t).dt/S_i \qquad (4)$$

where: S_i is the cross section area of each tank

5.2 Condition of simulation

The conductivity between the tanks are estimated from the slope gradient and the confusion of road network. The drainage ability was estimated from the river system and the scale of river.

The inflow to each tank are derived from actual precipitation (Fig. 3). The level of Ariake sea is fixed as a sea level. Tidal change was neglected because the drainage to the sea is controlled by manual. The condition of simulation model fix as follows.

1) The difference of actual water level and simulated water is smaller than 10 centimeters.

2) The peak time of flood and estimated time is smaller than 2 hours.

3) The difference of actual water level and simulated water level in tanks at end time of flooding is smaller than 2 centimeters and their time difference is less than 6 hours.

By these conditions, the parameters were adjusted in a little allowerance.

5.3 Simulated flooding

Simulated flooding is shown in Fig. 8. Water level of each tank and sequential changing are closely similar with the actual flooding. In this

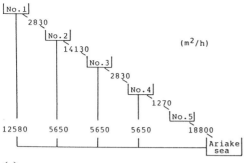

1) Actual drainage ability through river under the heavy rain ranges from 5650 to 12580 m²/h.

2) Actual flowing ability under the heavy rain ranges 1270 to 18800 m²/h.

3) Subsidence weaks the ability of drainage.

4) Subsidence elongates the period of drainage about 2 to 4 times.

(a)

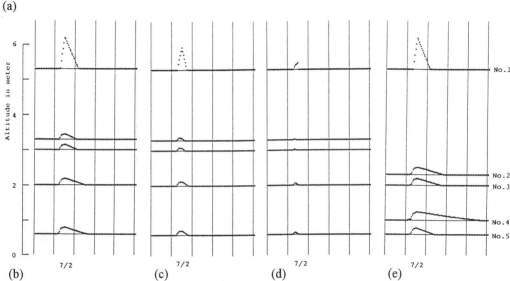

(b)　　　　　(c)　　　　　(d)　　　　　(e)

Fig. 8 Simulation results: a) conductivity, b) actual condition, c & d) twice and 4 times increase in drainage ability respectively, e) assuming one meter subsidence in area

simulation, the drainage ability and conductivity between the tanks are shown in Fig. 8a. The estimated flooding when the drainage ability was increased twice and four times are given in Fig. 8c & 8d respectively. When the central part of Saga area that simulated by tank 2, 3 and 4 was subsided about one meter, the flooding elongated about twice time. Especially, the flooding time in tank No. 4 elongated about three to four times (Fig. 8e).

6 CONCLUSION

The effect of drainage ability and land subsidence on flooding were presented by simulation of flood on July 2, 1990, in Saga lowland plain, Japan. The main results of simulation are noted:

REFERENCES

Mainichi Newspaper Pub. Co.1990. *Newspaper account in July 3*, (In Japanese).

Potak, A. J 1991, A study of the relationship between subsidence and flooding. *Proceeding of fourth symposium on land subsidence, Houston, Texas*. IAHS publication, No. 200:389-395.

Saga Meteorological Observatory 1990. *Record of the precipitation*.

Saga Newspaper Pub. Co. 1990. *Newspaper account in July 3*, (In Japanese).

Ditto 1990. Ibid. in 6th July, 1990.

Saga Prefectural Office 1989. *Record of bench mark*.

Ditto 1989. *Annual report of land subsidence*.

Togawa, H. 1977. *Numerical analysis and simulation*. kyoritsu zensho, Tokyo: pp.29-35

Geoscientific evaluation of geological and geotechnical barriers with respect to waste disposal projects

Évaluation géoscientifique des barrières géologiques et géotechniques pour les projets de décharges controllées de déchets

M. Langer

Federal Institute for Geosciences and Natural Resources (BGR), Hannover, Germany

ABSTRACT:

A site specific safety assessment is required for permanent waste respositories. The "waste-re-pository-rock" system has to be taken into consideration for this safety assessment. Since the "geotechnical barrier" in conjunction with "geological barriers" contributes considerably to long-term isolation of the harmful substances from the biosphere it is absolutely necessary to use engineering geology and hydrogeological methods for a quantitive assessment of the bar-rier effect of the host rock and the geological environment.

For a waste disposal mine, the load bearing capacity of the rock, the protective properties of the surrounding rock formations and the geological stability of the area are important factors in a safety analysis, of which the geotechnical stability analysis is an important part. Such an ana-lysis comprises an engineering geological study of the site, laboratory and in situ experiments and model calculations, long term monitoring, and special geological and geochemical investi-gations.

The existing geomechanical and hydrogeological modeling technique provides a useful tool to perform barrier integrity calculations and to design an underground disposal mine.

RÉSUMÉ:

Les dépôts de stockage final de déchets demandent une analyse de sûreté qui est spécifique pour le site. Pour cette analyse le système entier - les déchets, le dépôt et la roche - doit être pris en considération. Comme la barrière géotechnique contribue considérablement a l'isolation à long-term des matières dangereuses de la biosphére, il est indispensable d'utiliser la géologie de l'ingénieur et les méthodes de la hydrogéologie pour faire une analyse quantitive de l'effet de la barrière géologique de la roche d'accueil.

Pour le stockage de déchets dans une mine, la stabilité de la roche, les qualités protectives des formations géologiques de la région sont des facteurs importants dans une analyse de sûreté, dont l'analyse géotechnique est une part importante. Une telle analyse demande une étude de la géologie de l'ingeniéur du site, des essais in-situ et en laboratoire, des évaluations théoriques, une surveillance à long-terme et des recherches specials sur la géologie et la géochimie.

La technique des modèles géomécaniques et hydrogéologiques, qui est actuellement disponi-ble, se porte garant de réaliser des calculs sur l'intégrité de la barrière et de projeter une mine pour le stockage final en souterrain.

1. ENGINEERING GEOLOGY AND ENVIRONMENT

Many now-a-days tasks and works of engineering geology has resulted on the one hand from the development of modern construction techniques and large-scale technology and on the other hand from the continously increasing environment-consciousness of society.

Any application of a technical skill by society tends to grow into something "big". Modern technology, is fundamentally geared to large-scale use, and the question arises whether it does not become too big in particular cases for safety aspects. We put a mortgage on future life for example by building and operating nuclear power stations and by setting up waste disposal sites and permanent disposal mines. Perhaps we cannot completely avoid acting thus or similarily. But if that is the case then we must act in such a way that the chance of our descendants to manage the mortgage is not thrown away in advance.

The modern philosopher Hans Jonas calls this ethical category that is brought into the arena the "responsibility principle", meaning that the demands on responsibility grows in proportion to the actions of power. Modern engineering geology - as a servant of technical power (in the positive sense) - must face up to this responsibility.

The International Society of Engineering Geology therefore at its General Assembly of 1980 called for the attentiom of all experts in engineering geology and related fields to be directed, during the planning and building of the relevant projects, not only towards their reliability and effectiveness but also to the same extent towards the protection and meaningful use of the environment; quantitative forecasts should be made concerning the effects of human activities and natural processes on the geological environment with regard to place, time, kind and intensity.

Obviously the relation of mankind to the environment is at present so impaired and is so little supported from the philosophical point of view and from the point of view of social policy that many people regard refraining from all human intervention as the only acceptable and harmless way of treating nature. The possibility of shaping the environment according to criteria based on reason is thus considerably reduced. What arises from this attitude are unjustified definitions of and requirements on protection. Certainly it is legitimate and for ethical reasons also necessary to gear all measures taken by mankind today forwards avoiding damage to future generations. But it is equally legitimate to think today about what claims on protection the individual or groups of individuals will perhaps have 200 years hence on the basis of changed social structures.

Environment protection is an interdisciplinary cross-section task which encompasses a varied spectrum of human actions. The engineering geologist can certainly not judge, let alone solve, all the problems of environment protection. But it is precisely he who, in the interplay between the geosciences and the engineering sciences can point out and help to take the optimum steps to making provisions for existence (Langer 1989).

The activities of engineering geologists today are marked by the constant effort to face up to the great problems of our society which arise from the conflicts of the competing claims to the use of the earth: in underground and above-ground construction economic, safety and ecological aspects are taken into account. Their work therefore serves primarily to prevent damage in the course of technical interventions in nature (e.g. assessing the safety of embankments, dams, road and rail tunnels, caverns and disposal sites) and to reduce natural hazards. (e.g. landslides, earth subsidence and earthquakes) by using geological measures.

Specially in the assessment of geological barrieres at disposal sites and in underground storage facilities these activities are substantiated.

2. GEOTECHNICAL SAFETY ANALYSIS

The perception that assessing the safety of waste disposal repositories requires to consider all inherent uncertainties has led to the integration of probabilistic concepts into safety analysis. Therefore the demonstration of safety has to be based on a site specific safety plan. This safety plan has to be prepared for the entire plant in which the possible risks are described and any potential risks are defined (see SIA recommodations 260). The potential risks are subsequently to be introduced into a comprehensive overall assessment which is to cover all aspects, whereby interactive influences of the individual risks are to be borne in mind (Langer 1993).

Natural geological barriers are an important part of a multiple-barrier system. Thus, the loadbearing capacity of the rock (expressed, for example, through subsidence or cavern stability), its geological and tectonic stability (e.g. mass movement or earthquakes), and its geochemical and hydrogeological development (e.g. groundwater movement and the potential for dissolution of the rock) are important aspects of the safety analysis. Therefore safety cannot be assessed from a purely egineering point of view, but must include geological factors. A site-specific modelling of geological, hydrogeological, geochemical and geomechanical features and processes is needed.

The safety analysis must be based on a safety concept that includes the possibilities of failures (failures scenarios) that could occure during the excavation, operation and post-operation phases, as well as measures to avoid such failures. Monitoring is also a part of the safety concept (Fig. 1).

SAFETY ASSESSMENT by MODELLING

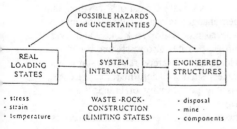

Fig. 1. Safety assessment by modelling

2.1 SAFETY PLAN

The site specific safety plan has to include the individual risk scenarios and possible contingencies for which again measures and/or verifications are required. The safety plan has to be updated as new experience become available (e.g. during the construction or during the operation of the plant).

The safety plan for an underground disposal plant (disposal mine) should, inter alia, describe the following (Langer et al 1993).
• measures to avoid or reduce risks
• possible actions to enhance stability of the plant based on monitoring systems
• acceptable residual risks.

The potential risks define the limiting situations which are to be avoided. These could be for example:
• locale fractures in mine openings
• failure of pillars and roofs
• rock burst
• rock mass loosening due to large cavity or shaft convergence, leading to a loss of integrity of the rock mass
• loss of functionality of seal structures (e.g. dams).

The safety plan should assess the following:
• system failure of load bearing structures
• long term integrity of barrier effect of rock mass and seal structures
• seal functions of seal structures.

To this end the following subsurface measurements and/or calculations are available:
• the short term and long term convergence of cavities
• large scale deformations in rock mass and neighbouring rock and overburden
• stress states in rock mass

2.2 SAFETY ENGINEERING ASSESSMENT BASED ON EXPERIENCE

In the case of usage or conversion of existing mines the safety assessments should include exposures, log data and in situ experience from the previous operations of the mine. It is often the case that many years of geotechnical records are available which can directly or indirectly, after renewed interpretation, provide important information. The safety rele-

vant knowledge and experience includes, inter alia:
- geological exposures and drilling results
- long term records of results of measurements of convergence and deformations
- cavity geometries which have proven their integrity,
- fracture formations and spalling
- visual reports on the behaviour of the rock mass in particular when driving cavities
- large scale observations of geotechnical and hydrogeological rock mass behaviour
- seismic logs
- gas and liquid ingress into mine,
- previous studies on integrity assessments of the mine, e.g. criteria for the selection of pillars and roofs, proof of rock impact safety etc.

The long term observations can also be used to check the selection of material laws and structure models for the theoretical verifications in that previous states are history matched with the results of in situ measurements.

In the evaluation of experience it must be born in mind that with respect to the use of existing mines or their conversion for underground disposal purposes other safety regulations or other mining situations may apply as compared to previous mining operations.

When transferring experiences from mines in neighbouring rock masses or mines cut in similiar rock mass situations it is vital to check to what extent the rock mass situation and the structures are comparable with the proposed disposal site.

2.3 SAFETY ENGINEERING ASSESSMENT BASED ON IN SITU MEASUREMENTS

When driving new cavities for the disposal facility and when operating the facility, in situ measurements of convergence, deformations, rock mass movements, stresses, micro seismic events etc. have both monitoring as well as safety relevant functions. These include, inter alia:
- checking the material laws selected for the rock mass and overburden

- monitoring the transfer of lab results on the deformation strength of the tested rock to the in situ conditions
- reviewing the structure models for theoretical verifications in that the states on which the measurements values are based are mathemathically checked
- direct safety statements, e.g. from determination of temporal development of stress situations in the cavity vicinity in pillars and in stopes
- investigation of the rock mass zones influenced by driving the cavity, loosening etc.

When evaluating in situ measurement results for safety considerations it must be observed that measurements only register actual status and that any statements concerning threshold or failure states (and hence also safety margins) based on measured actual status can only be arrived at by extrapolation

2.4 NUMERICAL PROOF OF SAFETY

The potential risks as described in section 2.1 which generally represents states of a hypothetical character and are hence not covered by the in situ measurements are to be checked mathematically. Relevant models have to be developed for each situation, and are to comprise the following component sections:
- presentation of the risk scenario under investigation
- effects such as primary state of rock mass, temperature, cavity driving, effects of waste, earthquake etc.
- structural calculation model which must cover rock mass formations, cavities and their changes as realistically as possible
- material models for rock mass and overburden, possibly also waste and backfill
- calculation of safety relevant status variables such as deformations, stresses, possibly permeability of liquids (and gases)
- checking and assessing calculation results
- safety concept which provides statements as to threshold values for risks

With respect to the effects it may be necessary to mathematically reproduce the long term influence of the heading of the mine workings and the individual time/space phases of the heading of the cavities, the backfill states and the long term states. When transferring the spatial rock mass section with the repository facility into a calculation structural model, particular attention must be paid to ensure that the section considered is large enough to satisfactorily include the extended reach of the rheological deformation influences within the rock mass. As a rule any neighbouring cavities and the overburden are to be included. The calculation model must be sufficiently realistic. They must be checked for their mathematical and programmed correctness (verification), their rock mechanical validity (validation) and their applicability to the specific site. It is recommended that test calculations are carried out on unequivocally defined rock mass states and, as far as available, mine working states for which measurement data are available are history matched.

The informative value of the theoretical proofs is decisively influenced by the expressions introduced into the calculation model. For this reason it is necessary to be aware of these influences when assessing safety matters. These include, inter alia:

- show sources of errors, e.g. in the structural description of the disposal facility, in the material laws, in the numerical calculation methods
- show the sensitivity of the results to changes in the input parameters by calculating with parameter variations (e.g. also to identify natural scatter).

In the practical performance of the proofs it may be useful to initially work with assumptions which are unequivocally conservative which produce simple verification methods and then move on to more complex expressions in places where the conservative calculations indicate possible critical states. When using this procedure of simplified verification the conservativeness of the expressions must be thorough presented.

2.5 SEQUENCE OF PROOFS

The more important proofs for the geotechnical integrity of the mine and also for satisfying the barrier requirements are to be produced during the planning and during the design of the underground disposal site, inter alia as preparation for the planning procedures and the environmental acceptability test. The safety assessment of this planning phase is based primarily on the evaluation of experience (2.2), the mathematical investigations (2.4) and, if necessary, on complementary geotechnical details and in situ and/or lab tests. The assessment must cover risk situations and contingency measures during:

- the construction of the underground disposal site
- its operation
- after completion of disposal.

When using existing mines the effects of the preceding mining engineering on the loading of the rock mass and on the integrity of the rock mass must be taken into account.

In cases where new openings are to be driven during the construction of the plant, control measurements are to be used with which the assumptions of the verification in the planning phase can be checked. It is necessary here to undertake structural calculations of the constructions states measured and then the interpretation of measurement and calculation results. If necessary the safety verifications during the planning phase are to be updated based on new knowledge.

Some experts may consider these requirements for a safety analysis too excessiv. It must be kept in mind, however, that a repository must provide a special protective function over a very long period of time. It must also be born in mind that uncertainties must be taken into consideration during planning, designing and construction. Uncertainties are for example:

- variation of material properties with respect to space and time,
- uncertainties in the determination of the load,
- inexactness of the model (simulation of the physical and geological conditions), and
- errors of omission and unexpected events.

Geological and geotechnical uncertainties can be mastered by a method of calculated risk ("Geoengineering Confidence Building", see Fig. 2). The main part of this method is the handling and validation of models.

SAFETY CONFIDENCE BUILDING

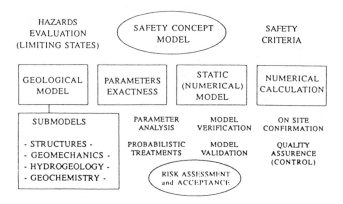

Fig. 2. Geo-engineering safety confidence building

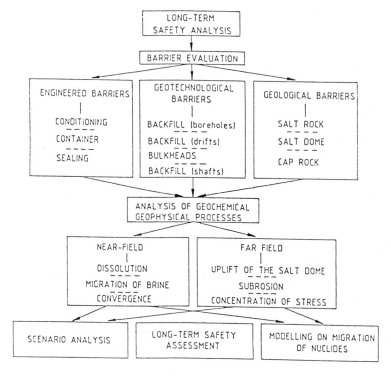

Fig. 3. -Long-term safety analysis (outline)

3. EVALUATION OF GEOLOGICAL BARRIERS FOR DISPOSAL SITES

A disturbance of the long-term stability of the ecosystem in the post operation phase must be ruled out. That is, the transport of hazardous levels of harmful substances into the biosphere by circulating groundwater must be prevented. In order to guarantee the health and safety of people over this long period, several independent technical and natural barriers are used in the system "waste/permanent deposit/geological medium" to prevent the release of harmful substances. These are the waste containers, the geological permanent disposal site barriers such as borehole filling, the packing of galleries and chambers, dams, shaft filling and plugging and the geological barrier, in other words the host rock in the immediate field and the further geological surroundings as a geohydraulic barrier. The development of a realistic and verifiable long-term safety concept to fulfil the protection requirements taking into account all the existing barriers is extremly difficult and is at present the subject of large-scale international research.

One possible methodological approach is shown in Fig. 3. It contains the separate analysis of the individual barrier systems (technical, rock-mechanical and geological), the analysis of the physical and geochemical processes in the immediate and wider surroundings of the permanent deposit and a summarized scenario and fault assessment. For judging technical barriers the probabilistic risk analysis is available. The evaluation of technical barrier is done by the geotechnical stability proof. The geological system is analysed by the prognosis of future geochemical, hydrogeological and tectonic events ("prognostic geology"). In the analyses of geochemical and physical processes in the immediate field solution processes and migration and the connected permeability changes must be examined. As the driving force for pollutant movement in certain fault scenarios in rock salt the convergence of mine workings through the creeping of the rock salt is a possibility to be considered. Relevant processes in the wider field are salt raising and subrosion of salt, by groundwater move-

ment, and phenomena through thermally induced stress concentrations. In the summarized fault analysis, the combined effect of all barriers in certain conceivable events (faults) are analysed, which could cause the danger of releasing pollutants into the biosphere (release paths).

The fault scenarios must be substantiated in detail and their boundary conditions ascertained. From the geological and geotechnical point of view these boundary conditions result from underground exploration. Exploration must also provide the data for setting up the models for the site-specific safety analyses. The data obtained must permit a sufficiently accurate estimate of the conservative assumptions made. Thus in underground exploration structural geological data in particular have to be obtained for fault identification, geochemical data for the analysis of the long-term geochemical process in the host rock and quantitative thermomechanical material data for the model calculation of pollutant mobilization.

Also the wider geological surroundings of the underground deposit (e.g. the overlying rock) can work as a barrier. For assessment hydrogeological model calculations are indispensable. They have to take account of factors which influence the possible transport of pollutants in the ground water. This includes, for example:

- convective transport through flow
- dispersive and diffusiv transport
- physical-chemical interaction with the rock along the path of flow (e.g absorption)
- solution rates.

In order to set a complete model for the propagation in the groundwater the availability of pollutants from the waste must be known. This availiability depends crucially upon the form of waste. In the simplest case of unidimensional groundwater movement in a porous medium the groundwater movement may be described by the Darcy Law.

In jointed rock special rock-hydraulic investigations are needed in order to assess the propagation processes of contaminated and toxic substances. Experiments must be performed to find out the flow paths in the form of joints and faults with a largely watertight rock mass.In addition to the natural jointing of the rock the mining working of cavities and the consequent loosening up of the rock creates an additional jointing and/or the originally existing joint structure is changed by loosening. It is necessary to find out the anisotropic permeability and to quantify the hydraulic properties of the individual joints sets or several major joints. The results of measurement may shown an anisotropic permeability for certain rock areas. Here the permeability depending on the rock and water pressure in the joints must be tested. A corresponding experimental arrangement (the system fracture flow test, Fig. 4) has been developed and used by the Federal Institute for Geosciences and Natural Resources (BGR) (Langer, Liedtke and Pahl 1989). The experiments have so far produced the following results in principle.

- The waterflow remains restricted to individual joints.
- Major joints make a quick and wide transport of liquid possible.
- The permeability of the joints is 100-1000 times greater in comparison to the permeability of the rock.
- The amount of the compression in the rock types investigated, sandstone and

granit, is almost eliminated in a radius of 50-100 m from the inflow point.
- Drive mechanisms for flow processes in jointed rock are:
  - gravitation (different hydraulic heads in the conductor pipes, density current due to temperature differences and the different specific density of possible lyes)
  - convergence of caverns, storage chambers galleries etc.
  - gasification process
  - pressure release of gas deposits
  - tectonic movements.
- The propagation speed in jointed rock under unfavourable geological conditions can be 100-1000 times greater than in loose rock with comparable permeability.
- The joint geometric size of the opening, the asperity and the degree of separation can vary widely from joint to joint even with homogeneous cleaving of the rock mass, so that big differences occur in the permeability of the individual joints.
- Boreholes which are made in the framework of construction create a variety of possible water routes in the rock mass. Here injection measures must be provided for.

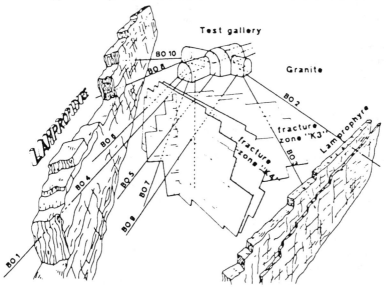

Fig. 4. Fracture system flow test

4. CONCLUDING REMARKS

In conclusion let us consider the outlook for engineering geology.

Society's demands on the state and politics to solve the steadily increasing problems of the environment are becoming more and more urgent. Engineering geology will be judged by its ability to help prevent damage in technical interventions in nature and to reduce natural disasters or avert the damage they cause.

The contributions of engineering geologists should concentrate on evaluation the predominant factors of the stability of the earth's crust and applying this knowledge in vulnerability maps and alarm plans, the development of prevention strategies on the basis of a better understanding of the mechanisms if natural disasters, in particular mass movements such as landslides and rock falls.

Geoscientific findings must be applied more than hitherto to carrying out engineering projects (like waste disposal projects) safely, precisely in endangered areas. The changes to the geosphere and biosphere by man has taken to alarming forms.

In the year 2000 fifteen per cent of the land surface of the earth will be occupied by engineering constructions. Man has become an outstanding longterm geological force, his influence on nature has become comparable with geological processes of the past, and it is uncertain whether this processes can be controlled.

Engineering geologists are therefore joining the ranks of those geoscientists who have recognized this danger and are prepared to contribute to mastering and controlling it.

REFERENCES

Langer, M. 1989: Engineering geology and environmental protection.- In: Demello Volume, Ed. Edg. Blücher Ltd.: 252-259, Sao Paulo

Langer, M. 1993: Safety concept and criteria for hazardous waste sites. Engineering Geology 34, 159-167, Elsevier Putb., Rotterdam

Langer, M. et al. 1993: Empfehlungen des Arbeitskreises "Salzmechanik" der Dt. Ges. für Erd- und Grundbau e.V. zur Geotechnik der Untertagedeponierung von besonders überwachungsbedürftigen Abfällen im Salzgebirge - Ablagerung in Bergwerken - Bautechnik 70, 12: 734-744, Berlin

Langer, M. Liedtke, L. and Pahl, A. 1989: Felshydraulische Tests zur Benutzung der Barrierewirkung von geklüftetem Fels.- Mitt. Ing. v. Hydrogeol., 32:467-484, Aachen

Sediment bound heavy metals and organic content – An example from Ganga river, Mirzapur, India

Métaux lourds et contenu organique dans les sédiments – Un exemple du Ganges, Mirzapur, Inde

Ajai Srivastava
Sedimentary Laboratory, Department of Geology, Banaras Hindu University, Varanasi, India

ABSTRACT: Sediment in an aquatic milieu acts as a 'voice-recorder' of a stream behaviour. The contamination of the sediment in an aquatic environment is not uncommon. The organic content in an aquatic sediments is contributed from the various sources. In an aquatic ecosystem, the presence of organic content and their decomposition products largely govern the behaviour and affinity of heavy metals in the sediments. Keeping this in view, sediments of Ganga; one of the important rivers of the world, in Mirzapur have been studied in the light of influence of organic content in the concentration of heavy metals. From the study and observation it has been inferred that the enrichment of heavy metals in the sediments of the area is partly due to a rather high organic content. Besides, location i.e. nearness to the outlets carrying industrial and domestic wastes and effluents, carbonate content, grain size and presence of clayey minerals have also played in the concentration of heavy metals in the sediments of the area.

RÉSUMÉ: Les sédiments dans un milieu aquatique agissent comme un magnétophone du comportement du courant d'eau. La contamination du sédiment dans un milieu n'est pas rare. Le contenu organique dans un sédiment aquatique provient de sources variées: de terre à cause de la corrosion par l'eau et l'air, de la colonne d'eau et du sédiment lui-même. Dans un écosystème aquatique, la présence du contenu organique et ses produits de décomposition déterminent en grande mesure le comportement et l'affinité des métaux lourds dans le sédiment. Les sédiments du Gange - un des fleuves plus importants du monde dans la région de Mirzapur - ont éte étudiés à la lumière de l'influence du contenu organique dans la concentration des métaux lourds. A partir de l'étude et de l'observation on a inféré que l'enrichissement des métaux lourds dans les sédiments de la région est partiellement du plutôt au contenu organique élevé. De plus, l'emplacement, c'est-à-dire la proximité des courants d'eau emportant les déchets domestiques et industriels et éffluents, le contenu carbonaté, la taille des grains et la présence des minéraux argileux ont aussi joué un rôle dans la concentration de métaux lourds dans les sédiments de la région.

INTRODUCTION

Sediments of different deposits contain varying amount of organic matter. The marine sediments may contain 10% or more, whereas the average near-shore sediments have 2.5% and those of the open ocean approximately 1% organic matter. According to Goldman (1924), the amount of organic matter preserved is a "product of an equation between the rate of supply of organic matter and the rate of decomposition". Besides, it also depends on the chemical composition, rate of growth of the organic materials, the rate of burial and the rate of decay after burial. Texture of the sediments, however, govern the amount of organic matter present therein e.g., clayey sediments contain twice as much organic content as the silty deposits does.

The amount of organic matter as occurring in the aquatic sediments is contributed from the land by water and wind erosion, from the water column, and from the sediment itself by biological activity (Eglinton et al. 1975). In addition to these, the cumulative effect of the various anthropogenic activities viz., agricultural practices, industrial processes, paper and sugar mills, oil refineries, slaughter houses, dairy companies, dumping of high nutrient wastes e.g., sewage sludge, waste disposal facilities lead to enhance the level of organic material in the aquatic sediments.

In an aquatic milieu, the organic substances and their decomposition products largely govern the behaviour and affinity of heavy metals. It is important to note that the trace metal (heavy metal) adsorption capacity of organic matter is oftenly between that for metal oxides and clays (Guy and Chakrabarti, 1976). A review of literature (Beck et al. 1974, Rashid 1974, Reuter and Perdue, 1977, Förstner and Wittmann, 1983) shows that the metal ions sorbed onto the humic substances become part of the particulate material (bottom sediments). However, metals may remain therein unless there is not a change in the physico-chemical conditions, viz., pH, salinity, redox condition etc.

Ganga sediments of the Mirzapur, an industrialized part of an eastern Uttar Pradesh, India have been studied in the light of influence of organic matter content in the concentration of heavy metals in the sediments. For the purpose, the sediment samples collected from both the bank of Ganga river in this sector (Fig.1) have been analyzed for heavy metals and organic content. The area under investigation is a plain tract and falls in the Middle Ganga Basin.

ENRICHMENT OF HEAVY METALS IN THE GANGA SEDIMENTS

The heavy metal content was determined by using Atomic Absorption Spectroscopy Pye Unicam SP 2900. Sediments of the area under investigation were found to be enriched in the heavy metals Pb, Zn, Ni, Cr, Cu, Co, Rb, Sr and Li. Of which Pb, Zn, Ni, Cr, Cu and Co are very toxic and relatively accessible whereas Rb, Sr and Li are noncritical (Wood, 1974). The toxicity and reactivity of a metal complex is largely governed by the nature of the ligands.

The distribution of heavy metals in the Ganga sediments is erratic and does not follow a particular trend (Table 1). The spatial distribution of the metals in the area is shown in Figs. 2, 3, 4 and 5. Metal concentration in the sediments has been compared with the average shale value (Turekian and Wedepohl, 1961), a geoche-

mical background value. It may be stated that the tolerance level of the toxic heavy metals in sediments has not been defined as are available with regard to water. Therefore, the value of average shale is taken for comparative study of the sediments of Ganga river (although it may not refer to the level of toxicity). The maximum concentration of metals Pb (226 ppm), Zn (717 ppm) and Cu (931 ppm) was recorded in the sediments collected near Oliar ghat while those of Ni (100 ppm) and Co (270 ppm) at Ghamhapur. Cr and Mn are present in its maximum concentration (150 ppm and 710 ppm respectively) to the downstream of Vindhyachal ghat ka nullah. The maximum concentration of Rb and Li (150 ppm and 64 ppm respectively) was found at Pipra Dand, while those of Sr (144 ppm) at the confluence of Ojhala nullah. When compared with average shale value (Turekian and Wedepohl, 1961), it is noted that at most of the places the concentration of Pb, Cu, Ni, Co and Cr are much higher than the corresponding average shale value (Srivastava et al., 1993).

ORGANIC CONTENT IN THE GANGA SEDIMENTS

The representative sediment samples were analyzed for their organic contents by rapid titration method (Walkely and Black, 1934). The result obtained is shown in the Table 2 and illustrated in Figs. 6 and 7. In the area, organic carbon and organic matter content range from 0.0312% to 0.660% and 0.0538% to 1.137% respectively. The maximum content of the both have been observed near Kutchery ghat, where the effluents of a Government hospital nullah meet the river. However, the recognition of organic pollutant might be assessed with the help of computerised gas chromatography.

In addition to this, samples were studied with the help of Infrared spectrographic method. For this purpose, sediment samples of minus 325 A.S.T.M. (American Society for Testing Materials) mesh size powder (i.e. less than 0.44 mm size) were taken. The I.R. absorption spectra of these powders have been recorded using nuzol technique with scantime 16 and source current 0.6 by Perkin Elmer infrared spectrophotometer model No. 783. The IR spectrum was scanned in the region 4000-200 cm^{-1} as the spectra characteristics of the important organic substances fall in this region. The shifting of peaks were corrected by comparing with the standard peaks of nujol. The absorption spectrum of nuzol comes at 2915 cm^{-1},

Name of the Nullahs

A Malhia ghat ka nullah
B Vindhyachal ghat ka nullah
C Diwan ghat ka nullah
D Baswaria nullah
E Ojhala nullah
F Nukkar ghat ka nullah
G Kachcha ghat ka nullah
H Churwa nullah
I Khandwa nullah
J Naar ghat ka nullah
K Pakka ghat ka nullah
L Sunder ghat ka nullah
M Oliar ghat ka nullah
N Hospital nullah
O Vijaypur kothi nullah
P Ghore Shahid nullah
Q Chhota Sakhaura nullah
R Bisundhar pur ghat ka nullah
S Pipra Dand nullah
T Belavan nullah
U Muwaiya ghat ka nullah
V Gahia ghat ka nullah
W Purwa nullah
X Dullapatti ghat ka nullah
Y Pachaura ghat ka nullah
Z Gigraon ghat ka nullah

Name of the Ghats

I Malhia ghat
II Vindhyachal ghat
III Diwan ghat
IV Bhatan ghat
V Ojhala ghat
VI Lohia Talab ghat
VII Lalla ghat
VIII Nukkar ghat
IX Kachcha ghat
X Chaube ghat
XI Naar ghat
XII Pakka ghat
XIII Sunder ghat
XIV Oliar ghat
XV Kuthery ghat
XVI Vijaypur Kothi ghat
XVII Fataha ghat
XVIII Chhota Sakhaura ghat
XIX Bisunderpur ghat
XX Company ghat
XXI Belavan ghat
XXII Newarhia ghat
XXIII Muwaiya ghat
XXIV Gahia ghat
XXV Dullapatti ghat
XXVI Akbarpatti ghat
XXVII Pachaura ghat
XXVIII Gigraon ghat

INDEX

Industrial waste nullah
Hospital waste nullah
Municipal waste nullah
Dry nullah
Seasonal nullah
Underground nullah
Kachcha ghat
Pakka ghat
Washing ghat
Cremation ghat
Sample location

Fig 1. Map showing location of Samples, Ghats and nullahs in the Mirzapur region

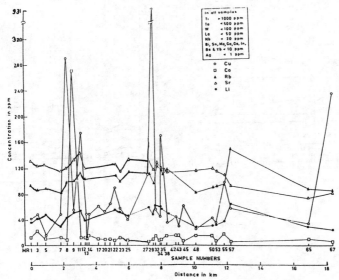

In all samples
Ti > 1000 ppm
Te < 500 ppm
W < 100 ppm
La < 50 ppm
Nb < 20 ppm
Bi, Sn, Mo, Ga, Ge, In,
Be & Yb < 10 ppm
Ag < 1 ppm

o Cu
□ Co
▲ Rb
△ Sr
● Li

Fig. 2. Metal (Cu, Co, Rb, Sr, & Li) concentration in the Ganga sediments, right bank ; Mirzapur region

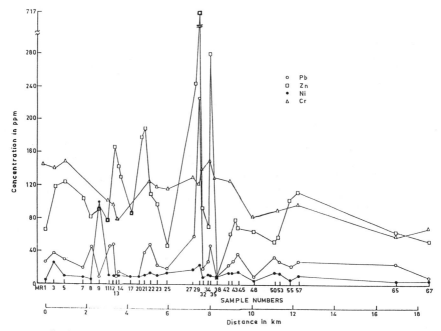

Fig. 3 Metal (Pb, Zn, Ni & Cr) concentration in the Ganga sediments, right bank ; Mirzapur region

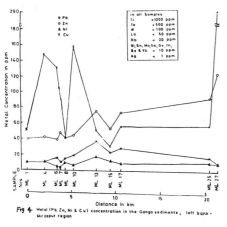

Fig 4 Metal (Pb, Zn, Ni & Cu) concentration in the Ganga sediments , left bank ;
Mirzapur region

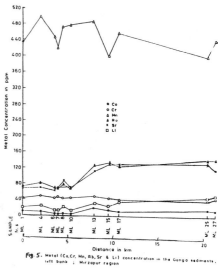

Fig. 5. Metal (Co, Cr, Mn, Rb, Sr & Li) concentration in the Ganga sediments,
left bank ; Mirzapur region

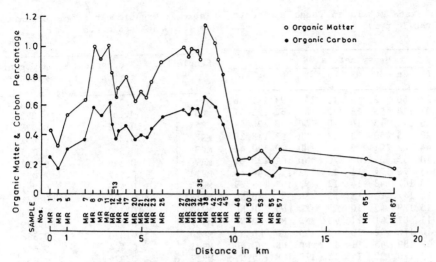

Fig 6 Plots of downstream Variation in Organic Matter and Carbon in the Sediments of Right bank of River Ganga

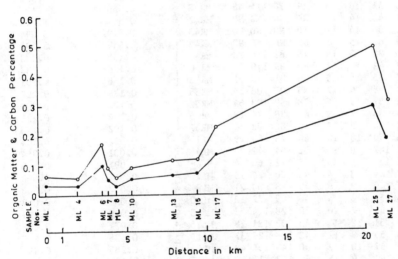

Fig. 7 Plots of downstream Variation in Organic Matter and Carbon in the Sediments of left bank of River Ganga

Table 1. Trace metals in Ganga sediments (all values in ppm).

Sample No.	Heavy metals								
	Pb	Zn	Cu	Ni	Co	Cr	Rb	Sr	Li
MR1	27	66	42	5	12	145	93	131	36
MR3	37	118	50	26	22	142	87	123	41
MR5	30	114	15	10	10	150	89	126	47
MR7	20	103	50	10	12	ND	83	118	29
MR8	44	82	291	6	8	ND	99	120	43
MR9	10	93	54	100	270	ND	102	133	51
MR11	45	78	174	11	12	102	112	144	55
MR12	48	167	48	14	12	97	103	119	39
MR13	10	141	10	13	10	80	ND	ND	ND
MR14	15	129	45	10	15	ND	ND	ND	ND
MR17	11	87	61	11	13	ND	ND	ND	ND
MR20	10	176	52	10	9	ND	ND	ND	ND
MR21	37	189	65	11	12	ND	ND	ND	ND
MR22	47	110	90	14	11	129	108	125	55
MR23	23	97	58	11	13	121	99	116	52
MR25	19	47	40	14	8	119	114	133	46
MR27	58	243	108	19	6	132	110	131	59
MR29	226	717	931	24	9	123	95	116	47
MR32	19	94	56	9	15	138	124	127	62
MR34	28	71	45	12	9	137	117	119	60
MR35	47	280	169	11	10	131	113	116	57
MR38	10	10	35	10	15	129	118	114	48
MR42	23	61	43	14	15	ND	ND	ND	ND
MR43	28	79	30	14	12	ND	ND	ND	ND
MR45	37	69	61	15	5	ND	ND	ND	ND
MR48	10	65	25	5	15	78	82	116	28
MR50	33	52	42	15	12	ND	89	119	29
MR53	28	58	35	14	5	86	91	113	31
MR55	21	103	98	6	17	ND	95	108	38
MR57	28	113	57	11	6	98	150	93	64
MR65	25	64	33	4	9	60	88	74	29
MR67	12	52	234	5	7	66	86	82	24
ML1	10	40	50	10	10	44	74	69	19
ML4	13	42	147	10	11	52	84	72	25
ML6	15	39	131	5	5	45	72	63	14
ML7	11	47	104	3	4	48	70	72	15
ML8	14	41	40	9	6	45	78	89	21
ML10	19	44	157	6	4	46	69	72	13
ML13	37	75	51	17	20	50	128	114	32
ML15	23	53	31	12	15	45	132	131	28
ML17	28	73	56	9	12	42	124	127	37
ML25	17	92	57	10	17	40	143	135	48
ML27	12	123	289	9	15	45	143	114	53
Average shale value for metals (Turekian & Wedepohl, 1961)	20	95	45	68	19	90	140	300	66

Samples of Right Bank — Samples of left Bank

ppm = part per million;
ND = not detected.

Table 2: Organic carbon and organic matter in the Ganga sediments.

Sample No.	% Organic carbon	% Organic matter
MR1	0.248	0.428
MR3	0.187	0.323
MR5	0.312	0.538
MR7	0.366	0.632
MR8	0.589	1.015
MR9	0.526	0.907
MR11	0.612	1.055
MR12	0.476	0.821
MR13	0.378	0.651
MR14	0.420	0.724
MR17	0.463	0.799
MR20	0.370	0.638
MR21	0.404	0.697
MR22	0.379	0.654
MR23	0.442	0.762
MR25	0.520	0.897
MR27	0.577	0.996
MR29	0.537	0.927
MR32	0.571	0.986
MR34	0.566	0.976
MR35	0.527	0.908
MR38	0.660	1.137
MR42	0.594	1.024
MR43	0.519	0.894
MR45	0.459	0.791
MR48	0.135	0.233
MR50	0.138	0.239
MR53	0.168	0.291
MR55	0.128	0.222
MR57	0.178	0.306
MR65	0.134	0.231
MR67	0.102	0.176
ML1	0.037	0.064
ML4	0.031	0.054
ML6	0.097	0.168
ML7	0.052	0.090
ML8	0.037	0.063
ML10	0.056	0.097
ML13	0.068	0.117
ML15	0.071	0.122
ML17	0.135	0.233
ML25	0.285	0.491
ML27	0.180	0.310

Samples of Right Bank — Samples of left Bank

1466 cm^{-1} and 1360 cm^{-1}. The wavelength of different absorption bands were noted and compared with standard chart for rock-forming minerals. The organic matter with their corresponding wave-number is shown in the Table 3.

Table 3. Organic matter inferred from IR spectral study.

Organic matter	Wave number cm^{-1}
Humic acid (tschernosem)	1379
Humic acid (Marsh soil)	1337
Humic acid (Podzol soil)	1515
Humic acid (artificial)	1370
Humic acid – Mycel (Stachybotrys-chartarum)	1715,1615,1330
Grey peat	1650
Doplerite peat	1380
Lignin (straw)	1660,1505,1360, 1030,770,740
Bitumen (gilsonite-albertite)	1610,1376

The organic matter content of the terrestrial sediment and/or soil predominantly consists of a series of high molecular weight, brown, nitrogenous polymers referred to as 'humus'. The 'humus' depositing on the ground would be decomposed thus leading to the formation of organic acids referred to as 'humic acid' rich in heterocyclic phenols. Its acidic nature is due to carboxyl (COOH) and phenolic OH (C$_6$H$_5$–OH) group.

In most of the sediments, lignin, various types of humic acid etc. are found as a organic content as inferred from IR spectral studies. The bands absent at 3650–3600 and 3500–2500 cm^{-1} confirm that there is no free –OH and no intermolecular and intramolecular hydrogen bonding due to phenolic or carboxylic group in the compound respectively. The IR band near 1715 cm^{-1} suggests the presence of C=O in carboxylic acid. The band at 1615 cm^{-1} confirms the presence of C=C in the benzene ring. The IR spectrum of humic acid shows band at 1370 and 1330 cm^{-1}. It is due to in-plane bending in homoxylic C-H bond (Rai and Shekhar, 1990). Lignin, another organic content in sediments is responsible for thickening and strengthening of plant cell walls. It cannot be defined chemically. It is a high cross linked, macromolecular branched polymer, formed irreversibly by the dehydrogenation and condensation of plants. It is amorphous, incrustation material found in wood, consisting of methoxylated phenol propane unit linked either by linkages and C-C bonds. In finger print region, the absorption band at 1030 cm^{-1} is due to C-O stretching in C-O-C group present in methoxylated phenyl group. The band at 770 and 740 cm^{-1} is due to out-of-plane bending of C-H bond (Silverstin et al., 1981) in benzene ring. The C=C present in benzene ring is assumed by the presence of IR band at 1660 cm^{-1}. The band at 1505 cm^{-1} is due to C-H bending of –CH$_2$ and –CH$_3$ group present in the propane. The C-C stretching due to linkage of one unit of lignin to another is confirmed by the IR band present at 1360 cm^{-1}.

Sources of organic matter in the Ganga sediments

Dumping of municipal sewage containing faecal matter, urine, human and animal excreta, domestic effluents, burnt wooden materials, filth, pieces of old clothes, solid wastes etc. are the prominent sources towards augmenting the organic matter content in the sediments. Besides, industrial activities prevailing in the region are also helping hands in the accumulation of organic matter in the sediments. Allochthonous sources of organic matter in the area are mainly of plant origin and forested and cultivated land. This source is significant over autochthonous production of organic matter. The industrial effluents and wastes coming from the various industries are equally responsible to increase the heavy metal budget of the Ganga sediments in the area. These industries viz., non-ferrous metal, dyeing and carpet industries and other enterprises like cold storage, dairy corporation, soap factories, electronic and electric industries, chemical and metal refinery industries and rice mills pour their effluent waste into the river through the effluent nullahs. The left bank of the river is frequently used by farmers to grow food crops. The agricultural activities e.g., the agricultural use of pesticides and the practice of spreading manure for its fertilizing properties along this bank lead to increase the organic content in the sediments. Consumption of nitrogenous fertilizers in the region is fairly high and is followed by phosphorous and potassium. Some amount of the nutrients added to the soil/sediment through

2843

fertilizers eventually goes to the river ecosystem (Srivastava,1990).

DISCUSSION

The distribution of heavy metals in the Ganga river sediments may be correlated with the corresponding high content of organic matter. The highest concentration of Pb, Zn and Cu (226 ppm, 717 ppm and 931 ppm respectively) near Oliar ghat ka nullah may be attributed to comparatively higher percentage of organic matter. Likewise, concentration of Ni and Co (100 ppm and 270 ppm respectively) around the Ojhala nullah and of Pb, Zn, Cu and Cr (48 ppm, 167 ppm, 48 ppm and 97 ppm respectively) near the confluence of Ojhala nullah and of Pb, Zn, Cu and Cr (47 ppm, 280 ppm, 169 ppm and 131 ppm) near the confluence of hospital nullah may also be related to comparatively fair content of organic matter in these sediments. Along the left bank of the river the highest content of organic matter was recorded little upstream of Girgraon ghat, which also records the maximum concentration of Cu i.e. 57 ppm.

The sorptional characteristic of organic matter is governed by the ligands, concentration of organic compounds and salinity of the aquatic milieu. According to Hunter (1980) and Tipping (1981), at least part of the organics absorbed onto the particulate matter in natural water has carboxylic and phenolic functional groups available for binding with trace metals. According to Beveridge et al. (1983), a significant part of the organic matter of the aquatic sediments consists of small colloidal aggregates, which in its turn are composed of highly cross-linked heteropolymeric materials of biological origin and are rather non-degradable. The polymeric materials of biological origin would interact with and sequester dilute metals from the water phase, ultimately depositing metals onto the sediments. The organic substance in the aquatic system form the organometallic complexes thus act as an anti-pollutant i.e. scavenger against toxic, obnoxious, hazardous substances etc. On the other hand, decomposition of organic matter mediated by bacteria creates anoxic conditions within a short distance beneath the sediment-water interface. In this partially restricted environment many heavy metals are mobilized. The process changes the intermingled organic matter in the sediment into dissolved organic matter. The heavy metals settled into the bottom sediments may be released from sediment to water phase owing to microbials activity and making the water unfit for drinking and other purposes.

CONCLUSION

It has been inferred that the enrichment of heavy metals in the sediments of the area is partly due to a rather high organic content. Besides, the other factors viz. location i.e. nearness to the outlets carrying industrial and domestic wastes and effluents, the flow behaviour of the river, the textural characteristics of the sediments, carbonate content and clay minerals together are considered responsible for enrichment of the heavy metals in the sediments. As the exchange of materials including heavy metals may take place at the sediment-water interface; even a small part of the bulk of heavy metals if remobilized from the contaminated sediments would effect the water quality of the river and the water may not only be unfit for drinking purposes but its enhanced pollution may cause health hazards. Keeping this in view, the prevention of sediment pollution is important. Various measures for checking the sediment pollution have been suggested. However, the simplest course and comparatively easily feasible method is to divert the effluent nullahs (outlets) and drive the dirty water elsewhere and/or to treat the harmful pollutants at the source.

ACKNOWLEDGEMENTS

Author is thankful to the Head, Department of Geology, Banaras Hindu University for providing necessary research facilities and Head, Department of Chemistry, Banaras Hindu University for providing facility for IR spectroscopic study. The author is grateful to Prof. (Retd.) M.N. Mehrotra, Department of Geology, Banaras Hindu University for his constant inspiration. Council of Scientific and Industrial Research (C.S.I.R.), New Delhi is gratefully acknowledged for awarding a 'Research Associateship' to the author.

REFERENCES

A.S.T.M., 1959. Symposium on particle size measurement. Am. Soc. Testing Materials. Sp. Tech. Pub. 234, 303.

Beck, K.C., J.H. Retuer and E.M. Perdue. 1974. Organic and inorganic geochemistry of some coastal plain rivers of the southeastern United States, Geochim. Cosmochim. Acta 38 : 341-364.

Beveridge, T.J., J.D. Meloche, W.S. Fyfe and R.G.E. Murray. 1983. Diagenesis of metal chemically complexed to bacteria: Laboratory Formation of metal phosphates, sulfides and Organic Condensates in artificial sediments. Appl. Environ. Microbial. 45: 1094-1108.

Eglinton, G., B.R.T. Simoneit, and J.A. Zoro. 1975. The recognition of organic pollutants in aquatic sediments. Proc. R. Soc. Lond. B. 189: 415-442.

Förstner, U. and G.T.W. Wittmann. 1983. Metal Pollution in the aquatic environment. 3rd Ed. Springer Verlag Berlin. P.486.

Goldman, M.I. 1924. Black Shale Formation in and about Chesapeake Bay. Bull. Am. Assoc. Petroleum. Geol. 8: 195-201.

Guy, R.D. and C.L. Chakrabarti. 1976. Studies of metal-organic interactions in model systems pertaining to natural waters. Can. Jour. Chem. 54: 2600-2611.

Hunter, K.A. 1980. Processes affecting particulate traces metals in the Sea Surface Microlayer. Mar. Chem. 9: 49-70.

Rai, U.S. & H. Shekher. 1990. Chemistry of organic eutectics: Phenanthrene-Benzoic acid and Phenanthrene-cinnamic acid system. Cryst. Res. Technol. 25: 771-779.

Rashid, M.A. 1974. Adsorption of metals on sedimentary and peat humic acids. Chem. Geol. 13: 115-123.

Reuter, J.H. and E.M. Perdue. 1977. Importance of heavy metal-organic matter interactions in natural waters. Geochim. Cosmochim. Acta. 41:325-334.

Silverstin, R.M.; G.C. Bassler & T.C. Morill. 1981. Spectrometric identification of organic compounds. John Wiley & Sons, Inc.

Srivastava, A. 1990. Study of sediments of Ganga with special reference to effects of pollution load in Mirzapur, U.P. Ph.D. thesis (Faculty of Science, Department of Geology, Banaras Hindu University) P. 192.

Srivastava, A.; M.N. Mehrotra & R.N. Tiwari. 1993. Study of pollution of the river Ganga in the Mirzapur region (India) and its impact on sediments. Intern. J. Environmental Studies. 43: 201-208.

Tipping, E. 1981. The adsorption of aquatic humic substances by iron hydroxides. Geochim. Cosmochim. Acta. 45: 191-199.

Turekian, K.K. and K.H. Wedepohl. 1961. Distribution of the elements in some major units of the Earth's Crust. Bull. Geol. Soc. Am. 72: 175-192.

Walkely, A. and D.C. Black. 1934. An examination of the Degtjureeff method for determining soil organic matter, and a proposed modification of the chromic acid titration method. Soil Science. 37: 29-38.

Wood, J.M. 1974. Biological cycles for toxic elements in the environment. Science. 183: 1049-1052.

Shear strength of dump soils with references to water saturation

Résistance au cisaillement des sols tenant en compte la saturation

S. Rybicki & H. Woźniak
University of Mining and Metallurgy, Kraków, Poland

ABSTRACT: The shear strength of a dump soil has been investigated with the special reference to water saturation of the soil. The shear strength of the water-saturated dump soil (lumps of a cohesive soil and their mixture with a sandy soil) decreases by ca. 3 - 30% depending on lithology and compaction.

RÉSUMÉ: On présente les resultats de recherches de la résistance au cisaillement du sol de décharge avec l'attention spéciale de sa saturation en eau. La résistance de sol (des boullets de sol cohérent et leur melange avec un sol sableux) se diminue 3 - 30% dépendant de la lithologie et du degré de consolidation.

1 INTRODUCTION

During open pit mining of mineral deposits their stripped overburden soil is crumbled, transported and redeposited as a dump soil, often inside the open pit. Reconstruction of the water level taking place when dewatering of the pit has ceased results in saturation of the soil with water, the easier the looser a structure of the soil. This affects physical and mechanical properties of the dump soil, and may lead to the loss of stability of dump slopes and even to liquefaction of a bigger volume of the deposited material. The changes depend mainly on lithology of the dump soil, that represents most often a mixture of lumps and fragments of various cohesive and non cohesive soils, and also on its compaction.

2 INVESTIGATIONS

2.1 Material tested

Cohesive and non cohesive soils of the Quaternary age from the overburden of a brown coal deposit, exploited by the open pit mine "Bełchatów" in the central Poland, were used for laboratory investigations. The overburden is built of intercalating sands, silts, boulder clays and clays. Because the soils are mixed during stripping and transporting by conveyor belts, the sandy soil (P), the cohesive soil (M), and their mixture (PM) in proportion 25+75, 50+50, and 75+25 were taken for testing. Such a selection was to simulate a composition of the dump material. The sandy soil (P) was formed by a medium grained sand mixed with a fine one. The cohesive soil (M) represented a varved silt (sand fraction 4.5%, silt fraction 87%, clay fraction 8.5%), predominant in the overburden. Lumps of the silt had dimensions 2 - 15 mm, corresponding to those occurring mainly in the dump.

2.2 Test procedure

All samples were examined for their shear strength, compressibility, and coefficient of filtration, with the aim at establishing correlation of these parameters with a type of the tested soil, its compactness and moisture content (water saturation). However, only the shear strength of the dump soils have been dealt with in the present paper.

The shear strength tests were carried out in the triaxial apparatus using samples with enlarged both diameters $D = 75$ mm and heights $H = 150$ mm. A rule that the maximum diameter of the lumps of the cohesive soil (d_{max}) should be five times smaller than the diameter of the sample (Schulze 1957, Fukuoka 1957) was observed. The standard and long term shear strengths were measured for samples with the natural moisture content (samples labelled NW) and after their saturation with water (samples SR). The shear strength at the increasing pore pressure, imparted through saturation of samples with water from inside, was also studied. Prior to investigations, all samples underwent the anisotropic drained consolidation at the horizontal to axial stress ratio $k = 0.8$. The vertical consolidation stress (σ_v) varied between 0.125 and 2.0 MPa depending on the type of testing. Shearing was being executed at the horizontal stress $\sigma_3 = 0.05 - 1.6$ MPa, altering accordingly to a value of the axial consolidation stress. The samples in question were overconsolidated (OCR = 1.25 - 16.0). Samples with the natural moisture content were studied without drainage and measurement of the pore pressure (type CAU), while the pore pressure at shearing was measured for samples saturated with water. When testing the shearing strength of samples saturated with water from inside, they were initially anisotropically consolidated and then saturated with water from their bottoms at the pressure of 0.02 - 0.03 MPa. An increase of the water pressure inside a sample after saturation was leading to its failure at the constant external load.

A preliminary consolidation and further saturation of some samples with water simulated natural processes taking place in the dump.

3 RESULTS

3.1 Standard shear strength

For most of the studied samples, composed of the sand alone P (100), the lumps of the varved silt alone M (100), and the mixture of the two

PM (50+50), both at the natural moisture content (samples NW) and after saturation with water (samples SR), plastic shearings were observed (barreled-shaped samples). The maximum value of the stress was being attained at axial deformations of ca. 5 - 23%, with smaller values corresponding with higher compression ratios the samples. No significant dependence of these deformations on the moisture content of the samples has been observed. The brittle type of shearing has been found only in samples with OCR ≥ 4, and values of the axial deformation in the critical state were ca. 3 - 15%.

Examples of the critical state envelope of the standard shear strength in the p_f - q_f system for the dump soils in question, consolidated at the load of 2 MPa, and for a natural undisturbed sample of the varved silt have been presented in Figure 1.

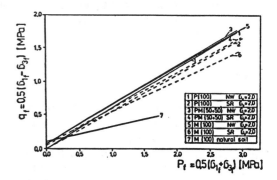

Fig. 1 Shear strength envelope of dump soils

A rectilinear shape of the envelope of all samples indicates that the preliminary compression at the load of at least 0.5 MPa eliminates empty spaces among the lumps in the silt, and in its mixture with the sand. As a result, the dump soil attains a stable structure and reacts as a natural soil. Changes of the porosity index during the preliminary consolidation proved that a fundamental compaction of the investigated dump soil took place at the load of 0.025 - 0.1 MPa. The change of the porosity index within this range of load amounted to ca. 80 - 97% of its total value reached at the load of 2.0 MPa (Fig. 2).

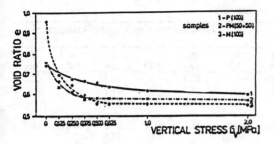

Fig. 2 Relation between void ratio and vertical stress

Saturation with water of the preliminary consolidated soil samples does not substantially influence under such conditions either soil structures (elimination of empty spaces among the lumps) or an increase of the moisture content. Curvilinearity of the strength envelope characterizes probably the range of small initial consolidation loads of up to ca. 0.2 MPa, and this depends on the type and consistency of the soil lumps. Values of the apparent angle of internal friction and cohesion obtained from measurements of the standard shear stress of the the sand samples P (100), silt M (100), and their mixture PM (50+50) have been shown in Table 1.

An increase of the standard shear strength of the dump soil due to a rise in its consolidation ratio is caused, as it is seen in Table 1, by an increase of cohesion of the soil. Saturation with water of the consolidated dump soil results in a relatively small decrease of its strength at ca. 7 - 10%. The shear strength of the 1:1 mixture of sand and lumps of varved silt is bigger than the strength of any of the single components. Cohesion of the consolidated lumps of the varved silt is clearly smaller than cohesion of this soil with an undisturbed structure, but an apparent angle of internal friction is bigger. This is caused by a lumpy structure of the dump soil in question and disturbance of structural bonds among soil particles.

3.2 Long term strength

The long term strength of the dump soils was measured according to Tan (Tan 1966) at a constant rate of loading. This strength is identified as such a value of stress that initiates a nonlinear dependence between the stress and a rate of strain. An example of this relation for the sand P (100) and varved silt M (100) samples has been shown in Figure 3.

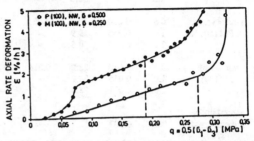

Fig. 3 Axial rate deformation versus shear stress

Values of the angle of internal friction and of cohesion determined in long term strength tests at the consolidation load of 0.5 MPa are almost equal for the all types of dump soils studied, being lower at respectively ca. 13 - 24% and ca. 30 - 36% from the values obtained for samples with the natural moisture content (Table 1). Saturation with water of the consolidated dump soil decreases a little its long term strength. The state of stress corresponding to the long term strength is being attained at smaller axial deformations (ca. 0.4 - 6.7%) than in the standard strength tests. The ratio of the total long term strength (q_T) to the standard one (q_t) averages 0.71 for the sand samples with the natural moisture content and 0.72 for their equivalents saturated with water. For samples of the varved silt the ratio is respectively 0.66 and 0.69, and for the mixture of the sand and lumps of the silt - 0.62 and 0.78.

3.3 Shear strength at internal water saturation of samples

The behaviour of the dump soil samples resulting from internal water saturation (the pore pressure increases) at the constant external load reveals more significant differences in various types of dump soils. An increase of the pore

Table 1. Parameters of standard and long term shear strength

Type of dump soil	Samples with natural moisture content (NW samples)					Samples saturated with water (SR samples)					
	σ_v MPa	c_u MPa	ϕ_u deg	c_T MPa	ϕ_T deg	c_u MPa	ϕ_u deg	c' MPa	ϕ' deg	c_T MPa	ϕ_T deg
sand P (100)	0.5	•	29.9	•	24.0	•	28.3	•	30.2	•	21.0
	1.0	•	30.0	•	•	•	27.7	•	31.7	•	•
	2.0	•	30.6	•	•	•	29.1	•	29.8	•	•
sand + varved silt PM (50+50)	0.5	0.030	30.5	0.021	23.0	0.022	23.4	0.022	29.6	0.017	20.0
	1.0	0.055	29.9	•	•	0.054	26.4	0.038	31.4	•	•
	2.0	0.070	30.5	•	•	0.053	28.3	0.032	31.8	•	•
varved silt M (100)	0.5	0.030	28.2	0.016	24.5	0.030	28.2	0.009	31.2	0.019	23.0
	1.0	0.055	27.9	•	•	0.070	25.2	0.042	27.8	•	•
	2.0	0.071	28.0	•	•	0.076	25.0	0.064	27.4	•	•
natural undis-turbed varved silt	0.5	0.107	11.7	•	•	•	•	•	•	•	•

c_u, ϕ_u, c', ϕ' - parameters of standard shear strength
c_T, ϕ_T - parameters of long term shear strength

pressure does not deform axially samples of the sand P (100), and of the mixtures PM (75+25) and PM (50+50) until the moment predecessing their failure. First deformations of only ca. 0.1 - 0.3% were noted at the pore pressure equal to ca. 90% of its critical value, causing a sudden failure and very big deformation of the samples. The density index for the sand samples P (100) amounted to 40 - 50%, that corresponds well with suggestions that liquefaction of sands takes usually place at the density index below 50% (Seed 1968). Figure 4 exemplifies a relationship between the vertical deformation and the pore pressure. In samples of the fragmented varved silt M (100) and of the sand/silt mixture PM (25+75), the first noticeable vertical deforma-tions originated at the pore pressure equal to ca. 76 - 80% of the pressure which resulted in big axial deformations (above 10%).

A further increase of the pore pressure in the soils was leading to a fast increment of the axial deformations but in none of the samples a sud-den failure took place. The presence of the varved silt lumps in the dump soil imparted it some cohesion and made difficult displacement of soil particles, preventing a sudden failure (liquefaction).

As it can be seen, the failure (liquefaction) of the soil resulting from an increasing pore pres-sure occurred not only in the sandy soil but also

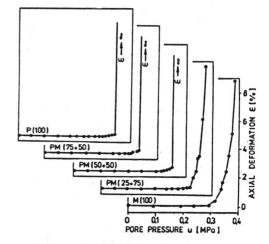

Fig. 4 Axial deformation versus pore pressure for samples consolidated at $\sigma_v = 0.5$ MPa

in its mixtures with lumps of the cohesive soil, making up to 50 wt. per cent. At such an amount of the varved silt lumps with diameters 2 - 15 mm (ca. 8 mm in average), their dispersion in the sand is not high enough to assure their con-tacts in the whole volume of the sample. As the earlier investigations revealed (Rybicki 1986), a mixture of sand with ca. 10 mm big lumps of the cohesive soil behaves as a cohesive soil. There-fore, the dump soil studied, containing only 25% wt. per cent of the silt lumps with diameters 2 - 15 mm, can be regarded as a non cohesive soil, that can be subject to liquefaction in the same way as the sand itself.

4 CONCLUSIONS

1. The dump soil represents a specific type of a soil considering its composing grains, lumps and fragments with different lithologies. They form at low loads a loose, unstable structure, that changes particularly at the increasing load or saturation. An adequate admixture of the sandy soil to the lumps of the cohesive soil affects properties of the whole mixture, but they also depend on diameters of lumps of the cohesive soil.

2. Curvilinear dependence of the standard shear strength on the consolidation stress (normal) holds true only in a relatively narrow range of loads between ca. 0.1 and 0.5 MPa. These values result from consistency of the soil lumps and a type of the soil. At bigger loads, the structure of the dump soil becomes more stable due to elimination of pores among lumps and - hence - is similar to a structure of the natural soil. The bulk density of such a dump soil also approximates the density of the natural soil.

3. An increase of the shear strength of the dump soil at a higher consolidation stress results mainly from the rising cohesion. However, despite attaining by the dump soil bulk densities similar to the natural soil, the cohesion of the former is lower and its angle of internal friction higher than the respective values of the latter. This can be attributed to disturbances of the structural bonding of dump soil particles (cohesion decreases) and to maintaining an original, lumpy structure of the consolidated dump soil in spite of elimination of pores among lumps (the angle of internal friction increases). A full homogenization of the structure of the compacted dump soil is reached probably after a longer time.

4. Increases of the standard and long term shear strengths of the dump soil resulting from its saturation with water can be limited to a significant degree by a preliminary compaction of the soil. In the studied dump soils compacted at loads of up to 2 MPa, such a saturation caused a fall not exceeding ca. 3 - 10% for the standard shear strength , and ca. 17 - 30% for the long term one . The smaller the preliminary consolidation load of the soil, the bigger decrease of the shear strength after saturation with water.

5. The standard and long term shear strengths of the studied dump soil increase with the growing consolidation load.

6. In dump soils, similar to natural soils, a method of loading affects parameters of the shear strength. In sands, the effective angle of the internal friction measured at an increasing pore pressure is smaller in comparison to the values obtained from the standard tests with an increasing axial stress, providing the same states of compaction.

7. A possibility that a structure will suddenly collapse (liquefy) as a result of a decrease of the pore pressure exists not only in the sandy dump soil but also in mixtures of sandy and cohesive soils, at contents of up to 50 wt. per cent of the latter.

REFERENCES

Fukuoka, N. 1957. Testing of gravely soils with large scale apparatus. *Proc. 4th ICSMFE*, vol. 1, London.

Rybicki, S. 1986. Structure and physico-mechanical properties of dump soils. *Proc. 5th Int. Congr. IAEG*, Buenos Aires.

Schulze, E. 1957. Large scale shear test. *Proc. 4th ICSMFE*, vol. 1, London.

Tan, T.K. 1966. Determination of the rheological parameters and the hardening coefficients of clays. *Rheol. Soil. Mech.* Berlin.

Seed, H.B. 1968. Landslides during earthquakes due to soil liquefaction. *Proc. I. SMFE, ASCE*, vol. 94, New York.

The engineering geological approach to purification of polluted rivers and protection of the surrounding groundwater conditions

Encadrement de la géologie de l'ingénieur sur la purification des cours d'eaux pollués et protection des conditions hydrogéologiques dans le voisinage

Yoshinori Tanaka
Toyo University, Japan

ABSTRACT: It is estimated that a polluted river is naturally purified to a certain extent through the process of the mutual exchange of water between the river and the groundwater. Based on this phenomenon, the engineering geological approach was examined to develop a method of purifying the polluted river by recirculating the groundwater via the infiltrated river water.

Through on-site investigation and laboratory testing, it was found that the absorption and decomposition directly under the river bed plays an important role in the reduction of total nitrogen.

In the case of the combination of leaking river water and the pumping of groundwater, it is desirable to plan to effectively use the infiltration process. It will be required in this case to reduce the disturbance of the groundwater environment accompanying the depression of water levels due to pumping, and to prevent the dissipation of polluted river water into the groundwater environment.

RÉSUMÉ: Il est estimé qu'un cours d'eau pollué est naturellement purifié dans une certaine mesure en passant le processus d'échange d'eau mutuel entre le cours d'eau et les eaux souterraines. Conformément à ce phénomène, l'approche de l'ingénierie géologique a été examinée en vue du développement de la méthode de purification du cours d'eau pollué par le retour des eaux souterraines dans le sens contraire au cours d'eau.

A la suite de l'enquête sur les lieux et l'expérience dans le laboratoire, il est certain que l'absorption et la décomposition directement sous le lit d'une rivière joue un rôle important pour la réduction de l'azote total. Dans le cas d'une fuite d'eau d'une rivière et d'un pompage des eaux souterraines, il est désirable d'utiliser efficacement le processus de l'infiltration. Il sera demandé, dans ce cas, de réduire la perturbation de l'environnement des eaux souterraines accompagnant la dépression du niveau de la rivière due au pompage, et de prévenir la dispersion des eaux polluées d'une rivière dans l'environnement des eaux souterraines.

1 INTRODUCTION

Frequently, rivers infiltrate underground in the midstream region and recharge groundwater. Most of the marginal zones of the principal lowlands of Japan consist of terraces, and in these areas through which rivers flow, the infiltration of river water is active and recharges the groundwater under the lowland plains. In the case of a polluted river, pollutants are brought underground and an area of pollution spreads along the river with the groundwater flow. In particular, it is to be noted that small rivers which have their sources located on the upper end of terraces are generally polluted with the development in the watershed. It is estimated that the pollution of groundwater via the infiltration of these polluted rivers must not

be disregarded when discussing the changes in groundwater conditions.

The soil layers have complicated functions purifying passing polluted water by mechanical, chemical and biochemical means and play a role in restricting the range of pollution. On the other hand, we may apply such functions for purifying a polluted river. It is expected that purified water, having been through soil layers, and then returned to the river, would be very useful in improving the water quality of the river. Simultaneously, this method would be effective to protect the mixing and the dissipation of pollutants into the groundwater.

The purpose of this paper is to mutually connect a series of problems; the pollution of the river originating from human activities, the pollution of groundwater caused by the

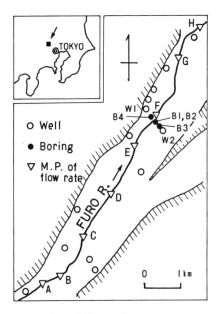

Fig.1 Location of Furo River

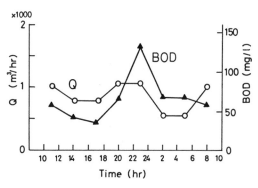

Fig.2 Hourly changes of flow rate and BOD in Furo River (after Saitama Pref. 1989)

leakage of river water, the purification of leaking water through soil layers and the purification of the river. For these purposes, a small river is taken and used as a sample.

2 HYDROGEOLOGICAL CONDITIONS OF THE SITE

Furo River, which flows on the western terrace of the Kanto District was chosen as the site of this study (Fig.1). This river is a small river in the suburban area of several cities. The flow rate and water quality of this river change over a day as shown in Fig.2. Both of these have higher values in the same

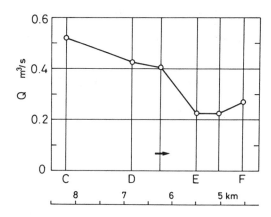

Fig.3 Change of flow rate along the Furo River

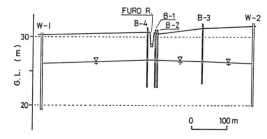

Fig.4 Profile of Furo River and its surroundings

hours, every morning and evening. Such a state of change shows obvious water pollution by the concentration of domestic waste water upstream. BOD value upstream shows that 78% of the flow rate originates from domestic waste water.

The terrace is underlain by sand and gravel and covered by a volcanic loam about 50cm thick. The permeability of this sand and gravel layer is 10^{-1}cm/sec in order in a shallower depth, but decreases to about 10^{-3}cm/sec with a clay mix where it is deeper.

The groundwater level is usually lower than the river bed level except in the case of an unusual rise when there is a typhoon. These hydrogeological conditions suggest the infiltration of river water into groundwater.

The flow rate of the river decreases in places downstream. One example of observation results in the flow rate shows an average flow loss of 0.1m³/sec/km downstream, which means a leakage of river water underground (Fig.3). Although the volume of leakage is different, depending on the location of the river bed and the season, it can affect the

2854

groundwater quality along the river.

3 METHODS OF INVESTIGATION

For the purpose of investigation of the hydrological and the qualitative relation between the river and groundwater, the observation line was set up by combining 4 boreholes and 2 wells in a cross direction to the river (See Fig.4. Refer Fig.1).

The hydrological investigation and the analysis of water quality were conducted for 3 years along this observation line. In the final year, thin tubes were set shallowly under the river bed and were used to extract the infiltrating river water. Via these tests, the data on the qualitative relation between the river and groundwater along the river was collected and the qualitative change of water over time was clarified.

These data were used to study a scale of qualitative effect of leaking river water on the surrounding groundwater, and the function of purification in soil layers on the process of infiltration and seepage.

4 INFLUENCE OF RIVER WATER ON THE GROUNDWATER QUALITY

On the observation line, boreholes B-1,B-2 and B-3 adjacent to the Furo River are different in water quality from borehole B-4, well W-1 and W-2 which are distant from the river. Among the long term qualitative observations (Temperature, pH, Electric conductivity, DO, Fluorescent matter and so on), the following matters are specially mentioned in comparison with the adjacent groundwater and distant groundwater.

(1) The temperatures of both groundwaters are different in a phase of seasonal change. The differences of temperature between the adjacent groundwater and the river are slight and show the same tendency of seasonal change. However the distant groundwater has a time lag of about 2 months on the measured yearly maximum temperature.

(2) The electric conductivity(EC) of river water is inversely correlated to the water level of the river. The EC of adjacent groundwater shows small fluctuations and does not have an obvious tendency of change (Fig.5). On the contrary, the EC of distant groundwater has a tendency to increase during the period of the rising of the rivers water level and because of that, it is roughly connected to the EC of the river with a reverse inter-relation. These results show that the distant groundwater is representative of the groundwater in the region which includes the observation line and the surrounding area, but the adjacent groundwater is affected by

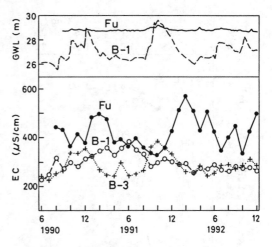

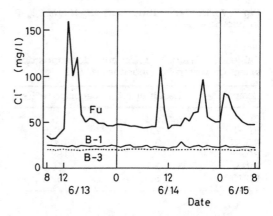

Fig.5 Seasonal variation of water level and EC
Fu : Furo River

Fig.6 Hourly change of Cl⁻ concentration

other factors adding to the groundwater conditions of the region.

(3) The concentration of fluorescent matter in adjacent groundwater is higher than the concentration in distant groundwater. Generally, fluorescent matter is contained in a cleanser, so its concentration is high in the river water.

(4) The quality of adjacent groundwater is more changeable with time than the quality of distant groundwater over a few days (Fig.6).

(5) The concentration of NH_4-N and PO_4-P are high in the boreholes B-1 and B-2 located near the river (See Fig.7). The concentrations of these components are higher in the river water and in groundwater of shallow depth directly under the water level. This condition shows that the concentrated water was supplied from above to the groundwater.

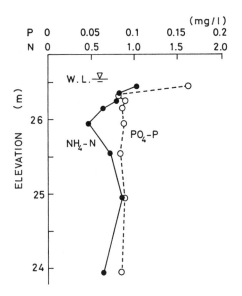

Fig.7 Vertical distribution of NH₄-N and PO₄-P in borehole B-1

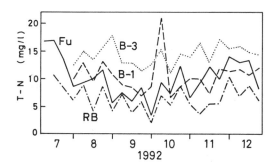

Fig.8 Weekly change of concentration of T-N
RB : River bed

Table 1. Residual ratio and decomposition effect on several components in B-1

Compo.	Residual R.	Decompo. effect
Cl^-	9.8%	0 %
NH_4-N	7.0	31.5
NO_2-N	0	100
PO_4-P	29.4	-188
EVAS	3.2	68.3
FLUOR.	6.9	32.4

It is considered that these characteristics of water quality in the site show the mixing of water leaked from the river into the adjacent groundwater.

5 PURIFICATION DURING INFILTRATION AND FLOW

If it is supposed that the difference of concentrations in borehole B-1 compared with the average water quality of the region originated in a mixing of river water and regional groundwater, the ratio of the qualitative difference in borehole B-1 to the concentration of river water can be calculated as a residual ratio on several components based on the long-term observation of water quality.

The left column of Table 1 shows the residual ratio. Usually, since Cl^- is regarded as an unabsorbed element, the value of residual ratio 9.8% of Cl^- means that the water leaked from the river was diluted with groundwater 10 times at the site of B-1. The values of the ratios of the other components are smaller than the value of Cl^-. The differences in values between each component and Cl^- show the effect of the decrease of each component mainly through absorption, oxidation, decomposition and so on underground, which does not include the effect of dissipation with flow. This effect is shown in the processes of both infiltration and groundwater flow.

The values given on the right column of Table 1 show that NO_2-N is mostly decreased and secondly, fluorescent matter and NH_4-N are in the same order. Anionic surfactant(EVAS) is relatively easily absorbed in soil and decomposed. NO_2-N and NH_4-N are joined to NO_3-N in the groundwater through the process of nitrification. The effect on P was uncertain. These confirm that the water quality is improved during the infiltration through the river bed and the groundwater flow, even though the effect of improvement through these processes is different depending on the components.

The above results include both processes of infiltration and groundwater flow as an action of purification. So, in order to observe the purification effect on the infiltration process, leaking water was directly taken at the depth of 1.5m under the river bed and analyzed on the same chemical components. In particular, the increase and decrease of nitrous components were noted.

Fig.8 shows that the concentration of total nitrogen(T-N) in leaking water in the infiltration process is 3.1mg/l lower on average than the concentration found in river water. This indicates the absorption of T-N or denitrification under the river bed.

On the other hand, the concentration of T-N in borehole B-1 near the river is almost the same or a little higher than the concentration of river water, except where there is an abnormal value, and is lower than the con-

2856

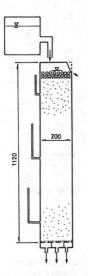

Fig.9 Apparatus of infiltration test

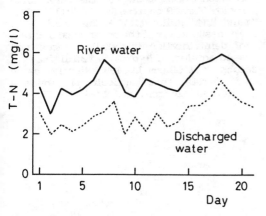

Fig.10 Concentration of T-N in the discharged water in the infiltration test

ground parts because its interstitial pores are filled with fine matter and slime, and is insufficient to effect the supply of gaseous oxygen.

The effective purification on the process of infiltration through soil layers under the river bed was also confirmed in the infiltration test, in which an apparatus as shown in Fig.9 was used. In the experiments, the river water flowed down in an unsaturated condition through the sand and gravel collected from the site. The discharged water from the lower end of the apparatus was smaller in value of electric conductivity and concentration of fluorescent matter, T-N and PO_4-P, than that of the river water (Fig.10). It is considered that the soil layer possessing the mixed condition of saturation and under-saturation is very effective in improving the quality of infiltrating water via its high physicochemical and biochemical activities.

6 PURIFICATION OF RIVER WATER AND PROTECTION OF GROUNDWATER CONDITIONS

Based on the results of the observation that the leaked river water is qualitatively improved through the process of infiltration from the river bed and groundwater flow, it becomes possible to investigate a method of purification of river water which uses the feed back of groundwater to the river following the infiltration and flow.

In the case of the combination of leaking river and pumping of groundwater, it is desirable to plan to use the infiltration process because the process gives a more efficient effect of purification than the groundwater flow process (See Fig.11). In that case, extra consideration will be needed in order to increase the number of reciprocal

centration in the regional groundwater as a background value. The effect of absorption and disintegration of T-N in borehole B-1 can be extracted by considering a mixing of river and regional groundwater. It was 3.7mg/l on average. This value shows little difference from the value directly under the river bed. It was found that the absorption and decomposition directly under the river bed plays an important function in the reduction of T-N.

The electric potential of oxidation and reduction of the leaking water is lower than the potential of river and groundwater. The gravel layer directly under the river bed possesses a condition in which denitrification is more likely to occur than in other under-

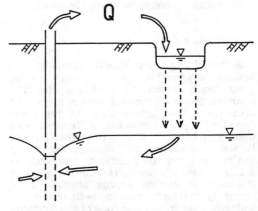

Fig.11 Schematic drawing of a purification method via the infiltration of river water

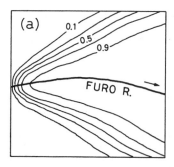 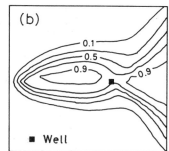 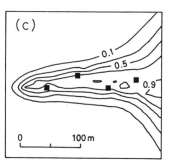

Fig.12 Distribution of pollutant compared with the number of pumping wells
(a) no well (b) 1 well (c) 4 wells

cycles of infiltration by setting a pumping well close to the river or to raise the capacity of treatment by increasing the water volume infiltrated from the vicinity of the river bed.

Although this method is fundamentally to return the water leaked from the river back to the river again by pumping the same volume of water, it is also required, when applying this method, to reduce the disturbance of the groundwater environment accompanying the depression of water levels, and, to prevent the dissipation of pollutant into the groundwater environment.

If this method is applied at the midstream site of the Furo River, it will be desirable to place several pumping wells dispersedly along the river as shown in Fig.12. The concentrated drawdown at the groundwater level and the dissipation of pollutant from the river to the groundwater can be minimized by increasing the number of pumping wells and reducing pumping volume per well.

In the case of no pumping well located at the site, the pollutant will dissipate far away from the river. However, the placing of pumping wells contributes in restricting the distribution of pollutant along the river.

7 CONCLUSIONS

The relation of water quality between the leaking polluted river and the surrounding groundwater depends remarkably upon the characteristics of the soil layers under the river bed. Soil layers play an important function in the formation of groundwater quality, so it is undesirable to disturb the soil layers located on the river conservancy works. Applying the purification method via soil layers for improving the water quality of the river is worthy of more studies. If we correctly distribute pumping wells at the site, we could make this method one of the environmental preservation plans which simultaneously improves both river and groundwater.

REFERENCES

Saitama Pref. 1989. Countermeasure against the pollution of Furo River, p.1–25.
Ronen,D. & Magaritz,M. 1988. Microscale haline convection -- A proposed mechanism for transport and mixing at the water table region. Water Resources Res. 24-7:1111–1117.
Trundell,M.R.,Gillham,R.W. & Cherry,J.A. 1986. An in-site study of the occurrence and rate of denitrification in a shallow unconfined sand aquifer. J. Hydrology 83:251–268.
Jacobs,T.C. & Gilliam,J.W. 1985. Riparian losses of nitrate from agricultural drainage waters. J. Environ. Qual. 14-4:472–478.

Engineering geology investigations on the behaviour of contamination of a new acrylate-based grout for making soil leakproof

Étude par la géologie de l'ingénieur du comportement à la contamination d'un coulis à base d'acrylates pour l'étanchéité des sols

H. Molek & J. Martens
Technical University, Darmstadt, Germany

ABSTRACT: A Ca- and Mg-acrylate-based grout used for making soils leakproof produced a tough-elastic gel. It contains no traces of acrylamide or acrylamide-derivatives. It can be reach a very low coefficient of permeability and increases strength. Further the behaviour of contamination of chemical grout injected into medium grain size sand were investigated through time under different storing conditions. More than 100 gel specimens were stored in H_2O, Na_2SO_4- and NaCl-solutions under constant conditions. Through time we took several water samples and analysed these for COD, pH, conductivity and several cations and anions. As a result considerable emissions were detected initially of the test. This should be eliminated in further tests by changing the grout composition.

RÉSUMÉ: Le moyen d'injection du sol examiné ici est basé sur des acrylates Ca et Mg et rend un gel visqueux-elastique, libre d'acrylamides et ses dérivates. Il est possible de pénétrer dans le sol du sable très fin. Le comportement du temps de ce moyen d'injection fut examiné sous des conditions de grisement différentes. Plus de 100 solides de gel restèrent dans l'eau destillée et des solutions de Na_2SO_4 et NaCl sous des conditions constantes. Périodiquement on prit des tests de l'eau selon la demande chimique en oxygène, le pH, la valeur de la conductibilité aussi bien que selon les cations et anions différents. Parfois, une contamination au début souvent importante a pu être constaté. On compte pouvoir l'éliminer dans des expériences ultérieures par un changement de la composition du produit.

1. INTRODUCTION

In the employment of a chemical grout for the sealing and solidification of the subgrade it is most important to check the enviromental impact. This complex of problems has gained in importance during the last few years and a lot of investigations have been done. For example on silicate gel injections (Müller-Kirchenbauer et al. 1985; Michalski et al. 1988), injections with acrylamids and ligno sulfates (Karol 1990).

In the following investigations a chemical grout based on monomers di- and tri-acrylates in water solution will be described. The grout is very good for sealing tasks especially for fine unconsolidated sediments.

The tests were performed on a highly uniform medium sand. The possible groundwater impact has been examined using boxes which do not pollute the test liquid. The used test liquids imitate natural ground water varieties in their chemical compositions.

The following results depend on the behaviour of contamination, permeability and strength of the grout. With later investigations the injections should be carried out in a stationary flow. Because the unharded grout comes in contact with the water the behaviour of contamination will be investigated.

2. GROUT CHEMISTRY

The tested chemical grout is composed basically of three components:
- Ca-/Mg-acrylates = base component
- 5 % $Na_2S_2O_8$-solution = hardener
- catalyst

The base component contents a mixture of unsaturated acrylderivatives dissolved in water. They could be polymerize with special redox systems. The reaction can be influenced by concentration and type kind of redox components. After the reaction we get polymer hydro gels. Hydro gels are specimens built from a polymer matrix in which water is stored homogenously without dissolving the polymer.

The test specimens are made from 445 g medium grain-size sand, 50 ml base component, 2 ml catalyst and 50 ml hardener.

3. ASSEMBLE OF THE INJECTED SAND

The sand was determined after DIN 18 196 to be well sorted (C_C = 1,2 ; U = 2,7) , medium sand. Grainsizes over 2 mm were removed to get a homogenous material.

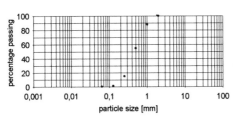

Fig. 1: Grainsize of the injected sand

The organogenic part of the sand has been investigated after DIN 4022 to 0.24 % and calcium was not demonstrable.

4. STRENGTH AND PERMEABILITY OF THE STABILIZED SAND

We performed four series of experiments. For each series we put two samples of stabilizied sand (5 cm in diameter, 10 cm in height and a volume of 196 cm^3) in a box (Volume = 1200 cm^3) made from HDPE. This procedure was performed twice. The samples were stored under folowing conditions:
- at air
- in 800 cm^3 pure water
- in 800 cm^3 Na_2SO_4-solution (200 mg/l Sodium sulfate purum p.a.)
- in 800 cm^3 of a NaCl-solution (5000 mg/l Sodium chloride extra pure).

The strength of stabilized sand has been analysed with an unconfined compression test after a storing time of 3 hours and 1, 6, 14 and 28 days. During the first 28 days of the investigation we reached in all four series of experiments an unconfined compressive strength between 5 and 6 N/mm^2. A dependence on storing time and storing conditions could not be shown during the first 28 days.

The permeability of the stabilizied sand has been ascertained using a permeameter after DIN 18130. We get a coefficient of permeability lower than 10^{-10} [cm/s]. The porosity of the wet test speciments range about 5 %. The porosity depends on the water content, because the gel is shrinking by drying and swelling by absorbing water. This process is fully reversible. Further investigations will detect the interdependence of interstices and permeability.

5. BEHAVIOR OF CONTAMINATION

After a storing time of 3 hours and 1, 6, 14 and 28 days we took several water samples. The following chemical parameters were examined:
- pH
- conductivity
- total hardness
- methyl-orange-alcalinity
- chemical oxygen demand (COD)
- sodium (Na^+)
- magnesium (Mg^{++})
- calcium (Ca^{++})
- potassium (K^+)
- sulfate (SO_4^{--})
- chloride (Cl^-)

5.1. pH-value

The pH-value has been fixed at all samples at 6.4 and there was no change during the investigation time of 28 days.

5.2. Conductivity

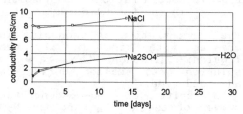

Fig. 2: Conductivity of the test liquids

After 28 days the sum of the soluted ions is constant for the Na_2SO_4- and H_2O-solutions. The curve of the NaCl-solutions is on a higher level because of the primary NaCl-concentration of 5 g/l. Conspicuous is the decrease of the conductivity after 3 days storing in NaCl-solution. After that the curve increases again. This depends on ions that build in the sample specimens (= reduction of conductivity). For this see the large decrease in Cl^--concentration in the NaCl-solution. The following increase of the curve could be explain by ions wich are soluted out of the test specimens (= increase of conductivity). See the

curves of concentration of Mg^+, Ca^{++}, Na^+ and K^+.

5.3. Total hardness

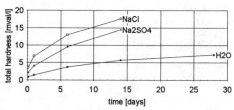

Fig. 3: Total hardness of the test liquids

Generally the total hardness increased during the investigation time. The increase was smaller by storing the test specimens in pure water than in the other two test liquids. The biggest values could be determined in the NaCl-solutions. This depends on the high Ca- and Mg-values (see fig. 6, 7).
After 28 days the total hardness did in all test liquids further increase.

5.4. Methyl-orange-alcalinity

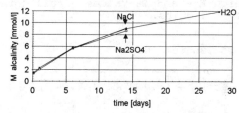

Fig.4: Methyl-orange-alcalinity of the test liquids

There is no difference between the three test liquids in M-alcalinity during the investigation time. The M-alcalinity increase and the inclination of the curves get flatter with passing time. This depends on the buffering of the hydronium-ions which increase with a longer storing time although the accumulation rate gets smaller in time. This depends on the balance between free available weak acids or alcalids of the test specimen and the storing liquid. The diffusion effects get smaller during the advanced time because of the balance in concentration.

5.5.Chemical oxygen demand (COD)

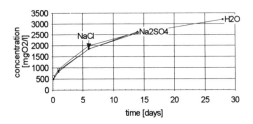

Fig. 5: Chemical oxygen demand of
the test liquids

There is no important difference
between the three test liquids in
COD during the investigation time.
The emission of organic components
was not stopped after 28 days.
However the rate of solution gets
smaller with advancing time.
Finally the emission of organic
components reached high values.
These organic components mainly
consist of Mg- and Ca-acrylate, not
connected with the hardener. An
indication of this is given by the
high Mg- and Ca-values of all three
test liquids.

5.6. Sodium (Na)

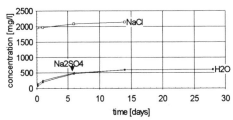

Fig.6: Concentration of sodium in
the test liquids

The concentration of sodium increases with time in all test
liquids. As a result of the higher
content of sodium we get bigger
values for the NaCl-solution. Be-

cause of this smaller concentration
difference it is impossible to
dissolve in the same time the same
rate of sodium as in pure water or
in the low concentrating Na_2SO_4-
solution. The increase of the concentration curve of the sodium
solution must be lower than that of
the other test liquids.

In spite of a higher primary concentration of sodium in the Na_2SO_4-
solution the concentration curve
has after 10 days the same values
as the curve from the pure water.
After 28 days we only saw a small
increase of sodium concentration.

5.7. Magnesium (Mg)

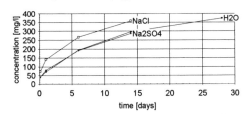

Fig. 7: Concentration of magnesium
in the test liquids

The concentration of magnesium increases with time in all test
liquids, although the rate of
solution gets smaller during the
advanced time.

In spite of Mg-concentration
curves of the Na_2SO_4- and the H_2O-
containing test liquids, which have
nearly the same values, the NaCl-
containing test liquid has about 60
mg/l higher values. The magnesium
belongs to the Mg-arylates of the
grout. One part belongs to Mg-
arylates that had not reacted with
the hardener and so were dissolved
in the test liquid. This has been
demonstrated in the high COD-
values.

Also there were reactions between
the grout and the sodium chloride
of the NaCl-solution. This has been
documented in higher Mg-values and
in the decrease of the chloride
values in the NaCl-solution during
advancing time.

5.8. Calcium (Ca)

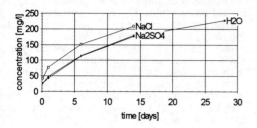

Fig. 8: Concentration of calcium in
the test liquids

The concentration of calcium
increases with time in all test
liquids, although the rate of
solution gets smaller with
advancing time. Like the concentration of Mg^{++} in the NaCl-
solution the calcium values are
much higher (about 70 mg/l) in
spite of the Ca^{++}-concentration in
the Na_2SO_4-solutions and pure
water.

The increase of the Ca^{++}-values
has not endet after the storing
time of 28 days. Because the grout
contains mostly Ca- and Mg-
acrylates dissolved in water and
other low concentrated additives,
the reasons for the emission of
Ca^{++} seem to be the same as that of
the Mg^{++}-concentrations.

5.9. Potassium (K)

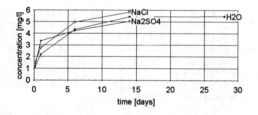

Fig. 9: Concentration of potassium
in the test liquids

In all test series the K^+-values
increase. After 28 days the
increase of K^+-concentration is at
a very low level. The emission of
potassium is very low in the three
test liquids. There is no important
difference between the three test
series. The source of the potassium
contamination will be determined in
later investigations. The source of
the potassium could be the sand
itself or small contaminants in the
grout components.

5.10. Sulfate (SO_4)

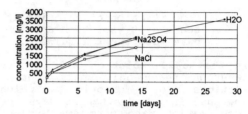

Fig. 10: Concentration of sulfate
in the test liquids

The curves of the sulfate
concentration increase during the
first 28 days and longer. The
accumulation rate gets smaller in
time.

The sulfate curve for the Na_2SO_4-
solution starts with a little bit
higher concentration than in pure
water because it contains sulfate.
But after 10 days the sulfate
values get smaller than that in the
pure water. How this continues we
could not say and this will be
shown in further investigations.

In spite of this the shape of the
sulfate concentration curve in the
NaCl-solution differs. The values
are visible at a lower level than
the values of the other two test
liquids. Therefore the distance
gets always greater in time.

The reason for the emission of
sulfate ions is not completely
reacted hardener (5 % solution of
$Na_2S_2O_8$ in water). In a NaCl-
solution it is not possible to
dissolve so much sulfate as in non-
chloride containing solutions,
because Cl^- is more reactive than
SO_4^{--}. So the sulfate concentration
curve of the NaCl-solution must be
on a lower level than in the other
two test liquids.

5.11. Chloride (Cl⁻)

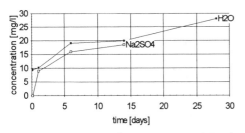

Fig. 11: Concentration of chloride in the Na_2SO_4-solution and pure water test liquids

In the Na_2SO_4-solution and in pure water we reach only small concentrations of chloride, which increase in time. After 28 days the acculumation rate gets smaller but had not reached 0.

In the Na_2SO_4-solution we get smaller values than in the other two test liquids. This depends on the higher concentration of anions (SO_4^{--}) and so a lower rate of chloride is in solution.

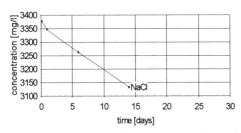

Fig. 12: Concentration of chloride in the NaCl-solution test liquid

In spite of the chloride concentration in pure water and Na_2SO_4-solution the chloride concentration in the NaCl-solution decreases linearly in time. This can explain the Ca- and Mg-acrylates of the grout, which have about 30% of free valances. Here it is possible to build in chloride anions and water get out.

6. CONCLUSIONS

We have investigated a chemical grout based on acrylate which has very good sealing behaviour. It is possible to penetrate soils with a grain size down to fine sand.

The grout contains no acrylamide and acrylamide derivatives.

The gel is shrinking by drying and swelling by absorbing water. This process is fully reversible.

In the contamination tests we detect emission of Ca- and Mg-acrylates. This was shown in the increase of the Mg^{++}-, Ca^{++}-ions and COD values.

The emission of sodium has been stabilized after 14 days storing time. The sodium comes from non-reacted hardener ($Na_2S_2O_8$). Equally the big increase of the sulfate values depends on non-reacted hardener.

Chloride emits in low concentrations in pure water and Na_2SO_4-solutions. In spite of this we reach in the NaCl-solution a decrease of Cl^--concentrations. In this case chloride is added on the Ca- and Mg-acrylates.

The investigation of the behaviour of permeability, strength and contamination with longer time will be continued. This will be nessessary for using this chemical grout in practice. To eliminate the harmful emissions it is necessary to optimize the composition of the grout.

7. REFERENCES

CLARKE, W.J. 1982. Performance characteristics of acrylate polymer grout. In ASCE/AIME (eds.), Proceedings of the Conference "Grouting in Geotechnical Engineering", 10.-12. Februar 1982 in New Orleans, p.418-432. New York, ASCE.

DIN 4022 (1987): Bennen und Beschreiben von Boden und Fels; Normenausschuß Bauwesen im DIN Deutsches Institut für Normung e.V.. Berlin: Beuth-Verlag.

DIN 18 130 (1988): Bestimmung des Wasserdurchlässigkeitsbeiwertes; Normenausschuß Bauwesen im DIN Deutsches Institut für Normung e.V.. Berlin: Beuth-Verlag.

DIN 18 196 (1988): Bodenklassifikation für bautechnische Zwekke. Normenausschuß Bauwesen im DIN Deutsches Institut für Normung e.V.. Berlin: Beuth-Verlag.

Karol, R.H. 1990. Chemical grouting. New York, Basel: Marcel Decker Inc.

Michalski, D. ; Lange, W.; Metzke, G. 1988. Löse und Sorptionsvorgänge bei der Um- und Druchströmung chemisch verfestigter Körper. In Freiberger Forschungshefte A 771:75-85. Leipzig.

Müller-Kirchenbauer, H.; Borchert, K.-H.; Friedrich, W. 1985. Veränderung der Grundwasserbeschaffenheit durch Silicatgelinjektionen. In Die Bautechnik 4, p. 130-142. Berlin, Ernst und Sohn.

Skeist, I. 1977. Handbook of Adhesives. New York: Van Nostrand Reinhold Company.

Model experiment on the aptitude of the clays from Wimpsfeld (Westerwald, Germany) as mineralic pollutant barriers – An experiment in the field

Simulation expérimentale in situ de l'aptitude des argiles de Wimpsfeld (Westerwald, Allemagne) comme barrières minérales contre les polluants

B. Carson, H. Molek & E. Backhaus
Technical University, Darmstadt, Germany

ABSTRACT: A clay from the Westerwald clay pits Wimpsfeld (Westerwald, Germany) was tested on its reactivity with various synthetic seepage-solutions.

A field experiment was composed, in which testbasins were built according to the common techniques for waste deposit construction and filled with synthetic seepage-solutions. With porous cups, which were arranged to screen the entire depth of the testbasins, the temporal change of the chemical composition of the clays subjected to seepage treatment was traced considering the ions Na^+, K^+, Ca^{++}, Mg^{++}, Al^{+++}, HCO_3^- and SO_4^{--}.

The progress of the percolation front in the clay can be determined by the temporal change of the element concentrations in the water samples from the porous cups.

There are indications for the dependance of the temporal order of mobilized elements and the velocity of percolation on the chemical character of the seepage solution.

Analogue clay samples are tested on the merging of selected petrophysical characteristics, which are proposed in the following, such as the consistency limits d'Atterberg throughout the duration of the experiment.

Conclusions are drawn concerning the chemical stability and its influence on the petrophysical characteristics of the clays, respectively the permeability and deformability in waste deposits.

RESUME: La réactivité de l'argile à des différents types d'eaux d'infiltration synthétiques est étudiée à l'exemple d'une argile provenant des glaisières de Wimpsfeld (Westerwald, Allemagne)

Dans un essai in situ, des bassins-tests étaient installés suivant les procédés habituels pour des constructions de dépotoirs, et remplis avec les eaux d'infiltration synthétiques.

A l'aide d'un réseau de lysimètres a bougie céramique le changement successif de la composition chimique des argiles sous l'influence d'infiltration est suivie par l'analyse les ions Na^+, K^+, Ca^{++}, Mg^{++}, Al^{+++}, HCO_3^-, Cl^-, NO_3^- et SO_4^{--}.

L'avancement du front d'infiltration dans la couche d'argile peut être déduit par les changements successifs des concentrations d'ions dans les échantillons d'eau prélevés.

Cette étude montre qu'il y a des rapports entre le type d'eau d'infiltration et la succession des éléments en solution, ainsi que la vitesse d'infiltration.

En plus, des échantillons d'argile sont prélevés et les changements de leurs caractéristiques physiques sont étudiés (granulométrie, limites d'Atterberg, capacité d'accumulation d'eau, contenu en carbonates, capacité d'échange de cations).

Des conclusions sont faites sur la résistance chimique des argiles et son effet sur les propriétés physiques, surtout par rapport à la perméabilité et la stabilité.

1 GENERAL INFORMATION

Clay is an important part of waste deposits. Through its quality as an aquiclude, its ability of adsorption and retardation and its plasticity it satisfies most technical demands of a waste deposit barrier. The workshop GDA (Geotechnik der Deponien und Altlasten) has defined the topical demands of waste deposit barriers. Yet there is still a lack of long-time investigations and field projects on the stability of clay under long-time chemical stress through aggressive substances.

In a field experiment the waste deposit situation is simulated: three testbasins were built and mantled with a clay lining according to the technical demands for waste deposits. The lined basins were filled with three chemically different synthetic seepage solutions.

After 17 months the first comprehensive investigations took place. The investigations intended to define the petrophysical, chemical and mineralogic effects of seepage solutions on the tested clay.

The following essay proposes the results concerning the selected parameters liquid limit w_L and plastic limit w_P and the resulting plasticity index I_P in context with the chemical composition of the pore water.

2 GEOLOGIC AND ENGINEERING CHARACTERISTICS

The tested clay comes from the clay pits Wimpsfeld near Mengerskirchen (Westerwald, Germany). The pits are run by the Schmidt KG company from Dornburg, Langendernbach (Westerwald, Germany). The clay was sedimended in Tertiary age (Eocene-Oligocene) as an erosion product of tropically weathered Devonian slates. Among the outcropping clay layers one clay named Wi 310, was chosen for the project.

The Wi 310 is a plastic, silty kaolinitic clay (Fig.1, Tab. 1). It's color is light brownish-gray with red streaks of varying intensity. About 1% of the clay consists of small red slaty pieces of up to 5 mm grain size. This material outcrops south of the clay pits.

The mineralogic composition of the material was determined by X-ray diffraction with the particles 2μ. The clay consists of 80% Kaolinite, 10% Illite and 10% mixed layer-clay of the Illite-Smectite-group

The grain size distribution was determined on a sedigraph 5100.

The grain size fractions of Wi 310 are: sand (3,7 . . . 6,3)%, silt (29,0 . . . 33,5)% and clay (48,0 . . . 63,0)%

According to its engineering properties the Wi 310 is a lightly plastic, silty, slightly sandy clay, TL after the plasticity diagram by CASAGRANDE.

3 TEST METHODS

3.1 Construction and arrangement

In a field project the Wi 310 is tested on its reactivity and stability in contact with three different synthetic seepage solutions in three testbasins.

The test basins have the shape of an inverted four-sided pyramid of approximately 90 m^2 basis area which are of app. 10 m length, 9 m width and 1,65 m depth (Fig. 2).

In order to achieve an enclosed system, and to

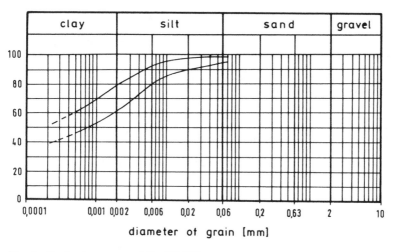

Fig. 1: Grain size range of the Wi 310

Tab. 1: The petrophysical characteristics of the Wi 310

Parameter		Value	Test Method
w_{nat}	natural water content	5 – 47%	DIN 18 121, 1
ϱ_S	grain density	2,57 g/cm^3	DIN 18 124
w_L	liquid limit	ca. 48%	DIN 18 122
w_P	plastic limit	ca. 25%	"
w_S	shrinkage limit	18 – 20%	"
I_P	plasticity index	19–21%	"
w_A	storage capacity	47,5%	ENSLIN & NEFF
w_{Pr}	proctor compaction	18,9%	DIN 18 127
ϱ_{Pr}	proctor density	1,84 g/cm^3	"

protect the environment the basins were lined and covered with plastic foil.

The Wi 310 was brought into the basins with the proctor water content of 19%. In each basin 7 layers were constructed, each layer is 10 cm thick. The basic layer is 15 cm thick. The thickness of the entire lining is 75 cm. In order to avoid the detouring of seepage solution around the clay lining the walls of the basins were also lined with clay.

Into the six top layers of each basin a set of 4 porous caps was integrated. This was accomplished by cutting a 5-cm-deep slit into the compacted clay. Porous cups and tubes were inserted, the slits filled with the original material and compacted again. The tubes were lain to the basin margin parallel to the layers.

The four porous cups of one set lie within 30 cm distance from one another. The six sets of porous cups lie step-wise 10 cm vertical and 50 cm horizontal distance, in order to avoid mutual influences (Fig. 3). The porous cups were connected with sample vessels and a suction pump by a tube system.

Analogue to the pore waters also the clay was sampled. The undisturbed samples were drawn by dynamic penetration with a sampler two-inch in diameter. The samplings were located with 50 cm distance in order to avoid influences by seepage and detouring of the test solution through former sampling cavities.

Pore water and the test solutions were sampled monthly, the clay was sampled once every three months.

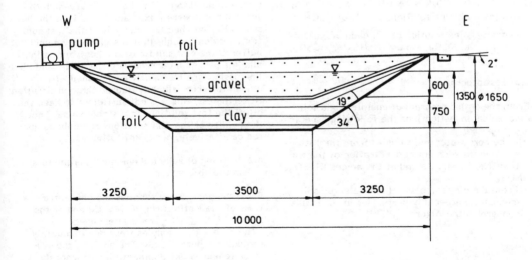

Fig. 2: Cross-section through a test basin

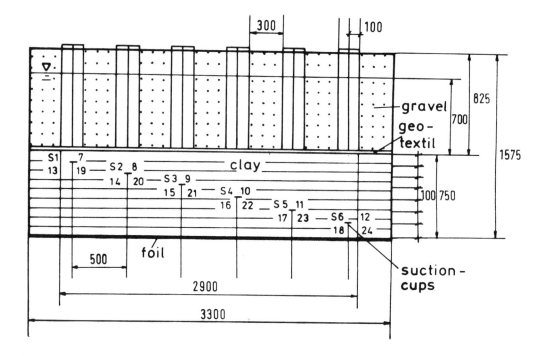

Fig. 3: Arrangement of the porous cups (S1 ... S24)

3.2 The test solutions

The three test basins were filled with three different synthetic solutions:

- acid: Mixture of three acids, pH3, 1/3 each of 1N HC1, 1N HNO$_3$, 0,5 N H$_2$SO$_4$
- alkaline: alkaline solution, pH 11, of NaOH
- neutral: saline solution of 1% each of Kations Na, K, Ca, Mg as: NaHCO$_3$, MGSO$_4$, KCl, NH$_4$Cl

3.3 Test programm

From the entire program of chemical and soil mechanical investigations the following aspects are proposed:
- In the pore water and samples from the test solution the element concentration of the Kations Na, K, Mg, Ca and of the anions Cl, SO$_4$, NO$_3$.
- From the clay samples pH and selected soil mechanics, especially properties in the results concerning the Atterberg limits.

4 ASSOCIATIONS

4.1 Reactions of the test solution

The element concentrations in the test solutions were intended to buffer the water of which the test solutions were mixed and to achieve the desired pH. The chemical analysis of the test solutions showed though, that the clay very quickly buffered the test solutions to a neutral or slightly alkaline pH. As a consequence the acidic and alkaline basin were frequently replenished with an amount of acids/alkalics figured through titration. Nevertheless, this was not sufficient to keep the pH of the test solutions at a stable value. Thus the acidity of the test solutions can only be quoted qualitatively, not quantitatively.

4.2 Merging of selected chemical parameters in the pore water

From the pore water samples the concentrations of the Kations Na, K, Mg, Ca and of the anions Cl, SO$_4$, NO$_3$ were determined.

The pore water samples were drawn monthly through the porous cups. The analysis shows a clear temporal development. Simultaneously the transport of material to the depth of the clay lining is clearly shown. The infiltration front can be traced by the different temporal deve-

2870

lopment in the porous cups distributed over six depths.

In the alkaline test basin there was an excessive supply of Na, which was easily mobilised. For Na a clear rise of the concentration could be found in all depths, while the concentration of K, Mg, Ca remained stable. The concentration of Mg and Ca even declined slightly, which indicates a fixation of these elements. The concentration of anions Ci, SO$_4$, NO$_3$ also remains stable in the alkaline test basin.

In the neutral saline test solution all alkaline kations and anions were present in excessive supply. Here for all analysed kations a time-dependant incline of the concentration is clear. The beginning of the incline is dependant on the natural mobility of the element. Fig. 4 shows by example of the results from 10 cm depth, how from the 5th month on successsively the concen-

trations of Na, K, Mg, Ca increase. Analogue the concentrations of Ci, SO$_4$, NO$_3$ rise.

In the acidic test solution the anions Ci, SO$_4$, NO$_3$ were present in excessive supply, while the basic kations were derived from solution out of the clay and are mobilised by excessive supply of H$^+$-ions. The temporal development showed the arrival of the analysed kations nearly simultaneously after 10 months, with exception of the extremely mobile Na and Ci. It is surprising that the anions derived from the test solution and the kations dissolved from the clay arrived in the depth at the same time.

In the acidic test basin the development of a depth profile can be traced (Fig. 5) by example of the incline of the Ca-concentration dependant on the depth in the profile. After 17 month the migration front had reached 20 cm depth, while in the layers below 30 cm the Ca-concen-

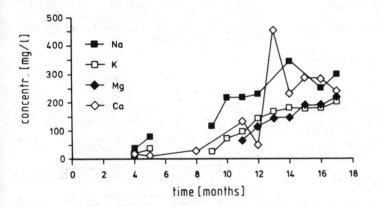

Fig. 4: Successive Mobilisation of Na, K, Mg, Ca in 10 cm depth, saline test basin. The sharp peaks of the Ca-Concentration in the 12th and 13th month are verified

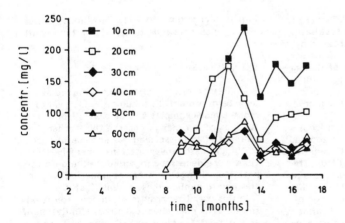

Fig. 5: Increasement of the Ca-concentration dependant on depth, acidic test basin

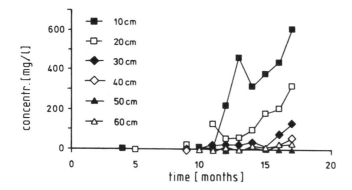

Fig. 6: Proceeding seepage of NO₃ acidic test basin

tration still lie at the normal level of app. 50 mg/l.

NO₃ migrates toward the depth in the acidic basin (Fig. 6). After 17 months the migration front had reached a depth of 40 cm, the concentration of NO₃ was still rising strongly in the overlying layers.

Generally from the changes of the chemical composition of the pore water in the clay it can be concluded, that in the alkaline milieu no mobilisation or material transport can be determined. In the neutral-saline and in the acidic test basin exchange processes as well as element migration take place.

4.3 Merging of the engineering properties

From the clay lining samples were taken once every three months from 6 depths. From these samples w_L and w_P were determined.

Dependant of the type of test solutions here also a definit temporal development can be seen. The intensity of the alteration decreases with increasing depth.

The most obvious changes of the Atterberg limits took place in the neutral saline test basin (Fig. 7). W_L after 17 months has decreased by 8% water content, calculated to be 17% of the normal value. W_P has decreased by 21% of the normal value. Thus results the decreasing of the plasticity index by 4% of the normal value.

In the acidic and the alkaline test basin no clear changes can be determined.

In Fig. 8 all data of the layers 0-2 cm from all three test basins are positioned in a zoomed section of the plasticity diagram by Casagrande. The data form groups which clearly differ. The alkaline milieu causes a merging toward less plastic clay within the range of the TL. The acidic milieu causes the merging toward the range of middle plastic clays, TM. In the neutral saline test basin the data merge toward the silty range.

The shifting toward less plasticity in the alkaline and saline milieu is a clear indication for the beginning hardening of the material caused by the influence of seepage solutions.

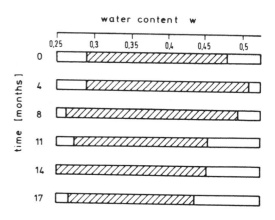

Fig. 7: Merging of the Atterberg limits in the neutral saline test basin, 0-2 cm depth

5 CONCLUSIONS

The discussed field project was designed as a long-term project. Exceeding the interval of 17 months more samples will be drawn from the test basins in order to observe the further development of the described phenomena. In addition to the discussed investigations other investigations were necessary to complete the determination of changes of other engineering properties and to explain these scientifically.

In this way first results about the alteration of the Wi 310 under chemical stress (similation of acidic, neutral-saline and alkaline waste seepage),

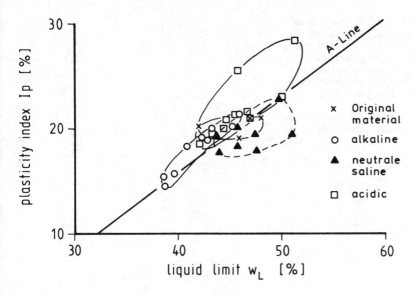

Fig. 8: Change of the soil mechanical properties under different chemical influences in the plasticity diagram by CASAGRANDE, borrow depth 0-2 cm

are discussed.

For the tested clay Wi 310 similar investigations had not yet been made, especially concerning the aptitude of the material as a mineralic waste deposit lining.

The parameters liquid limit w_L, plastic limit w_P and plasticity index I_P were chosen to define the changes as these are widely spread arguments in construction practice.

After a testing period of 17 months clear changes in the chemical composition of the pore waters were found. The alterations had reached depths of 20 . . . 30 cm in the clay lining. Alterations of the investigated soil mechanic properties could also be found in these depths.

Comparing Fig. 7 with Fig.s 4, 5 and 6 it can be stated, that the merging of w_L and w_P happen simultaneously with the increase of the element-concentrations in the pore water. It can be concluded, that massive exchange processes cause the changes of the engineering properties.

This confirms among many other investigations by USTRICH (1989), who observed the mobilisation of carbonatically bound kations after treatment with various test solutions. HASENPATT (1988) describes the formation of aggregates and the decrease of the clay fraction. Perhaps this can be an explanation for the loss of plasticity in the alkaline and saline milieu mentioned above. REUTER and MESECK (1986) observed changes of the swelling capacity of bentonites in contact with various test solutions. Here also the greatest changes were stated in the saline test solutions compared to the alkaline test solutions. KOMODROMOS and GÖTTNER (1986, 1988) found a hardening of clays in contact with

test solutions which was reversible in water. SCHNEIDER (1992) found a loss of material out of carbonatic clays up to 10%, when treated with acids. ECHLE et al. (1988) proved the decreasing of the Atterberg limits in clay from an authentic waste deposit lining run 8 years.

In the numerous publishments about the changes in mineralic clay barriers influenced by seepage all authors concordingly with the here discussed investigations conclude that clays are not at all stable in waste deposits and can only delay the extravasion of pollutants from waste deposits.

6 LITERATURE

Echle, W., Cevrim, M. & Düllmann, H. (1988): Tonmineralogische, chemische und bodenphysikalische Veränderungen in einer Ton-Versuchsfläche an der Basis der Deponie Geldern-Pont.- Schr. Angew. Geol. Karlsruhe 4, 99-121.

Hasenpatt, R. (1988): Bodenmechanische Veränderungen reiner Tone durch Adsorption chemicher Verbindungen.- Mitt. IGB Zürich, 134, 146 S.

Komodromos, A. & Göttner, J. (1986): Beeinflussung von Tonen durch Chemikalien, Teil 1: Durchlässigkeit.- Müll und Abfall, Berlin, 3, 102-107

Komodromos, A. & Göttner, J. (1988): Beeinflussung von Tonen durch Chemikalien, Teil 2: Gefüge- und Festigkeitsuntersuchungen.- Müll und Abfall, Berlin, 12, 552-562

Reuter, E. & Meseck, H. (1986): Die Beeinflussung der Quelleigenschaften handelsüblicher

Bentonite durch chemische Lösungen.- Wasser und Boden, Hamburg, 11, 563-568

Schneider, L. (1992): Der Einfluß von Prüfflüssig-
keiten (organische Säuren und chlorierte
Kohlenwasserstoffe) auf Deponiebasisabdich-
tungen (kalkhaltige mineralische Basisabdich-
tungen, Kombinationsdichtungen).- Diss. RWTH
Aachen, 310 S.

Ustrich, E. (1988): Tonminerale und ihre Wirksam-
keit in natürlichen und technischen Schadstoff-
barrieren.- Schr. Angew. Geol., 4, I-VII, 1-321,
Karlsruhe

Hydrogeology and water balance analysis in the underground dam region, Japan – Securing of subterranean water with underground dam

Analyse hydrogéologique et bilan hydrique dans la région de barrages souterrains, Japon

Toru Kuwahara, Kazuto Namiki & Kunioki Hirama
Technical Research Institute, Obayashi Corporation, Tokyo, Japan

Masatoshi Kushima
Civil Engineering Technical Division, Obayashi Corporation, Tokyo, Japan

Abstract : Underground dams are installed with cut-off walls to control ground water and develop new ground water. The new ground water is good in quality and can be utilized effectively, while ground water salinitization can be prevented near coastal region. This paper introduces the today's Japanese underground dams and relationship between civil engineering technology and hydrogeology. Especially, case studies of water balance analyses are shown in each step of underground dam construction such as planning, surveying, design, construction, and maintenance management.

Résumé : Les barrages souterrains sont dotés de murs parafouilles servant à contrôler l'écoulement des eaux souterraines et à rechercher de nouvelles eaux souterraines. Ces nouvelles eaux souterraines sont de bonne qualité et peuvent donc être utilisées efficacement pour les barrages souterrains. La salination de ces eaux à proximité du littoral peut également etre évitée. Le présent mémoire présente les barrages souterrains japonais actuels ainsi que la relation entre les technologies de génie civil et l'hydrogéologie. En particulier, des études par cas ou des analyses de bilan d'eau sont présentées pour chaque étape de la construction du barrage, à savoir planification, étude, conception, construction et gestion de la maintenance.

1 Outline of the Underground Dam

1.1 *Concept and effectiveness of underground dam*

The Underground dam is for the purpose of newly developing ground water by manually controlling the flow of ground water by construction a cut-off wall within the ground. Hydrogeological effectiveness of the underground dam can be summarized as follows.
(1) If the ground water flowing to lower reaches is shielded by cut-off walls, then the ground water table at the upstream side of the cut-off wall rises. As a result, it becomes possible to store the ground water in the boid in the ground at the upstream side and thus new good-quality water can be utilized (Fig.1, A ; storage type).
(2) Near the seashore, the reverse flow of seawater during pumping can be prevented by the cut-off wall. As a result, ground water can be stored at the upstream side of the cut-off wall and at the same time the salinitization of ground water can be prevented (Fig.1, B ; anti-salinitization type).
(3) Ground water table at the upstream side of the underground dam can be controlled by *1)* leaving required permeable layers between the ground

surface and the top of the cut-off wall and also by *2)* leaving permeable layers from the impermeable basement to the bottom of cut-off wall. By applying such an optimum design, it becomes possible to prevent the turning into marsh at the upstream side and the drying up of ground water at the downstream side. Also in the arid region, salt accumulation in the surface layer of the ground can be prevented.

Especially, as the water source development method in arid region, the underground dam has the following advantages;
(1) Utilization of surface water can be classified as follows: *(a)* above ground dam, *(b)* direct intank from river water, *(c)* direct intank from lakes and marshes. Compared to the underground dam, the surface water is creating some problems such as water rights *(a,b)* for relevant countries in the case of international rivers, a great amount of evaporation and accumulation of salinity*(a,c)*, worsening of water quality *(c)*, and great influence to environment *(a)*.
(2) Conventional forms of use of ground water can be divided into sub-surface water by shallow wells *(d)* and deep ground water by deep wells *(e)*. In this case, there are some problems such as drying up and worsening of water quality in shallow wells in

dry season *(d)*, and drying up of finite fossil water and a high salinity characteristics in deep ground water *(e)*.

(3) On the other hand in the underground dam, the rain water as circulating water is the main supply source, and surface water discharged without being used and river bed water is stored in the underground dam so that problems are less and minor compared to other water source developing methods.

1.2 *Steps of underground dam construction*

The step of underground dam construction comprises planning, surveying, design, construction, and maintenance (Fig.2). In the planning stage, the basic plan is made such as selecting construction site and determining the quantity of water to be developed. In the survey stage, various hydrogeology and ground water are surveyed and tested and research for environmental assessment is performed based on hydraulic, hydrological and geological data. In the design stage, water balance in the relevant area is analyzed, location of cut-off walls and cut-off wall method are selected, and the specifications of pumping wells (location, number of wells, aperture and quantity of pumping) are determined. After the construction of the cut-off wall, various survey, confirmation and verification are performed and the maintenance criteria are prepared in the maintenance stage.

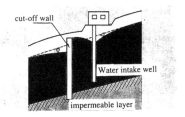

(A) Storage type underground dam in inland areas etc.

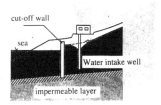

(B) Anti-salinitification type underground dam near seashore

Fig.1 Effectiveness of underground dam

1.3 *Japanese underground dam*

The concept of the underground dam is relatively old in Japan to be 1940's. As of February 1994, seven underground dams were completed or under construction and two underground dams are under

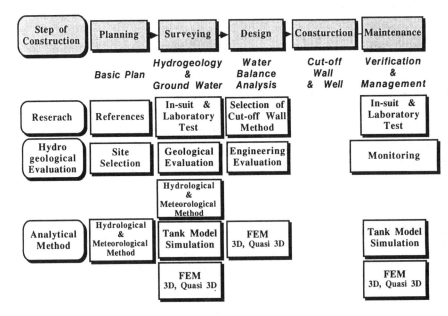

Fig.2 Steps of underground dam construction

execution and construction, including two experimental underground dams constructed under the guidance of the Government (Fig.3, Table 1).

The purpose of the construction of these underground dams can be roughly divided into the arrangement of waterworks in mountainous district, suburbs and remote islands and also the arrangement of the agricultural irrigation facilities in subtropical remote islands called Okinawa and Nansei Islands, Japan. Also, there are some examples of reviews by Japanese Government or companies for overseas projects in the west Africa (Sahel Green Belt Project), Gulf Coast area, and China. Especially, we discuss on the hydrogeology and methods of water balance analysis in the underground dam basins, Japan, during the construction of underground dam in this report [1),2),3)].

2 Hydrogeological potential for the underground dam construction site

Land suited to the underground dam should fulfill the following requirements for enhancing the storage effects; 1) impermeable basement should be distributed in shape of valley, 2) porous unconsolidated aquifer having a large porosity should be thickly and widely present over the said basement, and 3) intank area collecting rainfall should be sufficiently large. Also by considering the construction cost, it is topographically desired that the aquifer can be easily closed by the cut-off wall. This kind of topographic and geological conditions can be normally satisfied by alluvial sediment, buried rivers and varied valley tectonically. Also proper selection of the underground dam construction site and the optimum design cut-off wall are required for preventing environmental problems such as the turning into marsh at the upstream side and the drying up at the downstream side.

The underground dam in Japan are currently being built for alluvial sand gravel layers and porous limestone originated from coral reef in the Quaternary time as aquifer; and the shape of aquifer considered for construction is ordinary river topography and tectonically buried fault valley. Thickness of aquifer is about 10 m to 80 m. The impermeable basement is composed of Tertiary mudstone or more older basements.

The flow of research and hydrogeological evaluation is shown in Fig.2. Important points in the geological and ground water surveys and tests are as follows;
1) Shape, extension, thickness, and permeability characters of aquifer for ground water ;
2) Shape and hydraulic conductivity of impermeable layers as basement for aquifer ;

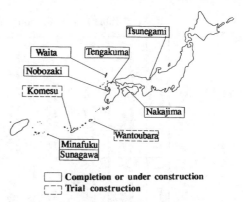

☐ Completion or under construction
⌐⌐⌐ Trial construction

Fig.3 Distribution of the Japanese underground dams

3) Hydrological and meteorological situation such as rainfall, air temperature, and evapotransipiration ;
4) Situation of fluctuation in ground water table and recharge relation between rivers and ground water ;
5) Characteristics of water quality (chemical components of ground water) ;
6) Engineering characteristics such as permeability and bearing capacity at the construction site for cut-off wall ;
7) Factors of environmental changes such as drying up of ground water or turning into marshes .

3 Ground water analysis of the underground dam basin

3.1 Purposes and methods of water balance analysis

Purposes of the ground water analysis in the underground dam construction are mainly to
1) establish new water quality to be developed,
2) figure out the fluctuation in ground water table, and 3) predict the possibility of turning into marshes. The result of ground water analysis are used for the design of the height to the top of cut-off wall, appropriate pumping discharge, recharge work and measures for preventing turning into marshes.

The analysis methods can be classified into
1) the water balance analysis by hydrological and meteorological methods, 2) water balance analysis by tank model simulation, 3) seepage analyses by Finite Element Method.   These should be properly chosen depending upon the purpose of analysis, stage of surveys and tests, and quality and quantity of analytic data.   The relationship between step of construction and analytical methods is shown in Fig.2.   Case studies of respective analytic methods will be described below in each step of construction.

Table 1 Specifications for the underground dams in Japan

Dam name	Location	Executing body	Completion year	Cut-off wall	Gross storage quantity	Purpose	Geology	Method
Nobozaki Underground Dam	Kabashima, Nobozaki Town, Nagasaki Pref.	Ministry of Construction, Nobozaki Town	1974	Length of levee: 60 m Max. levee height: 16 - 25 m	20,000 m³ (300 m³/day)	Waterworks	Clay with gravel, crystalline schist	Grouting method (double pipe double packer injection method)
Minafuku Underground Dam	Miyakojima, Okinawa Pref.	Okinawa General Affairs Bureau, Ministry of Agriculture, Forestry and Fisheries	1979	Length of levee: 500 m Max. levee height: 16.5 m	720,000 m³ (7,000 m³/day)	Water for agricultural use	Ryukyu limestone, Shimajiri mudstone	Grouting method (stage injection method)
Tsunegami Underground Dam	Mikata Town, Fukui Pref.	Mikata Town	1983	Length of levee: 202 m Max. levee height: 18.5m	73,000 m³ (300-420 m³/day)	Water for fishery processing	Clay with gravel, clay with rock pieces	Diaphragm wall method (slurry-hardened wall)
Amagakuma Underground Dam	Umi Town, Fukuoka Pref.	Umi Town	1988	Length of levee: 129 m Max. levee height: 12.5m	17,000 m³ (800-1000 m³/day)	Waterworks	Sand gravel, weathered granite	Grouting method (double pipe, double packer injection method)
Wantohara Underground Dam	Kikaijima, Kagoshima Pref.	Kyushu Agricultural Administration Bureau, Ministry of Agriculture, Forestry and Fisheries	From 1987	Length of levee: 2400 m Max. levee height: 35 m	Execution of trail construction (planned)	Water for agricultural use	Ryukyu limestone, Shimajiri mudstone	Various diaphragm wall methods, thin membrane method
Sunagawa Underground Dam	Miyakojima, Okinawa Pref.	Okinawa General Affairs Bureau, Agricultural Land Development Public Corporation	From 1989	Length of levee: 1853 m Max. levee height: 49 m	9,500,000 m³ (8,800,000 m³/year)	Water for agricultural use	Ryukyu limestone, Shimajiri mudstone	Diaphragm wall (Continuous pile type soil cement wall)
Komesu Underground Dam	Itoman City, Okinawa Pref.	Okinawa General Affairs Bureau	From 1990	Length of levee: 2600 m Max. levee height: 82 m	Execution of trail construction (planned)	Water for agricultural use	Ryukyu limestone, Shimajiri mudstone	Various diaphragm wall methods
Nakajima Underground Dam	Nakajima Town, Ehime Pref.	Chugoku Shikoku Agricultural Administration Bureau, Ministry of Agriculture, Forestry and Fisheries	From 1991	Length of levee: 87.7 m Max. levee height: 26.1 m	27,000 m³ (500 m³/day)	Water for agricultural use	Sand gravel, shale, sandstone	Diaphragm wall (continuous pile type soil cement)
Waita Underground Dam	Toyotama Town, Nagasaki Pref.	Toyotama Town	1992	Length of levee: 105.3 m Max. levee height: 7.5 m	34,000 m³ (280 m³/day)	Waterworks	Sand gravel, shale, sandstone	Diaphragm wall (slurry-hardened wall)

3.2 *Hydrological and meteorological method in the planning*

Regional site selection for construction is carried out by reference data such as topographic map (scale 1: 25,000) and geological map (scale 1: 50,000). Catchment and storage areas are picked up, and locations of cut-off wall are inferred. Among these preliminary selected sites, some suitable sites are selected for the next field survey by the social conditions such as arrangement of waterworks. Fig.4 is one of the examples on the regional site selection in the Tsushima Island, west Japan.

With respect to a basin separated by watersheds in mountainous region, that is, the ground water baisn, the balance of water inflow and outflow at the basins in a long period of time can be considered for each of rainfall, evapotranspiration, runoff of surface water, recharge and discharge of ground water, pumping discharge and ground water storage.

In this kind of water balance is valid, then the computation of approximate quantity of ground water to be developed becomes possible as follow; *[Quantity of potential ground water resources]* = *[Rainfall]* -*[Evapotranspiration]* -*[Runoff of surface water]*

For making this kind of analysis, required are the many data related to the air temperature, rainfall, surface runoff , infiltration rate of rainfall to the ground (runoff ratio of rainfall), and intank area of ground water for underground dam. Also from thickness and effective porosity of aquifer and the area of storage area, it is possible to calculate storage volume and the quantity of the rise of water level. These results will be approximate values but are good enough for the primary evaluation of aquifer when selecting the candidate site for underground dam.

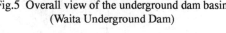

Fig.4 Case study of the regional site selection

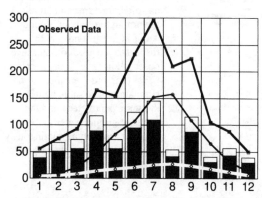

Fig.5 Overall view of the underground dam basin (Waita Underground Dam)

Regend of Data

- Rainfall(mm)
- Evapotranspiration(mm)
- Temperature (℃)
- Underground Flow(mm)
- Surface Flow(mm)

Fig.6 Case study of the results of water balance analysis by hydrological and meteorological methods (Waita Underground Dam, 1951 to 1980)

Fig.5 shows an example of the Waita Underground Dam, storage type, which is selected from the above mentioned sites in the Tsushima Island. The aquifer is composed of alluvial sand gravel layer, and impermeable layer is alternation of sandstone and shale of Paleogene age. Mean values for thirty years are used for the meteorological data, and the analysis of drought year is also possible as required. Shown in the Figure 6 are the surface runoff and ground water discharge (quantity of potential ground water resources) in addition to the meteorological data. The evapotranspiration is determined by the Thornthwaite Method based on air temperature. From these results, the status of surface water and ground water resources during drought season and wet season is determined and used as one of the items for engineeringly judging the properness of underground dam construction.

3.3 Tank model simulation in the surveying

The tank model is originally a flood prediction model for rivers[4] but has to be fitted to the analyses of ground water recharge by introducing the concept of effective porosity of ground[5]. In the tank model, the region to be analyzed is first divided into the intank area and storage area depending upon the topography and hydrogeology, and the layers to be considered are separated into surface layer and aquifer. In the tank model the layer in each section is considered as a tank, and water inflow and outflow at each tank are calculated. In this tank model quantity of outflow decrease exponentially to that of storage; the change of ground water level is shown by the following function in each tank.

$$h = h_0 * \exp[-(\alpha/Pa)t]$$

h: ground water level at time t, h0: initial groundwater level, α: runoff coefficient, Pa: effective porosity, t: time.

Required as analysis data are the air temperature, rainfall, evapotranspiration, ground water level, river discharge, and pumping discharge. For improving the accuracy of the analysis, daily observation data for at least one year are required. In the simulation, the ground water level repeatedly calculated based the previous mentioned exponential function while changing two parameters characterizing the tank (runoff coefficient α, and infiltration coefficient β). As a results, the parameter which gives the calculation results agreeing with the observation results for one year of ground water level is used as the tank model for the basin. The ground water discharge of the aquifer tank obtained at that time is used as the quantity of ground water developed. These results of analyses can be expressed by the type of water balance of monthly surface runoff and ground water discharge (quantity of newly developed ground water by underground dam) similarly to the case of water balance analysis by hydrological and meteorological method. Since tank model can be used for long-term simulation by utilizing the daily observation data, it is effective for data analysis not only for hydrological survey but also for preparing the maintenance criteria as later described.

Shown in Fig.7 to Fig.8 are the results of an example of analysis for the Waita Underground Dam. Fig.7 is the tank model for the relevant area, and Fig.8 shows the results of ground water level simulation (hydrograph simulation) for computing the quantity of ground water developed. The parameters of the tank model are considered to be corresponding to the macro-characteristics such as voids, fractures and faults governing the flow of ground water in the ground. With respect to the ground water level observation data for the next fiscal year and thereafter, simulation is possible by using the same parameters, and the status of ground water in the relevant area can be continuously simulated based on the original tank structure.

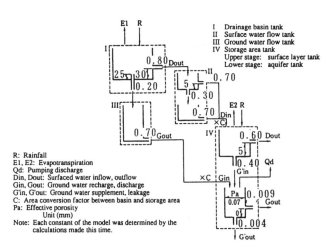

R: Rainfall
E1, E2: Evapotranspiration
Qd: Pumping discharge
Din, Dout: Surfaced water inflow, outflow
Gin, Gout: Ground water recharge, discharge
G'in, G'out: Ground water supplement, leakage
C: Area conversion factor between basin and storage area
Pa: Effective porosity
Unit (mm)
Note: Each constant of the model was determined by the calculations made this time.

I   Drainage basin tank
II Surface water flow tank
III Ground water flow tank
IV Storage area tank
Upper stage: surface layer tank
Lower stage: aquifer tank

Fig.7 Structure of tank model (Waita Underground Dam)

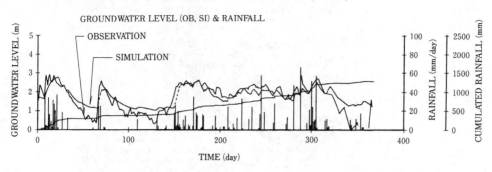

Fig.8 Case study of hydrograph simulation (Waita Underground Dam)

3.4 *Three-dimensional FEM analysis of ground water flow in the design*

The seepage analysis by the Finite Element Method (FEM) is able to analyze the flow of ground water under the conditions very close to the actual topography, hydrogeoloy, and detail design by utilizing a large scale computer. By the analysis, it could compute the ground water discharge, that is, the quantity of ground water developed, to predict the fluctuation in ground water table, and to forecast the flow current and velocity of ground water and probability of turning into marshes.

Fig.9 to Fig.12 are the examples of three dimensional flow analysis for the Waita Underground Dam. Fig.9 shows the analysis area modeled (three-dimensional element division) for FEM. This area is almost in the range of storage area plus the downstream portion of the cut-off wall. Distribution of aquifer and hydraulic parameters of the ground set for each section based on the results of surveys and tests. Also, the water level of rivers and the pumping discharge from the water source well were considered as the conditions for analysis.

The simulation was performed for three stage of
1) natural state, *2)* after constructing cut-off wall, and *3)* during pumping after taking measures against turning marshes, and the changes in the ground water flow in the storage area of the underground dam were analyzed. Results of analyses show how the ground water flowed through the basin, stored and finally lowed into the sea.

Fig.10 expresses the current direction and current speed by flux vector among the results of flow analysis seen on a plan. Arrow marks in the figure shows only the direction of flow (length of arrow is constant). Size of the current speed is expressed by the color in the original figure. We can judge the changes of flow

pattern and portions where ground water is stored from these results.

Fig.11 shows three-dimensionally the distribution of the values of hydraulic head among the results of flow analysis. In the saturated ground water zone, the hydraulic head generally increased as the elevation of ground surface becomes higher, however, if there is an unsaturated zone, the value of hydraulic head is expressed lower. The portions, where ground water is stored, almost correspond to not whole area but the more or less narrow area of upstream side of the cut-off wall. We can judge both the effectiveness of storage and portions where ground water is stored from these simulations.

Fig.12, based on a bird-eye-view, also the distribution of pressure head among the results of the analysis of ground water flow. That is, moisture state in the ground is indicated. If the value of pressure head is positive, the area is below the ground water table; if it is negative, the area is the unsaturated zone above the ground water table. Among these, the portion of the pressure head from 0.0 to minus 0.50 meters has the highest water content in the unsaturated zone. Therefore, this portion means that the water content is now high but still in unsaturated state, but saturation will occur if the ground water table increases slightly, thereby being placed below the ground water table. This portion has a very high possibility of turning into a marsh.

It can be known that extension of this portion in the storage zone will change depending on the condition of design and construction. That is, if the cut-off wall is designed and built by giving top priority to the storage effect, then the area having a high possibility of turning into marsh is widened compared to the natural state by the rise of ground water level (see *Natural state* and *After constructing cut-off wall* in the figure). However, by taking proper measures

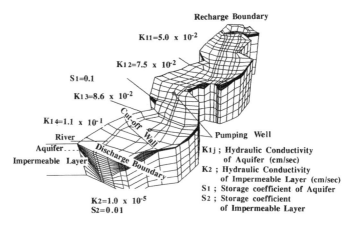

Fig.9 Model diagram for the FEM analysis

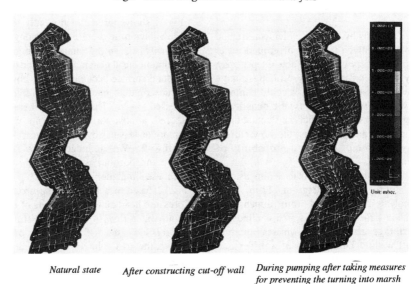

Natural state   *After constructing cut-off wall*   *During pumping after taking measures*
                                                        for preventing the turning into marsh

Fig.10 Flow of ground water (Flux vector, Waita Undergrouned Dam)

for preventing the turning marshes *(changing the height to the top of cut-off wall, crushed stone emplacement, earth fill, etc.)*, this area can be reduced almost the same as the present state while securing the quantity of water developed schemed in the original plan (see *During pumping after taking measures for preventing the turning into marsh* and *Natural state* in the figure). It was verified that the changes in the present storage sate and situation of turning into marsh on the site almost agree with the results of the simulation from the time before construction to the present after construction. Therefore, we can decide the detail condition of

design based on such as sensitive analyses using computer simulations.

3.5 *Tank model simulation in the maintenance*

Fig.13 shows the case study of the ground water monitoring after the underground dam construction. This case study was carried out in the Tsunegami Underground dam, central Japan, which is the type of anti-salinitification. The aquifer is composed of alluvial sand gravel and clay layer, and impermeable layer is sandstone, shale, chert of Mesozoic age. Fig.13 indicates the observation results during five

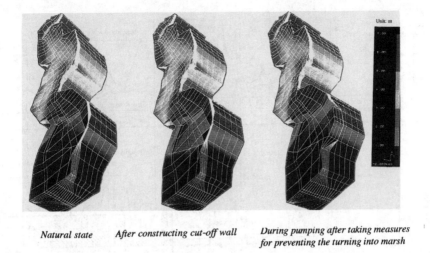

Natural state   *After constructing cut-off wall*   *During pumping after taking measures for preventing the turning into marsh*

Fig.11 Situation of distribution of ground water （Hydraulic head, Waita Underground Dam）

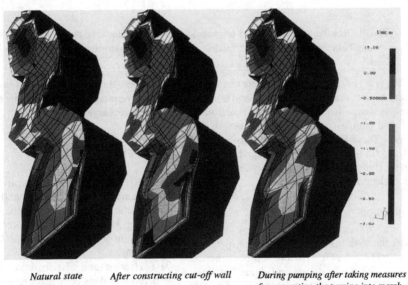

Natural state   *After constructing cut-off wall*   *During pumping after taking measures for preventing the turning into marsh*

Fig.12 Prediction analysis of the possibility of turning into marshes (Pressure head,Waita Underground Dam)

years maintenance period and the results of analysis by tank model simulation. This figure shows the correlations between the construction process (*cut-off wall , infiltration type artificial recharge system*) of the underground dam and the quantity of ground water developed, salinity density of ground water, and changes in monthly maximum and minimum ground water level. From these data, the increase in storage effect, decrease in the concentration of salinity, and fall in critical ground water level during pumping due to recharge are apparent. Also, another simulation was performed in order to further utilize the surplus ground water, and it was found that the pumping discharge can be increased by 10 % from the tank model simulation while maintaining the present concentration of salinity.

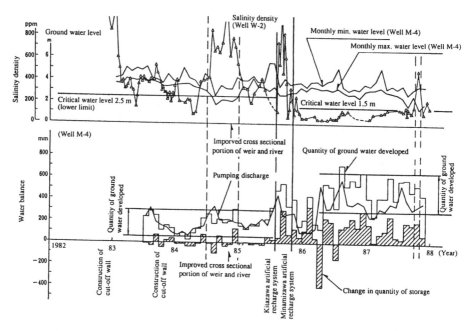

Fig.13 Results of ground water observation during maintenance period and the results of
water balance analysis by tank model (Tsunegami Underground Dam, 1983 to 1987)

After ten years of the construction, were carried out researches of both storage effectiveness and artificial recharge system. It is verified that function of underground dam is correctly acted at least during ten years. These data observed will also reflect to not only the further maintenance management of this Tsunegami Underground Dam but also the design of the new underground dams.

4 Conclusion

Under the hydrogeological condition in Japan, technical problems still exist but the construction technology for the underground dams has been almost established as a dependable method in this country. On the other hand, in the arid region, there are unresolved subject such as hydrogeology and engineering.

In the water balance analysis, we can use three types of analysis depending upon the purposed such as
1) hydrological and meteorological methods in the planning, *2)* tank model simulation in the surveying and maintenance, *3)* seepage analyses by Finite Element Method in the design. From these methods, it is enable to evaluate the water balance in the underground dam basin before and after construction of cut-off wall and also the quantity of water newly

developed, fluctuation of ground water table, change of environment, and effectiveness of storage and artificial recharge system.

REFERENCES

1) K.Hirama and T.Kuwahara: Water resource developments technology by underground dams, Symposium on the greeting of the G.C.C countries, Vol.1,pp.313-330, (1992), Execute Committee Symposium on the greeting of the G.C.C. countries, Tokyo, Japan
2) K.Hirama and M.Kushima : Examples of underground dam construction, Symposium on the greeting of the G.C.C countries,Vol.1, pp.331-347, (1992), Execute Committee Symposium on the greeting of the G.C.C. countries, Tokyo, Japan
3) Obayashi Corporation : Underground dam construction, p.28, (1993)
4) M. Sugawara: Runoff analysis method, Kyoritsu Publishing, Tokyo, Japan , (1972) *(in Japanese)*
5) M.Yoshikawa: A structure analysis of runoff and recharge of unconfined groundwater by storage model, Jour. Japan Soc. Engineering Geol., Vol.23, No.1, pp.1-6, (1982), Japan Society of Engineering Geology *(in Japanese)*

The effect of grouting on a loosely consolidated dam basement through which confined groundwater flows

Effets du traitement par injections d'une fondation peu consolidée d'un barrage sur l'écoulement des eaux souterraines

M. Ikeda & M. Furuta
Geotechnical Department, Hokkaido Engineering Consultants Co. Ltd, Tsukisamu-Higashi, Toyohira-ku, Sapporo, Japan

H. Yamashita
Pirika Dam Construction Office, Hokkaido Development Bureau, Japan

ABSTRACT : Grouting in dam construction is usually performed in order to consolidate ground deteriorated by excavation or to reduce the amount of seepage through cracks. However, for weakly consolidated rock with confined groundwater, precautions regarding piping of the confined aquifer and heaving of the confining layer must be taken into consideration. Some parts of the basement of Pirika dam, constructed in Japan, are comprised of a loosely consolidated Neogene rock through which confined groundwater flows. Therefore, the most important issue in constucting the dam was to prevent both piping and heaving actions. The effect of the grouting was scrutinized and assessed at four stages — 1) seepage prevention planning, 2) investigation of grouting methods and test grouting, 3) analysis of grouting and seepage prevention effects, and 4) dam management when the reservoir is filled. Three years have passed since the completion of Pirika Dam in 1991, and the safety of the dam in regards to piping, heaving and seismic actions has been confirmed.

RÉSUMÉ : L'injection est utilisée pour construire le barrage afin de consolider le sol détérioré par excavation et de réduire l'infiltration d'eau à travers des fissures. Mais, une assise rocheuse peu solide requiert des mesures particulières contre l'effet de renard. Surtout, quand elle comporte une nappe captive, son gonflement doit être évité. Le barrage Pirika dans du Japon a été construit sur le terrain tertiaire de Pliocène où existe certains sols médiocres comprenant la nappe captive. Donc la question primordiale était la sécurité à assurer de cettes assises contre le renard et le gonflement. Pour celà, on a évalué et analysé minutieusement des effets d'injection en 4 étapes— 1) Plan de prévention de l'-infiltration d'eau, 2) Examen de la méthode de l'injection et de son mise en oeuvre d'essai, 3) Examen du résultat de l'injection et de l'effet contre l'infiltration d'eau, 4) Plan de gestion et la surveillance du barrage après le réservoir rempli. Depuis l'achèvement de ce barrage en 1991, on constate, toujours sa qualité en sécurité relatives de renard, de gonflement et des actions séismiques.

1 INTRODUCTION

Pirika dam, constructed on the island of Hokkaido, Japan (Fig. 1), is a multipurpose combined type dam : a concrete gravity dam and a rockfill dam with a clay core. The height above the lowest foundation is 40m, the crest length is 1,480m (the longest in Japan), the reservoir capacity is 18,000,000 m³, and the active storage capacity is 14,500,000 m³.

Some parts of the basement are comprised of a loosely consolidated Neogene coarse sandstone through which confined groundwater flows. Therefore, the most important issue in constructing the dam was to prevent both piping and heaving actions.

The dam construction was divided into four stages—stage 1 : seepage prevention planning, stage 2 : investigation of grouting methods and test grouting, stage 3 : analysis of grouting and seepage prevention effects, and stage 4 : dam management when the reservoir is filled.

In stage 1, seepage analysis by means of the quasi-three dimensional finite difference method was performed in order to investigate the stability of the foundation layer (Ikeda and Furuta (1993)). The results are as follows : First, the groundwater velocity is the highest at the end of the cutoff zone, and the groundwater flowlines concentrate there. Second, a concrete cutoff wall would be more effective in cutting seepage than curtain grouting. However, a cutoff wall would concentrate groundwater flowlines more than grouting would. Therefore, the best way to cut the seepage would be a combination of cutoff wall and grouting. Third, it was found that the permeability of the grouting zone should be gradually increased to that of the aquifer to

prevent the concentration of flowlines.
 The length of cutoff zone was also investigated to ensure overall dam safety in regards
to the effect of piping of the basement layer,
heaving of the confining layer, and groundwater leakage through the aquifer.
 Based on the this seepage analysis, the
effects of the grouting were scrutinized and
assessed in the other three stages. This paper
mainly reports a summary of the grouting.

2 SUMMARY OF THE GEOLOGY

The dam basement is composed of Neogene layers.
The left bank of the river bed is underlain
by Miocene mudstone and shale, while the right
bank is underlain by Pliocene sandstone (see
Fig. 2). The sandstone varies from relatively
consolidated fine sandstone which shows almost
the same shear strength as that of mudstone
and shale, to weakly consolidated coarse sandstone that is susceptible to piping. Table 1
shows the various physical properties of the
different types of foundation materials identified during field surveys and laboratory tests.

3 IMPORTANT ISSUES FOR DESIGN AND FOUNDATION
TREATMENT

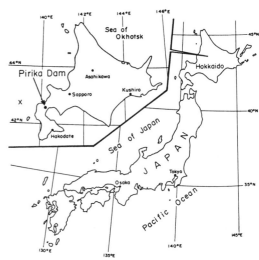

Fig. 1 Dam Location Map

(A cross (x) indicates the epicenter of
Hokkaido Nansei-oki Earthquake in 1993.)

A special design was not required for the
foundation which is to lie on the mudstone,
shale and siltstone (Shm, Sil) layers. However,
a special treatment was necessary for a foundation on the weakly consolidated coarse sandstone (Ssc(B): B refers to the brown color

Table 1 Physical Properties of Basement Layers

Geological formations	Main distribution	Physical Properties			
		Permeability			Elasticity (MPa)
		Hydro-geology	Coefficient of permeability (cm/s)		
sandstone	fine sandstone Ssf	right bank	confining layer	1×10^{-4} $\sim 1 \times 10^{-3}$	500
	coarse sandstone Ssc (B)	right bank	confined aquifer	3×10^{-3} $\sim (7 \times 10^{-4})$	100
	medium to coarse sandstone Ssm~c	right bank	confined aquifer	3×10^{-4}	250
mudstone, shale & siltstone Shm, Sil	from right bank to river bed	imperme-able layer	1×10^{-5}		$450 \sim$ 700

Fig. 2 Legend

Quater-nary		tr 6	terrace deposits
		tr 5	
Neogene	Setana Formation (Pliocene)	Ssf	fine sandstone
		Ssm	medium sandstone
		Ssc	coarse sandstone
		Pt	pumice tuff
	Miocene	Sil	siltstone
		Shm	shale

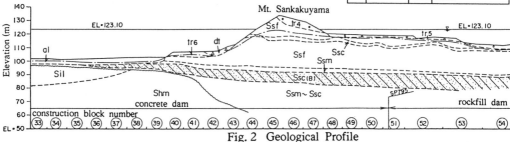

Fig. 2 Geological Profile

due to the oxidization), and medium to coarse sandstone (Ssm~c), excluding fine sandstone.

At Mt. Sankakuyama, especially, the analysis and design of the abutment and the foundation treatment are important issues, as the sound basement Shm layer falls off suddenly in the vicinity of Mt. Sankakuyama abutment (see Fig. 2). The relationship between Shm and Ssm~c to Ssc(B) is unconformed due to the geological time lag. The analysis and the treatment of the Ssm~c and Ssc(B) layers were conducted taking the following issues into account.

1. The location of the foundation of the concrete gravity dam needs to be changed from the Shm layer to the Ssf layer, avoiding an outcrop of Ssm~c or Ssc(B) layers. (Ssm~c and Ssc(B) layers have inadequate bearing capacity for a concrete dam foundation.)

2. An analysis of the deformation of the underlying Ssc(B) layer below the Ssf layer (approximately 5m thick).

3. Ssc(B) layer is a confined aquifer and has a slightly high permeability coefficient $(3 \times 10^{-3} \text{cm/s})$. Thus, groundwater streamlines are most likely concentrated at the edge of the grouting zone and immediately downstream of the dam. Therefore, preventative measures were nesessary against piping of the confined aquifer (Ssc(B), Ssm~c) and heaving of the confining layer Ssf.

4 INVESTIGATION OF GROUTING METHOD AND TEST GROUTING

The test grouting of the Ssc(B) layer was performed in order to improve the physical properties of the Ssc(B). Using the results of the test, the dam was designed aiming for both a reduction in the permeability and an improvement in the deformation properties.

Seven different grouting tests regarding both the grouting methods and the materials were conducted in the vicinity of Mt. Sankaku-yama, on the downstream side of the dam, and their individual characteristics were compared.

A summary of the results is given in Table 2. The most suitable method was the combination of superfine cement and the double tube double packer grouting method (Type A). The detail is described below.

4.1 Observation of Injected Ssc(B)

After the Type A test grouting, a shaft 3m in diameter was excavated in the test area in order to conduct various surveys including observation with the naked eye of the injeted Ssc(B) layer.

1. Situation of the shaft
Figures 4 & 5 show the developed view of the internal surface of the test shaft and a view of the bottom surface at different depths.

The groutmilk was injected alongside injection pipes and in the form of layered sheets. Figure 3 shows the grout coverage ratio. Although the ratio varied with the depth, the total coverage ratio was more than 60%.

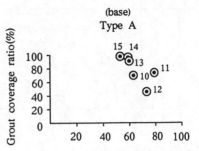

(base)
Type A

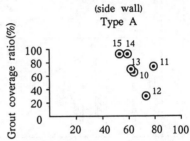

(side wall)
Type A

Total amount of cement injected (kg/V)
(coarse sandstone : depth 9 to 16m)

⊙ 10 : upper limit of depth
(It refers to 10 to 11m deep.)

Fig.3 Relationship between amount of cement injected per valve and grout coverage ratio of injected part

2. Boring Core Observation
After injection, check boring was carried out and core samples were obtained. Some parts with groutmilk injected showing the phenolphthalein reaction were observed and compared with the quantity of cement injected. It was found that when the injection amount exceeded 70kg/V (/V refers to different valves), the groutmilk in 1m long boring core exceeded 50% (Fig. 6).

4.2 Improvement of Elasticity and Permeability

1. Elasticity
The static elasticity of Ssc(B) was obtained by using lateral loading tests in check boreholes after injection(Fig. 7). The elasticity

Table 2 Grouting Tests for Ssc (B)

	Type A	Type B	Type C	Type D	Type E	Type F	Type G
Purpose	Check the suitability of double tube double packer grouting method	Check the suitability of stage grouting method	Compare the degree of improvement resulting from the combination of materials and grouting methods				
Grouting method	Double tube double packer	Stage	Stage	Stage	Double tube double packer	Double tube double packer	Double tube double packer
Material	Superfine cement	Superfine cement	Blastfurnace cement type-B	Colloid cement	Blastfurnace cement type-B	Colloid cement	1st borehole : Colloid cement 2nd & 3rd borehole : Superfine cement
Grouting hole array	Lattice pattern (1.0m)	Triangular pattern (3.0m)	Triangular pattern (4.0m)	Triangular pattern (4.0m)	Triangular pattern (4.0m)	Triangular pattern (4.0m)	Triangular pattern (4.0m)
Injection specification	· Injection pressure $P = Pd + \alpha H$ (Pd : Proportional limit of pressure, $\alpha = 0.5$) · Injected from the upper valve · Grouting using water	· Injection pressure 0.29 to 0.49 MPa · Stage Length 2.5m	· Injection pressure · Stage Length 2.5m	· Injection pressure 0.49, 0.98, 1.47 and 1.96 MPa · Stage Length 2.5m	· Injection pressure $P = Pd + \alpha H$ α (medium sandstone : 0.5 coarse sandstone : 0.8) · Valve interval 33cm	· Same as injection specification Type E	· Same as injection specification Type E
Results	· Average amount of cement injected was 129.6kg/V. · The reduction of permeability was excellent. · The permeability was 20 Lugeons at an injection amount of 10kg/V. · The improvement was considerable.	· Maximum injected amount was 50kg/m. · The method is not suitable for the proposed site.	· The improvement was poor.	· The improvement was poor.	· Injected amount was 30kg/V. · The improvement was poor.	· As the rate of displacement exceeded the allowable limit (20/100mm), the test was terminated. · The improvement was poor.	· The amount of grouting injected was 30kg/V for 1st hole, 145kg/V for 2nd and 3rd holes. · As a whole the injection amount was less than that of Type A. · The Lugeon value was usually around 30, with values of 10 at a few places.
Remarks	· The economy and simplicity of construction need to be considered.	· The improvement using the stage grouting method was poor.	· It is difficult to use this method.	· It is difficult to use this method.	· It is difficult to use this method.	· It is difficult to use this method.	· When both double pipe, double packer method and superfine cement are used, Ssc(B) could be improved. · In addition, Type G should be used in the places where Types B to F were tested.

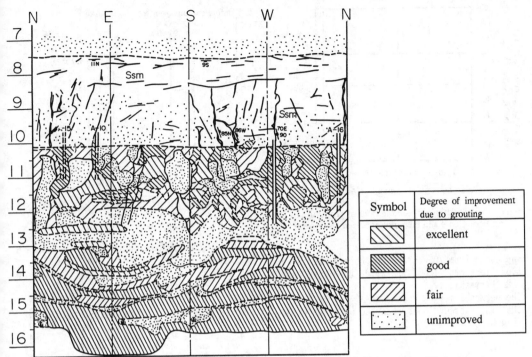

Fig. 4 Type A — Developed View of the Internal Surface of the Test Shaft

Symbol	Degree of improvement due to grouting
(hatched)	excellent
(hatched)	good
(hatched)	fair
(dotted)	unimproved

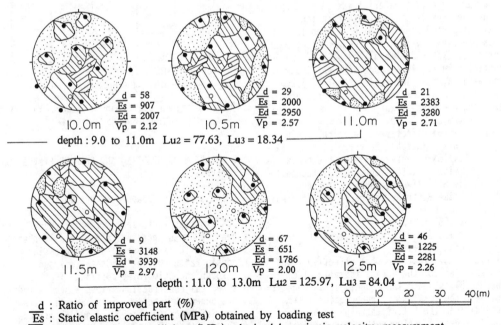

10.0m $\dfrac{d = 58}{\overline{Es} = 907}$ $\overline{Ed} = 2007$ $\overline{Vp} = 2.12$

10.5m $\dfrac{d = 29}{\overline{Es} = 2000}$ $\overline{Ed} = 2950$ $\overline{Vp} = 2.57$

11.0m $\dfrac{d = 21}{\overline{Es} = 2383}$ $\overline{Ed} = 3280$ $\overline{Vp} = 2.71$

—— depth : 9.0 to 11.0m Lu2 = 77.63, Lu3 = 18.34 ——

11.5m $\dfrac{d = 9}{\overline{Es} = 3148}$ $\overline{Ed} = 3939$ $\overline{Vp} = 2.97$

12.0m $\dfrac{d = 67}{\overline{Es} = 651}$ $\overline{Ed} = 1786$ $\overline{Vp} = 2.00$

12.5m $\dfrac{d = 46}{\overline{Es} = 1225}$ $\overline{Ed} = 2281$ $\overline{Vp} = 2.26$

—— depth : 11.0 to 13.0m Lu2 = 125.97, Lu3 = 84.04 ——

0   10   20   30   40(m)

  d : Ratio of improved part (%)
$\overline{Es}$: Static elastic coefficient (MPa) obtained by loading test
$\overline{Ed}$: Dynamic elastic coefficient (MPa) obtained by seismic velocity measurement
$\overline{Vp}$: Mean P wave velocity (km/s)
Lu2, Lu3 : Lugeon values of 2nd & 3rd bore holes, respectively

Fig. 5 Sketch of the Bottom Surface

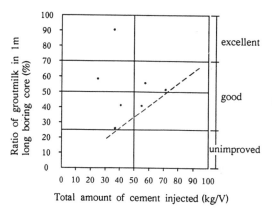

Fig. 6　Relationship between the ratio of groutmilk in 1m long boring core and the amount of cement injected

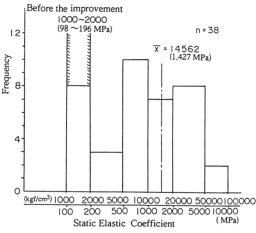

Fig. 7　Histogram of Static Elastic Coefficient in Check Boreholes

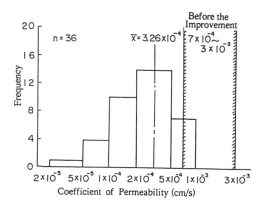

Fig. 8　Histogram of Permeability in Check Boreholes after the Improvement

increased from 100 to 200 MPa before the injection, to around 1,000 MPa after the injection.

2. Permeability

The permeability decreased to 3×10^{-4} cm/s, that is, 1 order lower than the original value (Fig. 8).

4.3 Groutability

A suitable grouting material was decided upon using the experimental formula from the U. S. Army Corps of Engineers which is used to assess groutability of a layer composed of soil particles (Table 3). When the ratio of small particles in the layer and large particles in the grouting material satisfy the follwing two formulae at the same time, effects of grouting are significant.

$$D_{15}/G_{85} \geqq 15, \quad D_{10}/G_{95} \geqq 8$$

where, D_{10}:10% effective size on the grain size distribution curve of the soil

D_{15}:15% effective size on the grain size distribution curve of the soil

G_{85}:85% effective size on the grain size distribution curve of the grouting material

G_{95}:95% effective size on the grain size distribution curve of the grouting material

Table 3 shows that the most effective grouting material for the Setana formation including Ssc(B) is superfine cement. It is consistent with the results of the grouting tests. In addition, the belief that the double tube double packer grouting method was the most suitable for the layer was strengthened, in that the method prevents boreholes from collapsing, and the method makes iterative injections in short stage lengths possible.

5 ANALYSIS OF GROUTING AND SEEPAGE PREVENTION EFFECTS

5.1 Elasticity Enhancement Design

The grouting tests showed that around 25 percent of Ssc(B) were not significantly changed, although the improvement due to grouting was considerable. Piping in Ssc(B) may occur through the unimproved portions. Therefore, a special construction method was required at the basement junction of the Shm layer and the Ssf layer (Chapter 3, issue 1). It was decided that the location of the basement should be changed from Shm to Ssf via a diaphragm wall. (It was the first of its kind in the world, and was called the "box-type" diaphragm wall(Fig. 9).)

Table 3. Groutability of the Setana Formation

	Superfine Cement		Colloid Cement		Blastfurnance Cement Type B	
	D_{15} / G_{85}	D_{10} / G_{95}	D_{15} / G_{85}	D_{10} / G_{95}	D_{15} / G_{85}	D_{10} / G_{95}
Sand and gravel D_{10} = 0.08mm D_{15} = 0.15mm	25	10	8	3	6	2
Coarse sandstone, Conglomerate Ssc2 D_{10} = 0.20mm D_{15} = 0.23mm	38	25	13	7	9	5
Coarse sandstone Ssc(B) D_{10} = 0.18mm D_{15} = 0.23mm	38	23	13	6	9	5
Medium to coarse sandstone Ssm ~ Ssc D_{10} = 0.10mm D_{15} = 0.15mm	25	13	8	3	6	3
Fine sandstone Ssf* D_{10} = 0.007mm D_{15} = 0.015mm	2.5	1	1	0.2	0.6	0.2
	G_{85} = 0.006mm G_{95} = 0.008mm		G_{85} = 0.018mm G_{95} = 0.030mm		G_{85} = 0.027mm G_{95} = 0.040mm	

* It is impossible to inject even if superfine cement is used.

The height of the dam above the Ssf basement is 25m, even when the box-type diaphragm wall is used. Thus, a dam embankment could be damaged from subsidence of Ssc(B). Stress analysis showed that the elasticity of Ssc(B) must be increased to around 500 MPa. On the other hand, the test grouting data showed that the mean elasticity of the Ssc(B) exceeds 1,000 MPa, and 75% of the layer's elasticity could be increased to 500 MPa. Depending on the analysis and the data, deformation prevention was planned.

5.2 Seepage Prevention Design

As briefly described in Chapter 1, to avoid groundwater streamline concentration, the permeability of both the cutoff and grouting zones should be gradually increased to that of the aquifer. Therefore, using the seepage analysis, it was decided to use, as the construction method, the combination of the box-type diaphragm wall(construction block No.40, upstream-downstream direction), a cutoff wall (block No.41 to 43, dam-axis direction), and a grouting zone (block No.44 to 52, dam-axis direction, the permeability is gradually increased as mentioned in Chapter 1). Fig.9 shows the schematic diagram. The target safety factors

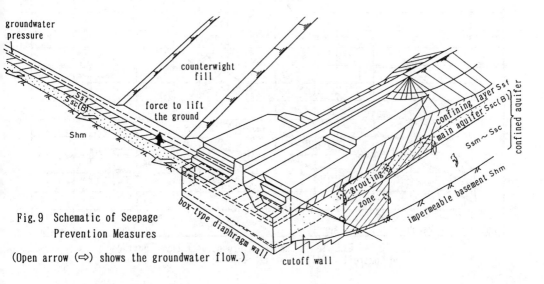

Fig.9 Schematic of Seepage Prevention Measures

(Open arrow (⇨) shows the groundwater flow.)

(Fs) regarding piping were as follows : Fs regarding the critical hydraulic gradient (ic) > 5, and Fs regarding the critical groundwater velocity > 20.

5.3 Evaluation of Dam Grouting and Seepage Prevention

Grouting of the dam was conducted until the safety factor Fs exceeded the target values (for ic, Fs>5). The average coefficient of permeability of Ssc(B) in check boreholes was K=8.32×10⁻⁵(cm/s). By means of both the borehole permeability tests and the seepage analyses, the overall safety of the dam was assessed. As the estimated safety factor sufficiently exceeded the target value (>5) regarding the critical hydraulic gradient), it is believed that the preventative measures regarding seepage are sufficient(Table 4). Thus, the controlled filling of the reservoir was carried out.

6 SAFETY CONSIDERATIONS DURING THE FILLING OF THE RESERVOIR

Thirteen observation boreholes were drilled to ensure dam integrity in regards to piping and heaving during the controlled filling of the reservoir. The safety criteria were as

Table 4 Increase in Safety Factor through by Seepage Prevention Measures

Seepage prevention measures	Safety factor regarding critical hydraulic gradient	
	Immediately downstream of the dam	Edge of the seepage prevention measures
(a): Nothing (Box-type diaphragm only)	3.4	—
(b): (a)+cutoff wall (No.41 to 43 block)	4.2	2.7
(c): (b)+curtain grouting(No.44 to 50 block)—planned	5.5	5.4
(d): (b)+curtain grouting(No.44 to 52 block) — constructed	7.7	6.1

(estimated by seepage analysis)

follows :
1. Caution-indicated groundwater level : When the measured groundwater level exceeded the values predicted by seepage analysis.
2. Dangerous groundwater level : when the safety factor regarding piping was lower than 5, or the factor regarding heaving was lower than 1.5.

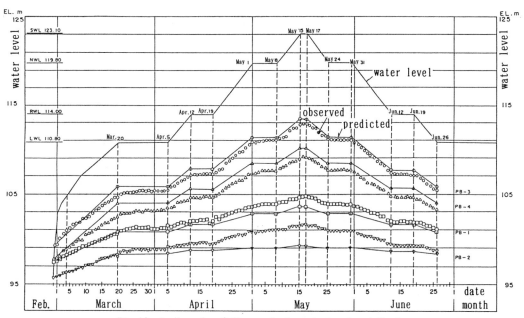

Fig. 10 An Example of Change in Groundwater Level During the Controlled Filling of the Reservoir

Fig. 10 shows an example of the groundwater
level during reservoir filling. The observed
groundwater level was consistent with the
predicted values, and the reservoir filling
was completed successfully.

7 CONCLUDING REMARKS

Pirika dam was first planned in the 1960s in
the upstream area of the Shiribeshi-Toshibetsu
river. It was completed in 1991, after more
than twenty years of survey and construction.
From engineering geology point of view, parts
of the basement have some of the worst sedimentary rocks in Japan to construct a dam on.
In order to overcome the geological problem of
weakly consolidated layers—Ssc(B) and Ssm~c,
many analyses including test grouting, and
both deformation and seepage analyses, were
conducted. This paper was written to provide a
reference regarding the foundation treatment
of soft sedimentary rock with confined groundwater, mainly stressing on grouting.
Three years have passed since the completion
of Pirika dam. On July 12th, 1993, an intensive
earthquake with a magnitude of 7.8 on the
Richter scale struck the island of Hokkaido,
and the southwest was terribly damaged.
Although the distance from the epicenter(Fig.
1) to Pirika dam was approximately 90km, no
damage to the dam, nor increase in the amount
of leakage, nor evidence of liquefaction of
the sandy basement was found. Thus, the safety
of the dam has been further substantiated.
Finally, we would like to express our thanks
to the staff members of the Hakodate Department of Development and Construction, the
Hokkaido Development Bureau, and Hokkaido Engineering Consultants Co. Ltd. for their help and
encouragement.

REFERENCES

Hokkaido Development Bureau 1992. Foundation
    Treatment of Pirika Dam. **
Ikeda M. & Furuta M. 1992. Seepage Analysis
    about a Dam Constructed on a Confined Aquifer. *
    Hokkaido Geotechnics '93. 4 : 49—55.
King J. C. & Bush E. G. W. 1961. Grouting of
    Granular Materials. Symposium on Grouting,
    American Society of Civil Engineers.
River Bureau, Ministry of Construction &
    Japan Dam Engineering Center 1987.
    Construction of Multipurpose Dam. **
Rushton K. R. & Redshaw S. C. 1979. Seepage
    and Groundwater Flow. New York : John wiley
    & Sons.

Terzaghi K. & Peck R. B. 1967. Soil Mechanics
    in Engineering Practice. New York : John
    Wiley & Sons.
Weaver K. 1992. Dam Foundation Grouting. New
    York : American Society of Civil Engineers.

    * in Japanese with English abstract
    ** in Japanese

7th International IAEG Congress / 7ème Congrès International de AIGI, © 1994 Balkema, Rotterdam, ISBN 90 5410 503 8

A rising groundwater level at Fawley, Hampshire

Élévation de la nappe à Fawley, Hampshire

J.J. Drake, D.P.Giles, W.Murphy & N.R.G.Walton
Department of Geology, University of Portsmouth, UK

ABSTRACT: Groundwater levels in the Fawley area of Hampshire have risen by up to 62m in the last 40 years, due solely to a decrease in the volume of groundwater abstracted from the confined aquifer underlying the oil refinery which dominates the area.

This paper describes the history of groundwater levels in the area since the 1930s, and concludes that despite a rate of groundwater rise of up to 2.2m/year, none of the problems commonly associated with rising groundwater levels appear to have affected the site.

RESUMÉ: Les niveaux de l'eau souterraine dans les environs de Fawley dans le comté de Hampshire sont montées d'un maximum de 62 mètres pendant les 40 dernières années. Ceci a pour seul cause une diminution du volume d'eau souterraine tirée de l'aquifère qui est confiné au-desous de la raffinerie d'huile de pétrole qui domine les environs.

Cette communication décrit l'évolution des niveaux de l'eau souterraine dans la région depuis les années 1930, et donne la conclusion que, en dépit d'une augmentation du niveau de l'eau souterraine de 2.2 mètres par an, le site étudié n'a pas éprové les problemes qui s'associaient normalement à ce genre d'augmentation.

1 INTRODUCTION

Groundwater levels are currently rising under many urban and industrial areas both in the U.K. and overseas, with London, Liverpool, and Paris being notable examples. Brassington (1990) briefly described many of the locations where groundwater levels are known to be rising, and noted that in all of these cases the rise is in response to a reduction in recent years in the amount of groundwater being abstracted from the underlying strata in. Rises in groundwater levels are of interest to both geotechnical and civil engineers, as they can lead to problems that can affect both the soil in which the rise is occurring, and any structures founded on, or in, the soil.

This paper describes the rise in groundwater levels observed in the Fawley area of Hampshire.

2 LOCATION

Fawley (National Grid Reference SU 460 031) and the adjoining villages of Blackfield and Holbury cover around 11km^2 and are located on the western coast of Southampton Water, a major estuary in the southern English county of Hampshire (Fig. 1).

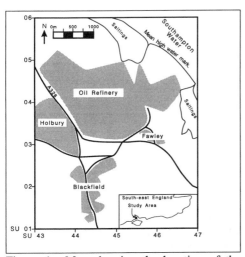

Figure 1. Map showing the location of the Fawley area. (Grid numbers are from U.K. national grid.)

The nearest major centre of population is the city of Southampton, an important international passenger and freight port, 10km to the north.

The area investigated (SU 450 040) is occupied by a major oil refinery, covering approximately 5km^2 and bounded by Fawley village to the south-east, Blackfield to the south, Holbury to the west, and Southampton Water to the east. There has been a refinery on the site since 1921, although it was not until the early 1950's, when the demand for petroleum products rose rapidly, that the present refinery, owned and operated by the Esso Petroleum Company Limited and the largest in the U.K., was built (Gilchrist 1991).

3 GEOLOGY

Fawley lies towards the centre of the Hampshire Basin; a synclinal structure filled with Palaeogene sediments which forms a large part of central southern England.

In the period 1935-1952, 19 deep boreholes were sunk at the Fawley refinery for water abstraction purposes. The availability of accurate logs from many of these boreholes has made it possible to construct a detailed geological map and cross-section for the site (Figs. 2 and 3).

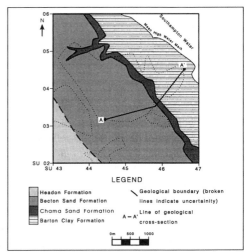

LEGEND

Headon Formation — Geological boundary (broken lines indicate uncertainty)
Becton Sand Formation
Chama Sand Formation — Line of geological A – A' cross-section
Barton Clay Formation

Figure 2. Geological sketch map of the oil refinery and adjoining areas (solid geology).

Up to 260m of Palaeogene strata has been proved in the boreholes, showing a sequence passing upwards from the London Clay Formation, through the Bracklesham and Barton Groups, and unconformably into Quaternary deposits (Fig. 4). These stratigraphic units comprise a range of lithologies from sandy clays to clayey sands and have been described in some detail by various authors (eg. Melville and Freshney 1982, Edwards and Freshney 1987a and Edwards and Freshney 1987b).

The strata at the site generally dip to the south-west at a little under 1° (Fig. 3), although both bearing, and amount, of dip are variable.

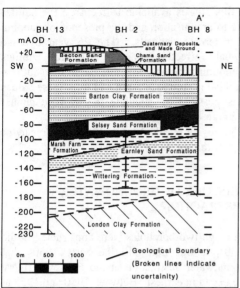

Figure 3. Geological cross-section through the refinery site.

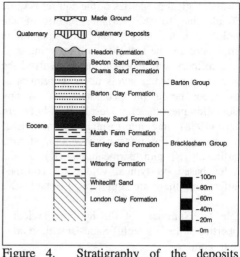

Figure 4. Stratigraphy of the deposits underlying the Fawley area.

4 TOPOGRAPHY

The topography of the area appears to be quite heavily influenced by the underlying geology, and consists essentially of:

1. A coastal fringe with an average elevation of less than 5m, in places consisting of saltings (SU 467 035 and SU 453 056). This fringe is around 1km wide.

2. A second, less regular, plain lying inland to the coastal fringe, with an average elevation of 25 to 30m. It is on this level that the majority of the refinery is sited.

The boundary between these areas is an irregular but marked slope, gaining 20m in height in around 250m distance; an average gradient of 4.5°.

The maximum elevation attained in the area occupied by the refinery and the adjacent villages is approximately 35m above Ordnance Datum (OD), and much of the land within a 20km radius of the site is below 60m OD.

5 HYDROGEOLOGY

Water is an important part of the oil-refining process, and hence the provision of an adequate, reliable supply of water is a factor which must be taken into account when a new refinery is planned.

At the Fawley refinery the majority of the water used was abstracted from the Whitecliff Sand, in the top 15 to 25m of the London Clay Formation, and from the sandy deposits of the overlying Bracklesham Group.

The main geological components of this sequence are sands, silts and clays in variable proportions, producing a series of lithologies from sandy clays to clayey sands. Due to lateral and vertical variations in lithology, however, the whole of the succession is considered to be in overall hydraulic continuity (Southern Water 1988), and hence can be considered as a single aquifer approximately 125 to 150m thick. This aquifer is confined by the Barton Clay Formation.

The calculation of the important

2897

hydrogeological properties of the aquifer is quite difficult due to the variable nature of the lithologies noted above, and this situation is complicated by the fact that well pumping tests at the refinery site appear to have involved the observation of water levels in the boreholes as they were being pumped, and not in any observation piezometers. Despite this, Southern Water (1988) were able to make estimations of the values of the transmissivity (T), storage coefficient (S) and permeability (K) of the aquifer based on typical values for confined aquifers, and these are summarised in table 1.

Table 1. Summary of the hydrogeological properties of the Whitecliff Sand-Bracklesham Group aquifer.

Property	Typical values
Transmissivity (T)	50 to 100m²/day
Storage Coefficient (S)	0.01 to 0.1%
Permeability (K)	0.1 to 1m/day

It should be noted that these values are for the aquifer as a whole, despite the fact that due to lithological variability only certain parts of the sequence are likely to contribute to the supply of water. Hence, variations in K, S and T are likely to be large over relatively short distances, both vertically and horizontally.

6 RECHARGE TO THE WHITECLIFF SAND-BRACKLESHAM GROUP AQUIFER

The outcrop of the deposits that comprise the Whitecliff Sand-Bracklesham Group aquifer in the area of interest is estimated to be around 200km², but it is probable that only around 50%

of this area is available to recharge the aquifer due to urbanisation and the presence of capping river terrace deposits. The 25 year average annual effective rainfall (rainfall minus evapotranspiration) for this area of southern England is around 325mm and from these values it is possible to calculate the average annual recharge to the aquifer:

325mm/year x 100 km² = 32.50 Mm³/year, or 32,500 million litres per year.

7 HISTORY OF GROUNDWATER ABSTRACTION

Although two boreholes were constructed in the 1930s, it was not until the early 1950s that groundwater abstraction at the Fawley refinery site began in earnest. Between 1950 and 1952 at least 14 boreholes were drilled, and in 1951 ten of these were producing a total of 5.8 million litres per day (Ml/d). By 1961 eleven boreholes were in use providing some 5 Ml/d, but by 1965 the abstraction had dropped to 4 Ml/d from the same 11 boreholes (Fig. 5).

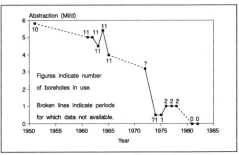

Figure 5. Rates of groundwater abstraction at the Fawley refinery in the period 1950-1985.

From 1965 data is again sparse, but by 1972 it was reported that only 3.2 Ml/d was being abstracted through an unknown number of boreholes. In 1974 and 1975 it appears that only 1 borehole was in operation, providing 0.5Ml/d, but by 1978 1.0Ml/d was being produced by 2 boreholes.

It is not clear when pumping ceased, but it was noted that no pumps were in use from 1981 onwards (Southern Water 1988). It is thought that the decline and eventual cessation of groundwater abstraction at the refinery site was in response to the increased provision of water through a reliable mains network.

8 GROUNDWATER LEVELS DURING THE PERIOD OF PUMPING

The first borehole at the Fawley site was drilled in 1935 and the original groundwater level in the borehole was noted at +7.3m OD. It can therefore be assumed that this was the original potentiometric level on the site.

In the early 1950s, after pumping of the new boreholes had begun, the groundwater levels in the boreholes was in the range +3m OD to -30m OD. By the mid-1950s this level had dropped to between -37m OD and -55m OD, although it appears that soon after this period a situation of approximate equilibrium was produced as groundwater levels remained relatively constant, even rising in some cases (Fig. 6).

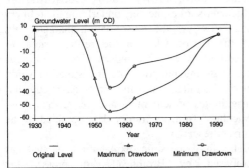

Figure 6. Groundwater levels at the Fawley refinery in the period 1930-1992.

By the early 1960s, the last period for which data is available, the groundwater level was between -21m OD and -45m OD.

From this data it can be seen that the maximum

drawdown was in the mid-1950s, with a value of 44 to 62m.

9 CURRENT SITUATION

During a reconnaissance visit to the refinery in September 1993, one of the few remaining boreholes, borehole No. 11, with a datum of +2.7m OD, was inspected. It was noted that artesian conditions were once again prevailing at this borehole, and hence the groundwater level in the borehole was above +2.7m OD.

It therefore appears that the groundwater level is approaching its original position of +7.3m OD, and this dramatic rise in groundwater level probably began in the early 1970s when abstraction of groundwater from the deep aquifer suddenly declined for operational reasons.

10 RATE OF GROUNDWATER LEVEL RISE

If it is assumed that the roughly constant abstraction rate of the 1960s and early 1970s allowed the groundwater level to find approximate equilibrium at the early 1960s level of -26 to -43m OD, and that this level only began to rise in the early 1970s, it can be seen that since this time the groundwater level has risen by around 30 to 45m, at a rate of 1.5 to 2.2m/year. However, it is unlikely that this rate was constant. More realistically, the early period of the rise was rapid, and the rate decreased with time, as shown in Fig. 6.

11 PROBABLE FUTURE SITUATION

It can now be considered that the groundwater level at the site is very close to, if not at, its original level of +7.3m OD. As no significant long term changes in the hydrogeological regime of the area (ie changes in rainfall, evapotranspiration, permeability etc) have occurred since the 1950s, it is unlikely that the groundwater level will rise any further. Hence it is reasonable to assume that the groundwater level at the site is now in equilibrium, and as long as abstraction does not restart there will be no further significant changes in this level, other than the usual annual hydrological fluctuations.

12 POTENTIAL GEOTECHNICAL EFFECTS

From the preceding discussion it is clear that groundwater levels in the Fawley area rose by up to 62m since the 1950s, and that this rise was solely due to a progressive reduction, and ultimate cessation, in the volume of groundwater abstracted in the area of the Esso oil refinery.

It is commonly feared that rises in groundwater level could have serious effects on the integrity of engineering structures through a variety of processes, which Wilkinson (1985) described as follows:
1. Flooding of basements and tunnels.
2. Hydrostatic uplift.
3. Chemical attack.
4. Reductions in foundation bearing capacity.
5. Heaving of clay soils.

It is the magnitude of ground movements likely to have occurred in the Fawley area due to the changes in groundwater level that is of prime interest in the current research. The results of preliminary work on the assessment of ground surface settlement and heave are described below.

13 SETTLEMENT AND HEAVE ANALYSIS

Preliminary oedometer testing has shown that ground settlement in the Fawley area, due to the lowering of groundwater levels in the Whitecliff Sand-Bracklesham Group aquifer, should have amounted to around 80mm, and that by 1995 nearly all of this settlement should have been recovered through ground surface heave.

Ground movements of these magnitudes would almost certainly have been noted and reported due to their potential effects on the integrity of engineering structures at the refinery. No such movements have been reported, however, and it is considered that there are three possible explanations for this.

Firstly, the period of time for which the groundwater level was depressed was relatively short when compared, for instance, to the situation in the London area (introduced below). This would have allowed only minor changes in the pore water pressure, and hence small magnitudes of settlement and heave, within the clay confining layer. However, both the low permeability of the Barton Clay and the limited time period for which groundwater levels were depressed, were taken into account during the settlement analysis, and, therefore, any variations between predicted and observed changes in the ground surface level due to these factors are considered to be minimal.

The second possible explanation for the discrepancy between expected and observed ground movements in the Fawley area, is that the samples of Barton Clay used in the preliminary tests were obtained from shallow excavation work some 30km to the south-west of Fawley. These samples, therefore, may not show the same permeability and consolidation characteristics as the clay underlying the refinery site. However, the consolidation characteristics of the Barton Clay in the refinery area, as measured by Soil Mechanics Limited (various dates) during previous site investigation work, predict similar, and in some cases greater, magnitudes of settlement and heave than those calculated from laboratory data obtained in the current research.

The third possible explanation is that, as proposed by Bromhead (in press), the permeability of a clay layer decreases with depth, due to increased overburden pressure. Thus, variations in the groundwater level of a confined aquifer may change the effective stress in the deeper levels of an overlying clay deposit by an amount which is substantially less than that predicted by basic soil mechanics principles.

This would lead to the actual magnitudes of settlement and heave caused by such variations to be smaller than those predicted using standard settlement analysis techniques.

The implications of this third point are quite wide ranging, and may have to be considered in a number of current rising groundwater level investigations, including that being conducted in London.

Groundwater levels in an aquifer comprised of the Cretaceous Chalk and Tertiary sands, and confined by the Tertiary London Clay, fell by up to 70m in the London area due to exploitation of water for industrial and public supply in the nineteenth and early twentieth centuries. Changing industrial practices in the last 50 years, however, have led to a decrease in the volume of water being abstracted, and since 1965 groundwater levels in the confined aquifer have been rising at a rate of around 1m/year in much of the area.

Topographic surveys in London suggest that up to 200mm of settlement of the ground surface level has occurred since 1865, and it is predicted that this settlement may reach an ultimate value of up to 250mm. If groundwater levels in the aquifer return to their original levels, ground surface heave may amount to up to 200mm (Simpson, Blower, Craig and Wilkinson 1989).

The London Clay is quite similar to the Barton Clay, which confines the Whitecliff Sand-Bracklesham group aquifer in the Fawley area, both in terms of geological history and geotechnical properties (Table 2). Hence, it is possible that if basic soil mechanics principles do not fully explain the response of the Barton Clay to rises in groundwater level, the same may be true for the London Clay. The magnitudes of heave predicted to occur as groundwater levels rise in the London Basin, therefore, may be far in excess of those that actually do occur.

Unlike in the Fawley area, topographic surveys are quite frequently conducted in London, and in the coming years these surveys should provide useful information on how the relative level of the ground surface is changing. By combining data from these surveys with that obtained from continuing groundwater level monitoring programmes, it should be possible to attain a greater understanding of the mechanical processes taking place within the London Clay as groundwater levels continue to rise. This should allow more accurate predictions of ground surface heave to be made not only for the London area, but also for a wide range of locations experiencing rises in groundwater levels in aquifers confined by similar clay deposits.

14 CONCLUDING REMARKS

This paper has described the changes in groundwater level that have occurred due to changes in groundwater abstraction rates at the Esso oil refinery in Fawley, Hampshire, in the period 1945-1995. It has been shown that since 1955 groundwater levels have risen by up to 62m and that, in theory, this rise in groundwater level should have produced a heave of up to 80mm at the ground surface. No such ground movements have been reported, however, and hence it appears that, in this case at least, basic soil mechanics principles are not totally applicable when the geotechnical effects of rises in groundwater level are being investigated.

Continuing work on the rise in groundwater levels in the London area should improve the current understanding of this problem.

15 ACKNOWLEDGEMENTS

The authors wish to thank Mr. T. Ford of the Esso Petroleum Company Limited for providing much of the borehole data included in this investigation.

This research is supported by a bursary provided by the Faculty of Science of the University of Portsmouth, U.K.

Table 2. Geotechnical properties of the Barton Clay, including those of the 'A3' horizon tested in the current research, and the London Clay.

Property	Barton Clay generally	'A3' horizon specifically	London Clay generally
Liquid limit, w_l, (%).	45 - 82[a]	66[a]	66 - 100[d,e]
Plastic limit, w_p, (%).	21 - 29[a]	23[a]	22 - 34[f,g]
Natural moisture content, w, (%).	17 - 28[a]	23[c]	23 - 49[h,i]
Clay fraction, (%).	35 - 65[a]	50[c]	42 - 72[d,f]
Bulk Density, p, (Mg/m³).	1.9 - 2.0[b]	2.0[c]	1.7 - 2.0[h,f]
Coefficient of volume compressibility, m_v, (m²/MN).	-	0.013 - 0.155[c]	0.05 - 0.18[j]
Coefficient of consolidation, c_v, (m²/yr).	-	0.159 - 1.466[c]	0.2 - 2.0[j]

References. [a] Kilbourn (1973). [b] Soil Mechanics Limited (various dates). [c] Current author. [d] Bishop, Green, Garga, Anderson and Brown (1971). [e] Skempton (1961). [f] Bishop (1966). [g] Skempton and Petley (1967). [h] Hutchinson (1967). [i] Hutchinson (1970). [j] Morton and Au (1974).

Note. The data contained in this table is from tests on weathered material, as unweathered samples of the Barton Clay have not been available in the current research.

REFERENCES.

Bishop, A. W. 1966. The strength of soils as engineering materials. *Geotechnique*: 16, 91-130.

Bishop, A. W., G. E. Green, V. K. Garga, A. Anderson & J. D. Brown 1971. A new ring shear apparatus and its application to the measurement of residual strength. *Geotechnique* 21: 273-328.

Brassington, F. C. 1990. Rising groundwater levels in the United Kingdom. *Proceedings of the Institution of Civil Engineers*, Part 1: 88, 1037-1057.

Bromhead, E. N. (In press). Interpretation of pore water pressure profiles in underdrained strata. *Proc. Int. Conf Groundwater Problems in Urban Areas*, London.

Edwards, R. A., & E. C. Freshney 1987a. Lithostratigraphical classification of the Hampshire Basin Paleaogene Deposits (Reading Formation to Headon Formation). *Tertiary Research* 8: 43-73.

Edwards, R. A., & E. C. Freshney 1987b. Geology of the country around Southampton. *Memoir of the British Geological Survey*, Sheet 315 (England and Wales).

Gilchrist, I. 1991. How it all began. Supplement to: *Insite Magazine* (Esso internal magazine), 66.

Hutchinson, J. N. 1967. The free degredation of London Clay cliffs. *Proc. Geotech Conf.*, Oslo, 1: 113-118.

Hutchinson, J. N. 1970. A coastal mudflow on the London Clay cliffs at Beltinge, north Kent. *Geotechnique* 20: 412-438.

Kilbourn, P. C. R. 1973. *Further studies of the Barton Clay coastal exposure at Highcliffe, Hampshire.* MSc. Dissertation, Southampton University.

Melville, R. V. & E. C. Freshney 1982. *The Hampshire Basin and adjoining areas* (4th Ed). London: Institute of Geological Sciences.

Morton, K. & E. Au 1974. Settlement observations on eight structures in London. *Proc. Conf. Settlement of Structures*, Cambridge, 183-203.

Simpson, B., T. Blower, R. N. Craig and W. B. Wilkinson 1989. *The engineering implications of rising groundwater levels in the deep aquifer below London.* London: Construction Industry Research and Information Association.

Skempton, A. W. 1961. Horizontal stresses in over-consolidated London Clay. *Proc. 5th Int. Conf. Soil Mech. Found. Eng.*, Paris, 1: 351-357.

Skempton, A. W. & D. J. Petley 1967. The strength along structural discontinuities in stiff clays. *Proc. Geotech. Conf.* Oslo, 2: 29.

Soil Mechanics Ltd. Various Dates. *Reports of site investigation works carried out at the Fawley refinery.* Unpublished reports.

Southern Water 1988. *Proposed groundwater abstraction by C.E.G.B. at Fawley.* Unpublished report.

Wilkinson, W. B. 1985. Rising groundwater levels in London and possible effects on engineering structures. *Proceedings of the 18th Congress of the International Association of Hydrogeologists*, Cambridge, 145-157.

Étude de l'écoulement dû à la marée dans un massif portuaire

Study of tidal flow in an harbour bank

Alain Alexis & Aïssa Rezzoug
Laboratoire de Génie Civil ECN, IUT de Saint Nazaire, France

ABSTRACT : The tidal flow occuring in an estuarine bank is a complex physical process located at the boundary between hydraulics and geotechnics, which creates undermining and instabilities on harbour slopes and structures. In this paper, this phenomenon is studied by means of a theoretical approach, followed by a numerical modelling and is compared with laboratory experiments and field measurments. We propose a theoretical modelization of the free surface in a bank, and the results of the numerical solving are compared with those of an original physical model (tide simulation on homogeneous sandy slice). This model simulates the behaviour of quays or slopes and gives the experimental cards of equipotential lines. The theoretical results and physical experimental model show a raising of water table level in the harbour slopes and a disturbed area which risks to alter behaviour of the material. The comparaison between theorical and experimental results shows good agrements between theory and laboratory experiments, specific contribution of field measurements.

RESUME : L'écoulement dans un talus portuaire soumis à la marée constitue un processus physique complexe à la frontière hydraulique-géotechnique. Nous proposons ici une approche théorique complétée par une résolution numérique comparée à des expérimentations en laboratoire et sur site. L'écoulement engendre dans un tel talus des perturbations dans sa stabilité. La modélisation théorique de la surface libre, basée sur la loi de Darcy dans la zone saturée, conduit à une équation aux dérivées partielles non linéaires. Sa résolution numérique est comparée aux résultats d'un modèle physique original (simulation de marée sur une tranche de sable fin homogène), simulant le comportement de talus ou de quais et permettant d'obtenir des cartes de lignes équipotentielles expérimentales. La comparaison des résultats théoriques et expérimentaux au laboratoire et en nature fait apparaître une concordance entre théorie et expérimentation complétée par un apport spécifique des mesures sur site validant ces résultats.

1 INTRODUCTION

L'écoulement dans un talus littoral ou estuarien soumis à la marée perturbe la stabilité des pentes de rives et de quais. Un tel phénomène constitue un important problème pour les aménagements portuaires. Lors d'un cycle de marée le talus se remplit et se vide alternativement. La surface libre de l'écoulement dans le sol oscille et il apparaît un amortissement de l'onde marée à l'intérieur du talus.

Dans cet article ce phénomène est étudié au moyen d'une approche théorique complétée par une résolution numérique. Les résultats de notre programme de calcul sont comparés aux mesures effectuées sur un modèle physique original dont nous décrivons le fonctionnement et l'exploitation. Ce modèle constitué d'un simulateur de marée relié à une tranche de sable fin homogène avec prises de pression, permet de simuler le comportement de talus et de quais.

Nous appliquerons ensuite cette étude à un cas réel en site naturel.

2 THEORIE ET MODELE NUMERIQUE

2.1 Modélisation

Nous cherchons ici l'équation de la surface libre s'établissant dans le talus sous l'action de la marée. Le talus est considéré comme un massif homogène, poreux, perméable, indéformable, et semi-émergé subissant une marée sinusoïdale de période T, d'amplitude A et de niveau moyen H_m constants. Le milieu poreux du massif est caractérisé par sa perméabilité k et sa porosité n et repose sur un substratum imperméable, (Fig.1). Nous supposons que l'écoulement suit la loi de Darcy. La ligne de surface libre suit l'oscillation de la marée sur le côté mer du talus et se stabilise progressivement en s'éloignant vers le côté terre pour donner un niveau constant que nous appelons ici le niveau d'équilibre Ne.

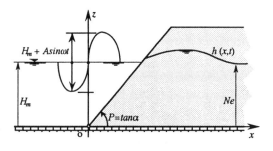

Fig.1. Modélisation de la surface libre induite par la marée dans un talus de rive.

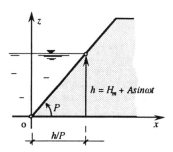

Fig.2. Modélisation de la condition sur la pente.

2.2 Formulation

L'équation, non linéaire, suivante a été obtenue par les travaux de Polubarinova (1962) sur les écoulements non permanents à surface libre. Elle s'écrit dans le cas plan,

$$\frac{\partial h}{\partial t} = \frac{k}{2n}\frac{\partial^2 h^2}{\partial x^2} \tag{1}$$

Dupuit ayant utilisé cette équation non linéaire pour les problèmes de rabattement en régime permanent, nous l'appelons donc équation de Dupuit en transitoire.

Dans le cas d'un talus présentant une pente P, la condition à la frontière eau-pente est (Fig.2) :

$$h\left(x = \frac{H_m + A\sin\omega t}{P}, t\right) = H_m + A\sin\omega t \tag{2}$$

La dépendance des variables dans la condition (2) rend la discrétisation de la pente plus délicate. Pour revenir à une condition plus simple aux variables indépendantes, nous proposons le changement de variables suivant :

$$X = x - (H_m + A\sin\omega t)/P$$
$$H(X,t) = h(x,t) \tag{3}$$

l'équation (2) devient alors :

$$\frac{\partial H(X,t)}{\partial t} = \frac{k}{2n}\frac{\partial^2 H^2(X,t)}{\partial X^2} + \frac{A\omega}{P}\cos\omega t\,\frac{\partial H(X,t)}{\partial X} \tag{4}$$

Des solutions analytiques partielles de cette équation non linéaire ont pu être établies, (Rezzoug et al. 1993 a).

2.3 Modèle numérique

2.3.1 Programme de calcul

L'obtention d'une solution numérique satisfaisante est possible en utilisant un schéma explicite centré en différences finies, (temps en différences avancées et espace en différences centrées) :

$$H_{i,j+1} = \frac{k}{2n}\frac{\Delta t}{\Delta X^2}[H_{i+1,j}^2 - 2H_{i,j}^2 + H_{i-1,j}^2]$$
$$+ \frac{A\omega}{P}\frac{\Delta t}{2\Delta X}\cos(\omega j\Delta t)[H_{i+1,j} - H_{i-1,j}] + H_{i,j} \tag{5}$$

Avec : $X = i\Delta X$ et $t = j\Delta t$

Le calcul débute à la frontière par :

$$H_{0,j} = H(X = 0, t = j\Delta t) = H + A\sin(\omega j\Delta t) \tag{6}$$

L'étude de stabilité des calculs donne la condition nécessaire, mais non suffisante, (Rezzoug 1990) :

$$\frac{k}{2n}\frac{\Delta t}{\Delta X^2} \leq \frac{1}{4A} \tag{7}$$

Notre programme MSDIF est basé sur cette résolution en différences finies. Les calculs sont initialisés (à $t = 0$) par une courbe voisine du niveau moyen, et poursuivis jusqu'à la périodicité des résultats.

2.3.2 Résultats

La figure 3 montre l'évolution de la surface libre dans un talus à chaque heure de la marée. La pente du talus est $P = 0,5$ (inclinaison 26 °). Le sol est un limon de perméabilité $k = 2,2 \cdot 10^{-5}$ m/s et de porosité 0,3. La marée est d'amplitude 2 m de niveau moyen $H_m = 8$ m. Ces courbes montrent un niveau d'équilibre Ne s'établissant 1 m au dessus du niveau moyen. Nous notons également la présence d'une "zone perturbée", s'étendant du niveau supérieur de la nappe jusqu'au substratum, (sa longueur horizontale est définie par une amplitude égale à 10 % de cette de la marée).

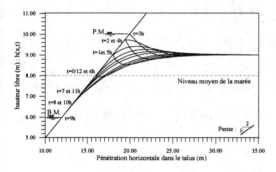

Fig. 3. Evolution de lignes de surface libre
(diagramme enveloppe).

Nous avons également entrepris cette étude comparative (Rezzoug 1994) avec d'autres modèles mathématiques comme :
 - la théorie générale des nappes aquifères qui traite des écoulements à surface libre dans les milieux saturés et utilise l'hypothèse de Dupuit ;
 - la théorie de Richard basée sur la prise en compte de la zone non saturée.
Les résultats des différents modèles se révèlent en bon accord.

Les résultats numériques montrent une onde progressive d'écoulement pénétrant à l'intérieur du talus et s'amortissant sur une distance fonction de la perméabilité du milieu. Les calculs mettent en évidence :
 - une surcote du niveau d'équilibre (Ne) dans le talus par rapport au niveau moyen de la marée (H_m).
 - une zone perturbée par des changements importants de vitesse d'écoulement et de pression interstitielle.

3 EXPERIMENTATION EN LABORATOIRE

3.1 Modèle physique

3.1.1 Description

Pour confronter les résultats théoriques à l'expérimentation, nous proposons un modèle physique bidimensionnel, original, qui permet d'effectuer un suivi de l'évolution de la surface libre.

La figure 4 montre d'une façon globale les deux parties principales du montage de cette installation : le système d'entraînement (simulation de la marée), et le bac de mesures.

Un mécanisme excitateur composé d'un moteur entraînant un entonnoir constamment en trop-plein, constitue le système d'entraînement. L'entonnoir tourne en décrivant un cercle dans un plan vertical, et permet d'obtenir une variation sinusoïdale du niveau d'eau simulant la marée.

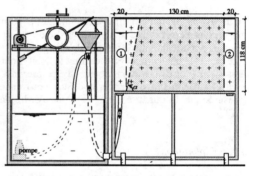

système d'entraînement   bac de mesures

→ sens de l'écoulement ① simulation du côté mer
+ prises de pression   ② simulation du côté terre

Fig. 4. Schéma du modèle physique expérimental.

Le domaine d'expérimentation est constitué, dans son ensemble, d'un bac de mesures de forme parallélépipèdique de dimensions 1,3 m x 1,18 m x 0,2 m. Ce bac repose sur une base horizontale imperméable et contient une tranche de sol comprise entre deux parois latérales de Plexiglas (rigidifiées par des profilés métalliques). Chaque paroi est également en contact avec un réservoir ayant la profondeur du bac, une section de 20 cm x 20 cm, et communique directement avec la tranche à travers des grilles. Nous avons appelé "réservoir côté mer", le réservoir amont où la simulation du phénomène de la marée est appliquée, et "réservoir côté terre" le réservoir du côté aval de la tranche où le niveau d'eau est laissé libre.

Ce dernier est exploité pour l'observation et la simulation de la limite infinie.

Les mesures de charges sont effectuées à l'aide de tubes piézométriques connectés aux prises de pressions réparties sur un maillage quadrilatère.

Le remplissage s'effectue par pluviation du sable dans le bac rempli d'eau, afin de s'assurer une bonne homogénéité du sol et minimiser les problèmes liés aux bulles d'air.

Nous pouvons faire varier l'angle de l'inclinaison α de l'interface eau-sol du côté mer de la tranche, dans le but de simuler la pente d'un talus.

Nous avons ainsi pu simuler les conditions aux limites naturelles en établissant une nappe à surface libre et en faisant varier le niveau pour représenter l'action de la marée, (Alexis *et al.* 1990). Nous avons ainsi pu suivre l'évolution de la surface libre et le champ de pression dans la zone saturée lors des oscillations périodiques de la nappe.

Nous avons également utilisé ce montage pour mesurer la perméabilité k de la tranche de sol en place en appliquant une différence de charge entre le côté mer et le côté terre et en mesurant le débit. Pour le sable fin utilisé nous avons obtenu une perméabilité $k = 8{,}77 \ 10^{-4}$ m/s, et une porosité efficace moyenne (mesurée par la quantité d'eau libre) de l'ordre de 0,12.

3.1.2 Exploitation

La réalisation des mesures nous a conduit à deux observations concernant :

- la périodicité : l'oscillation de l'onde incidente sinusoïdale (côté mer), est transmise à la surface libre à l'intérieur du massif, qui oscille avec la même périodicité ;

- le niveau d'eau : il apparaît une surélévation dans le réservoir côté terre par rapport au niveau moyen.

Pour chaque jeu de données expérimentales (pente, période, amplitude, niveau moyen...), les valeurs de pression recueillies sont représentées sous forme d'une carte de lignes équipotentielles, grâce à un programme d'interpolation. La position de la surface libre est calculée par la suite et mise en évidence sur la même carte.
Nous avons choisi, ici, de donner un exemple de nos résultats expérimentaux. La figure 5 montre une carte de lignes équipotentielles de l'écoulement dans la tranche de sol de notre modèle physique. L'axe

vertical donne la valeur de la charge hydraulique dans la zone saturée entre la base imperméable et le niveau de la surface libre, alors que l'axe horizontal donne la pénétration en cm dans la tranche. Cette carte est représentée pour les données : pente $P = \infty$, amplitude $A = 15{,}8$ cm et niveau moyen $H_m = 82{,}5$ cm pendant un cycle réduit $T = 120$s, à mi-flot.

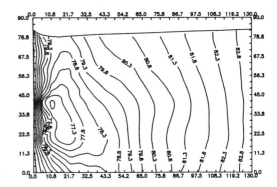

Fig. 5. Lignes équipotentielles et surface libre à mi-flot, (écran vertical).

Les différentes cartes réalisées montrent la zone perturbée qui est :
- en dépression à mi-flot,
- en surpression à mi-jusant,
- amoindrie pendant le reste du cycle.
En s'éloignant vers le côté terre de la tranche, nous constatons la quasi-verticalité des lignes équipotentielles.

3.2 Comparaison des modèles numérique et physique

Les figures (6a) et (6b) montrent deux exemples de comparaison entre le modèle mathématique basé sur l'équation de Dupuit en transitoire non linéaire et le modèle physique de laboratoire. Elles concernent une tranche de sable fin (§ 3.1.1) dans les conditions citées dans le tableau 1.

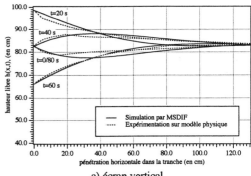

a) écran vertical

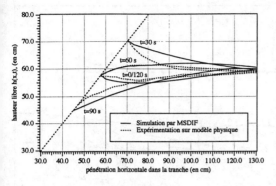

b) pente unité

Fig.6. Comparaisons de niveaux piézométriques et allure de surface libre pour une tranche de sable en cas de pente unité.

Nous remarquons une très bonne concordance entre les courbes de surface libre expérimentales et les résultats de MSDIF. Les écarts significatifs se situent dans la zone perturbée. Cette confrontation permet une validation de la modélisation tant pour la surélévation du niveau d'équilibre, que pour l'estimation de la longueur de la zone perturbée.

Tableau 1 Récapitulation et comparaison des grandeurs H_m, Ne, et Ne théorique pour 2 tranches de sable.

	Niveau d'eau dans le réservoir (cm)					
T	Côté mer				Côté terre	
(s)	H_{max}	H_{min}	H_m	A	Ne	Nethéor.
Ecran Vertical : Fig. 11.a.						
80	98,7	66,3	82,5	16,2	83,3	83,3
Pente = 1 : Fig. 11.b.						
120	70,3	44,9	57,6	12,7	58,7	59,9

4 EXPERIMENTATION EN NATURE

Compte tenu des hypothèses restrictives de toute modélisation (mathématique ou physique) et compte tenu de la complexité du phénomène naturel, il nous a paru important d'effectuer des mesures sur site naturel pour les comparer aux résultats de la modélisation.

4.1 Description du site

Le site des Sablières (figure 7), est un talus de remblai situé sur la rive nord de l'estuaire de la Loire, à proximité du Pont de Saint Nazaire. Le schéma de la coupe géologique montre des dépôts successifs de couches de sable et de vase. Le sol peut être qualifié de sable localement vasard de compacité moyenne. Pour stabiliser la pente qui est déjà assez douce (environ 1/5), le talus porte un enrochement en pierres naturelles (sécurité contre l'effondrement et l'entraînement par le courant).

La pleine mer (5 m cote marine, notée C.M.) couvre quasiment la pente d'enrochement. Alors qu'à basse mer (environ 1,5 m C.M.), le pied du talus est découvert, la hauteur d'eau est d'environ 1,5 m C.M.. La basse mer découvre un fond vaseux traversé par des trous d'affouillement d'environ 50 cm de diamètre, à travers lesquels le talus se vide pendant le jusant.

Nous avons réalisé 5 forages (F1 à F5), de 2,5 m de profondeur sur le site, forés le long d'une ligne normale à la rive et distants respectivement du point de la pleine mer de 13,00 m ; 20,95 ; 29,35 ; 38,89 ; 67,16 (Fig. 7). Ces forages ont permis la mesure du niveau d'eau, et donc d'effectuer le suivi de la surface libre pendant le cycle de marée.

4.2 Comparaison des mesures en nature et des résultats numériques

4.2.1 Mesures en nature

Les mesures ont été relevées le long d'un cycle d'une marée

- coefficient de marée 44,
- basse mer : 1,75 m C.M. à 8 h 17
- pleine mer : 4,20 m C.M. à 14 h 36

La figure 8 synthétise les mesures de niveau d'eau effectuées pendant le cycle de marée en joignant simplement tous les points de mesures. Le système d'axes est composé de la hauteur verticale mesurée en mètre C.M. et de la distance mesurant la pénétration horizontale dans le remblai. Les résultats sont représentés à chaque quart du cycle : mi-flot ($t = 0$), pleine mer ($t = 3$ h), mi-jusant ($t = 6$ h) et basse mer ($t = 9$ h), (Fig.8)

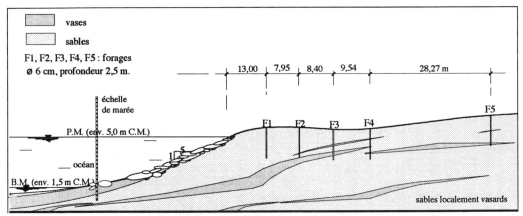

Fig.7. Site des Sablières, coupe du remblai.

Deux résultats importants sont à noter sur la figure :

- *la variation du niveau d'eau* est importante au voisinage de l'interface eau-sol et diminue e n s'éloignant de celui-ci. Au bout d'une distance horizontale d'environ une centaine de mètres comptée à partir du milieu de l'interface intertidale, le niveau se stabilise complètement pour donner ce que nous avons appelé le niveau d'équilibre.

- *le niveau d'équilibre* dans le remblai est voisin de celui de la pleine mer. Il est à noter que le niveau d'équilibre ($Ne = 4{,}57$ m C.M.) est supérieur à celui de la pleine mer du jour (4,20 m C.M.). Ce résultat semble indiquer que le niveau d'équilibre s'établirait autour d'une moyenne sur un cycle lunaire des niveaux de pleine mer.

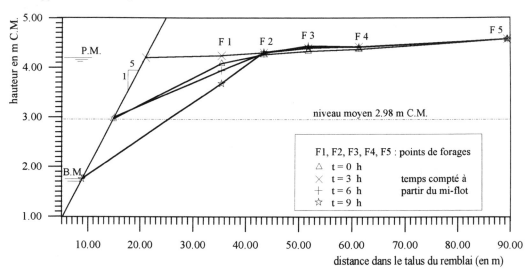

Fig.8. Résultats des mesures de l'évolution du niveau piézométrique dans le remblai des Sablières.

4.2.2 Simulation numérique du site naturel

Du fait de l'hétérogénéité du massif, la perméabilité et la porosité locales ne peuvent être représentatives. Nous avons estimé les valeurs globales à : $k = 2{,}6 \; 10^{-4}$ m/s et $n = 0{,}2$.

La figure 9 représente les résultats numériques obtenus par le programme MSDIF, dans la configuration du site naturel (marée citée au § 4.2.1). Nous constatons un relativement bon accord de l'allure des courbes de surface libre entre les mesures sur site et la simulation par MSDIF. Cependant la longueur de la zone battue calculée est inférieure à celle mesurée. Ceci est certainement dû à la sous-estimation de la valeur de perméabilité. D'autre part, le niveau d'équilibre mesuré (4,57 m C.M.) est supérieur à celui obtenu par la simulation (4,00 m C.M.) avec le coefficient de marée du jour (44).

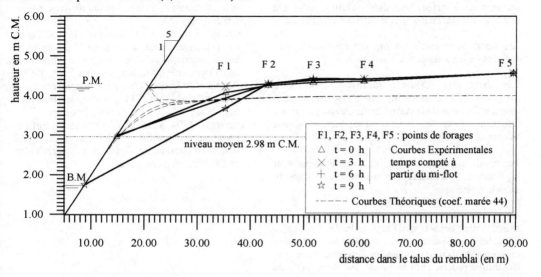

Fig.9. Résultats d'une simulation numérique du remblai des Sablières pour une marée de coef.44

La remarque du paragraphe 4.2.1 nous incite à procéder à un calcul avec un coefficient de marée plus important (74) dont les résultats sont représentés sur la figure 10. Nous remarquons un accord plus satisfaisant entre les niveaux d'équilibre mesuré et simulé. Ce résultat semble confirmer un établissement du régime d'écoulement fonction du cycle lunaire et non du cycle journalier de la marée.

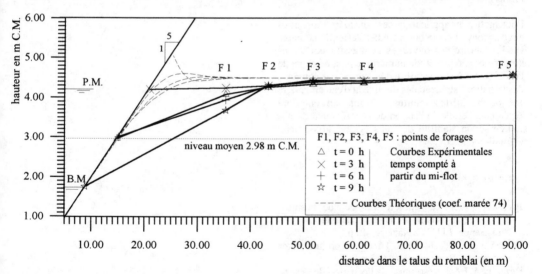

Fig.10. Résultats d'une simulation numérique du remblai des Sablières pour une marée de coef.74

5 CONCLUSION

Dans cet article nous avons abordé la modélisation de l'écoulement à surface libre dans un talus soumis à la marée à l'aide de l'équation de Dupuit en transitoire.

Les résultats numériques ont été confrontés aux mesures en laboratoire et en nature.

Les différents résultats montrent une onde progressive d'écoulement, qui s'amortit sur une distance fonction de la perméabilité du sol. Le niveau d'équilibre atteint dans le talus est supérieur au niveau moyen de la marée.

Les résultats du laboratoire, où les paramètres sont bien cernés, sont bien représentés par les résultats du modèle numérique .

En nature l'hétérogénéité du sol et la complexité de la marée réelle (déchet, revif) rendent la simulation plus délicate. Les résultats de la modélisation révèlent cependant satisfaisants et tendent à montrer un niveau d'équilibre proche de la moyenne des pleine mers du cycle lunaire.

Tous ces résultats montrent qu'il est certainement nécessaire de prendre en compte un niveau voisin de la pleine mer lors de l'évaluation de la stabilité des ouvrages portuaires (quais, gabions et talus) semi-immergés.

Pour mettre en application ces considérations, nous envisageons, dans le but d'étudier l'effet de la marée sur la stabilité des ouvrages et massifs portuaires semi-émergés, une étude numérique paramétrique de l'écoulement à surface libre dans un massif soumis à des variations sinusoïdales du niveau d'eau en variant les perméabilités, pentes... Nous envisageons également l'étude de l'action de l'écoulement dû à la marée sur la stabilité du talus (grand glissement, affouillement...).

REFERENCES

Alexis A. 1990. Réalisation d'un modèle de mesure des écoulements dans un massif en situation estuarienne, I.U.T St Nazaire, 46 p.

Polubarinova, P.Y.A. 1962. Theory of ground water movement, *Princeton Ed.*, 613 p.

Rezzoug A.1990. Résolution de l'équation de la surface libre de l'écoulement dans un talus soumis au marnage, *Mémoire de D.E.A.*, E.N.S.M. Nantes, 118 p.

Rezzoug A., Alexis A. 1992. Impact d'un écoulement cyclique sur un ouvrage semi-émergé, *Journées nationales de Génie Côtier-Génie Civil*, 26-28 février, Nantes, pp.222-231.

Rezzoug A., Alexis A. et Thomas P. 1993a. Ecoulement dans les talus intertidaux. *Xème Forum des Jeunes Océanographes*, 1 et 2 Avril, Wimereux. En cours de parution dans le *Journal de Recherche Océanographique*.

Rezzoug A., Alexis A. et Thomas P. 1993b. Theory and experimental validation of tidal seepage in banks. *International Conference on Hydro-Science and Engineering*, Washington June 7-11, 1993. *Advances in Hydro-Science and Engineering*, volume I part B, pp 1711-1716 ; edited by Sam S.Y. Wang. The University of Mississipi.

Rezzoug A. 1994. Modèles d'écoulement à surface libre dans un massif soumis à la marée. Application au comportement d'un talus portuaire. Thèse de Doctorat en préparation.

Design of groundwater preservation method by recharge-well system

Projet d'une méthode de préservation de l'eau souterraine par un système de recharge

M. Yanagida
Nippon Koei Co., Japan

M. Nishigaki
Okayama University, Japan

T. Uno
Gifu University, Japan

H. Nagai
Japan Highway Public Corporation

ABSTRACT: This recharging system is designed and executed, in order to restore the groundwater level on the downstream side close to the original condition, in view of the necessity to preserve the environment surrounding the groundwater, when the roads with the open–cutting structure continuously cut into the existing aquifers. This method applies the siphonic recharging as the basic system. Besides, various in situ tests and groundwater simulation analysis are carried out to evaluate the effectiveness of this engineering method and also to decide its suitable scale and basic form as well. Furthemore, fullscale test facilities are constructed to verify the actual effectiveness whereas efficiencies of various other recharge methods are reviewed in comparison with this system.

RÉSUMÉ: Ce système de recharge a été projeté et exécuté pour restaurer le niveau de la nappe du côte aval a fin de préserver des eaux souterraines quand les excavations routières traversent d'une forme continue les couches aquifères existantes. Basiquement, cette méthode applique la recharge par siphon. On exécuté aussi plusieurs essais in situ et des analyses de simulation hydraulique pour évaluer l'efficacité de cette méthode et décider sur l'échelle convenable et leur forme. En outre, on a construit des dispositifs en vrai grandeur pour vérifier l'efficacité réel et faire en même temps une étude comparative avec d'autres systèmes de recharge.

1 PREFACE

Nowadays, preservation of the environment surrounding the groundwater is observed as one of the most important assignments when execuiting construction projects. One measure to attain the above will the recharge system based on the artificial groundwater occurrence. However, almost all recharging works in Japan have been done as the temporary measures only during construction works to prevent land subsidence in the vicinity to be caused by the dewatering method. Besides, functional declines due to clogging have further obstructed the development of the above into a permanent preventive measure.

It was feared that shortage of water may result due to cutting into aquifers in connection with the planned road construction based on the continuous open–cutting structure (cov-

ering about 1 km). Accordingly, it had been decided to take necessary measures to preserve groundwater. The system is the groundwater conservation method which comprises of peri–style cut–off walls at the top of slopes on both sides of the road, which secure the stability of the slopes by cutting off the ground –water, and at the same time, artificially lot it occur and recharge by supplying groundwater on the downstream side, using dam–up effect on the upstream side and also applying the easy to control "reversal siphon system".

This report is to introduce the results of the study on the scale and structure of the recharge system, which reflects the road structure and the local hydraulic and geologic conditions, based on the results of simulation analysis and test construction.

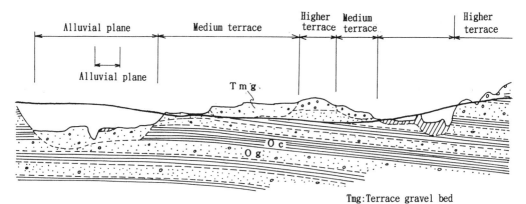

Fig.1 Topographic— and geological—situation in the open—cutting section

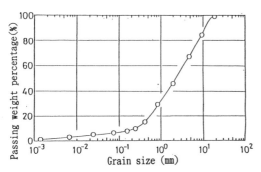

Fig.2 Grain size accumulation curve
(Terrace gravel bed)

2 PREPARATORY INVESTIGATION

2.1 *Present state of the area in the vicinity of the planned route*

Configuration of the ground in the vicinity of the planned route consists of high and low terraces built up by rivers on both sides, whereas its geological structure comprises of Osaka formations accumulated during Neogene and Quaternary periods and scattered terrace accumulations which cover the Osaka formations unevenly. (Fig.1)

The surfaces of the terrace accumulations vary substantially according to the location, however, sand and gravel bed form the shallow unconfined aquifers which serve as the intake source for some 250 shallow wells in the vicinity. On the other hand, the Osaka formations consist of unconsolidated clay, sand and grav-

el or alternation of these strata, while the sand and gravel bed form the confined aquifers. Test pumping indicates that the coefficient of permeability of each aquifers is 10^{-3} cm/sec, however, because of uneven soil layers, some locations seem to have higher permeability or seepage. (Fig.2)

Free groundwater in terrace gravel stratum is distributed among shallow places of GL— 1 to 2 meters and in general flows radiately to— ward rivers on both sides, just as to cut a— cross the planned route. Confined artesian head in Osaka formations are distributed 1 or 2 meters below the free groundwater level.

2.2 *Simulation analysis of groundwater*

In order to assess the influence of the const— ruction works upon the groundwater in the vic— inity as well as the effect of the preventive measure, simulation analysis of groundwater is carried out. The FEM quasi—three—dimensional aquifer model is used for this analysis. The study covers the area 700 meters from the road center on the upstream side and 800 meters on the downstream side, with the sideward limits up to the rivers on both sides. In our model, the topmost clay stratum of the Osaka forma— tions is considered as the relatively imperme— able layer, and the analysis was performed on the free groundwater in terrace accumulations supported by such clay stratum.

Based on the analysis on this model, it is estimated that the water level will decline by roughly 0.5 meter in the area within 300 me— ters on the downstream side of the road due to

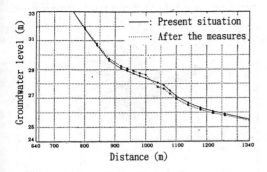

Fig.3 Analysis on measure's effect

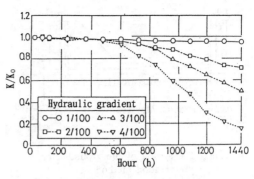

Fig.4 Change in permeability coefficient

cutting into aquifers, affecting some 40% of the existing wells and will further cause shortage of water for reservoirs and for irrigating rice paddies. For the above, the simulation analysis is carried out to verify the recharging effect of the artificial occurrence as a prevention measure. The study results indicate that the water level after the preservation measures will be roughly the same as the existing one both on upstream and downstream sides, in case the present volume of groundwater flow in the open–cutting area is replenished at 10 different places, meaning that the preservation measure is theoretically effective. (Fig.3)

2.3 Assessing the permeability characteristics of the ground

Indoor permeability test is performed on the samples of terrace gravel stratum obtained from the construction site. The test results indicate the decline in permeability due to movements of the fine–grain (advance of clogging) even with smaller hydraulic gradient than the existing groundwater flow gradient in the construction site. It is considered, however, that the disturbed test sample lacks the stable waterways which exist in the construction site, thus, even with smaller hydraulic gradient, the clogging takes place. In other words, the actual groundwater seems to be flowing through "local waterways", which may require additional attention to the irregular groundwater flows, when mapping out the preservation measure. (Fig.4)

3 THE SYSTEM OF THE GROUNDWATER PRESERVATION METHOD

3.1 The basic form

The necessary conditions of the groundwater conservation system is being able to replenish groundwater on the stable bases;both time–wise and area–wise. Limiting the clogging as little as possible is necessary to accomplish the above, and in our method, the basic policy shall be 1) to supply necessary volume of water in the form of groundwater. Moreover, 2) the clogging shall be avoided by replenishing groundwater without causing excessive hydraulic gradient, but through expanding the permeating cross sectional area, and 3) to improve the permeating efficiency taking the uneven groundwater flows into consideration. Whereas the technical side of the construction require 4) to attain stability of the cutting slopes and 5) to increase imperviousness against the groundwater. Furthermore overall plan shall be 6) workable on the construction site, 7) reliable, 8) easy for maintenance and 9) economically advantageous.

Having these conditions mentioned above 1) through 9) as the basic policy for the groundwater preservation method, it has been decided that the system based on the recharge method using "reversal siphon system" as shown in the Fig.5 shall be adopted. In this method, the following system shall be regarded as one unit, and multiple units will be installed in the site to achieve the necessary recharge volume.

The basic structure of one unit will be:

    1. The peristyle cut–off walls will be constructed to secure the stability of the cutting slopes of the road and to secure hydraulic gradient.

    2. Behind the peristyle cut–off walls, the drainage facility (drainage–well) shall be in–

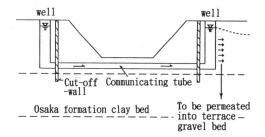

Osaka formation clay bed

To be permeated into terrace gravel bed

Fig.5 The system of the groundwater preservation method

stalled on the upstream side of the road and the occurrence facility (occurrence–well) will be built on the downstream side whereas the drainage facility and the occurrence facility will be connected by the communicating tube.

3. Devices to enlarge the effective radius of the wells, to overcome the ground irregularity and to expand the permeating cross sectional area shall be attached to this system.

3.2 *The way to expand the permeating cross sectional area*

When deciding the structures of the drainage facility on the upstream side and the occurrence facility on the downstream side, the nature of the soil layer in question and the probability of clogging shall be thoroughly studied. Furthermore, it is important to take the following items into consideration:

1. Area–wise continuous permeation shall be possible, so that the facility can cope with the changesin the soil profile of the ground to be recharged.

2. Recharge shall be performed in the situation as close to the natural hydraulic gradient as possible.

3. The facility shall be advantageous both in workability and cost. Particularly, enlarging the permeating cross sectional area may be the answers for 1. and 2. above. Therefore, three ideas as shown in the Fig.6 are discussed as the devices which can be attached to the drainage–well and the occurrence–well.

Type 1: The way to expand the effective radius of the well by installing horizontal drain holes which are situated radiately from the drainage–and occurrence–wells.

Type 2: The way to expand the effective radius of the well by installing trench–type drains radially close to the surface, which have bigger permeating cross sectional area than the

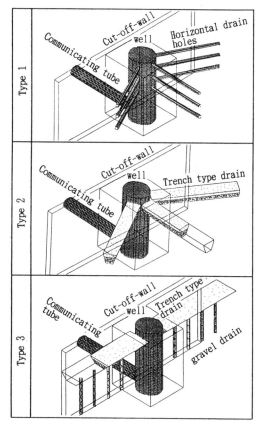

Fig.6 Various way to increase seepage

drain holes mentioned above.

Type 3: The way to expand the effective radius of the well both horizontally and vertically by installing trench–type drains along the cut–off walls and the gravel drains in vertical directions.

All devices aim at enlarging the effective radius of the drainage– and occurrence–wells, and the full scale test facilities of the Type 1 and the Type 3 are constructed to verify the actual effectiveness of the devices in the real construction site. The test is performed by installing 100 meters long peristyle cut–off walls on both sides of the road in the construction site, and also drainage– and occurrence–wells at the center of the site. During the predetermined test period of one month, groundwater levels are surveyed at 30 spots in the vicinity.

The results show that the groundwater levels rose as the test begun in both cases, meaning both devices have significant recharge effects, however, it has been further verified that the

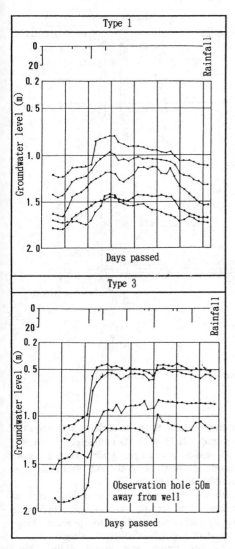

Fig.7 Results of test facility construction

Type 3 which recharges on the "Line" has better results than the Type 1 which recharge at
the "Spot", in such aspects as the proportion
of rising waterlevel, the range of effective
permeation and the volume of permeance. (Fig.7)

4 GENERAL STRUCTURE OF THE GROUN-
   DWATER PRESERVATION NETHOD

The structure of this recharge system is consisted of a "drainage–well" to collect groundwater on the upstream side, an "occurrence–
well" to recharge on the down–stream side and
a communicating tube to connect both wells.
"Trench drain" and "gravel drain" are attached
to each well, in order to improve their functions.

5 STRUCTURAL DESIGN OF THE FACILITY

5.1 *The shape of the facility*

1. Drainage–well and occurrence–well
The diameter of the well is 2 meters. Galva-
nized liner plate is used for durability, on
which $\phi 50$ millimeter holes are bored to realize 7.5% numerical aperture. Furthermore, the
well is covered with 3 millimeter mesh net to
prevent filter materials fall into the well
through drainage holes.
2. Trench drain
The trench drains which are attached to the
drainage– and occurrence–wells serve as the
"waterways" to drain (on the upstream side)
and recharge (on the downstream side) with
bigger cross sectional areas. Accordingly, its
base level is expected to be the "natural water level in dry season–0.5 meter", because 1.
it should always be flooding within the trench
(above the groundwater level), and 2. because
of the trenches both on upstream and downstream sides should be on the same level. The
shape is in principle 1.0 meter in base width
and1:0.5 slope gradient at the depth of 2.5
meters. After filling in the filter materials,
the excavated soil is refilled for the upper 1
meter, however, at the boundary between the
filter materials and refilled soil, cut–off
sheet is put in to prevent inflow of muddy water from the surface. (Fig.9)
3. Gravel drain
The gravel drains serve,together with trench
drains, to drain and recharge on infallible
and continuous bases. Crushed stones are
placed on $\phi 500$ millimeter auger holes on the
trench bottom. The distance between the holes
is 2 meters, because of the workability. The
depth of drains shall depend on the aquifers
in the construction site, however, in this
study, the result of test construction (0.26
penetration rate against the aquifer thickness)
is applied. (Fig.9)

5.2 *Selecting filter materials*

Filter materials are selected, taking the following conditions and grain distribution of
the aquifers in question into consideration:

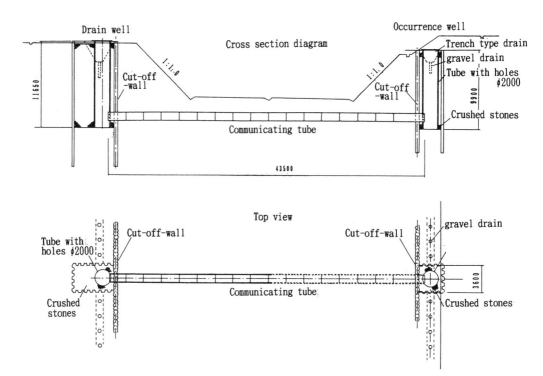

Fig.8 General structuer of the groundwater preservation method

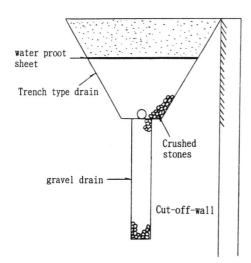

Fig.9 Trench and gravel drains

1. being equipped with sufficient permeability against the ground in the vicinity , 2. gener- ating no clogging, 3. no particle crushing in spite of compaction, and 4. no weathering al- teration after being submerged for long. Since

more than 5000m³ filter materials are required in this study, No.5 (13 −20 milimeter) ∼ No.7(2.5 −5 milimeter) simple grain crushed stones are used in view of cost and availability on the market. Furthermore,

5. As for the wells, No.5 and No.7 crushed stone are mixed at the rate of 4:6, applying a reportedly efficient method to prevent clog- ging by mixing stones of different sizes.

6. As for trench and gravel drains, No.6 crushed stones with the size in between are used.

5.3 The scale of the facility

According to the plan of this measure,the o- pen−cut site will be divided into 12 blocks, whereas 9 blocks out of them will be equipped with recharge facilities (in principle, one facility in each block). Each recharge system is required to have the drainage− and occur- rence−capacities to satisfy the necessary wa- ter supply volume in the area it covers (the distance from the center of the adjoining sys- tem). Thus, the construction scale is calcu- lated based on the following steps:

1. Necessary water supply volume of each system
Cross sectional, two dimensional seeping flow analysis is carried out per each construction block, and the present volume of the groundwater flow thus obtained is considered as the necessary water supply volume.

2. The capacity of the system to replenish (or to drain) the groundwater
Based on each shape, this recharge system can be disintegrated into three parts: the well itself, trench drains and gravel drains. Then, the theoretical capacity of each part is calculated as follow, based on the permeation result of the test construction and the hydraulic geological conditions of the tested area:

1) Wells. They are supposed to be located beside the impermeable boundary (peristyle cut-off wall), and the well formula by the "method of image" is applied.

2) Trench drains. They are considered to be equal to the wells which have similar cross-sectional shapes that of the trench drains, and the same well formula is applied. However, due to the fact that their actual depth is as shallow as 2.5 meters from the surface, the figures are adjusted by penetration rate into the aquifers, just like the case of imperfect penetrating well.

3) Gravel drains. They themselves are observed as the wells and the calculation is made for them as a cluster of wells situated close to the impermeable boundary.

3. Review on the construction scale
Based on the seepage volume of each part mentioned above, the construction scale of each block is decided based on the following procedures:

1) The average permeability of the construction block as a whole is calculated, based on the results of the site seepage test and the hydraulic gradient indicated on the groundwater contour chart. Then, the "permeation coefficient" is obtained, multiplying the average by the cross section of the aquifer in the construction site.

2) This "permeation coefficient" is compared with the one in each tested area to establish the ratio, based on which the designed permeation (or drainage) capacity of each facility in a certain block is decided.

3) The volume equivalent to the wells capacity is deducted from the necessary water supply volume in each block, then the remaining figure is divided by the "trench + gravel drains" capacity per unit length (2 meters) to obtain the required scale or length of the construc-

tion.

4) For further information, the actual construction scale is decided, taking the condition of the aquifers (effective length is zero for clay stratum) and the flowing direction of the groundwater into consideration.

5.4 *Maintenance*

The following maintenance facilities are installed in order to keep the recharge system functioning for longer period:

1. $\phi 40$ millimeter aeration pipes are attached to three peripheral places of the liner plates of the well, to make it possible to clean the filter materials from outside, by blowing in the compressed air.

2. $\phi 200$ millimeter vinyl chloride resin tube with holes is placed at the base of the trench drain in vertical position, to remove the clogging materials.

3. Observation holes are bored very close to both wells (about 1 meter from the filter zone) to investigate any development of clogging through changes in water levels in the well and the observation holes.

6 CONCLUSIONS

The results obtained from this study can be roughly summarized as follow:

1. The groundwater conservation system reviewed above has a distinctive feature to solve within one system the two contradictory problems; the one to deal with (drain) groundwater in the construction site and another to preserve the present groundwater.

2. The reversal siphon system is considered to be effective to be applied as the principal idea which determines the basic shape of the recharge system, because it makes preservation of the present situation and the maintenance works of the system easier.

3. In view of the facts that the groundwater seepage is uneven and that bigger hydraulic gradient near the occurrence facility is more apt to cause clogging, various ways to enlarge the permeation cross sectional area of the facility have been studied. The results indicate that using both trench drains and gravel drains together is effective.

4. When designing the facility, the conventional well theory is taken into consideration, on top of the results obtained from the actual test construction. Furthermore,

5. The test to verify the function of the completed facility shows bigger permeation volume than the design value.

6. The groundwater on both upstream and downstream sides of the construction site where the conservation method is executed are generally stable, thus, it is considered the stationary state is attained.

7. Two years after the conservation measure was applied, no changes in water level due to clogging have taken place in the vicinity of the well.

All the above goings are understood to imply that this facility has been functioning well up to this moment.

REFERENCES

Nagai, H. 1992. The recharge method by siphon system.
Proc. 47th Civil Engineering Academy Annual Meeting: 314–315. Japan

Nishigaki, M. 1988. The study on clogging characteristics of the gravel drain materials.
Proc. 43th Civil Engineering Academy Annual Meeting: 888–889. Japan

Changes in groundwater conditions at mining in connection with sulfide mineral oxidation

Modifications apportées à la circulation des eaux souterraines par les travaux miniers en relation avec l'oxydation des sulfures minéraux

Ivan L. Kharkhordin & Yuliya G. Vishnevskaya
St. Petersburg Mining Institute, Russia

ABSTRACT: The forecast the of artificial hydrochemical regime is very important for water protection at mines. The application of chemical modeling to a quick estimation of hydrochemical conditions at mines is difficult because a lot of initial information is needed. Proposed approach is based on the use of logical functions for description of the structure of hydrochemical systems at mines. It may be used as foundation of expert system for investigation, schematization and forecast of the hydrochemical regime at mines. Stability of hydrochemical system structure is determined as a probability of changes of values of logical variables.

RESUME: La prévision des changements de régime hydrochimique des eaux naturelles dans le courant d'exploitation des gisements sulfurés a beaucoup d'importance a la protection de l'environnement. L'emploi de la simulation chimique pour l'estimation rapide des conditions hydrochimiques n'est pas effectif, parce qu'il exige beaucoup d'information. La base de la méthode proposé est l'application des fonctions logiques pour schématiser le processus de la formation du chimisme des eaux naturelles. En employant ce méthode on peut créer un système experte pour étudier et schematiser des processus hydrochimiques aux gisements sulfures, et pour prevision hydrogeologique du régime technogenese d'eau superficielles et souterraines, pour l'analyse de l'effectivite de la protection d'eaux naturelles. La stabilité du régime hydrochimique se détermine comme la probabilité du changement des significations de variables logiques.

1 INTRODUCTION

The prognosis of artificial hydrochemical regime at sulfide containing ore deposits is very important and difficult. Importance of the problem is caused by high concentrations of heavy metals (to more than 10 g/L) and salinity (to 400 g/L) and low pH value (to 0 - 1) in drainage water. Main difficulties of hydrochemical forecast at mining are connected with a disequilibrium character of chemical processes in the system water - (air) - oxidizing sulfides and variety of mechanisms, which control kinetics of sulfide mineral oxidation. Different aspects of this problem were described in recent reviews by Lundgren and Silver (1980), Jowson (1982), Nordstrom (1982), Ferguson and Erickson (1987), Nordstrom (1991). Nevertheless there are no way to solve in a general case following questions that required for correct forecast of artificial hydrochemical regime:
   1. What data should be collected in a field?
   2. What laboratory experiments should be made?
   3. How hydrochemical conditions at mining may be schematized?
   This study is our first step to development of expert the system for investigation, generalization and forecast of changes in groundwater conditions at mining in connection with sulfide mineral oxidation. It's purposes are the analysis of preliminary study information and elaboration of conceptual foundation of the expert system.

2 PRELIMINARY STUDIES AND REVIEW

Preliminary studies include laboratory experiments and field investigations at two sites at the Far East of Russia (Fig.1). These results are partly published by Kharkhordin (1991), Abramov et al. (1992), Kharkhordin (1994). In this paper we present only summary of that investigation.

Fig.1 Location of Malomyr (1) and
Afonasyevskii (2) sites

Three main zones, with considerably different physical and chemical conditions may
be distinguished in the structure of hydrochemical fields of sulfide containing ore
deposits. Zone 1(ore body) is characterized
by presence of oxidizing sulfide minerals,
low pH value and high concentrations of
heavy metals and different sulfur oxyanions.
The pathways of sulfide oxidation occur
dominantly by microbially-mediated in well
aerated, unsaturated environments and
chemical oxidation in saturated, anaerobic
environments (Taylor et al., 1984). Hydrochemical conditions at zone 2 (groundwater
flow from an ore deposits to a stream) vary
from weak changes to large transformation of
groundwater chemical composition that are
connected with precipitation of secondary
minerals.

The discharge of groundwater into streams
conducts to strong changes of physical
conditions, oxidation of metastable sulfur
species, aeration of water. Metastable
sulfur oxyanions such as politionates and
thiosulfate have considerable influence on
solubility and speciation of a number of
metals (Goleva, 1977; Pogrebnyak et al.,

1989; Benedetti and Bouleogue, 1991;
Kharkhordin, 1991). The organic compounds
inhibit the metastable oxyanion oxidation.
Strong negative correlation between sulfate
concentration and dissolved organic content
in water was observed at the Afonasyevskii
site (Fig.2). The total concentration of
dissolved sulfur in broun water reach to
20 mg/l.

The pathway of chemical sulfur oxidation
is different in dependence from pH value.
Polythionates, in contrast to thiosulfate
and sulfate, are unstable in alkaline solution, but relatively stable under acidic
and neutral solutions (Goldhalber, 1983).
The rate of metastable sulfur oxyanion oxidation varied in a wide range.

3 APPROACH TO DESCRIPTION OF MINE WATER
  FORMATION

Chemical modeling is a wide-applied tool in
description of processes of the mine water
formation, processes of metal attenuation
and dispersion, speciating acid mine water
(Nordstrom, 1990). The main difficulties of
this approach are connected with redox dis-
equilibria of mine water (Nordstrom, 1990;
Kharkhordin, 1994) and a variety of control
mechanisms of the rate of sulfide mineral
oxidation. Moreover, it is very difficult
to use chemical modeling for the rapid qualitative estimation of hydrochemical conditions at mines and for the forecast of possible trends of changes of the hydrochemical
regime. Above-mentioned facts indicate that
chemical modeling is not suitable as a foundation of the quick expert analysis of hydrochemical condition sulfide mineral deposits.

We believe that to apply logical functions
to schematization of hydrochemical conditions is not only possible but could be very useful for prognosis of the artificial
hydrochemical regime, assessment of risk
and remediation action alternatives, development of groundwater monitoring strategies at mines. All mechanisms, that control
the pathway and kinetics of hydrochemical
processes , may occur in two states -
"included" or "excluded" ("1" or "0").
The schematization should be conducted
separately for each zone. Values of logical
functions may be changed as a result of human activities. For example, these changes
could be observed as a result of ore body
drainage:

1. Chemical pathway of sulfide mineral
oxidation from "1" to "0".

2. Oxygen diffusion rate of oxidation
control - from "1" to "0".

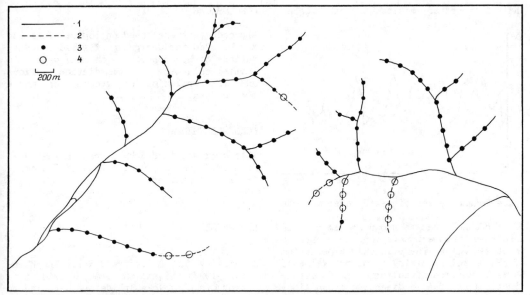

Fig.2 The map of sulfate distribution insurface water at Afonasyevskii site. 1 - streams with broun water (dissolved organic concentration more than 3 mg/l), 2 - streams with colourless water, 3 - sample points with sulfate concentration less 1 mg/l, 4 - sample points with sulfate ions concentration more 2 mg/l.

3. Microbiological pathway of sulfide mineral oxidation - from "0" to "1".

4. Microbiological reaction rate control - from "0" to "1".

It is important to estimate the stability of the structure of hydrochemical system (in our terms, it is the probability of changes of logical function values). Partly this task could be solved by the analysis of regime observation data. It is shown on a primitive example. The behavior of hydrochemical system was simulated for the case of influence of a random quality of the sulfuric acid mine water discharge on pH value of initially carbonate buffered stream water with numerical model. Quantity of sulfuric acid per liter of stream water is expressed by follow equation:

$$a = c + rq$$

where c, q are constants, r - random value (evenly distributed from 0 to 1). A few series of regime observation were generated for different values of c, q and concentration of hydrocarbonate ions in stream water by numerical modeling. Each series include the calculation of pH value for 50 time intervals. Typical results are shown on Fig. 3, 4, 5. For all of these situations q is the same. For the case, that is shown on Fig. 3 c is considerable less

than the concentration of hydrocarbonate ions in the stream water. A pH value is a function of carbonate equilibrium for all time intervals. Such hydrochemical system exhibit a small dependence from little fluctuations of the acid water discharge.

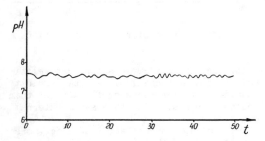

Fig.3 First series of regime observations

For the case, that is shown on Fig. 4, c is close to the hydrocarbonate concentration in the stream water. This system exhibits strong fluctuations of pH value as a result of small fluctuations of acid water discharge because there are two mechanisms of a pH control. They are the carbonate equilibrium when acid quantity less than hydrocarbonate ions concentration and the sulfuric acid concentration in solution, that form after mixing, in other case.

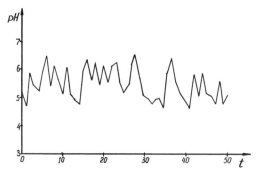

Fig.4 Second series of regime observations

For the case, that is shown on Fig.5, c is more than hydrocarbonate ions concentration in stream water. There is only a mechanism of pH control in mixing solution. Value of pH is a function of sulfuric acid. Little fluctuations of acid discharge cause little fluctuations of pH value of water stream.

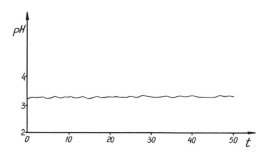

Fig.5 Third series of regime observations

This primitive example shown that relatively stable hydrochemical systems exhibit considerably less range of fluctuation than unstable systems in the same external conditions.

The same approach is applicable for more complicated hydrochemical system with more than two states with different mechanisms of rate control or pathway of hydrochemical processes. Such systems require more than one logical variables for describing.

Other approach to analysis of the hydrochemical system stability is the study of extreme cases. For extreme estimates of possible changes in groundwater conditions the expert system uses simple analitical and numerical models.

To collect initial information for expert estimation of groundwater conditions is the main purpose of field investigations.

CONCLUSIONS

Approach that are based on logical function may be used for designing of field investigations of hydrochemistry of mine water, in initial selection of remediation actions and interpretation of groundwater monitoring data.

ACKNOWLEDGMENT

This work was funded by a grant number 93-05-14091 from the Russian Fund of Fundamental Investigations.

REFERENCES

Abramov, V.Yu., A.A.rosapov & I.L.Kharkhordin, 1992. Migration forms of oxidation products of gold-sulphide ore deposits in surface water. Proc. 204th ACS National Meeting: p. 74.

Benedetti, M. & J. Boulegue 1991. Mechanism of gold transfer and deposition in a supergene environment. Geochim. et Cosmochim. Acta 55(6): 1539-1547.

Ferguson, K. & P. Erickson 1987. Will it generate acid ? An overview of methods to predict acid mine drainage. Proc. AMD Workshop, Environ. Canada, Halifax, Nova Scotia, March.

Goldhaber, M.B. 1983. Experimental study of metastable sulfur oxyanion formation during pyrite oxidation at pH 6-9 and 30 C. Am. J. of Sci. 283(3): 193-217.

Goleva, G.A. 1977. Hydrogeochemistry of ore elements. Moscow: Nedra.

Jowson, R.T. 1982. Aqueous oxidation of pyrite by molecular oxygen. Chem. Rev.82: 461-497.

Kharkhordin, I.L. 1991. Sulfur species in stream water of gold-sulfide ore deposit site. Proc. XII Meeting on Groundwater of Siberia and Far East: 133. Irkutsk-Tomsk.

Kharkhordin, I.L. 1994. Chemical model of forming of platinum-electrode measured redox potential in water solution in contact with oxidizing sulfide minerals. Proc. International Conference on Future of Ground-Water Resources at Risk, Helsinki (Prepared for publication).

Lundgren,D.G. & M.Silver.1980. Ore leaching by bacteria. Ann. Review Microbiol. 34, 263-284.

Nordstrom, D.K. 1982. Aqueous pyrite oxidation and the consequent formation of secondary iron minerals. Acid Sulfate Weathering (ed. D.K. Nordstrom): 37-56. Soil Sci. Soc. Amer., Spec. Publ. No 10.

Nordstrom, D.K. 1991. Chemical modeling of
    acid mine waters in the western United
    States. Proc. U.S. Geol. Survey Toxic
    Substances Hydrology Programm, Mallard,
    G.E. & D.A. Aronson, U.S. Geol. Survey
    Water-Resour. Invest. Report 91-4034:
    534-538.
Pogrebnyak U.F., Kondratenko L.A.,
    Laperdina T.G. et al. 1989. Ore elements
    in water of supergene zone of deposits a
    Zabaykalye region. Novosibirsk: Nauka.
Taylor, B.E., M.C. Wheeler & D.K. Nordstrom
    1984. Stable isotope geochemistry of acid
    mine drainage: Experimental oxidation of
    pyrite. Geochim. et Cosmochim. Acta 48
    (12): 2669-2678.

The movement mechanism of adsorbed water in saturated cohesive soil and its use in the engineering geology and hydrogeology

Mécanisme du mouvement de l'eau d'adsorption dans les sols cohésives et ses applications en géologie de l'ingénieur et hydrogéologie

X. L. Feng & T. Z. Yan
China University of Geosciences, Wuhan, People's Republic of China

ABSTRACT: The permeability tests on saturated cohesive soil composed of marketable kaolinite have been done with self-designed device. The changing laws of pore water pressure transferred by adsorbed water with time and distance have been obtained. The micromechanism of those laws has been explained rationaly. On the base of those, some phenomena existed in the field of hydrogeology and engineering geology has been discussed; Some important concepts such as influencing radius, critical hydraulic gradient have been proposed.

RESUME: les tests de perméabilité sur le sol saturé et cohésif, qui est composé par la kaolinite vendable, ont été faits avec le dispositif destiné par lui-même. Les lois du changement de la pression des eaux de pore, qui est transférée par léau dábsorption avec le temps et la distance, ont été obtenues. Les micromécanismes de ces lois ont été raisonnablement expliqués. Sur la base de ces données, quelques phénomènes de l'hydrogéologie et de la géologie civile dans ce domaine ont été discutés; Quelques concepts importants tels qul le rayon d'influence, la pente hydraulique et critique ont été proposés.

1 INTRODUCTION

In recent years, economics of the southeast region of China where the soft soil exists widely developed rapidly. The soft foundation improvement is necessary for meeting the need of industry and citizen architecture. For predicting the change regularity of soft soil property with the time, the Terzaghi's consolidation theory must be used. The precision of predicting result depends on the magnitude and distribution of pore water pressure existed in soft soil. Subsidence is related closely with the withdraw of underground water. Its rate not only depends on the magnitude and duration of withdrawing water, but also on the soil property and distribution of pore water prressure in the soil. In the process of with drawing underground water, it is necessary to investigate whether the adjacent acquifer induces leakage. When the drilling or excavation is done in saturated cohesive soil, the bearing water zone will form. All problemlems metioned above are related closely with the movement of pore water in soil. There are three kinds of pore water in the soil. They are gravity water, capillary water and adsorbed water. A great deal of work has been done for the former two kinds of water. The Key point is to seek the movement law of adsorbed water in the saturated cohesive soil.

Table 1 physical property index of kaolinite

specific gravity	density	water content	W_P	W_L	I_P	void ratio	saturation degree
(g/cm³)	(g/cm³)	(%)	(%)	(%)	(%)		(%)
2.58	1.54	71.0	35.7	75.0	38.3	1.865	98.3

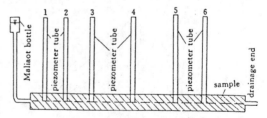

Figure 1 sketch of experiment device

2 EXPERIMENT AND THEORITICAL ANALYSIS

For the saturated cohesive soil, the movement of pore water is singed by the transference of pore water pressure. Typical experiment has been done to gain the movement regularity of adsorbed water.

2.1 Experimental condition

Table 1 indicates the physical property index of saturated marketable Kaolinite sample. Test device is shown as

Table 2   The function of waterhead with time

1	2	3	4
$H = 6.36t^{0.44}$ $R = 0.98$	$H = 1.63t^{0.57}$ $R = 0.98$	$H = 0.55t^{0.62}$ $R = 0.92$	$H = 0.58t^{0.46}$ $R = 70$
$H = 3.37t^{0.58}$ $R = 0.97$	$H = 2.34t^{0.53}$ $R = 0.81$	$H = 2.86t^{0.18}$ $R = 0.84$	$H = 2.45t^{0.18}$ $R = 0.99$

Fig. 1. The diameter of sample is 4 centimeter. The internal diameter of piezometer tube is 0.6 centimeter. The distance between each piezometer tube and supplied pressure point is 3, 7, 12, 19, 29, 39 centimeter respectively. The whole sample tube is 49 centimeter. The head of piezometer tube is netty. The applied constant water head difference is 150, 180 centimeter respectively. The observation lasts two months. During the observation, the test temperature is below one centigrade each day. Accumulated temperature difference is 5 centigrade. Some methylbenzene which has one percent concentration added to the distilled water, when the sample is remoulded. So that the growth of bacterium can be prohibed.

nent function are used to stimulate the variation of water head with time and distance at the different hydraulic differences. Table 2 is the stimulation result. The power function is the common form. On the base of stimulation results, following opinions can be obtained.

1 At the action of the same water head difference, the nearer the distance to the supplied pressure point, the higher the water head in the piezometer tube. At first, the ascending rate of water head in the first piezometer tube is greater than that of others. After a period of test, te second tube is greater than others. For example, 325 hours is needed for the occurence of this kind of phenomenon when the water head difference is 150 centimeter. Only 250 hours is needed when the water head difference is 180 centimeter. It is concluded that the larger the water head difference, the shorter the delayed time.

2 For the third and fourth pizometer tube, the water head increasse slowly whether high or low of the water head difference.

3 Figure 2 is the variation curve groups of water head changing with thime. It indicates that the H (x, t) curves have a tendency becoming to lines when the permeation time increases continuously. Meanwhile, each curve has a connection point with the abscissa at which pore water pressure is equal to zero. The distance to the supplied pressure point is the influenced scope of additional water head difference. The author defined it as inflencening radius, singned it with RI. The greater the

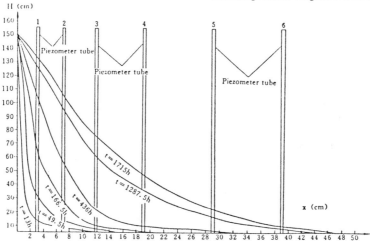

Figure 2   Variation curve groups of pore water pressure with distance and time

2.2 The treatment of data and analysis

Three functions they are hyperbola, power, and expo-

additional water head difference, the longer the permeation time, the larger the RI value. The maximum RImax will be reached at last.

2.3 *Study on the mechanism*

In order to research the internal mechanism causing the occurence of phenomena mentioned above, a model must be established. In light of properties of adsorbed water, following assumptions can be considered.

1 The sample is saturated which water content is equal to its liquid limit. Adsorbed water film existed at the surface of adjacent soil grain intersects each other Thickness of the film is equal to $D_0/2$ (D_0 is the diameter of the pore fulled of adsorbed water)

2 Bonded force between grains is large enough to prevent the movement of soil grains at the action of the permeating force.

3 Cohesive soil grain is flattened. The pore between grains is cylindrical

4 Shear strength of adsorbed water obeys the distribution of $\tau = \alpha e^{-\beta\lambda}$, in which α, β is constant. λ is the distance to the surface of grain ($0 \leqslant \lambda \leqslant D_0/2$)

Based on these assumptions, the model shown as figure 3 can be set up. where the dotted line is the permeating way of adsorbed water. Ho is supplied water head difference, do is permeating radius of adsorbed water[1]. With the increase of distance to the surface acted by water head difference, do decreases gradually. At last, do is equal to zero.

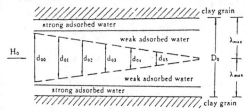

Figure 3   The sketch of adsorbed water transferring pore water pressure

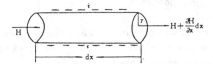

Figure 4   The stetch acted force on micro unit

In order to gain the equilibrium equation, a microunit must be taken out of the continuous pore (shown as figure 4). Following equation can be established at t time.

$$\rho_w H \pi r^2 - \tau \cdot 2\pi r dx - (H + \frac{\partial H}{\partial x}dx)\,\rho_w \pi r^2 = 0$$

$$\frac{\partial H}{\partial x}r = -\frac{2\tau}{\rho_w} \qquad (1)$$

The initial and boundary condition as well as the expression of $x = f(r)$ are needed to solve the above equation. It is difficult to obtain the equation of $x = f(r)$ theoritically. In light of the observation data, following re-

lation can be set up

$$r = -\frac{r_0}{RI}x + r_0 \qquad (2)$$

that is $\frac{\partial H}{\partial x}r = -\frac{2\tau_0}{\rho_w}\exp\,(\beta r_0 x/RI) \cdot \exp\,(\beta r_0) \qquad (3)$

In process of solving above equation, $\int e^x/x\,dx$ integral can't be avoided. This kind of integral only has numerical solution. Considering diameter of pores be small in general, r can be insteaded of average value $r_0/2$. At last, equation (4) can be obtained

$$H\,(x) = H_0 - \frac{4\tau_0 R_1}{\beta r_0 \cdot \rho_w}\exp\,(\beta r_0)\,[1 - \exp\,(-\beta r_0 x/RI)]$$

$$(4)$$

When $x = RI$, H (x) $= 0$

That is $\qquad RI = \frac{\rho_w \beta r_0}{4\tau_0}\frac{H_0}{e^{\beta r_0} - 1} \qquad (5)$

where $\tau_0 = \exp\,(-\beta D_0/2)$ shear strength at the point of adsorbed water film intersection

Ho——Applied water head difference

Ro——initial permeating pore radius

Formula (4) can be considered as the varying equation of pore water pressure with the distance (x). As $\tau_0 = 0$ (sandy soil), $RI \rightarrow \infty$; as $\tau \rightarrow \infty$ (rock), $RI \rightarrow 0$. This is coincide with the natural rule.

It is necessary to indicate water head value increases continuously with time for the same piezometer tube. This rule is complied. Creep of adsorbed water can't be avoided in the process of permeation. This means the shear strength of adsorbed water will decrease with time. This causes permeating radius increase with time for the same piezometer tube. The larger the permeating radius, the higher the water head. Because the relation $r = f(t)$ is difficult to decide, Formula (4) can't show the regulaity mentioned above directly. The micromechanism that adsorbed water transferes pore water pressure can be explained rationaly.

1 Derivating to the formula (4), $dH/dx < 0$. It indicates that H (x) decreases with the increase of distance x.

2 Shear strength of adsorbed water decays in exponent form with increasing of distance to surface of soil grain. For the first piezometertube, adsorbed water with smaller shear strength transforms into gravity water in a more fast rate at first. Then the transformation slows down because of the increase of shear strength of adsorbed water, Corresponding with this process, the permeating radius increases in a more fast rate at first and then slows down. on the other hand, although the water head difference acted on the surrounded soil of the second pizometer tube is relatively small, the initial permeation radius is relative small also. The percentage of adsorbed water with low shear strength is relatively high. permeation radius can increase with time in a long period. The ascending rate of water head at the second

piezometer tube can exceed that at the first pizometer tube after a long period, the absolute value of water head in the first tube is higher than that in the second tube.

With the increase of distance, the lose of water head increase. Although the thickness of adsorbed water with low shear strength is relatively large, the acting water head difference is too samll to overcome the shear strength. permeation radius increases slightly. the water head ascends slowly also.

3 USES IN THE PRACTICE

3.1 Bearing water zone in saturated cohesive soil

when the drilling or excavation be done in saturated cohesive soil. Phenomenon shown as fig. 5 occurs always. Professor zhang proposed a concept of bearing water zone in order to explain this kind of phenomenon. Experiences shows:

$$\frac{\Delta H_1}{L_1}=\frac{\Delta H_2}{L_2}=\frac{\Delta H_3}{L_3}=\frac{\Delta H_1+\Delta H_2}{L_1+L_2}=\cdots=Constant \quad [1]$$

The thickness of bearing water zone can be calculated with following formula.

$$T=H_0/ (I_0+sin\theta) \quad\quad [1]$$

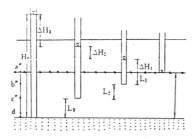

Figure 5   Bearing water zone in saturated cohesive soil

When the permeation occurs only in horizontal direction, $\theta=0$ $T=H_0/I_0$

Where   T——Thickness of bearing water zone

H_0——Pizometer water table of acquifer

I_0—— Initial hydraulic gradient of saturated cohesive soil

θ——Angle between permeating line and horizontal direction

Comparing phenomenon of bearing water zone with former experiment, it is concluded that figure 5 and figure 1 is equivalent. The influcening radius (RI) is equivalent with thickness of bearing water zone (T). So the initial

hydraulic gradient can be calculated with following formula

$$I_0=\frac{4\tau_0}{\beta\rho_w}(e^{A_0}-1)/r_0^2$$

It can be concluded that the regularity of bearing water zone formation is coincide with the law of adsorbed water transfering pore water pressure. It is easy to explain the micromechanism of bearing water zone formation.

3.2 Leakage supply of acquifer

In general, when water head difference in two adjacent acquifer exists, underground water in one acquifer will leaks to the other through the cohesive soil between them. The larger the water head difference and permeability coefficent of cohesive soil, the larger the leakage supply.

Key problem is what is the condition of occurence of leakage supply.

According to the discussion mentioned above, the permeation only occurs at the scope of influencing radius for a applied water head difference. If the thickness of cohesive soil is larger than that of bearing water zone, the leakage supply is impossible.

Assuming the thickness of cohesive soil be H (shown as figure 6), the critical water head difference, hydraulic gradient can be derived

$$H=RI=T=\frac{\beta\rho_w r_0 \cdot \Delta H_{cr}}{4\tau_0\ (1-e^{A_0})}$$

$$J_{cr}=\frac{\Delta H_{cr}}{H}=\frac{4\tau_0}{\beta\rho_w r_0^2}\ (e^{A_0}-1)$$

Where ΔH_{cr}——Critical hydraulic difference

J_{cr}——Critical hydaulic gradient

Only the practical hydraulic difference is greater than ΔH_{cr}, occurs the leakage supply.

3.3 Land subsidence

In general, subsidence is caused by withdrawing underground water from acquifer. In this process, pore water pressure in adjacent cohesive soil decreases, the effective stress increases. correspondingly. Terzaghi's consolidation theory will be used to predict the subsidence precisely

For the consolidation at the action of additional stress, it is assumed that total stress keep unchanged during consolidation. For the subsidence process, drainage consolizdation of saturated cohesive soil is caused by decreasing of water table in acquifer. The water table difference is equivalent with additional total stress. For the subsidence, consolidation of cohesive soil consists of two different process. One is the water table decreases continuously; the other is the water table keeps con-

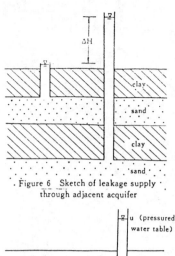

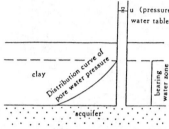

Figure 6 Sketch of leakage supply through adjacent acquifer

Figure 7 Sketch of drainage scope of cohesive soil

$$\therefore \quad S=\int_0^T \frac{a}{1+e_1}\left[u_0-\frac{4T\tau_0 e^{\beta r_0}}{\beta r_0^2}(1-e^{-\frac{\beta r_0}{T}x})-x\rho\right]dx$$

Where: a——compression coefficient of cohesive soil

e₁——initial void ratio of cohesive soil

Where: a——compression coefficient of cohesive soil

e_1——initial void ratio of cohesive soil

u_0——initial pore water pressure

ρ——natural density of cohesive soil

T——thickness of bearing water zone

Bacause the distrubution Curve can be measured, above formula can be used to calculate the subsidence of cohesive soil.

4 CONCLUSION

1 The changing law of pore water pressure transferred by adsorbed water with distance can be expressed using following formula

$$H(x)=H_0-\frac{4\tau_0 RI}{\beta r_0 \rho_w}e^{\beta r_0}(1-e^{\frac{\beta r_0}{RI}x})$$

2 The varying law of pressure transfrred by adsorbed water can be reflected with influencing radius rationly

3 Some practical phenomena can be explained with pressure transferring law. such as the formation of bearing water zone. the decision of leakage boundary.

References

1 Zhang, Z. Y, 1980. About the problem of dynamics of adsorbed water. Geology publishing house.

2 Shen, X. Y, 1985. Study of Envionmental engineering geology on subsidence of cities at yantze river delta and neighbouring sea shore plains of China. Geology publishing house.

3 Harold, W. O, 1966. Darcy's law in saturated kaolinite water resources research.

4 Childs, E. C, 1962. Darcy's law at small potential gradients, Journal of soil science.

5 Kemper, W. D, 1964. Mobility of water adjacent to mineral surfaces.

stant. In the first process, the compression of saturated cohesive soil will produce at the action of changing water table difference. It is contardictory with classic consolidation theory. In practice, this part of subsidence is always neglected, Difference between theoritical computation and practical measurement is caused.

On the other hand, it is considered only the saturated cohesive soil in bearing water zone produces consolidation corresponding to water table decending. A great deal of measurement must be done in order to decide the varying law of bearing water zone. Simulating test is done so that above opinion can be confirmed. Water table in all six piezometer tube is 135 centimeter (simulating the state before withdrawing). Let one end drain water (simulating the withdrawing process) and observe the water table in six tube. When the test lasts 50 days, water table in the tube adjacent to draining point decreases to 38 centimeter. the farthest tube decreases to 117 centimeter. If increasing the length of sample (to simulate increasing the thickness of saturated cohesive soil), water table in one part of sample will be unchanged at a given time.

Based on above analysis, calculating model on subsidence can be established (shown as Figure 7), compression can be computed with following formula

$$S=\int_0^T \frac{a}{1+e_1}u_x dx$$

$$\therefore \quad u_x=u_0-\frac{4\tau_0 T \cdot e^{\beta r_0}}{\beta r_0^2}(1-e^{-\frac{\beta r_0}{T}X})-x\rho$$

7th International IAEG Congress / 7ème Congrès International de AIGI, © 1994 Balkema, Rotterdam, ISBN 90 5410 503 8

Ground settlement and sinkhole development due to the lowering of the water table in the Bank Compartment, South Africa

Affaissement du terrain et développement de dolines causés par l'abaissement de la nappe dans le Bank Compartment (Afrique du Sud)

R.A. Forth
University of Newcastle upon Tyne, UK

ABSTRACT: The world's largest dewatering project took place in South Africa in the late 1960's and early 1970's following the catastrophic inflow of water into West Driefontein Gold Mine on 26th October 1968. At the time West Driefontein was the world's richest gold mine and developments had started on the adjacent East Driefontein mine, to the east of the Bank Dyke. In order to save West Driefontein and East Driefontein a massive pumping operation was initiated with pumping from shafts attaining a peak volume of about 340 megalitres per day in mid-1970. Pumping continued for several years diminishing to a roughly constant volume of about 75 megalitres daily by 1976. The effect of the pumping was to lower the water table about 450 metres close to the main pumps and by over 50 metres some 15 kms away within the Bank Compartment. This paper discusses the investigation of the subsurface geology of the Bank Compartment in an attempt to predict the location of areas of subsidence and sinkhole formation which resulted from the dewatering operation.

RÉSUMÉ: Le plus grande project d'épuisement d'eau a été éxécuté en Afrique du Sud a la fin des années 60 et au début de 70 a la suite de la venue d'eau catastrophique dans la mine d'or de Driefontein le 26 Octobre 1968. A cette époque West Driefontein était la mine d'or plus valable du monde dont le développement avait commencé dans la mine adjacent de East Driefontein, a l'Est du Bank Dyke. Pour préserver West Driefontein et East Driefontein on a éxécuté une large opération de dénoyage par pompage a partir de puits de mine jusqu'à atteindre un volume maximum de 340 megalitres par jour vers la moitié des années 70. Le pompage a continué pendant plusiers années en se stabilisant a 75 megalitres par jour pendant 1976. L'effet du pompage a été l'abaissement de la nappe d'environ 450 m près des pompes et d'environ 50 m à 15 km de distance dans le Bank Compartment. Cette communication discute les recherches sur la géologie de Sub-Surface du Bank Compartment en éssayant de prédire la localisation des zones de subsidence et de la formation de dolines résultantes de l'opération d'abaissement de la nappe.

1 INTRODUCTION

The Bank Compartment was the third compartment to be dewatered on the Far West Rand. Dewatering became necessary after the dramatic inrush of water into West Driefontein Gold Mine on 26th October 1968 in order to save the mine which had been flooded and to continue the development of East Driefontein Mine.

The result of the dewatering operation was, as in the previously dewatered Venterspost and Oberholzer compartments, the development of sinkholes and ground subsidence. The township of Bank was evacuated as sinkholes developed and buildings, roads and railways subsided. Comparison of the dewatering of the Bank compartment could be made with that of its westerly neighbour,

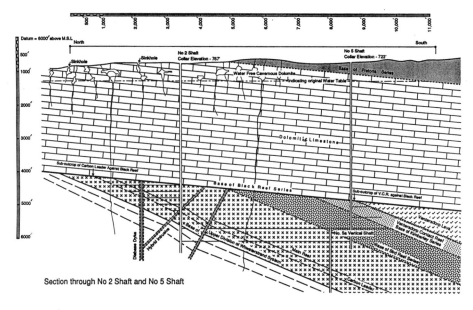

Fig. 1   Section - West Driefontein Gold Mine

the Oberholzer compartment, which was dewatered between 1962 and 1966. Indeed beneath the mining town of Carletonville the ground water had been lowered between 150 and 300 metres. This led not only to the well documented disaster at West Driefontein when the crusher plant was lost down a sinkhole 55 metres diameter (with the loss of 29 lives) but also large zones of subsidence, with subsidences of 8 metres in one case and 5 metres in another (Brink (1979)).

2 GEOLOGY

A generalised geological section north-south across the Far West Rand shows the Transvaal System progressively transgressing over the Ventersdorp System and the gold-bearing Witwatersrand System. In the northern part of West Driefontein Mine property the Dolomite Series and Black Reef Series of the Transvaal System, up to 4,000 feet thick and dipping at approximately 7° to the south, lies directly and unconformably over the Upper Division of the Witwatersrand System, which, in

its Main-Bird Series contains the gold-bearing Main Reef and Carbon Leader. Further to the south the Transvaal System directly and unconformably overlies the Ventersdorp Contact Reef (VCR). The Reefs, where they sub-outcrop against the Black Reef, dip at approximately 23° to the south (see Figure 1). Dips generally increase to the south. A series of approximately north-south trending dykes of Pilansberg age (1,290 million years) cut vertically through the complete geological section.

The Bank Compartment is defined by two such dykes; to the west is the Bank Dyke and approximately 8 miles to the east is the Ventersdorp Dyke. The dykes are impervious to water. To the north and south the Compartment is defined by the outcrop of the Dolomite which is limited by two low ranges of hills, the Gatsrande in the south and the Witwatersrand in the north.

This generalised geological picture is essential to an understanding of the dolomite water problem. Dolomite is a low porosity rock and within the solid rock itself the water content is negligible. It does, however, contain many joints and cracks and is faulted by post-Transvaal movement. Fissures develop in these areas. Furthermore, in the upper stratigraphic horizon of the Dolomite Series there are many chert bends, which are insoluble, but rain-water percolating through cracks in the chert and into the dolomite, erodes away the dolomite along bedding planes. Most of the solution takes place at the water table with the subsequent development of caves, often inter-connected, and open fissures. The water is stored in these cavities, continually enlarging them, but at depths much below the water table the acidic rainwater is neutralised by the alkaline dolomite and has little erosive power. Hence it is thought that most of the dolomite water is stored in fissures and caves which, when inter-connected, form "valleys".

For about 2,000 million years there has probably been continual percolation of rainwater into the rocks, and hence continual weathering and solution. But in that period, too, there have been variations in climate causing variations in rate of percolation (for example the Dwyka Glaciation), and tectonic events (for example the intrusion of the Pilansberg dykes) which have influenced the position of the water table. Quite what has been the effect of these events, and others, is problematical, but the extent of dewatering clearly shows that the erosive powers of the rainwater have been working in parts to several hundred metres below the water table before dewatering. This implies "valleys" in the upper stratigraphic horizons of the Dolomite Series to at least this depth, and can only be explained in two ways. Either the rainwater retains its acidic erosive power well below the water table, or the position of the water table must have been in the geological past at a lower level than at present. This second factor is likely to be the most significant but a combination of the two effects is not ruled out. Evidence of cavities well below the present

water table is shown by a number of boreholes. For example one borehole encountered shows cavernous ground between 60 and 70 m below ground level whereas the original water table before dewatering was at 27 m below ground level.

There is, then, a reservoir of limited volume in the upper stratigraphic zone of the Dolomite Series. This water only becomes a danger to mining operations when it seeps through between 1,000 and 2,000 metres of rock. There are ways in which this happens. A number of post-Transvaal faults cut through the Transvaal System and the Witwatersrand System, and the dolomite water will find its way, under force of gravity, into these cracks and lines of weakness right through into the mine workings. Another line of attack is along cracks caused by post-Transvaal uplift which strains the strata. Cracks may occur due to sagging of the hanging-wall caused by a weakening of the strata by stoping. In whatever way the cracks may develop the water irresistibly widens them into fissures.

3 EVENTS LEADING UP TO THE DEWATERING OF THE BANK COMPARTMENT

The Bank Dyke trending slightly east of north cuts through the eastern portion of the mine property of West Driefontein Gold Mining Company Limited. Prior to 1964 this mine concentrated on exploiting the western portion of its property and currently has Shafts No. 2, 3 and 5 for that purpose. The mining operation was hampered throughout by inrushes of water from the dolomite in the Oberholzer Compartment, defined by the Bank Dyke to the east and the Oberholzer Dyke to the west. Between 1962 and 1964 West Driefontein pumped out more than 30 million gallons of water a day, which was canalised into the Wonderfontein Spruit and hence to the Mooi River. The mine was eventually equipped to handle 63 million gallons a day.

In 1964 West Driefontein began development of the far eastern section, east of the Bank Dyke. This brought into the picture the unknown reservoir of water contained in the Bank Compartment. No problems were encountered, however, in mining the VCR east of the Bank Dyke, and the only inrush of water was of the order of 2 million gallons a day.

Furthermore, the dolomite water in the Bank Compartment had given comparatively little trouble to Libanon Mine, where pumping had never exceeded 3 million gallons a day. This mine is on the eastern margin of the Bank Compartment.

An advantage that the Bank Compartment had over the adjacent Oberholzer Compartment is that post-Transvaal faults here are known to be filled with mylonite (water-tight) and can only feed very limited quantities of water into the mine workings. With all this knowledge mining proceeded east of the Bank Dyke for over four years.

About 1,000 metres east of the Dyke, and striking almost north-south, runs a normal fault with an upthrow of about 170 metres to the east, the "Big Boy" fault. This pre-Transvaal fault has the effect of faulting up the gold-bearing VCR close to the base of the Dolomite Series. The position of stoping around the "Big Boy" fault coincides approximately with an anticlinal or doming structure in the Transvaal System vertically above it. This causes radial cracks to develop through the dolomite coinciding approximately with the sub-outcrop of the "Big Boy" fault against the Black Reef Series. Stoping out of the V.C.R. east of the fault caused movements and cracking of the Ventersdorp Lava, thus connecting the stopes with the dolomite above.

On the 26th October 1968, it is believed that cracks developed through the Lava to the dolomite and into a vast receptacle of water in the upper stratigraphic horizons of the Dolomite Series. This water poured through the cracks, was delayed temporarily by the impervious shale of the Black Reef Series, but under gravity broke through into the mine workings east of the "Big Boy" fault. The volume was quite unexpected, but it is estimated

that a single crack 3.5 m long and 25 mm wide, given the pressure overhead, was sufficient to pour 82 million gallons of water a day into the mine. (In fact many cracks, not just one, were probably responsible.) What happened then to West Driefontein mine is a story unto itself (Cartwright, 1969).

4 A REVIEW OF PUMPING IN THE BANK COMPARTMENT

It was decided after the catastrophic flooding that the Bank Compartment had to be dewatered, and a programme to this end was immediately instigated. In the early days the greatest problem was lack of knowledge both of the sub-surface topography of the dolomite, and of the amount of water in the Compartment.

Boreholes were immediately drilled over as wide an area as possible to obtain information. It was known from the rate of discharge of water at the Bank Eye (a spring) that the inflow into the Compartment from all sources was 10.6 million gallons per day. This is predominantly rainfall (about 25" a year), discharge from the Venterspost Compartment (4.5 million gallons a day) and a limited amount of underground seepage. There was a slight variation of about 2 million gallons a day depending on rainfall. In the rainy season, October to March, inflow, and hence discharge at the Eye, was greatest. The spillway at the Bank Eye is at 326 metres below datum (datum = 6,000 feet = 1,968.4 metres above mean sea level). This effectively marked the original water table prior to 26th October, 1968, the date of the water inrush into West Driefontein (see Fig. 2).

The Bank Eye immediately stopped discharging water into the Oberholzer Compartment for the first time in recorded history. The water table quickly dropped, in the case of borehole E2A, (see Fig. 2) which is near the breakthrough point, to 370 metres below datum, a drop of almost 50 metres. Immediately that the huge valves were closed on 19th

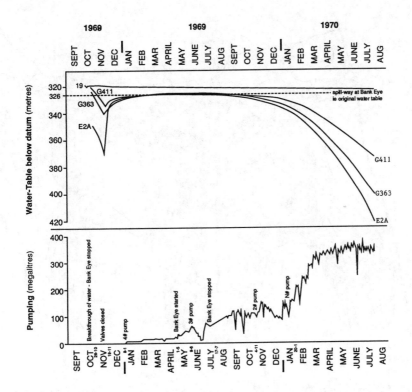

(See Fig. 4 for location of boreholes)

Fig. 2   Pumping and draw down - 1968-1970

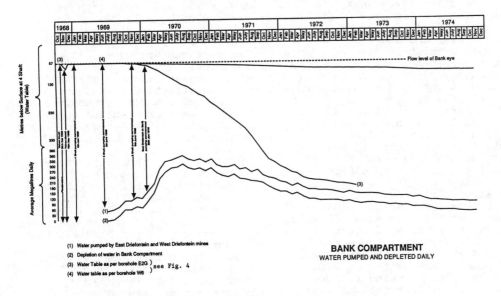

(1) Water pumped by East Driefontein and West Driefontein mines
(2) Depletion of water in Bank Compartment
(3) Water Table as per borehole E2G) see Fig. 4
(4) Water table as per borehole W6)

BANK COMPARTMENT
WATER PUMPED AND DEPLETED DAILY

Fig. 3   Pumping and drawdown - 1974

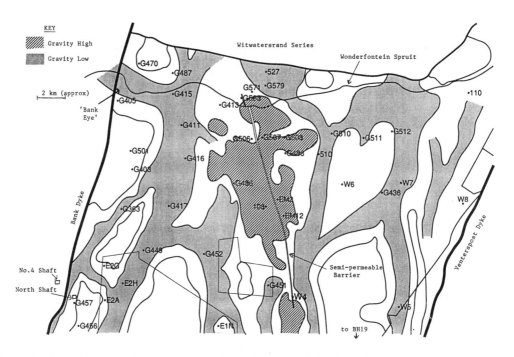

Fig. 4   Location of boreholes and semi-permeable barrier

November 1968, sealing 10 and 12
levels (or effectively sealing the
Bank Dyke) the water table shot up
again and in a matter of hours was
back near to its original level.
This operation saved West
Driefontein west of the Dyke, but
left No. 4 Shaft (West
Driefontein) and the area east of
the Dyke completely flooded.

Pumping began in No. 4 Shaft on
8th January 1969, at the rate of 2
million gallons a day increasing
to 4 million gallons a day by the
end of April. On 1st May 1969,
the Bank Eye began discharging
water once again, but the
installation of pumps in No. 3
shaft on 9th June meant a total
pumping of 12 million gallons a
day, and so on 17th July the Bank
Eye stopped flowing. It will
never flow again whilst mining
takes place in this area. Pumping
was increased to about 20 million
gallons a day until the pumps at
No. 2 Shaft and North Shaft were
introduced, the latter on 20th
January 1970. Pumping rose to a
peak of 75 million gallons a day
in 1970, decreasing gradually to
about 15 million gallons a day in
1976. (Fig. 3)

5   THE EFFECT OF THE DEWATERING
PROCESS

This is best illustrated by Figure
2 showing diagramatically how the
dewatering has progressed up to
and including July 1970, and this
in turn gives information on the
geology of the Dolomite Series.
Two features are especially
apparent:
a) The dewatering was initially
   only effective over about one-
   half of the Compartment, namely
   that west of a quartz outcrop
   striking slightly west of north
   down the middle of the
   Compartment.
b) There was a remarkably steep
   gradient of progressive
   dewatering running north-south
   along the west flank of this
   quartz outcrop. The steep
   gradient swings to an east-west
   trend around the Wonderfontein
   Spruit.
   Both these features are thought
to be related to the quartz
outcrop, and it is thought
possible that this outcrop was a
result of secondary post-Transvaal
movement on a fault. A quartz
vein filling the line of secondary

Plate 1   Subsidence effects from dewatering

Plate 2   Sinkhole (50 metres diameter) in maize-field

movement would then form a barrier which is seen to be semi-permeable.

6 THE SUB-SURFACE TOPOGRAPHY OF THE DOLOMITE AS INTERPRETED FROM THE DEWATERING PROCESS AND A DETAILED GRAVITY SURVEY

From experience in the Oberholzer Compartment a gravity survey was considered the best and most practical way of deploying geophysical techniques to determine the sub-surface structure. Interpretation from the Regional Gravity map was attempted and a simplified version is reproduced on Figure 4 to show main Bouger Anomaly "Highs" and "Lows".

7 SUBSIDENCE AND SINK-HOLE DEVELOPMENT

Surface subsidence (Plate 1) occurred over a wide area in the western section of the Bank Compartment and was a direct result of dewatering. Boreholes have shown that the "valleys" are filled in with weathered chert, clays, coal and wad (manganiferous clay), with a cover of red soil. The thicknesses varied considerably from nothing to tens of metres. Consolidation of this material occurred when the water table is lowered through it, for this unconsolidated material loses a certain amount of volume and tended to compact under its own weight. It caused cracking and subsidence at the surface (usually only a matter of a metre or two) and was obvious in places near Bank Station and elsewhere. This phenomena can only manifest itself where the water table was above the solid dolomite.

The process of dewatering was thought to be complete when the water table dropped below the level of the valley floors, and into solid dolomite. The percentage water content of the dolomite was steady at around 1.2%. This figure, however, can in no way be taken as an average for all the dolomite, because the water content of the "valleys" may be around 10%, whilst that of "high ground" is probably less than 1%. The whole dewatering operation, then, is arrived at lowering the water table below the "valley" floors, hence draining them of their high water percentage content.

Sink-hole formation (Plate 2), in contrast, occurs in all limestone and dolomite areas of the world, caused by underground erosion by rainwater and subsequent ground collapse. The erosion takes place from below the surface and gives no indication on the surface whatsoever, until a large, almost circular hole perhaps 30 metres in diameter and 20 metres deep, develops instantly. It was not possible, at an early stage to relate dewatering to sink-hole formation, but the predominance of sink-holes in the western sector of the Bank Compartment, and since dewatering was instigated, could not be overlooked. It seemed likely that the water provided a natural buoyancy in the caverns which was eliminated as the water was removed. On the other hand rock-bursts whilst mining, rain-storms, etc. can trigger off collapses of the roofs of cavities, thus forming sink-holes.

8 ACKNOWLEDGEMENTS

The author wishes to acknowledge the assistance of his former colleagues at Consolidated Goldfields Limited in the data acquisition exercise.

REFERENCES

Brink, A.B.A. (1979). Engineering Geology of Southern Africa. Building Publications, Pretoria, South Africa.

Brink A.B.A. (1984). A brief review of the South African sinkhole problem. Proceedings of the First Multidisciplinary Conference on Sinkholes, Orlando, Florida, p. 123-127.

Cartwright, A.P. (1969). West Driefontein - Ordeal by Water.

Use of environmental tracers and piezometer nest to elucidate origin and flow regime of hot springs

Emploi de traceurs et de piezomètres pour étudier l'origine et le régime d'écoulement des sources thermales

Kei Ichikawa
Public Works Research Institute, Tsukuba, Japan

Souki Yamamoto
Tokyo Seitoku University, Japan

Hiromichi Ishibashi
Suimon Research Inc., Chiba, Japan

ABSTRACT: To conserve environmental circumstances of hot springs, it is necessary to elucidate its origin and flow mechanism, as well as its interaction between surface water, precipitation and groundwater. The authors intend to assess probable influence which may be caused by construction of diversion works to guard hot springs several hundred meters downstream from flood. Some environmental tracers such as tririum, deuterium and oxygen-18 are chosen as reliable tool to evaluate groundwater flow regime, and piezometer nest to determine hydraulic gradient. This paper shows the interaction between groundwater and hot springs, hydrological characteristics such as origin and flow mechanism, as well as the result of assessment based on potential flow analysis.

RESUMÉ: Pout conserver l'environnement de l'eau thermale, it est nécessaire d'élucider leur origine, mécanisme de l'écoulement, l'interaction entre les eaux superficielle, souterraine et thermale. Les auteurs ont accompli un évaluation des effets possible à cause de la construction du déversoir pour défenser la station banéaire quelque cent mètres aval contre les crues. Dans ces recherches, quelques traceurs environnementals tel que tririum, deuterium et oxygène-18 ont été choisi comme un outil sûr pour évaluer le régime de l'écoulement de l'eau souterraine et aussi "niche" piézométrique pour le gradient hydraulique. Ce document presente ici que l'interaction entre l'eau souterraine et les sources thermales, traits hydrologiques ainsi que l'origine et le mécanisme de l'écoulement, et aussi le résulte de l'évaluation basée sur l'analyse potencielle.

1. INTRODUCTION

Studied area is the Asamushi Hot Springs located near to the estuary of the Asamushi River (Figs. 1, 2). Although it has been exploited densely and pumping became inevitable, it was an artesian spring. Most well are shallow (within 10m), but recently 4 deeper wells (40 to 50m) are of main yield (1.45m³ to the total of 2m³).

The purpose of this survey is to elucidate the origin and flow mechanism of thermal water, and relationship between surface water in the hot spring area.

1.1 *Physical condition*

Hot spring are in the downstream reach of the river within 200m from the seashore, under alluvium of less than 100m width. Elevation is about 3m above sea level, therefore spring is friable to sea-water instruction when excessively pumped, and self-flowing occurs when pumping stops. This is an artesian regime, and hot spring water is stored in conglomerate at a depth of 10 to 50m below surface.

1.2 *Geology and hydrogeology*

This area is located in the northeastern region of Honshu Island, and is consisted mainly of dacitic pyroclastics and lavas of Miocene age, slightly suffered hydrothermal alteration.

Surface deposit such as mud or muddy sand and gravel distributes along thalweg in about 20m thick or less. Basement is mainly consists of tuff of various facies from tuff-breccia to fine tuff, intercalating small amount of rhyolite. It generally dips towards north with a strike of E-W (Fig. 3).

Between the basement and surface deposit, there exists well-consolidated conglomerate and sandstone of Pleistocene age at the downstream reach, but not at the upper reach.

Surface deposit is largely low in permeability and thus artesian condition is preserved.

Hydraulic conductivity gained from pumping test is 8×10^{-4} cm/s in alluvium, whereas 1 to 4×10^{-3} cm/s in Pleistocene and Miocene rocks. Apparent hydraulic gradient is about 1/80.

Formerly there was a self-flowing spring, but over-exploitation caused severe descendance oh hydraulic

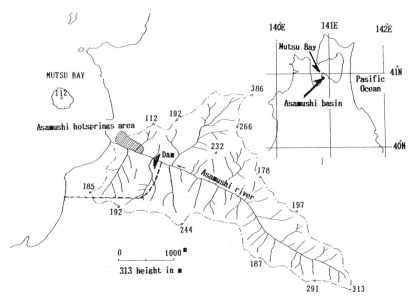

Fig. 1 Asamushi river basin: Asamushi dam and spillway

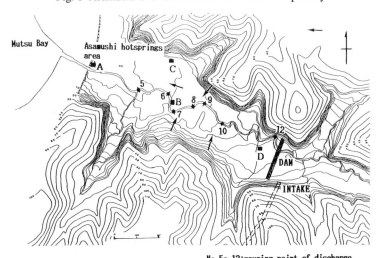

No.5~12:gauging point of discharge
A,B,C,D:bore holes & piezometer nest

Fig. 2 Asamushi hotsprings & study area

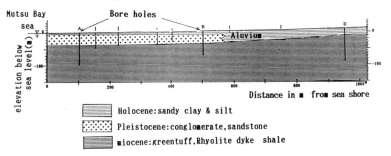

Holocene:sandy clay & silt
Pleistocene:conglomerate,sandstone
miocene:greentuff,Rhyolite dyke shale

Fig. 3 Geological cross section along the Asamushi river

head and sea-water intrusion thereafter. Nowadays, pumping wells are diminished from 106 to 23, and daily pumping is sustained to the maximum of 2m³ per day.

1.3 *Drainage basin*

Overall length of the main river is about 5km, and catchment area is 6.3km², the drainage basin consists a narrow river system. Flow regime of the river as shown in Fig. 4 have two times of high water, that is snow thawing in the spring and heavy rainfall of the typhoon in the autumn. The river channel at hot spring area is extremely narrow (less than 10m), and incapable to discharge flood water efficiently.

Diversion works and spillway are proposed to guard the hot springs, flood control scheme is shown in Figs. 5, 6).

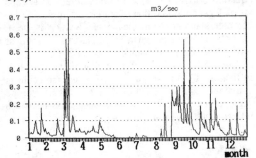

Fig. 4 Hydrograph of Asamushi river in 1989

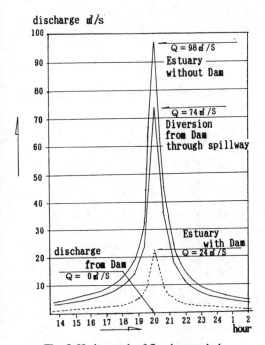

Fig. 5 Hydrograph of flood control plan

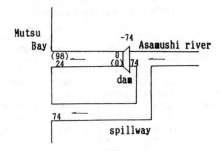

Fig. 6 Flood control system

2. METHODS AND MATERIALS

In-situ surveys are conducted during summer to autumn in 1992 and 1993. Its contents are:
1. 4 boreholes (50 to 100m deep).
2. Potential measurement by piezometer nest and electric logging.
3. Sampling of water at well, river and rain.
4. Pumping test near hot springs and upstream area.
5. Measurement of river water discharge.
6. Analyses of water quality and radioactive elements.

3. WATER BALANCE

3.1 *Surface water*

Calculation of water balance is based on following data:
1. Catchment area: 6.3km²
2. Annual rainfall: 1,040 to 1,290mm
3. Annual discharge through river: 269mm
4. Potential evapo-transpiration: 627mm to 705mm (by Thornswaite's method)
5. Average pumping rate from hot spring well: 1.45m³/min

Water balance can be calculated by a equation I=P-E-R (where, I: infiltration, P: precipitation, E: evapo-transpiration, R: river discharge). We got I=443mm/y, namely 1.2mm/day. This means an amount of 7.560m³ of infiltration can be expected from the overall catchment area.

Groundwater discharge in the form of hot spring pumping is about 2,100m³/day (1.45m³/min). Comparing these figures, fairly large amount of water runs off to the sea or outside of drainage basin. Thus, this rate of pumping is allowable and hot springs are supplied from rain water judging from water balance. Inversely, necessary surface area to supply to the well becomes about 1.74km².

Annual water balance is shown on Table 1.

2943

Table 1 Annual water budget

(mm/year)

	① P	② E	②×70%	①-②×70%
1983	1,252.0	640.7	448.5	803.5
1984	1,125.0	630.7	441.5	683.5
1985	1,163.0	657.1	459.9	703.1
1986	1,209.0	626.9	438.8	770.2
1987	1,207.0	647.6	453.3	753.7
1988	1,040.0	628.6	440.1	599.9
1989	1,177.0	663.9	464.7	712.3
1990	1,289.0	704.6	493.2	795.8
1991	1,223.0	672.8	471.0	752.0
Average	1,187.2	652.5	456.8	730.4

3.2 *Base flow in downstream reach*

As base discharge of a river is approximated to ground water discharge of the drainage basin (J. Boussinesq, 1905), base discharge around hot springs area and confluents are measured, as shown in Fig. 2. Measurement was worked out on the autumn of 1992, its result is shown in Fig. 7.

Knicks between Nos. 7 and 8, and Nos. 8 and 9 are due to an inflow of wastewater from the hot springs, and between Nos. 11 and 12 to a confluence. Stations down to No. 9 show higher temperature and increase of basic discharge due to resurgence of groundwater.

4. CHEMICAL COMPOSITION

Chemical composition of water sample is shown on Table 2, and stiff diagramme on Fig. 9. It shows that borehole A, C and hot springs contain high concentration of components, and least in rain water, river water next to it. It may be concluded that rain water runs off to the river in a rather short time, whereas hot springs, A and C move for a long time after infiltration to the ground, C is of lesser concentration and mixed with shallow groundwater.

From this result, water of this area can be divided into two groups: (1) hot spring, A and C; (2) rain water, river water and shallow groundwater of upstream reach.

Water of hot springs exists in the right hillside of the river and downstream of B, and flows in deep layers for a long time.

5. ISOTOPE COMPOSITION

Environmental isotopes generally used for water are tritium, deuterium and oxygen-18, as they behave all the same as hydrogen and oxygen-16. Degree of presence of such isotopes are described by a parameter δ, namely a deviation of isotope ratio from that of standard sample. This deviation is calculated by a equation $\delta = ((Rx/Rs)-1) \times 1,000$ (where, Rx: isotope ratio of a sample, Rs: that of standard sample, in case of water, standard mean ocean water SMOW is used).

Stable isotopes shown in Table 3 has a measurement error of $\pm 2.0‰$ for δD and $\pm 0.1‰$ for $\delta^{18}O$.

Both $\delta^{18}O$ and δD of rain water are intensely influenced by temperature when water molecule is converted rain water, through a process of isotope segregation. Therefore, natural water has specific δ reflecting its origin and circulation process.

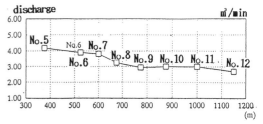

Fig. 7 Base flow of lower drainage in Asamushi river, (Nov.21, 1992)

Table 2 Chemical composition of Asamushi waters

1992

	No.1 G	No.2 G	No.3 r	No.4 r	No.5 r	No.6 H	No.7 H	No.8 H	No.9 R	No.10 R
pH	8.1	7.7	7.8	7.8	7.5	8.1	8.3	7.9	7.4	6.7
λ	283	171	106	132	168	1320	1360	1500	72.3	18.7
Mg	9.23	5.83	4.37	3.89	4.86	3.40	14.09	6.08	1.22	0.00
Ca	30.48	12.83	4.80	7.22	11.23	142.80	162.80	152.40	2.00	2.00
Na	23.40	18.10	16.10	19.00	21.00	211.00	188.00	257.00	2.50	1.80
K	0.68	0.86	0.80	0.97	1.05	5.10	10.92	5.79	0.55	0.20
HCO₃	90.18	55.70	45.09	39.78	47.74	47.74	122.00	31.83	10.61	5.30
SO₄	54.61	22.93	5.45	12.83	22.25	498.40	490.40	598.30	2.84	0.68
Cl	23.28	19.80	15.00	19.20	20.00	136.00	130.00	183.00	3.96	1.92
SiO₂	17.5	13.7	10.8	12.6	13.1	50.2	35.0	55.4	1.1	0.1

G: groundwater   r: river water

H: hotspring water   R: rain water

(unit : ppm)

λ : electric conductivity (μS/cm)

2944

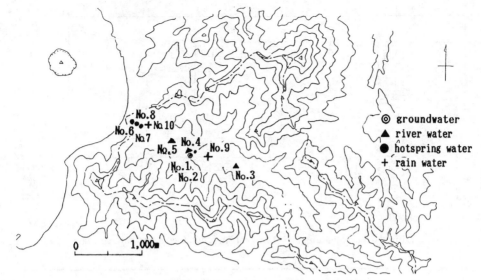

Fig. 8 Water sampling stations of Asamushi river basin, 1992

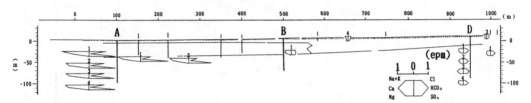

Fig. 9 Distribution of Stiff diagram of groundwater and hotsprings, 1993

Table 3 Isotopic compositions of Asamushi basin waters, 1992

Location	δD (‰)	$\delta^{18}O$ (‰)	^3H (TU)
NO.1	−56.6	−10.1	3.2 ±0.1
NO.2	−53.2	−9.8	6.4 ±0.1
NO.3	−51.8	−9.4	7.9 ±0.2
NO.4	−50.8	−9.3	7.5 ±0.2
NO.5	−53.6	−9.3	7.0 ±0.2
NO.6	−65.0	−11.2	0.90±0.09
NO.7	−61.3	−10.3	1.8 ±0.1
NO.8	−65.3	−11.2	0.3 >
NO.9	−61.4	−9.4	5.2 ±0.1
NO.10	−68.1	−10.0	4.7 ±0.1

Relationship between $\delta^{18}O$ and δD as shown in Fig. 11 says that all plots distribute over the meteoric line. α-parameter defined by $\delta D = 8\delta^{18}O + \alpha$ is bigger than 10, but α of rain water are quite near to meteoric line, whereas that of river, hot spring and groundwater are 20.8 to 25.2, and widely deviate from the meteoric line. This can be interpreted as a mixture of waters from different area and different time.

Water collected at higher altitude shows lighter isotope composition, and hot spring water next to this.

River water is heaviest.

Hot spring water is considered due to and infiltration of rain water at high altitude.

$\delta^{18}O$ at different depth is shown in Fig. 10. $\delta^{18}O$ of hot spring water is generally lower than −11.0‰, whereas that of groundwater is around −10‰ and surface water is higher than −10‰.

Temperature effect on $\delta^{18}O$ in the surveyed area is in Fig. 12, and distribution of $\delta^{18}O$ in Fig. 11. Temperature effect is quite clearly recognizable, also confirms that the origin of hot spring water is of high altitude.

Tritium ratio explains that Nos. 6 and 8 show low figure compared with 4.5 to 5.2 of rain water, whereas 7 to 8 of river water and groundwater must consider some effects of nuclear explosion in the atmosphere. Age of hot spring water is much older than other waters.

6. THERMAL PROCESS

Temperature logging at boreholes and some existing data is in Fig. 14. Isothermal lines of the upper section show a dome near A, lowering towards the surface and

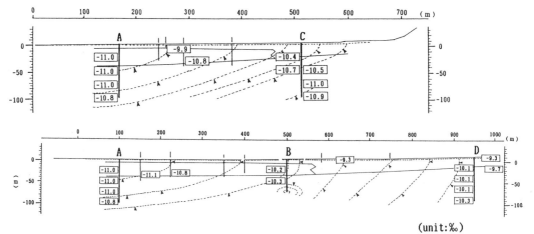

Fig. 10 Distribution of δ¹⁸O contents in groundwater and surface water, 1993

(unit:‰)

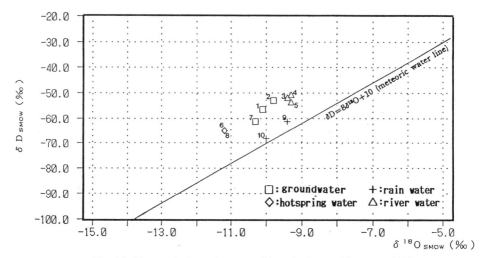

Fig. 11 Changes in isotopic compositions in Asamushi waters, 1992

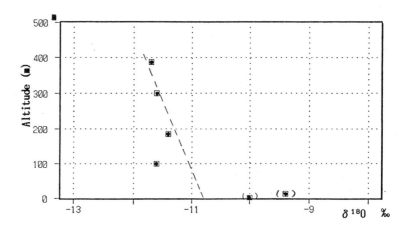

Fig. 12 Altitude effect in the slope of Asamushi river basin, 1993

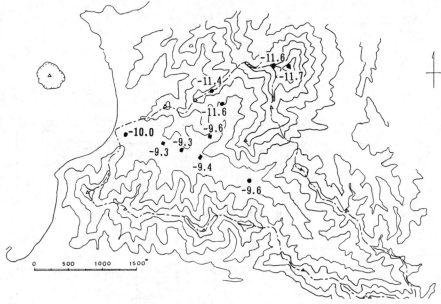

Fig. 13 Distribution of δ¹⁸O content in rain water, 1993

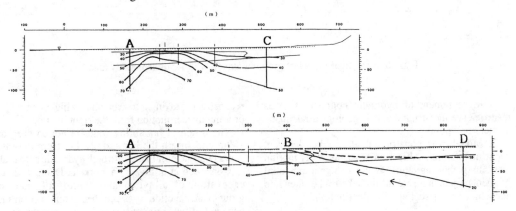

Fig. 14 Iso-thermal cross section

upstream. The lower section shows the same. At D, temperature is very low even at a depth of 100m, it is same as the surface of B. Cross section across the valley shows no horizontal difference, the deeper the higher and vice versa.

7. PIEZOMETER NEST

Piezometer nest (set of piezometers at depths of 10, 20, 30, 40, 50m or 25, 50, 75, 100m) was installed at the sites of boreholes. They allow to draw vertical potential head by water level measurement as shown on Fig. 15. Upper section shows an upward groundwater flow and surgence from the depth, on the other hand lower section shows another regime. Rate of increase of

hydraulic potential at upper reach is 40 to 50cm to 100m, whereas at lower reach 200cm to 100m. This also shows an upward flow.

8. ANALYTICAL MODEL

An analytical model of FEM was prepared, based on geological informations (Fig. 3) and hydraulic parameters (conductivity, water level, temperature, etc.). Result of calculation of potential and flow line corresponding to the lower section of Fig. 15 is shown on Fig. 16. They are quite different each other. This model cannot explain large increase of hydraulic potential especially towards the lower reach. Supposable cause of this discrepancy can be interpreted:

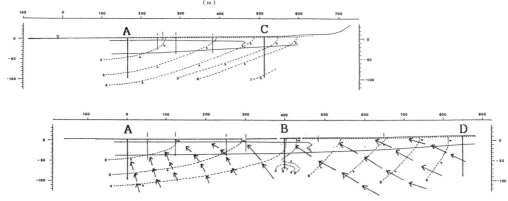

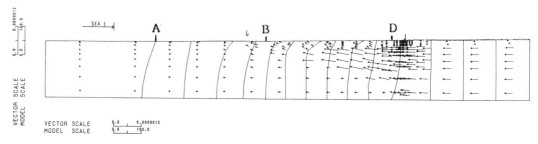

Fig. 15 Iso-potential diagram

Fig. 16 Analytical model of groundwater flow in Asamushi study area

1. Measurement of hydraulic conductivity was insufficient either in number or in geological adequacy of measured point.

2. Model was constructed along the thalweg, supposing that main direction of groundwater motion was along this direction. While a bigger potential gradient perpendicular to the model must be taken into account, as suggested by $\delta^{18}O$ distribution.

9. CONCLUSIONS

Groundwater (temperature of 16 to 20°C) near the dam site flows down along the thalweg for a distance of 400 to 500m, and ascend upwards being influenced by higher potential from the right hillside, also a rise of temperature around B is lower than the downstream reach.

Thus, the chemical or isotopic composition of B is almost identical with D (dam site), largely differs from A (hot springs).

The water of hot spring shows close resemblance with rain water fallen at the hillside of an altitude of 100 to 400m, as indicated by elevation effect of $\delta^{18}O$.

Analyses by 2-dimensional model along the thalweg cannot restore a rise in potential at the downstream reach. This also implies that modeling must be

reconstructed taking into account of a higher potential or groundwater motion from the right hillside.

The origin of hot spring water in not so deep as 100m underneath the hot springs, but from the rain water fallen at the hillside with high hydraulic potential.

Construction of diversion works at D site with an excavation of alluvium and shutoff of shallow groundwater motion by sheetpiles, cannot act on the behavior of hot spring water with inverse influences.

REFERENCES

Freeze, R.A. & Cherry, J.A. 1979. *Groundwater*. Prentice-Hall.

Hubbert, M.K. 1940. The theory of groundwater motion. *Jour. Geol.*, 48: 785-944.

Kayane, I. 1992. Methodology of field inverstigation on water cycle by environmental tracers in the humid tropics. In I. Kayane (ed.) *Water cycle and water use in Bali island*: 5-18. Japan Inst. Geosci., Tsukuba Univ.

Shimada, J. Shimmi, O., Tanaka, T., Nakai, N. & Itadera, K. 1992. The effect of Subak systems to the regional evaporation. In I. Kayane (ed). *Water cycle and water use in Bali island*: 175-188. Japan Inst. Geosci., Tsukuba Univ.

Sedimentology, engineering and environmental aspects in the River Gash basin, Sudan

Aspects sédimentologiques, techniques et de l'environnement dans le bassin du Gash, Soudan

O.M.Abdullatif

Department of Geology, University of Khartoum, Sudan

ABSTRACT: This paper investigates the various factors which have influenced and contributed to natural resources depletion and environmental degradation in the River Gash basin. The basin has been subjected to high rates of sedimentation, floods, desertification, diminision of water resources and loss of fertile agricltural lands. All above appear to be of major significance and adverse socio-economic effects in the area. Therefore, the situation necessitates the implementation of measures so as to preserve the natural resources and restore the environmental balance in the Gash basin.

RESUME: Ce document traite des différents facteurs qui ont influencé et contribué à la diminution des ressources naturelles et à la dégradation de l'environnement dans le bassin du Gash. Ce bassin a notamment subi une forte sédimentation, des inondations, une désertification, une diminution de ses ressources en eau, et a vu la surface de ses terres agricoles fertiles s'amenuiser. Tout ceci revêt une grande importance et engendre des effets socio-économiques negatifs sur toute la région. Cette situation necessite donc la mise en application de certaines mesures afin de préserver les ressources naturelles et de restaurer l'écosysteme dans ce bassin.

1 INTRODUCTION

The situation in the Gash basin represents a case study where both natural and man-made processes have resulted in floods, depletion of natural resources and environmental deterioration. The same situation applies to several other arid to semi-arid regions in Africa where population growth, unplanned agricltural developments and recurrent droughts have put great pressure on limited water resources used for domestic and irrigation purposes.

2 GEOLOGICAL SETTING

The River Gash basin is located in eastern Sudan (Fig 1). It is the only seasonal stream in the area which provides surface flood water and recharges the fluvial groundwater aquifer. The area has arid to semi-arid climate and the rainy season extends from July to Septe mber. The long term mean annual rainfall is 310 mm at Kassala town. Moreover, the River Gash is an ephemeral , braided stream which originates in the Eritrean highlands and ends forming a terminal fan delta in eastern Sudan (Fig. 1). The NW trend of the Gash fan suggests structural control by preexisting Basement shear zones. Gravity modelling in the central part of the fan shows a deep trough reaching a maximum of 2.5 Km filled with sediments. This most probably indicates a connection to the rift-related basins of central Sudan.

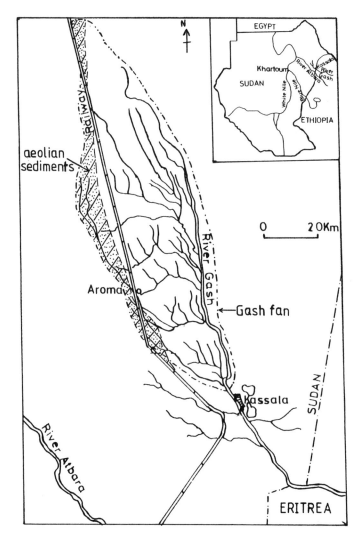

Fig.1. Location map of the River Gash basin

3 SEDIMENTOLOGY

The River Gash has a shallow sandy bed with wide floodplains. The main channel depth is between 1 and 2 m and a width which varies from 100 to 800 m (Waps). The river shows rapid discharge variation and the discharge loss is also characteristic, where about 50 percent loss is reported between the border and Kassala twon (Saeed, 1972; Ibrahim, 1980). Facies analysis has revealed two types of sequences consisting of channel-fill and sheet-flood. It seems that each sequence is produced by specific depositional event and both sequences represent two-end members for suite of mixed sequences. It appears that the fluvial sediments of the River Gash were produced in two phases, first, by braided channel aggradation and lateral migration and second by both channelized and sheet-flood deposition. The deposits of these two phases contribute to the fluvial sediments of the present day River Gash basin (Abdullatif,1989) North of Kassala most of the river

2950

water is lost by infiltration and evaporation and the river bifurcates into distributary channels forming the Gash terminal fan having characteristic conical shape (Fig. 1). These channels decrease in size and incision from the apex to the toe of the fan and here shallow sheet-floods dominate. Moreover, down fan there is progressive increase in the fine grained silt and clay facies. Summary of the morphological, sedimentological and hydrogeological aspects in the Gash basin is provided in Table (1).

FLOODS AND IRRIGATION PROBLEMS

Floods in the Gash basin usually result from heavy rain fall in the catchment area which causes high discharges above the channel capacity. During the last two decades Kassala area was flooded five times resulting in large damages to residential and agricltural areas. One of the causes of floods is the high rate of sediment aggradation in the river Gash. The river bed level at Kassala increased by 380 cm from 1936 to 1974 (Ibrahim,1980).

GENERAL	SEDIMENTOLOGICAL	HYDROGEOLOGICAL
morphology: braided to fan-like pattern	river: braided & ephemeral river bank: 1 - 2 m	mean annual flood:56 million m^3
tectonic setting: fan controlled by fault bounded-trough	river width: 500 m in medial part	min. annual flood:140 million m^3
river length: 280 Km to fan apex	depositional process: by both confined(channel) and unconfined flows (sheetflood)	max. annual flood:1260 million m^3
fan area: 2000 Km^2 fan gradient: gentle	discharge: variable and flashy	average, max. and min. days of flow: 90, 114 & 68 days.
catchment area: 21000Km^2 catchment width: 30 -90 Km	river subenvironments: channels(bars, sandflats) levees, overbank areas & floodplains.	saturated thickness: 15 -30 m
		discharge loss: 50 - 70 percent due to infilteration & evaporation
elevation: 2000 m a.s.l at upstream in Eritrea	facies types: St, Sp,Sh, * Sr, Ss, Fl	groundwater used in 1982 in basin: 174 million m^3
500 m a.s.l at Kassala 450 m a.s.l at mid fan	dominant grain size: coarse to fine sand	renewable potential in 1982: 229 million m^3
average rain fall: 645 mm in Eritrea 310 mm in Kassala 160 mm in fan area	maximum grain size: pebbles (1-7 cm) & 2.5 cm average sand facies percentage: 92-97 percent in median channel areas	Kassala area after 1988: groundwater used: 145 million m^3. renewable potential:110 million m^3
temerature: Winter (16-35° C) Summer (22-42° C) rainy season (23-36° C)	sorting: poorly-moderately sorted clast shape: subrounded-rounded	groundwater deficit:32 percent
relative humidity: 40 - 66 percent	facies thickness: 5 - 120 range c_m	groundwater drilling development in Kassala in 1982-1988 increases from 10.5 - 56 percent
mean evaporation: 5 - 16.5 mm/day	paleocurrent: unimodal facies lateral continuity generally greater than 10 m and a maximum of 20 m	flood surface water exploited at the Gash delta fan for irrigation ranges between 15-58 percent.
other: total forest area: 108000 acres	facies sequences: channel fill and sheet-flood	areas can be irrigated in Gash fan: 250000 acres.
total animal resources: 720000	facies vertical variation stacked fing upwards sequences	ratio of area cultivated to that can be irrigated is 32 percent

* codes (Miall,1977)

Table 1. River Gash basin morphological, sedimentological and hydrogeological data sheet.

Therefore, the sediment aggradation has chocked and raised the river bed leading to flash flooding, where dikes and spurs were insufficient to keep up with the rate aggradation and high peak discharges. This because the large sediment load exceeds the channel capacity and prevents easy flow of subsequent floods. This consequently forms middle channel bars, flow diversion and may lead eventually to banks erosion, avulsion and flooding. It seems that the predominance of the sand facies facilitates bifurcation, channel lateral migration and banks erosion (Abdullatif, 1990).

One of the main flood control measures used is spurs system which is built across the channel reach in Kassala area. The spurs are intendent basically to confine the flood within a narrow straight channel and to create a degradational process that helps removal of sediment load. Secondly, to prevent channel bifurcation, lateral migration, banks erosion and river avulsion. In addition, areas between the spurs, where wanning flow dominates, became sites for active sediment aggradation and vegetation and this more likely enhances banks stability and reduces chances for avulsion (Fig. 2a-d).

The existing spurs system, however needs revision particularly with respect to the time element in the designing under rapidly varied unsteady flow and unpredictable sediment transport condition.

Fig.2a. Excavation of silt from a pit in the river main channel.

Fig.2c. Thick vegetation cover in the inter-spur areas, river main channel.

Fig.2b. Silt aggradation on the spur side (1.5 m thick)

Fig.2d. The head of the spur is partly eroded by the flood.

The sediments aggradation in River Gash indicates that the rates of sediments depor .on exceeds that of sediments erosion. Although the spurs system enhances erosion, the change to active and effective degradational pattern is minimal because of rapid discharge loss and large sediments supply. Moreover the spurs system represents a bottle-neck situation which prevents easy flow, reduces storage capacity and eventually enhances avulsion and flooding. One more difficulty with the spurs is that the high rate of siltation in the inter-spurs areas (Fig.2a,b) has resulted in a laterally continuous silty layers. These layers have reduced the natural recharge of the fluvial aquifer in the Gash basin. One measure considered to improve the natural groundwater recharge is the removal of the upper silt layers from the bed of the river (WRM, 1993). Furthermore, the thick vegetation cover (Fig.2b,d) has caused physical and ecological changes and malaria-carrying mosquitoes are increasing during the flood period.

Recent observations indicate that after seven years in service most of the spurs in the city reach have been eroded away and the river restored its characteristic morphological pattern reflected in wide course with braided laterally shifting channels.

Therefore, to prevent catastrophic floods in the future, other flood control measures have to be introduced. For example, clearing the sediments aggradated from the main channel during the dry spills. This measure integrates very well with the irrigation system used in the Gash fan delta and both natural and artificial recharge measures now considered to augment the water supplies in the Gash basin (WRM, 1993). Moreover, non-constructional measures have to be considered so as to control man-made practices such as agriclture, building and industrial activities outside the banks and on flood plains areas (Ibrahim, 1988; Hamid, 1988; Abdella, 1988).

4.1 Irrigation difficulties

Irrigation by flood water in the Gash basin has become important especially after years of low rain falls and drought in the area. The Gash fan delta is irrigated using flush irrigation system through a network of canals connected to the river channel. The large sediments aggradation has resulted in heavy siltation in these canals and therefore reduced greatly their effectiveness. Consequently large areas of fertile lands which could no longer be irrigated are subjected to devegetation and soil erosion.

The sedimentary processes operating in the river with channel switching, lateral migration and high rates of siltation have influenced greatly the efficiency of water intake in the canals (Abdella,1991; Hamid,1991). Moreover, channel shifting causes change in river flow direction and consequently reduces the efficiency of water withdrawal. During the last ten years the data available show that the percentage of water used in flush irrigation ranges between 15 to 58 percent (Ahmed, 1991).

Comparison between annual flood discharges and cultivated areas in the Gash fan delta is shown in Fig(3). Maximum or minimum area irrigated does not relate to water available. The records als show that the ratio of the largest area cultivated to the area which could be irrigated is 32 percent (Table 1). Other factors contributed to inefficiency of the irrigation are poor supervision, operation and maintenance. Recently, the efficiency of the irrigation work declined 45 percent (Hamid,1991).

Therefore, any future expansion in agricltural developments require improvement of the irrigation system efficiency. Moreover, river training should be integrated with the irrigation system, since they are closely related.

5 DESERTIFICATION AND DEFORESTATION

The River Gash basin is subjected to devegetation and desertification which need rapid and efficient remedial measures. Both natural and man-made processes are responsible. For instance, collapse of

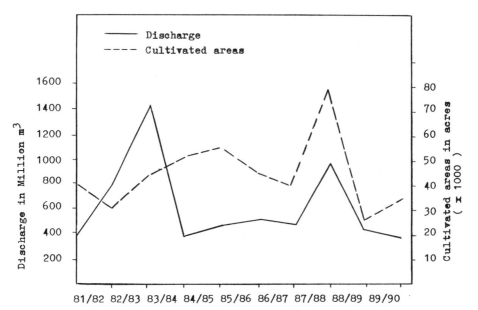

Fig.3. Comparison between annual flood discharges and
cultivated areas in the Gash fan delta (Ahmed,1991)

the irrigation on the western part of the fan delta resulted in devegtation and activated sand dunes migration (Fig.1), where more than 11 percent of the fan area has been covered by aeolian sediments (Mohamed, 1983). During the last two decades drought has aggravated the situation. Most of the natural tree cover in the Gash fan is economically important and useful for soil protection. The total area is estimated at 108000 acres(Mohamed and Seedi, 1991). The forests, however are threatend by low rain falls, sand dunes migration, fires, wood cutting and stock grazing. The preservation of forests is essential for sustainable development in the Gash basin. Therefore, by improving the irrigation efficiency large unused flood waters could be utilized in cultivation of productive forests in the Gash basin.

6. GROUNDWATER RESOURCES

The groundwater resources of the River Gash basin are diminishing due to extensive exploitation of these resources. This has concluded to the situation that the aquifer is exploited almost up to the maximum sustainable yield(Nur

El madina,1993). Many factors have contributed to the present situation and include unplanned agricltural development, heavy settlement of nomads, usual demographic increase and vast refugrees influx all have resulted in depletion of groundwater resources (WRM,1993). Declining of rain falls in the catchment are is considered to be a major factor. Comparison between the actual use and renewable potential groundwater is provided at three different river reaches (Fig. 4). For 1982 the groundwater reserve ranges between 14 to 29 percent. In Kassala area for 1988, however, an acute situation developed, where the actual use exceeded the renewable potential with a deficit of 32 percent. That is the renewable potential is best at 110 million m while the use has increased to 145 million m per year.This deficit is attributed to high drilling development which during the period 1982 to 1988, has increased from 10.5 to 56 percent (WRM, 1993). Moreover, the intensive development of groundwater in Kassala area has resulted in sharp decline in water level with drawdown reaching 7 m.(WRM,1993).

Some remedial measures have been suggested so as to tackle

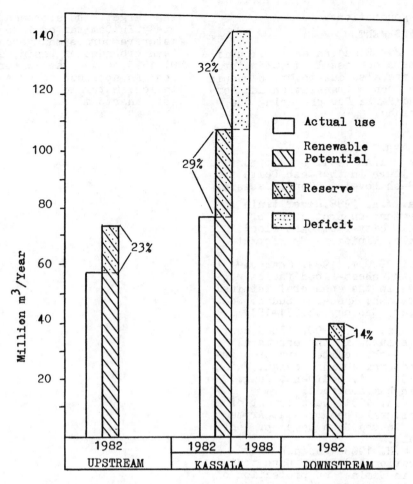

Fig.4. Comparison between the actual use of ground water and renewable potential at different river reaches (WAPS, 1982; WRM, 1993)

the critical groundwater situation (WRM,1993, Nur El Madina,1993), and these include (a) the mangament of both domestic and agriclture water demands, and (b) to augment the groundwater supply through natural and artificial recharge of the Gash basin aquifer.

7 CONCLUSION

Both natural and man-made processes have contributed to resources depletion and environmental deterioration in the Gash basin. Several aspects of water resources, floods, irrigation, agricltural development and desertification are closely related. This necessitates the implementation of measures so as

to preserve the resources and restore the environmental balance. For instance, the modernization of the irrigation system and the improvement of its efficiency and the integration with river training are essential. Better utilization of water resources by managing the demands and augmenting the supply has become a necessity. Moreover, the proper management of both surface and groundwater resources represents a critical element in a sustainable development in the Gash basin including preservation of native vegetation and fertile agricltural lands, reforestation and reach self sufficiency in food production. Such development can meet the needs of both present and future generations.

AKNOWLEDGEMENTS

I thank Nur Elmadina and Dr.Yousif for papers on Kassala groundwater. Thanks are also due to Mr. Hardouin for the french translation and for Wafa and Faiza for preparing the photographs.

REFERENCES

Abdella, I.A. 1991. Development of agriclture in the Gash Delta, in the Gash Develop. Conf. Kassala.

Abdella, M.A. 1988.River training in Seminar on protection of Kassala city from Gash flood, Kassala, Ministry of Information.

Abdullatif, O.M. 1989. Channel-fill and sheet-flood facies sequences in the ephemeral terminal River Gash, Kassala, Sudan. Sedimentary Geology,63,171-184.

Abdullatif, O.M. 1990. Flood control measures in the ephemeral River Gash: sedimentary processes and engineering design, Sudan The 13th Int. Sediment. Cong., Nottingham,abstract.

Ahmed, S.A. 1991. Development of agricltural resources at Aroma area, in the Gash Develop. Conf. Kassala.

Hamid, M.H. 1988. Kassala city protection and River Gash training, in Seminar on protection of Kassala city from Gash flood, Kassala, Ministry of Information

Hamid, M.H. 1991. Irrigation in River Gash, in the Gash Develop. Conf., Kassala

Ibrahim, A.A. 1980. River Gash behaviour and training. M.Sc. Thesis, Univ. of Khartoum

Ibrahim, A.A. 1988. Human influences and behaviour on River Gash dikes. in Seminar on protection of Kassala city from Gash flood, Kassala,Ministry of Information.

Mohamed, A.E. 1983. Some aspects of desertification in the Gash Delta region, M.Sc Thesis, Univ. of Khartoum.

Mohamed,A.E. and Seedi, A,A.1991 The forests in the Gash Delta, in the Gash Develop. Conf., Kassala.

Nur Elmadina, E.K. 1993. Water resources management in the Gash groundwater basin, the history, the target, the approach, case study.In Seminar on integrated water resources management. The Gash groundwater basin,Kassala.

WRM, 1993. River Gash water resources: An approach towards augmentation. Water Resources Management, Khartoum.

Water resources protection in the Baikal lake basin

Protection des eaux dans le bassin du lac Baikal

Ivan M. Borisenko & Albert A. Adushinov
Buryat Geological Institute, Russia

ABSTRACT: At the present time the technogene contamination of the Baikal lake and its basin water resources is taking place. Modern technologies of wastewater purification do not fully solve the problem of environmental protection. Engineering and geological methods of surface and underground waters protection from contamination are proposed. The results of experimental and practical works confirm their high efficiency for this region.

Resumé: A present on voit la pollution technogène des ressources d'eau du lac Baikal et de son bassin. Il est à noter que la technologie moderne de l'épuration des eaux des égouts ne résout pas des problèmes de la protection de l'environnement. On propose des méthodes d'ingénieur et de géologie de la défense des eaux superficielles et souterraines contre la pollution. Leur haute efficacité pour cette région est affirmée par des résultats des travaux expérimentaux ainsi que pratiques.

The Baikal lake is inserted in the first and foremost list of world heritage areas and its basin territory has particular conditions for the nature wealth use. Up to 1960 it has been one of the ecologically purest regions in the world. However, during the last decades the rapid development of industrial and agricultural production, the exploitation of mineral deposits and forest areas, the Baikal-Amur railway construction and other anthropogenic processes have markedly changed the geological environment in the Baikal lake basin which holds over 80% of Russia surface fresh waters.

Surface and underground waters contamination by industrial, agricultural and municipal wastes has got usual occurence. The construction of numerous treatment structures and hard wastes storages has not stopped the contamination of the Baikal lake basin water resources, though technogene pressing has been considerably decreased. It is necessary to eliminate throwing off wastewaters even treated into rivers and lakes for the protection of this unique region from subsequent contamination. The authors have carried out natural experimental research on the application of geological methods of wastewaters thorough final purification. For this purpose the areas with proper engineering-geological and hydrogeological conditions have been chosen.

Large contaminants of natural environment in the Baikal lake basin are the enterprises of pulp and paper industry. In spite of the perfect and expensive treatment technology these enterprises' wastewaters which are thrown off into rivers continue to contaminate the region water resources with specific wastes, i.e. tall oil, methanol, lignin, as well as sodium sulphide the high content of which is inadmissable in fishing reservoirs. To decrease the harmful influence of these wastes on water environment the solution has been taken to study the geochemical wastewater treatment in underground aquifers parallel with the working out of chemical, physical and biological treatment technologies.

An experimental and industrial ground has been created not far from the pulp and paper combine. In the alluvial water-bearing aquifer of the Selenga river, the largest tributary of the Baikal lake, an injection well of 100 m depth has been drilled for pumping treated wastewaters and 17 wells of 30-110 m depth have been drilled to observe the change of ground water level, chemical composition and the actual speed of contaminated water travel front. Observation wells have been located radially, i.e. along and across ground water stream.

The following depositions take part in the geologic structure of the ground territory (from top downwards), as:

1. Sandy and loamy depositions of Holocene - 0,0-2,0 m.

2. Gravelly and pebbly depositions of Holocene-Pleistocene - 2,0-104,0 m.

3. Greenish and blue clay of Miocene - 104,0-178,0 m.

4. Paleozoic granites - 178,0 m and deeper.

Ground water level lies at the depth of 2-3 m. The thickness of an alluvial water-bearing aquifer is 100 m. Filtration coefficient is 23,6 m/day, water conductivity is 1800-2000 sq. m/day, ground water stream slope is 0,003 and its natural travel speed is 0,1 m/day.

The treated wastewaters of the pulp and paper combine have been continuously injected into the well at 670 cub. m/day speed. The experiment has lasted for 7 months. The observations of water parameters and sampling for chemical analyses have been carried out daily.

As a result of this experiment the data on the effect of waste-water final purification when moving in groundwater stream have been got. It is found that organic substances formed by wood processing are extracted from wastewaters mainly by means of ground particles sorption on the surface, by their sedimentation in deadlock pores and biogeochemical decomposition. During the experiment the most ecologically dangerous wastewater components (lignin, tall oil and methanol) have been detained already at the first tens of metres from the injection well and they have not been observed in groundwaters in the remote observation wells (to 150 m). Sulphates migrate farther and their concentration lowering in groundwaters takes place owing to ion exchange reactions and dilution. The mathematical modelling method displays that wastewater mineralization and sulphate content will equal with the ones of groundwaters after moving to the 1200 m distance.

The real speed of moving the contaminated water front along a water-bearing aquifer at the present injection regime does not exceed 0,5 m/day. At this speed the contaminated water can reach the Baikal lake only in 270 years. Certainly, the danger of water-bearing aquifer contamination is real, but it is preferable than direct throwing off wastewaters into rivers and lakes. This method of wastewater final purification is a stop-gap measure till the development of highly effective and economically profitable technologies.

Filtration fields where municipal wastewaters are purified have got the world practice. This method is already used in the severe conditions of Transbaikalia. However, engineering-geological and hydrogeological conditions are often taken into consideration poorly, that is why water environment is contaminated by unpurified wastewaters.

For the effective application of filtration fields in the Baikal lake basin we have experimentally studied the work of filtration fields located in different geologic situation. Before the experiment we have carried out the detailed research of granulometric composition, physico-mechanical and water physical properties of grounds which form the aeration zone, hydrogeological parameters of a water-bearing aquifer lying under the section of filtration fields and also geocryological conditions. The territories composed by sandy grounds on the particles of which the main mass of contaminating components is sorbed are the most suitable for the arrangement of effectively working filtration fields. For increasing the time and length of wastewater infiltration way through the layer of the aeration zone rocks the thickness of the latter has been chosen to be not less than 5-7 m. The hydrodynamic parameters of a ground stream (filtration coefficient and water conductivity) must ensure the withdrawal of wastewaters infiltrated through the ground.

The results of the experiment show that the efficiency of final purification of municipal wastewaters from organic and bacterial contaminants can be very high when keeping to the stated conditions and load rates for filtration fields. Nitrogen compounds are fully detained in a ground layer already to 1 m and phenols and oil products are detained in the upper 30 cm layer. Wastewater infiltration through the ground of 2-3 m thickness is needed for the purification from synthetic surface active substances and for lowering to the rate of biological and chemical oxygen consumption. Hard metals do not react with ground and soil water substances and their concentration lowering occurs only owing to dilution in a ground stream.

A large contamination source is mine waters pumped when mining coal. They are highly mineralized (to 6,5 g/l) and contain sulphates and hard metals in the quantities exceeding maximum permissible concentrations. Their purification and utilization are connected with technological and ecological difficulties. After the thorough study of a mine geologic structure and fulfilling hydrodynamic calcu-

lations and hydrogeochemical prognosis we have suggested that mine waters should be injected into noncoal layers occuring lower than mine working. As a result throwing off mine waters into reservoirs will be eliminated. It is possible to pump to 1 mln. cub. m a year into the wells drilled in noncoal layers. The chemical composition and mineralization of underground waters contained in noncoal layers practically do not differ from the chemical characteristics of mine waters. When these waters are moving to the local centres of unloading and are interacting with coal layers, inclusion into a natural hydrogeochemical process takes place, as well as sulphate reduction to sulphides and dilution of hard metals. When reaching reservoirs these waters will possess the quality satisfying drinkable and fishing requirements.

Wastewater purification methods mentioned above are used "in situ" in geological environment, i.e. the natural properties of geological objects are used without their travel to the contamination source. At our Institute the method of industrial and municipal wastewaters is investigated by their filtration through the grain material of zeolite minerals the deposits of which are in sufficient quantity in Transbaikalia. These minerals are characterized by their high chemical stability, mechanical strength and the ability for active ion changing reactions and adsorption. The latter is conditioned by the presence of a vacuum and canal system in the crystal structure which is filled by the cations of potassium, magnesium, sodium and calcium, as well as by water molecules. The ions mentioned are able to exchange with the ones of wastewaters showing selective properties owing to which it is possible to carry out ion concentration and separation that favours wastewater purification.

There are also other aspects that negatively influence the state of water environment in the Baikal lake basin. The construction of a hydraulic power station in the head of the Angara river flowing out of the Baikal lake has resulted in more than 1 m level

rise of the lake water which in
turn has entailed intensive lakeside changing processes and flooding large shore areas. Expensive
engineering measures have been taken to protect the Trans-Siberian
railway road from destruction.
Drainage land-reclamation has been
used for the recurrence of land
bogged up into agricultural turn.

On irrigational territories adjoining the Baikal lake shores the
application of chemical weed-kil-
lers and mineral and organic fertilizers is forbidden to avoid the
contamination of lake waters. The
16 km long North Mujsk tunnel is
being built in the region of the
Baikal-Amur railway. The tunnel is
crossed by geologic faults along
which thermal waters circulate.
Drain works exhaust large water
resources which can be attributed
to irretrievable losses.

Special service for the control
of the environment and for the
prognosis of its changes has been
formed to increase the effectiveness of water resources protection
in the Baikal lake basin.

Permeability measurements of stressed and unstressed strata, Eastern Transvaal, South Africa

Détermination de la perméabilité des couches sous contraintes, Transvaal Oriental, l'Afrique du Sud

B.J.Venter, C.A.Jermy & F.G.Bell
Department of Geology and Applied Geology, University of Natal, Durban, South Africa

ABSTRACT: A coal mine situated in the eastern Transvaal, South Africa, experienced a number of serious roof falls. These falls were attributed to excessive methane and/or water build-up in the immediate roof strata of the coal seams being mined. As a consequence a laboratory investigation was initiated to try and establish the relationship of different roof facies types to the migration of the fluids responsible for the falls.

The focus of the investigation was therefore on permeability, which was measured under atmospheric and simulated underground stress conditions. For the tests under atmospheric conditions a modified Ohle cell was used with nitrogen and methane as the permeating fluids. The underground stress regime was simulated by the use of a modified Hoek triaxial cell, allowing permeability measurements under different stress conditions. Plotting the permeability at a range of gas pressures against the reciprocal mean gas pressures and fitting a straight line which intersects the y-axis to the points, the "liquid equivalent" permeabilities of the samples were obtained. The liquid equivalent permeabilities were used as a basis for comparison between the different facies types tested. Facies type was found to have an influence on the permeability. The permeability was also found to be very variable and unpredictable, even over short distances (centimetre scale), especially when methane was used as the permeating fluid. This was not entirely unexpected as methane is known to be adsorbed by carbon, which therefore can influence the permeability.

RESUMÉ: Une mine de charbon, dans l'Est du Transvaal, en Afrique du Sud, a subi de nombreux et sérieux affaissements. On les a attribués à un excès de méthane et/ou à l'augmentation de l'eau dans la couche immédiate de la voûte d'exploitation du charbon. En conséquence, un bureau d'enquêtes a entrepis d'essayer d'établir le rapport entre les différentes catégories de voûtes et le déplacement des fluides reponsables des affaissements.

Le but de cette enquête était de mesurer la perméabilité sous des conditions atmosphériques et des pressions souterraines simulées. Pour les tests sous conditions atmosphériques, on a utilisé une cellule d'OHLE modifiée, avec du nitrgène et du méthane comme fluides filtrants. Le système de pression du sous-sol fut simule avec l'utilisation d'une cellule d'HOEK triaxiale modifée, permettant de mesurer la perméabilité sous les différentes conditions de pression.

En faisant les graphiques de la perméabilité des differentes pressions du gaz et les pressions du gaz inverse, puis en traçant une ligne droite qui coupe l'axe Y en certains points, on obtient les perméabilités du "fluid équivalent" des expériences. On les a utilisées comme base de comparaison entre les différentes catégories de faciés analysées. On a également trouvé que le type du faciés avait une influence sur la perméabilité. On a ègalement conclu que la perméabilité est très variable et imprévisible, même sur de courtes distances (échelle en cm) particulièrement lorsqu'on utilise le méthane comme fluide perméable. Ceci n'était pas complètement inattendu ca on sait que le méthane est absorbé par le carbone ce qui peut donc influer sur la perméabilité.

1 INTRODUCTION

A coal mine in the eastern Transvaal, South Africa, experienced some roof falls with catastrophic results. These roof falls were identified as resulting from excessive gas pressure build up in the roof strata of a mined coal seam. Accordingly an investigation of the permeability of the roof rocks in the mine concerned to fluid flow was undertaken. In order to characterize the migration of methane into and around the mined seam the permeability of different lithologies in the immediate roof strata was measured in the laboratory. Twenty four separate facies types were identified in the roof strata concerned (Ward and Jermy, 1985) but only eleven of them were present in the borehole core sampled for the tests. The borehole core was obtained from exploratory holes drilled by the colliery through the roof strata of the mined seams. The permeability was first tested under atmospheric conditions with a modified Ohle cell (Venter, 1994). Secondly the permeability was tested under triaxial conditions, to simulate the *in situ* mine stress regime, with the aid of a modified Hoek cell (Daw, 1971). By plotting the permeability of a gas against the reciprocal mean gas pressure and then extrapolating the best fit line to intersect the y-axis, the permeability of the sample at high gas pressures could be deduced (ASTM D4525-85, 1990). This permeability is referred to as the "liquid equivalent permeability" (k_{le}) and was used as a method of comparison between the different tests in this study.

2 METHODOLOGY

2.1 Sample preparation

The samples were obtained from five boreholes drilled vertically into the roof of a mined coal seam experiencing roof falls. These yielded core of NX (54 mm) diameter. A borehole drilled from the surface to intersect the coal seam was also used for sampling and yielded TNW (60.3 mm) diameter core. The core samples were cut into appropriate lengths for the different tests. The Ohle cell required sample lengths of 12 mm to 20 mm irrespective of their diameter. The triaxial tests, to reduce "end" effects, required samples conforming to the 2:1 length to diameter ratio proposed by Obert and Duvall (1967). As a result the samples were cut into 100 mm or 120 mm lengths, depending on the diameter of the core. Carborundum paste was then used to grind the ends of the samples until they were parallel and smooth. This was done so that an even load could be applied perpendicular to the long axis of the sample in the triaxial case, and to assure a better seal between the O-rings and the sample when tested in the Ohle cell.

After the samples were sawn into their appropriate lengths and their ends ground with carborundum paste to the required smoothness, the samples were washed to remove all the grit and paste. This was done with care to avoid impregnating the pores of the sample with the grit and paste, which could affect the permeability of the sample. The samples were then oven-dried at 110°C to remove all the excess moisture and placed in desiccators containing silica gel to keep them moisture-free. Drying over a period of 12 hours was found to be sufficient in most cases for the sample to attain a constant weight, that is, to be assumed moisture-free.

2.2 Ohle cell

2.2.1 *Apparatus*

The original design of the Ohle cell as described by Ohle (1951) and Chakrabarti and Taylor (1968) was subsequently modified by the authors. These modifications mainly concerned the material used for its construction and the sample holder. Instead of stainless steel, brass was used because it allowed for a more flexible design. Brass was chosen because it does not react with any of the fluids used (eg. does not rust when testing with water) and is strong enough to withstand the pressures used. Modification of the sample holder allows it to accept a broader spectrum of samples with respect to diameter and length. The seal between the sample and the sample holder is achieved by two rubber O-rings pressed against the sample by means of a spacer.

The gas (either nitrogen or methane) was supplied by means of a commercial gas cylinder that was capable of delivering 20 MPa gas pressure. The supplied pressure was kept constant by means of a regulator and measured with a transducer and a digital readout. Flow rates were measured with a stopwatch and a series of bubble flow meters of 25, 100 and 250 cm³ capacity.

2.2.2 *Test method*

After the sample was installed in the chamber of the sample holder, the apparatus was assembled. Sealing was achieved by compressing an O-ring against the sample. The effectiveness of the seal could be tested by insertion of a steel disc of the same dimensions as the sample in the cell, applying a pressure and noting if any fluid reached the bubble flow meter. With a sample installed, a pressure of 150 kPa was applied. The sample was then left to reach equilibrium. Equilibrium is assumed when the measured rate of flow becomes constant. This also implied that the sample was fully saturated with the testing fluid. The flow rate then was recorded. Then the gas pressure was increased by 50 kPa. The system was again left to reach equilibrium and the flow rate recorded. This incremental increase in gas pressure was continued until the pressure reached 700 kPa. The permeability at each increment then was calculated using Equation 1 which is a derivative of Darcy's law (Stormont and Daemen, 1991):

$$k = \frac{2QP_0\mu L}{(P_1{}^2 - P_2{}^2)A} \qquad (1)$$

where: k = permeability in the direction of flow (Darcy); Q = volumetric rate of flow at inlet pressure (cm³/s); P_0 = reference pressure (atm); μ = viscosity of the fluid (cP); L = length over which the flow is measured (cm); P_1 = inlet gas pressure (atm); P_2 = outlet gas pressure (atm); A = cross-sectional area across which Q is determined (cm²).

The reference pressure, P_0, and the outlet pressure, P_2, are usually taken as equal to each other and equal to the atmospheric pressure, which is assumed to be 1 atm. A gas pressure of 700 kPa was chosen as termination point because the initial tests resulted in erratic permeability versus reciprocal mean gas pressure plots after a inlet pressure of 400 kPa was reached (Fig. 1). This behaviour was thought to be caused by either

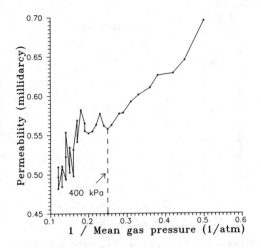

Figure 1. Graph showing irregular behaviour of the k_s versus $1/P_m$ beyond 400 kPa.

the Klinkenberg effect (Klinkenberg, 1941) or by the flow becoming turbulent so that Darcy's law was no longer valid (Desai, 1975). Subsequent experiments were continued up to 700 kPa inlet gas pressure to make sure that the cause could be identified as well as to provide enough data points for the later statistical manipulations required by Anon (1990).

Usually the sample was initially tested with nitrogen as the permeating fluid and directly afterwards with methane as the flowing fluid. Nitrogen was used first mainly for reasons of safety. If the geometry of the sample did not allow a seal to be effected and a large quantity of gas escaped into the atmosphere, it does not pose a hazard as does methane. To change over from one gas to another, the gas pressure was reduced to zero and the pressure released from the system without removing the sample from the cell. The procedure as outlined above then was repeated with the new gas.

2.3 Hoek cell
2.3.1 *Apparatus*

For the tests simulating *in situ* stress conditions a modified Hoek-Franklin triaxial cell was used (Daw, 1971). The modifications involve a series of interconnected circular grooves cut into the end platens. The grooves are designed to distribute the fluid over the whole cross-sectional area of the sample, at the high pressure end, and to collect the fluid passed through the sample over the whole cross-sectional area of the sample, at the low pressure end. In the case of coal it was found that these grooves pressed into the ends of the coal sample thereby effectively blocking any gas from penetrating the samples. Cintered discs of the same diameter as the platens were used to counteract this. Axial load was supplied by a hydraulic testing machine capable of applying a load of 1000 kN.

Confining stress was supplied by a hydraulic pump that could supply oil pressure up to 50 MPa. The gas was supplied and the flow rate recorded by the same equipment as used for the Ohle cell tests.

2.3.2 *Test method*

Before the sample was placed in the assembled Hoek cell, the annular space between the rubber sleeve and the cell was filled with hydraulic oil. The sample was then placed in the cell and a hydrostatic stress of 4 MPa applied (this was found to be the general stress at which no gas flowed between the sample and the sealing rubber sleeve). At the outset it was hoped to start the testing at the measured *in situ* stress conditions in the mine, but since these were estimated to be below 4 MPa, this could not be carried out. Only tests using methane were carried out, first because of the importance of the methane tests to this study and, secondly, because of the time aspect involved in reaching equilibrium. A full range of tests on one sample could take up to four days to complete. First, the gas pressure was increased in 50 kPa increments up to 700 kPa at a hydrostatic pressure of 4 MPa, similar to the Ohle tests. From this point a series of three tests were carried out:

1. The axial stress was increased in 2 MPa intervals (usually to 20 MPa) while the confining stress was kept at 4 MPa and the gas pressure at 200 kPa. The flow rate was measured after each load interval. Next the axial load was then decreased in 2 MPa steps, until hydrostatic conditions at 4 MPa were reached again.
2. The confining stress was increased in 2 MPa intervals while the gas pressure was kept constant at 200 kPa. This loading was also done until 20 MPa was reached at which time the hydrostatic load was decreased in steps of 4 MPa until the base 4 MPa hydrostatic conditions were reached again.
3. The hydrostatic stress was increased in 2 MPa intervals while the gas pressure was kept constant at 200 kPa. Similar to the previous method, the stress was increased until 20 MPa was reached. After this the confining stress was then relieved until the starting 4 MPa hydrostatic conditions were reached.

In each of the three cases the procedure was repeated two to three times. The effect of each cycle was determined by measuring the specific permeability at a series of gas pressures and determining the liquid equivalent permeability in each instance.

3 RESULTS AND DISCUSSION

3.1 Ohle cell

Plotting the permeability measured at each inlet gas pressure against the reciprocal mean gas pressure (P_m), in accordance with Anon (1990), it was found that the permeability decreased with increasing inlet gas

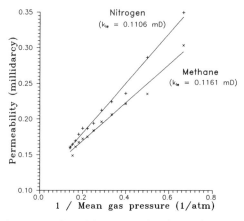

Figure 2. Plot of k_s versus $1/P_m$ for methane and nitrogen for the same sample.

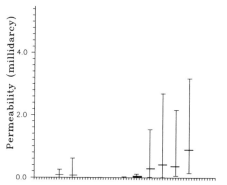

Figure 3. Plot showing the relationship between the k_{le} permeability ranges and of the different facies types obtained from the tests involving methane as the permeating fluid.

pressure. In the majority of the cases the nitrogen and methane plots for the same sample followed the same pattern (Fig. 2). The plot for nitrogen was found to plot higher with steeper slopes than the methane plots. This is as expected as nitrogen has a smaller molecular size than methane. Nearly half of the projected liquid equivalent permeabilities, however, did not follow this pattern. This is not important as the difference between the liquid equivalent permeability from nitrogen and the liquid equivalent permeability obtained from methane is what matters. Theoretically there should be no difference (Klinkenberg, 1941). Ohle (1951) concluded that the difference can be taken as experimental error. In this study the difference seldom exceeded 10% and the sample in Fig. 2 shows a difference of 5%.

Due to the extreme variability of the sediments tested, both over short distances (centimetre scale) and in the same facies type, the resultant liquid equivalent values obtained for each facies type is not confined to

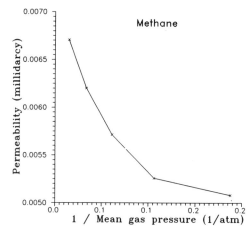

Figure 4. Plot of k_s versus $1/P_m$ showing a negative slope for a test involving methane as the permeating fluid.

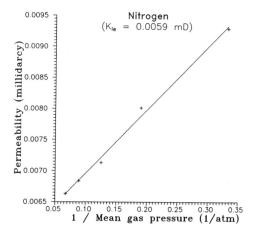

Figure 5. Plot of k_s versus $1/P_m$ showing a positive slope for a test involving nitrogen as the permeating fluid for the same sample as in Figure 4.

a single value, but rather to a range of values. This is illustrated in Table 1 which shows the ranges obtained from the tests involving methane as the permeating fluid. The relationship between the permeability ranges for the different facies types is better illustrated by Fig. 3. The main objective of this study was to determine if any correlation could be found between facies type and permeability. As illustrated in Fig. 3 a relationship, although not a straightforward one, does exist between facies type and permeability. In the coarser grained facies (facies type 8 and higher, see Table 1) an increase of average liquid equivalent permeability with increase in grain size can be observed. For the finer grained facies types (those below facies type 7) the average liquid equivalent permeability was found to decrease with increase in grain size. This could be the

Table 1. Liquid equivalent permeability ranges for the facies types tested using methane as the permeating fluid.

Facies type number and short description	Permeability Range (mD)			Number tested	Confidence (Average r²)
	Minimum	Maximum	Average		
2 Lenticular bedded mudrock.	0.0057	0.2590	0.1197	4	0.94
3 Alternating mudrock and fine grained sandstone.	0.0025	0.6146	0.0808	6	0.77
4 Flaser bedded sandstone.	0.0018	0.0040	0.0029	5	0.81
5 Ripple cross laminated fine grained sandstone.	0.0024	0.0070	0.0026	8	0.78
7 Massive fine grained sandstone.	0.0020	0.0247	0.0108	7	0.81
8 Cross laminated fine grained sandstone.	0.0001	0.1161	0.0282	8	0.92
9 Massive medium grained sandstone.	0.0213	1.5318	0.2788	22	0.80
10 Cross laminated medium grained sandstone.	0.0128	2.7132	0.4412	103	0.69
11 Massive coarse grained sandstone.	0.0681	2.1597	0.3902	34	0.80
12 Cross laminated coarse grained sandstone.	0.1567	3.1960	1.0546	11	0.69

result of many factors, but in the authors' opinion, the factor with the most significant influence is that of the number of samples tested. The more tests done on a specific facies type the more representative, statistically, the range of liquid equivalent permeabilities obtained. The coarser the grain size the larger the range of liquid equivalent permeabilities obtained because of the larger distribution in grain size present in the facies type. Another very significant factor that could influence the liquid equivalent permeability ranges is the choice of data points used in the extrapolation of the liquid equivalent permeability. Anon (1990) prescribes that a straight line be drawn through at least three data points at the lower values of reciprocal mean gas pressure. This best fit straight line then is extrapolated to the y-axis to obtain the liquid equivalent permeability for the sample. The points chosen for this best fit straight line may influence the value obtained in the extrapolation and the eventual liquid equivalent permeability ranges obtained. In this study all twelve data points were used when the straight line was fitted and the liquid equivalent permeability obtained.

In some cases (16 % of the tests involving nitrogen and 23 % of the tests involving methane) the plots of permeability versus reciprocal mean gas pressure gave plots with negative slopes for either methane alone (Figs. 4 and 5) or for both methane and nitrogen (Fig. 6). Such results were unexpected. The two main reasons that could be responsible for such results are the Klinkenberg effect and methane adsorption (Gawuga, 1979). Using Klinkenberg's method of analysis (or Anon, 1990) the effects of experimental error are negated (Klinkenberg, 1941). If the methane plot has a negative slope but the nitrogen plot for the same sample has a positive slope (Figs. 4 and 5) it can be assumed that adsorption is the dominant active phenomenon because methane is reactive towards the samples tested and not nitrogen. If both plots result in negative slopes (Fig. 6), gas slippage or the Klinkenberg effect must be the dominant mechanism present. In a small number of instances the extrapolated liquid equivalent permeabilities were found to have negative values which is a physical impossibility. Scheidegger (1974) stated that Darcy's law represent a straight line between quantity of flow and pressure differential and as such should result in a line through the origin. Experimental error could thus force the line to intersect the y-axis at a negative value.

Both the measured permeability (k_s) and the extrapolated liquid equivalent permeability was found to

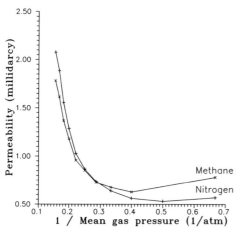

Figure 6. Plot of k_s versus $1/P_m$ showing negative slopes for both methane and nitrogen tests for the same sample.

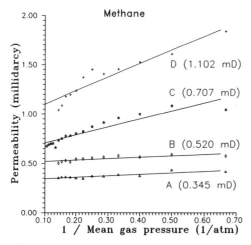

Figure 7. Plot of k_s versus $1/P_m$ showing the variation of permeability over a 10 cm interval. K_{le} is shown in brackets.

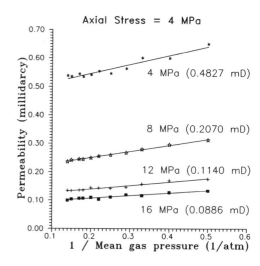

Figure 8. Plot of k_s versus $1/P_m$ showing the decrease of permeability with increase in confining stress from 4 to 16 MPa for the same sample.

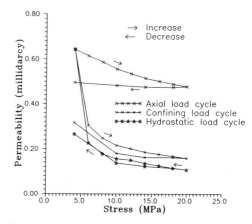

Figure 9. Plot of k_s versus stress showing the geometry and influence on the permeability of different stress load cycles on the same sample.

vary over very short distances. The plots represented in Fig. 7 are for 4 samples (each 15 mm in length) taken over a 10 cm distance. They all belong to the same facies type, but the grain size increases visibly from the medium grained sample A, to the more coarse grained sample D. It can be seen in Fig. 7 that k_s and k_{le} increase with increasing grain size. This shows that some relationship between facies type and permeability exists. The distribution of the data points also becomes less linear as the grain size increases and the gradient of the slopes declines as the grain size decreases. This means that the gas has a larger diameter path to flow through and some turbulent flow can occur, resulting in a non-linear plot. Fig. 7 also illustrates why a range of

liquid equivalent permeabilities, rather than singular values, were obtained for a specific facies type.

3.2 Hoek cell

The time needed for tests involving methane (or nitrogen) as the permeating fluid to reach equilibrium, was approximately 1 hour in the case of the Ohle cell tests. However, on applying a hydrostatic stress (i.e. $\sigma_1 = \sigma_2 = \sigma_3$) of 4 MPa, equilibrium was only reached after about 5 hours in the Hoek cell. The higher the facies number, that is, the coarser the grain size, the shorter the time needed to reach equilibrium. In order to obtain a basis for comparison with the tests under

Table 2. Liquid equivalent permeability ranges for the facies types tested with methane as permeating fluid under 4 MPa hydrostatic pressure.

Facies type number and short description	Permeability Range (mD)			Number tested	Confidence (Average r^2)
	Minimum	Maximum	Average		
7 Massive fine grained sandstone.	0.0061	0.0061	0.0061	1	0.92
8 Cross laminated fine grained sandstone.	0.0060	0.0666	0.0363	2	0.72
9 Massive medium grained sandstone.	0.3492	3.3127	1.3837	3	0.87
10 Cross laminated fine grained sandstone.	0.0096	2.9999	0.5852	14	0.69
12 Cross laminated coarse grained sandstone.	0.5280	0.5280	0.5280	1	0.99
14 Sandstone with carbonaceous drapes and slump structures.	0.0075	0.1316	0.0963	4	0.57
22 Mixed dull and bright coal.	0.4182	0.4182	0.4182	4	0.67

atmospheric conditions (Ohle cell), a series of permeability measurements at increasing gas pressures were made at differing hydrostatic loads. Table 2 shows the results at 4 MPa. Sixteen percent of the tests produced plots indicating a deviation from Darcy's law. The samples in these tests all contained some amount of carbon, suggesting that methane adsorption was the reason for the anomalies.

It was found that an increase in any of the applied stress regimes (axial stress, σ_1; confining stress, σ_3; hydrostatic stress, $\sigma_1 = \sigma_3$) led to a reduction in permeability. A increase in hydrostatic stress has the largest influence with confining stress the second largest and axial stress the smallest. Figure 8 shows the influence of a continued confining stress increase on the permeability. This decrease must be attributed to the closing of pore spaces under external pressure. The initial decrease in permeability which took place between 4 and 8 MPa, was found to be the largest in all cases, and in some this decrease was as much as a 50% reduction. Subsequent increases produced progressively smaller and smaller decreases in permeability as can be seen in Figure 8. Gawuga (1979) found that at some point there is no further decrease in permeability no matter how much the load is increased. The authors could not verify this, but after three consecutive load cycles the decrease in permeability due to any increase in stress was found to be very small.

Loading and unloading of the sample produced plots like Figure 9. The permeability decreases with increase in applied stress and then increases with decrease in applied stress. For the three plots in Figure 9 to start at the same initial permeability, the sample had to be relieved of the applied stress for two to three days. This means that some deformation of the internal structure of the sample takes place under load, but that if the load is

relieved the sample will gradual regain its initial internal structure. This is an-elastic behaviour. Testing the same sample on two occasions, a week apart, should therefore give the same results. In effect this suggests that the permeability of the samples are dependent on their stress history, at least for a time. The application of different load cycles for different stress regimes affect the samples similarly (Fig. 9).

4 CONCLUSIONS

Permeability is a physical property of a rock which can vary even over short distances and is dependent on many internal and external factors. It can be seen that the permeability of a sediment toward a gas, such as methane or nitrogen, will decrease with an increase in gas pressure. If carbon in one form or another is present in the sample being tested, it may adsorb methane and result in anomalous permeability versus reciprocal mean gas pressure plots. This mechanism can be shown to be more significant than gas slippage (Klinkenberg effect) by testing the sample with both nitrogen and methane and comparing the resulting permeability versus reciprocal mean gas pressure plots.

Application of an external stress results in a decrease in specific and liquid equivalent permeability. The amount by which the permeability decreases is controlled by the type of external stress applied. The largest decrease occurs when the applied stress is hydrostatic and the smallest decrease is obtained when the applied stress is axial. An increase in confining stress results in a larger decrease than an increase in axial stress alone, but results in a smaller decrease than an increase in hydrostatic stress. Permeability is dependent on the stress history of the sample, at least for a while. The influence of the stress history decreases with each application of a load cycle and will disappear after two to three load cycles.

Under a steady load application the permeability will decrease with time. This decrease continues until equilibrium is reached. Equilibrium is assumed when the flow rate is stabilized.

Tests involving gas as the permeating fluid requires substantially less time to complete than tests involving water as the permeating fluid. The liquid equivalent permeability obtained from the tests involving nitrogen and methane can be compared to the specific permeability obtained using water as the permeating fluid. In practice it is therefore not necessary that the lengthy tests involving water as the permeating fluid need to be carried out to obtain a significant correlation between permeability and facies type. It is only advisable to use water as the permeating fluid if it can be shown that the water might be reactive (eg. adsorbed) with the samples tested.

A relationship between increasing permeability with increasing grain size was found in the coarser grained facies (facies type 8 and higher). For the finer grained facies types the permeability decreased with increase in grain size. This is not detrimental if the results were to be used as a roof fall hazard indicator or to characterize the flow character for the extraction of coalbed methane. As facies type is closely related to grain size in the rocks tested, a graph of permeability versus facies type would offer a token idea of the permeability of samples taken from above the roof or in front of the advancing coal face.

ACKNOWLEDGEMENTS

The authors wish to thank Rand Coal (Pty) Ltd. for making the borehole core available for the tests and the University of Natal, Durban, for financial assistance.

REFERENCES

Anon (1990). *Standard test Method for Permeability of Rocks by Flowing Air,* ASTM D4525-85. American Society Testing Materials, Philadelphia. 730 - 733.

Chakrabarti, A.K. and Taylor, R.K. (1968). The porosity and permeability of the Zawar dolomites. *Int. J. Rock Mech. Min. Sci,* 5: 261 - 273.

Daw, G.P. (1971). A modified Hoek-Franklin triaxial cell for rock permeability measurements. *Geotechnique,* 21: 89 - 91.

Desai, C.S. (1975). Finite element methods for flow in porous media. *Finite elements in fluids - Volume 1: Viscous flow and hydrodynamics.* Galagher, R.H., Oden, J.T., Taylor, C. and Zienkiewicz, O.C. (Eds). John Wiley and Sons, New York. 157 - 182.

Gawuga, J.K. (1979). *Flow of Gas through Stressed Carboniferous Strata.* Unpublished Ph.D. thesis, University of Nottingham, 362p.

Klinkenberg, L.J. (1941). The permeability of porous media to liquids and gases. *Drilling and Production Practice,* Dallas. 200 - 213.

Obert, L. and Duvall, W.I. (1967). *Rock Mechanics and the Design of Structures in Rock.* John Wiley & Sons, New York. 337.

Ohle, E.L. (1951). The influence of permeability on ore distribution in limestone and dolomite: Part I. *Economic Geology.* 46: 667 - 705.

Scheidegger, A.E. (1974). *The Physics of Flow in Porous Media.* Third edition, University of Toronto Press., Toronto. 353 p.

Stormont, J.C., and Daemen, J.J.K. (1992). Laboratory study of gas permeability changes in rock salt during deformation. *J. Int. Rock. Mech. Min. Sci. & Geomech. Abstr,* 29: 325 - 342.

Venter, B.J. (1994). *Assessment of the Permeability of Vryheid Formation Sediments.* Unpublished M.Sc. thesis, University of Natal, Durban. 107 p.

Ward, J.R. and Jermy, C.A. (1985). Geotechnical properties of South African coal bearing strata. *Proceedings Symposium on Rock Mass Characterization, SANGORM,* Randburg, 57 -65.

Some aspects of waste deposit in the marl open-air casts

Quelques aspects de mise en dépôt des refus aux chantiers de marnes

Jan Jaremski
Technical University of Opole, Poland

ABSTRACT: The marl excavations in Opole region are located in very complicated hydrogeological conditions, i.e. above the aquiferous layer of shell limestone. In the paper the author found that it could be possible to use changes of water chemism caused by impurities in hydrogeological environment for evaluation of tightness of excavations. Changes of contents of nitrogen compounds and nitrification processes were analysed. It was found that contact between particular aquiferrous layers was possible by numerous faults and dislocations. Migration of impurities contained in eluates to particular aquiferrous layers in case of direct waste deposits was recognized as unquestionable.

RESUME: Les chantiers de marnes d'Opole sont localisés dans les très difficiles conditions hydrogéologiques - au-dessus de la couche aquifère de muschelkalk de Trias. On a montré la possibilité d'utilisation des changements du chimisme d'eaux, provoqué par des infections qui sont étrangères du milieu hydrogéologique dans l'évaluation de l'étanchéité de chantiers. On a analisé le changement de la teneur des composés azoïques, et on a observé les processus de la nitrification. On a constaté qu'un contact entre chaque couche aquifère est possible grâce aux nombreuses failles et dislocations. La migration des refus refermés en segments de chaque couche aquifère dans le cas de mis en dépôt direct des refus, on a reconnue comme incontestable.

1 INTRODUCTION

The excavations in inactive quarries of marls and limestones are more and more often used for waste depositing. In order to evaluate usability of such excavations it is necessary to determine
- hydrogeological conditions in the excavation,
- permeability of the walls (usually rock detritus),
- tightness of the layers occurring below the waste deposits.
In this paper these problems are discussed. The author chose excavations in Opole, a city in south-west Poland, as an example for his considerations. In the excavations wastes from the local power station are stored. In Opole marl excavations occupy about

230 ha and in this area any buildings are not built in spite of complete technical infrastructure. This fact strongly influenced localization of the Opole Power Station. The Opole Power Station is going to deliver 45 Tg of furnace wastes during 30 years of its operating. Municipal wastes will be stored in the excavations too. Triassic water intakes occur near the excavations and they are extremely efficient.

In the Opole region 3 water-bearing levels occur. From comparison of salinity of waters occurring in Turonian marls, Cenoman sandstones and Triassic shell limestones with composition of waters in the excavations it results that composition of the waters from the excavations is strongly influenced by Cenoman and Triassic waters.

Contact between particular water-bearing levels is possible by a complicated system of dislocations and faults which connects different stratigraphic water-bearing levels. This system is not well know yet. Owing to this system some impurities can get into underground waters. The system is a result of volcanic phenomena in the Tertiary period. The volcanic tuffs observed in sight-holes testify the above statement. Chemical constitution and solubility of toxic matters occurring in furnace wastes as well as their influence on environment have not been well know yet. We know, however, that these wastes contain toxic substances, for instance heavy metals and polycyclic aromatic compounds [1],[4],[6]. If a great amount of wastes from the power station are deposited in the excavations, below the water level and in complicated hydrogeological conditions, it will be a serious interference in nature.

The author was always opposed to ash depositing directly in the excavations. In his previous papers he postulated their tightening. He proposed untypical methods of evaluations of tightness of the layers above the Triassic water-bearing layer. In this paper migration of impurities was discussed and special attention was paid to changes in contents of nitrogen compounds, oxygen consumption demand of organic coal and nitrification processes. The results obtained can be an important source of information about permeability of tightening packages. It was found that decrease of the aeration zone strongly influenced secondary pollution caused by atrophy of the biological complex.

Marls and limestones occurring in Opole region and in some other regions of the world have similar properties and diagenesis and they were subjected to tectonic movements. Thus, remarks from this paper can be widely applied.

2 HYDROGEOLOGICAL CONDITIONS AND FILTER TERM TO REBUILD

The cretaceous deposits of the Opole region lie in the north-western slope of the Upper Silesian Anticlinorium. The Opole city is localized in the central part of the cretaceous formation on the disreate cretaceous horst.

The marl complex occurs on the surface on the right side of the Odra river, just beneath the soil (humus). This fact is important, because the open casts in the marl are situated on the area. Turonian marls are marl limestones and marls as well as their waste. Due to the numerous cracks and fissures the Opole marls display some specific filtration features. The marls and marl eluvium have been studied for many years. Numerous analyses have led to recognition of the phenomena occurring in the waste [2],[5]. These phenomena were tested experimentally in relation to the local hydrogeological conditions. Those conditions are as follows in the examined area of the Opole region. The level of ground water changes being dependent on precipitation. It rises to 1 m μ/g in the periods of the long-lasting rainfalls, whereas it sinks to the depth of 6 m drought time. The author registered it in the testing ground many times. Due to such the processes the soil mass begins to consolidate up to the cohesive soil with the filtration factor below 10 m per 24 hours in the extremal saturation stage.

Quarry wall after the ceasing of exploitation of the marl on all its is subjected to an intensive process of weathering. Under the influence of external elements the marl and especially its waste on the surface are under-going contant structural changes. In Opole region, the first water-bearing layer with free water is built up of Turonian marls. The second water-bearing layer as sub-artesion in the Cenoman sands and the third one occurs in the Trias limestone. According to the hydrogeological data included in many elaborations it has been stated that there are some enormously effective formations of shell-limestone in the Opole region. The designed water intakes near the excavations where wastes are going to be deposited are supposed to derive up to 100.000 m³/d of water. The limestone of the Górażdże stratum, Karchowice stratum and diploporous stratum are water-bearing formations. It should be stressed that the water occurring in shell limestone is of high quality.

3 POSSIBILITIES OF THE FLOW OF POLLUTED WATERS FROM THE COVERED EXCAVATIONS TO THE WATER-BEARING HORIZON

From the complex analysis of hydrogeological conditions of a given region with reference to the excavations, in which furnace wastes are to be storaged, we can draw the following conclusions:

1. The water-bearing horizon of Cenoman and waters occurring in shell limestone are tensed,

2. The Odra river drains the ground waters along all its course in the Opole area,

3. On the Piast excavations and Groszowice excavations and in their vicinity many recognized faults occur. The complicated system of dislocations and faults can be a reason why impurities enter into underground waters. The excavations occur in the region of tectonical movements of the Tertiary period and they border upon the outcrop of the Cenoman stage, so thickness of the Turonian is less, thus this region was more open to formation of tectonical dislocations. It results from volcanic effects of the Tertiary period and it is proved by results obtained from bore-holes near the excavation in Groszowice. In one of these bore-holes volcanic tuffs were found. However, their form and range are not know (they reached to the depth of 60 meters, i.e. depth of the bore-hole). There is an opinion that slightly permeable clayey marls and marl clays occur on the excavation bottoms and they isolate a Cenoman water level. According to the latest investigations this opinion is not right. From observation of bore-holes it results that there is a contact of the Turonian level with the Cenoman level through sandy marls having a greater filtration coefficient than clayey marls. According to the hitherto assumptions the clayey marls separate these two levels. Their thickness reaches 10 m. All those conclusions allow the author to construct the hypothesis that thanks to the faults there is a mutual contact between the described water-bearing layers [4]. The permeation of water from the higher-levels, which is rather scant (weak), is only an additional factor (Fig.1).

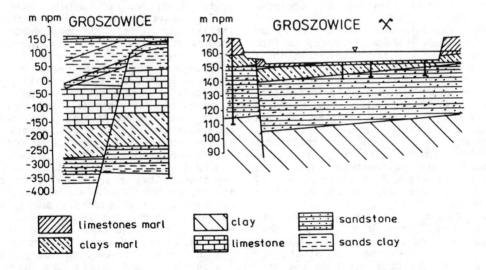

Figure 1. Examples of faults occurring in open casts and in their vicinity

The faults, however, complicate in many ways the hydrogeological relations and the movement of underground water, functioning as the pipers or channels which join different stratigraphic water-bearing levels, and leading out the waters from deeper levels to the upper ones. The topography of faults in Opole region area is still unknown, only the excavations in which the faults occur and their vicinity, are know. From all these consideration and from the settlements about the permeation of the surroundings covers of excavations, results that there is the contact of open casts waters with the waters drawn for the water supply service. Also the analysis of the saltness of the waters of the Odra river and of each water-bearing level speaks for this theory. The Piast and Groszowice excavations are partly flooded. The difference in pressures in the Cenoman sandstone waters and the Turonian marls waters is 9 m for "Groszowice" excavation and 7 m for "Piast" excavation. The chemical composition of the water from the Odra river differs from that of Turonian marls, Cenoman sandstones and shell limestone in particularly great saltness and pollution. The contents of dissolved substances (826 mg/dm^3 chlorides, 426 mg/dm^3 other elements) get very high values. The water from "Groszowice I" and "Piast" excavations includes 470÷520 mg/dm^3 of dissolved parts, 66.6÷81.0 mg/dm^3 of sulphates and 35÷36 mg/dm^3 of chlorides. The water from Cenoman sandstone includes 17.7÷125.3 mg/dm^3 of sulphate, 5÷30 mg/dm^3 of chlorides (also 54 mg/dm^3 were noticed) general hardness is from 2.72÷5.76 val/l. The water from shell limestone includes 454÷350 mg/dm^3 of dry remains, from 17÷25.5 mg/dm^3 of chlorides, 80÷69 mg/dm^3 of sulphate. The water taken from the Turonian marls on the opposite side of the town, i.e. on the premises of the ZOO, includes 64÷67 mg/dm^3 of chlorides, 285÷248 mg/dm^3 of sulphate. The results are comparable with the content of water from earthworks - 67 mg/dm^3 of chloride, to 285 mg/dm^3 of sulphate.

The Cenoman waters of Opole and waters from Groszowice excavations, which are near the river, are not salted. It proves the lack of the influence of the Odra river on the content of ground waters on the given water-bearing levels. Summing up the results the given analyses of chemical content of each level in the excavation area unmistakably the mutual contact of ground waters of every level [4].

4 CHANGES IN CHEMISM OF TRIASSIC WATERS CAUSED BY THEIR POLLUTION WITH OUTSIDE SUBSTANCES

Triassic waters can be polluted by organic substances which are foreign in the hydrogeochemical environment. Analysis of such pollutions may be a source of information about tightness of the layers occurring above Triassic waters. Taking into account age and conditions of formation of these waters and assuming that layers above the water-bearing layer are tight, we can say Triassic waters should not contain any organic matters or their derivatives. Thus, we should not observe increased oxygen consumption, ammonia, nitrites and the like.

4.1 Oxygen consumption

Increased oxygen consumption means that there are organic substances in water. Triassic waters in Opole region are slightly mineralized, so a low oxygen consumption should be expected. From the data collected during many years (see Fig.2) it results that oxygen content is increased and variable. Organic matters from the soil, dissolved in water, got to deeper layers owing to vertical infiltration. Next, in consequence of coagulation aided by calcium hydroxide from weathering of marls and limestones, they were deposited in fissures and free spaces, mainly on the walls of rock block and chips.

Reduction of oxygen consumption means that mineralization proceeds. Mineralization of organic matters is influenced by many complicated chemical and biochemical processes. Biochemical processes are particularly important. They proceed owing to the microorganism for which organic substances are a source of energy and building matter. For releasing of the energy microorganisms use their complex enzymatic apparatus acting in the presence of water. Microorganisms living in water change the complex organic matters into simpler mineral substances. Organic matter contains a large amount of potential energy. Its decomposition is connected with oxidation, i.e. processes releasing energy. Microorganisms oxidize organic carbon compounds and reduce them to

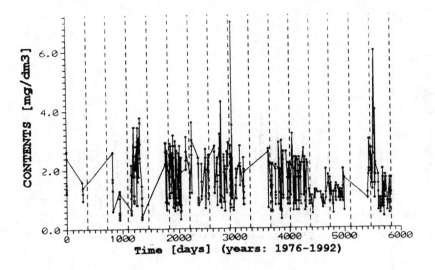

Figure 2. Oxygen consumption of the Opole Trias water

carbon dioxide and water; owing to that their potential energy releases.

4.2 Nitrogen compounds

Figs.3,4,5 shows content of nitrogen compounds in Triassic drinking water in Opole. The data were collected for many years. From the figure it results that content of nitrogen compounds is changeable.

The presence of ammonia and nitrites can be observed in the waters considered. It may be a result of penetration of impurities from waste deposits into drinking water of Triassic origin in Opole region. The author presented such suggestion in his previous papers [5]. Changes in content of nitrogen compounds are influenced by physical and chemical processes proceeding in Turonian marl eluvium (see the previous chapter).

Thus, we can say about periodical changes of chemism of water. In consequence of the processes mentioned, ground water level in Turonian marls raises from time to time (up to 1 m below the ground). It causes that the aeration zone decreases, a lack of oxygen is observed in this zone and, in consequence, all the biological complex decays. Thus, secondary pollution takes place. At the same time properties reducing pollution of the aeration zone change. Special ways of infiltration, i.e. dislocations and faults are of a great importance, too - they cause that the biological complex in the aeration zone is not able to work.

The data for Opole have been compared with the same data for industrial water used in Strzelce Opolskie. The town is on the outcrop of Opole Trias, 30 km from Opole, near Strzelce Opolskie shell limestone occurs just below the humus layer, without any tightening layers. In this town the water supply system collects Triassic water through the drilled wells situated not far from the melioration ditch for which municipal wastes are carried off after mechanical cleaning. A small part of the ditch is usually dry, so the wastes penetrate into the ground, i.e. Trias shell limestone. The region is situated on the outcrop and outside the town there are many animal farms. In Strzelce Opolskie the effects of biological complex in the aeration zone are really great and the water collected contains less ammonia and nitrites than the water in Opole region. Figs. 6,7,8,9.

Thus, we can say about periodical changes of chemism of water. In consequence of the processes mentioned, ground water level in Turonian marls raises from time to time (up to 1 m below the ground). It causes that the aeration zone decreases, a lack of oxygen is observed in this zone and,

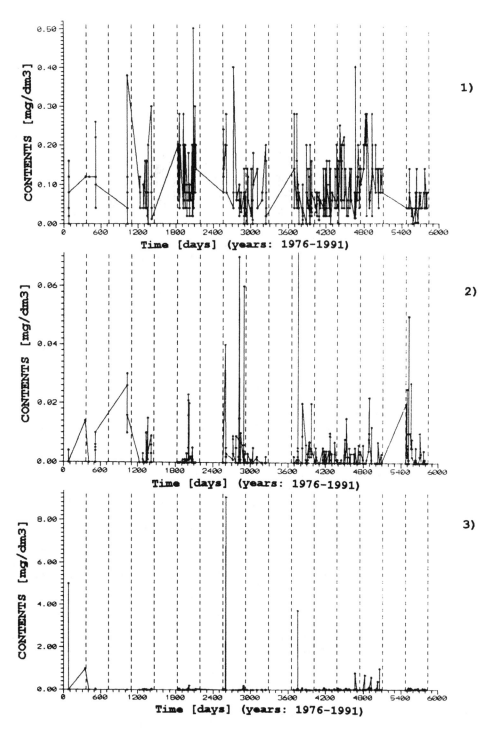

Figures 3,4,5. Contents of 1)ammonia, 2)nitrites, 3)nitrates Opole Trias water in consequence, all the biological complex decays. Thus, secondary pollution takes place. At the same time properties reducing pollution of the aeration zone change. Special ways of infiltration, i.e. dislocations and faults are of a great importance, too - they cause that the biological complex in the aeration zone is not able to work.

The data for Opole have been compared with the same data for industrial water used in Strzelce Opolskie. The town is on the outcrop of Opole Trias, 30 km from Opole, near Strzelce Opolskie shell limestone occurs just below the humus layer, without any tightening layers. In this town the water supply system collects Triassic water through the drilled wells situated not far from the melioration ditch for which municipal wastes are carried off after mechanical cleaning.

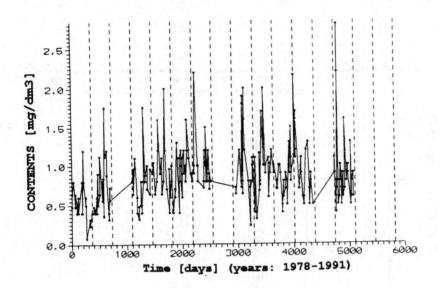

Figure 6. Oxygen consumption in the Strzelce Opolskie Trias water

Also oxygen consumption is less in Strzelce Opolskie but an amount of nitrates is incomparably higher. Such high content of nitrates is a result of work of the biological complex, namely nitrification processes of the first and second degrees which reduce impurities to nitrates.

These problems are discussed in the next chapter. Nitrification processes can be also taken into account while evaluating tightness of the deposit.

There are some differences between waters in Opole and Strzelce Opolskie. In Opole increased content of ammonia and nitrites can be observed but an amount of nitrates is rather low. In Strzelce Opolskie an amount of ammonia and nitrites is small and there is a great amount of nitrates, Figs.3,4,5,7,8,9.

Thus, we can say that in Opole the layers situated above the Triassic waters are not tight. It can be proved by the results of tests of water from the new drilled wells situated about 15 km from Opole, in Zimnice Małe. The new water intake is in nearly the same hydrogeological conditions as those in Opole. When the holes were made and during test pumping there were no ammonia and nitrites in water. They occurred after six-months operating.

The idea of direct ash depositing in the marl excavations in Opole was investigated in many research centres. From the tests it resulted that eluates from the excavations would not penetrate into water-bearing layers for at least 50 years.

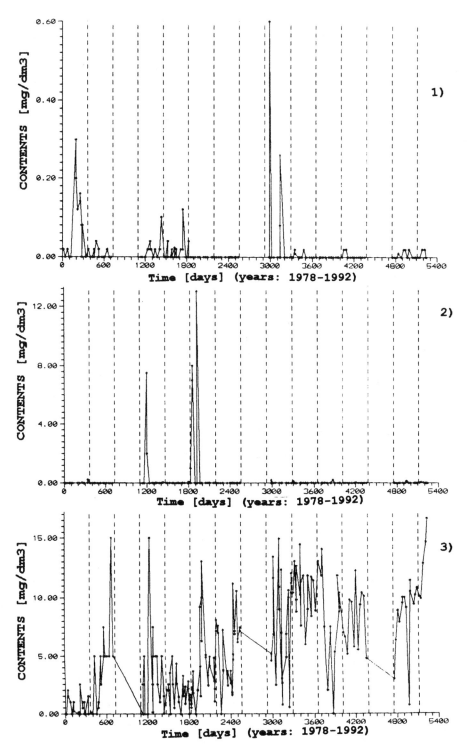

Figures 7,8,9. Contents of 1) ammonia, 2) nitrites, 3) nitrates Strzelce
Opolskie Trias water

5 REDUCTION OF ORGANIC COMPOUNDS UNDER HYDROGEOLOGICAL CONDITIONS IN WATER-BEARING LAYERS NEAR OPOLE AND STRZELCE OPOLSKIE

During vertical infiltration of the water-bearing medium we can observe not only physical and chemical processes but biochemical ones as well. They usually proceed under influence of bacteria and fungi. Apart from organisms decomposing organic materials there are also bacteria (nitrifying, sulfuric and some other types of bacteria) and fungi which process mineral matters. Intensity of microbiological life is reduced below the aeration zone. If materials which can be easily decomposed under influence of bacteria and fungi penetrate into the ground and if there is a sufficient amount of oxygen, we obtain carbon dioxide, ammonia, nitric acid, phosphoric acid and sulfuric acid. At the final stage of mineralization salts form on condition that an amount of base compounds in the ground is sufficient for acid killing.

Flora and the humus layer occurring near Strzelce Opolskie, a town not far from Opole, absorb or bond phosphates, nitrogen compounds and non-mineralized organic matters. Chlorides, nitrates, sulfates and any other compounds which are not absorbed or used by plants get to water. In Opole region water flows are disturbed by many dislocations and faults which form special ways of filtration and influence directions of water flows in Turonian marls. Action of the sorbtion complex of the aeration layer, in which impurities are reduced, is limited in consideration of a short flow time. The turbulent water movement and different flow velocities cause that acid waters penetrate into deep layers. These waters contain organic impurities being a results of fertilization of the fields. Triassic waters in Opole region cannot contain such compounds as ammonia, nitrites, taking into account condition of formation of the waters, their age and the system of geological strata. Similarly, in the soil particles of impurities are dispersed, i.e. they do not form aggregates. We can observe coagulation which is a result of treatment of organic materials dissolved in water with a little amount of solution of

calcium hydroxide being a product of rock weathering. Under influence of calcium base small particles form specific flocs. The particles formed are rather heavy and they settle. A similar phenomenon can be observed in the soil - it occurs spontaneously and is very slow. Fast coagulation of colloids saturated with hydrogen and calcium is a consequence of electrokinetic properties of the system. Under suitable conditions the ammonium ion can be subjected to nitrification. Nitrification is a process of enzymatic oxidation of nitrogen and it proceeds on two coordinated stages. In the first stage nitrosomans bacteria participate and it can be written as

$$2NH_4^+ + CO_2 \xrightarrow{\text{enzymatic oxidation}} 2NO_2^- + H_2O + 4H^+ + \text{energy}$$

The second stage with nitrobacters

$$2NO_2^- + O_2 \xrightarrow{\text{enzymatic oxidation}} 2NO_3^- + \text{energy}$$

In strongly basic environment oxidation of nitrites can be delayed up to reduction of concentration of NH ions. Nitrifying bacteria are always sensitive to surroundings. Weathered calcite on the walls of blocks and chips is a neutralizing agent during coagulation of impurities. Also illite being a remainder after weathering is a flocculent, i.e. it helps in coagulation. Nitrifying bacteria can act when humunication and biochemical transformation of organic matter is ending. This process goes together with synthesis of humus compounds.

6. RECAPITULATION

The results of investigations allow to make an explicit statement: infiltrated rain-water and the water from fluming of cinders, together with accumulated heavy metals polycyclic aromatic hydrocarbons and other pollution, will translocate to the water-bearing strata [1],[4],[6].

It is very important to storage cinders in the saturation zone eliminating, at the same time, the aeration zone and neglecting the sorptive complex in the reduction of pollution. The contact between particular water-bearing layers is possible through numerous faults which was proved by the comparison of composition of waters occurring in these layers. At present the conception of immediate storing cinders in excavations is good because it brings changes of the landscape and the possibility of making use of additional space for the residential buildings. The fact, that these areas are practically armed with the complete infrastructure and keeping this conception for many years in planning, designs, and economic profits, permits one to accept that storing cinders in excavations is inevitable. Depositing below the level of under-ground water is hardly possible, from of the technical point of view (Fig.10).

Figure 10. The example excavations visible ground water level

After analysing all the empirical investigations of the negative influence of furnace waste storage excavations the author proposed the conception of the maximum sealing system of the open cast section. The section has 3 ha of in excavations and it is separated from the local material (weathering marls) with the conception particular attention must be paid to the protection of ground water and the stability of the material dumped. The solution is under construction at present. The drain filter will be analysed. The proposed solution allows to investigate chemical constitution of the filtrate under real conditions of the Opole Power Station taking into account coal composition, combustion temperature in the kettles, solubility of compounds under real hydraulic conditions as well effects of emulgations. It also gives time for investigations on topography faults and it will be possible to realize the proposed solution with minimum costs and make the optimum decision on the basis of complate data concerning a quantity and composition of toxic substances contained in the ash.

Marls and limestones occurring in Opole region and in some other regions of the world have similar properties and diagenesis and they were subjected to tectonic movements.[3] Thus, the remarks from this paper can be widely applied.

REFERENCES

Dybczyński,R., et al. 1990. A comprehensive study on the contents and leaching of trace elements from fly-ash originating from Polish Hard Coal by NAA and AAS methods, Biol. Elem. Resc. 26. PUB-SET, INC.

Jaremski,J. 1990. Analysis of influence of swelling if marl eluvium of water infiltration. Proc., 6 th Int. Cong. IAEG, Amsterdam, A.A.Balkema.

Jaremski,J. 1992. The geotechnical properties of the marls eluvium occurring outside the Opole region. Zeszyty Naukowe WSI, Opole, No 179.

Jaremski,J. 1993. The influence of furnace waste from Power Station Opole deposited in the open casts upon the composition of ground water. Proc., 4 th Int. Symp. Rec. Treat. Utiliz. Coal Mining Waste, Cracow.

Jaremski,J. 1993. The influence of recognition of physical and chemical processes occurring in the weathering in the estimate of the extreme values of geotechnical parameters of the Opole marls eluvium . Proc., Int. Symp. Geot. Eng. of Hard Soils-Soft Rocks, Athens, A.A. Balkema.

Theis,T.I., et al. 1988. Physical and chemical characteristics of unsaturated pore water and leachate at a dray fly ash disposal site. Proc., 43 rd. Ind. Waste Conf., Purd. Univ. West, Laf. Indiana, Lewis Publishers, INC.

Surveillance du confinements de décharges à l'aide des traces et micro-traces minérales dans les eaux

Survey of the landfill confinement with water inorganic traces

M.-O. Looser, X. Dauchy, M. Bensimon & A. Parriaux
Laboratoire de Géologie (GEOLEP), École Polytechnique Fédérale de Lausanne (EPFL), Suisse

RESUME: Les eaux constituant souvent le véhicule principal des pollutions pouvant s'échapper d'un site contaminé, les traces inorganiques peuvent se révéler être de bons indicateurs pour surveiller la qualité d'un confinement. A ce jour, l'étude de ces traces a été limitée à quelques métaux, en raison des faibles concentrations dans les eaux, ou de problèmes d'interférences dans les percolats. L'apparition d'un outil analytique très performant, le spectromètre de masse à source à plasma à haute résolution (ICP-MS-HR), offre de nouvelles perspectives. Les analyses effectuées autour de sites contaminés ont permis de vérifier les avantages des traces inorganiques: facilités de prélèvements, d'analyses et d'interprétations des résultats.

ABSTRACT: Waters are the most important transport media for landfill contamination. Inorganic traces can be good pollution indicators for survey purposes of the confinement quality. Because of the low concentrations in the waters or interferences problem in leaches the study of the inorganic traces was limited to some metallic traces. Actually, an efficient method, the high resolution inductively coupled plasma mass spectrometry (HR-ICP-MS), offers new perspectives for inorganic traces analysis. Preliminary investigations near hazardous wastes confirmed the interest of inorganic traces: sampling, analysis and results interpretation are simple.

1- INTRODUCTION

Notre groupe de recherche en géologie appliquée s'occupe depuis plus de 10 ans du projet AQUITYP, qui consiste à caractériser une eau par sa composition chimique afin d'en déterminer son origine.

Sur la base de notre expérience de l'analyse des traces minérales dans les eaux souterraines, il nous est paru intéressant de développer cette approche à la détection ou à la surveillance des sites contaminés, car les eaux constituent bien souvent le véhicule principal des pollutions pouvant s'échapper de tels sites (Bensimon & al 1994).

Afin de vérifier l'intérêt de cette démarche, une série d'investigations préliminaires avaient été menées à proximité de plusieurs sites industriels et de décharges urbaines (Looser & al, 1993).

Ces premières investigations sur des eaux contaminées ont montrés qu'un certain nombre de traces inorganiques pouvaient être utilisées comme révélateurs de pollutions. Des analyses de percolats de décharges ont alors été effectuées, afin d'avoir un aperçu des pollutions inorganiques qui peuvent être libérés dans l'environnement par de tels sites. Cet article en présente les principaux résultats.

2- TECHNIQUE ANALYTIQUE

La spectrométrie de masse à source à plasma à haute résolution (ICP-MS-HR) est actuellement la méthode analytique la plus appropriée à l'étude des traces inorganiques dans le système hydrogéologique. Les avantages de l'utilisation des traces et micro-traces inorganiques et de l'emploi de cette technique analytique sont notamment:

- une bonne stabilité et représentativité des échantillons (par rapport aux traces organiques notamment),
- une procédure d'analyse rapide
- la quasi absence d'interférences sur le dosage,
- une interprétation simple des résultats,
- un seuil de détection bas pour une grande part des éléments du tableau périodique.

L'ICP-MS est le mariage de 2 systèmes performants: la source à ionisation à plasma (ICP) et l'incomparable pouvoir de résolution du spectromètre de masse (MS).

L'instrument est constitué d'une torche à plasma d'argon à pression atmosphérique, où règne une température de l'ordre de 7000 K. Sous ces conditions, l'essentiel des constituants du plasma sont des ions à charge unitaire avec un degré de ionisation supérieure à 80% pour la majorité des éléments. Les ions sont analysés en fonction de leur rapport masse sur charge (m/z) dans un spectromètre de masse à double secteurs et détectés par un multiplicateur d'électrons. C'est la configuration de notre instrument High Resolution Inductively Coupled Plasma Mass Spectrometer (HR-ICP-MS) commercialisé par Fisons VG Instruments.

2.1- Échantillonnage

Les échantillons sont prélevés dans des récipients en polyéthylène neufs. L'échantillon peut être stocké à une température de 4°C, après filtration à 45μ et addition d'acide nitrique Suprapur concentré (65%), de façon à ce que le pH de l'échantillon soit voisin de 2. Un pH bas inhibe l'adsorption des ions métalliques sur la surface du récipient et prévient la formation de précipités de métaux en traces, ou la coprécipitation des métaux en traces avec d'autres constituants majeurs. Le volume nécessaire pour l'analyse des traces est de 50 ml.

Des tests ont été effectués dans notre Laboratoire pour connaître l'influence des divers procédés de filtration (Dauchy, 1994). L'utilisation d'un dispositif de filtration en polysulfone est recommandé par rapport à un dispositif en pyrex comportant un élément de verre fritté, difficile à nettoyer. Par ailleurs, les tests comparatifs effectués sur différents filtres, mettent en évidence les performances du filtre en esters de cellulose qui permet de garantir l'absence de contamination ou d'adsorbtion significative.

2.2- Procédure d'analyse et précision de la technique

En général une première analyse semi-quantitative fournit des renseignements sur la composition globale de l'eau. Cette phase consiste à mesurer le spectre de masse pour chaque échantillon, dans un domaine de masse allant de m/z = 6 à m/z = 238, ce qui correspond respectivement au plus petit isotope du lithium ($^6Li^+$) et au plus lourd isotope de l'uranium ($^{238}U^+$). A partir des spectres de masse de chaque échantillon, on peut déduire la présence de traces et ultra-traces telles que Li, B, Al, Ti, V, Cr, Fe, Mn, Ni, Ga, Ge, Co, Cu, Zn, Ge, As, Br, Rb, Sr, Se, Zr, Rh, Mo, Pd, Ag, Cd, In, Sn, Sb, Te, Ta, La, Os, Ir, I, Cs, Ba, W, Tl, Hg, Pb, Bi, U et Sc.

Lors de cette première analyse, les concentrations des éléments peuvent aussi être estimées, ce qui permet de choisir les standards adéquats pour les mesures quantitatives.

A partir des résultats obtenus lors de l'analyse semi-quantitative, 5 standards sont utilisés pour établir les courbes de calibration. Il s'agit de standards de pureté élevée et dilués avec de l'eau déionisée Milli-Q dans 5% HNO3.

L'yttrium est généralement utilisé comme standard interne pour corriger la dérive de l'appareil, cet élément n'est donc pas mesuré dans les échantillons. Un échantillon blanc permet aussi de corriger les résultats afin de réduire les éventuelles interférences spectrales et les problèmes de contaminations dus aux réactifs.

Lors de l'analyse semi-quantitative la précision analytique varie en moyenne de 20 % par rapport à la réponse du spectromètre basée sur l'analyse d'un standard unique.

Lors de l'analyse quantitative, utilisant une courbe de calibration de 5 standards pour chaque élément, la précision est ramenée à 5% sur le dosage de chaque élément, ceci pour des concentrations de l'ordre du μg/l et sous des conditions analytiques normales. Le seuil de détection est alors d'environ 0.01 μg/l.

3- ANALYSES DES PERCOLATS DE DECHARGES

Douze percolats provenant de 10 décharges situées dans le Bassin Lémanique ont été analysés. Les sites échantillonnés sont des décharges urbaines de taille moyenne à importante (plus de 2 millions de m^3), contenant un large spectre de déchets qui peut être estimé dans les proportions suivantes:

- 30 à 50 % d'ordures ménagères
- 10 à 30 % de mâchefers (résidus d'incinération)
- 10 à 20% de boues de station d'épuration des eaux usées urbaines (STEP)
- 20 à 50% d'inertes (matériaux de démolition)
- moins de 1 % de déchets spéciaux

Toutes les décharges ont été ouvertes il y a une vingtaine d'années et plusieurs d'entre-elles sont encore en activité.

Ces décharges sont relativement sèches. Elles se situent toutes au-dessus de la nappe phréatique et les seuls apports d'eau sont constitués par les précipitations (de l'ordre de 1500 mm/an dans cette région).

Le substratum géologique est formé soit de terrains fluvio-glaciaire qui recouvrent des marnes et des grès molassiques, soit de terrains alluviaux épais. La plupart des sites ont été installés dans d'anciennes excavations de matériaux sablo-graveleux.

Les percolats ont été prélevés dans des piézomètres situés sur le site de décharge et qui en atteignent le fond, ou à la sortie des tubes de drainage des casiers, pour les trois sites qui en sont équipés.

3.1- Choix des percolats

Les échantillons ayant une résistivité inférieure à 1000 µS/cm ont été écartés afin de ne considérer que des percolats suffisamment concentrés (tab.1). La mesure du pH et l'analyse de la demande biochimique en oxygène (DBO5) et de la demande chimique en oxygène (DCO) ont permis de vérifier que tous les sites étaient dans une phase méthanogénique.

Tab.1. Variation (min-max) du pH, de la conductivité et de la demande biochimique (DBO5) et chimique (DCO) en oxygène.

pH		6.7 - 8.2
cond.	µS/cm	1089 - 12610
DBO5	mg O₂/l	5 - 173
DCO	mg O₂/l	117 - 1925

La distinction entre la phase acétogénique et méthanogénique a été basée sur les observations que Robinson et Gronow (1993) ont faites sur les percolats de décharges du Royaume-Uni: la phase acétogénique s'accompage de valeur en DBO5 supérieure à 2000 mg/l et d'une DCO généralement supérieure à 10000 mg/l. Cette phase acide ne dure généralement que deux à cinq ans . Les teneurs en DBO5 et DCO s'abaissent ensuite rapidement en-dessous de 2000 mg/l lors de la phase méthanogénique. La charge en métaux lourds est plus importante durant la phase acétogénique que durant la phase méthanogénique (Harmsen, 1983 in Borden & Yanoschak, 1989).

3.2- Analyses des éléments inorganiques

15 éléments inorganiques ont été analysés dans tous les percolats. Par la suite 27 autres éléments ont été recherchés dans au moins 6 des percolats les plus chargés (tab.2). Rappelons que seuls les éléments inorganiques sous forme dissoute dans les eaux ont été analysés. Les éventuelles fractions solides ou colloïdales (> 0.45 µm) retenues lors de la filtration des échantillons n'ont pas été analysées. Les éléments en solution nous intéressent tout particulièrement, puisque ce sont eux qui vont pouvoir franchir les barrières géotechniques et géologiques et atteindre les eaux souterraines L'origine de la présence des éléments a été établie sur la base de leurs utilisations courantes dans l'élaboration de produits manufacturés (Shreve &

Brink, 1977).

Quatre groupes d'éléments ont été distingués en fonction de leur fréquences et de leur teneurs moyennes:

3.2.1- Les micro-traces exotiques:

Dix éléments n'ont pas été décelés ou ont des concentrations à la limite du seuil de détection (0.01 µg/l). Il s'agit d'une partie des éléments du groupe du platine, soit osmium, iridium, paladium et rhodium, ainsi que du gallium, germanium, indium, thalium, tellurium, et tantale. Ce résultat n'est pas surprenant dans la mesure ou ces éléments sont plutôt rares dans la nature. Ils sont aussi très peu employés dans l'industrie, où ils interviennent surtout dans la fabrication de composés électroniques.

3.2.2- Les micro-traces rares:

Six éléments ont été détectés dans seulement 1/3 des échantillons analysés et à des teneurs moyennes inférieures à 100 µg/l. Il s'agit du sélénium, du scandium, du bismuth, de l'argent, du mercure et de l'antimoine.
Hormis l'antimoine qui est essentiellement utilisé dans les vernis, les autres éléments entrent dans la composition d'alliages ou de matériel électronique.
On pourrait s'attendre à ce que le mercure, qui est d'usage relativement fréquent (piles) par rapport aux autres éléments de ce groupe livre des teneurs plus importantes dans les percolats. Toutefois, c'est un élément peu mobile, qui de plus, a la capacité de se lier aux particules en suspension.
Le sélénium se distingue de ce groupe en présentant des valeurs assez élevées lorsqu'il est détecté. Son origine pourrait être liée soit aux composants électroniques, soit aux pigments utilisés dans les peintures et vernis.

3.2.3- Les micro-traces fréquentes

Il s'agit des éléments détectés dans plus de la moitié des échantillons et dont la teneur moyenne n'excède pas 100 µg/l. 18 éléments se trouvent dans cette catégorie.
La plupart d'entre eux sont des métaux qui entrent fréquemment dans les alliages ferreux et non ferreux: titane, vanadium, zirconium, molybdène, tungstène, cobalt, chrome et plomb, étain, zinc, nickel, cuivre et dans une moindre mesure l'arsenic.
L'arsenic se trouve le plus souvent dans des produits phytosanitaires. Le césium et le lithium sont liés, tout comme le cadmium et le nickel déjà cités, à la production de piles et batteries.

Tab.2. Synthèse des teneurs en éléments inorganiques de 12 percolats de décharges urbaines en phase méthanogénique. Valeurs en µg/l. (ec: ecart-type, n>0: nombre de valeur supérieure au seuil de détection nval: nombre d'échantillon analysés)

µg/l	min	max	moy	ec	n>0	nval
B	187.3	20015.0	4781.5	5563.5	12	12
Br	187.2	14897.0	4296.4	5068.2	12	12
I	72.94	3324.90	970.56	954.62	12	12
Mn	69.76	3047.00	617.46	904.42	10	10
Fe	46.95	2065.50	555.44	657.98	10	10
Ba	83.80	855.40	315.41	242.82	12	12
Rb	4.26	529.70	149.93	173.88	11	12
Al	7.43	537.00	112.60	174.82	9	10
Zn	14.97	386.00	86.50	115.86	10	10
Li	7.57	236.10	79.30	66.20	12	12
Ni	4.20	454.70	72.20	130.63	12	12
Cu	3.49	221.70	43.02	68.80	12	12
Cr	2.20	192.40	30.68	55.70	11	12
Ti	8.90	78.10	27.92	27.60	6	6
Mo	1.36	73.50	21.26	26.74	10	12
V	2.13	53.20	13.65	13.85	12	12
Pb	0.20	80.20	12.33	22.56	12	12
W	0.20	36.00	10.68	14.97	5	9
As	1.10	14.40	7.30	4.55	6	9
Co	0.90	26.10	7.12	7.70	12	12
Sn	1.00	11.00	4.25	4.72	4	6
U	0.12	9.10	3.59	2.56	12	12
Cs	0.58	7.20	3.51	2.74	5	9
Zr	1.00	9.00	2.68	2.73	8	9
La	1.00	2.00	1.40	0.55	5	6
Cd	0.03	2.08	0.96	0.88	6	12
Se	6.00	69.50	38.43	31.77	3	9
Sc	1.91	6.36	4.82	2.52	3	9
Bi	0.32	1.04	0.79	0.40	3	9
Ag	0.05	1.55	0.63	0.81	3	9
Hg	0.10	0.40	0.27	0.16	3	12
Sb	1.00	1.00	1.00	<0.01	2	6
In	0.01	0.01	0.01		1	9
Ga	<0.01	<0.01			0	6
Ge	<0.01	<0.01			0	6
Ir	<0.01	<0.01			0	6
Os	<0.01	<0.01			0	6
Pd	<0.01	<0.01			0	6
Rh	<0.01	<0.01			0	6
Ta	<0.01	<0.01			0	6
Te	<0.01	<0.01			0	6
Tl	<0.01	<0.01			0	6

La présence d'uranium pourrait trouver son origine dans certains colorants photographiques (nitrate d'U) et la présence du lanthane peut être attribuée à son utilisation comme additif dans la métallurgie.
On constate que le plomb, qui est par ailleurs aussi un polluant atmosphérique important, se trouve à de basses teneurs, probablement en raison de sa faible mobilité et des complexes peu solubles qu'il forme avec le soufre (PbS et PbSO$_4$).
Les basses teneurs observées pour le cadmium et l'arsenic indiquent que les risques présentés par ces éléments sous forme dissoute reste faible, contrairement aux formes solides sous lesquelles ils peuvent se trouver dans les sols ou les sédiments pollués.

3.2.4- Les éléments en traces

Les 8 éléments restant ont été détectés dans pratiquement tous les échantillons à des concentrations moyennes supérieures à 100 µg/l. Il s'agit de métaux courant comme l'aluminium et le fer (total), de non-métaux et métaux réactifs comme le brome, l'iode, le manganèse et le barium, un métal alcalin, le rubidium, et enfin un alcalin, le bore.
Fer, aluminium, chrome et manganèse proviennent essentiellement des nombreux composés métalliques qui sont mis en décharges. L'origine du bore est plus difficile a établir, mais cet élément entre dans la fabrication de plastiques, fibres de verre et détergents. Le brome a une origine diffuse, bien qu'il soit principalement utilisé comme composé anti-déflagrant dans les hydrocarbures. D'autres sources peuvent être évoquées (bains photographiques par exemple).
Le barium entre dans la composition des pigments (peintures), tout comme le chrome déjà évoqué. L'iode quant à lui, est utilisé pour la fabrication de composés organiques et d'iodure de potassium.
Sur les 32 éléments recherchés lors de cette étude, seuls As, Cd, Cr, Cu, Fe, Hg, Mn, Ni, Pb, Zn sont actuellement analysés couramment et cités dans la littérature.

4.- COMPARAISON AVEC LES EAUX SOUTERRAINES

4.1.- Référentiel

Les concentrations observées dans les eaux souterraines non contaminées, provenant de 5 aquifères molassiques et de 5 aquifères quaternaires du réseau AQUITYP (citées dans Mandia, 1991) sont utilisées comme référentiel pour mettre en évidence les principaux marqueurs anthropiques liés aux percolats de décharges (tab. 3).

4.2- Facteur de contamination

Pour chaque éléments, le rapport entre la concentration moyenne dans les percolats et la concentration moyenne dans les eaux naturelles permet d'obtenir un facteur de contamination noté FC (tab.3). Si ce facteur est élevé, l'élément considéré pourra être utilisé comme un marqueur de pollution.

Sur la base de ce rapport, 6 éléments en traces et micro-traces fréquents se détachent nettement: il s'agit du brome, du manganèse, du rubidium, du nickel, de l'arsenic et du césium.

Pour ceux-ci, le rapport FC est supérieur à 100 et la teneur minimale observée dans les percolats est toujours supérieure à la teneur maximale observée dans les eaux souterraines du quaternaire ou de la molasse. Leur présence dans les eaux souterraines au-dessus du seuil naturel devrait donc signaler une contamination anthropique.

Tab.3. Synthèse des teneurs en traces minérales de 10 eaux souterraines provenant d'aquifères quaternaires et molassiques faisant partie du réseau AQUITYP (en µg/l). FC désigne le facteur de contamination (moy. percolat / moy. aquityp).

µg/l	min	max	moy	ec	n>0	nval	FC
B	3.20	486.10	67.82	150.35	10	10	70
Br	7.00	51.30	17.96	13.56	9	10	239
I	0.90	161.50	21.72	49.38	10	10	45
Mn	0.10	3.20	0.88	1.20	6	10	>500
Fe	2.00	29.40	7.65	8.61	10	10	73
Ba	7.80	225.70	70.81	75.06	10	10	4
Rb	0.10	1.90	0.53	0.61	10	10	283
Al	0.40	9.60	2.09	2.79	10	10	54
Zn	1.70	107.90	29.33	52.39	4	10	3
Li	0.50	105.90	15.29	32.98	10	10	5
Ni	0.10	4.10	0.70	1.28	9	10	103
Cu	0.50	3.30	1.48	1.06	6	10	29
Cr	0.20	3.90	1.54	1.40	7	10	20
V	0.30	1.10	0.59	0.32	10	10	23
Pb	0.60	0.90	0.75	0.21	2	10	16
W	0.50	0.50	0.50		1	10	21
As	<0.01	<0.01	<0.01		0	10	>500
Co	0.10	0.30	0.17	0.07	10	10	42
U	0.10	1.50	0.61	0.44	9	10	6
Cs	<0.01	<0.01	<0.01		0	10	>500
La	0.10	0.10	0.10	0.00	4	10	14
Sc	0.60	2.10	1.27	0.51	10	10	
Bi	<0.01	<0.01	<0.01		0	10	

4.3- Corrélations

Des corrélations ont été effectuées pour les éléments analysés dans un nombre suffisant d'échantillons et ayant une valeur supérieure au seuil de détection. Le tableau 4 présente les résultats obtenus pour les éléments qui ont un facteur de contamination (FC) supérieur à 20. Les corrélations obtenues pour les percolats ont été reportées à droite du tableau, à gauche, figurent les valeurs obtenues pour les eaux de la molasse et du quaternaire.

Seuls deux couples d'éléments présentent des corrélations supérieures à 0.8 aussi bien pour les eaux naturelles que pour les percolats: il s'agit de Br / I et de Rb / Fe.

Dans les percolats, les principales corrélations concernent dans l'ordre, Ni, Co, Rb, Mn, Fe, V, Cu et Al. Les liens entre métaux prédominent donc, indépendamment de la teneur moyenne des différents éléments. Il semble que ces corrélations soient liées à la parenté chimique des éléments considérés.

Pour les eaux souterraines B, I, Mn, Rb et Br forment entre eux les principales corrélations. Par contre, les métaux Al, Co, V ne présentent aucune corrélation significative et il en est pratiquement de même pour Fe, Cu et Ni. Les facteurs de corrélations élevés doivent être influencés d'une part par les propriétés chimiques des éléments, et d'autre part par le milieu géologique.

Tab.4. Corrélations pour les principaux éléments détectés dans les percolats à droite et pour ceux détectés dans les eaux de la molasse et du quaternaire (Aquityp) à gauche (valeurs arrondies).

PERCOLATS →

	Al	B	Br	Co	Cu	Fe	I	Mn	Ni	Rb	V
Al		0.7	0.1	0.5	0.9	0.3	-0.3	0.2	0.7	0.6	0.9
B	0.0		0.3	0.5	0.9	0.1	0.1	0.0	0.6	0.5	0.6
Br	0.3	0.9		0.2	0.3	-0.1	0.9	0.0	0.0	0.3	0.4
Co	-0.3	-0.3	-0.2		0.4	0.8	0.1	0.8	1.0	0.9	0.7
Cu	-0.3	0.2	0.3	0.4		0.0	0.0	-0.1	0.5	0.4	0.7
Fe	0.1	0.5	0.3	0.1	0.1		-0.2	1.0	0.8	0.8	0.5
I	0.1	1.0	0.9	-0.4	0.2	0.4		-0.1	-0.1	0.2	0.1
Mn	0.0	1.0	0.8	-0.2	0.1	0.6	0.9		0.8	0.7	0.4
Ni	-0.2	-0.1	0.0	0.3	0.8	0.0	-0.1	-0.1		0.9	0.7
Rb	0.0	0.9	0.7	-0.1	-0.1	0.8	0.8	1.0	-0.2		0.8
V	-0.3	-0.2	-0.3	0.1	-0.4	0.0	-0.2	-0.2	-0.2	-0.1	

← AQUITYP

4.4- Graphiques XY

Sur la base des facteurs de corrélations des graphes XY ont été réalisés pour tenter de séparer des échantillons influencés par les décharges, d'autres échantillons non influencés. L'échelle logarithmique a du être utilisée en raison de l'amplitude des teneurs considérées (de 0.01 à 20000 µg/l).

Pour tester ces rapports entre éléments, 3 groupes d'échantillons ont été reportés:
- les eaux de la molasse et du quaternaire non polluées (a)
- les percolats (p)
- des eaux d'aquifères molassique et quaternaire prélevées en aval de décharges urbaines (n)

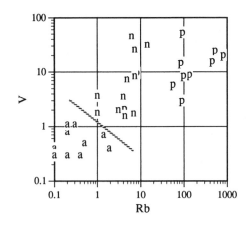

Fig.3. Graphe Rb / V. Facteur de corrélation élevé dans les percolats. Valeurs reportées en µg/l. Pour la légende se reporter à la fig.1.

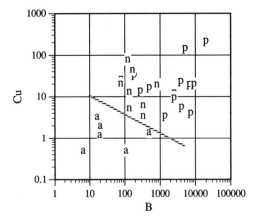

Fig.1. Graphe B / Cu. Facteur de corrélation élevé dans les percolats. Valeurs reportées en µg/l.
a: aquityp / p: percolat / n: aquifères contaminés par des décharges

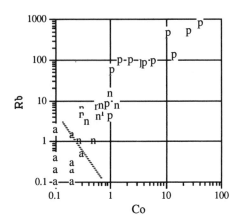

Fig.4. Graphe Co / Rb. Facteur de corrélation élevé dans les percolats. Valeurs reportées en µg/l. Pour la légende se reporter à la fig.1.

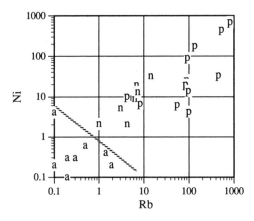

Fig.2. Graphe Rb / Ni. Facteur de corrélation élevé dans les percolats. Valeurs reportées en µg/l. Pour la légende se reporter à la fig.1.

Les graphiques XY permettant la meilleure discrimination entre ces groupes ont pour axes des éléments ayant des indices de corrélation élevés dans les percolats (fig 1 à 4).
Dans les autres cas, la distinction est plus difficile entre les eaux souterraines non polluées et les eaux souterraines polluées pour trois types de combinaisons:
- si les élément ont un coefficient FC trop bas, par exemple pour le rapport Li / Pb, ceci malgré un facteur de corrélation voisin de 0.8 pour les

percolats (fig.5).
- si des corrélations élevées existent aussi bien pour les percolats que pour les eaux souterraines (fig.6.).
- s'il s'agit d'éléments à bonnes corrélations dans les eaux souterraines et sans corrélation dans les percolats (fig. 7 et 8).

Dans ce dernier cas, l'influence du milieu géologique atténue probablement l'effet pollutif pour les éléments à fortes corrélations dans les eaux naturelles.

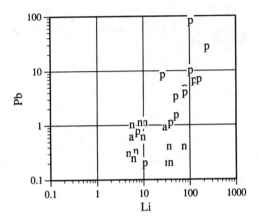

Fig.5. Graphe Li / Pb. Facteur de corrélation élevé dans les percolats, mais facteur de contamination (FC) bas. Valeurs reportées en µg/l. Pour la légende se reporter à la fig.1.

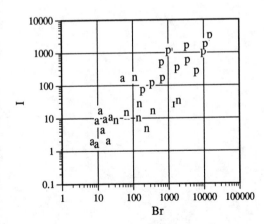

Fig.6. Graphe Br / I. Facteur de corrélation élevé dans les eaux souterraines et dans les percolats. Valeurs reportées en µg/l. Pour la légende se reporter à la fig.1.

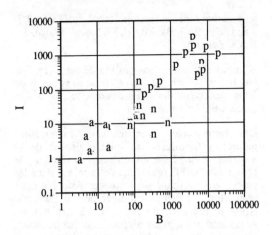

Fig.7. Graphe B / I. Facteur de corrélation élevé dans les eaux souterraines et bas dans les percolats. Valeurs reportées en µg/l. Pour la légende se reporter à la fig.1.

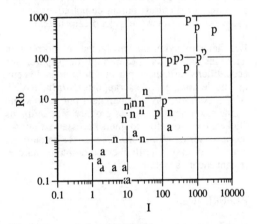

Fig.8. Graphe I / Rb. Facteur de corrélation élevé dans les eaux souterraines et bas dans les percolats. Valeurs reportées en µg/l. Pour la légende se reporter à la fig.1.

5. PERSPECTIVES

L'analyse de 10 percolats de décharges urbaines, à large spectre de déchets, en milieu sec et en phase méthanogénique à l'aide d'un ICP-MS-HR a permis de préciser la gamme des éléments en traces inorganiques qui s'y trouvent. En plus des éléments généralement analysés tel que As, Cd, Cr, Cu, Fe, Hg, Mn, Ni, Pb et Zn, 16 autres éléments ont été détectés dans au moins 60% des échantillons à des teneurs moyennes comprises entre 1 et 5000 µg/l. Il s'agit de B, Br, I, Ba, Rb, Al, Ti, Mo, V, W, Co, Sn, U, Cs, Zr et La.

D'autres éléments n'ont pas été détectés (In, Ga, Ge, Ir, Os, Pd, Rh, Ta, Te et Tl) ou dans moins de 30 % des échantillons et à des teneurs très basses (Se, Sc, Bi, Ag, Sb,

Une comparaison avec les eaux souterraines naturelles (aquifères du quaternaire et de la molasse) a été faite en utilisant un facteur de contamination FC (moyenne de la teneur dans les percolats / moyenne de la teneur dans les eaux naturelles). Ce facteur est élevé pour 6 éléments. Ainsi la présence de As et Cs à des teneurs supérieures à 1 µg/l est révélatrice d'une pollution. De même si les concentrations en Mn, Rb et Ni sont supérieures à 10 µg/l et à 100 µg/l pour Br, une pollution des eaux souterraines est pratiquement certaine. L'origine de la pollution ne peut toutefois pas encore être attribuée uniquement aux sites contaminés ou aux décharges. Des analyses complémentaires devront être effectuées pour s'assurer que d'autres sources de pollutions ne sont pas concernées (route, air, rejet d'eaux usées).

Les éléments ayant un bon facteur de corrélation dans les percolats, permettent de séparer les échantillons influencés par une pollutions de ceux qui ne le sont pas. Les rapports Ni/Mn, Rb/Ni, Al/Cu, Al/V et Al/Cu notamment ont permis une bonne discrimination sur les groupes d'échantillons étudiés. Il est toutefois toujours nécessaire d'utiliser des analyses de références, les courbes de partages entre le domaine pollué et le domaine naturel restant encore à établir.

REFERENCES

Bensimon M., Looser M., Parriaux A., Reed N. (1994). Characterization of groundwater and polluted water by ultra trace element analysis using high resolution plasma source mass spectrometry. - Eclogae Geologicae Helvetiae, à paraître 10p.

Borden R.C., Yanoschack T.M. (1989). North Carolina sanitary landfills: leachate generation, management and water quality impacts - Water Resources Research Institute, report 243, 52p.

Dauchy X. (1994), Usage de l'ICP/MS haute résolution dans l'étude de sites contaminés par des décharges - Rapport interne, GEOLEP, Ecole Polytechnique Fédérale de Lausanne, 28p.

Harmsen J. (1983) Identification of organic compouds in leachate from a waste tip - Water Research, 17 pp 699-705.

Looser M.-O., Parriaux A., Bensimon M. (1993). Surveillance des sites contaminés à l'aide des traces inorganiques. - Geoconfine 93, Montpellier, pp. 437-442.

Mandia Y. (1991). Typologie des aquifères évaporitiques du Trias dans le bassin Lémanique du Rhône (Alpes occidentales) - Thèse, Ecole Polytechnique Fédérale de Lausanne, 345 p.

Robinson H.D., Gronow J.R. (1993). A review of landfill leachate composition in the UK. - Procceedings Sardinia 93, Fourth International Landfill Symposium, Cagliari, Italy, pp. 821-832.

Shreve R.N., Brink J.A. (1977) Chemical process industries - McGrawHill, 814p.

The analysis of hydrogeological and hydrochemical factors affecting the spoil banks stability behaviour

Analyse des facteurs hydrogéologiques et hydrochimiques qui affectent la stabilité des terrains

Jan Moravec

Institute of Structure and Rock Mechanics of the Academy of Sciences of Czech Republic, Prague, Czech Republic

ABSTRACT: Hydrogeological and hydrochemical factors influence on clayey spoil bank bodies stability behaviour has been analysed. The problem has been solved on the spoil bank of Merkur brown coal opencast mine in the North-Bohemian basin related to the spoil bank construction, subsequent use and suitable integration into the landscape ecosystem. Spoil bank body, from the viewpoint hydrogeology hydrochemistry and engineering geology has been evaluated using in situ and laboratory measurements. The evolution of water-bearing system in spoil bank related to the stability of clayey spoil bank soils has been studied. The effect of the hydrological regime of a spoil bank on the occurrence of polluted effluents has been analyzed in connection with chemism of neighbouring groundwater bodies.

RÈSUMÈ: On analyse l'influence des facteurs hydrogéologiques et hydrochimiques sur les changements de la stabilité des corps des terrils d'argile. On étudie ce problème dans le territoire de la carrière du lignite "Merkur" qui se trouve dans le bassin houiller au nord de la Bohême en liaison avec l'edification l'exploitation et la bonne incorporation du terril dans l'ecosystème de la région. Le terril est estimé au point de vue l'hydrogéologie, l'hydrochimie et de la géologie d'ingénieur sur la base des mesures in situ et au laboratoire. On étudie aussi le developpement du systeme des couches aquifères à la stabilité de terril d'argile. On analyse l'influence de la régime hydrologique sur l'evolution du pollution de l'eau à l'interieur du terril au régard de la chimisme des couches aquifères environnates.

1 INTRODUCTION

The opencast mining in the North-Bohemian brown-coal basin (Czech Republic) calls fourth several problems with the construction and operation of spoil banks (waste dumps) and their subsequent re-introduction in the ecological system of the landscape.

Clayey spoil banks are man-made bodies characterized by the formation of water-bearing system and a special hydrological regime. The spoil bank water-bearing system is a system of water-bearing aquifers with variable hydraulic continuity rate and with specific boundary conditions.

Spoil banks are hydrogeologically anisotropic and heterogenous bodies. The clayey spoil bank soils have slight to very slight permeability with the coeficient of hydraulic conductivity

$$k = n \cdot 10^{-6} \text{ to } n \cdot 10^{-8} \text{ m.s}^{-1}.$$

2 HYDROGEOLOGICAL FACTORS

Spoil bank is a three-phase system. The system consist of solid phase, water and air. Each of these phases can be dominant under specific conditions. This means that spoil bank soils of the same modal composition may behave in quite a different maner. If the whole volume of voids (interstices) is filled up with water then spoil bank soil is saturated and forms a two-phase system. The shear strenght of soil will be unfavourably influenced by the two-phase system (the value of the angle of internal friction and cohesion will decrease).

The consolidation and stability clayey spoil bank soils are problems of engineering geology, difficult to solve so far.

One of the main reasons of selection the Merkur locality is that an extensive landslide of 140 million m^3 took place on the spoil bank of Merkur opencast mine in the North-Bohemian basin in October 1985. This is the largest landslide ever recorded in Czechoslovakia s opencast mining history. The similar problems exist in the Sokolov brown-coal basin (Western-Bohemia). The second largest landslide 50 to 70 million m^3 took place on outer dump at Vintiřov in June 1990. lts description is accompanied by an analysis of the causes of its inception. A reassesment of both previous and operating dumps is recommended with regard to their stability in the whole district (Rybář 1991).

It is found that the largest remaining difficulties in the prediction of settlement and stability of clayey spoil banks are associated with the estimation of permeability in nonhomogenous layers and determination of the drainage paths. Another problem is to define what the relevant "normal" spoil bank-
-water and groundwater conditions are and how to account for the change and fluctuations in groundwater level (Larson 1986).

Not so long ago. the spoil banks hydrogeology has been studied only in case the breakdown situations have to be solved. To improve the general status, the methodology has been developed to evaluate the water-bearing system in spoil banks, which is pointed at the realization of concrete solutions in the individual localities (Moravec 1987).

3 HYDROCHEMICAL FACTORS

The water flow velocity in spoil bank characterises the equation

$$v = (k/n_e) \cdot (dh/dl) \qquad (1)$$

where k is the hydraulic conductivity $|m.s^{-1}|$, n_e effective porosity, dh/dl is hydraulic gradient. From the equation follows that the water flow velocity depends on the spoil bank effective porosity.

The clay minerals swell during contact with flowing water (ion exchange). Therefore the effective porosity n_e depends on equilibrium relations between clay minerals and groundwater flow.

The groundwater flow in the spoil bank is expressively influenced by type and quantity of interstices filling. The interstices are more or less or even completely filled by clay minerals. The type of interstices filling depends on chemical and modal rock composition and chemical composition of flowing groundwater.

The clay minerals act as cation exchangers. The surface charge is mostly negativ. The clay minerals change for example Ca^{2+} in the solution for Na^+ ions in the mineral lattice. The specific surface of clay minerals occurs in order in decades $m^2.g^{-1}$ and the sorption capacity in hundres mmol H^+/kg^{-1} (Drever 1982).

Different types of exchange reactions characterise the empirical equations. The condition of equilibrium of exchange equation of one potentional ions of type

$$MA(s) + B^+ = MB(s) + A^+ \qquad (2)$$

characterises the relation obtained from the evaluation of experimental data according to the equation:

$$\frac{a(A^+)}{a(B^+)} = K_{AB} \frac{x(MA)}{x(MB)} \qquad (3)$$

when a is the activity of singular components in groundwater, x are molar fractions of A and B components in clay mineral n expresses the deviation from the condition of equilibrium K_{AB} is equilibrium constant. If the ion exchange takes place with different valence according to the equation

$$MA_2(s) + B^{2+} = MB(s) + 2A^+ \qquad (4)$$

the following relation is valid:

$$\frac{a^2(A^+)}{a(B^{2+})} = K_{AB} \frac{x(MA_2)^n}{x(MB)} \qquad (5)$$

When logarithming the equations
(3) and (5) the condition of equilibrium origines. This stright line
can be constructed from experimental data and afterwords the values
K_{AB} and n are calculated. The values of constant needed for calculations brings (Garrels – Christ
1965). On basis of obtained values
K_{AB} and n the degree of effective
porosity the affect rate condition
be judged and then the flow conditions in spoil bank. The acidity
of spoil-bank water is mainly caused by the oxidation of metal disulphides, most frequently pyrites. This process is rendered possible by the presence of oxygen
due the random deposition of dumped substances, the presence of
water of the own hydrological regime, and by the action of autotrophic bacteria. The acidic
spoil-bank effluents, thus formed,
react subsequently with the overburden minerals. The result is a
substantial alteration of the
chemical composition of the seepage water and formation of new
chemical compounds in the solid
phase. Similar reactions take place
also during the infiltration of
acidic rain water. Within the spoil
bank, highly polluted waters can be
found, with pH around 7, but with
very high concentrations of sulphates (2516 mg/l), iron (9,8
mg/l). aluminium (0,6 mg/l), manganese (9,7 mg/l), calcium (440
mg/), magnesium (460 mg/l). sodium
(340 mg/l), etc. Such water types
may be formed also in already consolidated spoil banks. The exchange
reactions between the clay minerals
and acidic effluents cause structural modifications of these minerals. Alterations of clayey minerals represent one factor affecting
also the stability of slopes of
spoil-bank bodies.

4 PROTECTION OF NEIGHBOURING
GROUNDWATER BODIES

The hydrological regime controls,
in the spoil bank, the continuously
occurring physico-chemical reactions between water, spoil-bank soil
(earth), and the atmosphere. These
reactions results in a heavily polluted spoil-bank effluent. The interaction between the hydrological
regimes of the spoil bank and the
neighbouring environment affects
mutually the chemism of spoil-bank
effluents and the water in neighbouring groundwater bodies.

The degree of pollution of the
spoil-bank water depends on the
chemical composition of the inflo-
wing water (in the spoil-bank body). such as the atmospheric precipitations, inflows from the subsoil, and on the mineralogical composition of the spoil-bank rocks
and soils. Further. it depends on
the residence time of water within
the spoil-bank.

The pollution of effluents and
its propagations into neighbouring
water-bearing collectors can be limited by a suitable regulation of
the spoil bank's hydrological regime by means of hydrotechnical measures.

The purpose of these measures is
to favour quick draining of atmospheric precipitation water from
the surface of the spoil-bank, thus
preventing the infiltration in its
deeper parts, futher the captation
of groundwaters and water springs
in the subsoil as well as the percolation and infiltration waters,
which penetrate into the subsoil
from the spoil-bank body. These
measures reduce the residence time
of water in the spoil-bank, thus
shortening the time of interaction
between water, dumping soils, and
the atmosphere. This results in the
occurrence of less polluted spoil-
bank effluents.

Heavy pollution of effluents is
encountered in spoil banks constructed on an undrained ground,
without any drainage system in the
subsoil of the spoil-bank. In these
spoil banks a hydrological regime
is established, which can be differentiated according to the residence time and degree of pollution of
water within the spoil-bank. The
pollution of these effluents is given, above all, by high concentrations of

Na^+, Ca^{2+}, Mg^{2+}, SO_4^{2-}, and the total mineralization.

For example, material deposited in the spoil bank of the opencast mine "Merkur" is formed, above all, by the overlying kaolinitic-illitic-montmorillonitic clays to claystones. The prevailing mineral is Ca-montmorillonite. Average mineralogical composition of these clays is: montmorillonite 38%, illite 26%, kaolinite 15%, quartz 8%, calcite+dolomite 6%, siderite 4%, accessorics 3%. The dumping soils contain also iron disulphides and recent sulphates. Basing on chemical analyses and measurements of the tritium activity in water samples from the spoil-bank of the opencast mine "Merkur", it could be established that the degree of pollution of effluents depends strongly on the residence time of water within the spoil--bank (see Tab. 1).

Effluents from spoil banks constructed on undrained ground (subsoil) are characterized by increased contents of Na^+, Ca^{2+}, Mg^{2+}.

SO_4^{2-} and by high overall mineralization.

The increase content of SO_4^{2-} is due the oxidation of iron disulphides. These reactions result in acidic effluents with pH < 4,5. Acidic waters are neutralized by circulation within the spoil bank. H^+ ions are consumed during the dissolution of carbonates yielding HCO_3^- ions.

The concentration of SO_4^{2-} in spoil-bank effluents is also increased by the dissolution of recent sulphates. In these reactions, an increased concentration of Na^+ ions is also encountered.

The weathering of the organic substance in the unsaturated zone of the spoil bank produces carbon dioxide, CO_2 which reacts with the infiltrating water from the atmospheric fallout, forming the carbonic acid. This acid dissolves calcite and dolomite, increasing

Table 1. Dependence of the chemical composition of spoil-bank effluents on the residence time of water within the spoil bank.

	sample 5	sample 12	sample 11
age	3-4 months mg/l	2 years mg/l	6 years mg/l
Ca^{2+}	86,77	393,69	440,35
Mg^{2+}	77,00	188,00	460,00
Na^+	87,00	720,00	340,00
K^+	12,90	53,00	51,00
NH_4^+	0,50	23,00	7,25
Fe^{2+}	0,17	1,70	9,89
Cl^-	49,28	86,80	87,57
Al^{3+}	0,002	0,07	0,06
Mn^{2+}	0,20	0,70	9,70
SO_4^{2-}	179,15	2594,52	2515,27
HCO_3^-	561,36	400,27	1266,10
mineralization total	1076,27	4815,62	5430,42
pH	7,1	7,8	7,2
tritium activity (T.U.)	61 ± 8	54 ± 8	36 ± 6

Note: T.U. = $3,2 \times 10^{-12}$ $Ci.l^{-1}$.

the concentration of Ca^{2+}, Mg^{2+} and HCO_3^- in the effluent.

5 CONCLUSIONS

Opencast mining in the North-Bohemian brown-coal basin calls forth problems with the construction and operation of spoil banks (waste dumps) and their subsequent re-introduction in the ecological system of the landscape.

The problem consist in determining of physico-mechanical and hydraulic properties of spoil bank soils which they have different water contents in different parts of the spoil bank. The prediction of the evolution of water-bearing system related to the stability of spoil bank is necessary. The results will be used in solving the stability of high spoil banks, in carrying out reclamation and foundation engineering on the spoil banks.

For evaluation of water-bearing system formation in the spoil banks in relation to their stability it is necessary to observe the water level course in the hydrogeological observation boreholes. The movement of spoil bank soils has been also observed by means of surface geodetical measurements, remote sensing and inclinometric boreholes. The observation has been performed for preventive danger signalling of spoil bank slope failure. The slope failure occurs during the critical increase of water level, and thus increase of pressure of the spoil bank water, in a certain location of spoil bank.

Study of water-bearing spoil bank has been focused at the research of the ratio between the

cations Na^+ and Ca^{2+} or Mg^{2+} in the structure of montmorillonite, illite and mixed clay minerals. The ratio of these cations in the clay mineral structure appears to influence the water-bearing system formation in the spoil banks. The quantitative representation and structure of clay minerals determine if either the zone of saturation will be created under the sufficient water inflow into the spoil bank, or almost all water be bound by these minerals, above all by montmorillonite and illite.

Special attention has been paid to the empirical approaches to the solution of stability problems of the spoil bank. Interdependence between the angle of the slope and its height has been searched and evaluated for various hydrogeological and hydrochemical conditions (Kudrna 1985, Zika-Rybár-Kudrna 1988). The hydrogeological results have been compared with instability phenomenas in the investigated area.

Within the spoil bank body, also

some physical and chemical processes take place, resulting in the acidic character of the drained spoil bank water. The acidity of spoil bank water is mainly caused by the oxidation of metal sulphides, most frequently pyrites. This process is rended possible by the presence of oxygen due to the random deposition of dumped substances, the presence of water of the own hydrological regime, and the action of autotrophic bacteria. The acidic spoil bank effluents, thus formed, react subsequently with the overburden minerals. The results is a substantial alteration of the chemical composition of the seepage water and formation of new chemical compounds in the solid phase. Similar reactions take place also during the infiltration of acidic rain water. Within the spoil bank highly polluted waters can be found. Such water types may be formed also in already consolidated spoil banks. The exchange reactions between the clay minerals and acidic effluents cause structural modifications of these minerals. Alteration of clay minerals represent one factor affecting also the stability of slopes of spoil bank bodies. These reactions will be investigated in detail

One of problems resulting from the insufficient evaluation of spoil bank hydrological regime is the interaction of effluents with the surrounding environment namely the contamination of neighbouring aquifers by polluted spoil bank effluents.

In order to accumulate sufficient data for the solution of this problem, systematic hydrological measurements and continuous water analyses should be effectuated in the spoil bank and in their relevant neighbourhood.

The solution of the formulated problems is very topical from the viewpoint of securing the optimum opencast working, and suitable integration of spoil bank into the landscape ecosystem with respect to groundwater regime affected by the mining activities.

REFERENCES

DREVER, J. Y. 1982. The geochemistry of natural waters. New Jer-

sey, Prentice - Hall, Inc.
Engl. Cli., pp. 82 - 85.
Garrels, R. M. - C. L. Christ
1965. Solutions, Minerals and
Equilibria. New York, Harper
and Row, pp. 1 - 450.
Kudrna, Z. 1985. Analysis slide
movements in open pit mine CSA.
Brno, Proc. Conf. Engng. Geology and Power Civ. Engng., pp.
201 - 206.
Larson, R. 1986. Consolidation of
soft soils. Linkoping, Swedish
Geotechn. Instit., pp. 174.
Moravec, J. 1987. The factors of
the water-bearing system formation in the spoil banks in the
North-Bohemian brown coal ba-
sims. Prague, IGG CSAS, pp. 106.
Rybář, J. 1991. The Vintiřov dump
near Sokolov - a landslip. Prague, Geol. Surv., 33, 5, pp.
133 - 134.
Zika, P. - J. Rybář - Z. Kudrma
1988. Empirical approach to the
evaluation of the stability of
high slopes. Lausane, 5th Int.
Symp. on Lanslides, 1, Balkema,
pp. 1273 - 1275.

Anticipated changes in groundwater conditions and environmental impact in a deltaic lowland due to a large-scale river diversion

Changements expectés dans les eaux souterraines et influences environnementales dans une région déltaique dûs à un projet majeur de dérivation d'un fleuve

P. Marinos & M. Kavvadas
National Technical University, Athens, Greece

V. Perleros
Athens, Greece

D. Rozos, E. Nikolaou, N. Nikolaou & J. Chatzinakos
Institute of Geological and Mineral Resources (IGME), Athens, Greece

ABSTRACT: The paper studies the anticipated changes in the seepage of a major river in Greece, towards the low-salinity ponds of its estuaries, through the deltaic alluvia and the underlying limestone. The study was triggered by the concern that changes in the salinity of the deltaic ponds due to the large scale river diversion would affect their rich ecosystems. It is shown that the contribution of river water seepage towards the deltaic ponds in the area is small compared to the discharge of excess irrigation water. Furthermore, the high salinity of the groundwater in the fine alluvia imposes a "barrier" on the flow towards the external parts of the delta, where the biotope has developed. It is concluded that even the small anticipated changes in the salinity of the influx of seepage to the deltaic ponds, can be easily controlled by monitoring the direct freshwater discharge towards the ponds.

RESUME: L' étude concerne des changéments possibles aux infiltrations d'un grand fleuve vers les étangs et lagunes périphériques de son délta, par ses alluvions et les calcaires sous-jacents. L' étude a été poussée par le souci que l'important biotope de la région serait affecté par la dérivation du fleuve, à la suite d'un changement de la salinité. Il est demontré que la contribution des infiltrations des eaux du fleuve est faible, notamment quand elles sont comparées avec les décharges des surplus des eaux d'irrigation. En plus, la salinité, très élevée, des eaux souterraines due à la nature des dépôts fins déltaiques, forme une importante "barrière" à l' écoulement vers ses parties externes, où se trouvent les biotopes. Il est conclus que les changements, finalement faibles, de la salinité des écoulements souterrains, peuvent être controllés par une gestion convenable de décharge vers les lagunes des eaux utilisées pour l'irrigation.

1 INTRODUCTION

The planned partial diversion of Acheloos river (a major river in Western Greece) may have important environmental consequences on the wetlands of its delta and their rich ecosystems which need to be investigated and corrected before the project is finalised. The proposed works, will diverge about 25 percent of the mean annual discharge of the river to the east for the irrigation of the Thessaly plains in Eastern Greece. The diversion will reduce the direct fresh water discharge in the vicinity of the river mouth. In addition to the direct discharge of river water, another possible source of fresh water to the deltaic lagoons and wetlands is the seepage of river water through the deltaic alluvia as well as the seepage of groundwater from the karstic aquifer underlying the river alluvia. The present paper assesses the significance of the last two sources of fresh water supply to the deltaic wetlands, compared to the direct supply of river water, and proposes remedial measures to compensate for any changes in the salinity of the lagoons due to the changes in the fresh water supply.

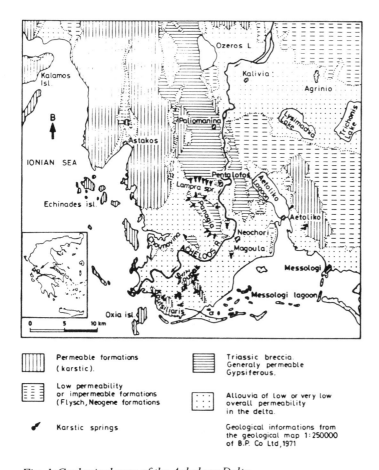

Fig. 1: Geological map of the Acheloos Delta.

Despite the obvious effects of any change in the amount of fresh water discharge on the salinity of the deltaic lagoons, it should be mentioned that the natural flow of the Acheloos river has already been artificially modified by:

1. the construction of a system of reservoirs for electric power generation in the river, which regulate the river flow by storing the flood waters,

2. the construction of an extensive irrigation network in the lower part of the Acheloos basin and the deltaic plains, consisting of water supply canals and discharge trenches which drain into the wetlands, and

3. the containment of the Acheloos river bed with lateral flood embankments, thus excluding the possibility of flooding in the alluvial plains.

Because of the previous activities, the annual flooding of the river is eliminated and thus the natural relationship of the river to its deltaic environment has been modified (the amount of fresh water supply has been decreased). However, the direct supply of significant quantities of irrigation water from the discharge trenches into the deltaic lagoons has maintained an amount of the fresh water supply and has thus influenced the salinity of the wetlands. Finally, sea water often advances into the river for a few kilometres upstream and increases the salinity of the river water when the hydroelectric plants are not operating and the river flow is practically stalled.

The hydrogeological conditions at the Acheloos delta were investigated via an extensive geotechnical investigation programme, designed by the authors and performed by the Hellenic Institute of Geological and Mineral Exploration (IGME) in the summer of 1993, at the southern part of the delta (see Fig. 1) in the area between the river estuaries and the Aetolikon lagoon. The programme consisted of geophysical measurements (electrical conductivity method) to establish the thickness of the alluvia and the morphology of the underlying bedrock, boreholes with continuous sampling and in-situ measurements of the permeability profile, installation of piezometers in the boreholes and monitoring of the groundwater table elevations, measurements of the conductivity, salinity and concentration of chlorides in the groundwater at various locations (boreholes, irrigation canals, discharge trenches, wells, springs, lagoons, etc.) and depths (in wells and boreholes).

Using the data collected from this investigation, the hydrogeological conditions at the deltaic alluvia were analysed using a mathematical simulation of the groundwater seepage regime driven by the river to the north, the deltaic lagoons to the south, the network of the irrigation discharge trenches covering the south section of the delta (see Fig. 2), and the karstic aquifer in the limestone underlying the deltaic alluvia (and outcropping mainly in the upstream area).

2 GEOLOGICAL STRUCTURE OF THE DELTA

British Petroleum (1971) investigated the geology of the Acheloos delta and issued a geological map in a scale 1:100,000. Recently, IGME (1986) issued a detailed geological map of the Acheloos delta in a scale 1:50,000. The following description is largely based on their findings as well as a re-evaluation based on the results of our 1993 site investigation programme.

The area of the Acheloos delta is covered with the alluvia deposited by the river in the region of its progressively advancing estuaries. These sediments are either deposited in freshwater (e.g. during the annual flooding of the river) or in water of increased salinity (marine sediments), such as fine grained sediments transported by the river and deposited in the sea. Most of the alluvia in the higher elevations are deposited in freshwater while the deeper sediments are mainly of marine origin, as evidenced by a consistent gradual increase in the salinity of the groundwater with depth measured in the boreholes. A detailed description of the

structure, geomorphologic conditions and bedding of these sediments is given by Psilovikos (1993). The sediments encountered in the boreholes are fine-grained, consisting of fine sands, silts and silty clays in random sequences both with regard to the lateral extent and the depth. Coarse-grained sediments (coarser than the medium sand size) were not encountered in any of the boreholes up to the depth investigated (35 meters).

The origin of the sediments and the methods of their deposition in the delta have contributed to a largely random structure with lack of continuity in the lateral extent and the creation of thin discontinuous sandy lenses in an otherwise impermeable matrix. The result of such a random interbedding of sandy layers with less permeable silts and clays is:

1. a significantly reduced average permeability of the composite material and

2. a permeability in the vertical direction less than that in the horizontal direction.

The bedrock underlying the deltaic alluvia to the east (towards the Aetolikon lagoon) consists of neogene formations and the triassic breccia of the Ionian zone (Fig. 1). The neogene deposits contain strongly cemented conglomerates, sandstones, marls and marly limestones which appear at the surface and form the hills between the delta and the Aetolikon lagoon. The triassic carbonate breccia, containing a significant amount of gypsum, appear near the surface to the north of Neochorion and at the Magoula hill inside the delta area. Finally, the bedrock at the western part of the delta consists of the Jurassic limestones of the Ionian zone overlying the triassic breccia. These limestones appear at the surface in various locations of the western part of the delta, such as the hills Chounovina, Koutsiliaris, Skoupas, Taxiarchis etc. The same limestones have formed the islands to the west of the Acheloos estuaries in the Ionian sea. The hydraulic conductivity of the neogene formations is relatively small while the limestones and the triassic breccia are more permeable by at least two orders of magnitude. Finally, it is worth mentioning that the water of springs in the triassic breccia has a high concentration of sulphates due to the existence of gypsum in the mass of the breccia.

The tectonic movements and faulting which occurred in the area have disturbed the geometry of the bedrock underlying the deltaic alluvia and have created large depressions in certain

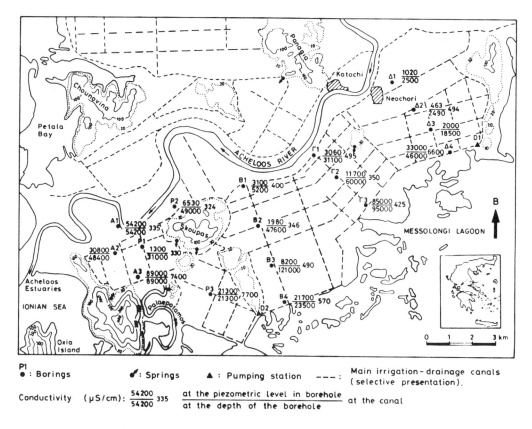

Fig. 2: Irrigation drainage system at the Acheloos river lowlands (south bank). The locations of the boreholes are also shown.

locations. In fact, a borehole drilled during a previous site investigation in the area of the Acheloos estuaries to a depth of 150 meters did not encounter the bedrock but penetrated solely through fine grained river and marine sediments.

The hydrogeological observations described above lead to the following conclusions:

1. The delta is essentially hydrogeologically isolated from seepage of river water originating at locations upstream of the delta area. Specifically, seepage of river water originating in the region Paliomanika-Pentalofos (to the north of the delta) towards the triassic breccia appear in the form of springs at the Lambra area to the north of the delta and do not contribute to the groundwater seepage in the area of the delta and the wetlands.

2. The karstic limestones appearing in the west part of the delta, form a karstic aquifer which is charged locally by the percolating rainfall and discharges along the periphery of the hills in the form of overflow-type springs at the interface of

the limestones with the surrounding alluvia. The annual discharge of these springs is compatible with the percolating rainfall and excludes any significant hydraulic contribution from the karstic masses to the north of the delta, thus indicating either complete isolation or, at least, the existence of appreciable obstacles in a hydraulic connection between the karstic masses below the deltaic alluvia.

3. The deltaic alluvia form a porous aquifer that discharges towards the river in the upstream part of the delta. In the downstream part of the delta, the aquifer has a relatively uniform gradient with a N-S direction (from the river towards the irrigation trenches of the delta).

3 THE PRESENT STATE OF THE DELTA

In the last decades, the natural evolution of the Acheloos delta has been disturbed by human activities consisting of the construction of a

2996

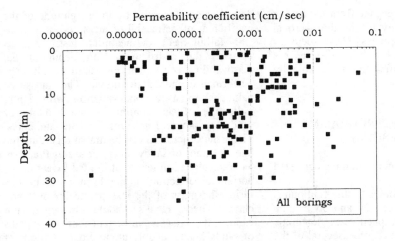

Fig. 3: Distribution of the permeability with depth in the deltaic alluvia.

series of reservoirs along the river, an extensive irrigation system of the lower delta plains and a containment of the river bed with flood embankments on its banks. The storage of flood water in the reservoirs and the complete regulation of the river flow dictated by the need for electric power generation has significantly altered the temporal distribution of fresh water discharge directly to the lagoons in the area of the estuaries, and consequently has modified their salinity. Furthermore, the reduction (or even complete stalling) of the river discharge during the periods that the power generation turbines are inoperative contributes to a corresponding increase in the salinity of the river water for a few kilometres upstream of the river mouth. As a result, any seepage of river water through the alluvia in the downstream part of the delta, towards the lagoons, involves (at least partly) water with increased salinity. This effect is further invigorated since most of the alluvia in deeper strata are marine sediments and the pore water has increased salinity; thus even fresh river water is bound to increase its salinity during seepage through the alluvia of the delta. As a result, the fresh water supply of the lagoons by seepage of river water is questionable regardless of the proposed partial diversion scheme.

The construction of an extensive irrigation network in the lower part of the delta (south bank of the river) has also significantly affected the hydraulic budget of the wetlands. The irrigated area (about 10000 acres) is supplied with fresh water via irrigation canals bringing

water from Lysimachia lake to the north-east (Machairas et al, 1988; NTUA 1992); the water of the river is not used for irrigation due to its variable salinity caused by the regulation of the river flow according to the needs for power generation as described above. The water supply for irrigation operates between April and September at an average rate of 3.5 m^3/sec (12500 m^3/hour). During the winter months, when the irrigation system is not in operation, the supply canals discharge the overflow of lake Lysimachia directly into the Acheloos river.

The groundwater table in the irrigated plains is maintained about two meters below ground surface by a regular grid of trenches collecting the irrigation overflows and discharging them into the lagoons via two pumping stations D1 and D2 (see Fig. 2) at an average rate of 1.5 m^3/sec (5400 m^3/hour). It is evident that the above average discharge of the irrigation trenches includes the discharge of excess rainfall during the winter months when the irrigation system in not in operation. Finally, is should be mentioned that part of the irrigation water supplied to the area is discharged directly into the trenches and finally to the wetlands since the irrigation network is not yet completed.

The network of the discharge trenches also maintains the groundwater table level in the deltaic alluvia 2-3 meters lower than the mean sea level and 3-4 meters lower than the water level in the adjacent part of the river. This piezometric head difference leads to seepage of river water towards the discharge trenches, seepage from the lagoons and wetlands towards the trenches and finally seepage from the karstic aquifer below the alluvia, in the vertical direction, towards the trenches. These seepages are added to the normal discharge of excess irrigation water and rainfall and are eventually pumped into the wetlands.

4 THE INVESTIGATION PROGRAMME - EVALUATION OF RESULTS

The 1993 site investigation programme consisted of:

1. Drilling 17 boreholes with continuous sampling to depths between 20 and 35 meters. The boreholes were equipped with standpipe piezometers at various elevations. Several of the boreholes were equipped with double piezometers to detect any hydraulic gradient in the vertical direction.

2. Measurement of the in-situ permeability coefficient with 188 constant and falling head tests performed inside the boreholes.

3. Geophysical investigation with the electrical resistance method along several axes in the area of the delta.

4. Measurement of the groundwater table elevation in springs, investigation and production boreholes, piezometers, wells, discharge trenches, as well as the level of the river (in several locations) and that of the wetlands.

5. Measurements of the conductivity and concentration of chlorides in the previous locations. Emphasis was given in the measurements at various depths inside boreholes to establish a profile in the vertical direction.

The main results of the hydrogeological and geotechnical investigation are as follows:

1. All boreholes encountered fine grained sediments consisting of alternating layers of fine sands, silts and clayey silts in random sequences with complete lack of continuity in both the horizontal and vertical direction. Up to the maximum depth investigated (35 meters), the bedrock was not encountered, neither any coarse grained sediments. The random variation of the granulometry of the sediments is a direct result of their depositional environment with sandy layers being deposited closer to the river and more fine grained sediments transported to larger distances and/or deposited during periods of flooding.

2. The random sequence of the alluvia in the delta is also evidenced by the magnitude of the coefficient of permeability measured during in-situ falling-head permeability tests. Fig. 3 presents the distribution with depth of the permeability coefficient measured in all tests performed in the deltaic alluvia. The measured values show a random variation with depth, indicative of the continuous interbedding of sandy layers with less permeable silts and clays. The same conclusion can be drawn regarding the lateral extent of sandy layers: it seems that they have developed in lenses of limited extent in the horizontal direction in the form of inclusions inside the mass of the less permeable silts and clays. These results should, however, not exclude the existence of certain continuous paths of sandy layers, mainly along older locations of meanders which have been cut-off from the present river bed and are buried by more recent alluvia. However, even if such routes of preferential hydraulic conductivity exist, their discharge capacity is most probably limited by the small magnitude of the piezometric head difference existing between the river and the wetlands. (1-2 meters) in conjunction with the long seepage path (a few kilometres).

3. The geophysical investigations indicate that the bedrock in certain parts of the delta is located at a depth of 30-80 meters and consists of limestones and triassic breccia. It should, however, be pointed out that the efficiency of the geophysical exploration is limited by the increased salinity of the groundwater, thus making more difficult the differentiation of the bedrock from the overlying alluvia. This conclusion is further supported by the fact that none of the boreholes reached the bedrock or any indication of its proximity up to the depth investigated (35 meters).

4. The measured piezometric levels in the alluvia indicate some "swelling" of the groundwater table (at least in the upper soil layers) in regions away from the river and the sea, resulting in seepage directions towards the river (in the northern part of the delta) and towards the sea (in the rest). This condition occurs especially in the upstream parts of the delta where the ground level is higher (at about +2.5 to +3.0 meters above mean sea level). On the contrary, in the downstream part of the delta, where the ground level is approximately at the mean sea level (or even lower than that) the piezometric regime is governed by the water level in the network of the discharge trenches

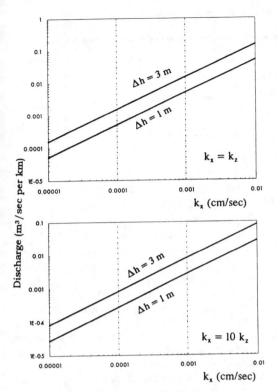

Fig. 4: Charts showing the computed seepage discharge of river water towards the irrigation discharge trenches as a function of the horizontal permeability of the deltaic alluvia.

which is maintained 2-3 meters below the mean sea level. This condition tends to create a groundwater seepage regime directed towards the discharge trenches. While this trend is in fact valid for the upper soil layers, the relatively small average permeability of the alluvia precludes the establishment of a hydraulic regime in the deeper soil strata with appreciable discharge capacity towards the irrigation trenches.

5. Measurement of the piezometric levels at various elevations in the alluvia indicates a small increase of the head with depth (15-35 cm of head difference in piezometers installed at depths of 5 and 25 meters). These results may be partially attributed to short-term effects, namely the lack of complete hydraulic equilibrium in the pairs of piezometers due to the short time between installation and measurement (about two weeks). It could, however, be a real effect due to the groundwater table lowering caused by the network of discharge trenches in the delta.

6. The measured conductivity of the water in small springs appearing at the periphery of the limestone hills in the western part of the delta is very high (1700-16800 μS/cm). These values are mainly due to a high concentration of chlorides (up to 3600 ppm) and indicate some mixing with sea water. The existence of springs with high chloride concentration allows to argue that these karstic springs are not hydraulically connected (in a appreciable degree) to the river and/or the limestone masses to the north of the delta.

7. The conductivity of groundwater measured inside the boreholes increased consistently with depth, with measured values up to 120000 μS/cm at the bottom of the boreholes (20-35 m), while the corresponding values near the surface in the vicinity of the river were in the range 1000-3000 μS/cm, but increased quickly to 10000-30000 μS/cm towards the wetlands. The conductivity measured at the bottom of the boreholes was significantly higher than that measured in the adjacent coast (where values about 50000 μS/cm are typical). In general, the increase of the conductivity with depth, all measurements indicate a dramatic increase of the conductivity in the N-S direction (from the river to the wetlands in the south). The increased conductivity in the southern part of the delta is also indicative of the low permeability of the alluvia which suppresses the tendency for leaching via seepage in the vertical direction (from the underlying limestone) or from the adjacent wetlands.

8. The conductivity of the water collected in the discharge trenches is significantly lower than that in the adjacent boreholes. This is mainly due to the fact that the discharge trenches collect excess irrigation water seeping through the upper soil strata which have been gradually leached (and thus the salinity of the porewater is relatively low).

5 SEEPAGE CALCULATIONS THROUGH THE DELTAIC ALLUVIA

Modelling coastal aquifers is significantly more difficult than modelling aquifers with homogeneous density, due to the obstacles in the diffusion and mixing processes caused by the differences of fluid density (Custodio, 1988). In

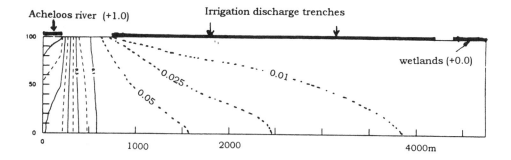

Fig. 5: Iso-piezometric contours corresponding to the computed seepage of Acheloos river water towards the irrigation discharge trenches (for unit head difference).

the present study, however, it is desirable to simply obtain an upper bound of the anticipated seepage of river water towards the wetlands; thus, the following analyses analyses assumed a homogeneous aquifer, which gives estimates on the safe side (larger than in reality).

The potential effects of groundwater seepages through the alluvia on the salinity of the deltaic wetlands were investigated quantitatively using an analytical model to describe the hydraulic relationships among the river, the deltaic alluvia and the adjacent wetlands to the south. The model was employed in the analysis of the following seepage routes (through the alluvia):

1. river water towards the discharge trenches,
2. water from the underlying karstic aquifer towards the discharge trenches.

The analytical model is a two-dimensional seepage model (in the vertical plane) based on the finite element method. It solves the seepage field equation in the time domain using a fully implicit scheme to advance the solution in time. The spatial discretisation of the domain is based on the Galerkin method of weighted residuals. The model can describe time-dependent seepages in porous media taking into account the storativity of the medium and/or boundary conditions variable with time. The model was developed for the study of Regional Flow problems in the horizontal plane (Marinos, 1993) and was modified appropriately for the study of the present problem.

The application of the finite element model in a parametric analysis of the seepage problem varying (i) the permeability of the alluvia, (ii) the ratio of the horizontal to the vertical coefficient of permeability and (iii) the thickness of the alluvia, allowed to conclude the

following:

1. Charts showing the computed seepage rate of river water towards the irrigation trenches (and finally to the wetlands) are presented in Fig. 4. For typical values of the geometry and the coefficient of permeability of the alluvia the computed discharge rate is 150-400 m^3/hour, values which are small compared to the direct discharge of the irrigation system to the wetlands. Fig. 5 presents the computed iso-piezometric contours corresponding to the seepage of river water towards the irrigation discharge trenches for a unit head difference and anisotropic permeability ($k_x = 10k_y$).

2. The computed seepage rate of river water towards the wetlands to the south can be approximated by the relationship:

$$Q = 1.552 \, k_x \tfrac{\Delta h}{L} H B$$

where k_x is the permeability in the horizontal direction, Δh is the mean head difference between the river and the wetlands, H is the thickness of the alluvia and B is the stretched length of the river along which the seepage takes place. It should be reminded that this relationship was obtained by disregarding the effect of different densities between the seeping river water and the water in the lagoons. Since such density differences tend to create a barrier and inhibit the free mixing of the two fluids (e.g. Custodio, 1988; De Breuck, 1991), disregarding density differences is conservative since the predicted seepage rates are higher than in reality. For typical values of the soil permeability parameters, the computed discharge is 30-300 m^3/hour. It should be mentioned that this rate is small compared to the rate of discharge of the irrigation trenches (about 5400 m^3/hour).

3. The computed seepage rate from the wetlands and lagoons towards the network of irrigation trenches is in the range of 50-150 m³/hour for typical values of the parameters involved in the model.

4. The computed seepage rate from the karstic aquifer underlying the alluvia towards the network of irrigation trenches depends on the piezometric head difference between the karstic aquifer and the trenches, the permeability in the vertical direction and the thickness of the alluvia. For a typical average thickness of the alluvia (50 meters), a head difference of two meters and a possible range of the vertical coefficient of permeability 10^{-3} - 10^{-6} cm/sec, the estimated seepage rate can be at most 150 m³/hour.

6 CONCLUSIONS

The paper studies the hydrogeological effects of the proposed partial diversion of Acheloos river (a major river in Western Greece) i.e., the effects caused by an estimated 25 percent reduction of its average discharge. Specifically, the paper analyses the seepage of river water through the deltaic alluvia towards the wetlands adjacent to its estuaries and towards the network of irrigation trenches (which eventually also drain into the wetlands). The study involved a review of the existing hydrogeological data, the collection of new data from an extensive site investigation programme and a parametric analysis of the seepage regime in the area of the delta using a finite element model.

The hydrogeological investigation revealed that the deltaic alluvia consist of fine grained soils (fine sands, silts and clayey silts) in random, generally discontinuous sequences, in both the lateral extent and the depth. Such profiles tend to reduce the average permeability of the composite material and decrease the seepage of river water towards the wetlands (both directly and indirectly). The estimates obtained via the computational model indicate that the anticipated seepage rates are small, especially if compared to the direct discharge of the excess irrigation water collected in the network of irrigation trenches existing in the deltaic plains. Typically, the seepage rates are expected to be in the range of 200-850 m³/hour (using conservative values of the governing parameters and disregarding the difference in

densities between the seeping river water and the water in the lagoons); these consist of direct seepage of river water towards the wetlands (30-300 m³/hour), seepage of river water towards the irrigation trenches (150-400 m³/hour) and seepage of water from the karstic aquifer towards the irrigation trenches (50-150 m³/hour).

These quantities are small compared to the discharge rate (about 5400 m³/hour) of the excess irrigation water which is collected in the network of trenches and pumped into the wetlands (in two locations). These quantities include the excess irrigation water, seepages of river water, seepages from the underlying karstic aquifer and finally water of the irrigation supply system discharged directly into the trenches. The relative portion of underground seepages in the total amount of the water discharged into the wetlands via the irrigation system is small (4-14 percent); it is thus evident that the river diversion scheme can only influence to some degree the magnitude and characteristics of that portion of the total discharge. It should, however, be realised that even that portion of the seepage will not be significantly affected for the following reasons:

1. The diversion will not reduce the water level in the river and thus the piezometric head difference will not be affected. The effect will be limited to a reduction of the river discharge rates and thus a somewhat increased salinity of the river water seeping through the alluvia.

2. The salinity of the river water in the present condition is not always low since the river flow is regulated by a series of reservoirs upstream. Thus, at present it is quite common that the salinity of the river water reaches levels that make the water inadequate for irrigation. Actually, this is the reason that at present the river water is not used for the irrigation of the deltaic plains but, irrigation water is supplied from lake Lysimachia.

3. The salinity of the water seeping through the deltaic alluvia is significantly increased since the pore water has a high concentration of chlorides. This became evident during the site investigation programme by the drastically increased salinity of the water in the boreholes at deep elevations. As a result, even the "fresh" river water will degrade as it seeps through the alluvia.

It is thus concluded that the proposed diversion of the Acheloos river will only have minor effects on the magnitude and the salinity of the seepage towards the wetlands. These effects are negligible compared to the effects of the direct discharge of irrigation water from the deltaic plains into the wetlands. Furthermore, even the small changes in the salinity of the seepage water can be compensated by regular measurements of the salinity levels in the wetlands and monitoring by adjusting the discharge rate of the irrigation system. In this way, the small anticipated increase in the salinity can be compensated even by slightly increasing the amount of fresh water discharging into the wetlands via the irrigation system.

ACKNOWLEDGEMENTS

The present paper is based on results of a study regarding the hydrogeological effects of the proposed diversion of Acheloos river, performed by the authors for the Greek Ministry of National Economy.

REFERENCES

British Petroleum Co Ltd, 1971. "The geological results of Petroleum exploration in Western Greece", *The Geology of Greece*, Vol 10, IGME, Athens, Greece.

Custodio E. 1988. "Present state of coastal aquifer modelling: short review", *Groundwater flow and quality modelling*, Reidel, 785-801.

De Breuck W. 1991. "Hydrogeology of salt water intrusion: a selection of SWIM papers", *Intern. Contributions to Hydrogeology*, Vol. 11, Int. Assoc. Hydrogeologists, Heise: 1-422.

IGME, 1986. "*Preliminary geothermal research in the region of the estuaries of Acheloos river*", Internal report prepared by M. Fytika and B. Kartalides (in Greek).

Marinos P., 1993. "*Report on the hydrogeological conditions of the Acheloos delta*" prepared for the Greek Ministry of National Economy (in Greek).

Marinos P., Frangopoulos J., 1973. "Systeme des sources karstiques de Lambra (Akarnanie-Grece)". *2 Convegno Int. Acque Sotterranee*, Palermo, Italie.

Machairas G. - HYDROEXYGIANTIKH, Lazarides L., 1988. "*Review of the Final design of the irrigation works in lower Acheloos (zones 8, 9B, 9C)*". Report prepared for the Greek Ministry of Agriculture (in Greek).

NTUA, 1992. "*Integrated management of the wetlands in Aetolikon-Messologgion*", research report by M. Bonazountas and M. Kallidromitou, prepared for the EC.

Psilovikos A., 1993. "*Report on the Geological conditions of the Acheloos delta*" prepared for the Greek Ministry of National Economy (in Greek).

Villas G.A., 1983. "*The Holocene evolution of the Acheloos river Delta, Northwestern Greece: Associated environments, geomorphology and microfossils*", MSc Thesis, Univ. of Delaware, USA. 201 p.

Géologie et ressources en sables et graviers dans la région d'Alger: Impact sur l'environnement

Geology, sand and gravel resources in Algiers region: Environmental impact

Belaid Alloul
Université des Sciences et Techniques (USTHB/IST), EL Alia Alger, Algérie

RESUME: Les formations géologiques qui ont fourni des matériaux sablo-graveleux dans la région d'Alger sont essentiellement des formations récentes .Les oueds qui descendent de l'Atlas Tellien et qui se jettent dans la mer représentent le plus grand gisement .Ces alluvions sont essentiellement calcaro-schisteuses dans la zone Ouest et légèrement siliceuses dans la zone Est. Les estuaires représentent une zone d'accumulation non négligeable.Les dunes côtières anciennes et récentes représentent le réservoir le plus facile à exploiter. Certaines sont consolidée par endroit et exploitées pour les remblais routiers alors que d'autres sont surtout des sables de plage qui sont les plus vulnérables et exploités pour le bâtiments.L'exploitation intense des alluvions d'oueds a eu un impact considérable sur les nappes alluviales dela région de Kabylie. Les prélèvements autorisés et non autorisés des sables de plages ont apporté un préjudice important à l'équilibre littoral.

ABSTRACT: Geological formations that supplied Algiers with sandy and gravely material are essentially recent. Wadi from Tellian to the sea are the most important deposits. These alluvions are essentially limestones and schist in west areas and slightly siliceous in the east. Estuaries are non negligeable zones of accumulation. Ancient and recent coastal dunes represent the easiest exploitable reservoir. Some are consolidated here and there and exploited for road embankmant whereas others are mainly beach sands , which are more vulnerable and used for construction. Intense exploitation of wadi alluvials have a considerable impact on alluvial aquifers of kabylie. Authorized and non authorized beach sand exploitation brought an important prejudice to littoral equilibrium. Deep geological research on other type of granulate will help to find new substitution ressources. This paper will be link between regional geology , granulate ressources of any kind and the impact of environnement.

1-INTRODUCTION.

Depuis l'indépendance de l'Algérie (1962) et particulièrement depuis une dizaine d'années toutes les villes d'Algerie connaissent une demande extremement forte en materiaux de construction. La province d'Alger,qui est soutenue par une géologie aussi riche que variée connait la même demande inassouvie en granulats de toutes sortes (sables graviers,concassés, argiles et marnes pour produits rouges et cimenteries etc). Si les granulats de roches massives necessitent le plus souvent une élaboration onéreuse qui ne permet pas une production trés importante de granulats ; les formations sablo -graveleuses des rivières et des dunes cotieres representent par contre des gisements faciles et peu coûteux à l'extraction. donc plus vulnérables à tout point de vue. Les formations géologiques qui sont porteuses de matériaux sablo-graveleux dans la région d'Alger sont essentiellement quaternaires. Parmi les travaux géologiques existant nous pouvons citer notamment ceux de :

L Glangeaud (1934), de A Aymée (1952,1954,1956) de Vesnine (1971) de A Saoudi(1982) et de O Betrouni (1983) etc.

2-LES CORDONS DUNAIRES

De couleurs ocres à rougeatres ces cordons d'épaisseur variant de quelques metres à des dizaines de metres sont visibles sur tout le littoral algerois depuis la ville de Tipaza à 80km à l'ouest jusqu'à l'embouchure de l'oued Sebaou à 70km à l'Est d'Alger.Ils occupent une grande partie du versant nord du Sahel de la baie d'Alger et de la zone cotière de la Mitidja.

Dans la baie d'Alger et à l'Est de celle ci le quaternaire marin est représenté par des dunes consolidées étagées du Pleistocène.Ces differents domaines correspondent à des terrasses marines consolidées qui débutent par un niveau de calcaire lumachellique à pétoncle qui se continue par une succession de terrasses qui sont soit des grès marins avec des galets soit des poudingues et grès grossiers .Le sommet de la série est constitué par des cordons dunnaires de sables fins récents qui se confondent parfois avec des sables de plage .

2.1 Géotechnique

Du point de de vue fondation ces dunes grésifiées représentent un bon sol d'assise pour tous les ouvrages d'infrastructure du batiment. Les terrains sont stables et portants.Les talus routiers tiennent à la verticale et ne necéssitent aucun soutènnement (RN 11 Ain benian-Bouharoun et RN 24 Bordj-el- kiffan). Du point de vue matériaux ces dunes présentent des niveaux grésifiés noduleux et des couches franchement sableuses.Ces conditions ne permettent pas l'exploitation sélective du sable fin .C'est plutot l'enssemble de la formation qui a été exploitée à differents endroits comme matériau de remblai routier (roccade sud Ben aknoun- Zeralda et autoroute de l'Est au niveau de Mohammadia et Bordj el kiffan). L'utilisation de ces dunes grésifiées comme granulats concassés pour béton pourrait s'avérer interressante.L'étude(Alloul,Aoudia,Hadibi1989) a montré que les résistances à l'écrasement d'eprouvettes confectionnées avec trois classes

granulaires 0/2,2/12,12 /20 à 7 jours sont Rc=163bars et à 28 jours 277bars .Les éprouvettes confectionnées avec du sable 0/2 donnent des Rc à 7jours de 150bars et à 28 jours de 204 bars. Ces resultats montrent que ces materiaux sont acceptables pour des bétons legers et mortiers de maçonnerie .

2-2 Impact sur l'environnement

Dans la région Algéroise les dunes grésifiées sont recouvertes soit par des forêts trés reposantes (domaine Bouchaoui à l'ouest et forêt de Beni mered à l'Est etc) soit trés urbanisées (complexes touristiques sur la côte ouest et le littoral Est algérois (Bordj el kiffan, Mohammadia) soit recouvertes par des terrains agricoles (terrasses de Cheraga, Ain benian et Zeralda). Il existe en vérité trés peu de sites exploitables où les impacts ne soient pas négatifs sur l'environnement. Les anciennes carrières ouvertes lors de la construction de la roccade Sud Ben aknoun -Zeralda represent des trous immenses et béants qui sont utilisés comme décharges sauvages. Il est donc bien entendu que cette étude de réutilisation des matériaux ne peut être prise en compte que dans le cas de terrassements routiers ou d'autres projets d'aménagement du territoire. L'ouverture d'autres carrières d'agrégats est inutile et ne peut apporter que nuisance à l'équilibre environnemental dans la région d'Alger.

3-LES DUNES ACTUELLES

Les dunes récentes ou actuelles sont présentes sur une grande partie du littoral Algérois .Elles sont particulièrement cartographiées dans les endroits suivants :

- dunes de sables jaunes du littoral ouest à Staouelli ,

-dunes de sables fin gris Sur la rive droite de l'embouchure de l'oued El Harrach au lieu dit "Lido".

-les formations dunaires éoliennes du Soltano- Tensiftien du littoral Boumerdès -Cap Djinet qui representent une large bande de sables fins gris (2km de large sur 15km de long .

-Des sables d'arènes granitiques sur le flanc nord du massif de Thenia (oued Boumérdès). Les quantités sont certes importantes mais non illimitées.Elles sont parfois urbanisées mais souvent nues ou recouvertes de forêts (Pins maritimes et Zemmouri)

Ces sables sont des materiaux qui présentent de bonnes caracteristiques géotechniques (sables fins propres) pour une utilisation en béton hydraulique, sauf pour les sables d'arènes qui nécessitent un lavage ou bien une utilisation comme couche de forme dans les chaussées. Les résistances obtenues sur un mortier de granulometrie 0/2 à 7jours sont de 180 bars et à 28 jours de 224 bars .

2.1Impact sur l'environnement

L'intense exploitation rarement autorisée par état dans certains secteurs induit un impact irreversible sur le rivage marin:

-une exploitation de sable de plage a réduit considérablement dans certains secteurs la zone de baignade (Sidi ferruch palm-beach, Club des pins , Zemouri etc.)

- l'exploitation des arènes granitiques de Boumerdès a contribué à la déforestation de certaines zones ,

-l'exploitation du cordon dunaire au niveau du lieu dit "Stamboul" a contribué à la formation d'une zone marécageuse très importante.

3 LES ALLUVIONS D'OUED

Les différents oueds qui ont charrié durant leurs histoires géologiques d'importantes quantités de sables et graviers prennent leurs sources soit dans les montagnes de l'Atlas soit dans la haute Kabylie. Ils traversent des zones de plaine et se jettent à la mer. D'Ouest en Est les grands oueds sont :

- L'oued Chiffa très exploité dans les gorges de la Chiffa,

-l'oued el Hachem à l'ouest de Tipaza fermé par le barrage de Boukourdane est intensément exploité à son embouchure, -

-L'oued El Harrach qui est intensément exploité à hauteur de Hammam elouane et Baraki par une dizaine de sabliéres et graviéres qui ont presque épuisé le lit mineur de l'oued .et qui s'attaquent au réservoir de graviers recouvert par les terres agricoles .

- L'oued Hamiz exploité au niveau du village de khemis El khechna lors de la reconstruction des pistes de l'aérodrome de Dar el beida . Le gisement est pratiquement épuisé. .

-L'oued Isser qui est exploité en amont des gorges de Lakhdaria mais qui présente des terrasses alluviales emboîtées fossiles situées sur toute la zone basse de l'embouchure.

-L'oued Sebaou qui est rejoint par l'oued Aissi au niveau de Tizi ouzou représente le réservoir en sables et graviers le plus important sur le parcours Tadmait -Tagdempt . Une dizaine de graviéres et sabliéres en exploitation a presque épuisé les terrasses actuelles du lit majeur . De rares gisements ont été préservés .

Du point de vue géotechnique ces alluvions sont des matériaux à dominante calcaro-argilitiques qui présentent souvent des caractéristiques suffisantes pour une utilisation dans les bétons hydrauliques et le corps de chaussée.Ils sont donc trés convoités.

3.1 Impact sur l'environnement

Nous citons les impacts les plus importants:

-L'exploitation intense des terrasses moyennes de l'oued El Harrach a considérablement contribué à la réduction de l'aquifère de la plaine de la Mitidja. Le rejet des boues de lavage des graviéres a entraîné un colmatage de la zone d'alimentation.

-l'exploitation des terrasses de l'oued Sebaou a contribué à l'abaissement du niveau piézométrique de la nappe alluviale qui est le réservoir principal de la ville de Tizi ouzou et de la plupart des villages de la haute Kabylie. Plusieurs exploitations ont été intérdites par les autorités régionales.

-formation d'une zone de marais sur l'embouchure de l'oued el hachem et pollution de la nappe alluviale côtière par les eaux de mer.

4 CONCLUSION

Les matériaux sablo-graveleux commencent à manquer sérieusement dans la région d'Alger. Beaucoup d'exploitations sont emmenées à être fermées. En raison du manque de matériaux et de l'impact négatif sur l'environnement il est urgent de penser à des solutions de ubstitution dont on peut annoncer quelques unes.

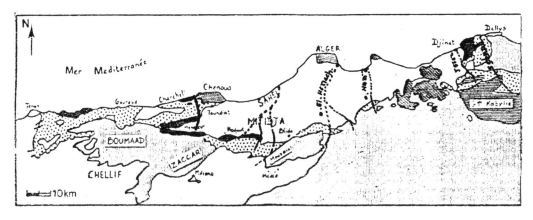

Figure 1:Carte des grands enssembles géologiques de l'Algérois.

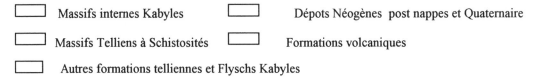

☐ Massifs internes Kabyles

☐ Dépots Néogènes post nappes et Quaternaire

☐ Massifs Telliens à Schistosités

☐ Formations volcaniques

☐ Autres formations telliennes et Flyschs Kabyles

1- La valorisation de granulats de roches massives , Les potentialités existent .

2- Dans la région de Zemmouri des exploitations de sables éoliens peuvent avoir lieu sans risque d'impact négatif sur l'environnement si les sites sont bien choisis et bien remis en état après exploitation.

3- recherche de nouvelles techniques de construction (type préfabriqué) ou bien briques en terre stabilisée qui ne nécessitent pas une grande quantité en sables et graviers .

4 - Les sables de dunes éoliennes sont très abondants dans les régions des hauts plateaux qui ne sont éloignés de l'Algérois que d'environ 200 km. Un transport par le rail serait plus judicieux dans ce cas.

5- lors des études routières qui nécessitent une grande quantité de matériaux d'emprunt pour corps de chaussées et remblais il faudrait éviter les alluvions sablo-graveleuses propre qui devraient être réservées à une utilisation en béton de ciment.

6- La revalorisation des produits rouges (briques ,tuiles ,hourdis,) est nécessaire en raison de L'intense utilisation de sables et graviers dans la fabrication d'agglomérés, particulièrement de parpaing (des centaines d'unités de fabrication) .Pour cela les marnes du Plaisancien très abondantes dans la région

seraient le matériau de base .

7-Les lois relatives à la protection de l'environnement devraient être appliquées avec une grande rigueur.

BIBLIOGRAPHIE

Alloul, Aoudia, Hadibi, Boukrara, Lazouni, Amir & Lethiet. 1992 et 1993 . Mémoires d'ingénieurs en géologie de l'ingénieur IST /USTHB Alger .

Ayme ,A. 1948 . Contribution à l'étude des terrasses entre Cap Matifou et l'oued Isser B.S.H.N.Af du Nord T 39 P 97.

Ayme ,A. 1952. Le Quaternaire littoral des environs d'Alger .Actes du IIème congrès panafricain de Préhistoire.

Ayme, A. 1956. Terrasses littorales entre Tipaza et le Sébaou.Histoire géologique de la province d'Alger.

Betrouni, M. 1983. Pleistocène supérieur du littoralOuest Algérois. Thèse de doctorat Université d'Aix -marseille II , faculté des sciences de Luminy.

Glangeaud, L. 1934 . Etude géologique de la région littorale de la province d'Alger. Bulletin de la carte géologique de l'Algérie.

Vesnine, V. 1971. Carte géologique de Thénia au 1/50 000ème accompagnée du rapport inédit (EREM).

The influence of geomorphological and hydrogeological structures upon the underground water pollution by the surface domestic waste deposits

L'influence des structures géomorphologiques et hydrogéologiques sur la pollution des eaux souterraines par les déchets ménagers

Aurel I. Harsulescu & Florian I. Pojar
Aquaproiect S.A., Bucuresti, Romania

ABSTRACT: One of the most noxious polluting source of the underground waters consists of the surface deposits of domestic waste.

The paper presents different criteria of waste deposits' classification.

We describe further, the influence of different geomorphological and hydrogeological structures upon the possibilities of polluting the underground waters by the deposits of urban garbage placed on the land surface.

Finally, we suggest some managing sollutions for surface domestic wastes deposits, underlying the importance of permanent monitoring of water quality, especially during and after the placing of noxious materials.

RESUME: L'une des plus nocives sources de pollution des eaux souterraines est réprésentée par les depôts ménagers de surface.

L'ouvrage presente des différents critères de classification des depôts de déchets.

En continuant on déscrit l'influence de differentes structures géomorphologiques et hydrogeologiques sur la possibilité de pollution des eaux freatiques par les depôts des ordures menagères urbaines situées a la surface du terrain.

A la fin, on fait des reccomandations sur l'aménegement de differents depôts des déchets ménagers de surface en insistant sur l'importance d'une vérrification permanente de la qualite des eaux souterraines spécialment pendant et après le déposement des matériaux nocifs.

1 INTRODUCTION

The environmental protection becomes more and more a priority, especially in the neighbouring of large cities, which -in societies with strong industry- seriously affect the ecological balance. The pollution is not a problem for only one country, that is why the necessity of underground water quality protection -one of the most important environmental factor- become a very important problem with a global approach.

In Romania, as well as in other Eastern countries, the problem of reducing pollution is more important due to the presence of the most polluting technologies in these regions. Moreover Romania is situated on the inferior sector of the Danube which is the second in length river in Europe, crossing eight countries and four capital cities, collecting an important part of the wastes on the continent. Fortunately Romania still has beautiful areas unaffected by the human activities.

2 WASTE DEPOSITS

The waste deposits are a very important source of environmental depreciation with noxious effects upon air, land and also with negative psychic, intelectual and moral effects upon people. The deposits of wastes are many time placed on the unfitted zones, such us pits, swamps, abandoned quarries, etc. which provide a reduced natural environmental protection.

Such preocupations concerning the organized collecting and deposits of garbage were recorded beginning with ancient times. The oldest known systems are those of collecting and removing of garbage from the cities of Mohenyo-Nard and d'Harappo on Indus valley 4000 years ago, as well as the systems of antic Egypt, antic China, Roman and Greek fortresse, etc

2.1 The classification of wastes deposits

Generally, such classifications are necessary for a better understanding of deposits effects and for establishing the best collecting, transport and laying methods. Out of the numerous criteria we consider as the most important ones, the followings:

1. The geomorphological structure on which the deposit is placed: plane, accidental, flood, sub-aquatic, impermeable lands, etc.

2. The way of deposits' achieving: organized and controlled, unorganized;

3. The origin of wastes: urban, industrial, mixed;

4. The way of transport: dry, hydraulic;

5. The composition of the deposited materials: soluble, organic, bio-degradable, radioactive, poisonous, etc

6. The degree of danger for: people animals, vegetation, etc;

Due to the specific situation, every placement can be considered according to distance to localities, the influence upon the landscape, available acces ways, etc.

2.2 Data about composition of the domestic' wastes deposits

A very characteristic of domestic wastes deposits is the non-homogeneous composition, which is different for every other country or region. Generally, the polluting substances coming from wastes may be: ideal miscible substances, uninfluencing the flowing features (nitrates, phosphates), unideal moving substances which affect the liquid density and viscosity (sea and industrial salts), organic or anorganic substances, with high or low degree of noxiousness, with long or short term effects, etc. Some of them are light and float, the heavier ones sink. For instance, 3 - 8 p/m of hydrocarbons is felt by human taste. As an illustration in Table 1 we present in a comparative manner the composition of some domestic wastes deposits in Europe and U.S.A.

After a concise analysis we notice that the noxiousness effect of domestic waste heaps upon the environment is felt through the following facts:

1. Not using within the social-economical circuit of large land surfaces close the big cities;

2. The presence in the air of fetid emanations and of dust spread in the atmosphere containing toxic and irritating substances;

3. The degradation of soil and subsoil, including underground water through the spreading and transport of polluting agents;

4. The proliferation of microbes, insects and rodents which may contribute to the transmission of different diseases either directly or through air, water, food, etc.

3 THE INTERACTION BETWEEN GEOMORPHOLOGICAL STRUCTURES AND URBAN WASTES DEPOSITS

Within the pollution of soil and subsoil made by the waste deposits, a

Table 1. The composition of domestic wastes deposits.

TYPE OF ELEMENTS (Percentage of weight)	COUNTRY			
	Germany	England	USA (California)	Romania (Bucharest)
Objects of leather or rubber.	1.3	1.8	1.6	2.3
Paper cardboards.	4.5	12.6	−	5.2
Organic scraps from kitchen.	18.7	13	68.4	−
Vegetable and animal scraps.	10	2.2	−	70.8
Metal wastes.	2.6	3.7	10.7	2.3
Glass and broken glass.	3	2.8	11.7	3.2
Ash, pottery, scraps, others.	60	63.9	7.6	16.2

very important role is held by the type of geomorphological and hydrogeological structures on which the deposits are placed.

3.1 Geomorphological considerations

In addition to the climate conditions, the presence of quaternary deposits and vegetation, and to the human activities, etc, the water flows -which facilitate very much the spreading of polluted substances are directly influenced by the geomorphological structures. Due to the great variety of these structures (for instance in Romania there are known 16 big geomorphological units) and for an easier understanding, the structures can be devided according to the stages principle into the three big groups:

1. The mountain stage of very high altitude which comprises a large variety of wavy and faulted rocks intercepted on the surface or at a little depth. In this region there are narrow valleys, abrupt slopes with high relief energy and transport speed interrupted in some places by flattened depressions even horizontal. As a rule the waters flows on the slope to the valleys, on the open joints near the surface or on the discontinuity planes. There are some special circulation cases of underground waters following the direction of permeable layers and faults dip, through carstic caverns, etc.

2. The hills and middle altitude plateaus stage can be often found toghether with mountain chain. This stage is not so abrupt and it has folded and monocline structures. The lithological composition of geological layers is very varied beginning with crystaline rocks and ending with types of soils covered with thicker quaternary deposits. The water circulation in this stage is higher than the first stage, being closely related with the presence of synclines and anticlines, the direction of joint systems and the discontinuity planes, the existence of permeable and soluble rocks, etc.

3. The third stage consisting of fields of low altitude is found after second stage; its general aspect is plane with small slopes. The lithological constitution is sedimental made up by uncohesive (permeable) or cohesive soils. The water circulation is facilitated by the porous soils and the flowing direction is, as a rule, towards the rivers' beds, these beds being numerous within this stage and having the role of natural drain.

In Fig. 1 you can see a hypothetical model of a slope with sectors of common characteristics, especially concerning the infiltrations and surface flowing of precipitation waters.

Fig. 1 Hypothetical slope model.

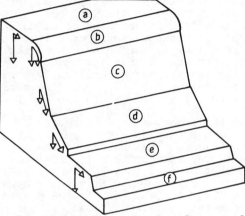

a -vertical movement of underground water, b -infiltrations and lateral flowing, c -superficial flowing, d -lateral flowing, infiltration, e -lateral flowing, infiltration, f -river bed

An important fact to be underlined is that the three stages above mentioned influence the water transport of polluters and depends on the thickness and permeability of surface quaternary deposits which affect especially precipitation water circulation in the soil.

3.2 Hydrogeological considerations

Within the distribution and configuration of surface and underground waters, very important were the paleological and geographical evolution of the earth crust, tectonic movements, the glacial and interglacial phases which contributed significantly to the existence and developing of the actual hydrogeological structures. In modern times the human activities like carring out the large water reservoirs, cannals for sailing, the exploitation of aquifer, oil and gas resources, etc. were added to these influences.

Due to the underground developing of the ground waters, it is difficult to determine their characteristics in a certain way. Nevertheless, there are a lot of estimation methods, sometime aproximate, of ground water hydrodynamics, although for this purpose there were necessary assimilations with flowings through porous homogeneous milieus (the case of macrogranular soils), through pipes (the case of fissured rocks). As a rule, the underground water movement((important factor of polluters spreading) is more slower than its movement at the land surface.

Generally, the underground waters are divided into deep ground waters ground waters of little depth, although there are some cases when these two types are interconnected.

1. Deep ground waters rarely arrives close to the surface zones so, they may spread the polluters only when they have an ascenting or artesian character. The frequently using in natural state of these waters in spas for medical purposes without being treated which could change their curative quallities, increases the possibilities of pathogen agents transmittion among people.

2. The ground waters of little depth are placed many times at a couple of meters depth under the soil level, frecuently being used as drinking water for people and animals. So, the protection of these type of resources against the pollution is of a high importance especially because in many areas of the world including Romania, their using is made in individual or agricultural farms without adequate equipments for treatment and purification. Due to the common characteristics on large zones and according to the hydrogeological features, the flowings of these waters follows preference directions imposed by hydraulic gradients and morphological aspect of the respective area.

3.3 The polluting way for underground waters

A well-known fact is that of existing forms of underground waters, namely motionless (chemically bond, hygroscopic, film water) and mobile (cappilary water and water from pores). For polluter spreading mobile water is very important. When we evaluated the spreading speed we had to take into account the effective porosity of watered layers which influence -depending on the total porosity- the piezometric height, the hydraulic slope, the flowing speed. As an illustration you can see in Table 2, some values of total and effective porosity for different types of soils.

Table 2. Some values of total and effective porosity.

	n_t (%)	n_e (%)
Clays.	40 — 70	1 — 10
Silts	35 — 45	15 — 25
Sands.	15 — 35	10 — 30
Gravels.	15 — 25	15 — 30
Volcanic tufas	30 — 40	5 — 15
Sandstones.	3 — 38	3 — 35
Limestones and dolomites.	0.5 — 12	0.1 — 10 *
Mica shists.	0.5 — 7.5	0.1 — 1 *
Quartzites.	0.8 — 1	0 — 2 *
Granites.	0.01 — 2	0.1 — 2 *

*Depending on the fissuration degree

In addition when we make analyses, we have to take into consideration the heterogeneousity of aquifers, their type (with free level or under pressure), the flowing type (laminar or turbulent), etc.

Fig. 2 Aquifer with free level, the capillary zone and saturation degree.

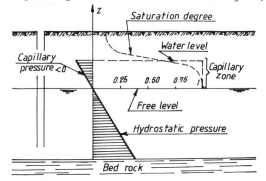

In Fig. 2 we present for exemple the saturation degree of a permeable layer which contains an aquifer with free level.

The pollution intensity is directly influenced by the permeability of geological layers in the foundation of the waste deposits, by the land morphology, by the depth of underground water, by the polluting agents characteristics, etc.

In Fig. 3 you can see the way of polluting an aquifer with free level by a surface wastes deposit.

Fig. 3 Pollution made by a wastes deposit.

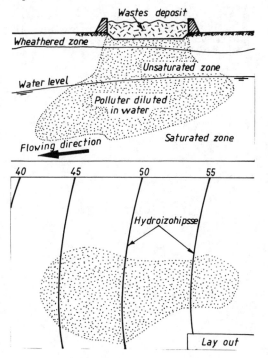

4 CONCLUSIONS

The vital importance of underground water resources imposes a more concise knowing of the geographical, geomorphological and hydrogeological conditions of the wastes deposit sites.

The complexity of polluting factors implies a multi-disciplinary studies made with the help of mathematics modellings which allow a very precise determination of desturbings produced by the polluting deposits

and by the protection solutions as well as their efficiency.

A special attention has to be paid to the understanding of geomorphological and hydrogeological conditions of the proposed sites which having unevident charcteristics, induces upon the analyses some uncertitude elements.

As a rule, before carring out a domestic wastes deposit, it is necessary to make some special studies which have to include also the assesment of works impact over the environment as well as the best solution for underground water protection.

Generally, when carring out urban wastes deposits, the following general principles have to be respected:

1. The depositing surface to be as horizontal as possible, and the foundation soils -preferable impermeable ones with adequate thickness- to have geotechnical characteristics comparable with the deposit load;

2. The slopes have to be stable, without landslides;

3. The geomorphological and hydrogeological condition of site have to be established as well as the original chemical conditions of ground waters;

4. It is compulsory to mention the types of wastes accepted in the deposit, the manipulation, transport and laying conditions, preventing their involving by atmospherical draughts;

5. The very noxious polluters will be recycled or treated at the source;

6. The heating and fertile features of wastes will be used through stamping or aerobe fermentation;

7. Depending of the real situation in the site and the type of deposited wastes, special precautions will be adopted for protecting the ground waters quality by making the impermeable foundation, the deep screening of the deposit site, the controlled draining of ground and precipitation waters, etc. For instance in Fig. 4 you can see a few scheme of the underground water protection measure against surface domestic wastes deposit pollutings;

8. All the technological operations concerning the collecting, transport and laying of the noxious materials will be made by specialized units;

9. The spaces in and around the deposits will be fixed-out in order

to harmonize with the natural landscape;

10. A permanent control of the wastes' influence upon the environment
has to exist before, especially upon
the underground water resources, during and after the using the deposit.

Fig. 4 A few scheme of ground protection measure.

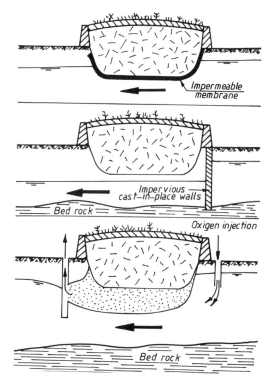

The adequate assurance of the gro-
und waters quality is very important
due to the high degree of noxious
wastes, the lasting effects with lower elimination possibilities and
the small distance between polluting
sources and consumers.

The point-like character of urban
wastes polluters facilitate the organizing of control and measurement
systems able to established at any
time, the quality of aquifers. These
systems, consisting of piezometric
drillings for instance, have to be
placed depending on the existing hydrogeological structure, especially
on the flowing direction of underground waters, in downstream and laterally to the polluting source.

Through periodical determination the
changings will be seen as well as the
sequences of adopted measures.

REFERENCES

Bancila, I. Geologia Inginereasca
    1980. Bucuresti: Ed. Tehnica.
Dessargues, A. Modelisation des re-
    servoir souterrais 1993. Liege
Marchidanu, E. Practica Geologica
    Inginereasca in constructii 1987.
    Bucuresti: Ed. Tehnica
Oncescu, N. Geologia Romaniei 1963.
    Bucuresti: Ed. Tehnica.
Pietraru, J. Halde pentru Depozita-
    rea Deseurilor 1982. Bucuresti: I.
    C.H.
Pojar, Fl. & Harsulescu, A. Masuri
    de Protectie a Surselor de Apa im-
    potriva Poluarii 1993. Bucuresti:
    Aquaproiect

Role of toe erosion in bluff recession along a section of Lake Erie's south shore near the Ohio-Pennsylvania border, USA

Le rôle de l'érosion du bas de l'escarpement dans la récession des falaises de la rive sud du Lac Erie, près de la frontière Ohio-Pennsylvania

Abdul Shakoor
Department of Geology and the Water Resources Research Institute, Kent State University, Ohio, USA

Shahalam M. Amin
Department of Geography, Kent State University, Ohio, USA

ABSTRACT: The role of toe erosion in causing bluff recession was investigated for four sites along a 3.5 km section of Lake Erie's south shore. The results showed that toe erosion occurred at a rate of 7-10 cm per month. The bluff response to toe erosion was found to be a function of bluff morphology and shape of the erosion zone within the toe area, the response being quicker where the notches had developed than where erosion had resulted in the formation of concave zones or negative slopes at the toe. The bluff recession was accompanied by the occurrence of numerous slumps, slab slides, earth falls, earth flows, and mudflows.

RÉSUMÉ: L'investigation à quatre localités à une section de 3.5 km de longueur de coté du sud du Lac Erie du rôle de l'érosion du bas de l'escarpement qui causé la récession des falaises a démontée que l'érosion se trouve à 7-10 cm par mois. La réponse des falaises à l'érosion du bas des escarpements est une fonction de la morphologie du falaise et la forme de la zone de la bas de l'escarpement. La réponse est plus vit où il y a des encoches, et plus longuement où l'érosion a causée la formation des zones concaves où des pentes négatives. La récession s'est accompagnée par beaucoup de dégringolades, de glissements des plaques, d'éboulis rocheux, d'écoulement de terre, et d'écoulement de boue.

INTRODUCTION

About 40% of the shoreline of the lower Great Lakes in the U.S.A. is formed in relatively weak Quarternary sediments of glacial, glacio-fluvial, and glacio-lacustrine origin. Bluff erosion in these areas has resulted in frequent loss of agricultural land, roads, and buildings. It has been postulated in many studies that erosion at the toe of the bluff plays a critical role in the overall recession of the bluffs (Quigley and Gelinas, 1976; Quigley et al., 1977; Davidson-Arnott and Askin, 1980; Emery and Kuhn, 1982; Buckler and Winters, 1983). However, our present understanding of the erosion mechanisms at the toe of the bluffs is quite limited. Except for the largely theoretical work of Sunamura (1977,1982), and field studies of Robinson (1977), McGreal (1979), and Carter and Guy (1988), there has been relatively little work on the mechanisms of toe erosion and the response of the bluff to toe erosion. Even fewer field studies have been done on the toe erosion processes operating along the costal bluffs of the Great Lakes. Consequently, most attempts at shore protection have been unsuccessful (Davidson-Arnott and Keizer, 1982). Thus, the objective of this study was to investigate the magnitudes and processes of bluff toe erosion, and to examine bluff recession in response to toe erosion.

Bluff toe erosion is a complex phenomenon and is dependent on a multitude of factors from different subsystems of the lakeshore environment. It is obvious, however, that the conditions favoring waves/swash to act upon the toe of the bluff constitute the primary prerequisite for any toe erosion to occur. The rate and nature of erosion, however, depend not only on the magnitude and frequency of wave/swash impact but also on the nature of toe material (strength, presence or absence of fractures, degree of weathering) and the type of wave erosion process (wave quarrying vs. swash abrasion).

METHODOLOGY

Selection of the Study Area

A 3.5 km long section of Lake Erie's south shore, near the Ohio-Pennsylvania border (Figure 1), was chosen for this study. The study area is located in the Springfield Township, Erie County, Pennsylvania.

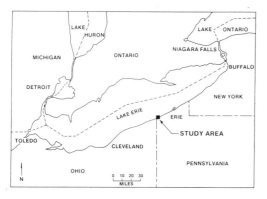

Figure 1: Location of the study area.

This segment of the shoreline is unique because no coastal protection measures have been undertaken here and it is off limits to the public. These attributes were important because this study required an unprotected natural shoreline where a number of measurement stations could be established and monitored regularly over a period of time without human interference. Accessibility of the shoreline was another important consideration in the selection of the study area. Within the study area, four specific sites were chosen for monitoring purposes (Figure 2).

Considerable attention was given to ensuring that the four sites were representative of the spatial and local variations of the bluffs within the study area with respect to bluff height and associated beach characteristics. Bluffs of different heights and beach widths were chosen to see whether response to toe erosion differed among bluffs of varying heights and beach characteristics.

Monitoring of Erosion Rates

In order to monitor the rate of toe erosion and changes in beach morphology, three peglines were established at each of the four sites. Peglines have been used successfully to measure erosion rates in several other studies (Bridges and Harding, 1971; Brunsden, 1974; McGreal, 1979). The distance between peglines at a given site varied from a meter to several meters. Each pegline consisted of 7 vertically spaced steel rods, 1.25 cm in diameter and 1 m in length, driven horizontally into the till material at 0.25 m intervals from the toe upwards (Figure 3), with the highest one being 1.75 m above the toe. Pegs were also installed across the beach, at 1.0-1.5 m intervals, to monitor changes in beach morphology. The exposed portions of the pegs were measured to 0.1 cm on a weekly basis from April to December,

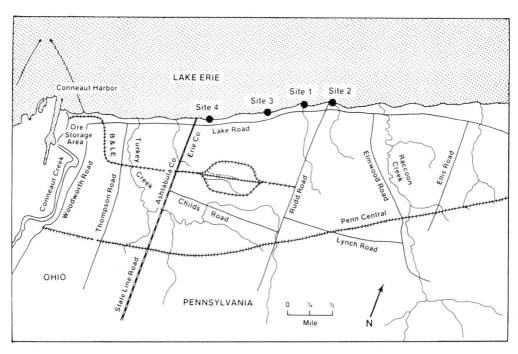

Figure 2: Location of the four study sites in West Springfield, Pennsylvania.

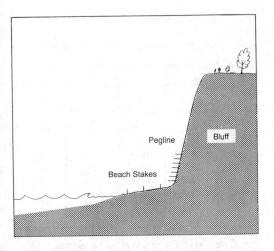

Figure 3: Schematic illustration of a pegline and beach stakes. Three such peglines were installed at each of the four study sites.

1986. The difference in subsequent measurements provided a measure of erosion or deposition for that particular period.

Continuous photographic records of various beach and bluff characteristics, and of the micro-erosion features at the toe, were maintained for all four sites throughout the study period. Various types of mass movement (slumps, falls, flows) affecting the bluff above the toe areas were also photographed. Recession of the crests of the bluffs at the four sites was also monitored periodically.

Laboratory Testing

Representative samples of the till material from the toe area were obtained from all four sites for determination of natural water content, Atterberg limits (liquid limit and plastic limit), and compressive strength. The natural water content, determined in accordance with the American Society for Testing and Materials (ASTM) procedure D2216, was monitored on a continual basis for the upper and lower portions of the pegged toe. The Atterberg limits were determined using ASTM procedure D4318 and were used to classify the till material. Because of the exceptional stiffness of the overconsolidated till material, a Schmidt hammer was used to measure the compressive strength according to the procedure described by Deere and Miller (1966). Compressive strength is considered to be a measure of resistance against wave erosion (Sunamura, 1983).

STRATIGRAPHY AND GOETECHNICAL PROPERTIES OF THE BLUFF TOE MATERIAL

The bluffs at the study sites are typically 5 to 15 m high and essentially vertical. The beach is relatively flat and has a variable width of 3 to 9 m. The general stratigraphy along Lake Erie's south shore is rather uniform and consists of four major lithologic units. These, from top to bottom, include: a) poorly-cemented lacustrine sand representing strand deposits; b) massive, lacustrine clay, silt, and sand; c) glacial till; and d) shale bedrock. Within the study area, however, the bluffs are primarily made of till material overlain by a thin deposit of lacustrine silt. Two distinct units of till material can be recognized in the study area, an upper unit of stiff to very stiff, yellow brown to gray, clayey silt to silty clay with trace amounts of sand and gravel, and a lower unit of very stiff to hard, highly fissured, gray silty clay with occasional cobbles and small boulders. The shale bedrock (Chagrin Shale) is not exposed in the study area.

The liquid limit values for the bluff toe material were found to range from 21.2 (site 1) to 25.5 (site 4), the plastic limit from 11.6 (site 1) to 18.0 (site 4), and the plasticity index from 7.5 (site 4) to 9.6 (site 1). Based on these properties, the till material was classified as a silty clay of low plasticity (CL). The liquidity index values for all four sites were determined to be less than zero, confirming the highly stiff and brittle nature of the till material. The average compressive strength of the till material was found to be 16 MPa (2320 psi). The shear strength, under undrained conditions, was assumed to be one half of the compressive strength, i.e., 8 MPa (1160 psi). A comparison of Atterberg limits and strength values for the four sites revealed that the till material was more or less uniform in nature from one site to the other.

BLUFF TOE EROSION: PROCESSES AND RATES

Processes of Erosion

Bluff toe erosion in the study area can be attributed to two distinct processes: a) direct wave attack and b) swash run-up. Direct wave attack occurs when the waves break on the bluff face as a result of high lake level, narrow beach, and wind set up (storm surge), or a combination of these. In direct wave attack, the erosion is due to hydraulic action of compression, tension, and cavitation; the abrasive action of material carried by the waves; and the wedging action of compressed air within the fissures present in the bluff material (Robinson, 1977; Sunamura, 1977, 1983). The erosional features associated with direct wave attack include concentrated erosion along joints, formation of potholes (selective erosion of weak material, plucking of boulders), and development of caves.

If the nearshore lake level is not high or the beach is wide, the waves break on the beach at variable distances from the bluff face. The resulting swash runup either reaches the toe of the bluff or is dissipated on the beach, depending on the size of the breaking waves and the width, height, and nature of the beach. In case of swash runup, erosion is due to a combination of hydraulic, wedge, and abrasion actions. The beach material in these processes serves as an abrasive, doubling the erosive power of the running water (Sunamura, 1982). In the study area, swash runup appears to be more dominant than direct wave attack in causing bluff toe erosion. Swash abrasion tends to erase small-scale erosional features (potholes, caves, etc.), resulting in smooth, concave erosional form of the toe.

Rates of Erosion

The average magnitudes of erosion at various peg levels for the four sites are shown in Table 1. At site 1, where the bluff is 8 m high, a maximum of 75.3 cm of erosion was recorded between April 19 and December 20, 1986, giving an average rate of 9.4 cm per month. The maximum erosion at site 1 occurred at peg 2, 0.5 m above the beach-bluff junction. Table 1 also shows that the absolute amount of erosion decreases with the height above the beach. This is not unexpected as both the direct and the swash wave impacts on the toe diminish with height, both in terms of magnitude and frequency.

At site 2, where the bluff is 5 m high, the erosion could be monitored only between July 15 and November 23, 1986. Frequent occurrences of slope movement and inhospitable shore conditions prevented data collection at other times. A maximum of 31 cm of erosion was recorded at the lowermost peg (Table 1), indicating a monthly erosion rate of 7.7 cm. The beach at this site was very narrow (about 1.5 m wide), allowing even small waves to reach the toe and erode. The amount of erosion for the next three pegs (2, 3, and 4) is more or less the same (Table 1), which indicates direct wave splash. The direct wave action resulted in the formation of notches.

The bluff at site 3 is about 17 m high, the highest among the four sites, and had a relatively wide (4.5-5 m) beach at the time of study. The maximum amount

Table 1: Average amounts of toe erosion at the four sites.

Peg No.*	Average Amount of Erosion **(cm)			
	Site 1	Site 2	Site 3	Site 4
7	29.9	6.5	14.6	3.9
6	39.4	11.8	24.2	7.9
5	38.6	13.6	27.7	16.2
4	51.8	19.6	35.1	23.8
3	68.7	17.4	37.6	24.8
2	75.3	18.1	47.8	26.8
1	55.5	31.0	49.0	25.6

* Pegs were vertically spaced at 0.25m interval, with peg 1 being the lowermost.

** The amounts listed are averages of 3 measurements for each height, representing 3 peg lines for each site.

of erosion for the five month period was observed to be 49 cm at peg 1 (Table 1), giving a monthly rate of 9.8 cm. This is the highest rate of erosion among the four sites. This can be attributed to the relatively wider beach at this site which allows the waves to break, roll, and pick up erosive material before reaching the bluff with considerable speed and energy. It should be noted that although wide beaches help to dissipate wave energy, the beach at site 3 is not wide enough to dampen moderate to big storm waves.

Site 4 with a 14 m high bluff exhibited a maximum erosion of 26.8 cm from August 5 to November 29 (Table 1), resulting in the lowest monthly rate of 6.7 cm among the four study sites. The erosion pattern at this site was quite variable from peg to peg, probably due to extensive fissuring of the bluff material.

Figure 4 provides a graphical comparison of erosion magnitudes at the four sites. It is clear from the figure that the maximum amount of erosion for three of the four sites occurred at 0.50-0.75 m height, rather than at the base. The figure also shows a continual decrease in erosion with increasing height. No erosion due to wave action was observed to occur beyond 2 m height above the beach level.

BLUFF RESPONSE TO TOE EROSION

The bluff response to toe erosion varied with bluff morphology, joint pattern, and shape of the erosion zone within the toe area, the response being quicker

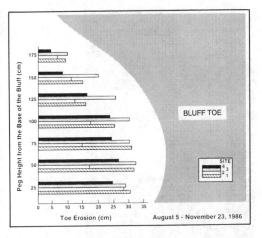

Figure 4: Comparison of the magnitude of toe erosion at the four sites.

time. The relatively narrow (2 m wide) beach at this site allowed swash and waves to reach the toe more frequently than the other sites, thereby promoting rapid development of notches. The notch development was followed by earth falls which frequently affected the entire bluff up to the crest line. Because of the presence of a prominent joint set parallel to the bluff face (Figure 8), planar slides, consisting of 2-3 m wide and 0.3 m thick slabs, were a prominent mode of failure at this site (Figures 9 and 10).

The bluffs at sites 3 and 4 were relatively high, about 17 m and 14 m, respectively, and were of sigmoidal shape, with the upper part being slightly concave and the lower part markedly convex. The beaches at these two sites were of moderate width (3-5 m). The cumulative erosion form at the toe could

where notches had developed (site 2) than where erosion had led to the formation of a concave zone (site 1) or a negative slope (site 3 and 4). At low bluffs (sites 1 and 2), nearly parallel retreat of the bluff took place, whereas at high bluffs (site 3 and 4), steepening of the bluff occurred with no significant recession of the bluff crest over the study period. The manifestation of bluff response to toe erosion was in the form of slope movements including slumps (rotational slides), slab failures (translational slides or plane failures), earth falls, earth flows, and mudflows. As stated previously, the frequency, distribution, extent, and nature (curved vs. planar) of soil joints played a major role in determining the mode of failure. Earth falls and slab slides were the predominant modes of failure immediately above the toe area, where the till was more jointed, whearas earth flows and localized slumps occurred more frequently near the crest area.

At site 1, where the beach was relatively wide and thick, a concave erosional zone, about 1.5 m high and 1 m deep, had formed at the toe, producing overhangs of the bluff above (Figure 5). The bluff did not respond immediately to toe erosion and the overhangs remained intact for some time. However, the overhangs started to yield in the form of earth falls between late fall and early spring, and by May, 1987, an almost vertical slope had formed. Although earth falls was the dominant mode of response at this site, a slump/ earth flow type of failure did occur in the upper portion of the bluff (Figure 6).

The development of notches was found to be a characteristic feature of toe erosion at site 2 (Figure 7). The notches were confined to the lowest 0.5 m of the toe and attained a depth of 35 cm in 4 to 6 weeks

Figure 5: Formation of a concave erosional zone due to wave abrasion, and the resulting overhang at site 1.

Figure 6: View of site 1 after the collapse of the overhangs and washing away of the failed material (earth falls). Notice the near vertical bluff face after failure on the right and a slump/earth flow type failure near the middle of the picture.

3017

Figure 7: Notch development at the base of the bluff at site 2.

Figure 9: Occurrence of plane failures (slab slides) promoted by the joint set that parallels the bluff face at site 2.

Figure 8: Three mutually perpendicular joint sets in glacial till at site 2. The set that parallels the bluff face was found to be the most well developed and most planar in nature.

Figure 10: Another example of a planar failure at site 2.

be best described as a negative slope with an arch-shaped zone, about 2 m high and 0.75 m deep, at the base (Figure 11). This caused a pronounced bulging of the lower till. The response of the lower part of the bluff to toe erosion was the failure of the bulge in the form of earth falls (Figure 12). The earth falls were quickly washed away and the whole process,

which constituted a yearly cycle, started all over again. Like site 1, the bluff responses at sites 3 and 4 were not immediate, rather a lag time of a few months was observed between the formation of the erosion zone at the toe and the failure of the convex bulge. Most of the earth falls occurred immediately after the spring thaw.

Figure 11: Toe erosion at site 3 resulting in the formation of a negative slope at the base of the bluff and the associated bulging of the bluff face above the toe area.

Figure 12: Failure of the bulge at site 3 by way of earth falls.

The upper parts of the bluffs at sites 3 and 4 remained relatively unaffected. Because the bluffs at these sites were relatively high, subaerial processes on the bluff faces, dominated by rain-wash and piping, caused the bluff crests to retreat independently, which ultimately produced the sigmoidal shapes of the bluff profiles. However, the failures of the bulges did cause an overalll steepening of the bluff faces.

Figures 13-16 show typical examples of how mass movement is causing recession of the bluff crest in the study area. In Figure 13, a slump has taken out a significant portion of the crest line, Figure 14 shows

Figure 13: Bluff recession caused by slumping and subsequent erosion.

Figure 14: Translational movement affecting the upper portion of the bluff including the crest line.

the effect of a translational slide on the bluff crest, Figure 15 shows the jagged appearance of the crest line and the presence of wide-open tension cracks, and Figure 16 is a good illustration of the extent of erosion over a short period of time. It should be pointed out that mudflows accompanied slumps and earth flows during wet periods. The sequence of bluff response to toe erosion over time at the four sites is summarized in Figure 17.

SUMMARY OF RESULTS

The results of this study can be summarized as follows:

1. Among the four sites studied, the erosion rate at the bluff toe varied from about 7 cm to 10 cm per month, with the total amount of erosion approaching 1 m over the 8-month monitoring period. The erosion

Figure 15: Irregular crest line and the development of tension cracks as a result of slope movement.

Figure 16: A remnant of the original bluff indicating the extent of bluff recession in the study area over a one-year period.

due to wave action was confined to approximately 2 m height from the bluff-beach junction and was most pronounced during the fall (October-December).

2. Spatial variation of the erosion was found to be related more closely to beach characteristics (mostly width) whereas temporal variation was observed to be a function of morphological conditions.

3. The three processes of erosion identified at the sites included wave quarrying, swash abrasion, and swash wash. Among these, swash abrasion was observed to be more frequent and more effective in eroding the in-situ toe material. The toe erosion

resulted in the formation of sharp notches, concave zones, negative slopes, and pronounced overhangs.

4. The bluff response to toe erosion varied with bluff morphology and shape of the erosion zone, the response being quicker where notches had developed than where erosion had lead to the formation of a negative slope at the toe.

5. The bluff recession was found to be the combined result of a variety of slope movements including slumps, translation or slab slides, earth falls, earth flows, and mud flows. Earth falls and slab slides, controlled by the presence of joints, dominated the lower half of the bluff whereas slumps, earth flows, and occasional mudflows were more dominant in the vicinity of the crest.

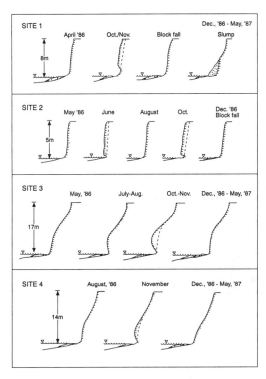

Figure 17: Schematic diagram showing bluff response to toe erosion over time at the four sites.

REFERENCES

American Society for Testing and Materials 1987. Soil and rock; building stones; geotextiles. Annual Book of ASTM Standards, vol. 4.08, Section 4, Philadelphia, 1189 p.

Bridges, E.M. & D.M. Harding. 1971. Micro erosion processes and factors affecting slope development in lower Swansea Valley. Inst. Br. Geogr. Spec. Publ. no. 3, pp. 65-79.

Brunsden, D. 1974. The degradation of coastal slope. Dorset, England. Inst. Br. Geor. Spec. Publ. no. 7, pp. 79-98.

Buckler, W.R. & H.A. Winters. 1983. Lake Michigan bluff recession. Annals, Association of American Geographers, vol. 73, no. 1, pp. 89-11

Carter, C.H. & D.E. Guy Jr. 1988. Coastal erosion: processes, timing and maginitudes at the bluff toe. Marine Geology, vol. 84, pp. 1-17.

Davidson-Arnott, R.G.D. & R.W. Askin. 1980. Factors controlling erosion of the nearshore profile in overconsolidated till, Grimsby, Lake Ontario. Proc. Canadian Coastal Conf., pp. 185-99.

Davidson-Arnott, R.G.D. & H.I. Keizer. 1982. Shore protection in the town of Stoney Creek, Southwest Lake Ontario, 1934-1979: historical changes and durability of structures. Journal of Great Lakes Research, vol. 8, pp. 635-647.

Deere, D.U. & R.P. Miller. 1966. Engineering classification and index properties for intact rock. Air Force Weapons Lab. Rep. AFWL-TR-65-116, Kirkland, New Mexico, 300 p.

Emery, K.O. & G.G. Kuhn. 1982. Sea cliffs: their processes, profiles, and classification. G.S.A. Bull., vol. 93, pp. 644-654.

McGreal, W.S. 1979. Marine erosion of glacial sediments from low-energy cliffline environment near Kilkeel, Northern Ireland. Marine Geology, vol. 32, pp. 89-103.

Quigley, R.M. & P.J. Gelinas. 1976. Soil mechanics aspects of shoreline erosion. Geoscience Canada, vol. 3, no. 3, pp. 169-173.

Quigley, R.M., P.J. Gelinas, W.T. Bou & R.W. Packer. 1977. Cyclic erosion-instability relationships: Lake Erie north shore bluffs. Canadian Geotechnical Journal, vol. 14, pp. 310-323.

Robinson, L.A. 1977. Marine erosive processes at the cliff foot. Marine Geology, vol. 23, pp. 257-271.

Sunamura, T. 1977. A relationship between wave-induced cliff erosion and erosive force of waves. Journal of Geology, vol. 85, pp. 613-618.

Sunamura, T. 1982. A wave tank experiment on the erosional mechanism at a cliff base. Earth Surface Processes and Landforms, vol. 7, pp. 333-343.

Sunamura, T. 1983. Processes of sea cliff and platform erosion. In P.D. Komar (ed.), Handbook of Coastal Erosion Processes, CRC Press Inc., Florida, pp. 233-265.

Land reclamation in Macau – Dikes between the islands

Reconquête de terres à Macao – Des digues parmi les îles

J.L.Tocha Santos
Hidroprojecto S.A., Lisbon, Portugal

ABSTRACT: This paper describes the design criteria and construction methods for the retaining dikes of a huge land reclamation project in Macau, Southeast Asia. The dikes are founded on very soft mud, 10 to 15 m deep and their construction included dredging activities, hydraulic filling, embankment reinforcement with geogrids and wick drains to accelerate the consolidation of the mud. References are made to some occurrences during the construction period.

RESUMÉ: Dans cette communication on décrit les critères de projet et les méthodes de construction adoptées pour les digues de soutènement d'une vaste zone à recupérer à la mer à Macau, Sudest de l'Asie. Les digues seront fondés sur la vase mou, avec 10 à 15 m d'épaisseur et dans leur construction sont utilisées des techniques de remblai hydraulique, renforcement avec des geogrids et drains verticaux pré-fabriqués pour accélerer la consolidation de la vase. On fera aussi référence à quelques occurrences survenues au cours des travaux.

1. INTRODUCTION

A grandiose project of land development has been launched by the Macau Government, the first phase of this project being the construction of the retaining dikes which have a total length of about 6.5 km. The dikes corresponding to the western reclamation area, totaling 3.7 km (the construction of a section of 0.6 km at the northern extremity of the dike A is still waiting for the final definition of the land-use in that particular area) have been concluded in February 1994 and the construction of the 2.2 km dike B for the eastern reclamation area is scheduled to start in April 1994.

2. LAY-OUT AND GEOMETRY OF THE DIKES

At present, the islands of Taipa and Coloane are linked by a 2.1 km long causeway, built in the 60's, bearing roughly North-South. This causeway divides the proposed land reclamation surface in two areas: the western and the eastern areas.

The western area will be enclosed by a 4.3 km long dike (dike A) which conventionally was separated in three sections, named the A1 (1.05 km), the A2 (2.40 km) and the A3 (0.85 km) dikes.

The project crest level of all dikes was established at 5.5 m, taking into account the maximum tide elevations and wave conditions. The crest width was initially 12 m supposing that it would bear a main road. Later it was reduced to 7.0

m when that idea was abandoned.

3. PHYSIOGRAPHIC AND GEOLOGICAL CONDITIONS

The western area, located in a quiet environment, protected from the action of waves and strong currents by the causeway, has been subjected since the completion of the link between the two islands to a high rate of sedimentation and consequently presents large zones partially emerged. The sea bottom along the dikes axis varied from +2.00 to -0.50 m referred to the MCD (Macau Chart Datum) which will be used as datum for the elevation throughout this paper. The higher ground elevations corresponded to the dike A1 location where mangrove type vegetation occurred.

The eastern area will be confined by the dike B, 2.2 km long. In opposition to the western area, this area has been opened to waves and currents and therefore the rate of sedimentation is smaller. Consequently the sea bottom along the East dike route is somewhat deeper, generally -1.0 m.

In both areas, the sea floor is covered by a 10-15 m thick very soft marine mud, overlaying diluvium deposits, the granitic bedrock being more than 30 m deep. The diluvium deposits consist of clayey and sandy soils. Between these deposits and the sound bedrock, an intermediate zone of completed decomposed (CDG) or weathered granite exists. This scenario can be observed in Fig.1 which

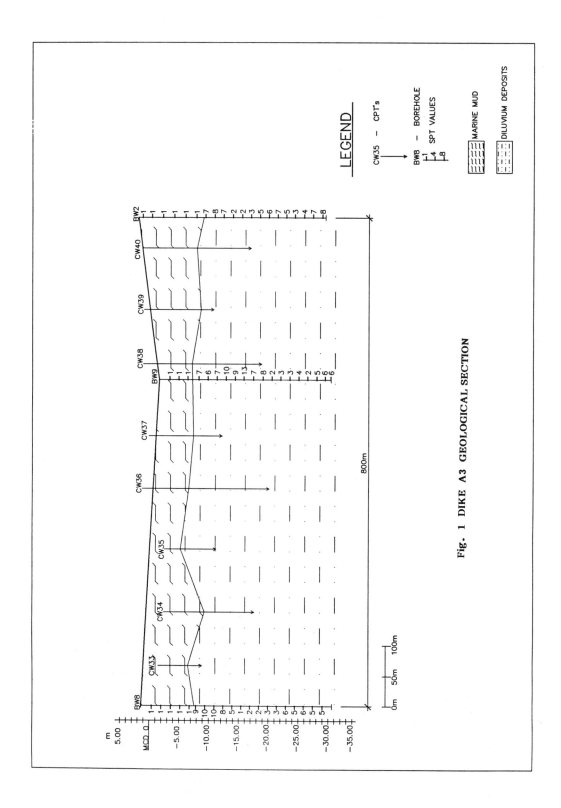

Fig. 1 DIKE A3 GEOLOGICAL SECTION

represents the geological section along the south dike where no bedrock was found up to the depth of investigation.

These adverse geological and physiographic conditions create geotechnical problems of stability and deformability of the embankments as well as difficult operational conditions during the construction.

4. CRITERIA USED FOR THE DESIGN

A very short period (barely one month) was allowed to the Consultant to carry out the design of the dikes, so no time was available to proceed to any site investigation. Consequently, a research and review of the existing geological and geotechnical data of other projects in the vicinity, namely of the airport project, was used to assume values for the geotechnical parameters as necessary for the design calculations.

A site investigation program was prepared to be executed at the beginning of the construction stage. The information then obtained would be used later to check the validity of the values assumed.

In the geotechnical model initially adopted for the design the parameters shown in Table 1 were assumed.

Table 1. Geotechnical parameters initially assumed

Soil Type	Depth (m)	Cu (kPa)	Ø (deg.)	T (kN/m3)	Cc/(1 + e0)
Fill (Dikes)	--	0	35	19	
Marine mud	≤5	5	0	16	
Marine mud	5-8	10	0	16	0.18
Marine mud	≥8	20	0	16	

Using this model, several solutions were analysed through computer stability analysis of cross-sections with several berm widths and different pre-dredging depths, with and without geogrids reinforcement.

For the same dike and foundation geometry and strength parameters, the embankment reinforcement with geogrids proved to be a solution with a factor of safety 10 to 20% higher than a solution without reinforcement. It was also assumed that the geogrid would perform a separating function in a certain degree, thus contributing to reduce somehow the penetration of the fill material into the mud.

The aim of all the analysis performed was to find the most cost effective solution which led to consider as acceptable a factor of safety as low as 1.1 (using the most severe condition of an instantaneous load of the mud by the entire dike cross-section, armour included), provided the construction would be phased, giving time for the mud to consolidate.

At first, when any dredging activities were discouraged, the envisaged method of construction was a step-by-step filling, the first stage being the displacement of the surficial very soft mud by truck end-dumping of fill over a geogrid layer, starting from both islands towards the centre of the area. This was carefully specified, but logistic and operational difficulties presented by the Contractor turned it unfeasible and the initial condition of avoiding dredging was discarded.

Consequently, the solution retained for all dikes, except the dikes A1 and B, consisted of pre-dredging down to -2.5 m level to remove the almost fluid mud, installing geogrid at the bottom, replacing the dredged mud by sand, and elevating the fill up to +3.0 m. At this elevation wick drains would be installed and the fill operations would be interrupted until at least 50% of the consolidation of the underneath mud had occurred. Finally, a second layer of geogrid would be placed and the fill elevated and compacted by layers up to the crest design level.

For the dike A1, the pre-dredging operation was not considered necessary due to the existence of a dessicated mud layer at surface or to the presence of vegetation which provided higher resistance.

For the dike B, the use of geogrid was avoided because the height of submergence is bigger than in the western area, where some difficulty was experienced in placing the geogrid in submerged conditions. To ensure a reasonable factor of safety without the reinforcement, the depth of pre-dredging was increased up to -5.0 m elevation.

The soil specified for the filling operations was sand with less than 30% of particles passing #200 ASTM sieve.

The seaward face of the dikes were protected from the waves with armour which was dimensioned taking into account the expected wave height, 1 m at dike A (A2 and A3 sections) and 2 m at dike B.

Geotextiles were specified wherever different graded materials were adjacent.

For quantity and quality control, the installation and monitoring of instrumentation was specified. This included bench-marks, settlement plates, inclinometers and piezometers.

In order to cope with the uncertainties at the date of the design, a trial section of the embankment with instrumentation and monitoring was specified.

5. CHANGES IN THE ORIGINAL DESIGN

Shortly after the construction of the dike A had started, the results of the specified site investigation became available. Comparing the parameters assumed for the design with those obtained, it was verified that they were similarly enough.

On the other hand, during the construction it was concluded that a 12 m wide crest for the dikes was not justified due to the alteration of the planned land-use and it was decided to reduce the width to 7 m in order to obtain a significant cost reduction.

For this reason a new series of computer calculations were carried out to find the most convenient solutions (Fig.2).

The stability calculations led to the design of

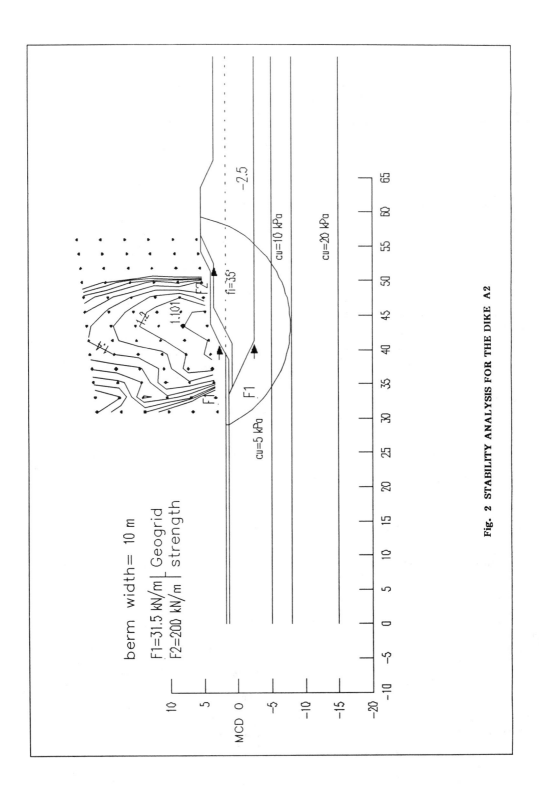

Fig. 2 STABILITY ANALYSIS FOR THE DIKE A2

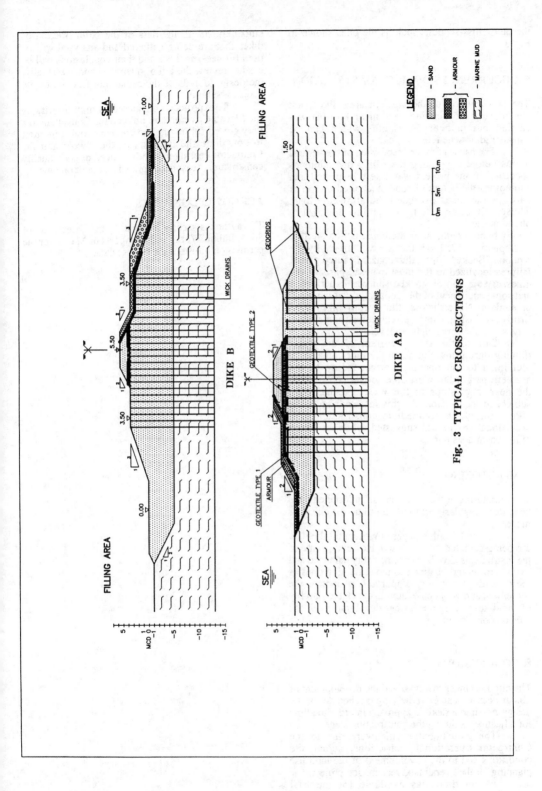

Fig. 3 TYPICAL CROSS SECTIONS

the new cross-sections such as the ones shown in Fig.3.

6. OCCURRENCES DURING CONSTRUCTION

The construction of the western area dikes (not considering the pending section of 800 m) was carried out in near 14 months which can be considered on schedule.

The pause after the installation of the vertical drains before continuing the filling operations as specified in the project was rarely respected and sometimes the height of sand stockpiled on top of parts of the dikes exceeded the allowable load. These situations led to minor instability phenomenae.

Nine months after the construction started, on September 17, 1993, the site was striked by the typhoon "Becky". This natural hazard caused partial failures localized in the most exposed parts of the dikes. Also a loss of stocked soil was experienced and some equipment of the Contractor was damaged or sunken. Nevertheless, the overall dike stood firmly against the joint effects of strong winds, waves and abnormal high sea levels.

Due to the strong current of the water flowing during low tide from the inner area to be reclaimed to the opening corresponding to the northern part of the West dike not yet constructed, the near inner slope of the West dike has been subject to toe erosion. As this fact endangers the dike stability, the immediate construction of a provisional groyne was suggested, while the filling of the basin does not start.

7. MONITORING

The monitoring of the instrumentation installed have been done regularly and will continue for one year more.

At the time this paper is written all the data are being gathered and analysed. However, some of the instruments have been irremediably damaged and extra care should be given to the remaining ones as their readings are very important to control the settlement and very convenient to check the stability of the dikes during the further development of the reclamation project.

8. CONCLUSIONS

The circumstances that involved the development of this project with an extremely tight schedule for the design required a flexible approach in order to allow for adjustments during the construction stage.

The adjustments made were due to the Contractor operational restrictions, actual site conditions and to the development of the land-use planning for the overall land reclamation project.

As no time was available for the trial embankment, all the data of the build fraction of dikes that are being gathered and analysed will be used for back-analysis and their conclusions will be used to review the bill of quantities and, eventually, the design details of the remaining fraction of the dikes to be built.

This project, considered as high priority by the Macau Government, obliged the Consultant to a very closed follow-up and frequent advice in order to conciliate the interests of the Owner and the Contractor bearing in mind the need of good quality construction, reasonably priced and built on time.

ACKNOWLEDGEMENTS

The author is greatly in depth to the Directorate of Soils, Public Works and Tranports of Macau for the permission of use of the project data.

Forecasting coastal cliff failure in jointed rocks

Prédiction du risque de rupture des falaises côtières dans les roches fissurées

A.T.Williams
Environmental Research Unit, Science and Chemical Engineering Department, University of Glamorgan, Pontypridd, UK

P.Bomboe
Geological Engineering Department, Faculty of Geology & Geophysics, University of Bucharest, Romania

P.Davies
Faculty of Science and Humanities, College of Higher Education, Bath, Avon, UK

ABSTRACT: To evaluate the cliff failure risk in Carbonaceus and Old Red Sandstone series, on West Wales coast, multiblock simulation models were developed as functions of cliff geometry, jointing configurations, joint friction angle and joint shear strength fluctuation. The distribution of the lateral loading on fractured blocks generates the structural loosening followed by differential block failure across the cliff face. Using appropiate input variables and numerical models, failure mode and failure risk can be forecasted and proper remedial actions selected.

RESUMÉ: Affin d'évaluer le risque au fracturation de la falaise de la série Old Red Sandstones sur la côte West Wales, on a dévelopé des modéles de simulation multiblock. Les modéles ont pris en compte la géométrie de la falaise, la distribution de fracturation, l'angle de frotement ainsi que la resistence au cissaillement des fractures. L'instabilité de la structure due aux charges latteraux a conduit finallement aux glissements et aux roulements differenciés des blocks. Par un bon calage du modéle ou peut analiser le processus et le risque de l'instabilité ainsi qu'évaluer les remédes possible.

INTRODUCTION

Litological environment of the forecasting models comprises limestones, rudaceous arenite, argillaceous sequences, dark - gray to black, compact siltstones, with thin beds of greyish - brown mudstone. It also contains many layers of grey, strong silty sandstones and fine - grained mudstones, which are exposed in large toppled masses, moderately to highly weathered.

Beadding planes and two principal sets of steeply dipping open joints define the structural environment. Discontinuities are in general regular and parallel to each other, almost in a grid - like pattern. The spacing of major discontinuities is mainly allong the bedding planes and the discontinuity separation is quite wide. Joint lengths is larger then 5m and continuity index approches 1, so that no intact rock bridges occurred on the stepped translation path. Minor joints lie irregularly on the rock surafce and are not particularly important for the simulation model.

The bedding and the two sets of perpendicular joints outline masses of structural blocks (fig.1) used as a system for the simulation model of the cliff failure forecasting.

Joints roughness friction angle (ϕ) and shear strength of the discontinuities were measured and representative values are avaiable as input variables for the forecasting models.

SIMULATION OF THE STEPPED SLIDING PATH

Although the actual position and shape of the stepped sliding surface in jointed rocks is uncertain, by a reasonable simulation method is possible to find out workable equivalent configurations as a result of succesive combinations of the fluctuating orientations, spacings and lengths of the two major joint sets.

Computer simulation begins with the identification of the dip angle (β) of the average slip surface (fig.1), computed as:

$$\beta = \alpha + \arctan [\lambda_m / t_m + \lambda_m \tan [90^0 - (\tau - \alpha)]]$$

where

$t_m = \lambda_s / \sin (\tau - \alpha)$

λ_s = average spacing of the steep joints,
λ_m = average spacing of the bedding planes,
α = dip angle of the bedding planes,
τ = dip angle of the steep joints.

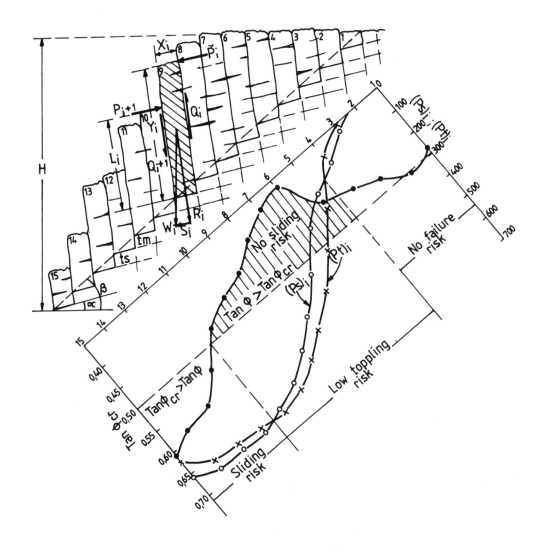

Fig.1. Multiblock model for cliff limiting equilibrium analysis
(site C1: $\alpha = 20^0$; $\beta = 38^0$; H = 24m; $\phi = 26^0$)

Knowing the dip angle β, the average sliding surface could be drawn from the toe to top of the cliff. The dip angle of the steep joints, the average spacing between the steep joints, the stepped sliding path and the cliff face define the multiblock model (fig.1) used to simulate the risk of sliding and toppling on the site.

LIMITING EQUILIBRIUM EVALUATION

Calculation begins with the upper blocks, to evaluate toppling forces $(Pt)_i$ and sliding forces $(Ps)_i$. If the equality $Yi / Xi = \cot \alpha$ is confirmed for the upper blocks, forces $(Pt)_i$ and $(Ps)_i$ equal zero.

For the next blocks the simulation program evaluate the necesary forces to prevent toppling failure:

$$(Pt)_i = [(Pt)_{i-1} (Y_i - X_i \tan \phi) + (W_i/2)]/L_i *$$

$$* (Y_i \sin \alpha - X_i \cos \alpha) / L_i$$

and forces to prevent sliding failure:

$$(Ps)_i = (Ps)_{i-1} + [W_i (\tan \phi \cos \alpha - \sin \alpha)]/(1 - \tan^2 \phi)$$

The components R_i and S_i are evaluated by the following solutions:

$$R_i = W_i \cos\alpha + [(Pt)_i - (Ps)_{i-1}] \tan \phi$$

$$S_i = W_i \sin \alpha + [(Pt)_i - (Ps)_{i-1}]$$

Limiting friction on the sides of the blocks are simplity computed by

$$Q_i = P_i (\tan \phi) \text{ and}$$
$$Q_{i+1} = P_{i+1} (\tan \phi).$$

The critical values of the joint friction angle $(\phi_{cr})_i$ is evaluated for each structural block as function of R_i and S_i forces. The rations between tan ϕ available on the joint and the critical value tan ϕ_{cr} required for limiting equilibrium condition, could be accepted as safety factor values for the stability of each structural block.

CLIFF FAILURE FORECAST

For a given site on the coast, the number and dimensions of the structural blocks are computed as functions of cliff heigth, distance from the cliff top to the occurrence point of the stepped sliding path, and the average spacing between the steep joints.
The competence of the simulation model was checked using, as input variables, field and laboratory measurements from various (sites) of the cliff, in Druidston Haven area. In station C1 the following values were avaiable:

Clifl height: 24 - 27m
Cliff face azimuth: $165^0 - 345^0$
Dip angle of cliff face: 60^0
Bulk density: 2,41 Mg / m³
Joint friction angle ϕ: 26^0
Strike of bedding plane: $153^0 - 333^0$
Dip angle of the bedding of main joints: 2 - 2,5m
Joint length: > 10m

By succesive iterations, the computing program evaluates proper values of Ps and Pt forces, and critical values of the friction coefficient, for every structural block, from the top to the bottom of the cliff (table 1)

Table 1. Computed values os sliding forces, toppling forces and critical friction coeficient.

Block nr.	Ps	Pt	Tan ϕ_{cr}
1	0	0	0.48
2	0	0	0.51
3	23	43	0.47
4	76	115	0.41
5	214	157	0.39
6	335	238	0.31
7	437	358	0.34
8	521	459	0.40
9	585	541	0.44
10	630	603	0.45
11	655	647	0.47
12	659	670	0.53*
13	642	663	0.57*
14	603	654	0.58*
15	541	610	0.62*

The computing program provides also values for perpendicular (R) and tangential (S) forces, and evaluates the available friction resistance (Fr) at the bottom of the structural blocks (Table 2)

Table 2. Computed R, S and Fr values.

Block nr.	R	S	Fr
1	73	48	55
2	220	172	166
3	335	244	252
4	461	330	347
5	588	418	442
6	298	179	221
7	283	179	213
8	272	178	205
9	262	178	197
10	251	178	189
11	241	178	182
12	224	168	168
13	231	192*	174*
14	176	133*	133*
15	246	239*	185*

There is an evident risk of sliding for the lower structural blocks. The tangential force (S) values are higher than the friction resistance available at the battom of the structural blocks.
The forecast is amphasized by graphical reprezentation of the computed values (fig. l), wich outlines that in the lower part of the cliff the critical friction cefficient is higher than the available friction coefficient.

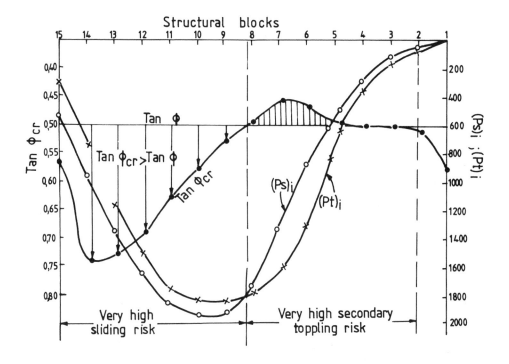

Fig.2. Structural block analysis of Druidston Haven (site C2: $\alpha = 20^0$, H = 50m, $\phi = 26^0$)

For a better understanding of the jointed cliff behavoir, it in necessary to analyse the sensitivity of the forecast to the variation of the cliff geometry, joint configuration and resistance parameters.The numerical model developed for cliff analysis by multiblock method demonstrates its potential to simulate failure modes and failure risk on hunchy uneven slopes in jointed rocks.In Druidston Haven area, site C2, computed forces configuration is very unfavorable (fig.2), necessary sliding friction coefficient for limit equilibrium being significantly higher than the actual joint friction coefficient. In the upper part of the cliff the Pt values are higher than Ps values, so there is a potential risk for local toppling of the blocks. Because the structural loosening by progressive sliding of the lower blocks there is a very high risk for secondary toppling in the upper part of the cliff.

CONCLUSIONS

The numerical model developped for cliff analysis

by multiblock method demonstrates its potential to simulate failure modes and failure risk on hunchy uneven slopes in jointed rocks.

The numerical and graphical results of the computing program outline the complex failure mechanisms and the affiliation between sliding and toppling risk in specific lithologic and structural environments.

REFERENCES

Davies, P., Williams, A.T. & Bomboe, P. 1991. Numerical modelling of Lower Lias rock failure in the coastal cliffs of South Wales. In Costal Sediments '91. Vol.2, p 1599 - 1612, N.Y.

Williams, A.T.,Davies,P. & Bomboe, P. 1993. Geometrical simulation studies of coastal cliff failures in Liassic strata, Souht Wales, U.K. In Earth Surface Processes and Landforms. Vol.18, p 703 - 720.

Les travaux de protection de la Côte des Basques à Biarritz

Protection works on the 'Côte des Basques' in Biarritz

G. Sève & M. Bustamante
Laboratoire Central des Ponts et Chaussées, Paris, France

G. Balestra
Temsol, Bordeaux, France

ABSTRACT : The so-called "Côte des Basques" in the city of Biarritz which is located on the south west coast of France, is a 50 meters high marly cliff overlooking the Atlantic ocean. Regressive movements of the cliff at an average rate of 1 meter per year jeopardized a lot of buildings as well as the scenic coastal road towards Spain.

These movements were due to the combined actions: erosion from the ocean at the bottom part of the cliff and the internal erosion of ground water circulation in the sandy alluvium at the upper part.

In order to save this scenic site, the French authorities and the European community have funded the stabilization and protective works planned for the site.

This paper presents the various coastal works which have been completed including the stability analysis, tests and constructions. Special actions were necessary such as earth works, drainage, wall nailing, prestressed anchorage, jetting and jet grouting.

The conclusions of observations made during the different stabilisation steps are also presented.

RESUME : La Côte des Basques à Biarritz est constituée par une falaise d'une cinquantaine de mètres de hauteur qui surplombe l'océan Atlantique. Formée d'une alternance de marno-calcaires, elle est soumise à une érosion régressive intense dont la vitesse, sur les cinquantes dernières années est de l'ordre du mètre par an.

Le phénomène résulte de l'action combinée de l'océan qui attaque le pied de la falaise et de celle de circulations d'eau lesquelles, par érosion interne, déstabilisent les dépôts récents de sables fins silteux situé en crête de la falaise.

Les désordres ont atteint une telle ampleur, qu'ils menaçaient de couper la route côtière menant à la frontière espagnole, et qui présente par ailleurs un grand intérêt touristique. Afin de préserver un site pittoresque ainsi qu'un axe routier important, les pouvoirs publics avec l'aide de l'Etat français et de la Communauté européenne, ont décidé d'entamer une campagne de travaux confortatifs.

Cette communication relate les différentes phases du projet : étude de la stabilité, travaux et essais de contrôles. En ce qui concerne les travaux côtiers proprement dits, ceux-ci ont donné lieu à l'application de différentes techniques confortatives telles que : drainage et reprofilage de la pente, installation de cordons d'enrochements, soutènements par ancrages précontraints, clouages, injections classiques et jet-grouting. On trouvera une description de la mise en oeuvre et des particularités de ces techniques dans la présente communication. On termine en faisant état des observations qui concernent l'évolution générale du site.

1 PRESENTATION GENERALE

1.1 *Le contexte d'aménagement urbain*

Les falaises de la côte des Basques s'étendent sur plus d'un kilomètre au sud de la ville de Biarritz, petite ville de villégiature au sud de la côte atlantique française. Elles s'élèvent à une cinquantaine de mètres au dessus de l'océan et sont soumises à une érosion régressive intense dont la vitesse moyenne sur les cinquantes dernières années est de l'ordre du mètre par an.

Plusieurs villas construites au siècle dernier et au début de ce siècle en bord de falaise, ont été détruite par les mouvements de terrains (figure 1). Dans les années 1930 la municipalité de Biarritz avait fait réalisé des contreforts de protection des falaises qui se sont progressivement effondrés.

Figure 1. Glissement de terrain en tête de falaise qui affecte une villa.

Il est devenu urgent de conforter et protéger ces falaises lorsque la route nationale conduisant vers l'Espagne s'est trouvée menacée. Un éboulement en masse aurait également endommagé plusieurs rues adjacentes et les réseaux d'adduction d'eau, d'assainissement, etc.. De plus, le risque de chutes de blocs sur la plage en pied de la falaise menaçait directement les baigneurs estivaux.

La municipalité, aidée du Conseil Régional, du Conseil Général, de l'Etat et de la Communauté européenne a mis en place, au début des années 1980, les financements nécessaires (répartis sur plusieurs années) pour conforter et protéger les falaises. Compte tenu de la longueur à traiter (près de 1200 mètres), les travaux sont réalisés en plusieurs tranches.

1.2 Le contexte géologique

Les falaises de la côte des Basques ont une hauteur moyenne de 50 mètres, elles sont constituées par des marnes de l'Eocène surmontées par des alluvions récentes d'une dizaine de mètres de puissance. Ces

alluvions sont le siège d'écoulements importants. Elles présentent de plus des instabilités qui provoquent une forte régression de la tête de la falaise (BRGM (1991)).

Les alluvions quaternaires sont composées pour l'essentiel de sables fins, avec des niveaux de granulométrie plus importante (grave à certains niveaux). De plus, de nombreux talwegs naturels ont été comblés récemment (au début du siècle) afin de permettre la construction. Ces remblais peuvent atteindre 5 à 6 mètres d'épaisseur. En partie basse des sables supérieurs, se trouve une couche décimétrique de sable limoneux, rouge (figure 2).

Figure 2. Les alluvions récentes et des argiles d'altération des marnes de l'Eocène supérieur.

Les marnes de l'Eocène, d'une trentaine de mètres d'épaisseur constituent l'essentiel de la falaise (figure 3). Dans la partie nord, qui est celle qui nous intéresse ici, cette formation se présente comme un ensemble marneux à marno-calcaire homogène, de couleur grise, dont la stratification générale, très peu visible est orienté au Nord-Est selon un pendage de 45 à 65°. Ce réseau de discontinuités favorise des écoulements d'eau privilégiés.

Le contact sable/marne se fait par l'intermédiaire d'une couche d'argile beige clair d'une épaisseur variant entre 50 cm et 1 m. La profondeur de ce contact varie de plusieurs mètres. Cette argile résulte de l'altération des marnes dont la surface a été une surface d'érosion longtemps exposée avant d'être recouverte par les alluvions sablo-graveleuse.

La partie des marnes qui forme la paroi des falaises présente un degré d'altération très variable du fait de la régression permanente. Ces marnes peuvent être saines lorsqu'elles viennent tout juste d'être découvertes par un glissement. Mais l'exposition aux intempéries et le fluage dû à la décompression réduisent rapidement leur résistance mécanique.

Les matériaux marneux éboulés forment le pied de la falaise. L'action conjuguée des eaux qui proviennent du massif et de l'océan a altéré ces matériaux qui présentent un aspect comparable à celui des marnes altérées en paroi.

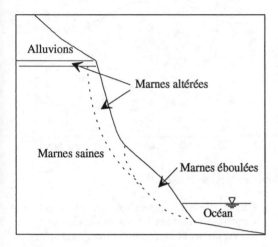

Figure 3. Schéma de la falaise initiale.

Les caractéristiques de résistance mécanique de la marne saine sont importantes. Des essais de compression simple, réalisés sur des échantillons prélevés en sondages mettent en évidence une valeur moyenne de résistance R_c = 1,44 MPa, ce qui correspond à des valeurs c' = 415 kPa et φ' = 35°. Cependant les études de géologie structurale et de stabilité à rebours sur des glissements dans ces matériaux ont conduit à prendre en compte dans les calculs de stabilité les valeurs suivantes : c' = 70 kPa et φ' = 27° le long des familles de discontinuités observées (discontinuités pentées vers l'ouest à 50° sur l'horizontale). Les pressions limites mesurées sont supérieures à p_l = 4 MPa.
Les caractéristiques résiduelles mesurées sur les marnes altérées sont c' = 0 et φ' = 30°.

Dans la nappe alluviale des sables, les caractéristiques mécaniques retenues sont c' = 0 et φ' = 30° avec des valeurs de p_l = 0,6 MPa.

1.3 *Les solutions de stabilisation et de protection*

Les différentes solutions de traitement des falaises ainsi que la description des premières phases de travaux (phases I, II et III) ont été décrites en détail par Lurenbaum et al. (1993). Nous présenterons ci-dessous uniquement la dernière phase de travaux réalisés : la phase IV, achevée à l'automne 1992.

La figure 4 présente le profil type de la solution adoptée : il consiste en un remodelage de la falaise en déblai/remblai. La partie en déblai, en tête de falaise a une pente très importante. C'est pour cette raison que des ouvrages de soutènement superposés ont été réalisés : le mur supérieur mur S, en encorbellement est un mur préfabriqué alors que le mur qui le supporte, mur I, est un mur de type berlinoise ancré par des tirants actifs. Trois murs cloués par des ancrages passifs ont été réalisés au dessus de l'enrochement de pied qui assure la protection de la falaise contre l'érosion due à l'océan.

La digue de pied, haute de 7 mètres au dessus de l'océan, est constituée par un remblai de matériaux filtrants recouverts, côté océan par un cordon d'enrochements. Elle est suffisamment large pour permettre le trafic d'engins, ce qui a servi pendant la phase travaux et autorise l'entretien. Elle est de plus utilisée comme promenade en période estivale.

La stabilité des marnes est assurée par les murs cloués, protégés par un parement en béton projeté.

La stabilité des alluvions est assurée par les murs de soutènement supérieurs (murs S et I). De plus un système de drainage à partir des ouvrages de soutènement par drains subhorizontaux assure l'écoulement des eaux supérieures.

Ces ouvrages de protection de la falaise ont été étudiés pour s'intégrer dans un aménagement paysager : végétalisation des risbermes, réalisation de promenades et escaliers piétons afin de permettre un accès agréable et aisé à la plage.

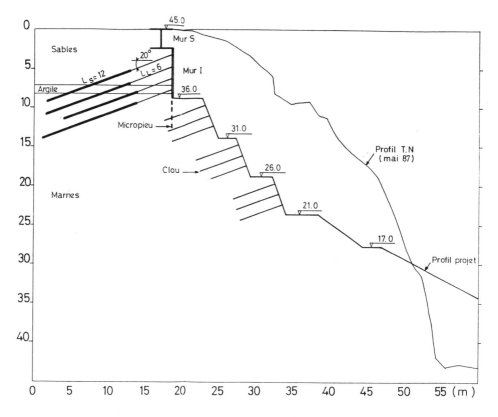

Figure 4. Profil type de l'aménagement de la falaise.

2 ETUDES PREALABLES, EXTRAPOLATIONS AU PROJET

2.1 Etudes de stabilité

2.1.1 Stabilité générale

L'étude de la stabilité générale du projet, dans la configuration décrite précédemment, a été réalisée sur la base des éléments fournis par l'étude structurale et l'étude rhéologique en laboratoire des marnes. En particulier, les valeurs de cohésion et d'angle de frottement dans les marnes ont été prises égales à c' = 70 kPa et φ' = 30° le long de discontinuités pentées à 50 - 60° sur l'horizontale. Cela impose un cercle de rupture très grand pour la vérification de la sécurité au grand glissement. Le coefficient de sécurité vis-à-vis de la stabilité au grand glissement du projet est F = 1,3. On exige généralement un coefficient de sécurité F = 1,5 pour la stabilité générale à long terme d'un tel projet. Le

Maître d'Oeuvre, cependant, compte tenu des nombreuses études réalisées a considéré que le site était suffisament connu pour accepter cette valeur.

2.1.2 Stabilité du mur de soutènement supérieur

Le mur de soutènement supérieur (mur S) étant un mur préfabriqué "classique", son dimensionnement a été fait en utilisant la théorie de Rankine pour la détermination de la poussée et en vérifiant la stabilité au renversement et au glissement sur la base. Le mur inférieur (mur I) assure la stabilité au poinçonnement du mur S.

2.1.3 Stabilité du mur de soutènement inférieur

Le dimensionnement du mur I, qui est un soutènement ancré par tirants précontraints de type berlinoise, a été réalisé en utilisant une méthode de calcul de type stabilité de pente. L'équilibre du massif est calculé en introduisant les efforts stabilisateurs apportés par les tirants (Delmas (1986)). Ceci a été adopté afin de rester cohérent avec l'ensemble des études de stabilité de la falaise sur lesquels ont été calés les paramètres de résistance des marnes. La méthode de calcul qui aurait été utilisée s'il n'y avait eu cette contrainte, est la méthode de calcul aux modules de réaction (Balay (1988)).

La stabilité de l'ouvrage a été vérifiée avec un coefficient de sécurité minimun F = 1,8 (figure 5) pour un cercle qui n'intercepte pas la semelle du mur S, avec un niveau de nappe à 41 NGF et des tensions en service dans les tirants valant respectivement de haut en bas 400 kN ; 400 kN ; 400 kN ; 500 kN.

assurer l'équilibre de la fouille. Ces tensions ainsi déterminées correspondent à la phase chantier. Les tensions définitives sont celles présentées plus haut qui prennent en compte un niveau de nappe permanent à une cote supérieure et également, la charge apportée par le mur S, qui a été réalisé après le mur I.

Tableau 1.
Coefficients de sécurité
dans les différentes phases de terrassement.

Cercle	T1 kN	T2 kN	T3 kN	T4 kN	F_{glisst}	$F_{butée}$
1	100	0	0	0	3,6	2,1
2	100	200	0	0	2,6	2,5
3	100	200	200	0	1,5	3,5
4	100	200	200	200	1,6	3,6

Pour le calcul du parement, réalisé en béton projeté sur un ferraillage en treillis soudé, il est nécessaire de faire une hypothèse sur la répartition des pressions derrière le parement. Schlosser (1990) préconise de prendre une distribution uniforme des efforts appliqués par les tirants. Aussi, pour le dimensionnement des armatures des nervures et du voile, on a supposé d'une part une répartition uniforme des efforts induits par les tirants et d'autre part une distribution trapézoïdale équivalente. C'est l'enveloppe des moments fléchissants maximum obtenus pour les deux distributions qui a permis de déterminer le ferraillage nécessaire.

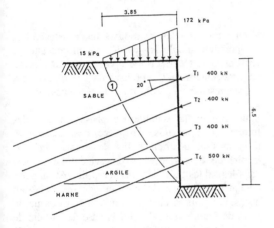

Figure 5. Profil de calcul de la stabilité du mur I.

Les tensions dans les tirants en phase d'exécution ont été calculées en cherchant sur des cercles de rupture qui passent en pied de la fouille pour chacun des niveaux de terrassement, un coefficient de sécurité vis-à-vis du glissement F > 1,5 et un coefficient de sécurité vis-à-vis de la mise en butée des terres F > 2. Ce dernier coefficient de sécurité est la valeur de l'effort de butée $\frac{1}{2}K_p\gamma h^2$ rapportée à l'effort apporté par les tirants (K_p, coefficient de butée des terres est $K_p = tg^2(\frac{\pi}{4}+\frac{\varphi'}{2})$). Les calculs ont été réalisés en supposant une absence de nappe. Le tableau 1 présente les tensions dans les tirants calculées pour

2.1.4 Stabilité temporaire de la fouille dans les sables

La réalisation du mur I nécessitait d'excaver les alluvions sableuses et il était donc nécessaire d'évaluer la stabilité à court terme des sables non soutenus. Contrairement à d'autres techniques de soutènement (paroi moulée par exemple), l'exécution d'un mur cloué est une phase critique vis-à-vis de la stabilité locale ou globale. En particulier, la stabilité locale de l'excavation lors des terrassements dépend directement de la hauteur de sol excavé. A titre d'exemple, on peut citer l'expérience que l'on a du sable de Fontainebleau (Schlosser (1990)) pour lequel φ' = 38° et c' = 4 kPa qui met en évidence une hauteur critique de fouille de 2 mètres (figure 6).

Aussi un certain nombre de précautions avait été préconisées à l'entreprise :
- limiter à 1 m - 1,2 m la hauteur de fouille,
- mettre en place des inclusions avant terrassement,
- ouvrir par plots de faible largeur entre inclusions,
- projeter un voile mince de béton de protection immédiatement après terrassement afin de limiter les risques d'éboulement locaux dus à la décompaction et à l'altération des terrains,
- protéger la fouille contre les eaux de ruissellement,
- rabattre la nappe en amont et mettre en place un drainage de l'ouvrage.

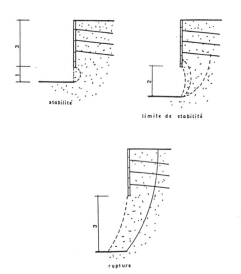

Figure 6. Stabilité d'une fouille dans le sable de Fontainebleau.

De plus et comme il était exclu de terrasser sous le niveau de la nappe sans précautions particulières, les derniers mètres de terrassement ont été réalisés après avoir rabattu la nappe et après injection de gel de silicates.

Les caractéristiques mécaniques des sables nécessaires pour assurer la stabilité à court terme des terrassements du mur I, hors nappe, ont été évaluées et comparées à celle mesurées dans le sable en place, en considérant une largeur de plots de terrassement de 1,8 m (ce qui correspond à la largeur entre les poteaux de la berlinoise) et un poids volumique du sable $\gamma = 19$ kN/m^3. La résistance au cisaillement d'un sable est représentée habituellement par un angle de frottement et une cohésion nulle. Cependant une légère cohésion de quelques kPa peut

être observée. Cette cohésion apparente provient habituellement d'une légère cimentation, d'une éventuelle présence de fines et de phénomènes capillaires. La façon dont la cimentation se développe au cours du temps par circulation d'eaux chargées (calcite, silice, oxydes de fer, etc.), conduit à une hétérogénéité importante de la cohésion résultante.

L'état actuel des connaissances ne permettant pas de calculer de façon fiable la stabilité d'une fouille dans des sables du fait de l'hétérogénéité de la cohésion et du peu d'expérience que l'on a de ce type de travaux. Il a été imaginé de se référer à l'expérience que l'on a des terrassements de fouilles dans des matériaux argileux. La hauteur critique d'une fouille dans un massif purement cohérent est :

$$h_c = \frac{2c_u}{\gamma}$$

Si on extrapole à des sables (avec une résistance à la compression simple $R_c = 2C_u$ et en prenant un coefficient de sécurité F = 2), pour la première passe de terrassement de 2 m de hauteur, une résistance à la compression simple $R_c = 75$ kPa est nécessaire.

Pour évaluer la stabilité en phase travaux des fouilles pour la mise en place des deuxièmes, troisièmes et quatrièmes lits de tirants, on s'est inspiré de l'expérience du creusement de tunnels dans les sables qui, bien qu'encore limitée, apporte quelques éléments d'évaluation de la cohésion nécessaire pour assurer la stabilité à court terme du front de fouille verticale. Sur la base de l'article de Leca & Dormieux (1990) dans lequel le problème de la stabilité de front de taille de tunnels est considéré à l'aide de l'analyse limite, on a proposé d'approcher la valeur de la résistance à la compression simple R_c nécessaire à la stabilité locale, par la valeur de la pression de confinement nécessaire pour équilibrer le front de taille ; cela, en tenant compte d'un effort vertical dû aux tirants supérieurs et en prenant un coefficient de sécurité F = 2. Pour une profondeur de 3,6 m avec une ouverture de 1,8 m, la valeur de la pression de confinement d'équilibre 55 kPa. On a pu considérer que la valeur de R_c nécessaire dans les sables était 110 kPa.

2.1.5 Stabilité des murs cloués

La stabilité de chacun des murs cloués a été calculée en faisant l'hypothèse d'une courbe de rupture circulaire qui passe en pied de mur et est tangente à un coin incliné à 50 ° sur l'horizontale. La méthode de calcul utilisée est la méthode proposée par Delmas et al. (1986) pour la vérification du dimensionnement de ce type d'ouvrage. Elle présente l'avantage de tenir compte des efforts de flexion par le biais d'un calcul au module de réaction de pieu sollicité latéralement par un déplacement de sol. Cette méthode permet de calculer :

- F_s : le coefficient de sécurité de l'ouvrage vis-à-vis de la stabilité,
- F_{anc} : le coefficient de sécurité du clou en ancrage,
- F_{CA} : le coefficient de sécurité du clou sollicité en flexion composée,
- F_{CB} : le coefficient de sécurité du clou sollicité en traction-cisaillement,
- $F_{s/c}$: le coefficient de sécurité vis-à-vis de l'interaction latérale sol/clou.

Il est habituellement demandé pour chacun de ces coefficients de sécurité une valeur supérieure à 1,5 exception faite de l'interaction latérale pour laquelle on demande $F_{s/c} > 2$.

Pour satisfaire aux différentes conditions de sécurités sur les armatures et sur le sol, les clous mis en oeuvre ont été forés en 100 mm de diamètre et les armatures utilisées varient de 32 mm à 40 mm selon la hauteur du mur (5, 6 et 7 mètres) et la position du clou (les clous du bas étant les plus chargés). Les hypothèses de frottement latéral dans les marnes ont été vérifiées par le biais d'essais d'arrachement.

2.2 Plots d'essais préalables

L'importance du projet exigeait que les hypothèses de calcul concernant le dimensionnement des tirants et des micropieux fassent obligatoirement l'objet d'essais préalables en vraie grandeur. Ceux-ci réalisés de décembre 1991 à février 1992, avaient pour objectifs :
- de vérifier les hypothèses du dimensionnement (Bustamante & Doix, (1985)) ;
- d'apprécier la faisabilité en offrant la possibilité, avant d'entamer les travaux définitifs, de prendre la mesure des difficultés que pouvaient présenter la mise en oeuvre dans les conditions de chantier ;
- d'optimiser la méthode de forage du point de vue des rendements mais aussi de manière à garantir les capacités d'ancrages ou portantes les plus élevées (Bustamante & Le Roux, (1993)).

En vue de l'extrapolation à l'ouvrage définitif, les emplacements des plots d'essais, qu'il s'agisse des tirants des clous ou des micropieux, furent choisis de manière à ce qu'ils correspondent à des conditions de sols, caractéristiques et représentatives de celles que l'on pouvaient rencontrer sur l'ensemble du site.

2.2.1 Les tirants d'essais.

Il fut décidé de vérifier la capacité d'ancrage de trois tirants, à raison de deux tirants avec scellement dans les marnes, TE.1 et TE.2, et un tirant TE.3 avec scellement dans les alluvions quaternaires.

Les deux premiers tirants TE.1 et TE.2, d'une longueur totale $L_t = 17$ m, étaient scellés sur $L_s = 6$ m dans les marnes compactes, de pression limite p_l moyenne = 4 MPa. Le projet exigeait que leur traction admissible T_a soit au moins égale à 550 kN.

Le tirant TE.3 scellé à la fois dans les alluvions et les marnes avait une longueur totale $L_t = 10$ m, pour une longueur scellée $L_s = 5$ m. La capacité d'ancrage admissible exigée valait 165 kN. Les 3 essais ont été réalisés conformément aux directives du mode d'essai statique des LPC. La figure 7 montre la réalisation des essais sur les tirants TE.1 et TE.2 en décembre 1991. Les capacités d'ancrages effectives des 3 tirants d'essais se sont avérées tout à fait satisfaisantes, aucun déchaussement du scellement n'ayant été observé pour les tractions maximales d'épreuves, à savoir 950 kN pour la série TE.1 et TE.2, et 665 kN pour le tirant TE.3.

Figure 7. Réalisation des essais de tirants

Par la suite, l'ensemble des essais de mise en tension, effectués sur les tirants définitifs, ne révélèrent de défaut d'ancrage confirmant par leur fluage les fluages mesurés lors des essais préalables.

2.2.2 Les micropieux d'essais.

Les essais préalables, réalisés en février 1992, ont concerné 2 micropieux d'essais, M.1 et M.2, d'une longueur totale de 8 m et 12 m, scellés sur 6 m dans les deux cas, dans les marnes compactes caractérisées par des pressions limites moyennes au niveau des essais p_l de l'ordre de 4,75 MPa. Il s'agissait d'essais à la traction. L'expérimentation a été notablement valorisée par l'instrumentation des micropieux qui ont été équipés d'extensomètres amovibles (Bustamante et al., (1990)), afin de connaître la distribution des efforts le long des fûts. Le recours aux extensomètres amovibles a permis de mesurer les efforts le long du scellement et leur loi de mobilisation sur 8 niveaux.

Figure 8a. Essai de micropieu en cours.

Après les essais, les micropieux ont été excavés afin de pouvoir relever la géométrie exacte des scellements. Les figures 8a et 8b montrent dans l'ordre : l'essai en cours (on remarque

l'instrumentation du micropieu) ainsi que l'aspect du scellement après sa mise à nue. Les figures 9 et 10 présentent pour le micropieu M.2, la distribution de l'effort le long du scellement et les courbes de mobilisation du frottement latéral unitaire pour différents niveaux de mesure. Les résultats des essais d'arrachement ont confirmé la justesse du dimensionnement adopté.

Figure 8.b. Vue du scellement du micropieu d'essai.

2.2.3 Les clous d'essais.

Ceux-ci ont été au nombre de six. Ils ont été réalisés sur des clous de différentes longueurs, de 4 à 12 m, scellés gravitairement dans les marnes. Les essais ont permis de vérifier que les capacités prévisionnelles étaient systématiquement garanties.

3 REALISATION DES TRAVAUX

L'ensemble des travaux proprements dits de la phase IV qui ont débuté en novembre 1991 furent achevés en novembre 1992 (figure 11).

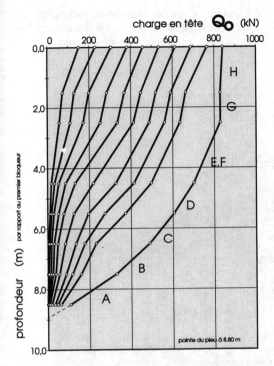

Figure 9. Distribution de l'effort le long du scellement du micropieu.

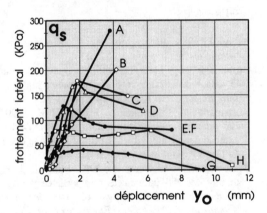

Figure 10. Courbes de mobilisation du frottement latéral unitaire le long du micropieu.

3.1 Le traitement provisoire des terrains

On rappellera qu'il était nécessaire de consolider les sols avant de pouvoir réaliser le mur de couronnement supérieur (mur I) . Cela a consisté en l'injection au gel de silice de 50 m3, soit environ 15 % du volume des sables.

Pour réaliser la fouille dans les alluvions sableuses, l'enteprise a procédé par phases d'excavation de faible hauteur avec mise en place immédiate par fonçage de drains courts et d'une mince couche de béton projeté.

3.2 Les ouvrages de soutènements

Le mur inférieur I, en béton projeté armé et nervuré d'une longueur de 100 m, représentant 397 m^2 de surface, a été ancré par 104 tirants, ce qui représente un linéaire de 2216 mètres. Ils ont été finalement précontraints entre 450 et 550 kN.

Le mur cloué MC, également en béton projeté mais non nervuré a été ancré par un total de 470 clous passifs soit un linéaire de 2788 m. Leur capacité variable, était comprise entre 125 et 157 kN. La surface de cet ouvrage, qui se développait sur 90 m, valait 1616 m^2.

Le système de drainage

Il était principalement constitué par 120 drains subhorizontaux, de 3 à 8 m de longueur chacun, soit un linéaire de 1316 m.

Les travaux de terrassement

Ils ont donné lieu à quelques 35000 m^3 de terrassements.

Le montant de la totalité des travaux a été estimé à 11 MF.

4 LE CONTROLE DES TRAVAUX

L'importance des travaux de la phase IV, au même titre d'ailleurs que ceux des phases précédentes, justifiait l'instrumentation des différents ouvrages afin de connaître leur comportement dans le temps.

C'est ainsi que furent installés :
- 4 profils inclinométriques,
- 4 piézomètres ouverts,
- 4 cellules de mesure des pressions interstitielles,

- 16 cales dynamométriques annulaires pour le suivi de l'évolution des tractions des tirants (figure 12) et des clous.

Le suivi de l'ensemble des ouvrages n'a donné lieu à ce jour, c'est-à-dire plus d'un an après l'achèvement des travaux à aucune observation particulière. La publication détaillée de l'ensemble des observations fera l'objet d'une communication ultérieure.

Figure 11. Vue des murs S, I et MC après achèvement.

Figure 12. Vue d'une cale annulaire pour le contrôle de la traction d'un tirant précontraint.

REFERENCES

Balay, J. 1988. Parois moulées et ancrages. *Les techniques de l'ingénieur.* C-252.

Bustamante, M.., Doix, B.,1985. Une méthode pour le calcul des tirants et des micropieux injectés. *Bull. liaison Labo P. et Ch.* **140**. pp 75-92.

Bustamante, M.., Gianeselli L., Jézéquel J.-F., 1990. La mesure des déformations à l'aide des extensomètres amovibles LPC. *Méthode d'essai LPC n°34, octobre 1990*, pp. 1-16.

Bustamante, M., Le Roux, A., 1993. Minéralogie et texture des sols indurés, leur importance pour le géotechnicien. Proceedings of the *Internat. Symposium Geotechnical Engineering of Hards Soils - Soft Rocks, Athens, 20-23 September*, pp.939-945.

BRGM 1991. Rapport d'étude géologique des falaises de la côte des Basques. Dossier de consultation des entreprises.

Delmas, Ph., Berche, J.-C., Cartier, G., Abdelhedi, A., 1986. Une nouvelle méthode de dimensionnement du clouage des pentes : programme PROSPER. *Bull. liaison Labo P. et Ch.* **141**. pp 57-66.

Leca, E. & Dormieux, L. 1990. Upper and lower bound solutions for the face stability of shallow circular tunnels in frictional material. *Géotechnique* **40**.No 4. pp 581-606.

Lunrenbaum, N., Page, .A., Combe, M., Riondy, G. Confortement des falaises de la côte des basques à Biarritz (Pyrénées-Atlantiques), bilan des travaux réalisés entre 1983 et 1992. *Revue Travaux.* Janv. 1993. pp 15-25.

Schlosser, F. 1990. Bilan du Projet National CLOUTERRE. *Communication écrite.*

Étude du comportement de dépôts terrestres de vases portuaires

Study of the behaviour of harbour muds deposited on land

S.Gallois, A.Alexis & P.Thomas
Laboratoire de Génie Civil ECN, IUT Saint Nazaire, France

B.Gallenne
Port Autonome, Nantes-Saint Nazaire, France

RÉSUMÉ: Le dépôt à terre de vases de dragage pose de nombreux problèmes en raison, notamment, de la grande quantité d'eau contenue dans ces sédiments et de la lenteur avec laquelle cette eau s'évacue. Après une étude bibliographique des travaux antérieurs sur la caractérisation du matériau et la recherche de solutions pratiques, nous présentons un outil expérimental permettant de modéliser, à échelle réduite, le tassement d'un matériau de dragage sous diverses conditions de dépôt. Cet outil, appliqué à l'étude de l'influence d'adjuvants mélangés à la vase au moment du dépôt, fournit un ensemble de données originales et particulièrement intéressantes pour tout organisme portuaire.

ABSTRACT: On land disposal of dredged muds raises several problems mainly due to the large amount of water contained in these sediments and to its very slow draining off. After a bibliographical study of earlier researches aiming to characterize the material and to find practical solutions, we present an experimental device allowing to modelize, on a reduced scale, the settlement of a dredged material under various deposit conditions. This tool, applied to the study of the influence of additives mixed with mud during the filling, provides a set of original datas specially interesting for harbour authorities.

1 ÉTUDE BIBLIOGRAPHIQUE ET DISCUSSION

Nous présentons ici une synthèse de notre étude bibliographique (Gallois, Alexis & Thomas 1993) portant sur plus de soixante articles scientifiques internationaux.

1.1 *Caractérisation et dépôt des vases de dragage*

Caractérisation et comportement du matériau

Les vases estuariennes se caractérisent par une grande variabilité spatiale et temporelle de leur structure, de leur granulométrie et de leur composition bio-chimique.

Des mesures par sédimentométrie sur des vases prélevées en différents points du port de Hamburg (Werther 1986) , montrent une fraction sableuse d'importance très variable: 10 à 80 % en masse (50% en moyenne).

D'autre part F. Ottmann & G. Lahuec (1972) soulignent les variations de composition d'une même vase au cours de son transport. Ces différences sont dues au lavage subi par le matériau au cours des différentes phases d'aspiration, clapage et refoulement.

L'étude de la distribution granulométrique dans la fraction fine pose aussi des problèmes spécifiques (Kinnaer 1991):
- la défloculation, préalable à une analyse sédimentométrique standard, détruit les agrégats constituant le matériau,
- dans le cas d'une sédimentométrie sans défloculation, le principe même de la mesure (loi de Stokes) n'est plus applicable aux flocons (particules poreuses),
- les poids volumiques de toutes les particules ne peuvent être considérés comme équivalents notamment du fait de la teneur en métaux lourds des fines pouvant donner des densités voisines de 3,6.

Une méthode basée sur la diffraction de Fraunhofer d'un rayonnement Laser donnerait de meilleurs résultats.

L'étude des fines (Werther 1986 et Van Mieghem, Smits & Jeuniaux 1991) fait apparaître, pour des vases estuariennes de la Mer du Nord, une forte concentration en métaux lourds dans cette fraction: 95 à 98 % de la concentration sur l'ensemble du matériau. Ces métaux sont susceptibles de migrations gravitaires à travers des dépôts de vase, indépendamment du mouvement d'ensemble du dépôt (Mondt, Laus, Leermakers & Bayens 1991).

Comme les métaux lourds, les matières organiques se concentrent essentiellement dans la fraction fine (Werther 1986). Cependant des composés organiques lourds (huiles, hydrocarbures,...) sont fréquemment décelés dans des sables ou dans la fraction grossière de vases, qui agissent alors comme des filtres (Van Gemert, Quakernaat & Van Veen 1986).

L'étude du comportement mécanique de la vase avec les moyens classiques d'un laboratoire de géotechnique (œdomètres, cellules triaxiales...) nécessite certaines adaptations permettant d'appliquer et de mesurer des contraintes très faibles (Kumbasar & Özaydin 1977 et Stepkowska et al. 1991). L'essai triaxial, dont la mise en oeuvre est difficile avec des matériaux très mous, peut être remplacé par des essais au scissomètre de laboratoire et des essais de pénétration (Dieltjens 1991).

Les mesures sur sites restent très sommaires et ne permettent pas une caractérisation précise du matériau. Le tassement, la densité (Van Mieghem, Smits & Jeuniaux 1991) et la teneur en eau (Thomas, Fleming & Riddell 1991) ont été étudiés sur divers types de dépôts.

Etudes pratiques, applications

Trois types de procédés de traitement fondamentalement différents ont été mis en oeuvre, en fonction de la qualité prioritairement recherchée pour le matériau transformé.

Le premier procédé s'appuie sur les deux propriétés suivantes:
- les mixtures de dragage sont constituées d'en moyenne 50 % de sable,
- les polluants sont concentrés dans les fines.
Le principe de ce procédé est donc de séparer sables et fines réduisant ainsi le volume de matériaux à traiter.

Le second procédé au contraire utilise la plus grande perméabilité du mélange sable-fines pour obtenir une consolidation plus rapide.

Le troisième procédé consiste à ajouter à la mixture un réactif chimique qui, en consommant l'eau interstitielle, crée un liant qui solidifie très rapidement le matériau.

Abondamment étudié autour des grands ports de Belgique, des Pays-Bas et d'Allemagne où la pollution est un problème crucial, le procédé par séparation sables-fines comporte quatre étapes principales:

- séparation sable-fines, par hydrocyclonage (Werther 1986) ou par décantation sélective (Van Mieghem, Smits & Jeuniaux 1991),
- traitement de la vase destiné à éliminer ou réduire les divers polluants (Calmano 1986),
- évacuation de l'eau interstitielle par centrifugation (West & Pharis 1991, Werner Ihle & Sparrow 1980, Hamburg Port Authority 1987 et Werther 1986), ou dans des bassins de consolidation (Van Mieghem, Smits & Jeuniaux 1991 et Claessens & Smits 1991),
- dépôt ou réutilisation de la boue consolidée, en remblais paysagers (Claessens & Smits 1991), en comblement de fosses littorales (Hamburg Port Authority 1987) ou pour d'autres usages (Kleinbloesem & De Gast 1980).

Les techniques de traitement global de la mixture (Thomas, Fleming & Riddell 1991 et Kleinbloesem & De Gast 1980), beaucoup plus simples, recherchent en priorité un drainage efficace:
- dépôt par couches avec couches de sables intermédiaires,
- altération de la surface afin de favoriser l'évaporation.
Elles donnent des résultats intéressants en quelques mois.

Nous abordons, enfin, une troisième catégorie de méthodes issues directement des méthodes d'amélioration des sols en technique routière. Ces méthodes utilisent des réactifs chimiques qui consomment l'eau interstitielle pour former un liant. Le matériau ainsi obtenu est très rapidement solide. On voit immédiatement qu'un défaut de ces méthodes est de conduire à une augmentation globale du volume des matériaux. Cependant elles présentent l'avantage de fixer leurs divers composants chimiques et d'éviter ainsi, après mise en dépôt, les risques de détérioration du sous-sol.

Les principaux produits ayant cette capacité de solidification et stabilisation de matériaux à grande teneur en eau (Calmano 1986) sont: ciment, water glass (alkali-silicate), CHEMFIX (Mukaï 1985), SOLIROC (deux produits brevetés à base d'alkali-silicates), cendres volantes, chaux, gypse. Des associations de ces produits peuvent donner de bons résultats.

Études théoriques, modélisation

Les tentatives de mise en équation de la consolidation de matériaux cohésifs sous poids propre, ont été nombreuses. Cependant, les hypothèses prises en compte par la plupart des auteurs ne permettent pas de représenter de manière satisfaisante le comportement particulier des mixtures de dragage, par nature, très hétérogènes. Nous avons présenté par ailleurs (Alexis, Thomas & Gallois 1993) l'ensemble de ces différentes approches et une abondante bibliographie constituant un état des connaissances dans ce domaine.

Quelques propositions de modélisation sont axées plus spécifiquement sur les matériaux de dragages:
- sédimentation et consolidation sous l'eau d'un un matériau composé de deux types de grains: les éléments cohésifs fins et les éléments grossiers non cohésifs (Toorman & Berlamont 1991) à partir de la théorie de Kynch,
- prise en compte d'une grande variété de conditions aux limites dont certaines s'appliquent bien au cas de dépôts de dragage à terre: fond du dépôt drainant, évaporation à la surface du matériau...(De Smedt, Sas & Buelens 1991),
- modèle de Gibson complété par des méthodes originales de détermination des paramètres des lois de comportement (Van Heteren & Greeuw1991).

Aspects économiques du traitement des vases

Très peu d'auteurs donnent des éléments de coût permettant de comparer entre elles, d'un point de vue économique, les diverses méthodes de traitement proposées.

On peut cependant estimer les coûts suivants (Van Wijck et al. 1991), en francs français (FFR) par tonne de matière sèche (t.DM):
- dragage seul: 45 FFR/t.DM,
- traitement complet par séparation et dewatering mécanique puis dépôt dans des zones protégées: 600 à 1500 FFR/t.DM,
- méthodes semi-naturelles associées à une mise en dépôt paysager: 200 à 280 FFR/t.DM.

1.2 Études complémentaires

Les méthodes appliquées et les résultats obtenus sur des vases de dragage, peuvent être complétés par des travaux sur deux matériaux posant des problèmes similaires:

Boues résiduelles du traitement des eaux usées

L'évacuation de l'eau interstitielle peut être accélérée par:
- adjonction de floculants (Kang et al. 1990),
- adjonction de chaux ou de cendres volantes (Zall, Galil & Rehbun 1987).
La stabilisation et la solidification peuvent être obtenues par traitement à l'aide de:
- chaux (Bevins & Longmaid 1984),

- mélange ciment-cendres volantes (Faschan et al. 1991),
- procédé CHEMFIX (Barth 1990).
Toutes ces améliorations se font au détriment de la réduction de volume.

Cas des argiles très molles

La consolidation de ce type de matériau peut être accélérée par:
- application du vide dans des drains verticaux, sous une membrane étanche (Kotera, Sakemi & Matsui 1991),
- électro-osmose (Pilot 1977).
Une solidification sans évacuation de l'eau interstitielle peut être obtenue par:
- l'hydroxy-aluminium (Bryhn, Loken & Aas 1982),
- combinaison de chaux avec divers déchets industriels (Kamon & Nontananandh 1991).

1.3 Discussion, perspectives

Nous avons effectué une synthèse des connaissances sur les vases de dragage et les méthodes de traitement des dépôts de dragage à terre et donné des éléments d'appréciation et de comparaison de ces différentes méthodes.

Nous nous sommes également intéressés quelques expériences dans des domaines voisins dont les résultats sont susceptibles de contribuer à la compréhension ou au développement de techniques équivalentes pour les vases.

Nous constatons tout d'abord que:
- très peu d'auteurs se sont intéressés au comportement et au traitement de la mixture brute de dragage,
- aucun outil expérimental intermédiaire entre les appareils classiques de laboratoire et les techniques de mesures sur site n'est proposé,
- quelques pistes d'utilisation de matériaux de récupération sont présentées mais les informations concernant ces méthodes et leur efficacité sont insuffisantes,
- les méthodes qui donnent un accroissement rapide de la résistance mécanique du matériau engendrent une faible diminution de volume,
- aucune solution universelle n'a été proposée car chaque site avec son matériau et son environnement géo-économique impose ses contraintes propres.

Toute étude d'un procédé de traitement des dépôts de dragage à terre doit donc commencer par l'évaluation objective des contraintes (et avantages) de site, permettant d'orienter les recherches vers des méthodes acceptables du point de vue de l'environnement économique, écologique, géographique...

Ceci nous conduit à définir le cadre géo-économique, propre à l'estuaire de la Loire:

- une densité de population au voisinage de l'estuaire relativement réduite, laissant disponible des surfaces assez importantes,

- un surcoût admissible pour le traitement faible,

- un objectif idéal: l'obtention d'un matériau suffisamment résistant pour, à moyen terme ($\approx$ 3 ans), supporter un remblai constructible.

Ces quelques remarques, la volonté généralisée aujourd'hui de réduire ou de valoriser au maximum tous les types de sous-produits de nos industries, ainsi que l'abondance et le coût réduit de certains déchets, ont alors orienté nos recherches vers des méthodes peu coûteuses et susceptibles de donner rapidement un matériau consolidé de caractéristiques mécaniques satisfaisantes. Nous proposons donc d'étudier le comportement de mixtures de vase et sous-produits industriels en reproduisant, à échelle réduite, en laboratoire, les conditions de dépôt en vraie grandeur.

1.4 *Récupération, recyclage et géotechnique*

Dans l'optique d'une utilisation de déchets pour le traitement de vases de dragages, il est intéressant de rappeler les principales utilisations courantes ou possibles de déchets dans la fabrication de liants hydrauliques et de matériaux de structure (Murat 1981):

- laitiers (déchets de l'industrie sidérurgique, production: environ 10 Mt / an), dans la fabrication du ciment,

- cendres volantes (résidu de la combustion du charbon dans les centrales thermiques, production: environ 3Mt / an), dans la fabrication du ciment,

- laitiers mousse, expansé ou granulé, cendres volantes expansées ou frittées, ordures ménagères frittées... dans la fabrication de bétons légers.

- sulfates de calcium résiduaires pour la fabrication du plâtre,

- carbonates de calcium résiduaires pour la fabrication de la chaux,

- goudrons, brais et bitumes (résidus de la distillation de la houille et du pétrole) en technique routière,

- laitiers et scories en technique routière,

- cendres volantes dans les couches de base et de fondation de chaussées,

- cendres volantes pour la stabilisation d'argiles très plastiques (Canada) ou de sables de dunes (Brésil),

- déchets de matières plastiques (PE, PVC, PS)

incorporés au bitume et goudron dans les couches de roulement des chaussées.

Dans le cas du traitement de la vase, la priorité est l'évacuation de l'eau interstitielle (effet drainant) obtenue de trois manières (Gallois, Alexis & Thomas 1993-b):

- modification de la granulométrie et donc de la perméabilité globale de la vase,

- création de chemins de drainage préférentiels ,

- consommation de l'eau interstitielle par une réaction d'hydratation.

Un effet de renforcement dû, soit à des efforts de frottement ou de cohésion de la vase sur un matériau d'apport moins déformable, soit à l'apparition au sein du matériau de cristaux solides fortement liés entre eux est souhaitable.

2 NOUVELLE ÉTUDE EXPÉRIMENTALE

Nous présentons succinctement les méthodes et résultats de cette étude. Le détail de nos travaux est décrit par ailleurs (Gallois, Alexis & Thomas 1993-b).

2.1 *Définition des objectifs*

Outre les objectifs généraux définis au §1.3, plusieurs objectifs originaux justifient cette recherche:

- étude de procédés de traitement nouveaux, n'ayant jamais été appliqués à un matériau de teneur en eau si élevée,

- reproduction à échelle réduite du comportement de dépôts de vases à terre.

Notre dispositif doit simuler des conditions limites correspondant à un dépôt émergé en vraie grandeur, tout en autorisant un suivi de la consolidation (mesure du tassement, des débits de drainage, de la cohésion) simple et précis.

2.2 *Sélection des matériaux d'apport*

Les matériaux d'apport utilisés ont été choisis en fonction des critères suivants:

- rapidité et simplicité d'approvisionnement,

- représentation des différents types d'action drainante,

- possibilité d'un approvisionnement en grande quantité.

Nous avons, donc, sélectionné un matériau correspondant à chacun des trois types d'action drainante: modification de la granulométrie (sable), création de chemins de drainage (papier) et consommation de l'eau interstitielle (cendres volantes).

Nous avons utilisé un sable de Loire de granulométrie très étroite (0/4). C'est un sable à béton dont le coût, livré en grande quantité, peut être estimé entre 30 et 40 F par tonne.

Le papier provient d'une entreprise de recyclage. La qualité de papier choisi est la plus abondante et la moins chère. Le papier est broyé et réduit à une dimension de l'ordre de 10 à 30 cm puis compacté en balles d'environ 600 kg. Le coût hors livraison de ce produit est, selon la quantité, de l'ordre de 200 à 300 F par tonne.

Les cendres volantes, résidus de combustion de charbon pulvérisé (Grandjean 1989) proviennent de la centrale thermique EDF de Cordemais, site de production le plus proche des zones de l'Estuaire concernées par cette étude. Leur coût, hors transport, ne devrait pas être supérieur à 50 F/t pour de grandes quantités.

Les copeaux de bois sont les déchets d'une menuiserie. La disponibilité de ce matériau pour un traitement de la vase à une échelle industrielle n'a pas été vérifiée.

2.3 Conception et réalisation de l'outil expérimental

Cet outil n'est pas conçu dans le but de quantifier l'efficacité réelle d'une méthode, mais de permettre la comparaison de plusieurs méthodes appliquées à la même vase, dans les mêmes conditions de dépôt. Cette comparaison s'effectue sur trois critères mesurables: la variation de volume du dépôt, la quantité d'eau drainée et la résistance mécanique de la mixture. Il se compose (fig.1) de:
- cinq fûts de 200 litres avec dispositifs de drainage,
- un portique de mesure.

Afin d'obtenir des mesures de résistance significatives dans un matériau très peu portant, nous avons réalisé un pénétromètre très léger. Il se compose d'une pointe en P.V.C. surmontée d'un tube du même matériau et d'éléments de tige d'acier ronds servant de lest.

Les résistances de pointe mobilisables sont:

Figure 1. Schéma de principe de l'outil expérimental. Schematic view of the experimental device.

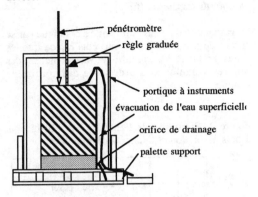

- pointe + tube PVC seul   Rp = 4,3 kPa
- avec un élément de tige   Rp = 7 kPa
- avec deux éléments   Rp = 10 kPa
- avec trois éléments   Rp = 15 kPa
- avec quatre éléments   Rp = 20 kPa

(Cette dernière valeur correspond au poids d'un homme sur ses deux pieds)

2.4 Préparation des essais et mesures

En l'absence d'éléments bibliographiques permettant d'évaluer les proportions d'adjuvants à mettre en oeuvre, nous avons choisi (tab.1):
- pour les matériaux inertes: 10 % du volume des grains de vase,
- pour les cendres volantes: même proportion massique grains/cendres que granulats/ciments dans un béton standard.

Les mesures sont faites à intervalle de temps croissant sur une période de deux mois.

Tableau 1. Proportions et volumes des différentes mixtures. Proportions and volumes of each mixture.

N° du culot	Avant addition			Adjuvant			Après mélange		Remarques
	Concentration des solides		Volume total (l)	Type (Masse totale)	Proportion		Volume total (l)	Masse vol. (kg/m3)	
	g/l	cm3/l			g/l	cm3/l			
1	621	230	120	/	/	/	120	1392	
2	541	200	127,5	bois (2,5 kg)	20	20	130	1334	200/20=10
3	617	229	119,5	Cendres (17 kg)	140	52	126	1453	617/140=4,4≈5
4	760	281	121,5	Papier (2,5 kg)	20	20	124	1468	281/20=14>10
5	831	308	119,5	Sable (10 kg)	80	30	123	1557	308/30≈10

2.5 Synthèse des résultats

Mesures de tassement (fig.2)

La vase naturelle, sans adjuvant, se distingue par sa vitesse de tassement beaucoup plus rapide que les autres mixtures. Au contraire, la vase contenant des cendres volantes a la plus lente diminution de volume.

Figure 2. Evolution de l'épaisseur des différents dépôts. Evolution of thickness of each layer.

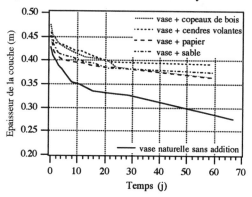

Mesures de résistance mécanique (fig.3, 4 & 5)

Afin de pouvoir tenir compte des disparités de concentration initiale des différentes mixtures lors de l'interprétation des mesures de résistance

Figure 4. Pénétration en fonction de la concentration moyenne pour Rp =10 kPa. Penetration versus average concentration for Rp =10 kPa.

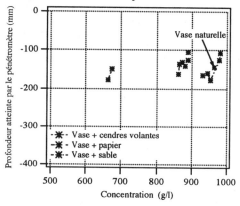

mécanique, nous choisissons de tracer les courbes d'enfoncement de la pointe du pénétromètre en fonction de la concentration moyenne du dépôt en éléments solides non rapportés (grains de vase seulement).

Ces courbes sont des obliques montantes qui traduisent l'accroissement logique de la résistance d'un matériau avec sa concentration. Elles sont sensiblement parallèles et leur position relative permet de comparer la qualité des mixtures. Les mixtures caractérisées par des courbes situées en haut et à gauche du graphe sont, à concentration équivalente, les plus résistantes, tandis que celles situées dans le secteur inférieur droit sont les moins résistantes.

On peut observer les phénomènes suivants:

Figure 3. Pénétration en fonction de la concentration moyenne pour Rp =4,7 kPa. Penetration versus average concentration for Rp =4,7 kPa.

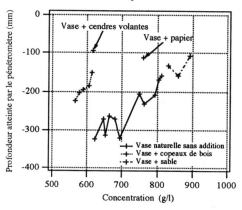

Figure 5. Pénétration en fonction de la concentration moyenne pour Rp =15 kPa. Penetration versus average concentration for Rp =15 kPa.

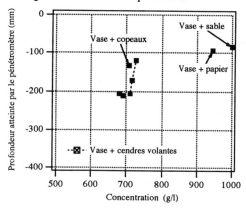

Les courbes (ou leurs extrapolations) correspondant à la vase naturelle et à la mixture vase-sable sont confondues. L'addition de sable, dans les proportions utilisées ici, ne semble donc pas avoir un effet de renforcement sur le matériau naturel. Vase naturelle et mixture vase-sable sont les deux dépôts les moins résistants.

Les courbes correspondant aux mixtures vase-copeaux et vase-cendres volantes sont toutes situées dans la partie gauche des graphes et traduisent une résistance mécanique plus élevée que celle des autres mixtures. Il est difficile d'établir la supériorité de l'un ou l'autre de ces deux matériaux d'apport qui ont un effet de renforcement certain.

La mixture vase-papier semble avoir une résistance intermédiaire. L'effet de renforcement dû à la présence de papier reste limité.

Mesures de volume drainé (fig.6)

On remarque qu'à moyen et long terme, la vase naturelle est beaucoup plus drainante que toutes les mixtures.

L'adjonction de cendres volantes semblent au contraire réduire considérablement le drainage, notamment au cours des premiers jours, durant lesquels les cendres consomment une partie de l'eau interstitielle.

On constate, à court terme aussi, une vitesse de décharge très grande dans la mixture vase-sable. Nous pensons que ce phénomène est dû au poids volumique des grains de sable qui entraînent vers le bas la structure fragile de la vase et accélère la remontée de l'eau. Ce phénomène prend fin, rapidement, lorsque les contraintes effectives développées fixent les grains de sable.

Figure 6. Evolution de la vitesse de décharge de l'eau à travers la surface supérieure de chacun des dépôts. Evolution of vertical discharge water speed

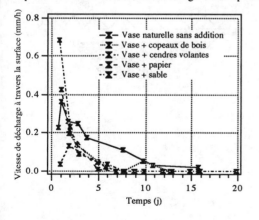

2.6 *Conclusion*

L'ensemble de ces résultats, valables seulement pour les dosages en matériau d'apport choisis, place aux premiers rangs en matière d'efficacité les cendres volantes et les copeaux de bois. Le papier broyé, avec une efficacité un peu plus limitée, devance le sable qui, dans les proportions choisies, reste pratiquement sans effet sur la vase.

Nous faisions l'hypothèse, dans la première partie de cet article, que l'amélioration d'un dépôt de vase par mélange avec un matériau d'apport pouvait être due à deux types d'action:
- effet drainant, qui accélère l'augmentation de la densité de la mixture,
- effet de renforcement, qui rigidifie, à concentration constante, la structure de la vase.

L'étude rigoureuse des résultats de ces essais met en évidence les points suivants:
- l'effet de renforcement est primordial: la structure formée dès mélange (ou dans les jours qui suivent, pour les cendres volantes) fixe l'ordre de grandeur de la résistance du dépôt,
- la structure trop fragile de la vase naturelle remaniée (par dragage, par exemple) ne peut, sous le seul effet du drainage, évoluer rapidement pour lui permettre d'acquérir une résistance suffisante,
- l'évacuation de l'eau interstitielle est ralentie par la présence de matériaux d'apport (sans doute par leur effet de renforcement, qui en constituant une structure plus solide au sein de la vase, diminue les pressions interstitielles et donc diminue le drainage), il n'y a donc pas réellement d'effet drainant.

3 PERSPECTIVES

3.1 *Méthodes de traitement*

La qualité des résultats obtenus incite à multiplier les essais avec d'autres matériaux d'apport ou avec des dosages différents des mêmes matériaux, afin de constituer une base de données comparative sur l'efficacité des méthodes de traitement des vases de dragage par adjonction de déchets ou sous-produits industriels.

Il sera en particulier intéressant de comparer les activités respectives des cendres volantes, de la chaux et du ciment.

La qualité d'une méthode ne se mesure pas uniquement à son efficacité selon les critères définis dans cet article. Le coût du matériau d'apport, la facilité d'approvisionnement, la simplicité de mise en oeuvre à grande échelle et la stabilité bio-chimique du produit sont des paramètres à prendre en compte avant toute application.

3.2 *Essais en vraie grandeur*

L'interprétation de résultats expérimentaux de laboratoire, en vue de comprendre et de prévoir des phénomènes de très vaste échelle, est difficile. C'est pourquoi nous estimons nécessaire la création d'un outil d'expérimentation permettant de tester et de comparer, en vraie grandeur, les différentes solutions techniques envisagées pour l'amélioration de dépôts de vase à terre.

Les objectifs spécifiques de ce nouvel outil sont:

- conditions de mise en dépôt (forme du dépôt, refoulement...) les plus proches (ou représentatives) possible des méthodes habituelles des organismes de dragage,

- vase identique (provenance, prélèvement...) à celle habituellement refoulée à terre,

- suivi (type de mesures, fréquence...) comparable à celui des essais en laboratoire, complété de carottages et d'analyses granulométriques.

Le chantier expérimental (fig.7) se composera de:
- une souille en Loire,
- deux bassins tampons,
- une série de douze bassins d'essais carrés, à section trapézoïdale, de 5 X 5 m au fond et de 3 m de profondeur, alignés en bordure de Loire et d'une piste existante.

L'ensemble du site d'expérimentation aura pour dimensions 100 x 250 m.

Le principe de fonctionnement suivant est retenu (fig.8): la vase prélevée dans le chenal de navigation par la drague aspiratrice en marche est déversée dans une souille en Loire préalablement préparée à cet effet par une drague aspiratrice stationnaire (DAS). De là, elle est reprise et refoulée en direction

Figure 7. Avant projet d'outil expérimental en vraie grandeur. Pilot study of a field scale experimental device.

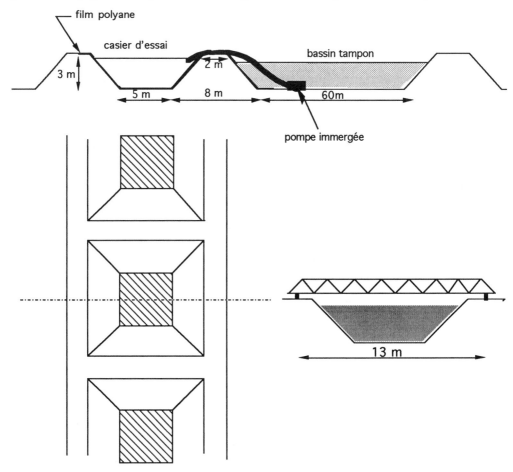

Figure 8. Schéma de principe du remplissage. Schematic view of the filling.

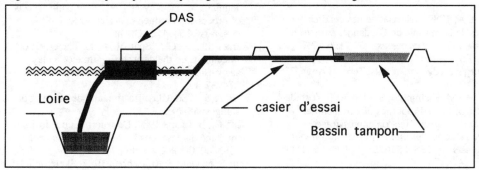

d'un bassin tampon à terre qu'elle remplit en quelques heures. La vase est alors pompée du bassin tampon vers les casiers d'essai à l'aide d'une pompe à boues.

4 CONCLUSION

La première partie de cet article, essentiellement axée sur l'analyse bibliographique, met en relief l'ampleur d'un problème commun à tous les grands ports mondiaux, ainsi que différents aspects scientifiques, expérimentaux et économiques de sa résolution jusqu'alors mal connue (tant dans les milieux portuaires que dans les milieux scientifiques).

Dans la seconde partie, nous proposons un ensemble de méthodes et d'outils originaux prenant en compte les spécificités naturelles et économiques du domaine d'étude: conception et instrumentation de culots de laboratoire et premières expérimentations.

Cette procédure a permis d'effectuer une étude comparative de l'impact de différents adjuvants sur le comportement mécanique des vases portuaires de l'estuaire de la Loire et de sélectionner expérimentalement (selon des critères à la fois scientifiques et économiques) des matériaux d'appoint qui accroissent la résistance mécanique des vases en laboratoire, avec une efficacité prometteuse.

Dans la dernière partie, nous ouvrons à ces nouvelles méthodes des perspectives de validation en Nature, avant application en site réel.

L'avant-projet d'un outil adapté à cette validation est présenté. Il sera réalisé et exploité grâce à l'excellente coopération entre ingénieurs, techniciens et chercheurs du Port Autonome de Nantes - Saint-Nazaire et de l'Institut Universitaire de Technologie de Saint-Nazaire.

RÉFÉRENCES

Alexis, A., P. Thomas, S. Gallois 1993. *Tassement des sédiments cohésifs*. Saint-Nazaire, Institut Universitaire de Technologie.
Barth, E.T. 1990. The site demonstration of the CHEMFIX solidification/stabilization process at the portable equipment salvage company site. *JAPCA*, ISSN 0894-0630, 2: 166-170.
Bevins, R.E. & F.M. Longmaid 1984. Stabilization of sewage-sludge cake by addition of lime and other materials. Water pollution control, ISSN 0043-129X, Vol. 83, 1: 9-22.
Calmano, W. 1986. Stabilization of dredged mud. 80-98.
Claessens, J. & J. Smits 1991. Riverbottom sanitation and landscaping beneficial use of fine grained dredged material from the antwerp region. *Proc. CATS congress:* 4.35-4.42. Ghent.

De Smedt, F., M. Sas, J. Buelens 1991. Consol : a computer simulation model for the consolidation of dredged material. *Proc. CATS congress:* 2.19-2.24. Ghent.
Dieltjens, W., 1991. Mechanical strength determination of cohesive sludges. *Proc. CATS congress:* 2.65-2.69. Ghent.
Faschan, A., M. Tittlebaum, F. Cartledge , H. Eaton 1991. Effects of additives on solidification of API separator sludge. *Environmental monitoring and assessment*, ISSN 0167-6369, 2: 145-161.
Gallois, S., A. Alexis & P. Thomas 1993-a. *Amélioration des dépôts de vase émergés: I (mars 1993)*. Nantes: Port Autonome - I.U.T. de Saint-Nazaire.
Gallois, S., A. Alexis & P. Thomas 1993-b. *Amélioration des dépôts de vase émergés: II (dec.*

1993). Nantes: Port Autonome- I.U.T. de Saint-Nazaire.

Grandjean, A. 1989. Valorisation des cendres volantes de la centrale de Gardanne. *Industrie minérale, mines et carrières,* vol. 71: 68-71.

Hamburg Port Authority 1987. How to dispose of Hamburg's dredged materials. Terra et Aqua, 33: 13-15.

Kang, S. M., M. Kishimoto, S. Shioya, T. Yoshida & K. Suga 1990. Properties of extracellular polymer having an effect on expression of activated sludge. *Journal of Fermentation and Bioengineering,* ISSN 536253, Vol. 69, 2: 111-116.

Kinnaer, L.R.J. 1991. Particle size characterization. *Proc. CATS congress:* 2.1-2.10. Ghent.

Kleinbloesem, W.C.H. & J. De Gast 1980. Dredged silt in Rotterdam looks valuable. *Proc. of WODCON IX:* 273-285. Vancouver.

Kumbasar, V., Özaydin I.K. 1977. Les caractéristiques de la consolidation des sédiments maritimes pollués: 1159-1162.

Mondt, W., G. Laus, M. Leermakers, W. Baeyens 1991. The spatial distribution of pollutants (inorganic and organic) in a dredged sludge pumping ground. *Proc. CATS congress:* 2.71-2.85. Ghent.

Mukaï, H. 1985. Research on solidification of sea bed mud for control of red tide. 253-264.

Murat, M. 1981. *Valorisation des déchets et des sous-produits industriels.* Paris: Masson.

Ottman, F. & G. Lahuec 1972. Dragages et géologie. *Travaux,* juin-juillet 1972: 1-7.

Stepkowska, E.T., J.L. Perez-Rodriguez, A. Justo, C. Maqueda, D. Boels & J. Van Heteren 1991. Study in behaviour and micro-structure of a dredged sludge. *Proc. CATS congress:* 2.51-2.58. Ghent.

Thomas, B.R., G. Fleming, J.F. Riddell 1991. Topsoil from dredgings: dewatering procedures required to rapidly form a topsoil product for land reclamation.*Proc. CATS congress:* 4.9-4.19. Ghent.

Toorman, E.A., J.E. Berlamont 1991. Prediction of settling and consolidation of cohesive sediment. *Proc. CATS congress:* 1.49-1.54. Ghent.

Van Gemert, W.J.T., J. Quakernaat, H.J. Van Veen 1986. Methods for the treatment of contaminated dredged sediments: 45-64.

Van Heteren, J., G. GREEUW 1991. Consolidation behaviour of dredged slurries. *Proc. CATS congress:* 1.63-1.66. Ghent.

Van Mieghem, J., J. Smits, F. Jeuniaux 1991. Large scale separation and consolidation techniques for on land disposal of fine-grained dredged material. *Proc. CATS congress:* 1.55-1.62. Ghent.

Van Wijck, J., J. Van Hoof, M. De Clercq, F. Jeuniaux, C. De Keyser 1991. Economic aspects of processing dredged material. *Proc. CATS congress:* 4.71-4.82. Ghent.

Werner Ihle, S., G.J. Sparrow 1980. The effect of flocculants on the de-watering time of a clay suspension. *Proc. Australas. Inst. Min. Metall.,* 274: 37-40.

Werther, J. 1986. Classification and dewatering of sludges: 65-79.

West, G., B. Pharis 1980. Dewatering cuts drilling mud and disposal costs. *Oil and Gas Journal,* ISSN 0030-1388, Vol. 89, 39: 84-88.

Zall, J., N. Galil & M. Rehbun 1987. Skeleton builders for conditioning oily sludge. *Journal - Water Pollution Control Federation,* ISSN 0043-1303, Vol. 59, 7: 699-706.

Movement patterns and coastal land loss associated with landslides along the Atlantic Coast of Trinidad

Caractéristiques du mouvement et pertes de terrain côtier associés aux glissements de terre sur la côte Atlantique de Trinidad

R.J. Maharaj

Institute of Marine Affairs, Chaguaramas, Trinidad

ABSTRACT: A geotechnical investigation of landslides developed along a segment of a 1.8 km long shale coastline is presented. Failures are by sliding with secondary flows at the toe. The landslide varies from earth to debris slides. Soils contain highly plastic and potentially expansive clays. Landslides are rainfall induced and demonstrate sudden failure, usually following an increase in passive earth pressure due to undrained loading, increase in pore water pressure and decrease in effective stress at approximately 60-65 m from the toe of the landslide. Movement is primarily in response to increase groundwater levels, while basal shear surfaces are near-parallel to seepage zones, which exhibit high seepage rates. Coastal land loss was as much as 9066 m^3 within seven months, 66% of this within the rainy months from June to December.

RÉSUMÉ: Une investigation géotechnique des glissements de terre qui se sont développés à l'intérieur d'un segment de 1,8 km de longeur d'un littoral d'argile chisteuse, se présente ci-dessous. Les ruptures sont de deux types, les glissements principalement, et les coulées sécondaires et progressives au front. Le glissement peut être soit de terre, soit de débris. Les sols contiennent des argiles très plastiques potentiellement expansibles. Les glissements de terre sont provoqués par les précipitations. L'intensification de la pression interstitielle et l'affaiblissement des contraintes effectives à 60-65 m environ du front du glissement de terre mènent à la rupture subite. Le mouvement répond principalement à l'augmentation des niveaux d'infiltration pendant que les surfaces des cisaillements basaux sont presque paralèlles aux zones d'infiltration que manifestent des niveaux augmentés d'infiltration. La perte de terrain côtier était autant que 9066 m^3 dans sept mois, 66% du quel se perdaient dans les mois pluvieux, Juin à Decembre.

1 INTRODUCTION

This paper presents results of a geotechnical investigation conducted along Mayaro Bay on the Atlantic coast of Trinidad. The purpose of this study was to examine the nature of slope deformation mechanisms operating within a segment of low mudrock/clay coastline and to measure the loss of hillslope material to the marine environment.

To the best of the authors' knowledge, there has never been any field geotechnical studies conducted on deformation mechanisms in mudrock/clay slopes in Trinidad, nor has there been any estimation of the volume of terrestrial material lost to the marine environment from landslide sites. Previous studies conducted on coastal erosion in Trinidad have highlighted the erosion problem (Deane 1971; Bachew et. al 1983 and Bertrand et. al. 1991). However, coastal land loss associated with landslides was never addressed.

In the context of geotechnical studies of tropical soil and mudrock slopes, the Mayaro Bay sites are very useful field laboratories, as they can provide first hand data on the movement patterns and deformation mechanisms of soil slopes, a subject which there is no information for Trinidad. The information gathered from this investigation can be used as a baseline for evaluating coastal landslide hazard along other segments of coastline in geologically similar materials and can be useful for engineering of clay slopes in these areas e.g. in the construction of coastal seawalls and shoreline protection structures.

From an environmental standpoint, this study can give some indication about the amount of coastal slope material lost to the marine environment and the rate of erosion of this material. It therefore can be useful in assessing the importance of coastal landslides as a contributor to marine sediments along Trinidad's coastline.

2 THE STUDY AREA

The study area is located at the northern end of Mayaro Bay on the Atlantic coast of Trinidad (Figure 1a and 1b). The area lies at approximately $10°19'$ N latitude and $60°59'$ W longitude and forms part of a 1.80 km long segment of low to high mudrock and clay cliff coastline. From previous geotechnical investigations of this area, (Maharaj and Pfister 1992) the study area can be described as characterized as a landslide prone coastline, characterized by earth topples, slides and flows.

Slopes along this coast rise to a maximum of 75 m above mean sea level (msl) but varies between 65 - 75 m above msl. Slopes are generally gentle to moderately steep, between $15°$ - $30°$, but can be up to $41°$ especially at their toe and head. Slopes are generally concave in their long profile, but with very steep toe segments, maintained by erosion during high water tide.

The shoreline is subject to continuous wave attack, especially during extreme high water spring tide, storms and hurricane events. These result in considerable erosion and removal of basal slope material and subsequent shoreline retreat. The study site usually experiences waves and breaker between 0.20 - 1.00 m high, but can be subject to larger waves during storm events.

At the base of the coastal cliffs is a sandy beach and Holocene wave-cut platform. The beach sediment are subject to seasonal offshore and onshore movement ,sometimes exposing the wave cut platforms (shale and silt sequence).

The coastline receives a mean annual rainfall of 435-500 cm, with a mean annual wet season rainfall of 375-435 cm and mean annual dry season rainfall of less than 38 cm (Berridge 1981).

3 METHODOLOGY

Desk studies were conducted which involved a review of existing literature and preliminary analysis of the topography, geology, geotechnics, hydrology, soils and climate of the study area. Preliminary analysis of the topography was made using black and white, panchromatic stereopairs of the study site (scale 1:10,000, 1976 edition) obtained from the Lands and Survey Division. This was supplemented by a study of topographic maps of the site, namely 1:10,000 (1972 edition) and 1:25,500 (1977 edition) sheets obtained from Lands and Survey Division, Trinidad.

Some geotechnical, geological, soils and surface water data were also extracted from black and white stereopairs.

Field studies formed the body of the data acquisition means. On-the-ground surveys were necessary to establish ground control for instrumentation and mapping to obtain topographic details of the landslide. A set of three benchmarks were installed on stable ground at the toe of the landslide and on the beach (Figure 1c). The heights of these benchmarks above sea level were transferred by precise levelling loops from the established national benchmarks in the area, while the benchmarks were tied together by triangulation using a precise level and theodolites. From these points the slide topography was surveyed and slope profiles were mapped. All slickensides, cracks, seeps, bulges, water ponds, scarps and other major landslide surface features were mapped using these reference points.

Profiles were surveyed along the center line of the landslide and across the flow-slide track, using a clinometer and Sokkia theodolite and levelling instrument. Surface displacement was monitored by installation of two flexible (PVC) plastic tubings (T1 and T2) and four survey stakes (S1 - S4) within the moving parts of the landslide.

A landslide map was prepared by surveying the surface of the failure along three survey lines, from each of the three benchmarks. For all surface profiles which were done, measurements (distance and slope) were taken at major slope breaks. Where a straight segment of slope was encountered, such as along the flow track, readings were taken at 1 m intervals.

Sub-surface exploration consisted of two large boreholes (BH1 and BH2) and four logged, hand dug test pits (TP1 - TP4, Figure 1c). This allowed the collection of sub-surface soil samples, mapping

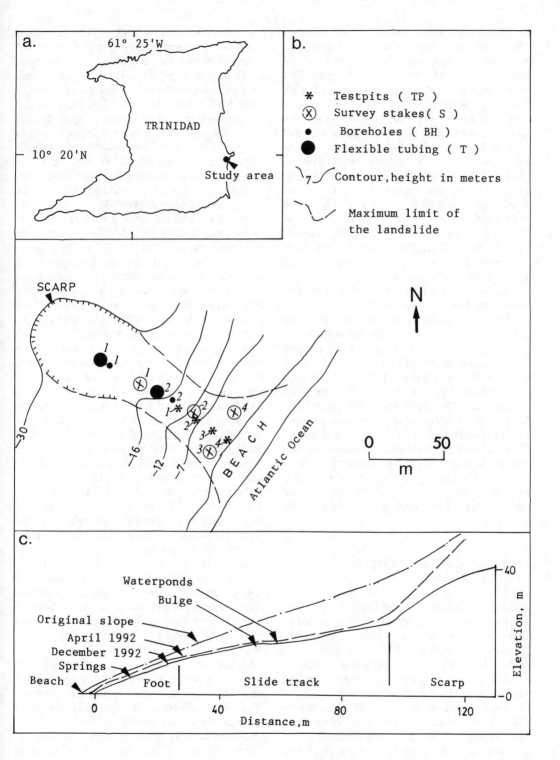

Figure 1. Location and morphology of the landslide

of groundwater levels and sub-surface flows. Several other hand dug test pits were also done to map field soil properties within the body of the landslide. Soil properties noted were grain size, color, moisture, plasticity, dilatancy, organic content, stiffness and weathering. A screw-type hand auger and thin wall Nalgene plastic tubes were used to collect soil samples. Both instruments had an inner diameter of 50 mm. In-situ shear strength was measured using a Soiltest hand pentrometer.

Soil samples were also collected from the toe, body and upper flow track of the slide. All soil samples were immediately sealed in plastic bags to prevent moisture loss. Sampling procedures follow those of the American Society for Testing Materials (ASTM 1982). These were taken for laboratory determination of their grain sizes, plasticity and natural moisture (ASTM 1982).

Groundwater levels were monitored by installation of open standpipe piezometers and manned monthly by using a wetted tape. Thoroughflow rates were measured in test pits by collecting sub-surface flow across seepage zones from a vertical face of a test pit. A plastic funnel with an area of 0.0211 m^2 (diameter of 15 cm) was placed across the seepage zone, with the center of the funnel at the main seepage line. The funnel was lightly pushed into the soil at an angle to the horizontal (inclined downslope) and parallel to the seepage direction. The water draining/flowing into the funnel was then collected in plastic measuring cylinders for at least 5 hours. The soil type/s along the seepage zone was also described, especially its textural characteristics. The discharge was subsequently calculated.

4 LANDSLIDE MORPHOLOGY AND CHARACTERISTICS

Based on the classification of landslides proposed by Varnes (1978), the landslide can be described as a complex failure, in which slope movement is a combination of sliding and flow. However, from field observations of movement patterns, translational-slab sliding is the dominant deformation mechanism.

Figure 1 shows the morphology of the failure for different times of the year, and the original slope. The original slope was reconstructed from extrapolation of the upper slope gradient down to the shoreline, as well as field measurement of slope angles on similar mudrock/clay slopes in the area.

From these methods, a slightly concave original profile was produced, becoming more concave with continued erosion and landsliding. The lowest part of the original slope would more likely have been steep to vertical at the toe, maintained by continued wave erosion during high tide. Just above the toe, the original slope might have been slightly concave due to sliding following removal of material from the toe of the slope. This would have eventually caused the development of an initial point of rupture and failure along the mid-slope, which with continued retrogressive activity, led to the development of the present profile.

The present landslide surface is characterized by slopes ranging from 6-26°. The present scarp measures between 18° at the head of the slope, to 44° at its' base. Field surveying have shown that there is a continual degradation of the main scarp, with a continual decrease in slope angles. However, the slope angles along the flow track and at the foot of the slide is basically constant all year, with only very small angular variations.

The slope along the flanks of the landslide measures 28° - 32°, with very little degradation, thereby maintaining the lateral limits of the slide track. Gentle slope segments along the slide track are characterized by ponding of water, especially in the upper flow track. The head of the flow track sometimes show bulging in response to failure of the immediate main scarp (Figure 1).

A pronounced secondary scarp is found at the head of the zone of deposition, with an approximate vertical displacement of 0.65 m, but can be up to 2.00 m. The surface of the flow track is characterized by pronounced slickensides along its' flanks and parallel to the direction of failure. These are found in the clayey soils and sometimes also cause stretching of plant roots following displacement. The surface of the flow track material contains numerous fissures and tension cracks up to 30 cm wide and 40 cm deep. These are particularly common during the rainy periods and following removal of toe material by high water tide.

The zone of deposition is characterized by small, frequent mudflows and debris flows especially at the points of exit of groundwater (springs). Numerous mudflows are also formed in response to heavy rainfall events. Sliding is also common at the toe, following wave erosion by high water tide.

The landslide measures between 150 to 157 m in total length (along the long axis of the failure) and has an average width of 28 m at the scarp, 10 - 15 m along the slide track and 28 m at the toe of the deposit. The scarp measures approximately 62 m in length, while the deposit and slide track varies between 88 - 95 m long. The landslide, as well as the length of deposit and slide track is dependant on failure and further movement of the toe unto the adjacent beach. Where the toe is eroded, as during high water tide, smaller lengths are recorded.

The maximum height at the top of the scarp was approximately 52 m above sea level, while the maximum height of the head of the deposit was 24 m above mean sea level. The scarp has an average slope of 30°, but varies between 18° - 44°, steeper at the base and gentler at the toe. As a result of continued degradation of the scarp, there has also been a decrease in the angle. The slope on the slide track varies between 5° -18°, while the toe of the deposit measures between 20° - 30°. The landslide has an estimated thickness of 2 m in the slide track area and about 3 m - 4.5 m in the foot.

The landslide has a steep scarp or feeder zone which supplies slope material to a gentle and generally concave temporary accumulation zone, which ultimately feeds into the foot. The foot is wider than the body of the slide track and therefore the landslide demonstrates a divergent slide/flow pattern. The landslide has a generally concave form with steeper and topographically higher right and left flank. The failed upper slope material is therefore bounded within these high flanks. Marginal slickensided shear and failure surfaces also help confine the moving mass of soil.

5 BEDROCK PROPERTIES

The study area is characterized by interbedded siltstone and shale sequences of late Tertiary age belonging to the Mayaro Formation and Moruga Group (Barr and Saunders 1965). These sequences are thinly bedded and are usually dark grey to black, but occasionally brown. Beds are well defined and measure between 2 to 5 cm thick.

Mudrocks are highly fractured and disintegrate into angular and trapezoidal shaped blocks. These fractures represents oblique to normal faults dipping southwest. Stress relief cracks are also widespread within these units, and may be due to removal of overburden stress (probably associated with initial deposition).

Bedrock are particularly prone to wetting and drying effects, especially in response to daily tidal and seasonal rainfall changes. This causes rapid crumbling and disintegration of these lithologies into angular fragments. Natural water content of shale samples collected within the zone of permanent saturation range from 18 -20%. These samples are usually soft and plastic, especially on their outer surfaces. Water absorption tests conducted in the laboratory revealed water absorption of 35%.

Discontinuities within shale are less than 60 mm wide and usually contains clay infillings. These are especially plastic and are potentially expansive. The infillings are usually grey/black to brown and are usually moist, sometimes wet, even during dry periods. Discontinuities are usually planar and smooth, with minimal surface asperities.

Bedrock are moderately to highly weathered. Between 35-70% of the rock mass have been converted to soils. Excavation of test pits and augering of boreholes within the body of the landslide revealed that similar weathering grades are present within the surface 2 m of landslide material. Below this depth, less weathered bedrock is present.

6 SOILS

Soils from the flow track and accumulation area of the landslide are generally coarse and can be classified as debris. Most particles are within the sand size category, with clayey fines and with angular, gravelly shale fragments. The largest particle sizes are within the coarse gravel fraction. These soils represents a mixture of surface highly weathered clays with less weathered sub-surface mudrocks.

Soils are dark brown to dark gray, depending on the original color of the parent mudrock (gray or brown). Surface samples are generally dry to moist, but always wet during the rainy season with deeper samples have higher water content. Fines are highly plastic, while plastic threads are stiff. Within the accumulation zone these soils are soft when wet, forming a hard outer surface crust when dry. These usually produce desiccation cracks down to 40 cm deep and as much as 15 cm wide. Fine soils also demonstrate slickensides following failure of slope material along the flow track.

Measurements of unconfined compressive strength (Uc) along the walls/sides of test pits 1 and 2 revealed that soils located near to and just above seepage zones are of low strength. Unconfirmed strength decreases with depth, from a moist to dry surface soil to soils along seepage zones at 80 cm deep. Soils above failure surfaces have Uc values of less than 2.00 kg/cm^2 but greater than 0.25 kg/cm^2. Below this surface Uc values are greater than 2.00 kg/cm^2 with a peak of 3.25 kg/cm^2. Failure surfaces like those identified in test pits are characterized by yellow/brown, highly plastic, slickensided clays of low Uc values and along or above which seepage zones are present. It is significant to note that soils within the flow track which are sliding and are above failure surfaces still possesses relatively high unconfined shear strength, with values between 0.25 - 1.25 kg/cm^2.

Measurements of thoroughflow rates along seepages zones revealed that these rates can be very high. This suggests that the effective porosity and permeability of these soils are high, even though plastic clays and silts are present. This is due to the high sand and gravel content of these soils. This can also explain why high seepage pressure and thoroughflow rates can develop within the deposit.

Excavation of test pits revealed that the surface 60 cm of soils in the flow track and surface 100 cm soils in the depositional zone are of low density. Seventeen soil samples collected had between 0 - 35 gravel, 21-62% sand, 10-29% silt, 6-30% clay and 22-59% fines. Gravel and sand fractions were highly angular, with planar, smooth surfaces and composed entirely of shales. Plastic limits vary between 20-25% and liquid limits between 65-72%.

7 SURFACE AND GROUNDWATER FLOW

Due to the concave nature of the ground surface created by past and present landslide activity, surface flow and runoff from the upper slope is channelled along the length of the landslide. However, these flow are only short term and usually infiltrates into the ground via tension cracks, secondary scarps, slickensides and desiccation cracks to become thoroughflow. In some flatter segments of the body of the landslide, rainfall and surface runoff accumulate, producing water ponds (Figure 1). This usually occurs only during and following heavy rainfall events, more likely after field capacity has been satisfied, or surface tension and desiccation cracks have been closed up. In such instances, the walls/sides of cracks usually crumble and fill in the lower sections of these discontinuities (which act as a deterrent to further ground water infiltration).

In the zone of accumulation of the landslide and near the toe, several small springs a represented with flow parallel to the dimensional length of the slide. Here, very significant discharges are common, especially following rainfall. Measurement of this surface flow during October, 1992 (rainy season), revealed that the discharge can be greater than 60 cm^3/min. It is significant to note that when this measurement was taken there had been a relatively long dry period at the study site (more than 2½ weeks). Consequently, higher discharge should be expected for more rainy periods.

Groundwater flow is particularly confined to the surficial 1 - 2 meters of material in the slide track and depositional zone. Flow is near-parallel to the ground surface and occurs as definite shallow seepage zones. These zones can vary from 20-25 cm deep to 100-115 cm deep. The shallower zones are mainly found at the base of the landslide, while the deeper zones are confined to the body or slide track.

Measurement of groundwater flow along these seepage zones revealed that flow varies from 1.68 cm^3/min in shallow, near-surface, highly plastic clays, with very few sand and gravel shale fragments, to 42.69 cm^3/min in slightly deeper plastic clays, with significantly more gravelly shale fragments. These correspond to discharges of 79 cm^3/min/m^2 and 202cm^3/min/m^2 of slide area. It is significant to note that these measurements were taken following a period of 2½ weeks without rainfall. Much higher rates can therefore be expected during more rainy periods.

Natural moisture content in boreholes 1 showed that this parameter increases from 31% at the surface to a maximum of 43% at approximately 1 m deep, then decreases to 38% between 1.5 - 2.0 m deep. In borehole 2, surface moisture was 26%, increasing to 39% at 1.0 m deep and then decreasing to 33% between 1.5 - 2.0 m deep. Measurement of the piezometric levels in these two boreholes in the slide track revealed that there was a steady rise in groundwater levels from the dry season into the latter part of the rainy season of the year. The water levels at the start of monitoring (in April 1992) of borehole 1 was 115 cm below the surface. In borehole 2, the depth was 100 cm. In borehole 1, the water level reached the surface at the start of September, until it was subsequently displaced downslope. In borehole 2, the water level reached the surface in July and maintained that level throughout the following month until

September. During mid-September 1992, boreholes 1 and 2 were displaced downslope. Therefore, further reliable measurements could not be taken at this site. Borehole 1 was also drained of all its water following displacement.

It should be noted that although the water level was at the surface of the slide track from as early as June 1992, measurement of subsurface flow in the same area of the foot of the slide revealed relatively low discharge. It is believed that this may have been due to diversion of sub-surface flow to the left lateral flank of the slide, as well as the retention of some of this water within the soils of the slide track. This may also be partly responsible for the development of undrained loading and the development of positive pore water pressure within the slide track material and therefore may be an important contributory element to subsequent sudden failure or surging within the body of the slide.

Laboratory measurement of the natural moisture content of soils from small debris flows at the toe of the slide and in the vicinity of a line of springs (within the foot), was done for three times periods, June, August and December 1992. The result of this analysis showed that the water content in flowing material (9 samples each time) varied between 36 and 45% with a mean of 42% in June, 40 and 45% with a mean of 42% in August and 26 and 39% with a mean of 33% in December. This suggests that fluidisation of the toe material can take place at low water contents (less than the liquid limit of the fines). This process may be facilitated by the poorly graded nature of the deposit.

8 COASTAL LAND LOSS

Because the topography of the slide changes significantly with continuing movement, the failure was surveyed several times of the year. This has allowed the calculation of coastal land loss between successive periods. Approximately 23,262 m^3 of sediments were lost to the marine environment prior to the commence of this study in April, 1992. Between May to December 3, 1992, a further 9,066 m^3 were lost from the landslide site. Of this volume, 2,521 m^3 were lost between May 27 to June, 1992 and 6,545 m^3 between June 26 to December 3, 1992. The land loss during the latter period was more so during the months of July and August, when higher rainfall was experienced and significant failure and movement occurred. This

represents a total loss of 32,328 m^3 of hillslope material from this landslide site.

Of the total land loss during the survey period, at a total of 9,155 m^3 of material was degraded from the landslide scarp. This material entered the slide track with 9,066 m^3 of it (99%) being transported downslope and lost to the marine environment and 89 m^3 (1%) being retained in the slide track.

Between April to June 1992, 3217m^3 of material entered the slide track (from failure of the scarp), but only 2521m^3 was lost to the marine environment. This represents a retention of 685m^3 during this time. Between June to December 3, 1992, a further 5937m^3 entered the slide track (from failure of the scarp), while 6545m^3 was lost to the marine environment during this time.

9 MOVEMENT PATTERNS

9.1 *Upslope Displacement*

Major displacements occurred between April to May and August to September, 1992. Tubing 1 was displaced 21.50 m downslope between May 9 to December 3, 1992, while Tubing 2 was displaced 15.50 m during the same period. The most significant displacement of each plastic tubing took place between April to May, 1992, with 8.60 m for Tubing 1 and 8.20 m for Tubing 2. However, there was greater displacement of Tubing 1 due to its' proximity to the foot of the failure and therefore its' subject to greater shear stresses from the upslope sliding mass. Larger displacements were recorded during rainy periods, while during dry periods June to July 1992, no displacement was recorded. Three weeks without rainfall during October 1992 also caused a reduction in movement to only 0.50 m for that month.

Tubing 1 was deflected 22° from the vertical on October 14, 1992, 30° one month after and then to 40° up to December 3, 1992. This deflection was concentrated at approximately 1.0 m deep, causing the tubing to become concave upslope. Following December 1992, the tubing broke and was deposited with the soil mass at the foot of the landslide. An interesting pattern of upslope displacement was recorded for both tubings. Tubing 1 and 2 maintained an equal spacing of 31 - 31.70 m from the start of the survey in April 1992 until August 1992 (even while both were in motion). After August, the spacing increased by about 6 meters to 37.10 m. During the initial period of the survey, the groundwater level was at 1.15 m. In August 1992, when the spacing between tubing 1 and 2 increased, the level rose to o.40 m near tubing 1 and was already at the surface at tubing 2 from as early as June 1992. This period of significant displacement, was also accompanied by a continual rise in the piezometric surface. From the movement pattern observed, it appears as though there is a slow continuous and constant rate of movement of the material in the slide track between April to August. After August, the rates are increased, ultimately culminating in surges. It is also significant to note that such surging behavior seems to be taking place within the slide track, approximately 60 - 65 m from the toe survey stake. This suggests that passive earth pressure from upslope loading may be very high in this part of the landslide. This may also be the area where undrained loading and positive pore water conditions (within the sliding mass) may be at their peak.

9.2 Surface Vector Movement

Within the upper section of the flow track, there was a consistent sliding direction towards the 140^0. This was determined from displacement of plastic tubing 1 and 2 placed within the slide track. Within the lower section of the slide track, movement direction was measured using four stakes placed at equal spacing (3 m), along a bearing of 148° and along the center line of the foot and within the soil. Analysis of field measurement data shows that there is a net displacement of slope material to the left flank of the slide from within the vicinity of stakes 1, 2, and 3. This is followed by movement more southeasterly, until it reaches the toe. Material just above the toe (Stake 4) does not show any appreciable change in movement direction. Material upslope of stake 1, moved towards 148^0. In the vicinity of stakes 1 and 2, the direction shifts further east, towards 108^0. Further downslope movement is accompanied by a further directional change, towards 060^0. This causes a significant diversion of upslope displaced soil to the left flank of the foot, towards the area a the line of springs and secondary debris flows are found. Further downslope movement here reverts to the southeast, to 131^0 and eventually towards 150^0 at the toe of the deposit on the beach. This movement pattern shows that much of the upslope material is diverted to the left flank of the deposit and therefore is not uniformly distributed over the beach area. Although

some material is deposited on the other flank and center of the foot, this is much less. This flow direction is also consistent with thoroughflow direction and the location of seepage zones.

10 MECHANISM OF FAILURE

The mechanism of failure operating within this landslide is extremely variable and can be complicated. At this point, enough information is not available to conclude on the same, as this is still being investigated. However, several field observations indicates certain aspects about the failure mechanisms.

It is apparent that sliding is the major form of movement associated with this landslide. This is especially so for the slide track where the main body of soils are being displaced. Here slickensides can be seen immediately after failure along the flanks of the landslide, with stretched plant roots.

The foot of the failure is characterized by both flow and sliding. Between the slide track and depositional zone, where the slope changes from 8^0 - 12^0 to a steeper 30^0 slope, flows are common. This area is also where sub-surface flow intercepts the ground surface (springs). Therefore, the supply of water from these springs here causes further saturation of the deposit leading to secondary debris flows. Based on the water content from samples taken here, it was found that flow can take place when the soil has only 27% moisture (a water content between 26 - 45% was recorded for soil samples taken here). Based on measurement of the thoroughflow rates here (between 79.62- $202.32 cm^3/min/m^2$) slurification of the deposit can take place easily.

Within the slide track, loading by deposition of failed scarp material (between 60 - 70 m from the toe) may cause the generation of high passive earth pressure especially where water is incorporated within the soil. This may become a major driving mechanism for movement, leading to bulging in the lower part of the slide track, and ultimately to catastrophic or sudden failure by surging/bursting. Surging can cause an effective decrease in passive earth pressure, at the point of loading and the creation of a concavity within the upper reaches of the slide track. This failure mechanism may also be partly responsible for the development of the numerous tension crack on the surface of the landslide, oriented almost perpendicular to the direction of mass transport. Surging and sudden failure, can also cause the development of high active earth pressure, accompanied by longitudal stretching along the length of the landslide and the center of the slide track. Consequently, many large tension cracks can develop in this part of the landslide.

From analysis of the deformation of flexible plastic PVC tubings within boreholes augered within the slide track, it is apparent that the rate of movement is greatest at about 1 m deep and therefore, is not uniform with depth. This movement is greatest along and just above seepage zones within the landslide. This non-uniform movement with depth and near to seepage zones further suggest that the surface of sliding is dependant on the location of the seepage lines. The measurement of the unconfined compressive strength of soils along the walls of test pits has showed that this area also represents the juxtaposition of soils of two different strengths. The surface sliding mass has strength values less than half of the underlying material.

11 CONCLUSIONS

Geotechnical investigations have revealed that landslides at the study site are mainly translational debris to earth slides, which moves along planar shear surfaces, near-parallel to seepage zones. Movement appears to be in response to high passive earth pressure following rainfall and increase in pore water pressure. Undrained loading also appears to be partly responsible for generating high shear stresses within the sliding mass. As a result, high volumes of sediment are deposited in the coastal and nearshore environment.

Further investigations are aimed at detailed examination of the mechanics of failure of this landslide based on earth pressure theory. This can provide useful information for engineering these types of slopes in Trinidad.

12 ACKNOWLEDGEMENTS

The author is grateful to the Institute of Marine Affairs, Chaguaramas for supporting this research and to Mrs. Charmain Pontiflette-Douglas for carefully typing this manuscript.

REFERENCES

Bertrand, D., O'Brien-Delpesh, C., Gerald, L. and Romano, H. 1991. Coastlines of Trinidad and Tobago - a coastal stability perspective. In G.Chambers (ed). New York *Coastlines of the Caribbean*: 1-16. New York : American Society of Civil Engineers.

Bachew, S., Hudson, D., and Gerard, A. 1983. Analysis of the coastal problems at Los Iros Bay, Trinidad, West Indies. *Trans. 10th Carib. Geol. Conf.* Cartagena, Colombia.

Deane, C. 1971. *Coastal erosion - Point Fortin to Los Gallos*. Second Interim Report. Trinidad and Tobago : Ministry of Planning and Development and Ministry of Works.

ASTM, 1982. *Annual book of ASTM Standards, Part 19, Soil and rock and aggregrates*. Philadelphia: American Society for Testing Materials.

Barr, K.W., and Saunders, J.B. 1965. An outline of the geology of Trinidad. *Trans. 4th Carib. Geol. Conf.*: 1-10. Trinidad.

Berridge, C.E. 1981. Climate. In: G.E. Cooper and P.R. Bacon (eds). *The natural resources of Trinidad and Tobago*:2-12. London: Edward Arnold.

Varnes, D.J. 1978. Slope movement types and processes. In: R.L. Schuster and R.J. Krizek (eds). *Landslides, analysis and control*: 11-33. Washington: National Academy of Sciences.

Maharaj, R.J., and Pfister, M. 1992. The stability of coastal cliffs on Radix Point, Trinidad, W.I. In:

M.Hermelin (ed). *Proc. II Latin American Symposium on Geological Risks in Urban Areas*: 225-246. Colombia.

Sea level rise, climatics, morphodynamics and delay in land processes

Élévation du niveau de la mer, climat, morphodynamique et retard sur la conquête des terres

J.Javier Díez
Politechnical University of Madrid, Spain

ABSTRACT: Since the Bruun's rule, linking shoreline regressions and coastal erosions due to it, sea level rise has increasingly drawn the attention in different fields of knowledge. But it has been likely taken into consideration for future estimations more than for past time analysis. Climatic changes had to have a strong influence on the development of sailing routes; they had to affect also land accessibility, by their effect on beach slopes and configuration of bays, estuaries, lagoons an inlets (harbour areas much principally being linked to them); and they could be meaning factors on landing process, in addition to other climatic and environmental considerations affecting living conditions, especially in low and wet lands. Columbus' looking for a pass towards "Cipango" and China cannot explain the great delay in european settling (not exploration) on the whole east coast of North America. This paper also extends to different types of climatic change impacts (and not only sea level ones) on coastal evolution, critically analizing the possibilities of discerning the impacts really due to the "sea level rise" and their consequences.

RESUMÉ: L'élévation du niveau de la mer est un principal problème côtière et litoral depuis que BRUUN établi sa relation entre les niveaux de la mer et les procès d'erosion des plages; mais il n' a pas été considéré sufissament dans l'analyse des temps passées, de manière que le problème est en train d'être mauvaisement interpreté. Dans ce travail les changes du niveau, relationnées avec les changes du climat, et aussi avec l'évolution historique des routes maritimes et de la morphodynamique côtière son analisées. Et tout celá est utilizée pour traiter d'expliquer le grand et paradoxal retard des colonies europeennes sur la côte Est de l'Amérique du Nord.

1. INTRODUCTION.

Climatic changes along "human era" are fairly well known, though not totally well defined. Until recently, only "great changes" have been taken into account, assuming that climate (not weather) could be considered as a constant, for all practical purposes. From then on, other minor climatic changes are receiving increasing attention, particularly those having affected both protohistoric and, over all, historic times. The generally assumed relation between the carbon dioxide emission rates and some present climatic changes, through the "greenhouse effect", has led to the meaning perception of the climate variability and its impacts on environment and other human life conditions. And although special alarm was produced by its effect on sea level rise rates, it also drew to the "climatic determination" of the History. Little Ice Ages and genial warm periods in the past have been noticed from literary and other artistic descriptions (Lamb, 1982), but they are being analized and quantified now in a more accurate and comprehensive way.

Since the Bruun's rule (1962), which linked the shoreline movements, and the coastal erosions/accretions related to them, with sea level changes, the present "sea level rise" has increasingly attracted the attention. Although obvious, the climatic influence on sea level changes has only recently been emphasized in short term analysis, and related to current a) increasing carbon dioxide emissions and rates, b) warmer climate and c) sea level rise. On the other hand, the wellknown influence of maritime climate on coastal morphology and morphodynamics appears not to have been fully considered until recently: longshore and onshore-offshore litoral transports, erosive and sedimentry processes and genesis and migration of barrier islands and other forms of sedimentary deposits have been more or less widely studied, though primarely in relationship with current environmental impacts; however, other longer term morphodynamics and pattern changes seem not to have been so extensively considered. A time-correlation was found (DIEZ, 1992) between different post-Würm orientations of a restricted shoreline stretch and the contemporary average latitudes of the paths of extratropical cyclones, which seem also to be linked to the Saharian succession of pluvial and interpluvial periods (M. de PISON); it supposes an "immediate answer" of "short enough" coastal stretches to average wave direction. And maybe the same hypothesis can explain the perceivable change of orientation of Long Beach Island (N.J.) during the last two centuries (C.E.R.C. pp. 4.7-4.9). This kind of topics seems not to be fully incorporated to the analysis.

Climate, affecting the oceanic pattern of winds and currents, had a strong influence on sailing routes. Climate put also constraints on accessibility, through their effect on beach slopes and coastal configurations (harbours laying out depending on them); and these have been meaning factors, with other climatic and environmental, for settling, especially in low and wet coastal areas. The analysis of the problem with a certain accuracy is only possible with the documents of the last two or three centuries, and it is easier in Eastern and Gulf Coasts of North

America than in Europe for the greater documentation availability about its pre-anthropic coastal conditions. When europeans arrived to North America were about of being able of mapping accurately, but at that time coastal lowlands in Europe were notoriously anthropized already. Unfortunately the little Ice Age had practically concluded in the middle of eighteenth century, when the first detailed maps and charts were available, but the poor older coastal cartographic information itself may be related to the special difficulties to settle in coastal areas in all Gulf and Atlantic Northamerican coasts. Columbus' hypothetic aim and insistence in looking for a pass towards "Cipango" and China can explain a certain delay in spanish exploration and even landing of more northernly latitudes, though it can not explain why the whole east coast of North America remained practically unoccupied so long.

2. RECENT AND LATE QUATERNARY CLIMATIC CHANGES.

The process of global warming seems to have been more or less persistent and relatively fast since the last glaciation (15000-30000 yrs., depending on different authors, though some extend it up) until well into the Neolithic, though it must have suffered meaning fluctuations, in agreement with marine transgressions and regressions (RABAN, 85 7 GALILI, 85 - CFR SIVAN, 90). Würm glaciation reached its top with an average temperature of about 5°C less than at present. Late Paleolithic ends in Eurasia with a far march away after "deere", while ices retired, and the warmer climate may have been the principal factor arising the Neolithic in the middle-east (11000 yrs b.p. about). But the transgression atenuated about 8000 years ago and could even stop and be followed by a meaning regression during the cooler next milenium (RABAN, 85); Egyptian and Mesopotamian empires flourished in that period. A very stabilized (likely fluctuating) warmer period ("atlantic"), but still cooler than today, followed for nearly two milenia (emerging indoeuropean peoples), before coming a new 2000 years cool period ("subboreal") which rose the Sea peoples and the golden Greek world; in the following increasingly warmer milenium the Roman empire emerged and died. The warmest moment of all this postglaciar period seems to have happened about 5000 years ago, with 1°C approximately above current average temperature.

So a certain historical determinism seems to relate cooler periods with emergence and domination of nations from mediterranean and southern and/or lower lands, whilst warmth carried nations on from northern and or higher areas. Warmth weaked Rome and extended "barbarian". The cold of ther former "Middel Age" flourished the mediterranean societies up to the establising of muslim and roman-germanic empires. And the new warmth of the later Middle Age ("little climatic optimum", 1000-1450) led to their decadence, the emergence of Hansa, The Romanic and Gothic artistic periods and the expansion of wikings. The beginning of the epysode known as "little ice age" is widely placed around 1450; the life conditions hardened in all central and northern Europe, becoming extremely cold and dry except in its west end, softed by the Gulf Stream, and have been showed in contemporary documents (LAMB, 1975); on the contrary they were mild around the Mediterranean, becoming even better (in agreement with the hypothesis of african "pluvials" coinciding with glacial epysodes, wealthier and much healthier (decreasing the epydemies, so frequent in the previous period), which led to the "Renaissances" -christian and islamic - and permited the othoman and spanish empires to establish, among many other meaning events which characterize the Modern Age. The coldness must have been very significant to permite the increase of ice surface in the polar and montainous areas.

2.1 Sea Level Changes

The marine transgression has followed the climate evolution as a general rule, with its minor fluctuations, though modified by other isostatic and sedimentary factors; as a matter of fact the sea level has been namely used to approximate the quaternary climatic changes. According to Sivan, the transgression atenuated in Calcolithic (10000-7000 b.p. aprox.), and the Flandriense transgression could likely reach its top between 5000 and 3000 yrs. ago, with between 2 and 4 meters above p.s.l. (present sea level), falling down until 1 meter under p.s.l. 2000 years ago. Close after the first milenium nederlander Frisia required some dique shore protection, which might indicate the beginning of a marine transgression (corresponding to the "little climatic optimum") whose last episode (the last but one of the more comprehensive Flandrien transgression), the Dunquerquian, begun in the XIIIth century. On the contrary, a very important generation of "polders" (Beemster, Purmer, Vermer, Shermer...) was dried in the Nederlands along XVI and XVII centuries, taking advantage of the strong regression during the "little ice age". The average temperature must have decreased rather more than 1°C in that period. since it was enough to induce not only volumetric changes (as in the top of Holocene very principally happened, with 1°C) but important eustatism as well.

Leaving apart all other factors, and taking into account the relative sea levels corresponding at both the bottom of the last glaciation (80-130 m., depending on authors and places) and the top of the Holocen (2-7 m), the last regression had to be very important; the generally agreed band of aproximately 3 m. above /under current sea level has to have been underpassed just then, especially in strongly subsident coastal areas as mostly in the middle east are. Nevertheless, we must consider that in long term thermic changes, their eustatic and volumetric variations only can be hidden/amplified by tectonic and primarely isostatic effects, but in shorter thermic changes even diferred processes of subsidence (glaciar and sedimentary) and consolidation (sedimentary) can notably interfere. Therfore the undertaken variations of the last centuries need a strong depuration.

3. COASTAL MORPHODYNAMICS INVOLVED.

As a primary general thought we must notice that transgression / regression movements are not the only climatic effect on littoral morphology. A primary characteristic of the climate is the average and extreme latitudes of the extratropical cyclons paths; it affects the fetch and, consequently, the effective direction of the wind waves causing littoral processes. If the "answering inertia" of a coastal stretch is under a certain value (barrier islands, pocket beaches,...) it can move in plant and change its orientation becoming more perpendicular to the climatic average wind wave direction (in according with the well known behaviour of the groin-artificial beaches under variable weather). The cases refered in the Introduction - in Javea, Spain, and Long Beach Island, N.J. - are supposed to obey this kind of morphodynamic evolution.

Concerning to northamerican east coast, between Yucatan and Terranova, three wide type of coastal zones can roughly be distinguished: 1) northern of Long Island, of glaciar origin, rock-morrenic morphology and roughly negative subsidence; 2) Florida platform, karstic, stable and no subsident; and 3) Gulf and Mid-atlantic coastal zones, sandy and with positive subsidence.

The "rigidity" of the first region keeps relatively invariable its structural morphology and its rocky and tortuous bottoms (notice their bathymetries) though transgression has led to coastal accretion of marine sands and gravels, produced from cliffs and bottom erosion. Landing on this coast without charts or previous knowledge is extremely risky; and it had to be worse with a lower sea level.

Florida may be used as the best reference in northamerica for considering the sea level movements and their effects; meanwhile the noticeable reefs had to

have remained unaffected but for erosion and relative depth, both related, we must question the existence or, at least, the character of the present barrier islands, for the unavailability of enough sandy materials, which seem to have increased during the last three centuries for anthropic deforestation in Appalachian region; yet supposing their existence, even in their present magnitude roughly, they had to have been generated by litoral drift, as spits (WILLIAM & BUILDING), so that, in any case, their growth -as the dunes' and other continental sand deposits'- had to increase during the cold period -greater litoral dynamic- and its following time of anthropic -european- appalachian deforestation, and their questionable landward migration -their growth could compensate the sea level rise effects, as it has happened in other hyperstable (BORES) coasts -did not affect the morphology nor reduce the dimensions of lagoons and bays; and since tidal prisms favour, against longshore drift, inlet steadiness sailing and landing conditions then had to have been much worse than nowadays.

In within the third coastal type, a) the Gulf coast, with much smaller tides and extraordinary abundant continental sands, is principally deltaic, hyperstable, subsident and fully spit-barriered, so that the significance of barrier island migration is also uncertain; therefore the accesibility to mainlands had to have been also difficult. b) the atlantic coast, with greater tides and estuaries, much less continental sediments, but in Georgia and part of South Carolina, where most of estuaries have become deltas, can be subdivided in two general regions, depending on subsidence and amount of sediments, approximately separated by Cape Hatteras. In the northern stretch the barrier migration under sea level rise has been shown (LEATHERMAN & AI) and may be related to the lack of natural nourishment; the subsidence is moderated; the estuaries -in great glaciar valleys- have become greater, salter, and less "estuarine"; shorelines retreated and

3068

wetland reduced. In the southern stretch the deltaic evolution of the estuaries -in smaller and fluvial valleys- has led to shoreline advance and renewed front islands, and more extended wetlands with fresher waters. There are appreciable differences in offshore and shelf slopes whose influence on morphodynamics of both zones is worthy of considering, but we only notice here: a) the existence of several multibarrier stretches the southernmore along each the narrower its bay or lagoon; it gives a continuity along every stretch as if its islands shaped one only and long spit. And b) the inversion in the sense of the longshore transport in the northern part of each island for the wave refraction on ebb tidal deltas. Both circumstances affect the unsteadiness of inlets, and more during the "little ice age" and following, what decreased the accessibility of mainlands.

4. HISTORICAL REVIEW.

The first permanent settlement of europeans in the Atlantic coast of North America was San Agustin, Fl. (1565) (Quebec cannot be considered a coastal place), and it had a nearly only military function, to protect the spanish convoys in their return to the mother country. After that, considering enough protection, spanish looked for the expansion into the country towards the Tennessee valley, while they only established one more permanent settlement in the coast (in Perrin Is. 1566), which was left in 1587, after Saint Agustin destruction by Drake (1586); by then, precisely, the misteriously disappeared english Roanoac had been settled (1586) in Albermale sound, N.C. The several known wreks up and down of Cape Hatteras show however that spanish tried unsuccesfully to establish along the coast up to at least Chessapeaque bay. And the first french and english settlements do not happen until 1605 (Port Royal, N. Scotia) and 1606 (Jamestown, V.). The accurate map of the bay area drawn by then (1616) by Harriot & White shows a cape Henry morphology perceptibly different to

nowadays, with its headland hooked northeastward instead of southwestward. The National Geographic Soc. confers to this map a high reliability, therefore it must have a great morphodynamic meaning. The differences have to be related to the average wind wave direction, which must be more oblique from the northeast, and with more capability for longshore littoral transport, at present, which is consistent with bigger path latitudes of the extratropical cyclons and with warmer climates.

As a matter of fact it is amazing the delay in establishing settlements on the atlantic coast, specially in relation with its early exploration. Explanations based on obsessive asiatic aims of Columbus cannot be used not even before his death happened, taking into consideration the wider objetives of the spanish empire, as legal, organizative and chartographic documents show. And others based on piracy practices or in lack of interest of europeans in Northamercia itself, have been shown (DIEZ, 1992) inconsistent. The explanation contained in next point is based on the specially dramatic climatic conditions and, meaningly, on their littoral morphodynamic effects.

5. HYPOTHETICAL CONCLUSION

Although this paper does not pretend any detailed research, the opportunity of the proposed question has been proved, especially as the present climatic character seems to be dramatic. This short review shows that major climatic changes have happened along relatively recent episodes and some of them can be related to some historic problems, directly and through their influence on coastal morphology. This hypothesis can be supported on the following inductions about Atlantic and Gulf coasts (though the present lowland appearance of the whole stretch between Yucatan and Long Island shows similarities, there are important genetic and morphodynamic differences which were likely more notorious during XV-XVII centuries, with a 2-5 m under present sea level, and which affected differentially to ship accessibility and maneouvrability under much less knowledge):

- The general feature of Gulf coast could seem to have been siilar to nowadays but the lower sea level, probably increased by subsidence, has to have accentuated the port and landing advantages of some punctual singularities as old (Florida, Yucatán) and younger (Veracruz, quaternary?) littoral reefs or structural hadlands (Alabama). These singular points have been, precisley, the landing areas to settle and penetrate the continent westward from the Appalachians. Florida, dangerous for its abundant reefs and unsteady inlets, poor, karstic, wet and not yet so eolically infilled with the apalachian and calcareous sands from littoral drift, only was used to protect convoys and as platform to cross the mountains towards the Tenessee valley.

- In New England the conditions were likely more similar, not being clear the sense of the subsidence since then; but it suffered much worse climate and sailing route (north) conditions (for ice and storms). Besides, the referred nature of the whole coastal region maked it very unaccessible, being Plymouth one of the very scarce accessible points from the sea. The succesful Cape Cod landing and survival was beleived to be miraculous and even providential (Providence, R.I.) and it remains as a first mark of the american character, what reveals the difficulties of the venture even in 1616. And the other meaning early coastal settlements ere selected from the land.

- The greater changes must have happned in front of Apalachians, where the accesses to mainlands, drylands, had to be specially difficult for not very adequate ships and under extremely variable morphodynamic conditions (notice these coasts were usual refugee for privateers and corsairs). The sea level could be 3-5 m under present, depending on subsidence; the difference was smaller

northern of Cape Hatteras, with less subsidence and continental deposits, where precisely english firstly settled. The greater subsidence southern of the cape could be partially compensated with shoreline advance by continental -fluvial-deposits, but it had to occur with a very great variability of inlets and shoals, and of sailing and landing conditions. A similar process could happen between Cape Henry and Long Island, but with much less deposits and subsidence (diferred); its three big estuaries make easily accessible the mainland in this stretch at present; and no morphologic factors can explain the first landing having happened in Cheassepeaque bay, the most infilled and less fluvial of them; therefore, climate itself must have been first factor as in New England.

ACKNOWLEDGEMENTS

To the spanish universitary authorities and to the spanish-northamerican ensemble commitee for their respective grants.

REFERENCES

BORES, P.S. Clasificación de formas costeras simples. Interior report. E.T.S. Ingenieros de Caminos, Canales y Puertos. Madrid 1975.

DIEZ,J.J. Estudio de los procesos litorales en la Bahía de Javea (Alicante), 1984. (Restricted document)

DIEZ, J.J. Cambios climáticos y geomorfológicos: su influencia en el proceso de asentamientos españoles en Norteamérica. Rev. Obras Públicas. Oct. 1992. Madrid.

E.P.A. Climatic changes effect. Washington, 1990.

GIEGENGACK,R. Personal comment. Philadelphia, 1991.

LAMB, H.L. Climate History and the modern World. Methuen & Co. N. York, 1982.

LEATHERMAN, S. & al. Virginia barrier island configuration. A reappraisal. Science, 1982.

MARTINEZ DE PISON,E. Personal lecture. Fuerteventura, 1986.

SIVAN, D. Personal communication. Sandy Hoock, 1990.

WIALLIAM, S. & BUILDING, K. Barrier island shorelines: an assesment of their genesis and evolution. Restricted paper. Fort Belvoir, 1982.

Artificial beach nourishment and offshore dredging effects on the Alicante coastal area (Spain)

Alimentation artificielle des plages et les effets des emprunts offshore sur la zone côtière d'Alicante (Espagne)

V. Esteban Chapapría
Laboratorio Puertos y Costas, Universidad Politécnica, Valencia, Spain

ABSTRACT: Artificial beach nourishment requires borrowed materials. Demand for offshore borrow areas consisting of sand is increasing. Comprehensive studies in order to select borrow areas have been promoted in Spain in recent years, including site investigation. A borrow area was identified in the Mediterranean area located in front of Sierra Helada close to Benidorm. This area has already been used to implement some restoration projects in large beaches in the Alicante area. More than 3,000,000 m³ has been exploited using trailing suction hopper dredges. Several studies on littoral processes were previously carried out to test potential effects of offshore dredging on the stability of beaches and coasts. They were especially assessed in the Benidorm area. The paper focuses on the results of both the preliminary geological studies on the borrow area and the impact assessment on the Benidorm beaches. The conclusions highlighted the need of monitoring the borrow area and some morphological features.

RESUMÉ: L'alimentation artificielle des plages requiert des matériaux d'emprunt. La demande de ces materiaux est en augmentation. Des études destinés a sélectionner des zones d'emprunt se sont développées en Espagne ces dernières années, incluant le site de recherche. Une zone d'emprunt a été identifié dans une zone mediterraneénne située en face Sierra Helada, près de Benidorm. Ce lieu a été déjà utilisé pour implanter des projets de restauration sur les larges plages de Alicante. Plus de 3,000,000 m³ ont été exploités en utilisant des dragues aspiratrices en marche. Plusieurs études sur les processus littoraux ont été realisés auparavant pour tester les effets du dragage au large sur la stabilité des plages. Ces tests ont été principalement centrées sur la zone de Benidorm.

1 INTRODUCTION

Coastal zone is in most cases an area of conflicts. The transition between a marine and terrestrial environment presents special dynamical problems. Physical processes on the coast are many and varied. Furthermore, time-scales of these processes are very different. The identification of coastal changes is essential for coastal planning and management. For many years people have been increasingly moving towards the coastal zone. Development and usage of coastal zones are increasing. Then, many interests are involved in this area.

Man's intervention in the shoreline processes has been considerable. Effects have resulted from coastal developments, specially harbour infrastructures. Coastal defences and selective protection of developed areas have changed the coastline. In many

cases problems derived from man's ever increasing proximity to the sea. All around the world coastal erosion is frequently the final effect. Coastal engineering has considered many options depending upon the final objective: (1) land reclamation, (2) coastal protection, or (3) beach restoration. Alternatives usually considered either singly or in various combinations are:

1. Retreat and set back.

2. Coastal structures: sea walls, rock groynes, offshore breakwaters and reefs, artificial headlands and artificial beach nourishment.

Measures and methods of coastal protection and land reclamation have been widely described and compared. In recent years soft coastal protection works, especially artificial beach nourishment, has become increasingly considered as an effective solution to coastal erosion. Nowadays environmental arguments have far greater consideration.

2 THE ALICANTE COASTAL AREA.

The Alicante coastal area is located on the Spanish Mediterranean coast (fig 1). In recent years, aproximately since the sixties, Alicante has undergone great economic changes. The most significant have arisen from holiday and tourist resort development. Benidorm and many other places along this coastline have international recognition as major holiday resorts. The Alicante area has been, probably more than other any area in Spain, under pressure from a growing population in

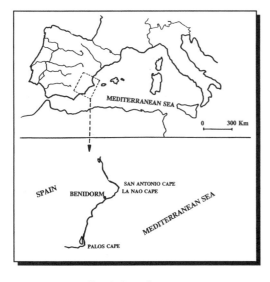

fig. 1 Location map

recent years. This population has become increasingly mobile and has more leisure time. Natural environment and very good climatic conditions are much appreciated in Alicante, especially in the coastal zone.

2.1 Geological and geomorphological general frame

The central Spanish Mediterranean coast is morphologically divided by the Betic alignments. The general geological and geomorphological frame differs on both sides. The Nao and San Antonio headlands are located at the end of the Betic alignments in the Iberica peninsula and are located in the Alicante area. The coastal area is

morphologically varied. Denia is located in the north of the San Antonio headland, in the Valencian gulf. The Denia coastline is comprised of large fine sandy beaches. Southward at the San Antonio and Nao capes the coast becomes structural due to Betic influence. The coastline which extends to Calpe and Altea areas contains important cliffs and pocket beaches. Most of the existing beach materials are gravel and sand. Benidorm Bay (fig 2) comprises of two large sandy beaches, Levante and Poniente beaches. The southern part of the Alicante coastal area includes extensive sandy coastal formations (San Juan, Guardamar, etc.).

Geologically speaking materials in the Alicante coastal area are mesozoic and cenozoic together with an underlying bed of Keuper materials. In many cases these materials appear as diapirs. During Triassic a limestone series was deposited, from the Jurassic to the upper Cretaceous. During the Palaeocene and the Neogene periods tectonic stability of the region underwent important changes. In the Oligocene, basins evolved quickly registering important local accumulations. Cenozoic deposits show many unconformities. In some areas there is a transgressive Pliocene. Plioquaternary sediments are many, both continental and marine.

2.2 Coastal actions undertaken

For two decades coastal erosion has been induced in many significant sandy beaches in the Alicante area. The Ministry of Public Works decided to implement artificial beach nourishment in the following:

1. San Juan beach restoration, close to Alicante city. Artificial beach nourishment consisted of recharging the beach with 2,000,000 m³ of sand over a length of 5.7 km.

2. Poniente beach in Benidorm. A total amount of 700,000 m³ of sand was recharged in a 1.5 km long stretch of coast.

3. Vila Joiosa beach restoration: two rock groynes were constructed to control and reduce longshore transport potential. In addition 350,000 m³ of sand was added.

4. Calp beach nourishment: some 850,000 m³ of material was supplied for this scheme.

3 BORROWED MATERIALS FOR BEACHES

Sand replenishment and other coastal restoring actions require borrowed materials. Usually a number of options can be considered for the supply

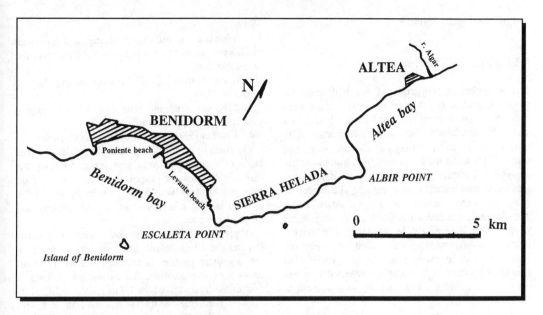

fig. 2 Sierra Helada and Benidorm bay

of material for artificial beach nourishment. Nevertheless, quantities involved, transport considerations and the basic nature of the material make marine sediments the preferred option in most cases (Murray 1992). As a result of this, demand for offshore borrow sand areas is increasing. Both ecological reasons and other interests may put a constraint on the availability of material from sea bed locations. Furthermore, materials replaced on beaches are subject to littoral drift processes and could therefore be displaced. It is necessary to compensate for longshore and offshore losses by periodically feeding new material to the nourished beach. New recharges projects are therefore required for regular renourishments to be undertaken.

The size of the material for artificial beach nourishment has to be tested according to beach profile characteristics, littoral dynamics and conditions (Graaff, J & Koster, M.J. 1.990). The items to be considered in detail are: (1) selection of sediment size, (2) development of profile/volume, (3) analysis of sediment budgets, and (4) long and cross shore transport rates.

4 BORROW SITE INVESTIGATION

Since 1987 the Spanish Ministry of Public Works has been promoting comprehensive studies in order to locate and select these borrow areas (M.O.P.U.

1987). These studies were carried out in areas between 6 and 35 meters in depth. On the Spanish central Mediterranean coast, studies were carried out between the Huertas and Albir headlands, a coastal stretch of 45 km. In order to achieve the objectives mentioned above, studies carried out included the following research:

4.1 Site investigation

This included bathymetric and geophysical research. Bathymetric surveys were implemented with an Atlas Deso-20 echosound. Seismic profiles were carried out with Uniboom E.G.G. and microprofiler Pinger. A Klein mod 531 side scan sonar was used to identify morphological features. Topographic works had been previously implemented. Sea level was recorded continously during the investigation period and a Maxiran integrated radiopositioning system was used.

Maps of scale 1:5,000 of the area were obtained. They represented: (1) isopachs of non consolidated sediments to the depth of the first potent regional reflector under the sea bed, (2) bed thicknesses of non consolidated sediments to the depth of the first discontinuous mapped reflector, (3) bed thicknesses of non consolidated sediments until the first discontinuous reflector below the first discontinuous mapped reflector, (4) isobaths each 5 m, and (5) sea bed morphological features.

4.2 Geotechnical characterization

The geotechnical properties of sea bed materials were also evaluated. A total of 260 samples were obtained in the study carried out between Vila Joiosa and Albir headland. They were granulometrically and mineralogically analysed. Mean size and standard deviation were obtained from granulometric analysis. Materials were classified as mud, fine, mean and coarse sands, and gravel. The places of sampling were located on maps. After being granulometric determined mineralogical components were further determined by counting. Carbonated fractions in sands were also evaluated. In general the quartz component was low but the carbonated fraction was very high. Samples collected in areas deeper than 20 meters were usually mud and clays.

4.4 Results obtained

An important borrow area was identified in front of Sierra Helada, between the bays of Altea and Benidorm. Sea bed materials are adequate for beach nourishment in this area. The mean size of the sea bed sand is between 0,30 and 0,45 mm. In this extensive area sediments of more than 30 m thick exist. Vibrocore drilling was carried out in the recomended area to confirm the obtained results. The results showed that sand under the sea bed was also adequate for utilization in coastal protection schemes. Due to their depth, situation materials could be exploited by using trailing suction hopper dredges. Total amount of sand volume to be exploited was quantified at 80,000,000 m³.

5 COASTAL EFFECTS OF OFFSHORE DREDGING

The results obtained in the preliminary studies recomended the exploitation of the marine bed in front of Sierra Helada, for the implementation of the above mentioned coastal nourishment schemes in the Alicante coastal area. Several studies on littoral processes were previously carried out to test potential effects of offshore dredging on the stability of adjacent beaches and coasts. They were especially assessed in the Benidorm area, given its touristic importance and nearness to the borrow area. Questions to be addressed are the following (Parrish

1989):
1. Whether the area of dredging is far enough offshore so as to prevent beach draw-down into the deepened area.
2. Whether dredging will interrupt material supply to the beaches.
3. Whether dredging will reduce banks which provide coastal protection.
4. Whether dredging will change wave patterns.

Variables to be considered must be selected on the basis of them either providing information on the direct influences and responses of the coast, or on their implications with respect to the impact of the erosion and any coastal restoration strategy that may be implemented.

In general, it is assumed that dredging in areas beyond the 10 m isobath has a negligible effect on the nearshore profile. In the Netherlands, to avoid erosion of the coastline the extraction of sand is prohibited landward of the 20 m isobath (C.C.E.R. 1.987). Nevertheless this is assumed in complete beach profiles. Exploitation in front of Sierra Helada was undertaken in front of a significant cliff, at depths between 15 and 30 meters. The volumes planned to be used in beach nourishment were the above mentioned, with a total amount of almost 4,000,000 m³. In the following paragraph, the studies on littoral processes carried out to test potential effects of this planned offshore dredging on the stability of Benidorm beaches, will be analysed.

6 SIERRA HELADA AND BENIDORM BAY

6.1 Description of the area

Sierra Helada separates the bays of Altea and Benidorm (fig 2). Altea bay has its extreme north in the Toix point, and situated in it is the mouth of the river Algar. It forms an important deltaic front. It is a plane coastal area with a peneplain with a large drainage frame. In Altea bay there are many ophitic outcrops. Beach materials consist of gravel, mostly calcareous. Sierra Helada is a classic accident on a northeast-southwest trend. In the extreme northeast lies Albir point. Sierra Helada is formed for the most part by material belonging to the lower Cretaceous, the Island of Benidorm being connected to it geologically. The landward facing slope is a potent quaternary glacis, and the seaward facing slope is a series of cliffs made up of cretaceous materials. The cretaceous series of Sierra Helada (substantially Neocomian-Aptian) is formed from bottom to top by: (1) a red sandstone series of some

100 m thickness, (2) some calcareous banks, with abundant fossils, of 200 m thickness, and (3) an alternate series of calcareous and well stratified marls.

The Escaleta point is in the extreme sothwest of Sierra Helada is. Benidorm bay houses two sandy beaches of great significance, Levante and Poniente beaches. The Levante beach of Benidorm is occupied by quaternary sand deposits. It is limited by Canfali point, an eocen outcrop of marls. Marls continue along the Poniente beach until Finestrat.

6.2 Sedimentological characterization

Sea bed materials in front of Sierra Helada (M.O.P.U. 1987) consists of sands sized 0,32 to 0,36 mm. Carbonated fraction is up to 30%. Sediments in front of Sierra Helada are mineralogically rich: quartz (transparent, white and orange), rounded tourmalines, zircon, garnets, opaque, andalusites, rutiles and some feldspar. Beach sand samples were analyzed (Esteban 1987). Granulometric analysis was conducted with sieves A.S.T.M. 16 - 30 - 40 - 50 - 80 - 100 - 200 - < 200. Carbonate rates were determined in the held fractions. Mineralogical composition was also analyzed in non carbonated fractions. The degree of roundness of white quartz was established. In general, fractions of 50 - 80 - 100 - 200 were analyzed. These sedimentological studies show that the beaches of Benidorm bay exhibit material clearly provided by the river Algar in favor of net littoral transport southward. This conclusion clearly demonstrates that a connection exists between the sediments in front of the Sierra Helada and those of Levante beach: longshore transport carries materials provided by the river Algar to Levante beach by means of the profile beach. This question explains the genesis of the submarine spit that starts at Escaleta point. Mineralogical results were the main criteria that led to the previous conclusion, especially the presence of red, blue and black quartz. Its origin can only be explained by the intervention of the river Algar, in particular the fact that this source of materials is located slightly north of Sierra Helada. The tourmaline is another significant mineral, with its high presence in fine fractions. Its rounded character indicates a distant source. This mineral is also present in beaches to the North of the Sierra Helada, while in beaches to the South of Benidorm it has not been found. The matter of its presence in the beaches surrounding Benidorm is a good indication confirming the relationship that exists between the sediments on both sides of Sierra Helada.

6.3 Bathymetrical characteristics

One of the most important morphological features is the submarine spit that begins in Escaleta point and extends in a southwesterly direction. It is some 200 m long and is found between depths of 6 and 9 m. The materials that form this submerged spit are not consolidated sands. Benidorm beaches have complete profiles. Close to the coastline in the extreme northeast of Levante Beach there is a relict beach. In its central section there are two small bars between depths of 0,5 and 3 m. The pending mean is 4,4 % and 2,6 % in their two sections. Poniente beach displays similar characteristics. Detail bathymetrical site investigation was undertaken to gain information about the whole area. The results of the projects indicate differences in the profiles obtained in different zones along Sierra Helada. Near Albir point, in the extreme northeast of Sierra Helada, there is an emerged headland of some 100 m with a height of 14 and 15 m below sea-level. Somewhat more to the south, there is a smaller headland of similar profile below sea-level. Moving southward from this headland, where the coastline turns inwardly, there is a section of rocks at a depth of 4 or 5 m. The borrow area is located at the center of this zone. Towards the southwest of this area, a significant submarine morphological change was detected. Inmediate to the coastline depth is of 3 m. Depth falls gradually to - 10 m over a distance of some 150 - 200 m from the coastline. Existing material is sand. It is a sand wedge either from a transverse and longitudinal point of view, with depth decreasing close to the coastline both in a northeast and southwest coastal direction. Morphology is constantly maintained: - 2 or - 3 m near the coastline and a zone close to it where the accumulation of sedimentary non consolidated materials is greater than at Albir point. This general form extends towards Escaleta point and even further in the form of the submerged spit.

6.4 Wave climate and littoral transport rates

The coastal dynamics in the area were analyzed. Visual sea and swell records were analyzed. The most intensive of which were of ENE (0.6 % - H > 3.5m). Most frequent were NE (7.6 % - H > 1m), ENE (4.7 % - H > 1m) and SW (4 % - H > 1m).

Littoral transport rates were evaluated (M.O.P.U. 1988) in 40,000 - 150,000 m3/ year according to CERC and BIJKER formulations. This must be considered as a theoretical potential case for the existence of a complete beach profile, that it does not exist in reality. However, longshore transport at the bottom of cliffs is qualitatively different to that of complete profile beaches. Consequently, effected calculations can only suggest a certain imbalance, not quantified, from northeast toward southwest.

6.5 Littoral morphodynamics

The results of the previous paragraph qualitatively establish coastal processes in the area. Concerning the general state of the beaches of Benidorm a number of points must firstly be outlined in order to determine its degree of balance. Generally, references and existing information tend toward a general balance in its behaviour. And this independently of given variable phenomena which could on occasions lead to the conclusion that erosion takes place in some zones of these beaches. The existing morphology and the evaluation of the coastal dynamics indicate that the deposits in front of the Sierra Helada tend to be displaced toward the southwest and the interior of Benidorm Bay and its beaches. This coincides with the existence of greater depths near the headland at in Albir point, where littoral drift is generated. Throughout the Sierra Helada fret decrease together to the headland supposes a certain deposit process of the circulating materials, tobe stabilized the currents in the zone. But net transportation, in this sense, is maintained. This is verified by the presence of the submarine spit at the Escaleta point. In general in all the submarine profile in front of Sierra Helada, sand material is of greater size than it is in any other area the surrounding submarine environment, which makes it suitable for utilization in beach regeneration. This can be explained because it is the only material, by its size and weight, whose presence is compatible with the existing currents.

6.6 Dredging effects

An artificial depression of the natural profile in the submerged beach in front of Sierra Helada will tend to be filled when storm waves generate a crossshore transport to the depth of dredging. A depth limit in complete profile beaches can be fixed in 3Hs. In front of Sierra Helada the effects could be greater due to the higher intensity of longshore currents, local irregularities and reflections. As the storm waves having a 10 years period of return are within the order of Hs=6 m, it is prudent to consider that crossshore transport can mobilize sediments and distribute the deficit (in 10 years) to the depth of 20 meters. This depth coincides with the prohibition limit of dredging used in the Dutch coast. That has also been suggested as acceptable for Mediterranean coasts, although not sufficiently proved.

Furthermore there are many other uncertainties (bathymetry, wave climate, calculation methods,etc.).

7 CONCLUSIONS

Results establish the existence of littoral transport betweeen Altea and Benidorm bays, along Sierra Helada. Morphological features explain littoral processes. Nevertheless, it was not possible to quantify littoral transport rates. There are many uncertainties that do not allow precise conclusions. Dredging in front of Sierra Helada might affect Escaleta submarine spit and Benidorm beaches. However, it is not possible to quantify these effects.

Conclusions highlight the need of monitoring the borrow area and some morphological features. Best results for defending coasts can be achieved by combining careful designs with monitoring and analysis. The growing use of beach nourishment has brought an increase in coastal surveying. Surveying and analysing the coast clearly indicates real coastal erosion problems. Monitoring could also provide useful information on the relative performance of different coastal protection works. Furthermore, data collected will be invaluable in the development of coastal engineering.

REFERENCES

Brampton, A.H. 1992. Beaches - the natural way to coastal defence. *Proc. Coastal management '92: integrating coastal zone planning and management in the next century*: 221-229. London: Thomas Telford.

C.C.E.R. 1987. Manual on artificial beach nourishment. Codes and Specifications. Rijkswaterstaat. Delft Hydraulics. Report 130.

C.E.R.C. 1984. Shore Protection Manual. Department of the Army, Waterways Experiment Station, Vicksburg, Miss. U.S.A.

Esteban, V. 1987. Procesos litorales en las costas

valencianas al sur del Cabo de San Antonio. Thesis Doctoral. E.T.S.I. C.C.y P.. Universidad Politécnica de Valencia.

Graaff, J. & Koster, M.J. 1.990. Coastal Protection. *Proc. of the short course on coastal protection.* Pilarczyk, K.W. Ed. Delft University of Technology. Rotterdam: Balkema.

Horikawa, K. 1.988. Nearshore Dynamics and Coastal Processes. University of Tokyo Press.

M.O.P.U.. 1987. Estudio geofísico marino en la costa de la provincia de Alicante. Tramo : Puerto de Villajoyosa - Punta Albir. Subdirección General de Costas y Señales Marítimas. Dirección General de Puertos y Costas. Madrid.

M.O.P.U. 1988. Extracción de arenas del fondo marino entre Punta de la Escaleta y Punta Albir. Benidorm - Alfaz del Pí (Alicante). Estudio del transporte sedimentario e influencia del dragado entre calados de 15 y 30 metros en dicho transporte". CONSOMAR S.A. Subdirección General de Costas y Señales Marítimas. Dirección General de Puertos y Costas. M.O.P.U. Madrid.

Murray, A.J. 1992. The provision of marine sediments for beach recharge. *Proc. Coastal management '92: integrating coastal zone planning and management in the next century*: 283-291. London: Thomas Telford.

Parrish, F. 1989. Marine resources. *Proc. Coastal management. ICE. 1989*: 187-196. London: Thomas Telford.

Littoral morphodynamic and sedimentological properties in Alicante coast (Spain)

La morphodynamique littorale et les caractéristiques sédimentaires sur la côte d'Alicante (Espagne)

V. Esteban Chapapría
Laboratorio Puertos y Costas, Universidad Politécnica, Valencia, Spain

ABSTRACT: Some comprehensive investigations in the Alicante coastal area were performed in order to define littoral morphodynamic characteristics. Works were extended to every beach between San Antonio and Cervera Capes. Sedimentological characteristics of all beaches were related to the littoral dynamic conditions. Coastal protection works were additionally studied. The general evaluation of their results was considered. General discussion was based on all this data. On the completion of this investigation, final coastal zoning was established. Coastal zoning divided the area into coastal morphodynamic units. This paper describes the investigations carried out in the Alicante coastal area and focuses on their results. They allowed the implementation of coastal engineering zoning and the accurate planning of coastal protection works. The conclusions highlighted the need for the creation of an oceanographical data base.

RESUMÉ:Quelques recherches ont été realisées sur la côte de Alicante afin de définir les caracteristiques morphodinamiques du littoral. Les caractéristiques sedimentaires de toutes les plages ont été reliés aux conditions dinamiques du littoral. Les travaux de protection côtière ont été etudiés ensuite. Ses résultats ont été pris en considération. A la fin on a établi des unités morphodinamiques côtières. Les études realisés ainsi que les résultats obtenus ont permis la planification des travaux pour la protection de la côte. Les conclusions ont mis en évidence la nécessité de créer une base de données oceanographique.

1 INTRODUCTION

Coastal areas are of particular importance in Spain. It is accepted that for any definition of the coast interacting environments, both continental and marine, are incorporated. Human intervention produces a great amount of pressure on this changing environment. Coastal management has to take into account: (1) natural resources, and (2) coastal uses, both potential and current. It is necessary to establish littoral morphodynamic conditions. Natural conditions and coastal processes must be recognised. From this coastal erosion can be established in order to implement coastal works and to assess development proposals. In recent years, the Spanish Mediterranean coast has undergone high touristic development. The Alicante area is located on the Spanish Mediterranean coast (fig 1) and has recently been under pressure from growing coastal develpoment. Man's intervention on shoreline processes has been considerable. Coastal planning (Cox 1992) is one of the activities concerning

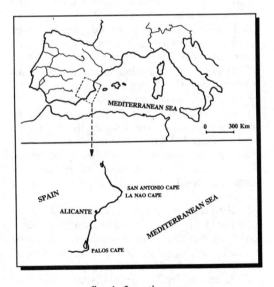

fig. 1 Location map

coastal engineering and other disciplines. Preliminary studies are necessary to enable adequate planning. In order to define littoral morphodynamic conditions in the Alicante area, studies carried out included the following research:

2 GEOLOGICAL AND GEOMORPHOLOGICAL CHARACTERISTICS

The studied area has its northern extreme in the San Antonio cape, with the Valencian gulf lying to the north. The coastline is almost totally comprised of a continuous large beach of fine sand. However, southward at the San Antonio and Nao capes the coast becomes structural (Sanjaume 1983) due to Betic influence (fig 2). The studied area is situated within the Betic zone, especially southern Prebetic and Subbetic. Materials in the Alicante coastal area are mesozoic and cenozoic. The most significant are the cenozoic and the cretaceous ones, together with the plioquaternary deposits. Betic structures have a tipically northeast-southwest trend. Morphological features were identified by field work. Iniatially the study established in the whole area nine units according to geological and geomorphological conditions:

1. San Antonio cape - Nao cape: Xàbia bay extends from the cretaceous alignment of the San Antonio headland to marine palaeocen materials of La Nao cape. The natural depression of Xàbia was filled by quaternary deposits. Beach materials consist of gravel, mostly calcareous. There is just one sandy beach in this area.

2. Nao cape - Moraira point: this stretch of coast consist of cliffs. Materials are limestones and marls belonging to the upper Cretaceous period.

3. Moraira point - Ifac: in this unit there are some sandy beaches. Quaternary and miocen deposits are present. Ifac is an eocen limestone rock.

4. Ifac - Sierra Helada: bay of Altea is situated on this stretch of coast, where the mouth of the river Algar is located. There are many ophitics outcrops. Beach materials consist of gravel, mostly calcareous. Sierra Helada is an accident on a northeast-southwest trend. Materials belong to lower Cretaceous.

5. Sierra Helada - Amadorio river: Benidorm bay is situed in this unit. It houses two sandy beaches of great significance. Levante beach is limited by Canfali point, an eocen outcrop of marls. West of Canfali point is Poniente beach.

6. Amadorio river - Barranco Aguas: here there are only pocket beaches of gravel and coarse sand.

Materials in this area are from Palaeocene-Miocene.

7. Barranco Aguas - Huertas cape: the significant deltaic front of the Seco river is located on this coastline. Beach materials consist of coarse sand and gravels. The large sandy beach of San Juan lies to the south of the river's mouth.

8. Huertas cape - Santa Pola cape: this unit contains some beaches between cenozoic outcrops. The southern part of the coastline consists of a large sandy beach with significant dunes and an old lagoon beyond them.

9. Santa Pola cape - Cervera cape: this is the most extensive large sandy formation in the whole area. Guardamar bay houses the mouth of the Segura river, one of the most important rivers in the Spanish Mediterranean coast. The coastline consist of a continuous beach with notable dunes and lagoons.

3 SEDIMENTOLOGICAL PROPERTIES

A total of 175 samples were obtained in the study carried out between the San Antonio and Cervera capes. To verify the existence of different types of sand, sampling was conducted on every beach by digging. In some cases samples were obtained from the sea bed and from dunes where they existed. On beaches where the materials are coarse sand and gravels, samples were also taken. On large beaches samples were taken at regular intervals.

Granulometrical analysis was conducted with sieves A.S.T.M.. Firstly, organic matter was eliminated. Carbonate rates were determined in the held fractions. Mineralogical composition was analyzed in a total of 462 non carbonated fractions. The degree of roundness white quartz was established by distinguishing four types of morphology (very rounded, rounded, less rounded and non rounded). By determining sedimentological properties of sand samples, relationships (or not) between materials of different beaches could be established, as well as their sources, crossed distance, etc.. Carbonated fractions of sandy samples also define their origin. It was also possible to calculate the general degree of roundness of the white quartz as a final result of the classiffication established.

Witha afew exceptions the mineral white quartz appeared with greatest frequency. Great variety of mineals were found in smaller size fractions. In these, a high percentage of round white quartz was determined. Granulometrical results can be related to the type of accumulation or even to the intensity of littoral processes. A detailed analysis of the

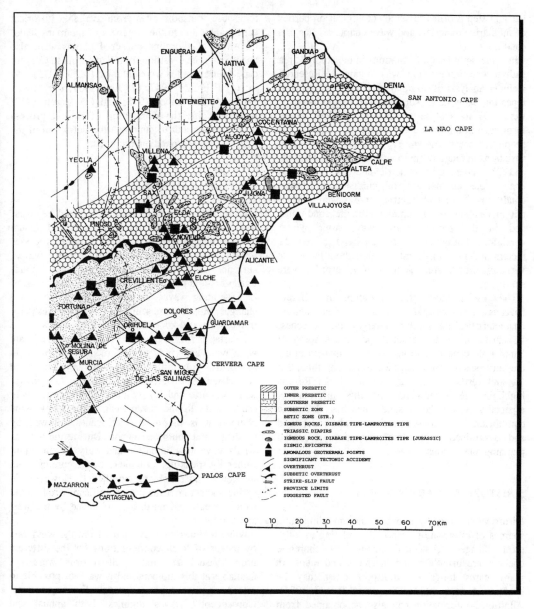

fig. 2 Geological pattern

results at every beach was established, relating them when continuity in the characteristics was demostrated. The northern units are isolated, especially the first and the structural stretch. In the large sandy beaches in the southern part of the Alicante coastal area material has greater classification. They also have higher general degrees of roundness but lower mean sizes.

4 BATHYMETRIC CONDITIONS

Bathymetric and morphological conditions of beaches and coasts relate to wave climate, material properties, and long and cross shore transport. The form of coastal profiles and their variations are mainly governed by cross-shore transports (C.C.E.R. 1987). Different types of coastal profile are distinguished. Research usually attempts to relate bar and step profiles to sedimentological properties of submarine materials and wave characteristics and conditions.

In some large beach formations in the studied area bathymetric information was obtained. Research includes analysis of seasonal variations by means of repeating bathymetric surveys. The main objectives were: (1) to analyse profiles morphology (step, form,etc.) in different areas where sedimentological properties were known, (2) to establish seasonal variations in relationship to wave climate conditions, and (3) to create a bathymetric data base that would enable the comparison of current and future conditions. Most bathymetric surveys were carried out in October, after summer climatic conditions, and in the period March-May, after winter conditions. Bathymetric surveys extend to areas 12 meters in depth to gain information about the whole variation profile area. A total of 79 profiles were surveyed.

The results show great variation in different beaches and coastal stretches. Nevertheless, characteristics are similar in large sandy beaches. For instance, in the coastal open areas south of Santa Pola cape conditions were maintained along the unit: beach slopes are, near to the coastline, 2.5 %, and further from the coastline 1 %. Results highlight the significance of this information, especially in the large sandy beaches. Dynamic equilibrium of the coastal profile must be understood and determined in the implementation of coastal planning and actions.

5 HISTORICAL CARTOGRAPHY

Historical cartography is one of the most important sources of information, especially relating to long-term changes. Historical records of shoreline changes are usually found in charts and maps. In many cases longshore transport rates may be evaluated from coastline changes (C.E.R.C. 1984). Additional information can also be obtained from local newspapers, local residents, etc.. Local people usually describe special factorsthat are to be considered and provide qualitative information.

Available historical cartographic archives and documentation concerning the Alicante coastal area were reviewed. Recent temporary morphological changes on historical maps were analyzed. Historical cartography provided information about changes in the lagoons situated in the south of the Alicante. Morphological changes in inlets were detected. In the studies carried out was not possible to evaluate

longshore transport rates from available historical cartography due to the largeness of the maps scale.

Position of shoreline on aerial photos were also analyzed in the coastline from Santa Pola to Cervera cape. Aerial photos were obtained (M.O.P.U. 1979) in: (1) February 1947, (2) June 1957, (3) August 1965, (4) November 1972, and (5) March 1977. Results enable the completion of littoral process determination and to approach the definition of the final coastal morphodynamic units.

6 LITTORAL DYNAMICS

Littoral dynamics were analyzed. Wind regimes, wave regimes, littoral transport rates and annual mean resultings were determined. Littoral processes result from the interaction of winds, waves, currents, tides, etc. in the littoral area. Winds transport sand on the dry beach and dunes, as well as generating waves, some currents and sea level changes. Most intense sediment transport takes place in the nearshore zone due to breaking waves.

To establish wind regimes in the area several data wind were analyzed. Finally records contained in the British Admiralty Charts were considered the best (Díez 1988). According to these: (1) general wind regimes, and (2) seasonal wind regimes were established. In the mean year the easterly wind component is the most significant, followed by northeast and southwest winds. During the autumn winds have low intensity and vary in direction. Generally speaking, in winter, winds coming from the fourth quadrant are significant. However, in spring and summer the most significant winds come from the east and northeast, foloowed by the west and southwest.

Wave regimes were evaluated from the wind data by means of fetch configurations for the different units. Visual sea and swell records were not available at this moment and it was not possible to afford another type of wave climate analysis. Consequently waves regimes, both annual and seasonal, were obtained. For wave generation, due to fetch configuration, significant winds come from east and southeast. Determined wave regimes highlight the greater significance in the northern units of waves coming from the east and northeast, followed by southeast, south and southwest. South of Huertas cape significant annual wave regimes are the east, south, southeast and northeast. Therefore, some differences may be noted between northern and southern units in the Alicante coastal areas. Seasonal variations could also be observed in each case.

Littoral transport rates according to CERC formulation were evaluated. Seasonal variations in the longshore transport rates were also analysed. Longshore transport is throughout the area in average southward direction. However the relevance of some partial transports was noted, especially northward. Rates obtained were considered too high. Nevertheless, they have an orientative value if they are used to make relative comparisons.

7 GENERAL DISCUSSION

Littoral processes in the area were defined through the general analysis and synthesis of the obtained results. They have permitted to establish . In some cases it has also been possible to determine genesis and evolution of beaches and large coastal formations. Finally coastal morphodynamic units of different ranges were determined by the different degrees of connection between coastal formations. Limits in the initial units were removed in some cases as a consequence of general discussion.

8 CONCLUSIONS

The purpose of coastal planning is to solve spatial conflicts, interests and environmental impacts by assessing and balancing uses, resources and processes. Studies carried out allowed the implementation of a method of coastal engineering zoning. They establish a methodology of analysing properly littoral processes. They can also allow for the accurate planning of coastal protection works. A geological and geomorphological approach make planning possible. Results highlighted the need for the creation of an oceanographical data base.

REFERENCES

Aguilar, J. 1984. Naturaleza y distribución de las corrientes producidas por la rotura del oleaje.Doctoral Thesis. E.T.S.I. C.C. y P.. Universidad Politécnica de Valencia.

C.C.E.R. 1987. Manual on artificial beach nourishment. Codes and Specifications. Rijkswaterstaat. Delft Hydraulics. Report 130.

C.E.R.C. 1984. Shore Protection Manual. Department of the Army, Waterways Experiment Station, Vicksburg, Miss. U.S.A.

Cox, T. M. 1992. Coastal zone planning. *Proc. Coastal management '92: integrating coastal zone planning and management in the next century*: 181-195. London: Thomas Telford.

Díez, J. 1988. Metodología para la determinación de los datos para el Planeamiento en Ingeniería de Costas. Curso de Planeamiento y Diseño de Obras Marítimas. 21st ICCE, 18-19 Junio de 1.988.

Esteban, V. 1987. Procesos litorales en las costas valencianas al sur del Cabo de San Antonio. Doctoral Thesis. E.T.S.I.C.C.y P.. Universidad Politécnica de Valencia.

Horikawa, K. 1988. Nearshore Dynamics and Coastal Processes. University of Tokyo Press.

Madoz, P. 1845-50. Diccionario geográfico-estadístico e histórico de España y sus posesiones en ultramar. Madrid.

M.O.P.U. 1975. Plan Indicativo de usos del dominio público litoral. Provincia de Alicante. Dirección General de Puertos y Costas. Madrid.

M.O.P.U. 1979. Estudio de la dinámica litoral en la costa peninsular mediterránea y onubense. Madrid.

Sanjaume, E. 1983. Las costas valencianas: sedimentología y aspectos de morfología litoral. Doctoral Doctoral Thesis. Facultad de Geografía e Historia. Universidad de Valencia.

Soil physical and soil mechanical aspects of reclamation in the Rhenish lignite mining area

Propriétés physiques et mécaniques des sols récultivés dans la zone de lignites du Rhenish

K. Winter

Rheinbraun AG, Abt. Gebirgs- und Bodenmechanik, Cologne, Germany

ABSTRACT: The current state of knowledge in the field of reclamation is reported from the soil physical and soil mechanical point of view. The engineering processes involved in reclamation (excavation, transport, application) destroy the original structure of the soil. The soil material becomes more sensitive to pressure and settlement and exhibits a greater tendency to consolidation. Moreover, it reacts highly sensitively to any changes in moisture content. Even a slight increase in the water content will cause the soil material to change from a semi-solid or stiff consistency to a pasty state. Due to these soil mechanical properties, the levelling work performed after dumping of the material has to be reduced to the inevitable degree in order to obtain new land with high-quality soil. Apart from the use of low ground pressure machines and minimization of the necessary levelling work by dumping the material with low ridge heights, the moisture conditions of the soil and the selection of the optimum levelling time are of great importance to the future quality of the new land.

RESUMÉ: L'état de l'art actuel en matière de recultivation du point de vue de la mécanique et la physique du sol fera l'objet de l'étude de la présentation. Le processus technique de la recultivation (excavation, transport, mise en oeuvre) détruit la structure initiale du loess inexploité. Le sol est ainsi plus sensible à la pression et au tassement, c'est-à-dire qu'il tend plus à se tasser. Le sol réagit également d'une manière très sensible aux variations d'humidité. Même une faible augmentation de la teneur en eau se traduit par une transition d'une consistance semi-rigide/rigide à une consistance pâteuse.Vu ces propriétés mécaniques du sol, l'objectif visant à obtenir des nouveaux sols d'excellente qualité, suppose la réduction au minimum inévitable des travaux d'égalisation après le déversement. Outre l'application d'un matériel exerçant une faible pression sur le sol, une minimisation des travaux d'égalisation par le déversement pour obtenir des faibles hauteurs des talus, la teneur en humidité du sol ainsi que la sélection du moment le plus avantageux pour l'égalisation sont d'une grande importance pour la qualité ultérieure de ce nouveau sol.

1. INTRODUCTION

By 1993, opencast mining activity in the Rhenish lignite mining area had affected a total 25,314 ha of land in the triangle between Aachen, Cologne and Mönchengladbach. The interference that such operations involve in the existing landscape with its agricultural, forestry, ecological areas, communication routes and settlements will meet with public acceptance only if it can be ensured that, once mining operations are completed, a multifunctional new landscape can take the place of the old.

This being so, the prime objective in the reutilization and reclamation of mining areas is to restore the soil and, in this way, to lay the basis for the development of new agricultural land or new forests.

	Total	Agriculture	Forestry	Building/ traffic	Water resources
Land in use	25314	16825	6805	1599	84
Reutilization	16314	7506	6884	1117	807

Table 1: Land area in ha (position: 31 Dec. 1993) Rhenish lignite mining area

By 1993, 16,314 ha had been made reusable. Agricultural reclamation accounted for 7,506 ha and forestry reclamation for 6,884 ha (Table 1). Success in reclamation depends very much on the soil material available for reclamation and how it is deposited. Since the mid-1970s, numerous studies have been carried out to examine the soil-physical and soil-mechanical properties of reclamation material, and trial series have tested the spreading and working of the reclamation layer.

The following report deals with the present level of our knowledge in the reclamation field from a soil-physical and soil-mechanical point of view.

2. RECLAMATION MATERIAL USED

The soil materials used for reclamation purposes are **loess/loess-loam** and the so-called **forestry gravel**.
Loess is a yellowish-brown to greyish-brown, more or less lime-containing silt sediment, mostly with a low, varying content of clay and fine sand, the result of drifting in the periglacial dry climate toward the end of the Pleistocene period. The loess or its disintegrated product, loess-loam, form the - in places very thick - Quarternary top layer in the Rhenish lignite mining area.

Present knowledge (v.d. HOCHT and WINTER 1994) indicates that only loess and loess-loam, which form the upper layers of overburden, are suitable material for agricultural reclamation. For forestry reclamation, a mixture of sandy-pebbly overburden layers and loess or loam, so-called forestry gravel, can be used as well.

By contrast, the sand, gravel and clay layers of the Tertiary period underlying the Quarternary covering layers are completely unsuitable for reclamation purposes. The sands and gravels, which are extremely rich in quartz with very small quantities of decomposable feldspar (feldspar content 2-3 %), are low in nutrients and weak in sorption capacity. The clays consist mainly of sorption-weak kaolinite. The Tertiary material, owing to its embedded pyrite and marcasite, which release SO_4 ions when they disintegrate, yields an acid and sterile soil. Even when mixed with terrace gravel, loess and loess-loam, therefore, the Tertiary material is not used for forestry reclamation.

Based on the breakdown of the loamy covering layers in the Rhenish lignite mining area, Table 2 shows the range of soil types and the main soil ratios, such as are employed at present in reclamation, with a breakdown by suitability class.

Suitability rating	Top layer structure	Soil type	Grain size distribution Clay %	Grain size distribution Silt %	Density g/cm³	Porosity %	CaCO₃ content %
A	Recent silty-loamy loess deposits	slU - uL	12-25	65-84	1,39-1,82 mean 1,6	37-52 mean 44	12-19
B	Older silty loamy loess deposits	ulS - uL - lU	12-30	>70			<5
C	Silty clayey basic loam	utL - uL - sL Fine silty sandy layers and pebbly bands	20-45				calcium-free

A = very suitable, B = quite suitable, C = not suitable for agriculture, but suitable for forestry recultivation

Table 2: Range of soil types and characteristic values per suitability level

3. GEOTECHNICAL PROPERTIES OF THE RECLAMATION MATERIAL

On the basis of its grain distribution, the loess/loess-loam soil material is classified among the fine-grained, and forestry gravel among the mixed-grained unconsolidated materials (Table 3).

Depending on water content and/or compactness, both the constitution of the loess or forestry gravel material and its soil-physical properties vary. Based on the physical demands that plants place on the soil, the following discusses the geotechnical properties of soil material and the special features of each specific reclamation.

		Unit of measure	Agriculture sites "loess/loess-loam"	Forestry sites "forestry gravel"
Range of soil-physical rations				
Soil type :			medium-loamy silt	low to medium-loamy sand
grain size :	clay	%	14 - 17	5 - 10
	silt	%	70 - 80	20 - 40
	sand	%	2 - 5	25 - 48
	gravel	%		15
Humus content:		%	0,4 - 0,5	0,3 - 0,5
average increase :		%	ca. 0,03/a	
Density :		g/cbm	1,55 - 1,75	1,40 - 1,80
Porosity :		Vol. %	35 - 52	32 - 48
Share of air-conducting pores (> 30 µm) :		Vol. %	4 - 12	4 - 15
Usable water capacity relative to 1 m root area :		l/m^3	high 160 - >200	90 - 180
Liquid limit :		Weight %	26,0 - 29,0	
Plastic limit :		Weight %	18,0 - 22,0	
Plasticity index :		Weight %	5,5 - 8,0	
Shrinkage limit :		Weight %	5,0 - 10,0	
Range of chemical rations				
pH value :			weak alkaline 7,5 - 8,0	very acid to very weak alkaline
Carbonate content :		%	4,5 - 11,5	4,5 - 7,5
Nitrogen :	Tot.-N	%	0,02 - 0,05	0 - 2
Potassium :	Tot.-K$_2$O	mg/100g soil	300 - 700	0,02
	CAL-K$_2$O	mg/100g soil	6 - 12	
Phosphorus :	CAL-P$_2$O$_5$	mg/100g soil	3 - 6	3 - 5
Magnesium :	CaCl$_2$-Mg	mg/100g soil	10 - 15	2 - 3
Cation-exchange-capacity :		mval/100g soil	8 - 10	3 - 7

Table 3 : Range of soil physical and chemical rations (according to DUMBECK and WINTER 1993)

3.1 Soil-Physical Correlations between Soil and Plant

Plants place two central demands on the soil:

- an adequate supply of water and
- rootability, to permit an intake of water and nutrients.

From a soil-physical point of view, the supply of water for the plant and the rootability depend entirely on the structure of the soil.
Soil structure is a matter of the way the solid soil components are arranged in space, and stratification of the soil components relative to one another organizes the void volume in a complex system of pores. Pore size and distribution correlate with the water-binding quality of the soil. This has a crucial impact on the water and air content and on rootability.

The field capacity of the soil is used as a parameter for the water supply to the plant. This is equivalent to the water content of the soil, which follows gravity when fully saturated with water. The plant can only use the water stored in the rootable soil area. On the one hand, this presumes a sufficiently high storage of plant-available water. On the other hand, there must be good soil rootability, since poor rootability makes actual benefit from these water quantities impossible.

Assessing the rootability of the soil means considering all those soil parameters that describe the inter-relationships of the soil's 3-material system (soil particles, interstitial water and interstitial air) and, hence, the properties of the soil structure. The chief soil-physical characteristics in this respect are

- dry density
- porosity
- water content.

The dry density of a soil has a crucial impact on its aeration and on its resistance to mechanical penetration. The growth of roots can be hindered both by excessively high resistance to penetration and by a deficient exchange of gas.
The exchange of gas depends on the volume of the air-filled pore content. What matters is the air-filled pore space in field capacity and the links between the pores, expressed in terms of air and water permeability.
The degree of soil saturation can be derived via the water content from the dry density and porosity and indicates the extent to which the overall pore area is filled with water and what air volume is available. High water content means that coarse pores, too, are filled with water and that low soil moisture tension suffices

for drainage. This can give rise to deficient aeration, which makes the plant's exploitation of the water more difficult or impossible. The denser the soil structure, the higher must be the soil moisture tension and the lower the water content at which there is still an adequate air volume.

3.2 Soil-Physical Properties of Natural Soil/Deposited Soil

Grown soil is marked by a dynamic equilibrium between soil-climate-organisms that has evolved in the course of soil formation and development. The soil horizons that have emerged, the grown soil structure and the water, soil and nutrient balance are indicative of a stable state in soil-mechanical terms. The technical process of reclamation disturbs the original effective structure and changes the soil. After discharge, we have a more or less thoroughly mixed, unstructured spread soil. Its soil-mechanical behaviour and specific initial soil structure depend on soil-physical properties and reciprocal action. In soil-mechanical terms, this is an unstable state (Fig. 1).
The crucial features of the reclamation material, the loess, are its low plasticity and the loss of strength that the loess material sustains from the destruction of its structure when it passes from unworked to a disturbed state (WINTER 1990). Typical of this is the destruction of the macro and coarse pores compared with the natural soil structure. The share of draining pores in total porosity falls in favour of medium pores. Capillarity and permeability are diminished (Fig. 2).
The loess material is more sensitive to pressure and settlement, so that it tends to become compressed. Loess also responds very sensitively to changes in moisture. Even a slight rise in the water content can turn the loess material from a semi-solid or rigid consistency into a pulpy state, so that it completely loses its strength.

The soil-physical properties of forestry gravel are to a great extent dependent on the share of loess that is added. Its density and, hence, its porosity and permeability as well as its strength correlate closely with its loess content. Unfavourable conditions are preprogrammed if the mixed soil layer of forestry gravel has the grain size distribution known in earthworks as soil-cement. In addition to the precondition of this specific grain composition, the highest compressions can always be obtained if the soil is worked with an optimal water content (> earth-damp, < complete saturation) (WINTER 1990).

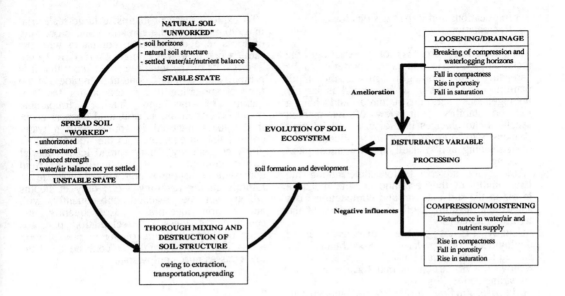

Fig:1: Natural soil - spread soil system

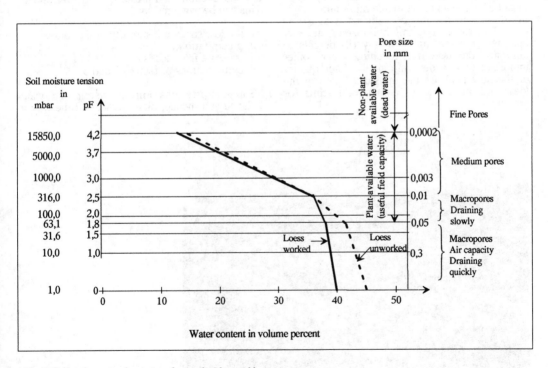

Fig. 2: Soil moisture tension curve of natural and spread loess

3.3 Implications and Repercussions for Reclamation

The technical process of extracting the reclamation material and its transportation to the spreading site thoroughly mixes the input material. The material to be spread is largely homogeneous and unstructured and has a reduced stability. In view of its marked sensitivity to changes in moisture and its high propensity to compact, there are implications for the spreading action itself and for subsequent working.

Among the adverse factors determining the later quality of the new land are the moisture levels in the soil material and compaction from subsequent levelling work on the surface of the terrain or spread soil.

Usually, it is not possible to dispense with the deployment of graders. Specifically for subsequent agricultural use, it is necessary to remove the rib structures that form during the spreading action (Fig. 3).

Likewise, grading work may be necessary to offset or fill hollows caused by differences in subsoil settlement or by other operational imponderabilia.

To obtain good new soils, therefore, the levelling work that follows the spreading stage must be kept to an unavoidable minimum.

In addition to the use of vehicles with a low ground pressure (bogland bulldozers) and any options for spreading with low rib heights to minimize the need for levelling work, other important factors determining the later quality of the new land are the moisture levels in the loess and the choice of a favourable time for grading.

To explore these problems, a large-scale trial in 1990 examined the spreading and processing of loess in a dry climate, compared with the same work in a wet climate (WINTER 1992). The results of the levelling trials show that it is primarily the moisture content in the loess at the time of spreading that is crucial for the later quality of the new land. Irreparable compressions in the subsoil (and deeper) occur if the loess material is spread with a water content that is equivalent to the liquid limit (= pulpy consistency). Dry spread loess material (water content ≤ the plastic limit = rigid consistency) becomes compressed in spite of rainfall and/or levelling work only at depths which can be reached subsequently with amelioration measures. Consequently, the spreading of loess for reclamation purposes must be stopped when, owing to precipitation levels, e.g. strong rainfall, the consistency of the loess passes into a pulpy state.

4. FINAL REMARKS

Owing to the soil-physical properties, reclamation material loess or forestry gravel tend to become compressed or to consolidate. This can be triggered by

1. the low structural stability of the loess,
2. precipitation,
3. pressure from loads or
4. combinations of factors 1 and 3.

Soil compressions impair the air and water balance and, hence, plant growth all the way to

Fig. 3: Rib structure of freshly spread loess

total failure.

The results of soil-mechanical studies show that, in order to obtain good soils in the new land, it is necessary to reduce soil working after the spreading action to the unavoidable minimum. Starting from this, the aim should be to use machines with low ground pressure, to minimize working processes and to perform the spreading and soil working in suitable weather. The low structural stability and the dead weight of the soil, both of which adversely affect the quality of the reclaimed soil, can only improve in the course of a longpedological development process and/or require suitable cultivation measures, the final state being an intact set of interactions between soil, soil organisms, flora and fauna.

REFERENCES

v.d. Hocht, F. and Winter, K. (1994): Die Lößlagerstätte im Rheinischen Braunkohlenrevier, ihre Verwendungsmöglichkeiten und besonderen Eigenschaften bei der Rekultivierung (The Loess Deposit in the Rhenish Lignite Area, its Use Options and Special Properties in Soil Reclamation) - published in the reclamation manual "Braunkohlentagebau - Landschaftsökologie Folgenutzung, Naturschutz" (Lignite Mining - Landscape Ecology, Follow-Up Use, Conservation) -, ed. Prof. Pflug (in preparation)

Dumbeck, G. and Winter, K. (1993): Bodenphysikalische und bodenmechanische Aspekte bei der Rekultivierung im rheinischen Braunkohlenrevier (Soil-Physical and Soil-Mechanical Aspects of Reclamation in the Rhenish Lignite Area) - published in: Mitteilungen der Deutschen Bodenkundlichen Gesellschaft (Bulletin of the German Pedological Society), 71, 29-32

Winter, K. (1990): Bodenmechanische und technische Einflüsse auf die Qualität von Neulandflächen (Soil-Mechanical and Technical Influences on the Quality of Reclaimed Soil) - published in: Braun-kohle 42,
Vol. 10, p. 15-23

Winter, K. (1992): Neue bodenmechanische Untersuchungen zum Kippen und Planieren von landwirtschaftlich zu rekultivierenden Flächen (New Soil-Mechanical Studies on the Depositing and Levelling of Agricultural ReclamationAreas) -published in: Braun-kohle 44, Vol. 9, p. 12-17

The 'Mal Pas' and 'Del Puerto' beaches from Benidorm (Alicante, Spain) both examples of artificial beaches

Les exemples des plages artificielles de 'Mal Pas' et 'Del Puerto' à Benidorm (Alicante, Espagne)

José Serra
Laboratory of Ports and Coasts, Universidad Politécnica de Valencia, Spain

José J. Díez
Universidad Politécnica de Madrid, Spain

ABSTRACT: The "Mal Pas" and "El Puerto" beaches are examples of haw stable artificial beaches are formed through direct and indirect intervention of man.

RÉSUMÉ: Les plages de "Mal Pas" et "El Puerto", sont un exemple de plages artificielles stables due direct et indirectement à l'intervention de l'homme.

1. INTRODUCTION

The Port of Benidorm are the most significant works carried out on the Poniente beach.

The works were stopped due to the inmediate effects they had such as sand depositing in breakwater sheltered areas.

This gave rise to what is now know as *"The Port Beach"* and a second beach, *"Mal pas"*, between the former point and *"Punta de Canfali"*.

2. LOCATION

The beaches are those within the municipality of Benidorm, Alicante.

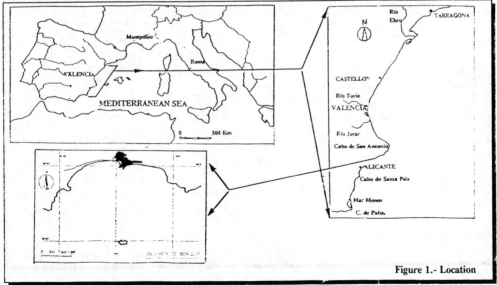

Figure 1.- Location

The Benidorm beaches are located between *"Punta de la Escaleta"* or *"El Cavall"*, the southwestern end of *"Sierra Helada"*, and a calcareus rock outcrop situated on the western end of the bay known as *"El Tossal"*.

The bay of Benidrom has two long beaches, the well known *"Levante"* and *"Poniente"* beaches. The Poniente beach lies on the western side of the bay of Benidorm between *"Punta de Canfali"* and *"El Tossal"*.

The Port of Benidorm is located on the eastern end of this beach.

3. HISTORY OF THE PORT OF BENIDORM

The port works began in 1919 and were finally carried out by mid 1929. The initial project included a 280 meter mound breakwater from *"Punta de Canfali"* oriented W-1/4SW, a stone wall, a secondary road connecting the port to the main Murcia-Valencia road, and complementary works such as bollards and staircases.

The port works were interrupted when the mound breakwater reached a length of 190 meter, the armor stone, stone wall and complementary works had not been completed. This was due to the inmediate sand depositing next to the breakwater.

Beach dynamics (i.e. sand displacement and redisplacement along the beach), was already known by local fisherman with the arrival of West-Southwest and Southeast strons.

The secondary road was the first works to be completed, in 1919. Simultaneously cliff erosion at the town's perimeter is halted due to the new road which acts a barrier against wave action.

In order to attempt to recuperate the sands movement. A channel was opened at the breakwater base thus permiting currents to flow through the port. This had a twoflod ain, frist to recuperate original sand displacement along the beach and second to avoid sand depositing within the port.

Unfortunately this channel had negative effects with the arrival of the previously mentioned strons, due to agitation to generate within the port.

The channel due to natural causes also began to close itself. In 1953 the channel was completely closed by the port authorities.

In the 1940's the concern over the port grows. The two main problems were, frist the disprotection against strons specially S-10ºW. The second problem was the low breakwater height which allowed free-movement of waves over the actual breakwater.

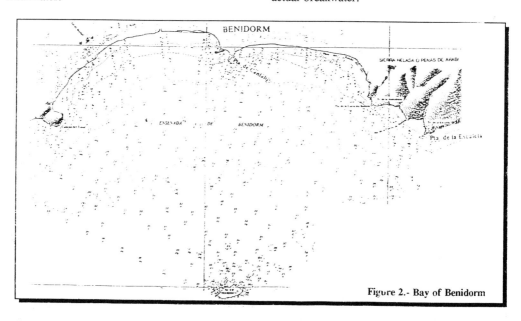

Figure 2.- Bay of Benidorm

Solutions provided to increase the shelter and to halt sand invasion les to the construction of a counter-dike. The best proposed projet was presented in 1946 which included a counter-dike built over the rocky sea-bed of *"Fontanelles"*.

Currently the port of Benidorm belongs to the *Generalitat Valenciana*, the autonomous government of the Valencian Community. The port is classified as a *marina* with only sports and touristic services.

After the last works carried out on the Poniente beach, artificial sand nourishment, a rubble-mound groin in the form of a small letter *"i"* was built. This serves as a counter-dike to reduce sand advancement in the *"El Puerto"* beach, which menacing with annulling the moorning surface withing the port.

4. HUMAN INTERVENTION ON THE PONIENTE BEACH

Human intervention on the Poniente beach can be grouped into three: Benidorm harbour, road infrastructure and urban development.

sea-wall becomes a potential risk factor when reached by waves, thus creating wave erosioneffecting the sea-wall.

* The over-urban developing has occupied large sandy extensions, destroyed dune formations, public beach occupation, and interfered with wind dynamics.

These man-made works and others of less importance have led to a modification of beach dynamics, loss of sand reserves, reduction of beach mobility as well as other negative effects.

An important consequence of these works is the recession which has been induced in the centre of the Poniente beach, concretely Fontanelles beach, solved by artificial nourishment that in its turn has heavily increased the accumulation of sands in the harbour's anchoring place, and consequently a strong advance of *"El Puerto"* beach.

5. COASTAL DYNAMICS

The Poniente beach can be considered as a

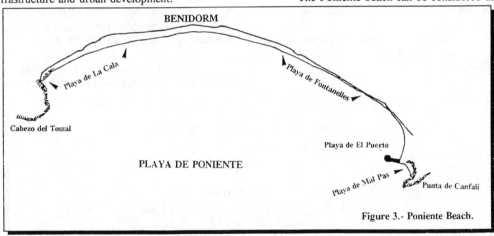

Figure 3.- Poniente Beach.

* The port of Benidorm is the frist and most significant works carried out on the Poniente beach. As previously exposed.

* The main road (N. 332, Murcia - Valencia) running parallel to the coast line is modified into a dual-carriageway. In order to protect the dual-carriageway a sea-wall is constructed, thereby substituting the low-cliff. The

morphodynamic subunity with a dynamic much more active internally than the contiguous subunities.

The shape of the Poniente beach is submitted to seasonal regular movements to accommodating itself to the action of the waves. The eastern waves (SE) carry sediments towards *"El Tossal"* reducing the width of the Fontanelles beach. On the other hand, the souhtwestern waves produce the opposite effect.

Greater potential of eolic transportation ofthe western and southern winds generated an eolic-dune transport west-east with resulting net towards the east and with natural renourishment from land in the eastern side of the beach.

In the strict balance hypothesis, the losses of materials opposite of Finestrat were compensated with the net income opposit the *"Punta de Canfali"*, and the exits from *"La Cala"* towards the dunes were compensated with the income in Fontanelles from the dunes.

The result is a closed circuit, coastal towards the west and eolic towards the east, with an exit in *"La Cala"* and entrance in Fontanelles, complete with a coastal entry flow from the Levante beach and an exit at Finestrat.

The various human interventions that have been accomplished in Poniente have led on one hand to a balanced readjustment af beach positioning. On the other hand this has meant immobility af large volumes of sand trapped within the breakwater or leveled by constructions.

Due to these conditions the sands have no effect in seasonal regular movements, nor in nourising itself.

Finally, the road infraestructure and the buildings near the beach have limited the oscillation margin of the coast line, so that seasonal local erosions that were acceptable before, now cause serious damage, even though the artificial nourishment carried out this risk.

6. CURRENT SITUATION

The last intervention accomplished in the Poniente beach has been an artificial nourishment, with sands originating from the sea-bed in front of the Sierra Helada.

Artificial nourishment that in its turn has increased a heavy accumulation of sands in the harbour's anchoring place, and consequently a strong advance of *"El Puerto"* beach.

Due to this action, rubble-mound groin was built to limit sand entering. The groin acts as a counter-dike for the port and indicates the port limit.

The first part of the groin is not overcome by wave action this therefore limits the dry-beach zone. The second part of the groins does allow sand flow and the third part is a samall island which acts as a buoy for a local traffic.

The groin is shaped as a small letter "i" togetter with the dot thus adopting its name from

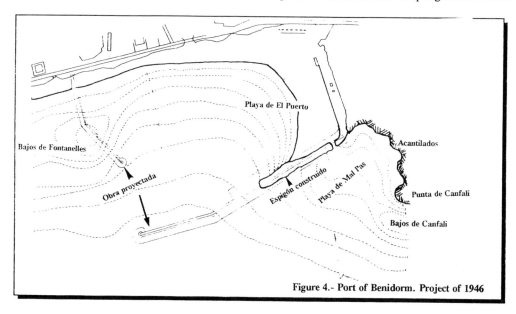

Figure 4.- Port of Benidorm. Project of 1946

3098

its appearance

Untill now both beaches, *"El Puerto"* and *"Mal Pas"*, are stable. The *"Mal Pas"* beach maintains its dimensions, and *"El Puerto"* beach advances or reduces according to wave and wind direction.

7. CONCLUSIONS

The modifications carried out on contour conditions in the Poniente beach by the construction of the port induced changes to the natural seasonal regular movements, the sands became trapped within the breakwater shelter. This beach progressively grew from sands originating from the Poniente beach.

This artificial beach is now know as *"El Puerto"* beach, due to its large size a prak was built on it. After artificial nourishing the dry-beach has become too large.

The *"Punta de Canfali"* before the construction of port of Benidorm was a passing point for the sands betwen Levante and Poniente beachs.

After the construction of port of Benidorm, the breakwater and the *"Punta de Canfali"* become obstacles which shelter part of the sands that originatied from the Levante beach.

This artificial beach, created in the shelter of the breakwater and of the *"Punta de Canfali"*, is now know as *"Mal Pas"* beach.

The orientation of the breakwater and the *"Punta de Canfali"* permit the free flow of sands betewn the Levante and the Poniente beaches. Due to this the beach remains stable and is only subjected to seasonal beach-profile changes.

In conclusion this is an example in the which some port works have not meant the classic accumulation-erosion effect, but given the particular conditions of the bay, and above all its dynamic, the works have created two stable artificial beachs, considering the effect of the works as beneficial.

ACKNOWLEDGEMENTS

Authors extend their grateful thanks to José R. García, town-council Civil Engineer in Benidorm.

BIBLIOGRAFIA

Benidorm recopilación fotografica. Ayuntamiento de Benidrom. Benidorm, 1985.
Datos de las obras que precisan en el puerto de

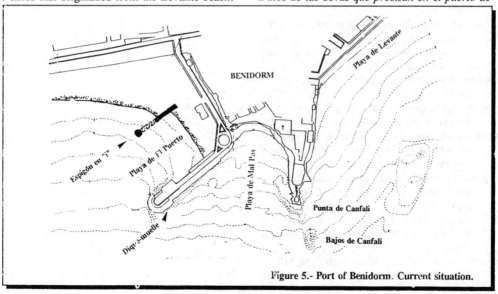

Figure 5.- Port of Benidorm. Current situation.

Benidorm. Abril 1941. Ing. D. Vic-nte Laporta Pérez.Esteban, V. 1987. *Procesos litorales en las costas valencianas al sur del cabo de San Antonio*. Tesis doctoral. E.T.S.I.C.C y P. Universidad Politécnica de valencia.

Estudio de las obras que precisan en el puerto de Benidorm para conseguir un abrigo eficaz y evitar la invasion de arenas. Abril 1.941. Ing. D. Vicente Laporta Pérez.*Estudio puerto de Benidorm*. Abril 1.941. Ing. D. Vicente Laporta Pérez.

Estudio del impacto de las obras de abrigo del Puerto de Benidorm sobre la playa de Poniente y de soluciones para la regeneración de la misma. Laboratorio de Puertos y Costas, 1991. Universidad Politécnica de Valencia.

Estudio de la influencia sobre las playas de Benidorm de las extracciones de arenas previstas en el proyecto de regeneración de la playa de Villajoyosa en los fondos marinos ante la Sierra Helada. Laboratorio de Puertos y Costas, 1991. Universidad Politécnica de Valencia.

Estudio de la influencia del nu vo Puerto Deportivo de Benidorm en las playas próximas. Centro de estudios y Experimentación de Obras Públicas "Ramón Iribarren". M.O.P.U. Madrid, 1984.

Horikawa, K. 1988. *Nearshore Dynamics and Coastal Processes*. University of Tokyo Press.

Manual on artificial beach nourishment, 1987. Centre for Civil Engineering Research, Codes and Specifications. Rijkswaterstaat. Delft Fhydraulics. Report 130.

Proyecto del dique de abrigo de la playa de Benidorm. 1916

Proyecto del dique de abrigo de la playa de Benidorm. Proyecto de liquidacion. 1.936.

Proyecto de terminacion del dique de abrigo en la playa de Benidorm. 1.941. Ing. D. Vicente Laporta Pérez.

Proyecto reformado para la terminacion de un dique de abrigo en la playa de Benidorm. 1.941. Ing. D. Vicente Laporta Pérez.

Proyecto de terminacion de las obras del puerto de Benidorm. 1.946. Ing. D. Vicente Vicioso Vidal.

Proyecto de morro del espigon del puerto de Benidorm. 1.981. Ing. D. José luis Campello.

Reparación de las averías causadas por los temporales del invierno de 1.947 - 1.948. 1.948. M.O.P.U. 1948

Cierre del Portillo del Espigón del Puerto de Benidorm. 1.953. M.O.P.U. 1953

Rectificación del espigón del Puerto de Benidorm. 1.958. M.O.P.U. 1958

Reparación del espigón de Benidorm. 1.971. M.O.P.U. 1971

Prolongación del Espigón de Benidorm. 1.972. M.O.P.U. 1972

Flandrian cliff lines in the Severn Estuary

Les falaises du Flandrian dans l'estuaire de la Severn

A. B. Hawkins, R. W. Narbett & D. R. Taylor
Engineering Geology Research Unit, University of Bristol, UK

ABSTRACT: Tidal cliffs developed in the soft Flandrian sediment are important in that they may result in differential settlement to structures constructed over them and are significant in that they indicate where erosion is actively taking place.The tidal cliffs within the Flandrian sediments in the middle part of the Severn Estuary where a new large bridge (the Second Severn Crossing) is being constructed are described.

RESUME: Les falaises de marée qui se développent dans le sédiment mou Flandrian sont importantes parce qu' elles ont pour résultat d'un affaissement différentiel des constructions au-dessus, et elles sont significatiffes parce qu'elles montrent où l'érosion se passe activement. Les falaises de marée dans le sédiment mou Flandrian qui se trouvent aux bords de l'estuaire de la Severn, près d'où on construit un grand pont nouveau, sont décrites

INTRODUCTION

Many authors have produced sea level curves for the Flandrian period. Although some of the variations between these can be accounted for in part by sample reliability and localised effects, there is a significant difference between the curves of Fairbridge (1961) and Morner (1969) and those of other workers who have produced curves similar to that of Jelgersma (1961). Nevertheless, all have concluded that sea level rose at a fast rate until some 6,000 years ago, since when the eustatic rise has been much slower.

During the period of fast sea level rise, it is likely that the coastal lowlands and estuaries were areas of steady accretion with little if any erosion taking place. Although slower than the 1.5 m per century rise during the main Flandrian Transgression (Godwin, 1940), in the period between 6,000 BP and Roman times, sea level rose by some 5 m (0.1 m per century) Consequently, during this period accretion was slower and there was the opportunity for the formation of peat horizons and even the development of forests, both of which subsequently became submerged under younger alluvium. As a result, the late Flandrian sediments contain peat bands, some of which have been dated. Devoy (1979) is fortunate in having obtained dates from both the top and bottom of a number of peat bands in the Thames Estuary while Tooley (1978) has undertaken a similar radiometric exercise in North West England. In the Severn Estuary there is little detailed data and only rare examples of radiometric dates for the top and bottom of individual peat bands. One such peat band near Clevedon, at a depth of approximately 6 m, has been

dated between 6100 and 4145 BP and a 0.4 m band at Avonmouth between 4960 and 4110 years BP (Hawkins, 1971).

In the area of the Caldicot Levels, some of these peat bands are now exposed outside of the sea walls. The cliff lines exposing these organic-rich deposits and the stumps of the submerged forest show evidence of the erosion of what must be considered recent sediments. In addition to this undisputed evidence that sediments deposited within the last five thousand years are now being eroded, there are numerous minor cliff lines around the Estuary which demonstrate active erosion. These cliff lines reach a height of over 2 m at the edge of the present estuary and from the stepped morphology inland of the salt marsh, it is likely that other cliffs have existed in the past which are now buried and difficult to distinguish. From a geotechnical point of view, such cliffs are very important. They indicate the temporary nature of the active coastline and hence the stability of structures placed on recent alluvium but also are the cause of differential settlement when structures are placed across the buried and topographically almost insignificant features.

This paper describes the cliff lines which are present for a limited part of the Severn Estuary, concentrating on the area between Caldicot and Chepstow to the west of the area and between Portishead and Aust to the east of the Estuary. It considers the nature of the cliff lines and their origins and discusses the likely development of such features should there be a faster rate of sea level rise in the future. Whilst at present most of these cliff lines occur outside of the sea walls, erosion could develop so that the protective tidal embankments lose their stability and hence may result in significant flooding.

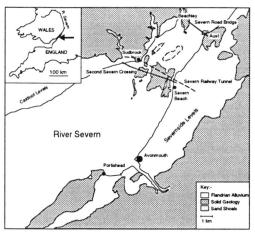

Figure 1. Map showing distribution of Flandrian sediments in the middle area of the Severn Estuary.

BACKGROUND GEOLOGY

Figure 1 shows the shape of the relevant part of the Estuary, delineating the solid rock outcrops from the more extensive deposits of Flandrian alluvium along the estuary margin. In the case of Aust, Severn Beach and Sudbrook, the surface solid geology is of Triassic age while Carboniferous rocks are exposed near Portishead, the Beachley Peninsula and the line of the Severn Road Bridge. Whilst the Triassic cliff line at Sudbrook is only 15 m high, at Aust cliff it reaches a height of 42 m and hence formed the ideal springing level for the Severn Bridge. Figure 1 also indicates the Levels on which the industrial development of Severnside has taken place in the form of large docks and major industrial complexes. The site of the M4 Severn Road Bridge is indicated, as is the alignment of the Second Severn Crossing which is currently being developed by a joint Anglo/French Consortium.

Hawkins (1990) published a map showing the likely subsurface topography present before the Flandrian deposits were placed. From this it is clear that the surrounding river valleys had deep bedrock features eroded at times of lower sea level and which are now buried below the Flandrian deposits. To the eastern side of the estuary (Severnside), a major river valley passed beneath the alluvium, entering the area between Aust and Severn Beach and passing out through the coastal flats in the vicinity of Avonmouth. The significance of this was first noted by Lloyd Morgan (in Richardson, 1887) when he recorded the then exposed Quaternary section along the line of the approach cutting to the Severn Railway Tunnel. Although Lloyd Morgan recorded striated pebbles

within the gravel horizons at the base of the channel, at that time he was not aware that the area had been glaciated and that this channel was the original course of the River Severn (Hawkins, 1990). When this river was blocked by glacial material, probably in the penultimate glaciation (Wolstonian), the Severn was diverted westwards and merged with the channel of the Mouton Brook which at that time had probably eroded a small feature in the area now known as The Shoots. The Shoots form the main low tide flow channel of the River Severn, being 25 m below Chart Datum (31.5 m below Ordnance Datum) and approximately 350 m wide. This channel, through which the tidal scour can pass at a speed of up to 8 knots (Admiralty Chart 1166) , is the main feature to be spanned by the bridge currently being constructed.

The Severn Estuary is unusual in having the second highest tidal range in the world; the variation between high and low water at Avonmouth being 14.8 m at astronomical tides, 12.3 m at spring tides and 6.5 at neap tides. As a consequence of this extreme tidal range, the Flandrian alluvium in the Severn Estuary has accreted to an unusually high level for modern coastal flats (see Hawkins, 1994). The alluvium has accreted naturally to a level of about 7 m above mean sea level, the actual height increasing upstream due to the north east as a result of tidal constriction. Thus the tidal marshes reach heights of 6.5 m, 7 m and 7.5 m above mean sea level at Caldicot, Aust and Oldbury respectively.

Since the construction of the sea walls, probably some 500 years ago (Cole, 1912) the ground outside of the tidal embankments has accreted by up to 1 m in

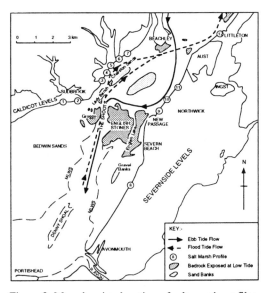

Figure 2. Map showing location of salt marsh profiles together with the main tidal flow directions in the middle Severn Estuary.

3102

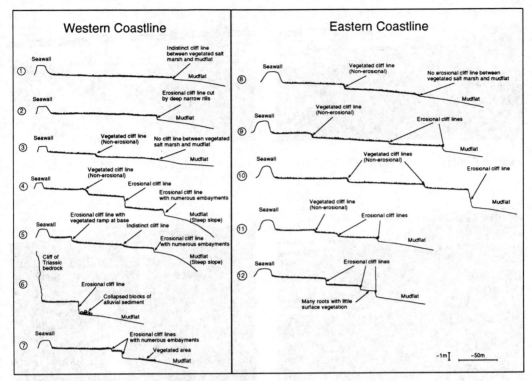

Figure 3. Morphology of recorded salt marsh profiles

level above that of the protected land within the sea walls. The reason for this is the constriction of area of the inundation, such that the level to which the water reaches is higher than would otherwise be the case. This not only results in a thicker layer of soil but also, because of the height, a prolific growth of salt marsh vegetation which assists accretion by trapping further sediment. The significance of this higher accretion outside of the sea walls to the formation of the tidal cliffs is not readily apparent. It is believed, however, that the tidal cliffs are more related to erosion by directed currents in either flood or ebb conditions.

DESCRIPTION OF THE SALT MARSH MORPHOLOGY

Figure 2 shows twelve localities around the Estuary where profiles of salt marshes have been described. The morphology of each of these profiles is illustrated in Figure 3.

South of Sudbrook

South of Sudbrook there is an unstepped salt marsh surface. Profile 1 is an area where the coast is protected from the direct attack of the flood tide by the recent deposits which form the Bedwin Sands and the

rocky outcrops of the Black Bedwins. A combination of the wide expanse of the Bedwin Sands and their elevation close to high water spring tide means that current and wave attack at the adjacent coastline is minimal. There is an indistinct cliff between the grass marsh and the mudflats in Profile 1 but in Profile 2, closer to the Sudbrook headland, an erosional cliff has developed. In this area the cliff has been cut by deep narrow rills (Figure 4) which penetrate the spartina marsh which appears to merge with the grass marsh in front of the tidal barrier with no perceptible cliffline. Shelly debris in the rills appear to have been transported although buried shelly lenses show limited colonisation in low energy current conditions in the past. These rills have been described by Allen (1987) as wave eroded ridge and furrows and seem to be formed by the action of waves and coarse debris cutting into drainage channels from the marsh levée.

North of Sudbrook

Between Sudbrook Point and the outlet of the Mouton Brook at St Pierre Pill, three profiles are included. Profile 3 has a minor vegetated cliff line (0.3 m) with many embayments. Between the tidal marsh and the mudflats there is no distinct cliffline. Profile 4, however, immediately north of the Shoots/Black Rock Channel, shows three distinct cliffs. Away from

Figure 4. Profile 2 near Sudbrook (ST 501 874).

Figure 5. Profile 4 (ST 517 886).

Figure 7. Profile 6, north of St. Pierre Pill (ST 524 896).

Figure 8. Erosional cliff line at Littleton Warth (ST 586 912).

Figure 9. Erosion of sea wall between Profiles 4 and 5 due to the lack alluvial cliff protection.

Figure 10. Remedial protection of sea wall (background) and alluvial cliff line near Profile 6.

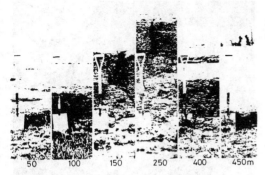

Figure 6. The changing amplitude of the grass /Spartina cliff at Mathern Wharf (after Hawkins, 1979).

the tidal bank is the distinct levee feature which probably results from material thrown up from the erosion zone by the strong flood tide inundation. Seawards, the three cliffs reach 0.5, 1.2 and 0.6 m in height. The upper cliff is generally vegetated, but is now showing evidence of some erosion. The middle cliff is almost vertical and exposes well laminated sediments above which a mature grass salt marsh vegetation/soil has developed. Below the main cliff the sediment is covered by spartina grass, the roots of which bind the soil and hence this is clearly helping to maintain the topographic ledge. The lower cliff is rounded and broken with significant embayments (Figure 5). The cliffs in this area change markedly in height along the distant coastline (Figure 6), as described by Hawkins (1979). As seen in Profile 5 the same upper terrace is present but has more evidence of erosional activity while the main cliff effectively dies out. Between the spartina marsh and the mudflats the 0.4 m cliff is similar to that in Profile 4.

North of St Pierre Pill (Profile 6), the Triassic bedrock is only some 50 m from the coastal margin and here there is only one pronounced cliff of some 1.5 m in height (Figure 7). On the surface of the mudflats adjacent to the cliff are many fallen blocks, typical of those referred to by Allen (1993). The erosional resistance and thus preservational potential of these blocks is proportional to their size, degree of vegetational binding and current/wave effects at the foot of the cliff. Towards the mouth of the River Wye (Profile 7), the main upper marsh is wider. Between this grassed area and the mudflats are two erosional clifflines totalling 1.2 m. The lower, some 0.9 m at the line of the profile, diminishes towards the north. Both cliffs have deep embayments. It is of note that here vegetation is beginning to colonise the upper mud flats, showing a clear balance between the erosional cliffs and the accreting mudflats, particularly where the tidal waters are deflected by the Beachley Peninsula.

Avonmouth to Severn Beach

Only one profile has been presented for this area as south of this the coast has been influenced by the development of industrial Severnside. At the position of Profile 8 there is only one distinct cliffline whis is heavily vegetated, indicating little if any erosional activity is now taking place. There is no pronounced topographic break between the vegetated marsh and the mudflats.

Severn Beach to Aust

Profiles 9 to 11 show three cliffs are present in this area. They are seen to occur at different distances from the tidal embankment and are very different in character. The most spectacular is the middle one (Profile 10) where below two vegetated cliffs there is a single one, reaching over 2 m in height. This cliff is seen to have some slight embayments. Again, a laminated structure is seen which is sufficiently bound by vegetation that it forms erosional blocks resting on the upper mudflats. In the case of Profiles 9 and 11, only the upper cliff is vegetated while seawards there are two smaller erosional cliff of 0.8 and 0.3 m.

A study of the salt marshes in this area has indicated that the southern part is accreting rapidly at present hence the coastline immediately north of Severn Beach is changing rapidly. The influence of the Second Severn Bridge construction has not yet been determined; the bridge is designed to have only a minimal impact on the tidal flow and hence in this accreting area, some 1 km north of the bridge construction, the coastline may not be unduly affected.

North of Aust

Along Profile 12 three cliffs are again present and each one is erosional. It is interesting that between the middle and lower cliff the flat surface is effectively unvegetated in winter and would appear to be part of the mudflats. The stems and roots of Spartina appear to have bound the sediments and hence in the summer period of low energy conditions, Spartina grass becomes established again. Figure 8 shows the upper cliff line together with a water filled erosional scour at its base.

Origin of the cliffs

It is clear that the most prominent cliff lines occur to the south of the Caldicot Levels, between Sudbrook and the Beachly peninsula and in the area of Northwick Oaze between the rock outcrops of Aust and Severn Beach.

South of the Caldicot Levels the main flood tide of the Severn Estuary is bifurcated by the sand banks of the

Middle Grounds. Whilst the main flow may pass to the north east towards the River Severn, a significant flow passes northwards towards the mouth of the River Usk and the adjacent Flandrian flats via the Newport Deep. It is the fetch of this tidal flow which has been observed in the past to be a significant erosional feature, as there is little refraction of the flood tide flow after the flow leaves the restricted marine channel south of Uskmouth. As a consequence, in this area erosion is seen to have occurred over a vertical height of 5 m exposing the peat bands and submerged forest.

In the area of Sudbrook, the flood tide is directed between the Gruggy, Ladybench and Charston Rock to the west and the English Stones to the east by the main channel of The Shoots. It is not surprising therefore that a significant tidal scour occurs directly ahead of this channel feature where floodwaters at the rate of 6-8 knots attack the soft sediments. It is of interest that the main tidal cliff, reaching a height of 1.2 m, is relatively short in length decreasing to 0.4 m between Profiles 4 and 5. The middle and upper clifflines effectively disappear over a reach of about 50 m. In this area the sea wall has been left vulnerable and is currently being attacked such that remedial works are urgently required (Figure 9). Similarly, in the region of Profile 6, there is only a small distance of salt marsh from the cliff to the sea wall. As a result the sea wall has been eroded and is now protected by rock blocks. The cliffline is also being protected by a rock bank to prevent further inland erosion (Figure 10).

The most prominent tidal cliff, however, is that between Severn Beach and Aust. Here the cliff lines are formed above the deepest part of the buried valley which runs beneath this area. Clearly the very high cliffline is related to the main tidal ebb flow which passes south from Slime Road across the estuary to abut the Flandrian deposits at Northwick (Figure 2). This tidal flow can be seen clearly in the infra red air photograph (Plate 6) which shows the nature of the river flow at low water spring tide conditions.

CONCLUSIONS

The paper has described the geological setting of the Severn Estuary which experiences the world's second highest tidal range. Such a large tidal range in a conical shaped estuary results in very fast currents. The geological structure and the more resistant lithologies of the Triassic and earlier formations form a number of headlands compared with the coastline of the Flandrian alluvium. These headlands and the sand banks which have accumulated within the estuary,have a major effect on the direction of the flow regime at both flood and ebb tide. The flood tide is seen to cause the significant cliffs south of the Caldicot Levels, north of Sudbrook Point and north of Aust while the swing of the ebb tide (Figure 11)

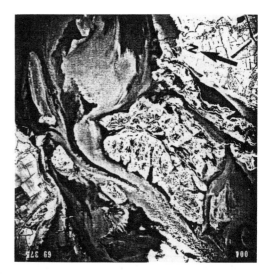

Figure 11. Infra red aerial photograph of the Severn Railway Tunnel region.

has been responsible for the high cliff line in the middle of the area between Sudbrook Point and Aust.

In general the salt marshes can be characterised as actively erosional or non-erosional. The actively erosional salt marshes have a number of embayed soft sediment cliffs bounding definite stepped salt marsh plateaux. Each plateau level is leveed away from the cliff towards the sea bank and can be vegetated by grass and/or spartina. There can be noticeable vegetational changes from one step to another. The cliffs within the salt marsh may be bare and being actively eroded, or may be vegetated and becoming buried and indistinct. Although the main cliff bounding the tidal mudflats is invariably eroding the salt marshes are simultaneously accreting vertically. Non-erosional mudflats form a wide levee from the mudflats to the sea wall and show no obvious buried cliffs. The front drainage apron towards the estuary is often heavily rilled, truncated by a small cliff bounded together by spartina. The vegetation of these marshes is predominantly spartina, becoming grassy inland. The margins of coastal salt marshes are in dynamic equilibrium with the prevailing tidal conditions. Slight adjustments in the direction of tidal currents brought about by the movement of channel bedforms may have a significant influence on where erosional forces affect the alluvial coastline.

The effect of a "relatively sudden" rise of sea level resulting from the greenhouse effect would be to increase the height to which the high spring tides/high astronomical tides reach and hence raise the level of the marsh outside of the sea wall. A rise in sea level would necessitate a heightening of sea defences. The potential for current erosion may well increase with the greater volume of water entering the estuary, which could result in the erosion of the tidal embankments.

It is essential that detailed studies of these topographic features are undertaken and their formation understood so that changes which occur in the future can be adequately recorded and their significance appreciated.

REFERENCES

Allen, J R L, 1987. Reworking of muddy intertidal sediments in the Severn Estuary, southwestern U.K. - a preliminary survey. *Sedimentary Geology*, 50, 1-23.

Allen, J R L, 1993. Muddy alluvial coasts of Britain: field criteria for shoreline position and movement in the recent past. *Proceedings of the Geologists' Association*, 104, 241-262.

Cole, S D, 1912. The Sea Walls of the Severn. Bristol. Printed for Private Circulation.

Devoy, R J N, 1979. Sea level changes in the Thames Estuary. *Phil. Trans. Roy. Soc.* Ser. B. Vol 285. 355 - 410

Fairbridge, R W, 1961. Eustatic changes in Seal Level. *Physics and Chemistry of the Earth*, 4, 99-185.

Gowdin, H, 1943. Coastal peat beds of the British Isles and North Sea, *Journal of Ecology*, 31, 199-247.

Hawkins, A B, 1971. Sea level changes around south-west England, pp 67-88 in Colston Papers, 23: *Marine Archaeology*, London Butterworths.

Hawkins, A B, 1979. Estuary evolution - with special emphasis on the Severn Estuary, pp 151-161 in Colston Papers, 31: *Tidal Power and Estuarine Movements*, London, Butterworths.

Hawkins, A B, 1984. Depositional characteristics of estuarine alluvium: some engineering implications. *Quarterly Journal of Engineering Geology*, 17, 219-234.

Hawkins, A B, 1990. Geology of the Avon Coast. *Proceedings of the Bristol Naturalists' Society*, 50, 3-27.

Hawkins, A B, 1994. Construction on Recent alluvial deposits: the importance of a correct interpretation of the Quaternary geology. In E F J DeMulder (ed.), *Special Volume: Engineering Geology of Quaternary Sediments*, Elsevier (in press).

Jelgersma, S, 1961. Holocene sea level changes in The Netherlands. *Mededelingen van de Geologische Stichting* (Ser.C), 6, 7.

Morner, N A, 1969. The Late Quaternary History of the Kattegat Sea. *Sver. geol. Unders.*, Ser C NR 640, Arsbok 63, NR3.

Richardson, C, 1887). The Severn Tunnel. *Proceedings of the Bristol Naturalists Society*, 5, 49-81.

Tooley, M.J., Torstensson and Davies. 1978. *Sea level changes in north west England during the Flandrian stage*. Clarendon Press, Oxford.

Sea cliff evolution and related hazards in miocene terranes of Algarve (Portugal)

L'évolution des falaises côtières et les risques associés dans les terrains du miocène (Algarve, Portugal)

Fernando M.S.F. Marques
Department of Geology, University of Lisbon, Portugal

ABSTRACT: In this paper are presented results of the use of a cost-effective procedure for the detection of slope mass movements, in a coastline with 46km of sea cliffs cut in miocene terranes of Algarve.

The data acquisition was mainly based on a detailed and systematic comparative study of aerial photographs of different dates completed by field surveys and comparison of large scale topographic maps.

This procedure allowed the identification of a large number of mass movements in the period 1947-1983, the assessment of their relevant dimensions and geomorphologic features, approximate dating and also the recognition of the types of the movements and the detection of some mechanisms involved.

The systematic character of the set of data obtained makes it a solid basis for the subdivision of the cliffs in sections with homogeneous behavior and for the assessment of the probability of occurrence of mass movements in each section.

The approximate dating of the mass movements also provides a basis for the analysis of the influence of the major factors in the evolution of these sea cliffs.

In the paper are also referred the implications of the data obtained in the assessment of the hazard associated to the evolution of these cliffs and in the definition of planning policies of cliffed coastal areas.

RESUMÉ: Dans ce travail sont présentés les resultats de l'aplication d'un procedé economique pour la detection de mouvements de versants, sur un secteur côtier avec 46km de falaises taillés en terrains miocénes de l'Algarve.

L'aquisition des donnés a eté basée sur un étude comparatif systematique de photografies aéreénnes de diferentes dates, completé par travail de champ e par des comparations de cartes topographiques em échele grande.

Ce procedé a permis la detection de nombreux mouvements de versant dans le période 1947-1983, la determination de ses dimentions et caracteristiques geomorphologiques, sa datation aproximative et aussi l'identification du type de mouvement et de quelques mécanismes associées.

Le caracter systématique des donnés permet la subdivision objective du secteur côtier em domaines avec comportement homogéne et la détermination de la probabilité d'occurrance de mouvements de versant dans chaque sub-secteur.

La datation aproximative permet aussi un essai d'avaliation de l'influence des facteurs externes sur l'évolution de ces falaises.

Dans le travail sont aussi présentés implications des donnés obtenus sur la détermination des risques associés à l'évolution de ces falaises et sur la définition de normes pour l'utilization de zones côtiéres avec falaises.

1 INTRODUCTION

Slope mass movements are the most evidént geomorphologic events in sea cliff evolution, and may constitute a major geological hazard in cliffed coastal areas.

In central Algarve, where an increasing number of coastal resorts are located on top of, or nearby sea cliffs, these movements may threaten life and property.

On the other hand, the economic value of land located in the coastal area is high. Thus, coastal management and planning policies, namely the definition of protection zones, must rely on solid basis, in order to minimize conflicts between authorities and land owners.

In spite of the obvious interest of this subject, the number of studies published about portuguese cliffs is very limited and most of them portray a qualitative and non systematic approach.

The necessity of obtaining a precise insight into this problem led to the development of a simple but precise procedure for monitoring the evolution of sea cliffs using a quantitative basis, mainly based on comparisons and measurements performed on aerial photographs of different dates.

Fig. 1 - Location of the study area.

Aerial photographs have been widely used to monitor cliff evolution, following two main perspectives that, in particular cases, may be complementary: qualitative identification of instability phenomena and, on the other hand, photogrammetric rectifications adapted to the problem to be analyzed. Each of the perspectives has shortcomings namely, the difficulty in obtaining quantitative data with the former, and the cost of the latter.

The procedure outlined in the paper combines qualitative comparison for detection of mass movements with precise measurements to assess their relevant dimensions.

Same preliminary results of the application of this procedure have already been published in Portugal (Marques & Romariz, 1991; Marques, 1991a, b).

The continuation of these studies in the sea cliffs build up of miocene terranes of central Algarve provided a data set that allows a precise insight of the evolution rates and processes involved. It can constitute also a solid basis for the assessment of the probability of occurrence of slope mass movements.

2 GENERAL ASPECTS

In the study area (Fig. 1), the exhumation of extensive karst features by sea action is the main

cause for the relatively complex morphology of the coastline.

The sea cliffs are mainly composed of biogenic calcarenites included in the Lagos - Portimão Formation, of miocene (Burdigalian) age (Antunes et. al., 1981), karstified, horizontal or sloping gently to south or southeast. These calcarenites are partially covered by reddish silty-clayey sands, of controversial age, usually referenced as Plio-pleistocene (PQ).

The Lagos - Portimão Formation is composed of alternate layers of fine grained calcarenites and calcarenites with high content in macro fossils. The latter are more resistant to surficial erosion, and in consequence, the alternate layers are visible in cliff profiles not affected by recent mass movements.

The calcarenites present two main types of morphology.

In the central sector (Fig. 2), extending from the mouth of the river Arade to Levante Hotel, the plio-pleistocene cover is restricted to the infilling of few karst depressions, usually small sinkholes one or two tens of meters wide. The sea cliff is moderately high (approximately 80% of its length between 20m and 40m), generally vertical, frequently notched at the toe and profusely perforated by karstic caves enlarged by wave action. In several zones, the top of the cliff and the adjacent area is indurated by calcite re precipitation, defining remnants of an older topography.

Surficial induration, karstic caves and wave erosion (specially in the intertidal zone) are the main causes for the presence of natural bridges, thin cave roofs and wide overhanging benches in the cliff face.

The coast line is irregular and mainly composed of short almost rectilinear stretches, forming small bays and headlands, frequently very narrow.

In the remaining sectors, the plio-pleistocene cover is almost always present. Karst features are extremely frequent and dominated by small (typically one or two tens of meters wide) sinkholes whose walls are indurated by calcite precipitation. The coastline is very irregular with frequent arch shaped stretches resultant of the exhumation of karst depressions. Stacks are also very common.

The general profile of the cliffs in eastern sectors is composed of a lower bench of calcarenites, usually notched at the toe, topped by one or several concave sections slope (displaying variable average slope, usually from 12° to 26°, but reaching 45° to near vertical in the steepest portions), extensively gullied, cut in the plio-pleistocene sands. In western sectors, the cliff profile, particularly in headlands, is near vertical, notched at the toe and composed of calcarenites. In the inner part of the small bays the cliffs are frequently composed of silty-clayey sands that correspond to the fill of karst depressions whose bottom lies below present sea level.

In terms of geotechnical properties, the fine grained calcarenites have uniaxial compressive strength typically from 1 to 3MPa and the calcarenites with macro fossils from 5 to 15MPa. The indurated calcarenites can produce higher values, up to 50MPa.

Discontinuities are generally very widely spaced, except near faults and in zones with extremely developed karst. Bed thickness is quite variable from place to place, ranging from medium to very thick. However bedding planes do not usually constitute discontinuities because there is a gradual transition between fine grained calcarenites and calcarenites with macro fossils.

The plio-pleistocene sands are dense, coarse grained, frequently pebbly, non plastic, with clay+silt content from 10 to 24% and clay content usually lower than 10%. Dry unit weight ranges from 18 to 23kN/m^3. Drained direct shear tests provided values of c' around 18kPa and ø' around 36° with minor variations from place to place.

3 METHODOLOGY

In terms of processes and mechanisms involved, toe erosion by waves is responsible for removal of debris of previous mass movements, by carving notches mainly in the intertidal zone and by widening of former karst caves. The notches cause the formation of tension fractures parallel to the cliff face, that ultimately lead to slumping or toppling (according to Varnes, 1978). The debris of these movements are then reworked and removed by waves, starting a new cycle of evolution.

Simultaneously, in calcarenites, the cliff face is affected by disintegration due to raindrop impact and alternate dry and wet conditions, while in the plio-Pleistocene sands the slope is dissected by gullies due to surficial runoff. However, these processes have a subordinate importance in the evolution of these cliffs, being responsible for a minor portion of erosion.

On the contrary, mass movements, although triggered by external causes as heavy rainfall, wave impact during storms or cyclic loading of seismic origin, and prepared by the toe erosion, are the most visible and potentially hazardous events in sea cliff evolution.

In order to assess the general evolution of the cliffs a procedure to detect mass movements was devised, mainly based in aerial photographs comparison performed with a mirror stereoscope with 8x binoculars.

Two overlapping pairs of photographs of the oldest and the more recent surveys corresponding to the same zone were selected. The older photos were fixed to the table in position for proper stereographic view. The more recent photographs are then positioned over the older pair, fixed by only one border, and in such a way that the selected zone is viewed. This arrangement allows easy and repeatable comparison of the same area in both sets, by raising partially the top photographs.

Selecting each time a very small area it is possible to detect modifications on the cliff top contour, and frequently, if a movement has occurred, to watch the corresponding top scar and toe debris in the more recent photographs.

Considering that frequently the cliff top borderland is not very far from horizontal and that in this particular coastline mass movements are not deep seated, it is possible to measure directly the dimensions of the horizontal area lost at the top of the cliff that corresponds to the depletion zone (IAEG, 1990). In fact, as movements detected were falls and rapid topples and slumps, each one corresponds to a neat and very quick loss of land at the cliff top, that can be fairly accurately measured.

The measurement can be performed with a parallax bar using nearby fixed references that exist in the two sets of photographs. Alternatively, measurements can be made with a 35mm photographic slide of a millimetric grid, taken with a low distortion lens and with careful alignment between lens axis and the plane of the paper.

Working with 8x magnification the minimum clearly visible size of the smallest division is approximately 0.08mm. With 1:30,000 photographs distances can thus be measured with accuracy in the order of half division, i. e., one and an half meter in the field.

It should be noticed that a particular care must be exercised when detecting movements and performing measurements in order to minimize errors. In consequence the user of this or similar procedures must be aware of the principles of aerial photography interpretation (see for ex. ASP, 1960) and of photogrametry (A.S.P., 1980).

In this particular case, as there was no previous experience of the procedure, the comparison was repeated twice: first only qualitative, the second with preliminary measurements and the third with very careful measurements. In the last comparison movements were detected using aerial photographs of 1947 (1:30,000) and 1983 (1:30,000). When a movement was detected photographs of 1958 (1:30,000), 1964 (1:15,000), 1972 (1:15,000), 1980 (1:15,000) and large scale maps (1:5,000 and/or 1:2,000) were used in order to define an approximate dating. Geomorphologic and dimensional features of each movement were also recorded in a purpose made registration form.

A large number of movements have been confirmed in the field, although no systematic confirmation was attempted, mainly because of the size of the study area and of the difficult access in some zones. Yet, no movement identified by means of photographs comparison had no supporting evidences in the field.

A critical review of the procedure suggests the following limitations, shortcomings and sources of error:
1- The procedure is quite dependent of the skill and patience of the user. In our experience, the 2nd and the 3rd comparison provided similar results both in terms of number of mass movements as in area lost at the top of the cliffs. It seems reasonable that for a user with experience in aerial photographs

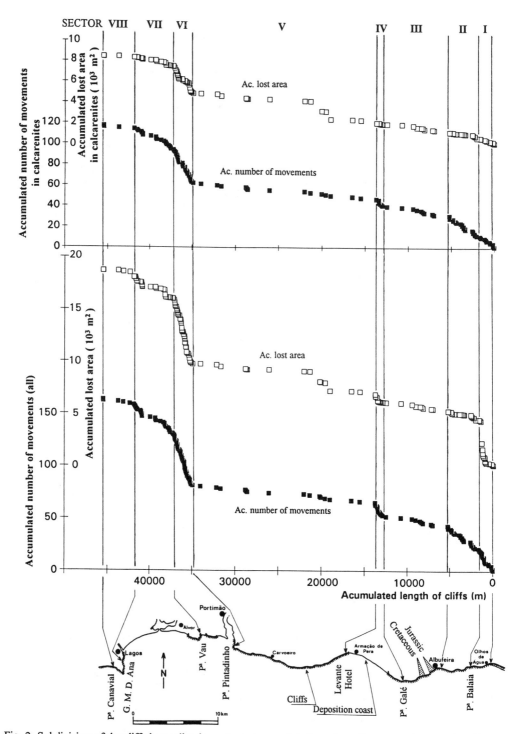

Fig. 2. Subdivision of the cliffed coastline in sectors. Accumulated number of events and accumulated lost area versus accumulated length of cliffs (from East to West). Movements in calcarenites (top) and all movements (middle).

interpretation two comparisons will be enough to obtain fairly accurate data.

2- Usually, movements can only be detected and the area lost measured at the top of the cliff. The detection and measurement of movements that only affected the cliff face must not be considered because the photographs are not orthorectified. Retreat at the toe must not be considered also, because usually there are no fixed references on the ground at the same height and common to both sets of photographs.

3- The retreat at the top of the cliff is assumed to be representative of the global evolution. This implies a parallel retreat model that, may not be strictly adequate under particular local conditions.

4- The accuracy of the identification and measurement in 1:30,000 photographs can reach 1 to 2m in the field, in conditions of near horizontal topography of the cliff top borderland and of fixed references located nearby. In cases where these conditions are not present or where ground surface was modified by man, the resultant accuracy is lower.

5- Movements smaller than those limits could not be detected. These can be important in particular lithologies, requiring that in all cases processes and mechanisms should be studied in the field.

6- The scale and photographic quality of distinct surveys is quite variable. This may introduce variations in accuracy of the data obtained for distinct time lags included in the global period of analysis, specially in the identification of movements.

On the other hand, this procedure has clear advantages over other methods of sea cliff evolution monitoring:

1- It is much more accurate than the comparison of topographic maps of different dates. In the case of Portugal, almost all the country is covered by two editions of the military map in 1:25,000 scale, witch is completely inadequate for this kind of studies. In fact, for Praia da Falésia, located immediately at East of Olhos de Água, with cliff evolution rates at least one order of magnitude higher than those detected in the cliffs in miocene rocks, Andrade et. al., 1989, based in maps comparison presented cliff retreat rates of 0.8 to 2.1m/year, with an average of 1.3m/year. This data has been revised with a photographic comparison that yielded a mean retreat rate of 0.25m/year (Marques & Romariz, 1991). This latter value has been confirmed by Bettencourt (1991) using a very precise (1:1,000) photogrammetric survey performed with purpose made photographs and land marks, that suggested mean retreat rate always much lower than 1m/year.

In the case of the study area the comparison of maps in scales 1:5,000 and 1:2,000 was undertaken but results were not generally consistent because the criteria used in maps production was not always the same. In consequence, the cliff top and face was not represented the same way in the two series of maps and there were cases where cliff had "grown" between the older and the new survey.

2- It allows at least a partial identification of processes and mechanisms involved.

3- It can be extended for time periods much larger than those possible with topographic surveys. Also, the method can be carried out in a systematic way along all the stretch of coast, what would not be feasible or, at least, extremely expensive with classical topographic surveys. These two points constitute a great advantage because retreat phenomena is discontinuous both in space as in time.

4 DATA OBTAINED AND DISCUSSION

In the study area, were detected 163 individual or groups of mass movements that occurred from 1947 to 1983 (Table 1). The space distribution of these movements was plotted in a graph of accumulated length of cliffs (from East to West) versus accumulated number of movements and, accumulated lost area at the top of the cliffs (Fig. 2).

This graph shows the contrasting behavior of distinct sections of these sea cliffs and allows the objective division of the study area in sectors with a near homogeneous mass movement space distribution.

The plot of the movements occurred only in calcarenites shows a similar pattern (Fig. 2), suggesting that the former subdivision reflects variations of properties of the miocene rocks along the whole coastal area.

Because of inherent limitations of the procedure applied, movements could only be assigned to time periods defined by the dates of the documents used in the comparative study. The shortcomings of this approach are twofold: the time distribution of the events frequency is artificially smoothed and, the relation between movements and triggering causes can not be precisely drawn. This relation can only be discussed in a general way and in a qualitative basis.

The available data set about external factors is also limited because wave action and particularly storm events were not systematically recorded in the southern coast of Portugal (Anon., 1992).

Bearing in mind these limitations and also the fact that for distinct zones the dates of the documents are not the same, the time distribution of frequency of events is generally not uniform and does not have a consistent pattern of variation (Fig. 3a). This can be due to the limited number of events in most sectors, statistically non significant.

The frequency of events for all sectors (Fig. 3b) shows a large variation along time, although clearly narrower than one order of magnitude. It should be noted that even in a 3 years long period (1980-83) movements did occur, suggesting that the retreat phenomena in these cliffs may be treated as a continuous variable at intermediate or long term time scales.

It is also clear that in sectors more sheltered in terms of wave action (VI and VII) mass movements were more frequent during the sixties. This could be explained by higher activity of external factors, namely rainfall and seismic activity during that period (Fig. 4). During the seventies, in sectors VII and VIII

3113

Table 1. Synopsis of the mas movements detected between 1947 and 1983.

SECTOR	Lithology	Length of cliffs (m)	Number of events	Length of cliff affected (%)	Mean width of movements (m)	Maximum width of movements (m)	Mean retreat rate (m/year)
I - Olhos de Água - Pª. da Balaia	Calcarenites	1.174	12	11	6	11	0,02
	Sands	396	8	≈100	11	22	0,3
	All	1.570	20	34	8	22	0,08
II - Pª. da Balaia - INATEL (Albufeira)	Calcarenites	3.596	21	5	4	8	0,005
	Sands	49	2	≈100	7	8	0,12
	All	3.645	23	6	4	8	0,006
III - INATEL (Albufeira) - Pª. da Galé (E)	Calcarenites	7.612	9	1,6	7	14	0,003
	Sands	8	1	≈100	N.D.	5	0,11
	All	7.620	10	1,7	7	14	0,003
IV - Pª. da Galé (E) - Levante Hotel (Armação de Pêra)	Calcarenites	750	8	6	4	5	0,005
	Sands	120	4	≈100	7	11	0,14
	All	870	12	19	5	11	0,02
V - Levante Hotel - Pª. do Pintadinho (Point of Altar)	Calcarenites	21.171	15	1,4	11	45	0,004
	Sands	19	1	≈100	N.D.	5	0,08
	All	21.190	16	1,5	10	45	0,004
VI - Pª. do Pintadinho - Pª. do Vau (W)	Calcarenites	1.787	30	27	7	25	0,04
	Sands	488	19	≈100	10	20	0,2
	All	2.275	49	43	8	25	0,08
VII - Pª. do Vau (W) - Geod. M. Dª. Ana	Calcarenites	4.516	21	5	5	11	0,005
	Sands	144	7	≈100	21	36	0,23
	All	4.660	28	8	6	36	0,01
VIII - Geod. M. Dª. Ana - Pª. do Canavial	Calcarenites	3.833	3	0,8	N.S.	6	0,001
	Sands	87	2	≈100	N.S.	6	0,16
	All	3.920	5	3	5	6	0,005

N.D. - Not determined   N.S. - Not significant

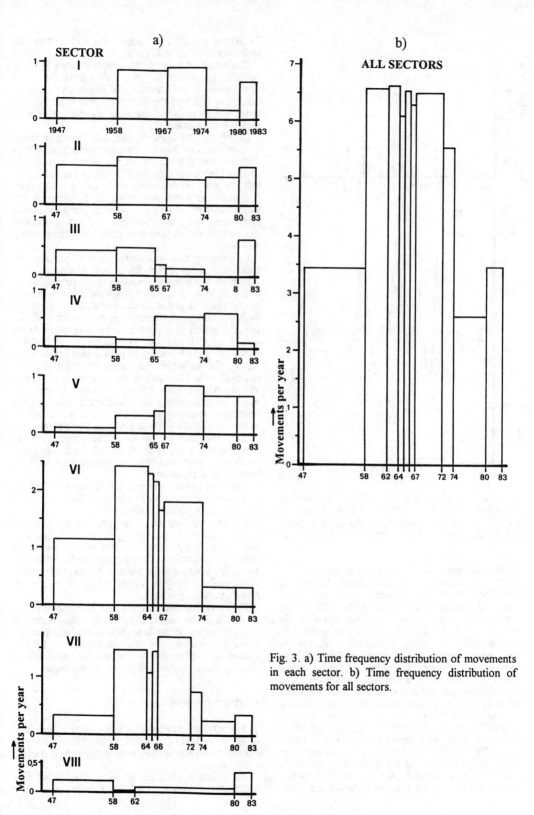

Fig. 3. a) Time frequency distribution of movements in each sector. b) Time frequency distribution of movements for all sectors.

3115

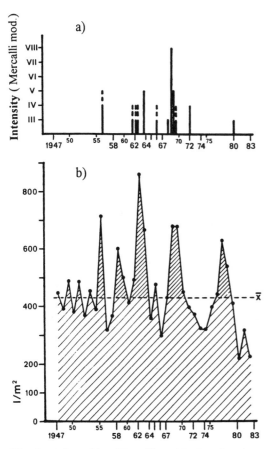

Fig. 4 - a) Intensities (Mercalli mod.) of earthquakes (data adapted from IGN, 1982 and LNEC, 1986).
b) Annual rainfall at Praia da Rocha (data adapted from Anuário dos Serviços Hidráulicos, 1948-1984).

retreat rates were significantly lower although no global explanation can be ascertained. In fact, in sector VI, beach nourishments were performed during that decade (Psuty & Moreira, 1990; 1992), causing a temporary reduction of sea action at the toe of the cliffs. At the same period, the sector VII, not susceptible of being significantly affected by beach nourishments, had a similar evolution. These observations suggest that other major external factors are involved in a complex combination and that particular features of the pattern of evolution can not be adequately precised without a complete data base about these factors.

However, it is clear that the earthquake of 23 November 1969, with local intensity of VIII (IGN, 1982; LNEC., 1986) did not produce an increase of the retreat rates observed during the first half of the sixties. This observation is also supported by the fact that in the hydrological years of 1968-69 and 1969-

70 annual rainfall was about 50% higher than the average for the zone (Fig. 4b).

The recorded size of the depletion zone is quite variable for different types of movements. No relation was found between cliff height and mean or maximum width of the depletion zone (Fig. 5). The same was found when analyzing the data relative to each type of movement.

In the case of toppling, the data (typically mean width from 2 to 5m and maximum width from 2 to 7m) indicates that tension cracks form in a relatively narrow band parallel to the cliff face and at a distance that is not related with cliff height. This can constitute an interesting point to consider in the study of the relations between tension crack formation and the strength of the miocene calcarenites and its variations along the whole coastal section.

Considering that time distribution of movements, although uneven during the period of analysis, had variations much lower than one order of magnitude and, that space distribution is almost uniform in each sector, the data presented can constitute an adequate basis for forecasting future evolution of the cliffs. In fact, the period of analysis includes very high relative annual rainfall and an VIII intensity earthquake, which are good examples of peak activity of external factors. On the other hand, cyclic evolution of active sea cliffs occurs because of the complex combination and interplay of the different factors. An extreme event of one factor can accelerate failures, and produce momentary increase of evolution. However, as further evolution is also controlled by the intensity of the other factors, this interdependence seems to preclude that a single event of very high magnitude can cause large variations in intermediate or long term rates of evolution.

Knowing space, time and size distributions of the movements it is possible to assess the local frequency and the return period of the events.

No attempt was made to produce estimates of these parameters, because geology and lithology have often sharp variations and results would be rather site dependent. This is particularly obvious in sector V, where a large number of caves occur and very detailed surveying is needed in order to provide reliable hazard zonations.

However, it is felt that the data presented in table 1, that includes separate values for movements in calcarenites and in sands in each sector, constitutes a valuable guide for the site assessment of hazards related to cliff evolution and for the definition of sound planning policies of cliffed coastal areas.

5 CONCLUSIONS

The procedure outlined is a cost-effective and fairly accurate method of monitoring sea cliff evolution. It allows systematic studies even in remote or difficult access zones.

Its careful application provides quantitative data sets that constitute a reliable basis for the assessment

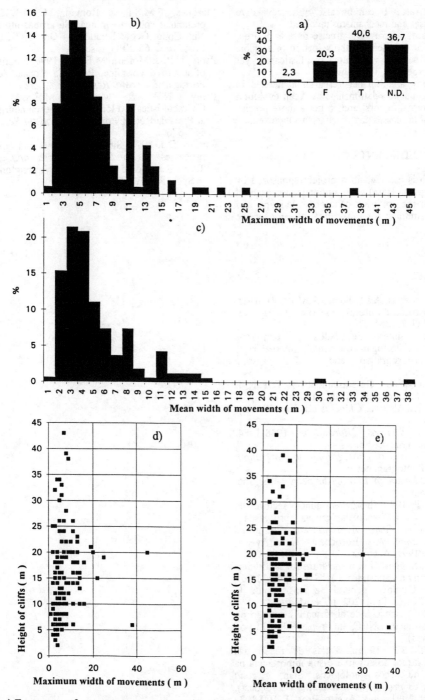

Fig. 5. a) Frequency of mass movements types (C - collapse of sinkholes fill; F - fall; T - Toppling; N.D. - not determined. b) Frequency of maximum width of the depletion zone of movements. c) Frequency of mean width of depletion zone of movements. d) Cliff height vs. maximum width of movements. e) Cliff height vs. mean width of movements.

3117

of evolution rates. It can be also quite revelative about processes and mechanisms involved.

Even in cases where very precise data is desired, the method can be useful in the selection of sensible areas to be studied with more sophisticated, yet much more expensive, techniques.

In the study area further research is needed in order to obtain solid relationships between evolution rates and external factors, and to get a more precise knowledge of processes and mechanisms involved.

6 ACKNOWLEDGMENTS

This research is included in a project sponsored by JNICT (Programa Base). The author is also grateful to DGOT for the provision of large scale maps, and to Prof. C. Andrade for the critical revision of the manuscript.

REFERENCES

Andrade, C., Viegas, A.L., Tomé, A.M.B., Romariz, C. 1989. Erosão do litoral cenozóico do Algarve. *Geolis*, III (1-2): 261-270.

Anon. 1992. Síntese de dados de temporais ocorridos em Portugal continental. Internal rep., Instituto Hidrográfico and LNEC, Lisbon (unpublished).

Antunes, M.T., Bizon, G., Nascimento, A. & Pais, J. 1981. Nouvelles donnés sur la datation des depôts miocènes de l'Algarve. *Ciências da Terra* (UNL), 6:153-168.

Anuário dos Serviços Hidráulicos 1948-1984. Instituto da Água, Lisbon.

ASP 1960. *Manual of photographic interpretation*. Am. Soc. Photogrammetry.

ASP 1980. *Manual of Photogrammetry*. Am. Soc. Photogrammetry.

Bettencourt, P. 1991. Erosão em quatro sectores da costa algarvia: avaliação das causas , processos e consequências. Sem. *A zona costeira e os problemas ambientais* (Eurocoast). Univ. Aveiro, 18-20 Set. 1991. Abstract.

IAEG, 1990. Suggested nomenclature for landslides. Commission on Landslides. *IAEG Bull.*, 41:13-16.

IGN, 1982. Catalogo general de isosistas de la Península Ibérica. Publ. 202, IGN, Madrid.

LNEC, 1986. A sismicidade histórica e a revisão do catálogo sísmico. LNEC, int. rep. 99/86, Lisbon (unpublished).

Marques, F.M.S.F. 1991a. Taxas de recuo das arribas do litoral do Algarve e sua importância na avaliação de riscos geológicos. Sem. *A zona costeira e os problemas ambientais* (Eurocoast), 18-20 Sept. 1991, Univ. Aveiro. Proc. 1: 100-108.

Marques, F.M.S.F. 1991b. Importância dos movimentos de massa na evolução de arribas litorais do Algarve. *Memórias e Notícias, Publ. Mus. Lab. Min. Geol.*, Univ. Coimbra, 112: 395-411.

Marques, F.M.S.F. e Romariz, C. 1991. Nota preliminar sobre a evolução de arribas litorais. IV Nat. Cong. Geotechnique, 2-4 Oct. 1991, Lisbon. Proc. 1: 57-66. SPG, Lisbon.

Psuty, N.P. & Moreira, M.E.S.A. 1990. Nourishment of a cliffed coastline, Praia da Rocha, Algarve, Portugal. *J. Coastal Res.*, 6, s. i.:21-32.

Psuty, N.P. & Moreira, M.E.S.A. 1992. Characteristics and longevity of beach nourishment at Praia da Rocha, Portugal. *J. Coastal Res.*, 8 (3): 660-676.

Varnes, D.J. 1978. Slope movements and types and processes. In *Landslides: analysis and control*. Trans. Res. Board Nat. Ac. Sci. Washington, Spec. Rep. 176: 11-33.

The relevance of sea cliff retreat assessment in hazard estimation:
An example in the western coast of Portugal

L'importance de l'évaluation du recul des falaises côtières dans l'analyse de risque:
Un exemple de la côte ouest du Portugal

Frederico G. Sobreira
Federal University of Ouro Preto, Brazil

Fernando M. S. F. Marques
Department of Geology, University of Lisbon, Portugal

ABSTRACT: A cost-effective procedure was used to assess sea cliff retreat in a 10km long coast located near "Lagoa de Albufeira", about 25km south of Lisbon. This procedure is based in the detailed and systematic comparison of aerial photographs of different dates (1958 to 1989).

In the study area, sea cliffs are cut in cretaceous marly limestones, sandstones and marls, in miocene calcarenites, and miocene to pleistocene sands.

Results reflect major variations in lithology and strength of the sea cliffs materials but also suggests strong influence of the adjacent beach dynamics.

In this paper are presented the more relevant results of this study and are discussed processes and mechanisms involved. Are also referred both the hazard estimation and planning implications of the data obtained.

RESUMÉ: Un procedé economique a eté utilisé pour determiner l'évolution des falaises existantes sur un trait litoral avec 10km de longeur, localisé prés de "Lagoa de Albufeira", 25km au Sud de Lisbonne. Le procedé a eté essenciellement basé sur un étude comparatif, detaillé et sistematique, de photographies aereénnes de diferentes dates (1958 à 1989).

Ces falaises sont taillés sur des calcaires marneux, arenites et marnes, du cretacé, des calcarenites du miocéne, et des sables du miocéne au pleistocéne.

Les résultats obtenus reflexissent les variations majeures de lithologie et de résistance des materiaux composant les falaises et aussi une forte influence de la dinamique des plages adjacentes.

Dans cet travail sont présentés les résultats plus importants de cet étude et aussi les processus et méchanismes d'évolution des falaises. Sont aussi discutés les implications sur l'analyse des risques géologiques et sur la gestion de de la zone côtiére.

1 INTRODUCTION

Sea cliff evolution is controlled by a complex combination of a large number of factors, frequently difficult to assess. Besides its academic interest, the study of cliff evolution becomes a fundamental task as a mean to provide basic data for adequate planning policies and for the reduction of associated geological risks in areas of increasing human occupation.

This subject has been studied under various perspectives comprehended in the fields of Geomorphology, Geology and Engineering (see for example Sunamura, 1983; Trenhaile, 1987; Anderson & Richards, 1987). Although processes and mechanisms have been studied and described to a great extent, much remains undone in terms of the assessment of evolution rates.

In order to obtain quantitative data of cliff evolution a cost-effective monitoring procedure,

mainly based in aerial photographs comparison, completed by field surveys, has been applied to cliffed coastal areas in Portugal. These studies provided a set of basic data and allowed the identification of some problematic zones (Marques & Romariz, 1991; Marques, 1991a, b; Marques & Andrade, 1992).

Extension of these studies to the sea cliffs located in the western coast of the Setúbal Peninsula, led to the identification of a high erosional section, where local retreat rates of up to 1m/year were detected. These rates have obvious implications on land use of the cliff top borderland and constitute a fine example of previously unrecorded high erosion zone.

This paper is intended to present the data set, results of a preliminary survey, and the description of processes and mechanisms involved. Implication of the data on existing models of shore dynamics and on coastal land use are also referred.

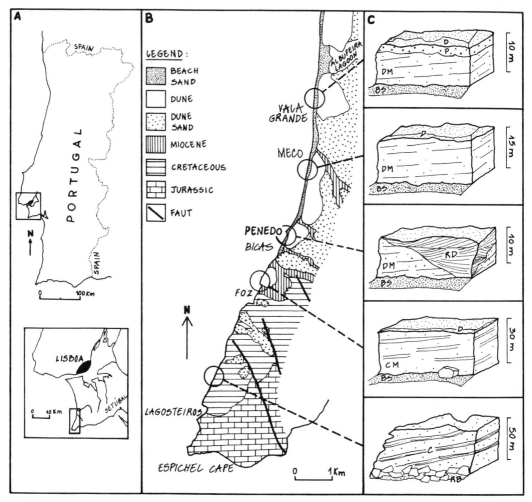

Figure 1 - (A) Location of the study area (B) Geological sketch of the area. (C) Typical cliff profiles where D-dune, P-pliocene sandy deposits, DM-miocene sandy deposits, BS-beach sands, RD-cemented dunes and recent deposits, CM-calcarenites miocene, C-interbedded limestones, sandstones and marl, RB-rock blocks and debris.

2 GEOLOGY AND GEOMORPHOLOGY

The study area is located in the Sesimbra county, about 25km south of Lisbon, and include a 10km long coastline dominated by sea cliffs (Fig. 1). Geographic limits of the area are, at north, the Albufeira Lagoon and, at south, the small bay of Lagosteiros.

These cliffs are cut in cretaceous, miocene, pliocene and pleistocene terranes, successively outcropping northwards.

From Lagosteiros Bay to Foz, sea cliffs are composed of cretaceous interbedded (Berriasian to Cenomanian) limestones, sandstones and marls (Rey, 1972). General dip of strata is NNW, their tilt decreasing progressively from south to north.

Further north, cliffs are cut in miocene (Burdigalian) terranes (Zbyszewski et. al., 1965), mainly composed of fine grained calcarenites, fine silty sands and limestones. Northwards of Bicas Beach miocene terranes (Helvetian and Tortonian) are more sandy, including sands, fine to coarse silty-clayey sands, very fine silty, sands and silty clays.

Near the lagoon, the miocene basement is covered by sands of pliocene age (Azevedo, 1982).

Pleistocene weakly cemented dunes are also present in the beaches of Bicas and Penedo. These also fill paleo valleys of the miocene basement.

Holocene dunes cover part of the sequence and constitute the landwards limit of the beaches at the mouths of the creeks of Lages, Amieira and Vala Grande.

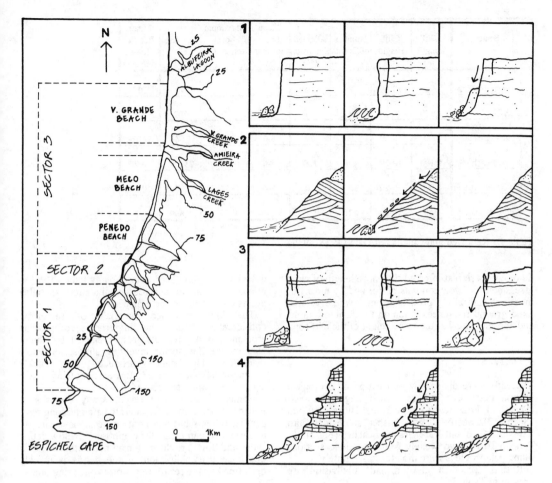

Figure 2 - Sectors defined by lithology and typical evolution for each sector. (1) Sector 3, (2) Cemented dunes filling paleo valleys at Penedo Beach, (3) Sector 2 and (4) Sector 1.

Morphology of the cliffs is quite variable from place to place and reflects lithology. Figure 1 represents a geological sketch of the area and main cliff types.

In cretaceous terranes cliffs are higher (up to 50m), profiles are irregular and debris at the toe is a constant feature. Mass movements are typically falls.

In miocene calcarenites, cliffs are 20 to 50m high displaying a near vertical profile. Mass movements are usually topples.

From Bicas Beach to north, a continuous beach exists, bordered by 10 to 20m high cliffs. Mass movements are produced by toe wave erosion during storms and higher spring tides, and are usually of the earth slump type. Surficial erosion is also important, and cliff face is more or less dissected by small gullies.

At Penedo and Bicas beaches weakly cemented dunes and recent sandy deposits filling paleo valleys constitute cliffs sloping around 60°.

3 METHODS

The evolution of these cliffs has been assessed using comparison and measurements performed on aerial photographs of different dates with a procedure described by Marques & Romariz, 1991 and Marques, 1994.

Aerial photograph surveys used were the following:
- 1958 - 1:30 000 (USAF);
- 1967 - 1:15 000 (ARTOP);
- 1981 - 1:15 000 (Cartoprojecto);
- 1989 - 1:15 000 (FAP).

The evolution of the cliffs has been monitored in two different ways. In cretaceous rocks and miocene calcarenites, as most part of the cliff top remained unchanged during the period of analysis, the comparison of the aerial photographs allowed the detection and measurement of relevant dimensions of single mass movements. In the less resistant

Sector	Cliff Length (m)	Cliff Height (m)	Number of Events	Affected Extension (%)	Mean Local Retreat (m)	Maximum Local Retreat (m)	Lost Area (m²)	Mean Retreat Rate (m/year)
1 Lagosteiros/Foz	3,750	20-50	8	2.5	3	5	90	10^{-3}
2 Foz/Bicas	1,100	20-50	8	9.1	3	6	302	10^{-2}
Penedo Beach	600	10-20	7	59	5-6	11	1,040	6×10^{-2}
Meco Beach	1,100	10-20	>20	100	5-6	34	25,250	0.7
3 V. Grande beach	1,000	10-20	5	30	5-6	14	2,800	0.1
Cemented Dunes	350	10-20	>20	100	-	26	8,800	0.8

Table 1 - Reported and Calculated Data (1958-1989).

lithology, as retreat affected the entire length of cliffs assessment of evolution was performed by means of measurements along profiles located where existed fixed ground references common to all photographs and approximately at the same height of the cliff top.

4. DATA SET

The cliffs were divided by application of geological criteria into three sectors named Lagosteiros/Foz (Sector 1), Foz/Bicas (Sector 2) and Bicas/Albufeira (Sector 3). Sector 3 was further subdivided into Penedo, Meco and Vala Grande beaches (Fig.2). At Penedo Beach two weakly cemented dune sections filling paleo valleys were monitored separately.

Results for each division and subdivision are present in Table 1.

To North of Albufeira Lagoon evolution of the cliffs is only sub aerial, mainly by gulling. The application of these procedures indicated only detectable erosion at the head of the major gullies.

5. PROCESSES

Lithological and structural features have strong influence on the types of mass movement. Typical movements have been remarked for each sector (Fig. 2). Variations in evolution rates at sector 3 can be explained by the dynamics of the adjacent beaches.

In sector 3, cliffs are mainly composed of interbedded layers of limestones, marls and sandstones. The weaker strata suffer superficial disintegration leaving unsupported limestone benches that evolute usually by fall. Sub aerial processes are more important than marine action because of the sheltering effect offered by a shore platform that promotes early dissipation of breaking wave energy.

At sector 2 waves erode directly the cliff base

inducing topples and subordinate falls. Tension cracks are common and toe accumulations of fallen debris also occur on some places.

The low resistance materials of sector 3 considerably favour the importance of marine erosion. Parallel retreats mainly caused by slump characterize this sector involving almost its whole extension. The debris produced are quickly disintegrated and removed by waves allowing a new cycle of evolution to succeed.

Cemented dunes and recent sandy deposits at Penedo Beach also suffer intermediate and long term parallel retreats but sub aerial process have also an important role in local cliff evolution. Shallow debris slides and flows are the dominant mass movements in this zone. Remobilisation of toe debris is quickly performed by the sea during storms on spring tides.

6. DISCUSSION

A general survey of the study area indicates an increase of erosion intensity from south to north reaching maximum values at the Meco Beach. The relative importance of marine erosion decreases to north and near Albufeira Lagoon it affects only the cliff base. Further north the active processes are only sub aerial.

The low erosion rates found southwards of the Bicas Beach can be explained by lithological constraints. Furthermore the debris accumulations at the toe of cliffs in miocene and cretaceous terranes constitutes an effective protection against wave action.

The change of marine erosion importance between Bicas Beach and Albufeira Lagoon does not reflect lithological differences. These variations can be explained by the particular pattern of the marine circulation cells of the whole coastal sector located to south of the Tagus River. A discussion of the existing circulation model for this zone is beyond of the scope of this paper. However, erosion data provided by this study suggests that this model need some adjustments, namely in the definition of the boundaries separating distinct coastal cells.

Cliffs cut in cretaceous rocks exhibit low retreat rates. However, as its evolution is essentially produced by mass movements of long return periods, they constitute a source of geological hazard in the area. In this case, retreat rates are not the most important parameter in hazard estimation but maximum local retreat.

Cliffs cut in miocene calcarenites a similar situation occur, with low global mean retreat rates but, the even larger size of mass movements indicates that maximum local retreat is by far, the most important indicator for hazard estimation and definition of criteria for protection zones.

The cliffs cut in the upper miocene terranes recorded the higher retreat rates, local values exceeding 1m/year. The filled paleo valleys of this sector also reach similar rates. In spite of the decrease in erosion rates found near Albufeira Lagoon this whole sector has been classified as highly hazardous. Due the concentration of activities on this area a potential risk situation must be considered.

It must be reminded that all values reported to each sector are average rates. Exceptional hazards due to local particular conditions or to the occurrence of extremely high intensity events of the external factors, can produce significant acceleration of retreat rates and must not be excluded in forecasting future evolution.

7. CONCLUSIONS

The detailed and systematic study of aerial photographs of different dates is a cost-effective and not excessively time consuming method for monitoring cliff evolution. It allows the identification and measurement of slope mass movements and is sufficiently accurate to compute mean retreat rates, that are fundamental tools to the assessment of geological hazards.

Despite the inherent shortcomings and limitations of the method, it is useful in preliminary evaluation and in the detection of more active sectors. These can be lately studied in more detail using, if very high accuracy is needed, more sophisticated techniques.

The rates of evolution reported from Setúbal Peninsula southwest coast show strong lithological control but also the influence of the shore dynamics in lower resistance cliffs.

The high retreat rates at Meco Beach, which exceed locally 1m/year show a geological hazard situation that may constitute a high risk zone due to the increasing use of that sector.

The data set presented provides a good basis for hazard and planning applications. However, it is felt that further work is needed in order to get a better insight into the complex combination of factors that control the evolution of these cliffs, and the processes and mechanisms involved.

REFERENCES

Anderson, M.G. & Richards, K.S. 1987. Slope stability. John Wiley & Sons. New York.

Azevedo, T. M. 1982. O sinclinal de Albufeira. PhD. Thesis. University of Lisbon.

Marques, F.M.S.F. & Romariz, C. 1991. Nota preliminar sobre a evolução das arribas litorais. *IV Cong. Nat. Geotechnique*. Vol.1: .57-66. Lisboa.

Marques, F.M.S.F. 1991a. Taxas de recuo das arribas do litoral sul do Algarve e sua importância na avaliaçãodos riscos geológicos. Sem. *A Zona Costeira e os Problemas Ambientais*, Univ. Aveiro. Proc. 1: 100-108.

Marques, F.M.S.F. 1991b. Importância dos movimentos de massa na evolução de arribas litorais do Algarve. *Memórias e Notícias, Publ. Mus. Lab. Min. Geol.*, Univ. Coimbra, 112:395-411.

Marques, F.M.S.F. 1994. Sea cliff evolution and related hazards in miocene terranes of Algarve (Portugal). *7th. Int. Cong. IAEG*, 5-9 Sept., 1994, Lisbon (in press).

Marques, F.M.S.M. & Andrade, C.F. 1992. Evolução das arribas litorais e actividade humana: um caso particular no Algarve ocidental. *Geolis*, Vol VI, (1 e 2) p.111-120. Lisboa.

Rey, J. 1972. Recherches géologiques sur le Crétecé inferiéur de l'Estremadura (Portugal). *Mem. Serv. Geol. Port.*. n° 21. Lisboa.

Sunamura, T. 1983. Processes of sea and plataform erosion. In: *CRC Handbook of Coastal Processes and Erosion*. CRC Press, p. 233-235. Florida.

Trenhaile, A.S. 1987. The geomorphology of rock coasts. Oxford Research Studies in Geography. Claredon Press. Oxford. 384p.

Zbyszewski, G.; Ferreira, O.V.; Manuppella, G.; Assunção, C.T. 1965. Notícia explicativa da folha 38-B, na escala 1:50.000. Serv. Geol. Port. Lisboa.

Geological-engineering aspects of the coastal zone development in Poland

Aspects de la géologie de l'ingénieur dans le développement de la zone côtière en Pologne

W. Subotowicz
Technical University of Gdańsk, Poland

ABSTRACT: The antropopression in relation to the coast has a consequence that building settlement and towns, and the development of this exceptionally attractive region calls for knowledge of complicated processes modelling the coast. This knowledge is necessary taking into account the hazards that the coast causes with respect to its developed interland. The destructive effect is observed mainly in form of abrasion and generated by it geodynamic phenomena occurring in the coastal zone. Hence there is a need to determine the boundary of safe investment and the protective zones along the coastline. Under the Polish coastal conditions, this boundary runs at a distance of approximately 100 m from its edge.

RESUMÉ: Concernant la côte de la mer l'anthropopression cause que la construction des villes et l'aménagement de cette zone excessivement attractive exije la connaissance des processus géologiques, très compliqués qui forment la côte. Cette connaissance est indispensable parce que la côte de la mer est une ménace incessante pour les terrains aménagés près de la mer. Il s'agit principalement de la délimitation où on peut investir sans perte dans la zone de côte. Quant à la côte en Pologne la ligne de sûreté parcourt dans une distance à peu près de cent mètres de son bord.

1 MORPHO-GEODYNAMIC CHARACTERISTIC OF THE COAST

Within the coast zone there can be distinguished its three elements: the undercoast (the sea-bottom), the beach, and the overcoast. Regardless of the occurring in the undercoast mostly two submerged bars, the character of the undercoast is determined by the shape of the abrasion profile (Subotowicz 1989). It is shaped in the matrix bed generally made up of till and intermoraine deposits, i.e. sand and gravel, and clay.

The coast whose overcoast had also been formed by the above sediments is known as cliff shore. Its overall length is 100 km.

The concentration of the sandy material originating from abrasion, primarily nearly exclusively from the cliff shores had accumulated on some other concave lengths of the coast. In this way were created sandbars and dunes. Hence is derived the term-the dune coast. Its occurrence is conditioned by the presence of postglacial dune sediments in the overcoast. However, one should not exclude

the existence of older matrix bed in their undercoast. The total length of the dune coast is 400 km.

The overall coastline in Poland is 500 km.

Fig. 1 The coast in Poland

1.1 The undercoast

On the basis of present knowledge in the undercoast there has been proved the occurrence of dynamic sand bed resting on the matrix bed (Fig.3). The bed is mainly represented by till, and intermoraine clay, sand and gravel. In the case of dune coasts the matrix bed can also be composed of contemporary organic sediments (gyttja and peat).

The dynamic bed consists of sand on the base of gravel, pebbles and boulders. The maximum thickness of this bed is 5 m and generally extends down to the izobath of 10 m, which corresponds to the distance of 800 m from the waterline to the mean sea level. Within this bed there are usually found two submerged bars of which the internal one is situated at the distance of up to 100 m from the waterline, whereas the other bar, the external one-at the distance of approximately 200 m.

The internal bar is characterized by a great variability of form and small thickness of the rubble. The other bar, the external one is definitely larger, of several metres rubble thickness, and presents a significantly stable form of the undercoast. As it follows from the current investigations the bar was formed between the XIX-th and the XX-th century during an extreme storm surges (Majewski, Dziadziuszko and Wiśniewska 1983). The pebbles, which are found in the bar base, and the boulders were brought here just during the great storm. Only the energy of high waves and the intensive currents were able to transport the pebble and boulder rubble followed next by sandy material. During the indirect and yearly storm surges of much weaker strength of interaction upon the bottom, only the surface part of the bar not exceeding the thickness of 1,5 m has been affected by the wavy motion and currents.

It should be noted that in the matrix bed in the area of the second bar ridge there occurs a buried abrasion escarp (Fig.3). It might have also been formed during an extreme storm surges, when the whole bar was activated, and the uncovered matrix bed was here mostly affected by abrasion.

The repeated each year exposures and abrasion of the bed take place only within the bar area and beyond the dynamic sand bed range towards the sea.

The sand rubble, of which the dynamic bed with the bars is made up, transports from west to east (Fig.1). The resultant direction and form of transporting the rubble is known as the rubble is known as the rubble stream. The rubble balance withinn the stream reveals some negative tendencies, i.e. its volume is subject to gradual reduction. As a proof of that can be its decreasing thickness and the occurrence of greater and greater areas of uncovered matrix bed.

1.2 The beach

The width of the beach with respect to the mean sea level ranges from 10 to 30 m. Narrower beaches are situated at the foot of the cliffs, while wider ones along the dune coasts. The beach rubble thickness varies between 0.5 and 1.5 m. Under the sandy-pebble ruble, similarly to the undercoast, there is the matrix bed. In time of high storm levels when the waterline is displaced towards the foot of the overcoast, the beach represents an extension of the undercoast.

1.3 The cliff and dune overcoast

The height of the overcoasts oscillates between several and tens of metres above sea level.

The overcoast dunes are composed of postglacial sandy formations. Sometimes there occur also peat and humus sand interbeddings, as an evidence of fossil soil levels.

Within the cliffs of about 30 m high it is possible to distinguish two till levels and intermoraine mainly sandy and clayey formations. The intermoraine sands are usually accompanied by underground water, which can be found in the cliff slopes in form of seapage springs. At places where water appears together with clays there is an intensive occurrence of landslide phenomena. It is worthwile pointing out that the phenomena are closely connected with abrasion resulting in the steepness of the cliff and its permanent recession. The land factors following from some favourable geological structure and hydrogeological conditions of the cliff, as well as relatively disadvantageous strength parameters of the cliff subsoil are subsoil are responsible only for the form and spatial range of the landslide.

The cliff abrasion rate is for (Fig.1):
1. Jastrzębia Góra - 0.94 m/year (1977-1990),
2. Dębina near Ustka - 0.90 m/year (1978-1990),
3. Niechorze-Trzęsacz - 1.10 m/year (1974-1983).

The mean cliff destruction rate in Poland in the last sixteen-year period (1974-1990) it is 1.0 m/year.

A question appears-which can be the extention of expected landslides towards the land? The answer to this question is to be found in the geodynamical researches of the cliff and in the estimation of its stability.

2 GEODYNAMICS AND ANALYSIS OF CLIFF STA- BILITY

Three geodynamical types of cliffs have been distinguished of cliff coast in Poland (Subotowicz 1982), (Fig.2). Those are:
1. The fallen ground type.
2. The fall down type and
3. The landslide-flow type.

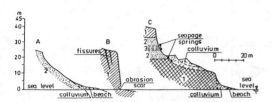

Fig. 2 Geodynamical types of cliffs in Poland A - fallen ground type, B - fall down type, C - landslide-flow type; 1 - till, 2 - sand and gravel, 3 - clay

The fallen ground type is characteristic for the cliffs built of fluvioglacial sand and gravel formations. They decide of the slight declination of cliff slope.

The fall down type is characteristic for the cliffs built of stiff till or clay giving them a vertical profile of the slope. In the case of increasing of humidity of cohesive formations, a change of their state follows. Landslide phenomena which develop then as flows and landslides quality the given cliff as another geodynamical type.

The landslide-flow type as a rule is characterised by a great lithological differentiation and by the presence of underground water. Apart mostly two till levels, intermoraine dividing sand, gravel, silt and stangnant clay occur. Among intermoraine formations usually underground water appearing on the slope in shape of seepage springs occurs.

The above accepted generalisations concerning the classification of the cliffs in Poland have a direct influence on the method of the analysis of their stability. The greatest importance, because of the frequency of occuring and complication of problem has among the mentioned geodynamical types, the analysis of stability of the cliff type three (the landslide-flow type).

Dependently on concrete geological and water conditions as well as physical and mechanical properties of soils, different sliding surfaces can appear. In the calculations of stability of slopes of different types, mostly a sliding surface in cylindrical shape or any sliding surface is accepted (a broken surface or a combination of a cylindrical and a broken surface). Every time the choice of accepted in calculations sliding surface will be depend on detailed geological and geotechnical analysis of subsoil of cliffs.

Independantly of considered method of calculation of stability and accepted sliding surface, the corresponding definition of geotechnical parameters has the essential importance for the proper estimate of cliff stability. It is abvious that carrying out of full geotechnical researches for all soil layers and all cliffs of coasts in Poland, or even distinguished sections is practically imposible. The full analysis requires a statistical estimation of geotechnical parameters for all soils from which cliffs are built.

In this connection it is indispensable to accept some simplifications. In every interesting profile (representative for considered section of cliff coast) layering of subsoil kind of soils and their state as well as water conditions in the cliff are known. On the ground of abovementioned information the average values of physical and mechanical parameters can be accepted for deterministic estimation of stability.

In farther analysis the method of estimation of stability in using probabilistic elements has been proposed. It has been accepted that one essential layer on which or in which appear sliding surfaces deciding of its stability, influences on the global stability of cliff. Such an assumption is based on the geological analysis of cliffs and long observations of real landslides on cliffs in Poland (Subotowicz 1982, 1990). For such indicated layer of soil widened researches of mechanical properties and their statistical estimation of expected values, standard deviation and standard variations. The coeficient of stability calculated by using a statistical estimation of geotechnical parameters for the layer deciding of cliff stability is treated as a random variable. Then it can except any values from determined range with some probability. The probability of failure is a measure of risks (Tejchman, Subotowicz and Gwizdała 1992):

$$P_f = p\,[\text{failure}] = p\,\left[F < F_D\right]$$

where: F and F_D are coefficient of stability correspondently calculated and

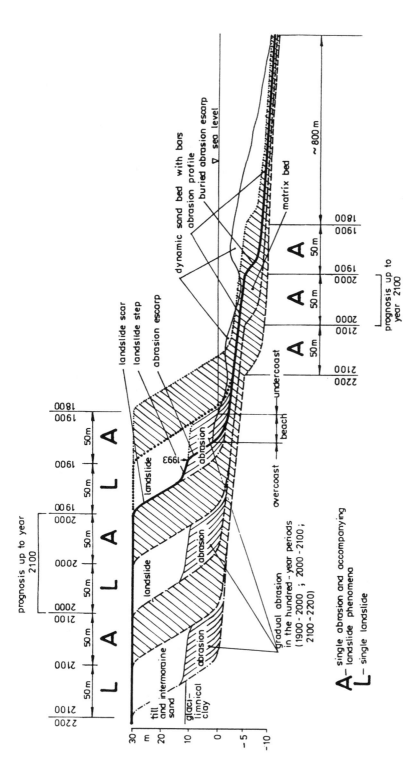

Fig. 3 Transformation of the cliff coast at Jastrzębia Góra in Poland

A — single abrasion and accompanying landslide phenomena
L — single landslide

admissible.

For example asuming a probability of failure $Pf = 10^{-3}$, the following reaches of slip zones are obtained (Fig.3):

p $[F < 1.3]$, reach 37 m
p $[F < 1.5]$, reach 49 m

3 TRANSFORMATION OF THE COASTAL TRANSVER-SE PROFILE

During an extreme storm surges that took place about the year 1900, a single intensive abrasion process A (Fig,3) caused the displacement of the cliff slope towards the land by about 50 m. As a result, a large landslide L, or a series of landslides towards the cliff's hinterland by another 50 m happened. The cliff retreated totally by about 100 m.

Assuming an analogy of a single abrasion of the cliff A it should also be taken for granted that today's buried escarp in the sea-bottom had also been dislocated by 50 m in the direction to the land. At that time also, the whole matrix bed of the undercoast and the beach underwent remodelling in a way of a single event. There was shaped a characteristic abrasion profile of the zones, which in principle, has not changed for ninety years.

In the vicinity of the buried abrasion escarp there is situated the second external bar. Its origin is also connected with the extremal release of the sea power, and as a matter of fact, it followed later when the storm was quieting down. However, during the yearly storms only the surface part of the bar mentioned is subject to remodelling.

4 FINAL REMARKS

Recurring to the presented transformation of the crosswise profile of the coastal zone there can be made an assumption that the process will take a similar form during the next extremal storm. Theoretically it is expected to occur in the year 2000 (Fig.3). By an analogy to the earlier considerations reconstructing the process of the year 1900, there is going to take place abrasion in the coastal zone of a catastrophic character. The whole abrasion profile modelled within the undercoast matrix bed, the beach, and the cliff overcoast will be at one time disclocated along by approx.50 m towards the land.

The steep cliff of the height 30 m a.s.l. dislocated in the year 2000 towards the land will create conditions for landsliding activation of the cliff on a large scale. The range of the landsliding phenomena, similarly to the year 1900 will also amount to 50 m (Fig.3).

Thus, it is probable that the cliff edge will recede by about 100 m.

However, it should be pointed out once again that the process of modelling the coastal zone has a catastrophic character. It occurs at a frequency of once in one hundred years.

The probablility of occurrence of such a process urges us to establish some safe boundaries of investment in the hinterland of the cliff, which should follow the line at a distance of at least 100 m from the cliff edge. It is a prognosis for the nearest one hundred years.

REFERENCES

Majewski, A., Dziadziuszko Z.and Wiśniewska, A. 1983. Monography of extreme storm surges in 1951-1975 (in Polish). Wydawnictwa Komunikacji i Łączności.

Subotowicz, W. 1982.Lithodynamics of the cliff coasts in Poland (in Polish). Ossolineum.

Subotowicz, W. 1989. Investigation of the cliff coast of Jastrzębia Góra (in Polish). Technika i Gospodarka Morska 7, 336-338.

Subotowicz, W. 1990. Engineering Geological Problems of the Coast in Poland. Proceedings Sixth International Congress International Association of Engineering Geology, Vol.4, Amsterdam: 2869-2872.

Tejchman, A., Subotowicz, W. and Gwizdała, K. 1992. Calculation and forecast of coastal cliff stability. Proceedings of the Sixth International Symposium "Landslides", New Zealand: 579-583.

Shoreline evolution of the South Bank of the Tagus River between Cacilhas and Cova do Vapor

L'évolution de la rive gauche du Tage entre Cacilhas et Cova do Vapor (Portugal)

M. Alexandra Chaves, J. A. Rodrigues-Carvalho & J. S. Cruz Rodrigues
New University of Lisbon, Faculty of Science and Technology, Geotechnical Department, Quinta da Torre, Monte de Caparica, Portugal

ABSTRACT: The south bank of the Tagus River, facing Lisbon, is a high sensitive area in the Geological and Geotechnical point of view, and so calls for a comprehensive study to support adequate land use planning. To help this, the Geotechnical Department of the New University of Lisbon has been developing the project "A Geological-Geotechnical study of the South Bank of the Tagus River between Cacilhas and Cova do Vapor". This project includes a particular study on the "Shoreline evolution of the South Bank of Tagus River between Cacilhas and Cova do Vapor, during the last hundred years".
The results based on a comparative analysis of some ancient maps (1871) and aerial photographs taken from 1938 to 1991 are presented in this paper. According to this investigation, the authors concluded that the shoreline evolution was, directly or indirectly, influenced by man's activity.

RÉSUMÉ: L'aménagement projecté de la rive gauche du Tage, région défavorable sur le point de vue géologique et géotechnique, imposa la necessité d'entreprendre sa reconaissance. Donc, la Section Géotechnique de l'Université Nouvelle de Lisbonne a entrepris un project de recherche "Étude Géologique et Géotechnique de la Rive Sud du Tage" qui a aussi envisagé "L'évolution de la Rive Sud du Tage de Cacilhas à Cova de Vapor depuis la dernière centaine d'années".
On présente, dans ce rapport, les résultats obtenus en faisant la comparaison des anciennes cartes topographiques (1871) et des photographies aériennes de plusieurs dates (1938-1991). On a remarqué que, directe ou indirectement, l'intervention humaine a été la grande réponsable de l'évolution de la rive.

1. INTRODUCTION

The south bank of the Tagus River is a high sensitive area in the Geological and Geotechnical point of view. Hence, the knowledge of its geological and geotechnical features is essencial to achieve an adequate land use planning. In order to help this, the Geotechnical Department of the New University of Lisbon has been producing different kind of documents within the project "Geological-Geotechnical study of the South Bank of the Tagus River between Cacilhas and Cova do Vapor". This includes a specific study on the "Shoreline evolution of the South Bank of Tagus River between Cacilhas and Cova do Vapor during the last hundred years".

One of the most remarkable conclusions provided by this investigation is that the shoreline evolution along the time was, unquestionably, directly or undirectly related to man's activity.

2. METHODOLOGY

2.1. *Selection of information sources*

The search of available maps and aerial photos of the area, concerning different periods was the first phase of work. The comparison study started, based upon an available map dated 1871 named "Planta do Rio Tejo e suas margens (Plan of River Tagus and its banks)", in a scale of about 1/10 200, covering the area between Cacilhas and Porto do Bucho (Fig. 1). This map is supposed to be the oldest quantitative cartographic representation of the region. However it was possible to do just a qualitative comparison analysis with this map. In what concerns aerial photos, the oldest (1938) and the most recent one (1991), were selected together with others taken in different years and similarly spaced in time.

2.2. *Processing*

The work involved five major steps:
1. establishement of a large number of reference points both on the aerial photos and on a topographic reference map in the scale 1/10 000;
2. assessment of the approximate scale of each photographs set ;
3. stereoscopic viewing and interpretation of each set of photographs and drawing out the interpreted shoreline that was discriminated as a natural shoreline

- beach and scarp - and as artificial shoreline - quays, walls, piers and embankments;

4. conversion of the different drawings to the reference scale 1/10 000;

5. examination, analysis and comparison of the consecutive drawings, and trial of dissimilarities.

The different quality of the aerial photographs studied caused some problems due to technical take off conditions and processing methods. As a mater of fact the sharpness and variability of scale, even in each photograph, is remarkable in the oldest set. This obliged a carefull work of adjusting and correction. The better quality of the more recent sets of photographs points out the evolution of aerophotographic methods in the last decades.

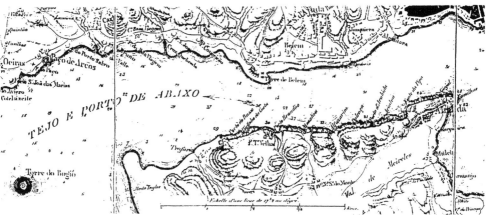

Fig. 1 - "Plan de Lisbonne, son Port, ses rades et ses environs". Ch. Calmet-Beauvoisin, 18th century.(Excerpt from a reduced copy)

3. RECORD AND ANALYSIS OF THE SHORE LINE EVOLUTION

In order to make easier the description of the shore line evolution, the studied region was subdivided in three sections according to its morphology and human occupation.

1 - Cacilhas - Bridge over the Tagus River: a shore scarp section where the natural shoreline outcrops just along small areas due to the high density of coastal constructions - dominantly quays and walls.

2 - Bridge over Tagus River - Trafaria: a mostly scarped shoreline carved by streams whose mouths form small re-entrances which give place to small beaches. This section reveals an increasing occupation by important coastal constructions through time.

3 - Trafaria - Cova do Vapor: a shoreline made up by a continuous line of beaches interrupted by several piers and embankments. The most remarkable changes within all the studied shore line, during the period under study, was noticed in this section's embankments.

3.1.Cacilhas - Bridge over the Tagus River

The morphology of this section corresponds to an abrupt scarp built up in miocene sandstones, siltstones, clays and calcareous rocks, whose toes are now intensively occupied by man, and so they crop out only along small areas.

From 1871 to 1991 a significative change in this section was not found. In 1871 the occupation of the riverine zone was restricted to some edifications on the border line between Cacilhas and Trafaria and almost all the existing harbour structures were already built: a quay along the riverside between Cacilhas and Ginjal and a still existing protection wall from Olho de Boi, till Arialva. By that time the shoreline extended about 100 m in relation to the wall (COSTA,1986) due to a fan deposit. This shoreline extension is no longer visible in the photographs taken from 1938 on.

The more relevant change is the disappearance, between 1871 and 1938 of an almost continuous line of beaches.

In 1938 some remains of beach were preferently located in small bays. From this year, till now, these beaches have been disappearing as they are being occupied by coastal constructions.

Between 1938 and 1958 there were no changes worthy of note, according to the available information concerning the Olho de Boi - Bridge over the Tagus river section.

In the interval 1958-1964, it should be mentioned the construction done in Cacilhas of an embankment that covered an area of nearly 8500 m^2 to remold the landing pier and enlarge the bus terminal (Fig. 2).

In 1964 the work on the Brige over the Tagus was in progress. These operations caused a large embankment at the south end wich pushed out the shoreline between 30m to 50m in a stretch of approximately 700m.

Between that date and 1984, the only mention to do relates to a shoreline advance of nearly 40 m along almost 400 m, caused by the construction of an embankment (1978-1984) in the area between Cacilhas and Olho de Boi.

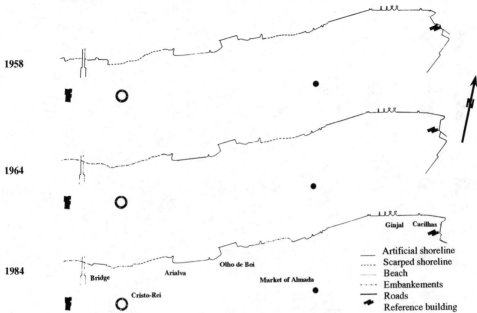

Fig. 2 - "Shoreline evolution between Cacilhas and Bridge over the Tagus River"

3.2 *Bridge over Tagus River- Trafaria*

Morphologically, this sector differs from the last because the scarp is shaped in degrees and because it's cut by perpendicular valleys to the shoreline of the river. The main modifications of the coastline are related to the increasing occupation of port structures. In 1871 the continuous line of beach, referred in the last section, proceed, till Portinho da Costa. To the west of this place, till Trafaria, the shoreline is in scarp whose evolution till 1938 consisted in a general retreat of the toe (talus), enlargement of the Portinho da Costa recess and the pronoucement of the Porto do Bucho re-entrance. From this year, till now, the main modification of this section were related with the increasing occupation by harbour structures.

This section is devided into two zones to simplify the description of change:

1. Zone between the Tagus Bridge and ETC boat terminus (Boliden);

2. Zone between ETC and Trafaria.

3.2.1. *Zone between the Tagus Bridge and ETC boat terminus*

In 1938, east of Forno de Tijolo (brick ovens), the presence of a large deposit, possibly of waste materials discarded from a clay pit farther up, can be verified in the scarp (Rodrigues-Carvalho and Costa, 1986).

In the years between 1938 and 1958, only four alterations, all important ones, can be registered (Fig. 3):

1. Erosion along 150 m at the base of the landfill of rejected material from the old clay pits, located just east of the Bridge (referred to above).

2. Enlargement of the Banatica pier to the west some 1300 m².

3. Enlargement of the Petrogal pier by some 3400 m².

4. Disappearance of the old wall that stretched 150 m, just east of the current day Tagol pier.

Between 1958 and 1972 there were no further relevant modifications on the shoreline, except a slight change due to the reconstruction of the Porto Brandão pier after 1964.

From 1972-1975, another embankment was built next to Tagol, lengthened to the east in nearly 2500 m². The situation was stable until 1978. The scale of the work documents used register nothing of note.

Between 1978 and 1984, the Tagol pier was enlarged, extending the shoreline between 30 m and 40 m, in a zone of 700 m. The Petrogal pier was also amplified also some 50 m to the east.

Between 1984 and 1991, the Banatica pier wall was extended to the west, covering over the small re--entrance that was there. In the same period, the Tagol pier was extended west via construction of an

3133

embankment and supporting structures. This new area, with a wideness of 10-30 m, stretches along 150 m of coastline.

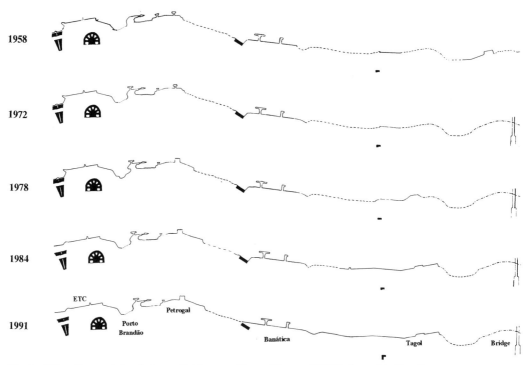

Fig. 3 - "Shoreline evolution between Bridge over Tagus River and ETC" (legend in Fig. 2)

3.2.2. *Zone between ETC and Trafaria*

Between 1938 and 1958, the only construction to name in this work was a protective wall, immediately east of the Trafaria landing pier and along nearly 70 m (Fig. 4).

Between 1958 and 1964, a new landing pier was built 70 m east of the old one, wich helped form a small beach between them. In the same period a pier was built in Portinho da Costa, west of the re-entrance, with an extension of about 200 m along the shore, and a wideness of 15-40 m. An embankment was then built on the western part of the bay to enable automobile access to the pier.

Between 1964 and 1972, the Portinho da Costa pier was widened by 20 m along its extension. On the other hand, due to the Porto do Bucho constructions, the small re-entry was covered up. A small landing bridge, about 80 m west of the old entry was built in connection with these structures.

In the period 1972-1975 there were no changes.

Between 1975 and 1984, the following took place:

1. covering up of the bay where the ETC Maritime Terminus exists today with back fill, creating a narrow stretch of ground (60-70 m) along 200 m in a western direction;

2. beach line advance of 20 m in Portinho da Costa.;

3. pier construction, west of the Portinho da Costa pier, possibly for protecting the scarp against the waves.

(The 1978 information only covers the area between Porto do Bucho e Trafaria where no modifications were detected).

In 1991, the small beach between the landing bridges of Trafaria disappears, and the Portinho da Costa beach looses almost 10 m. However, these signs may be related partially or totally with tide dynamics, making it difficult to determine if there is a true variation (and what its value is).

3.3. *Sector Trafaria - Cova do Vapor*

This section presents the most intense shoreline variations, in the period studied. Here are the factores involved:

1. nature of the shoreline ground: the formations are incoherent - recent alluvium and dune formations which are easily erodible and easily transported;

2. Nature of the hydraulic processes: this section is located in the mouth of the Tagus River, and is thus characterized by a strong interaction between river and maritime currents, as well as by the effect of

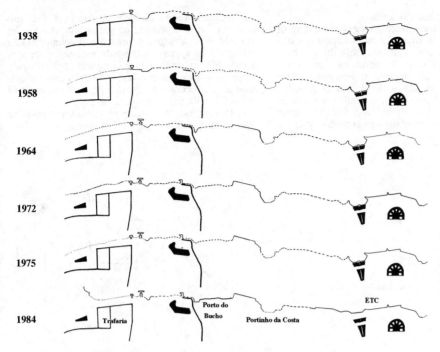

Fig. - 4 - "Shoreline evolution between ETC and Trafaria" (legend in Fig. 2)

waves;

3. anthropic intervention: construction of port structures - docks, land fills and piers - which affect the hydraulic system in a particularly sensitive area.

This third section can be divided into two zones in a similar way and for the same reasons that were used for the preceeding section:

1. Trafaria - Cova do Vapor;
2. Westernmost end of the South Margin.

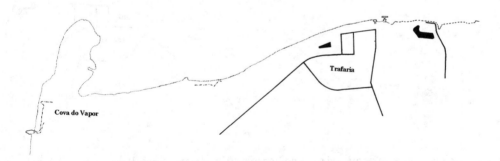

Fig. 5 - "Shoreline between Trafaria and Cova do Vapor in 1964" (legend in Fig. 2)

3.3.1 *Trafaria - Cova do Vapor*

Between 1938 and 1964, the Trafaria beach's widening of 60-80 m can be detected.

In 1964 (Fig. 5), north of Cova do Vapor, there used to be a long spit, with a rounded edge and a basic north-to-south grouth, that reached almost 600 m in length and had a medium wide of 250 m. At this date, the western and specially the northern edge of the village of Cova do Vapor were protected by supportive structures whose nature cannot be defined by simple observation of the aerial photographs. It is interesting to notice that the spit looks like the represented in the mentioned map from the 18th century (Fig. 1) rotated clockwise.

In 1972 (Fig. 6), there were no signs remaining of the area's spit. At this time, the small village of Cova do Vapor was already protected from the sea by a levy that surronded it on the north, east, and west. Immediatly east of the place and its protective levy, a inlet had grown because of the diminishing of the beaches shoreline whose maximum value was 150 m.

About 500 m east of Cova do Vapor and at the same time, a spit began developping - most probably due to

the migration and change of form of the spit described in the preceeding paragraph. As can be verified in the attached figures, it was composed of a cord with an approximate north-northeast direction, nearly 600 m long, ending in a barb edge (roughly eliptic) whose largest axis, with a northwest-southeast direction and 350 m long, made an angle of about 75° with the direction of the cord.

In this period, there were no important shoreline variations up to Trafaria.

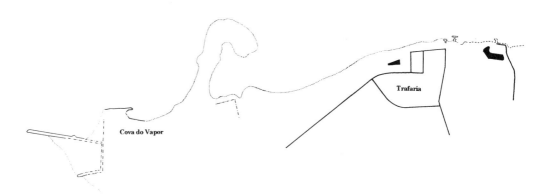

Fig. 6 - "Shoreline between Trafaria and Cova do Vapor in 1972" (legend in Fig. 2)

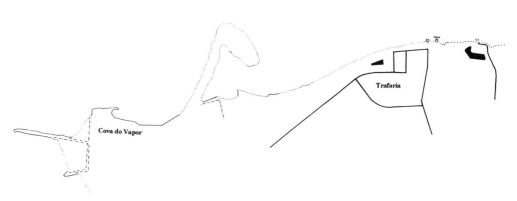

Fig. 7 - "Shoreline between Trafaria and Cova do Vapor in 1975" (legend in Fig. 2)

In 1975 (Fig. 7), the above-mentioned spit dislocated nearly 100 m to the east, with a rotation of 2° in the clockwise direction. The angle between the axes of the edge and of the cord decreases to a value close to 67°. It can be also noted the thinning out of the cord and a lengthening of the edge up to 400 m.

To the west of the spit, the shoreline suffers a great loss - in the order of 130 m. This accentuate the new bay grown up between Cova do Vapor and the spit.

The building of levies to stop the erosion that threatened some beach constructions should be referred to here.

The shoreline to the east of the spit until Trafaria diminished with values oscillating between 30-50 m.

Between 1975-1978, the eastward movement of the spit continues - in the order of 250 m - with the opening up of the cord-edge angle to nearly 105°. There was no apreciable rotation during this period, and the proportions of the two parts remained constant.

This last register of this structure dates to 1978 (Fig. 8) because, in the photographs of 1984 (Fig. 9), it appears to have coalesed in the consequence of its eastern migration into the huge embankment built between 1978-1984 in front of the Trafaria beach. A sandy extension was thus formed adjacent to the embankment that reached almost 11 ha. The small lagoon found in the middle was probably the remains of the bay formerly located east of the spit.

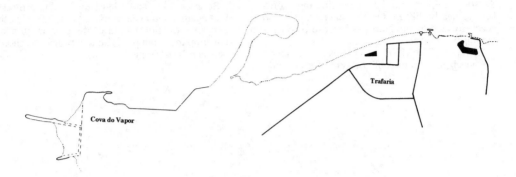

Fig. 8 - "Shoreline between Trafaria and Cova do Vapor in 1978" (legend in Fig. 2)

In the 1984 photographs, sedimentation actions to the west of the above-mentioned sandy area are notoriously responsible for favoring its growth. There are no alterations registered in the area between the sandy zone and Cova do Vapor due probably to the protection afforded by the levies (as mentioned, between 1975-1978). It can be observed, however, the lengthening of the northern pier of Cova do Vapor, to the east-northeast in about 130 m.

With the building of the forementioned embankment, whose area covers up to 20 ha, the Trafaria beach is reduced to the eastern area, where it develops along a strip of sand 300 m long.

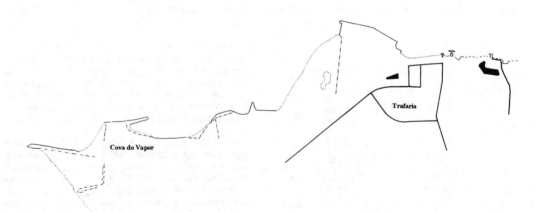

Fig. 9 - "Shoreline between Trafaria and Cova do Vapor in 1984" (legend in Fig. 2)

In 1991, changes can only be observed in one area, next to the coast of the sandy zone and west of the embankment, due to erosion, with deposits in a similar area but further to the North in the same line.

Fig. 10 is a sinthesis of the spit evolution repported above.

3.3.2 Westernmost End of the South Margin

In 1964 (Fig. 5), the protective structure for the small village of Cova do Vapor partially bordered the northern end of the village, and extended along the western end where it formed two branches perpendicular to the coast: the north pier and central pier. A third pier, also perpendicular to the coast existed to the south.

In the time span up to 1972 (Fig.6), the central and southern piers (located in the western end) were rebuilt. The lengthening of the central pier by nearly 500 m originated deposit formations between it and the southern pier, and thus increasing the shoreline here to nearly 300 m. On the other hand, the tiny beach located to the North suffered erosion and a diminishing of its northern end.

Between 1972-1975 (Fig. 7), erosion continued in the beach to the north of the central pier, and the beach located to the south of the same lost nearly 100 m. At this time, the small village of Cova do Vapor had already formed well-defined protuberances on her shoreline, being hemmed in by protective levies in all directions except south.

From 1975-1978 (Fig. 8), no alterations worthy of note were observed.

Between 1978 and 1984 (Fig. 9), the central pier was again lengthened nearly 100 m. Intensive sandy deposit formations south of the pier accelerated to cover up the south pier, and to advance the shoreline between 100-150 m.

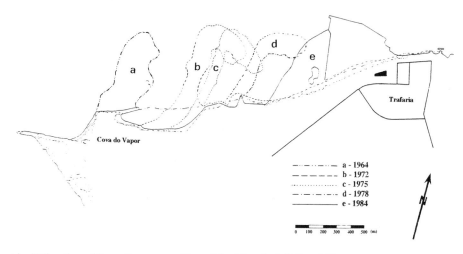

Fig. 10 - "Migration of the spit ; superposition of the Fig.s 5, 6, 7, 8 and 9"

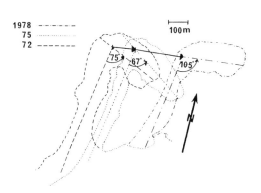

Fig. 11 - "Detail of Fig. 10"

4. FINAL CONSIDERATIONS

As defined early on, attention was paied exclusively on shoreline variations within the time period considered. In addition to the referred situations, others occurred eventually, of a rather efemerous nature between the dates when information was obtained and analyzed, due to the natural evolution of the scarp. This is the case of current and frequent accumulations of earth masses in the riverside area, which rapidly disappear due to the erosive action of the river.

From this analysis it is very clear that the most significant modifications were caused by anthropic interventions, or had these as their origin. This fact underlines the need for previous and adequate studies in the area, proven in its high sensibility, before any intervention in order to predict the type and magnitude of the alterations which may take place.

REFERENCES

Calmet-Beauvoisin, Ch. (18th century). Plan de Lisbonne, son Port, ses rades et ses environs.

Costa, C. (1986). Nota preliminar sobre a evolução geomorfológica das escarpas de Almada nos últimos cem anos. *Geotécnico* 1:127-131, SAGT, Monte de Caparica.

Planta do Rio Tejo e suas margens (1871). Direcção Geral dos Trabalhos Geodésicos, Lisboa.

Rodrigues-Carvalho, J. A., Chaves, M. A. and Rodrigues, J. C. (1994). Evolução da linha de costa entre Cacilhas e Cova do Vapor, Esc. 1:10.000, SAGT, Monte de Caparica.

Rodrigues-Carvalho, J. A., Chaves, M. A. and Silva, A. P. (1990). Nota descritiva da carta geológica dos taludes da margem Sul do Tejo (Cacilhas-Trafaria), Folhas 1 a 6, Esc. 1:2000, SAGT, Monte de Caparica.

Rodrigues-Carvalho, J. A., Costa, C. (1986). Modificações nas escarpas da margem sul do Tejo entre 1958 e 1975 - Um exemplo de utilização da detecção remota para o estudo da influência antrópica nos processos geológicos e naturais. *Geotécnico* 2:175-177, SAGT, Monte de Caparica.

Rodrigues-Carvalho, J. A., Silva, A. P. and Lamas, P.(1991). Nota descritiva da carta geológica dos taludes da margem Sul do Tejo (Trafaria-Cova do Vapor), Folhas 7 e 8, Esc. 1:2000, SAGT, Monte de Caparica.